Elementary Linear Algebra
Fourth Edition

Stanley I. Grossman
University of Montana and University College London

Saunders College Publishing
Philadelphia   New York   Chicago   San Francisco
Montreal   Toronto   London   Sydney   Tokyo

Text Typeface: Times Roman
Compositor: York Graphic Services, Inc.
Acquisitions Editor: Robert Stern
Developmental Editor: Alexa Barnes
Managing Editor: Carol Field
Production Management: York Production Services
Manager of Art and Design: Carol Bleistine
Art Director: Christine Schueler
Text Designer: C. J. Petlich, Hunter Graphics
Cover Designer: Lawrence R. Didona
Director of EDP: Tim Frelick
Production Manager: Robert Butler

Cover Credit: Cover photographed by Susan Robins

Printed in the United States of America

ELEMENTARY LINEAR ALGEBRA, 4th ed.

ISBN 0-03-031193-4

Library of Congress Catalog Card Number:
90-8877

0123   016   987654321

THIS BOOK IS PRINTED ON **ACID-FREE, RECYCLED** PAPER

To Kerstin, Aaron, and Erik

Preface

As recently as thirty years ago, the study of linear algebra was largely confined to mathematics and physics majors and to those who needed a knowledge of matrix theory to work in technical areas such as multivariate statistics. Linear algebra is now studied by students in many disciplines due to the invention of high-speed computers and a general increase in the application of mathematics in traditionally nontechnical areas.

Prerequisites

In writing this text I have had two goals in mind. I have tried to make a large number of linear algebra topics accessible to a wide variety of students who need only a good knowledge of high school algebra. Because many students will have had a year of calculus, I have also included several examples and exercises involving topics from calculus. These are indicated by the symbol $\boxed{\text{Calculus}}$. One optional section (Section 6.7) requires calculus, but otherwise *calculus is not a prerequisite*.

Applications

My second goal was to convince students of the importance of linear algebra in their fields of study. Thus, examples and exercises are drawn from a variety of disciplines. Some of these examples are short, like the application of matrix multiplication to the spread of a contagious disease (page 47). Others are quite a bit longer. These include the Leontief Input-Output Model (pages 16–18 and 74–78), Graph Theory (Section 1.12), Least Squares Approximation (Section 4.11), and a Model of Population Growth (Section 6.2). In addition, the booklet *Applications Supplement to Elementary Linear Algebra* contains interesting applications of linear algebra to linear programming, Markov chains, and game theory.

Theory

For many students linear algebra is their first real *mathematics* course. In this course students are required not only to carry out mathematical computations but also to create proofs. In this book I have tried to strike a balance between technique and theory. All the important techniques are described in great detail with many examples illustrating their use. At the same time, all the theorems that can be proved using the results given in the text are proved in the book. The more difficult proofs are placed at the ends of sections or in separate sections, but they *are* given. The result is a book that I am confident will give students all the algebraic skills necessary to solve linear algebra problems that arise in their areas of study and a greater appreciation for the beauty of mathematics.

FEATURES

Many of the features of the third edition have been retained in the fourth edition and are described below. New or enhanced features are indicated in italics.

Examples

Students learn mathematics by seeing clearly worked out examples. The fourth edition contains over 350 examples and *there are 80 more in the Applications Supplement*. Each example includes all the algebraic steps needed to complete the solution. In many instances explanations are highlighted in color in order to make the steps easier to follow. *In addition, the examples are now titled so students grasp more easily the essential concept each example illustrates.*

Exercises

The text contains approximately 1900 exercises *(with an additional 455 in the Applications Supplement)*. As in all mathematics books, these are the most important learning tool in the text. Problems are graded in order of increasing difficulty, and there is a balance between technique and proof. The more difficult problems are marked with * and a few exceptionally difficult ones with **. These are supplemented by review exercises at the end of each chapter. Answers to *all* odd-numbered problems, including those requiring proofs, are given at the back of the book.

The Summing Up Theorem

A major feature of the text is the frequent appearance of the Summing Up Theorem which ties together seemingly disparate topics in the study of matrices and linear transformations. The theorem is first encountered in Section 1.2 (p. 5). Successively more complete versions of this theorem are found in Section 1.9 (p. 78), Section 1.11 (p. 91), Section 2.4 (p. 146), Section 4.5 (p. 237), Section 4.7 (p. 263), Section 5.4 (p. 370), and Section 6.1 (p. 398).

Chapter Summaries

At the end of each chapter, a detailed review of the important results of that chapter appears. It includes page references and is a new feature of this text.

Geometry

Some of the important ideas in linear algebra are better understood by observing how they can be interpreted geometrically. *I have pointed out geometric interpretations of important concepts in several places in this edition.* These include

- The geometric interpretation of the 2×2 determinant (pp. 116, 187)
- The geometric interpretation of the scalar triple product (p. 188)
- Another geometric way to look at a plane (p. 201)
- The geometric interpretation of linear dependence in $\mathbb{R}^3$ (p. 233)
- The geometry of linear transformations from $\mathbb{R}^2$ to $\mathbb{R}^2$ (pp. 355–363)
- Isometries of $\mathbb{R}^2$ (p. 376).

Figures

The fourth edition contains *over 50% more figures than the previous edition* to give students as visual and geometric an understanding of linear algebra as the subject allows.

Historical Emphasis

Mathematics becomes more interesting if one knows something about the historical development of the subject. To stimulate this interest I have included a number of small historical notes scattered throughout the text. In addition, I have written eight "focuses" which are longer and somewhat more detailed. *All but three of these essays are new to this edition.* They cover:

- Carl Friedrich Gauss (p. 19)
- Sir William Rowan Hamilton (p. 33)
- Arthur Cayley and the Algebra of Matrices (p. 53)
- A Short History of Determinants (p. 139)
- Joseph Willard Gibbs and the Origins of Vector Analysis (p. 190)
- Carl Gustav Jacobi (p. 490)
- History of Mathematical Induction (p. A-6)
- George Dantzig and the History (and Future) of Linear Programming (p. 69 in the *Applications Supplement*)

Numbering

Numbering in this book is fairly standard. Within each section, examples, problems, theorems, and equations are numbered consecutively starting with 1. Reference to an example, problem, theorem, or equation outside the section is by chapter, section, and number. Thus, Example 4 in Section 2.5 is called Example 4 in that section but outside the section it is referred to as Example 2.5.4. In addition, page numbers are frequently provided to make referenced items easier to find.

ORGANIZATION

The approach I have used in this text is gradual. Chapters 1 and 2 contain the basic computational material common to most elementary linear algebra texts. Chapter 1 discusses systems of linear equations, vectors, and matrices. It has been reorganized to cover all systems of equations material before introducing matrices. This presentation provides more motivation for the student and follows a more common course order. There is also a new section (1.12) applying matrices to graph theory. Chapter 2 provides an introduction to determinants and includes a new historical essay on the contributions to linear algebra of Leibniz and Cauchy (Section 2.3).

Even in this early material there are optional sections to challenge the student a bit more. For example, Section 2.3 provides a complete proof that det AB = det$A$det$B$. The proof of this central result, using elementary matrices, is not often included in introductory texts.

Chapter 3 discusses vectors in the plane and in space. Many of the topics in this chapter are covered in a calculus sequence and may be familiar to students already. However, since much of linear algebra is concerned with a discussion of abstract vector spaces, students need a storehouse of concrete examples, which are most easily provided by the study of vectors in the plane and in space. The more difficult and abstract material in Chapters 4 and 5 is illustrated with examples from the concrete Chapter 3. Section 3.4 now includes a historical essay on Gibbs and the origins of vector analysis.

Chapter 4 contains an introduction to general vector spaces and is, necessarily, more abstract than the earlier chapters. I have tried, however, to present the material as a natural extension of properties of vectors in the plane, which is really how the subject evolved. The fourth edition discusses linear combination and span (Section 4.4) before linear independence (Section 4.5) to motivate these topics more clearly. Chapter 4 also now includes a new applications section (4.11) on least squares approximation.

At the end of Chapter 4 I have added an optional section (4.13) in which I prove that every vector space has a basis. In doing so I discuss partially ordered sets and Zorn's lemma. The material here is more difficult than anything else in the book and can easily be omitted. However, since linear algebra is often considered the first course in which proofs are as important as computations, I feel strongly that a proof of this fundamental result should be available to the more motivated student.

Chapter 5 continues the discussion begun in Chapter 4 with an introduction to linear transformations from one vector space to another. The chapter begins with two examples showing how such transformations can arise in a natural way. In Section 5.3 I have added a detailed description of the geometry of transformations from $\mathbb{R}^2$ to $\mathbb{R}^2$, including expansions, compressions, reflections, and shears. And Section 5.5 now includes a more detailed discussion of isometries of $\mathbb{R}^2$.

Chapter 6 describes the theory of eigenvalues and eigenvectors. These are introduced in Section 6.1, and a detailed biological application to population growth is given in Section 6.2. Sections 6.3, 6.4, and 6.5 all involve the diagonalization of a matrix, while Section 6.6 illustrates, for a few cases, how a matrix can be reduced to its Jordan canonical form. Section 6.7 discusses matrix differential equations and is the only section of the book that requires a knowledge of freshman calculus. This section provides an illustration of the usefulness of reducing a matrix to its Jordan canonical form (which is usually a diagonal matrix). In Section 6.8, I introduce two of my favorite results from matrix theory: the Cayley-Hamilton theorem and the Gershgorin circle theorem. The latter result, rarely discussed in an elementary linear algebra text, provides an easy way to estimate the eigenvalues of any matrix.

In Chapter 6, I had to make a difficult decision: whether or not to discuss complex eigenvalues and eigenvectors. I decided to include them because it seemed to be the only honest thing to do. Some of the "nicest" matrices have complex eigenvalues. To define an eigenvalue as a real number only may at first make things seem simpler, but it is certainly wrong. Moreover, for many applications involving eigenvalues (including some in Section 6.7), the most interesting models involve periodic phenomena, and these require complex eigenvalues. Complex numbers are not avoided in this book. For students who have not encountered them before, the few properties they need are discussed fully in Appendix 2.

In this era when students have access to computers and handheld calculators, it is important that a linear algebra text describe how computations are done in "real-life." In Chapters 1 through 6, many computations are done using a calculator. Chapter 7 introduces several numerical techniques for solving systems of equations and computing eigenvalues and eigenvectors. Section 7.1 discusses some of the problems that can arise when solving problems with a computer. This section also now contains a discussion of computational complexity. The techniques in Sections 7.1, 7.2, and 7.3 can be covered any time after Chapter 1. The material in Section 7.4 (on computing eigenvalues and eigenvectors) uses material presented in Section 6.1.

This book has two appendices, one on mathematical induction and one on complex numbers. Some of the proofs in the book use mathematical induction, so Appendix 1 provides a brief introduction to this important technique for students who have not used it before.

A word on chapter interdependence: The book is written sequentially. Each chapter depends on the ones that precede it, with two exceptions. Chapter 6 can be covered without most of the material in Chapter 5. Chapter 7 (except Section 7.4) can be covered after Chapter 1. Sections marked "Optional" can be omitted without any loss of continuity.

ACCURACY

The success of a mathematics textbook depends to a large extent on its accuracy. A few badly placed typographical errors, particularly in the answers to problems, can turn a good teaching tool into a source of frustration. This is a fourth edition and most (I'd like to say all) of the errors in earlier editions have been found and corrected. However, we have gone to considerable additional lengths to ensure that the text is error free and that the answers at the back of the book are correct.

Step 1   All odd-numbered problems were solved by Richard Spoonts, a student at Harvard University.

Step 2   I solved each problem, compared my answers with those provided by Mr. Spoonts, and resolved any discrepancies.

Step 3   After the answers were typeset in galleys, Dennis Grantham at East Texas State University and Marvin Zeman at Southern Illinois University checked the solutions to all the examples and the answers to all the odd-numbered exercises to make sure they were accurate.

Step 4   I corrected all errors found by Professors Grantham and Zeman. Then all three of us separately checked the answers again in the page proof stage.

The result is a text and an answer section that is as clean as it is within human ability to compile. However, if you do find an error in an answer or in the text, please send it to the publisher or to me; it will be corrected in the next printing.

SUPPLEMENTS

Applications Supplement for Elementary Linear Algebra contains two applications chapters on linear programming, Markov chains and game theory with numerous examples and exercises. It is available free of charge with new copies of this book.

A **Students Solutions Manual** prepared by Rick Miranda at Colorado State University contains complete detailed solutions for all odd-numbered exercises including review exercises and proofs.

An **Instructor's Manual,** also prepared by Rick Miranda provides complete solutions for all the exercises in the text and is available free to adopters.

RealMaTrices is a linear algebra computer software package written by Rich Miranda for use with IBM PC/XT/AT computers. This software is an extremely easy-to-use yet powerful tool for performing computations with real matrices, i.e., matrices with real entries. There are dozens of functions from simple matrix operations to solving systems of linear equations, finding eigenvalues and eigenvectors, and operating on polynomials and orthogonal matrices. Full documentation is included with the software. A preliminary disk with demonstration problems and a limited version of the software is available for review by potential adopters.

Student Manual for RealMaTrices keys the software to the fourth edition. It provides additional examples and exercises to give students practice using the software and it discusses the advantages of using software to solve certain problems in linear algebra.

ACKNOWLEDGMENTS

I am grateful to many people who helped as this book was written. Some of the material here first appeared in *Mathematics for the Biological Sciences* (New York: Macmillan, 1974) written by James E. Turner and myself. I am grateful to Professor Turner for permission to use this material.

A great deal of this book was written while I was a research associate at University College London. I wish to thank the Mathematics Department of UCL for providing office facilities, mathematical suggestions, and, especially, friendship during my annual visits there. Richard Spoonts, Dennis Grantham at East Texas State University, and Marvin Zeman at Southern Illinois University played a critical role in ensuring that the final product be accurate.

Special thanks go to the editorial and production staffs at Saunders College Publishing for the care and skill they brought to this product. I am especially grateful to Bob Stern and Alexa Barnes, my editor and developmental editor at Saunders, for their many helpful suggestions and constant encouragement. Thanks too to Kirsten Kauffman at York Production Services for her skill at helping to turn my manuscript into a bound book.

I also appreciate the numerous suggestions for improving this text from the reviewers of this and earlier editions. Their contributions to the reliability and teachability of this book have been invaluable.

Fourth Edition Reviewers

Thomas Cairns
University of Tulsa

Steven Kent
Youngstown State University

Gerald Leibowitz
University of Connecticut

John Ratcliffe
Vanderbilt University

Vance Underhill
East Texas State University

Marvin Zeman
Southern Illinois University
    at Carbondale

Third Edition Reviewers

John Chapman
Southwestern University

Thomas D. Forrest
Middle Tennessee
    State University

Henry Helson
University of California,
    Berkeley

William Jacob
Oregon State University

R. Bruce Lind
University of Puget Sound

Bryan Smith
University of Puget Sound

R. Van Enkevort
University of Puget Sound

Second Edition Reviewers

P. A. Binding
University of Calgary

Joseph M. Cavanaugh
East Stroudsburg State College

Louis Florence
University of Toronto

Eugene Johnson
University of Iowa

Graham A. Chambers
University of Alberta

Ronald Eldringhoff
St. Louis Community College

John Sawka
University of Santa Clara

Keith Stumpff
Central Missouri State University

First Edition Reviewers

Donald F. Bailey
Cornell College

Richard W. Ball
Auburn University

Sandra A. Bollinger
Longwood College

James Bradley
Roberts Wesleyan College

Joel V. Brawley
Clemson University

Paul Bugl
University of Hartford

Gary Chartrand
Western Michigan University

Richard J. Easton
Indiana State University

Gary L. Eerkes
Gonzaga University

Garret J. Etgen
University of Houston

William R. Fuller
Purdue University

John D. Fulton
Clemson University

Lillian Gough
University of Wisconsin

Douglas D. Smith
Central Michigan University

Barbara Ann Greim
University of North Carolina
    at Wilmington

Charles H. Haggard
Transylvania University

David Kinsey
Indiana State University

Thomas Lupton
University of North Carolina
    at Wilmington

J. J. Malone
Worcester Polytechnic Institute

Edwin L. Marsden
Norwich University

Martin E. Nass
Iowa Central Community College

John Petro
Western Michigan University

William B. Rundberg
College of San Mateo

William Rundell
Texas A & M University

David Schedler
Virginia Commonwealth University

Jack C. Slay
Mississippi State University

Cary Webb
Chicago State University

T. W. Tucker
Dartmouth College

Marcellus E. Waddill
Wake Forrest University

James Wall
Auburn University

James J, Woeppel
Indiana University Southeast

Dennis G. Zill
Loyola Marymount University

Stanley I. Grossman

Contents

The Applications Supplement to accompany Elementary Linear Algebra, Fourth Edition, contains the following material:

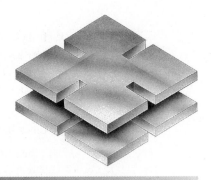

1

Systems of Linear
Equations and Matrices

1.1 INTRODUCTION

This is a book about linear algebra. If you look up the word "linear" in a dictionary, you will find something like the following: lin-e-ar (lin′ ē ər), adj. 1. of, consisting of, or using lines.† In mathematics, the word "linear" means a good deal more than that. Nevertheless, much of the theory of elementary linear algebra is in fact a generalization of properties of straight lines. As a review, here are some fundamental facts about straight lines:

i. The **slope** m of a line passing through the points (x_1, y_1) and (x_2, y_2) is given by

$$m = \frac{y_2 - y_1}{x_2 - x_1} = \frac{\Delta y}{\Delta x} \qquad \text{if } x_1 \neq x_2$$

ii. If $x_2 - x_1 = 0$ and $y_2 \neq y_1$, then the line is vertical and the slope is said to be **undefined**‡

iii. Any line (except one with undefined slope) can be described by writing its equation in the slope-intercept form $y = mx + b$, where m is the slope of the line and b is the y-intercept of the line (the value of y at the point where the line crosses the y-axis).

iv. Two distinct lines are parallel if and only if they have the same slope.

---

† Taken from the pocket edition of *The Random House Dictionary*.

‡ In some textbooks a vertical line is said to have "an infinite slope."

v. If the equation of a line is written in the form $ax + by = c$ ($b \neq 0$), then, as is easily computed, $m = -a/b$.

vi. If m_1 is the slope of line L_1, m_2 is the slope of line L_2, $m_1 \neq 0$, and L_1 and L_2 are perpendicular, then $m_2 = -1/m_1$.

vii. Lines parallel to the x-axis have a slope of zero.

viii. Lines parallel to the y-axis have an undefined slope.

In the next section we shall illustrate the relationship between solving systems of equations and finding points of intersection of pairs of straight lines.

1.2   TWO LINEAR EQUATIONS IN TWO UNKNOWNS

Consider the following system of two linear equations in the two unknowns x and y:

$$a_{11}x + a_{12}y = b_1$$
$$a_{21}x + a_{22}y = b_2 \tag{1}$$

where a_{11}, a_{12}, a_{21}, a_{22}, b_1, and b_2 are given numbers. Each of these equations is the equation of a straight line. The slope of the first line is $-a_{11}/a_{12}$; the slope of the second line is $-a_{21}/a_{22}$ (if $a_{12} \neq 0$ and $a_{22} \neq 0$). A **solution** to system (1) is a pair of numbers, denoted by (x, y), that satisfies (1). The questions that naturally arise are whether (1) has any solutions and if so, how many? We answer these questions after looking at some examples. In these examples we make use of two important facts from elementary algebra:

Fact A If $a = b$ and $c = d$, then $a + c = b + d$.

Fact B If $a = b$ and c is any real number, then $ca = cb$.

Fact A states that if we add two equations together, we obtain a third, valid equation. Fact B states that if we multiply both sides of an equation by a constant, we obtain a second, valid equation. We shall assume that $c \neq 0$ for although the equation $0 = 0$ is correct, it is not very useful.

EXAMPLE 1   **A System with a Unique Solution**   Consider the system

$$x - y = 7$$
$$x + y = 5 \tag{2}$$

Adding the two equations together gives us, by Fact A, the following equation: $2x = 12$ (or $x = 6$). Then, from the second equation, $y = 5 - x = 5 - 6 = -1$. Thus the pair $(6, -1)$ satisfies system (2) and the way we found the solution shows that it is the only pair of numbers to do so. That is, system (2) has a **unique solution.**   ■

EXAMPLE 2   **A System with an Infinite Number of Solutions**   Consider the system

$$x - y = 7$$
$$2x - 2y = 14$$

(3)

It is apparent that these two equations are equivalent. To see this multiply the first by 2. This is permitted by Fact B. Then $x - y = 7$ or $y = x - 7$. Thus the pair $(x, x - 7)$ is a solution to system (3) for any real number x. That is, system (3) has an **infinite number of solutions.** For example, the following pairs are solutions: $(7, 0)$, $(0, -7)$, $(8, 1)$, $(1, -6)$, $(3, -4)$, and $(-2, -9)$.   ■

EXAMPLE 3   **A System with No Solution**   Consider the system

$$x - y = 7$$
$$2x - 2y = 13$$

(4)

Multiplying the first equation by 2 (which, again, is permitted by Fact B) gives us $2x - 2y = 14$. This contradicts the second equation. Thus system (4) has **no solution.**   ■

It is easy to explain, geometrically, what is going on in the preceding examples. First we repeat that the equations in system (1) are both equations of straight lines. A solution to (1) is a point (x, y) that lies on both lines. If the two lines are not parallel, then they intersect at a single point. If they are parallel, then they either never intersect (no points in common) or are the same line (infinite number of points in common). In Example 1 the lines have slopes of 1 and -1, respectively. Thus they are not parallel. They have the single point $(6, -1)$ in common. In Example 2 the lines are parallel (slope of 1) and coincident. In Example 3 the lines are parallel and distinct. These relationships are all illustrated in Figure 1.1.

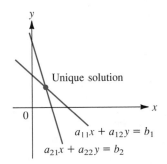

(a) Lines not parallel; one point of intersection

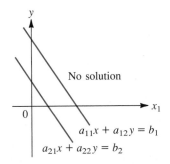

(b) Lines parallel; no points of intersection

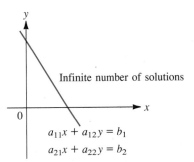

(c) Lines coincide; infinite number of points of intersection

Figure 1.1   Two lines intersect at one point, no points, or (if they coincide) an infinite number of points

Let us now solve system (1) formally. We have

$$a_{11}x + a_{12}y = b_1$$
$$a_{21}x + a_{22}y = b_2 \tag{1}$$

If $a_{12} = 0$, then $x = \dfrac{b_1}{a_{11}}$ and we can use the second equation to solve for y.

If $a_{22} = 0$, then $x = \dfrac{b_2}{a_{21}}$ and we can use the first equation to solve for y.

If $a_{12} = a_{22} = 0$, then system (1) contains only one unknown x.

Thus we may assume that neither a_{12} nor a_{22} is zero.

Multiplying the first equation by a_{22} and the second by a_{12} yields

$$a_{11}a_{22}x + a_{12}a_{22}y = a_{22}b_1$$
$$a_{12}a_{21}x + a_{12}a_{22}y = a_{12}b_2 \tag{5}$$

Equivalent systems   Before continuing we note that system (1) and system (5) are **equivalent.** By that we mean that any solution to system (1) is a solution to system (5) and vice versa. This follows immediately from Fact B assuming that the c in Fact B is not zero. Next we subtract the second equation from the first to obtain

$$(a_{11}a_{22} - a_{12}a_{21})x = a_{22}b_1 - a_{12}b_2 \tag{6}$$

At this point we must pause. If $a_{11}a_{22} - a_{12}a_{21} \neq 0$, then we can divide by it to obtain

$$x = \frac{a_{22}b_1 - a_{12}b_2}{a_{11}a_{22} - a_{12}a_{21}}$$

Determinant   Then we can substitute this value of x into system (1) to solve for y, and thus we have found the unique solution to the system. We define the **determinant** of system (1) by

$$\boxed{\text{Determinant of system (1)} = a_{11}a_{22} - a_{12}a_{21}} \tag{7}$$

and we have shown the following:

$$\boxed{\begin{array}{c} \text{If the determinant of system (1)} \neq 0, \text{ then} \\ \text{the system has a unique solution.} \end{array}} \tag{8}$$

How does this statement relate to what we discussed earlier? In system (1) we see that the slope of the first line is $-a_{11}/a_{12}$ and the slope of the second is $-a_{21}/a_{22}$. In Problems 31, 32, and 33 you are asked to show that the determinant of system (1) is zero if and only if the lines are parallel (have the same slope). So, if the determinant is *not* zero, the lines are not parallel and the system has a unique solution.

We now put the facts discussed above together in a theorem. It is a theorem that will be generalized in later sections of this chapter and in subsequent chapters. We shall keep track of our progress by referring to the theorem as our "Summing Up Theorem." When all its parts have been proved, we shall see a remarkable relationship among several important concepts in linear algebra.

THEOREM 1   **Summing Up Theorem—View 1** The system

$$a_{11}x + a_{12}y = b_1$$
$$a_{21}x + a_{22}y = b_2$$

of two equations in the two unknowns x and y has no solution, a unique solution, or an infinite number of solutions. It has:

i.   A unique solution if and only if its determinant is not zero.

ii. No solution or an infinite number of solutions if and only if its determinant is zero.   ∎

In Section 1.3 we shall discuss systems of m equations in n unknowns and shall see that there is always either no solution, one solution, or an infinite number of solutions. In Chapter 2 we define and calculate determinants for systems of n equations in n unknowns and shall find that our Summing Up Theorem—Theorem 1—is true in this general setting.

PROBLEMS 1.2

In Problems 1–12 find all solutions (if any) to the given systems. In each case calculate the determinant.

1. $x - 3y = 4$
   $-4x + 2y = 6$

2. $2x - y = -3$
   $5x + 7y = 4$

3. $2x - 8y = 5$
   $-3x + 12y = 8$

4. $2x - 8y = 6$
   $-3x + 12y = -9$

5. $6x + y = 3$
   $-4x - y = 8$

6. $3x + y = 0$
   $2x - 3y = 0$

7. $4x - 6y = 0$
   $-2x + 3y = 0$

8. $5x + 2y = 3$
   $2x + 5y = 3$

9. $2x + 3y = 4$
   $3x + 4y = 5$

10. $ax + by = c$
   $ax - by = c$

11. $ax + by = c$
   $bx + ay = c$

12. $ax - by = c$
   $bx + ay = d$

13. Find conditions on a and b such that the system in Problem 10 has a unique solution.

14. Find conditions on a, b, and c such that the system in Problem 11 has an infinite number of solutions.

15. Find conditions on a, b, c, and d such that the system in Problem 12 has no solutions.

In Problems 16–21 find the point of intersection (if there is one) of the two lines.

16. $x - y = 7$; $2x + 3y = 1$ **17.** $y - 2x = 4$; $4x - 2y = 6$

18. $4x - 6y = 7$; $6x - 9y = 12$ **19.** $4x - 6y = 10$; $6x - 9y = 15$

20. $3x + y = 4$; $y - 5x = 2$ **21.** $3x + 4y = 5$; $6x - 7y = 8$

Let L be a line and let $L_\perp$ denote the line perpendicular to L that passes through a given point P. The **distance** from L to P is defined to be the distance† between P and the point of intersection of L and $L_\perp$. In Problems 22–27 find the distance between the given line and point.

22. $x - y = 6$; $(0, 0)$ **23.** $2x + 3y = -1$; $(0, 0)$

24. $3x + y = 7$; $(1, 2)$ **25.** $5x - 6y = 3$; $(2, \frac{16}{5})$

26. $2y - 5x = -2$; $(5, -3)$ **27.** $6y + 3x = 3$; $(8, -1)$

28. Find the distance between the line $2x - y = 6$ and the point of intersection of the lines $2x - 3y = 1$ and $3x + 6y = 12$.

*29. Prove that the distance between the point (x_1, y_1) and the line $ax + by = c$ is given by

$$d = \frac{|ax_1 + by_1 - c|}{\sqrt{a^2 + b^2}}.$$

30. A zoo keeps birds (two-legged) and beasts (four-legged). If the zoo contains 60 heads and 200 feet, how many birds and how many beasts live there?

31. Suppose that the determinant of system (1) is zero. Show that the lines given in (1) are parallel.

32. If there is a unique solution to system (1), show that its determinant is nonzero.

33. If the determinant of system (1) is nonzero, show that the system has a unique solution.

34. The Sunrise Porcelain Company manufactures ceramic cups and saucers. For each cup or saucer a worker measures a fixed amount of material and puts it into a forming machine, from which it is automatically glazed and dried. On the average, a worker needs 3 minutes to get the process started for a cup and 2 minutes for a saucer. The material for a cup costs 25¢ and the material for a saucer costs 20¢. If $44 is allocated daily for production of cups and saucers, how many of each can be manufactured in an 8-hour workday if a worker is working every minute and exactly $44 is spent on materials?

35. Answer the question of Problem 34 if the materials for a cup and saucer cost 15¢ and 10¢, respectively, and $24 is spent in an 8-hour day.

36. Answer the question of Problem 35 if $25 is spent in an 8-hour day.

37. An ice-cream shop sells only ice-cream sodas and milk shakes. It puts 1 ounce of syrup and 4 ounces of ice cream in an ice-cream soda, and 1 ounce of syrup and 3 ounces of ice cream in a milk shake. If the store used 4 gallons of ice cream and 5 quarts of syrup in a day, how many ice-cream sodas and milk shakes did it sell? [*Hint:* 1 quart = 32 ounces; 1 gallon = 128 ounces]

---

† Recall that if (x_1, y_1) and (x_2, y_2) are two points in the xy-plane, then the distance d between them is given by $d = \sqrt{(x_1 - x_2)^2 + (y_1 - y_2)^2}$.

1.3   *m* EQUATIONS IN *n* UNKNOWNS: GAUSS-JORDAN AND GAUSSIAN ELIMINATION

In this section we describe a method for finding all solutions (if any) to a system of *m* linear equations in *n* unknowns. In doing so we shall see that, like the 2×2 case, such a system has no solutions, one solution, or an infinite number of solutions. Before launching into the general method, let us look at some simple examples. As variables, we use x_1, x_2, x_3, and so on instead of $x, y, z, \ldots$ because the subscripted notation is easier to generalize.

EXAMPLE 1   **Solving a System of Three Equations in Three Unknowns: Unique Solution**
Solve the system

$$2x_1 + 4x_2 + 6x_3 = 18$$
$$4x_1 + 5x_2 + 6x_3 = 24 \qquad \textbf{(1)}$$
$$3x_1 + x_2 - 2x_3 = 4$$

Solution   Here we seek three numbers x_1, x_2, and x_3 such that the three equations in (1) are satisfied. Our method of solution will be to simplify the equations as we did in Section 1.2 so that solutions can be readily identified. We begin by dividing the first equation by 2. This gives us

$$x_1 + 2x_2 + 3x_3 = 9$$
$$4x_1 + 5x_2 + 6x_3 = 24 \qquad \textbf{(2)}$$
$$3x_1 + x_2 - 2x_3 = 4$$

As we saw in Section 1.2, adding two equations together leads to a third, valid equation. This equation may replace either of the two equations used to obtain it in the system. We begin simplifying system (2) by multiplying both sides of the first equation in (2) by -4 and adding this new equation to the second equation. This gives us

$$-4x_1 - 8x_2 - 12x_3 = -36$$
$$\underline{4x_1 + 5x_2 + 6x_3 = 24}$$
$$ - 3x_2 - 6x_3 = -12$$

The equation $-3x_2 - 6x_3 = -12$ is our new second equation and the system is now

$$x_1 + 2x_2 + 3x_3 = 9$$
$$ - 3x_2 - 6x_3 = -12$$
$$3x_1 + x_2 - 2x_3 = 4$$

We then multiply the first equation by -3 and add it to the third equation:

$$x_1 + 2x_2 + 3x_3 = 9$$
$$ - 3x_2 - 6x_3 = -12 \qquad \textbf{(3)}$$
$$ - 5x_2 - 11x_3 = -23$$

Note that in system (3) the variable x_1 has been eliminated from the second and third equations. Next we divide the second equation by -3:

$$\begin{aligned} x_1 + 2x_2 + 3x_3 &= 9 \\ x_2 + 2x_3 &= 4 \\ -5x_2 - 11x_3 &= -23 \end{aligned}$$

We multiply the second equation by -2 and add it to the first and then multiply the second equation by 5 and add it to the third:

$$\begin{aligned} x_1 \phantom{{}+2x_2} - x_3 &= 1 \\ x_2 + 2x_3 &= 4 \\ -x_3 &= -3 \end{aligned}$$

We multiply the third equation by -1:

$$\begin{aligned} x_1 \phantom{{}+2x_2} - x_3 &= 1 \\ x_2 + 2x_3 &= 4 \\ x_3 &= 3 \end{aligned}$$

Finally, we add the third equation to the first and then multiply the third equation by -2 and add it to the second to obtain the following system [which is equivalent to system (1)]:

$$\begin{aligned} x_1 \phantom{{}+2x_2+3x_3} &= 4 \\ x_2 \phantom{{}+3x_3} &= -2 \\ x_3 &= 3 \end{aligned}$$

This is the unique solution to the system. We write it in the form $(4, -2, 3)$. The method we used here is called **Gauss-Jordan elimination.**† ∎

Gauss-Jordan
elimination

Before going on to another example, let us summarize what we have done in this example:

i. We divided to make the coefficient of x_1 in the first equation equal to 1.

ii. We "eliminated" the x_1 terms in the second and third equations. That is, we made the coefficients of these terms equal to zero by multiplying the first equation by appropriate numbers and then adding it to the second and third equations, respectively.

iii. We divided to make the coefficient of the x_2 term in the second equation equal to 1 and then proceeded to use the second equation to eliminate the x_2 terms in the first and third equations.

iv. We divided to make the coefficient of the x_3 term in the third equation equal to

---

†Named after the great German mathematician Karl Friedrich Gauss (1777–1855) and the German engineer Wilhelm Jordan (1844–1899). See the biographical sketch of Gauss on page 19.

1 and then proceeded to use the third equation to eliminate the x_3 terms in the first and second equations.

We emphasize that, at every step, we obtained systems that were equivalent. That is, each system had the same set of solutions as the one that preceded it. This follows from Facts A and B on page 2.

Before solving other systems of equations, we introduce notation that makes it easier to write down each step in our procedure. A **matrix** is a rectangular array of numbers. We shall discuss matrices in great detail beginning in Section 1.6. For example, the coefficients of the variables x_1, x_2, x_3 in system (1) can be written as the entries of a matrix A, called the **coefficient matrix** of the system:

Matrix

Coefficient matrix

$$A = \begin{pmatrix} 2 & 4 & 6 \\ 4 & 5 & 6 \\ 3 & 1 & -2 \end{pmatrix} \tag{4}$$

$m \times n$ matrix

A matrix with m rows and n columns is called an **$m \times n$ matrix.** The symbol $m \times n$ is read "*m* by *n*." The study of matrices will take a large part of the remaining chapters of this text. We introduce them here for convenience of notation.

Augmented matrix

Using matrix notation, we can write system (1) as the **augmented matrix**

$$\begin{pmatrix} 2 & 4 & 6 & | & 18 \\ 4 & 5 & 6 & | & 24 \\ 3 & 1 & -2 & | & 4 \end{pmatrix} \tag{5}$$

We now introduce some terminology. We have seen that multiplying (or dividing) the sides of an equation by a nonzero number gives us a new, valid equation. Moreover, adding a multiple of one equation to another equation in a system gives us another valid equation. Finally, if we interchange two equations in a system of equations, we obtain an equivalent system. These three operations, when applied to the rows of the augmented matrix representation of a system of equations, are called **elementary row operations.**

To sum up, the three elementary row operations applied to the augmented matrix representation of a system of equations are:

Elementary Row Operations

  i. Multiply (or divide) one row by a nonzero number.

  ii. Add a multiple of one row to another row.

  iii. Interchange two rows.

Row reduction

The process of applying elementary row operations to simplify an augmented matrix is called **row reduction.**

Notation

1. $R_i \rightarrow cR_i$ stands for "replace the ith row by the ith row multiplied by c."
2. $R_j \rightarrow R_j + cR_i$ stands for "replace the jth row with the sum of the jth row and the ith row multiplied by c."
3. $R_i \rightleftarrows R_j$ stands for "interchange rows i and j."
4. $A \rightarrow B$ indicates that the augmented matrices A and B are equivalent; that is, the systems they represent have the same solution.

In Example 1 we saw that by using the elementary row operations (i) and (ii) several times we could obtain a system in which the solutions to the system were given explicitly. We now repeat the steps in Example 1, using the notation just introduced:

$$\begin{pmatrix} 2 & 4 & 6 & | & 18 \\ 4 & 5 & 6 & | & 24 \\ 3 & 1 & -2 & | & 4 \end{pmatrix} \xrightarrow{R_1 \rightarrow \frac{1}{2}R_1} \begin{pmatrix} 1 & 2 & 3 & | & 9 \\ 4 & 5 & 6 & | & 24 \\ 3 & 1 & -2 & | & 4 \end{pmatrix} \xrightarrow[R_3 \rightarrow R_3 - 3R_1]{R_2 \rightarrow R_2 - 4R_1} \begin{pmatrix} 1 & 2 & 3 & | & 9 \\ 0 & -3 & -6 & | & -12 \\ 0 & -5 & -11 & | & -23 \end{pmatrix}$$

$$\xrightarrow{R_2 \rightarrow -\frac{1}{3}R_2} \begin{pmatrix} 1 & 2 & 3 & | & 9 \\ 0 & 1 & 2 & | & 4 \\ 0 & -5 & -11 & | & -23 \end{pmatrix} \xrightarrow[R_3 \rightarrow R_3 + 5R_2]{R_1 \rightarrow R_1 - 2R_2} \begin{pmatrix} 1 & 0 & -1 & | & 1 \\ 0 & 1 & 2 & | & 4 \\ 0 & 0 & -1 & | & -3 \end{pmatrix}$$

$$\xrightarrow{R_3 \rightarrow -R_3} \begin{pmatrix} 1 & 0 & -1 & | & 1 \\ 0 & 1 & 2 & | & 4 \\ 0 & 0 & 1 & | & 3 \end{pmatrix} \xrightarrow[R_2 \rightarrow R_2 - 2R_3]{R_1 \rightarrow R_1 + R_3} \begin{pmatrix} 1 & 0 & 0 & | & 4 \\ 0 & 1 & 0 & | & -2 \\ 0 & 0 & 1 & | & 3 \end{pmatrix}$$

Again we can easily "see" the solution $x_1 = 4$, $x_2 = -2$, $x_3 = 3$.

EXAMPLE 2   **Solving a System of Three Equations in Three Unknowns: Infinite Number of Solutions**   Solve the system

$$2x_1 + 4x_2 + 6x_3 = 18$$
$$4x_1 + 5x_2 + 6x_3 = 24$$
$$2x_1 + 7x_2 + 12x_3 = 30$$

Solution   To solve, we proceed as in Example 1, first writing the system as an augmented matrix:

$$\begin{pmatrix} 2 & 4 & 6 & | & 18 \\ 4 & 5 & 6 & | & 24 \\ 2 & 7 & 12 & | & 30 \end{pmatrix}$$

We then obtain, successively,

$$\xrightarrow{R_1 \rightarrow \frac{1}{2}R_1} \begin{pmatrix} 1 & 2 & 3 & | & 9 \\ 4 & 5 & 6 & | & 24 \\ 2 & 7 & 12 & | & 30 \end{pmatrix} \xrightarrow[R_3 \rightarrow R_3 - 2R_1]{R_2 \rightarrow R_2 - 4R_1} \begin{pmatrix} 1 & 2 & 3 & | & 9 \\ 0 & -3 & -6 & | & -12 \\ 0 & 3 & 6 & | & 12 \end{pmatrix}$$

$$\xrightarrow{R_2 \to -\frac{1}{3}R_2} \begin{pmatrix} 1 & 2 & 3 & 9 \\ 0 & 1 & 2 & 4 \\ 0 & 3 & 6 & 12 \end{pmatrix} \xrightarrow[R_3 \to R_3 - 3R_2]{R_1 \to R_1 - 2R_2} \begin{pmatrix} 1 & 0 & -1 & 1 \\ 0 & 1 & 2 & 4 \\ 0 & 0 & 0 & 0 \end{pmatrix}$$

This is equivalent to the system of equations

$$
\begin{aligned}
x_1 \quad - \quad x_3 &= 1 \\
x_2 + 2x_3 &= 4
\end{aligned}
$$

This is as far as we can go. There are now only two equations in the three unknowns x_1, x_2, x_3 and there are an infinite number of solutions. To see this let x_3 be chosen. Then $x_2 = 4 - 2x_3$ and $x_1 = 1 + x_3$. This will be a solution for any number x_3. We write these solutions in the form $(1 + x_3, 4 - 2x_3, x_3)$. For example, if $x_3 = 0$, we obtain the solution $(1, 4, 0)$. For $x_3 = 10$ we obtain the solution $(11, -16, 10)$. ∎

EXAMPLE 3 **An Inconsistent System** Solve the system

augment.

$$
\begin{aligned}
2x_1 + 4x_2 + 6x_3 &= 18 \\
4x_1 + 5x_2 + 6x_3 &= 24 \\
2x_1 + 7x_2 + 12x_3 &= 40
\end{aligned}
\tag{6}
$$

Solution We use the augmented-matrix form and proceed exactly as in Example 2 to obtain, successively, the following systems. (Note how in each step we use either elementary row operation (*i*) or (*ii*).)

$$\begin{pmatrix} 2 & 4 & 6 & 18 \\ 4 & 5 & 6 & 24 \\ 2 & 7 & 12 & 40 \end{pmatrix} \xrightarrow{R_1 \to \frac{1}{2}R_1} \begin{pmatrix} 1 & 2 & 3 & 9 \\ 4 & 5 & 6 & 24 \\ 2 & 7 & 12 & 40 \end{pmatrix}$$

$$\xrightarrow[R_3 \to R_3 - 2R_1]{R_2 \to R_2 - 4R_1} \begin{pmatrix} 1 & 2 & 3 & 9 \\ 0 & -3 & -6 & -12 \\ 0 & 3 & 6 & 22 \end{pmatrix} \xrightarrow{R_2 \to -\frac{1}{3}R_2} \begin{pmatrix} 1 & 2 & 3 & 9 \\ 0 & 1 & 2 & 4 \\ 0 & 3 & 6 & 22 \end{pmatrix}$$

$$\xrightarrow[R_3 \to R_3 - 3R_2]{R_1 \to R_1 - 2R_2} \begin{pmatrix} 1 & 0 & -1 & 1 \\ 0 & 1 & 2 & 4 \\ 0 & 0 & 0 & 10 \end{pmatrix} \xrightarrow{R_3 \to \frac{1}{10}R_3} \begin{pmatrix} 1 & 0 & -1 & 1 \\ 0 & 1 & 2 & 4 \\ 0 & 0 & 0 & 1 \end{pmatrix}$$

The last equation now reads $0x_1 + 0x_2 + 0x_3 = 1$, which is impossible since $0 \neq 1$. Thus system (6) has *no* solution. In this case the system is said to be **inconsistent.** ∎

Inconsistent system

Let us take another look at these three examples. In Example 1 we began with the coefficient matrix

$$A_1 = \begin{pmatrix} 2 & 4 & 6 \\ 4 & 5 & 6 \\ 3 & 1 & -2 \end{pmatrix}$$

In the process of row reduction A_1 was "reduced" to the matrix

$$R_1 = \begin{pmatrix} 1 & 0 & 0 \\ 0 & 1 & 0 \\ 0 & 0 & 1 \end{pmatrix}$$

In Example 2 we started with

$$A_2 = \begin{pmatrix} 2 & 4 & 6 \\ 4 & 5 & 6 \\ 2 & 7 & 12 \end{pmatrix}$$

and ended up with

$$R_2 = \begin{pmatrix} 1 & 0 & -1 \\ 0 & 1 & 2 \\ 0 & 0 & 0 \end{pmatrix}$$

In Example 3 we began with

$$A_3 = \begin{pmatrix} 2 & 4 & 6 \\ 4 & 5 & 6 \\ 2 & 7 & 12 \end{pmatrix}$$

and again ended up with

$$R_3 = \begin{pmatrix} 1 & 0 & -1 \\ 0 & 1 & 2 \\ 0 & 0 & 0 \end{pmatrix}$$

The matrices R_1, R_2, and R_3 are called the *reduced row echelon forms* of the matrices A_1, A_2, and A_3, respectively. In general, we have the following definition.

DEFINITION 1   **Reduced Row Echelon Form** A matrix is in **reduced row echelon form** if the following four conditions hold:

i. All rows (if any) consisting entirely of zeros appear at the bottom of the matrix.

ii. The first nonzero number (starting from the left) in any row not consisting entirely of zeros is 1.

iii. If two successive rows do not consist entirely of zeros, then the first 1 in the lower row occurs farther to the right than the first 1 in the higher row.

iv. Any column containing the first 1 in a row has zeros everywhere else.

EXAMPLE 4   **Five Matrices in Reduced Row Echelon Form**   The following matrices are in reduced row echelon form:

i. $\begin{pmatrix} 1 & 0 & 0 \\ 0 & 1 & 0 \\ 0 & 0 & 1 \end{pmatrix}$ **ii.** $\begin{pmatrix} 1 & 0 & 0 & 0 \\ 0 & 1 & 0 & 0 \\ 0 & 0 & 0 & 1 \end{pmatrix}$ **iii.** $\begin{pmatrix} 1 & 0 & 0 & 5 \\ 0 & 0 & 1 & 2 \end{pmatrix}$

iv. $\begin{pmatrix} 1 & 0 \\ 0 & 1 \end{pmatrix}$ **v.** $\begin{pmatrix} 1 & 0 & 2 & 5 \\ 0 & 1 & 3 & 6 \\ 0 & 0 & 0 & 0 \end{pmatrix}$

■

DEFINITION 2   **Row Echelon Form**   A matrix is in **row echelon form** if conditions (*i*), (*ii*), and (*iii*) hold in Definition 1.

EXAMPLE 5   **Five Matrices in Row Echelon Form**   The following matrices are in row echelon form:

i. $\begin{pmatrix} 1 & 2 & 3 \\ 0 & 1 & 5 \\ 0 & 0 & 1 \end{pmatrix}$ **ii.** $\begin{pmatrix} 1 & -1 & 6 & 4 \\ 0 & 1 & 2 & -8 \\ 0 & 0 & 0 & 1 \end{pmatrix}$

iii. $\begin{pmatrix} 1 & 0 & 2 & 5 \\ 0 & 0 & 1 & 2 \end{pmatrix}$ **iv.** $\begin{pmatrix} 1 & 2 \\ 0 & 1 \end{pmatrix}$ **v.** $\begin{pmatrix} 1 & 3 & 2 & 5 \\ 0 & 1 & 3 & 6 \\ 0 & 0 & 0 & 0 \end{pmatrix}$

■

Note. The row echelon form of a matrix might not be unique.

Remark 1. The difference between these two forms should be clear from the examples. In row echelon form, all the numbers below the first 1 in a row are zero. In reduced row echelon form, all the numbers above and below the first 1 in a row are zero. Thus reduced row echelon form is more exclusive. That is, every matrix in reduced row echelon form is in row echelon form, but not conversely.

Remark 2. We can always reduce a matrix to reduced row echelon form or row echelon form by performing elementary row operations. We saw this reduction to reduced row echelon form in Examples 1, 2, and 3.

As we saw in Examples 1, 2, and 3, there is a strong connection between the reduced row echelon form of a matrix and the existence of a unique solution to the system. In Example 1 the reduced row echelon form of the coefficient matrix (that is, the first three columns of the augmented matrix) had a 1 in each row and there was a unique solution. In Examples 2 and 3 the reduced row echelon form of the coefficient matrix had a row of zeros and the system had either no solution or an infinite number of solutions. This turns out always to be true in any system with the same number of equations as unknowns. But before turning to the general case, let us discuss the usefulness of the row echelon form of a matrix. It is possible to solve the system in Example 1 by reducing the coefficient matrix to its row echelon form.

EXAMPLE 6   **Solving a System by Gaussian Elimination**   Solve the system of Example 1 by reducing the coefficient matrix to row echelon form.

Solution   We begin as before:

$$\begin{pmatrix} 2 & 4 & 6 & | & 18 \\ 4 & 5 & 6 & | & 24 \\ 3 & 1 & -2 & | & 4 \end{pmatrix} \xrightarrow{R_1 \to \frac{1}{2}R_1} \begin{pmatrix} 1 & 2 & 3 & | & 9 \\ 4 & 5 & 6 & | & 24 \\ 3 & 1 & -2 & | & 4 \end{pmatrix}$$

$$\xrightarrow[R_3 \to R_3 - 3R_1]{R_2 \to R_2 - 4R_1} \begin{pmatrix} 1 & 2 & 3 & | & 9 \\ 0 & -3 & -6 & | & -12 \\ 0 & -5 & -11 & | & -23 \end{pmatrix} \xrightarrow{R_3 \to -\frac{1}{3}R_3} \begin{pmatrix} 1 & 2 & 3 & | & 9 \\ 0 & 1 & 2 & | & 4 \\ 0 & -5 & -11 & | & -23 \end{pmatrix}$$

So far, this process is identical to our earlier one. Now, however, we only make zero the number (-5) below the first 1 in the second row:

$$\xrightarrow{R_3 \to R_3 + 5R_2} \begin{pmatrix} 1 & 2 & 3 & | & 9 \\ 0 & 1 & 2 & | & 4 \\ 0 & 0 & -1 & | & -3 \end{pmatrix} \xrightarrow{R_3 \to -R_3} \begin{pmatrix} 1 & 2 & 3 & | & 9 \\ 0 & 1 & 2 & | & 4 \\ 0 & 0 & 1 & | & 3 \end{pmatrix}$$

Back substitution

Gaussian elimination

The augmented matrix of the system (and the coefficient matrix) are now in row echelon form and we immediately see that $x_3 = 3$. We then use **back substitution** to solve for x_2 and then x_1. The second equation reads $x_2 + 2x_3 = 4$. Thus $x_2 + 2(3) = 4$ and $x_2 = -2$. Similarly, from the first equation we obtain $x_1 + 2(-2) + 3(3) = 9$ or $x_1 = 4$. Thus we again obtain the solution $(4, -2, 3)$. The method of solution just employed is called **Gaussian elimination.** ∎

We therefore have two methods for solving our sample systems of equations:

> i. **Gauss-Jordan Elimination**
>   Row-reduce the coefficient matrix to reduced row echelon form.
>
> ii. **Gaussian Elimination**
>   Row-reduce the coefficient matrix to row echelon form, solve for the last unknown, and then use back substitution to solve for the other unknowns.

Which method is more useful? It depends. In solving systems of equations on a computer, Gaussian elimination is the preferred method because it involves fewer elementary row operations. We shall discuss the numerical solution of systems of equations in Sections 7.2 and 7.3. On the other hand, there are times when it is essential to obtain the reduced row echelon form of a matrix (one of these is discussed in Section 1.9). In these cases Gauss-Jordan elimination is the preferred method.

We now turn to the solution of a general system of *m* equations in *n* unknowns. Because of our need to do so in Section 1.9, we shall be solving most of the systems by Gauss-Jordan elimination. Keep in mind, however, that Gaussian elimination is sometimes the preferred approach.

The general $m \times n$ system of *m* linear equations in *n* unknowns is given by

$$
\begin{aligned}
a_{11}x_1 + a_{12}x_2 + a_{13}x_3 + \cdots + a_{1n}x_n &= b_1 \\
a_{21}x_1 + a_{22}x_2 + a_{23}x_3 + \cdots + a_{2n}x_n &= b_2 \\
a_{31}x_1 + a_{32}x_2 + a_{33}x_3 + \cdots + a_{3n}x_n &= b_3 \\
\vdots \qquad \vdots \qquad \vdots \qquad \vdots \qquad \vdots \\
a_{m1}x_1 + a_{m2}x_2 + a_{m3}x_3 + \cdots + a_{mn}x_n &= b_m
\end{aligned}
\tag{7}
$$

In system (7) all the *a*'s and *b*'s are given real numbers. The problem is to find all sets of *n* numbers, denoted by $(x_1, x_2, x_3, \ldots, x_n)$, that satisfy every one of the *m* equations in (7). The number a_{ij} is the coefficient of the variable x_j in the *i*th equation.

We solve system (7) by writing the system as an augmented matrix and row-reducing the matrix to its reduced row echelon form. We start by dividing the first row by a_{11} [elementary row operation (*i*)]. If $a_{11} = 0$, then we rearrange† the equations so that, with rearrangement, the new $a_{11} \neq 0$. We then use the first equation to eliminate the x_1 term in each of the other equations [using elementary row operation (*ii*)]. Then the new second equation is divided by the new a_{22} term and the new, new second equation is used to eliminate the x_2 terms in all the other equations. The process is continued until one of three situations occurs:

---

i. The last nonzero‡ equation reads $x_n = c$ for some constant *c*. Then there is either a unique solution or an infinite number of solutions to the system.

ii. The last nonzero equation reads $a'_{ij}x_j + a'_{i,j+1}x_{j+1} + \cdots + a'_{i,j+k}x_n = c$ for some constant *c* where at least two of the *a*'s are nonzero. That is, the last equation is a linear equation in two or more of the variables. Then there are an infinite number of solutions.

Inconsistent system

iii. The last equation reads $0 = c$, where $c \neq 0$. Then there is no solution. In this case the system is called **inconsistent.** In cases (*i*) and (*ii*) the system is called **consistent.**

Consistent system

---

†To rearrange a system of equations we simply write the same equations in a different order. For example, the first equation can become the fourth equation, the third equation can become the second equation, and so on. This is a sequence of elementary row operations (*iii*).

‡The "zero equation" is the equation $0 = 0$.

EXAMPLE 7   **Solving a System of Two Equations in Four Unknowns**   Solve the system

$$x_1 + 3x_2 - 5x_3 + x_4 = 4$$
$$2x_1 + 5x_2 - 2x_3 + 4x_4 = 6$$

Solution   We write this system as an augmented matrix and row-reduce:

$$\begin{pmatrix} 1 & 3 & -5 & 1 & | & 4 \\ 2 & 5 & -2 & 4 & | & 6 \end{pmatrix} \xrightarrow{R_2 \to R_2 - 2R_1} \begin{pmatrix} 1 & 3 & -5 & 1 & | & 4 \\ 0 & -1 & 8 & 2 & | & -2 \end{pmatrix}$$

$$\xrightarrow{R_2 \to -R_2} \begin{pmatrix} 1 & 3 & -5 & 1 & | & 4 \\ 0 & 1 & -8 & -2 & | & 2 \end{pmatrix} \xrightarrow{R_1 \to R_1 - 3R_2} \begin{pmatrix} 1 & 0 & 19 & 7 & | & -2 \\ 0 & 1 & -8 & -2 & | & 2 \end{pmatrix}$$

This is as far as we can go. The coefficient matrix is in reduced row echelon form—case (*ii*) above. There are evidently an infinite number of solutions. The variables x_3 and x_4 can be chosen arbitrarily. Then $x_2 = 2 + 8x_3 + 2x_4$ and $x_1 = -2 - 19x_3 - 7x_4$. All solutions are, therefore, represented by $(-2 - 19x_3 - 7x_4, 2 + 8x_3 + 2x_4, x_3, x_4)$. For example, if $x_3 = 1$ and $x_4 = 2$, we obtain the solution $(-35, 14, 1, 2)$.   ■

As you will see if you do a lot of system solving, the computations can become very messy. It is a good rule of thumb to use a calculator whenever the fractions become unpleasant. It should be noted, however, that if computations are carried out on a computer or calculator, "round-off" errors can be introduced. This problem is discussed in Section 7.1.

We close this section with three examples illustrating how a system of linear equations can arise in a practical situation.

EXAMPLE 8   **The Leontief Input-Output Model**   A model that is often used in economics is the **Leontief input-output model**.† Suppose an economic system has *n* industries. There are two kinds of demands on each industry. First there is the *external* demand from outside the system. If the system is a country, for example, then the external demand could be from another country. Second there is the demand placed on one industry by another industry in the same system. In the United States, for example, there is a demand on the output of the steel industry by the automobile industry.

Let e_i represent the external demand placed on the *i*th industry. Let a_{ij} represent the internal demand placed on the *i*th industry by the *j*th industry. More precisely, a_{ij} represents the number of units of the output of industry *i* needed to produce 1 unit of the output of industry *j*. Let x_i represent the output of industry *i*. Now we assume that the output of each industry is equal to its demand (that is, there is no overpro-

---

†Named after American economist Wassily W. Leontief. This model was used in his pioneering paper "Qualitative Input and Output Relations in the Economic System of the United States" in *Review of Economic Statistics* 18(1936):105–125. An updated version of this model appears in Leontief's book *Input-Output Analysis* (New York: Oxford University Press, 1966). Leontief won the Nobel Prize in economics in 1973 for his development of input-output analysis.

duction). The total demand is equal to the sum of the internal and external demands. To calculate the internal demand on industry 2, for example, we note that industry 1 needs a_{21} units of the output of industry 2 to produce 1 unit of its output. If the output from industry 1 is x_1, then $a_{21}x_1$ is the total amount industry 1 needs from industry 2. Thus the total internal demand on industry 2 is $a_{21}x_1 + a_{22}x_2 + \cdots + a_{2n}x_n$.

We are led to the following system of equations obtained by equating the total demand with the output of each industry:

$$
\begin{aligned}
a_{11}x_1 + a_{12}x_2 + \cdots + a_{1n}x_n + e_1 &= x_1 \\
a_{21}x_1 + a_{22}x_2 + \cdots + a_{2n}x_n + e_2 &= x_2 \\
\vdots \qquad \vdots \qquad \qquad \vdots \qquad \vdots \quad & \quad \vdots \\
a_{n1}x_1 + a_{n2}x_2 + \cdots + a_{nn}x_n + e_n &= x_n
\end{aligned}
\tag{8}
$$

Or, rewriting (8) so it looks like system (7), we get

$$
\begin{aligned}
(1 - a_{11})x_1 - a_{12}x_2 \qquad - \cdots - \qquad a_{1n}x_n &= e_1 \\
-a_{21}x_1 + (1 - a_{22})x_2 - \cdots - \qquad a_{2n}x_n &= e_2 \\
\vdots \qquad \qquad \vdots \qquad \qquad \qquad \vdots \qquad & \quad \vdots \\
-a_{n1}x_1 - a_{n2}x_2 \qquad - \cdots + (1 - a_{nn})x_n &= e_n
\end{aligned}
\tag{9}
$$

System (9) of n equations in n unknowns is very important in economic analysis.

∎

EXAMPLE 9   **The Leontief Model Applied to an Economic System with Three Industries**
In an economic system with three industries, suppose that the external demands are, respectively, 10, 25, and 20. Suppose that $a_{11} = 0.2$, $a_{12} = 0.5$, $a_{13} = 0.15$, $a_{21} = 0.4$, $a_{22} = 0.1$, $a_{23} = 0.3$, $a_{31} = 0.25$, $a_{32} = 0.5$, and $a_{33} = 0.15$. Find the output in each industry such that supply exactly equals demand.

Solution   Here $n = 3$, $1 - a_{11} = 0.8$, $1 - a_{22} = 0.9$, and $1 - a_{33} = 0.85$. Then system (9) is

$$
\begin{aligned}
0.8x_1 - 0.5x_2 - 0.15x_3 &= 10 \\
-0.4x_1 + 0.9x_2 - 0.3x_3 &= 25 \\
-0.25x_1 - 0.5x_2 + 0.85x_3 &= 20
\end{aligned}
$$

Solving this system by using a calculator, we obtain, successively (using five-decimal-place accuracy and Gauss-Jordan elimination),

$$
\left(
\begin{array}{ccc|c}
0.8 & -0.5 & -0.15 & 10 \\
-0.4 & 0.9 & -0.3 & 25 \\
-0.25 & -0.5 & 0.85 & 20
\end{array}
\right)
\xrightarrow{R_1 \to \frac{1}{0.8}R_1}
\left(
\begin{array}{ccc|c}
1 & -0.625 & -0.1875 & 12.5 \\
-0.4 & 0.9 & -0.3 & 25 \\
-0.25 & -0.5 & 0.85 & 20
\end{array}
\right)
$$

$$
\xrightarrow[R_3 \to R_3 + 0.25R_1]{R_2 \to R_2 + 0.4R_1}
\left(
\begin{array}{ccc|c}
1 & -0.625 & -0.1875 & 12.5 \\
0 & 0.65 & -0.375 & 30 \\
0 & -0.65625 & 0.80313 & 23.125
\end{array}
\right)
$$

$$\xrightarrow{R_2 \to \frac{1}{0.65}R_2}\begin{pmatrix} 1 & -0.625 & -0.1875 & 12.5 \\ 0 & 1 & -0.57692 & 46.15385 \\ 0 & -0.65625 & 0.80313 & 23.125 \end{pmatrix}$$

$$\xrightarrow[R_3 \to R_3 + 0.65625R_2]{R_1 \to R_1 + 0.625R_2}\begin{pmatrix} 1 & 0 & -0.54808 & 41.34616 \\ 0 & 1 & -0.57692 & 46.15385 \\ 0 & 0 & 0.42453 & 53.41346 \end{pmatrix}$$

$$\xrightarrow{R_3 \to \frac{1}{0.42453}R_3}\begin{pmatrix} 1 & 0 & -0.54808 & 41.34616 \\ 0 & 1 & -0.57692 & 46.15385 \\ 0 & 0 & 1 & 125.81787 \end{pmatrix}$$

$$\xrightarrow[R_2 \to R_2 + 0.57692R_3]{R_1 \to R_1 + 0.54808R_3}\begin{pmatrix} 1 & 0 & 0 & 110.30442 \\ 0 & 1 & 0 & 118.74070 \\ 0 & 0 & 1 & 125.81787 \end{pmatrix}$$

We conclude that the outputs needed for supply to equal demand are, approximately, $x_1 \approx 110$, $x_2 \approx 119$, and $x_3 \approx 126$. ∎

EXAMPLE 10   **A Problem in Resource Management**   A State Fish and Game Department supplies three types of food to a lake that supports three species of fish. Each fish of Species 1 consumes, each week, an average of 1 unit of Food 1, 1 unit of Food 2, and 2 units of Food 3. Each fish of Species 2 consumes, each week, an average of 3 units of Food 1, 4 units of Food 2, and 5 units of Food 3. For a fish of Species 3, the average weekly consumption is 2 units of Food 1, 1 unit of Food 2, and 5 units of Food 3. Each week 25,000 units of Food 1, 20,000 units of Food 2, and 55,000 units of Food 3 are supplied to the lake. If we assume that all food is eaten, how many fish of each species can coexist in the lake?

Solution   We let x_1, x_2, and x_3 denote the numbers of fish of the three species being supported by the lake environment. Using the information in the problem, we see that x_1 fish of Species 1 consume x_1 units of Food 1, x_2 fish of Species 2 consume $3x_2$ units of Food 1, and x_3 fish of Species 3 consume $2x_3$ units of Food 1. Thus $x_1 + 3x_2 + 2x_3 = 25{,}000 =$ total weekly supply of Food 1. Obtaining a similar equation for each of the other two foods, we are led to the following system:

$$\begin{aligned} x_1 + 3x_2 + 2x_3 &= 25{,}000 \\ x_1 + 4x_2 + x_3 &= 20{,}000 \\ 2x_1 + 5x_2 + 5x_3 &= 55{,}000 \end{aligned}$$

Upon solving, we obtain

$$\begin{pmatrix} 1 & 3 & 2 & 25{,}000 \\ 1 & 4 & 1 & 20{,}000 \\ 2 & 5 & 5 & 55{,}000 \end{pmatrix}$$

$$\xrightarrow[\substack{R_2 \to R_2 - R_1 \\ R_3 \to R_3 - 2R_1}]{} \begin{pmatrix} 1 & 3 & 2 & 25{,}000 \\ 0 & 1 & -1 & -5{,}000 \\ 0 & -1 & 1 & 5{,}000 \end{pmatrix} \xrightarrow[\substack{R_1 \to R_1 - 3R_2 \\ R_3 \to R_3 + R_2}]{} \begin{pmatrix} 1 & 0 & 5 & 40{,}000 \\ 0 & 1 & -1 & -5{,}000 \\ 0 & 0 & 0 & 0 \end{pmatrix}$$

Thus, if x_3 is chosen arbitrarily, we have an infinite number of solutions given by $(40{,}000 - 5x_3, x_3 - 5{,}000, x_3)$. Of course, we must have $x_1 \geq 0$, $x_2 \geq 0$ and $x_3 \geq 0$. Since $x_2 = x_3 - 5{,}000 \geq 0$, we have $x_3 \geq 5{,}000$. This means that $0 \leq x_1 \leq 40{,}000 - 5(5{,}000) = 15{,}000$. Finally, since $40{,}000 - 5x_3 \geq 0$, we see that $x_3 \leq 8{,}000$. This means that the populations that can be supported by the lake with all food consumed are

$$x_1 = 40{,}000 - 5x_3$$
$$x_2 = x_3 - 5{,}000$$
$$5{,}000 \leq x_3 \leq 8{,}000$$

For example, if $x_3 = 6{,}000$, then $x_1 = 10{,}000$ and $x_2 = 1{,}000$.

Note. The system of equations does have an infinite number of solutions. However, the resource management problem has only a finite number of solutions because x_1, x_2, and x_3 must be integers and there are only 3001 integers in the interval [5000, 8000]. (You can't stock 5237.578 fish, for example.)   ∎

Focus on . . .

Carl Friedrich Gauss, 1777–1855

Carl Friedrich Gauss
(Library of Congress)

The greatest mathematician of the nineteenth century, Carl Friedrich Gauss is considered one of the three greatest mathematicians of all time—the others being Archimedes and Newton.

Gauss was born in Brunswick, Germany, in 1777. His father, a hard-working laborer who was exceptionally stubborn and did not believe in formal education, did what he could to keep Gauss from appropriate schooling. Fortunately for Carl (and for mathematics), his mother, while uneducated herself, encouraged her son in his studies and took considerable pride in his achievements until her death at the age of 97.

Gauss was a child prodigy. At the age of 3, he found an error in his father's bookkeeping. A famous story tells of Carl, age 10, as a student in the local Brunswick school. The teacher there was known to assign tasks to keep his pupils busy. One day he asked his students to add the numbers from 1 to 100. Almost at once, Carl placed his slate face down with the words "There it is." Afterword, the teacher found that Gauss was the only one with the correct answer, 5050. Gauss had noticed that the numbers could be arranged in 50 pairs, each with the sum 101 (1 + 100, 2 + 99, and so on), and 50 × 101 = 5050. Later in life, Gauss joked that he could add before he could speak.

When Gauss was 15, the Duke of Brunswick noticed him and became his patron. The duke helped him to enter Brunswick College in 1795 and, three years later, to enter the university at Göttingen. Undecided between careers in mathematics and philosophy, Gauss chose mathematics after two remarkable discoveries. First, he invented the method of least squares a decade before the result was published by Legendre. Second, a month before his nineteenth birthday, he solved a problem whose solution had been sought for more than two thousand years. Gauss showed how to construct, using compass and ruler, a regular polygon with the number of sides not a multiple of 2, 3, or 5. On March 30, 1796, the day of this discovery, he began a diary, which contained as its first entry rules for construction of a 17-sided regular polygon. The diary, which contains 146 statements of results in only 19 pages, is one of the most important documents in the history of mathematics.

After a short period at Göttingen, Gauss went to the University of Helmstädt and, in 1798 at the age of 20, wrote his now famous doctoral dissertation. In it he gave the first mathematically rigorous proof of the fundamental theorem of algebra—that every polynomial of degree n has, counting multiplicities, exactly n roots. Many mathematicians, including Euler, Newton, and Lagrange, had attempted to prove this result.

Gauss made a great number of discoveries in physics as well as in mathematics. For example, in 1801 he used a new procedure to calculate, from very little data, the orbit of the planetoid Ceres. In 1833, he invented the electromagnetic telegraph with his colleague Wilhelm Weber (1804–1891). While he did brilliant work in astronomy and electricity, however, it was Gauss's mathematical output that was astonishing. He made fundamental contributions to algebra and geometry. In 1811, he discovered a result that led to the development of complex variable theory by Cauchy. We encounter him here in the Gauss-Jordan method of elimination. Students of numerical analysis study Gaussian quadrature—a technique for numerical integration.

Gauss became a professor of mathematics at Göttingen in 1807 and remained at that post until his death in 1855. Even after his death, his mathematical spirit remained to haunt nineteenth-century mathematicians. Often it turned out that an important new result was discovered earlier by Gauss and could be found in his unpublished notes.

In his mathematical writings, Gauss was a perfectionist and is probably the last mathematician who knew everything in his subject. Claiming that a cathedral was not a cathedral until the last piece of scaffolding was removed, he endeavored to make each of his published works complete, concise, and polished. He used a seal that pictured a tree carrying only a few fruit together with the motto *pauca sed matura* (few, but ripe). But Gauss also believed that mathematics must reflect the real world. At his death, Gauss was honored by a commemorative medal on which was inscribed "George V. King of Hanover to the Prince of Mathematicians."

PROBLEMS 1.3

In Problems 1–20 use Gauss-Jordan or Gaussian elimination to find all solutions, if any, to the given systems.

1.
$$\begin{aligned} x_1 - 2x_2 + 3x_3 &= 11 \\ 4x_1 + x_2 - x_3 &= 4 \\ 2x_1 - x_2 + 3x_3 &= 10 \end{aligned}$$

2.
$$\begin{aligned} -2x_1 + x_2 + 6x_3 &= 18 \\ 5x_1 \quad\quad + 8x_3 &= -16 \\ 3x_1 + 2x_2 - 10x_3 &= -3 \end{aligned}$$

3.
$$\begin{aligned} 3x_1 + 6x_2 - 6x_3 &= 9 \\ 2x_1 - 5x_2 + 4x_3 &= 6 \\ -x_1 + 16x_2 - 14x_3 &= -3 \end{aligned}$$

4.
$$\begin{aligned} 3x_1 + 6x_2 - 6x_3 &= 9 \\ 2x_1 - 5x_2 + 4x_3 &= 6 \\ 5x_1 + 28x_2 - 26x_3 &= -8 \end{aligned}$$

5.
$$\begin{aligned} x_1 + x_2 - x_3 &= 7 \\ 4x_1 - x_2 + 5x_3 &= 4 \\ 2x_1 + 2x_2 - 3x_3 &= 0 \end{aligned}$$

6.
$$\begin{aligned} x_1 + x_2 - x_3 &= 7 \\ 4x_1 - x_2 + 5x_3 &= 4 \\ 6x_1 + x_2 + 3x_3 &= 18 \end{aligned}$$

7.
$$\begin{aligned} x_1 + x_2 - x_3 &= 7 \\ 4x_1 - x_2 + 5x_3 &= 4 \\ 6x_1 + x_2 + 3x_3 &= 20 \end{aligned}$$

8.
$$\begin{aligned} x_1 - 2x_2 + 3x_3 &= 0 \\ 4x_1 + x_2 - x_3 &= 0 \\ 2x_1 - x_2 + 3x_3 &= 0 \end{aligned}$$

9.
$$\begin{aligned} x_1 + x_2 - x_3 &= 0 \\ 4x_1 - x_2 + 5x_3 &= 0 \\ 6x_1 + x_2 + 3x_3 &= 0 \end{aligned}$$

10.
$$\begin{aligned} 2x_2 + 5x_3 &= 6 \\ x_1 \quad\quad - 2x_3 &= 4 \\ 2x_1 + 4x_2 \quad\quad &= -2 \end{aligned}$$

11.
$$\begin{aligned} x_1 + 2x_2 - x_3 &= 4 \\ 3x_1 + 4x_2 - 2x_3 &= 7 \end{aligned}$$

12.
$$\begin{aligned} x_1 + 2x_2 - 4x_3 &= 4 \\ -2x_1 - 4x_2 + 8x_3 &= -8 \end{aligned}$$

13.
$$\begin{aligned} x_1 + 2x_2 - 4x_3 &= 4 \\ -2x_1 - 4x_2 + 8x_3 &= -9 \end{aligned}$$

14.
$$\begin{aligned} x_1 + 2x_2 - x_3 + x_4 &= 7 \\ 3x_1 + 6x_2 - 3x_3 + 3x_4 &= 21 \end{aligned}$$

15.
$$\begin{aligned} 2x_1 + 6x_2 - 4x_3 + 2x_4 &= 4 \\ x_1 \quad\quad - x_3 + x_4 &= 5 \\ -3x_1 + 2x_2 - 2x_3 \quad\quad &= -2 \end{aligned}$$

16.
$$\begin{aligned} x_1 - 2x_2 + x_3 + x_4 &= 2 \\ 3x_1 \quad\quad + 2x_3 - 2x_4 &= -8 \\ 4x_2 - x_3 - x_4 &= 1 \\ -x_1 + 6x_2 - 2x_3 \quad\quad &= 7 \end{aligned}$$

17.
$$\begin{aligned} x_1 - 2x_2 + x_3 + x_4 &= 2 \\ 3x_1 \quad\quad + 2x_3 - 2x_4 &= -8 \\ 4x_2 - x_3 - x_4 &= 1 \\ 5x_1 \quad\quad + 3x_3 - x_4 &= -3 \end{aligned}$$

18.
$$\begin{aligned} x_1 - 2x_2 + x_3 + x_4 &= 2 \\ 3x_1 \quad\quad + 2x_3 - 2x_4 &= -8 \\ 4x_2 - x_3 - x_4 &= 1 \\ 5x_1 \quad\quad + 3x_3 - x_4 &= 0 \end{aligned}$$

19.
$$\begin{aligned} x_1 + x_2 &= 4 \\ 2x_1 - 3x_2 &= 7 \\ 3x_1 + 2x_2 &= 8 \end{aligned}$$

20.
$$\begin{aligned} x_1 + x_2 &= 4 \\ 2x_1 - 3x_2 &= 7 \\ 3x_1 - 2x_2 &= 11 \end{aligned}$$

In Problems 21–29 determine whether the given matrix is in row echelon form (but not reduced row echelon form), reduced row echelon form, or neither.

21. $\begin{pmatrix} 1 & 1 & 0 \\ 0 & 1 & 0 \\ 0 & 0 & 1 \end{pmatrix}$

22. $\begin{pmatrix} 2 & 0 & 0 \\ 0 & 1 & 0 \\ 0 & 0 & -1 \end{pmatrix}$

23. $\begin{pmatrix} 1 & 0 & 1 & 0 \\ 0 & 1 & 1 & 0 \\ 0 & 0 & 0 & 0 \end{pmatrix}$

24. $\begin{pmatrix} 1 & 0 & 0 & 0 \\ 0 & 0 & 1 & 0 \\ 0 & 0 & 0 & 1 \end{pmatrix}$

25. $\begin{pmatrix} 0 & 1 & 0 & 0 \\ 1 & 0 & 0 & 0 \\ 0 & 0 & 0 & 0 \end{pmatrix}$

26. $\begin{pmatrix} 1 & 0 & 1 & 2 \\ 0 & 1 & 3 & 4 \end{pmatrix}$

27. $\begin{pmatrix} 1 & 0 \\ 0 & 1 \\ 0 & 0 \end{pmatrix}$
    28. $\begin{pmatrix} 1 & 0 & 0 \\ 0 & 0 & 0 \\ 0 & 0 & 1 \end{pmatrix}$
    29. $\begin{pmatrix} 1 & 0 & 0 & 4 \\ 0 & 1 & 0 & 5 \\ 0 & 1 & 1 & 6 \end{pmatrix}$

In Problems 30–35 use the elementary row operations to reduce the given matrices to row echelon form and reduced row echelon form.

30. $\begin{pmatrix} 1 & 1 \\ 2 & 3 \end{pmatrix}$
    31. $\begin{pmatrix} -1 & 6 \\ 4 & 2 \end{pmatrix}$
    32. $\begin{pmatrix} 1 & -1 & 1 \\ 2 & 4 & 3 \\ 5 & 6 & -2 \end{pmatrix}$

33. $\begin{pmatrix} 2 & -4 & 8 \\ 3 & 5 & 8 \\ -6 & 0 & 4 \end{pmatrix}$
    34. $\begin{pmatrix} 2 & -4 & -2 \\ 3 & 1 & 6 \end{pmatrix}$
    35. $\begin{pmatrix} 2 & -7 \\ 3 & 5 \\ 4 & -3 \end{pmatrix}$

36. In the Leontief input-output model of Example 8 suppose that there are three industries. Suppose further that $e_1 = 10$, $e_2 = 15$, $e_3 = 30$, $a_{11} = \frac{1}{3}$, $a_{12} = \frac{1}{2}$, $a_{13} = \frac{1}{6}$, $a_{21} = \frac{1}{4}$, $a_{22} = \frac{1}{4}$, $a_{23} = \frac{1}{8}$, $a_{31} = \frac{1}{12}$, $a_{32} = \frac{1}{3}$, and $a_{33} = \frac{1}{6}$. Find the output of each industry such that supply exactly equals demand.

37. In Example 10 assume that there are 15,000 units of the first food, 10,000 units of the second, and 35,000 units of the third supplied to the lake each week. Assuming that all three foods are consumed, what populations of the three species can coexist in the lake? Is there a unique solution?

38. A traveler who just returned from Europe spent $30 a day for housing in England, $20 a day in France, and $20 a day in Spain. For food the traveler spent $20 a day in England, $30 a day in France, and $20 a day in Spain. The traveler spent $10 a day in each country for incidental expenses. The traveler's records of the trip indicate a total of $340 spent for housing, $320 for food, and $140 for incidental expenses while traveling in these countries. Calculate the number of days the traveler spent in each of the countries or show that the records must be incorrect because the amounts spent are incompatible with each other.

39. An investor remarks to a stockbroker that all her stock holdings are in three companies, Eastern Airlines, Hilton Hotels, and McDonald's, and that 2 days ago the value of her stocks went down $350 but yesterday the value increased by $600. The broker recalls that 2 days ago the price of Eastern Airlines stock dropped by $1 a share, Hilton Hotels dropped $1.50, but the price of McDonald's stock rose by $0.50. The broker also remembers that yesterday the price of Eastern Airlines stock rose $1.50, there was a further drop of $0.50 a share in Hilton Hotels stock, and McDonald's stock rose $1. Show that the broker does not have enough information to calculate the number of shares the investor owns of each company's stock, but that when the investor says that she owns 200 shares of McDonald's stock, the broker can calculate the number of shares of Eastern Airlines and Hilton Hotels.

40. An intelligence agent knows that 60 aircraft, consisting of fighter planes and bombers, are stationed at a certain secret airfield. The agent wishes to determine how many of the 60 are fighter planes and how many are bombers. There is a type of rocket carried by both sorts of planes; the fighter carries six of these rockets, the bomber only two. The agent learns that 250 rockets are required to arm every plane at this airfield. Furthermore, the agent overhears a remark that there are twice as many fighter planes as bombers at the base (that is, the number of fighter planes minus twice the number of bombers equals zero). Calculate the number of fighter planes and bombers at the airfield or show that the agent's information must be incorrect, because it is inconsistent.

41. Consider the system

$$2x_1 - x_2 + 3x_3 = a$$
$$3x_1 + x_2 - 5x_3 = b$$
$$-5x_1 - 5x_2 + 21x_3 = c$$

Show that the system is inconsistent if $c \neq 2a - 3b$.

42. Consider the system

$$2x_1 + 3x_2 - x_3 = a$$
$$x_1 - x_2 + 3x_3 = b$$
$$3x_1 + 7x_2 - 5x_3 = c$$

Find conditions on a, b, and c such that the system is consistent.

*43. Consider the general system of three linear equations in three unknowns:

$$a_{11}x_1 + a_{12}x_2 + a_{13}x_3 = b_1$$
$$a_{21}x_1 + a_{22}x_2 + a_{23}x_3 = b_2$$
$$a_{31}x_1 + a_{32}x_2 + a_{33}x_3 = b_3$$

Find conditions on the coefficients a_{ij} such that the system has a unique solution.

44. Solve the following system using a hand calculator and carrying five decimal places of accuracy:

$$2x_2 - x_3 - 4x_4 = 2$$
$$x_1 - x_2 + 5x_3 + 2x_4 = -4$$
$$3x_1 + 3x_2 - 7x_3 - x_4 = 4$$
$$-x_1 - 2x_2 + 3x_3 = -7$$

45. Do the same for the system

$$3.8x_1 + 1.6x_2 + 0.9x_3 = 3.72$$
$$-0.7x_1 + 5.4x_2 + 1.6x_3 = 3.16$$
$$1.5x_1 + 1.1x_2 - 3.2x_3 = 43.78$$

1.4 HOMOGENEOUS SYSTEMS OF EQUATIONS

The general $m \times n$ system of linear equations [system (1.3.7), page 15] is called **homogeneous** if all the constants $b_1, b_2, \ldots, b_m$ are zero. That is, the general homogeneous system is given by

$$
\begin{aligned}
a_{11}x_1 + a_{12}x_2 + \cdots + a_{1n}x_n &= 0 \\
a_{21}x_1 + a_{22}x_2 + \cdots + a_{2n}x_n &= 0 \\
\vdots \qquad \vdots \qquad\qquad \vdots \qquad &\ \ \vdots \\
a_{m1}x_1 + a_{m2}x_2 + \cdots + a_{mn}x_n &= 0
\end{aligned}
\tag{1}
$$

Homogeneous systems arise in a variety of ways. We shall see one of these in Section 4.4. In this section we solve some homogeneous systems—again by the method of Gauss-Jordan elimination.

For the general linear system there are three possibilities: no solution, one solution, or an infinite number of solutions. For the general homogeneous system the situation is simpler. Since $x_1 = x_2 = \cdots = x_n = 0$ is always a solution (called

Trivial solution
Zero solution

the **trivial solution** or **zero solution**), there are only two possibilities: Either the zero solution is the only solution or there are an infinite number of solutions in addition to the zero solution. Solutions other than the zero solution are called **non-**

Nontrivial solution

trivial solutions.

EXAMPLE 1   **A Homogeneous System with Only the Zero Solution**   Solve the homogeneous system

$$2x_1 + 4x_2 + 6x_3 = 0$$
$$4x_1 + 5x_2 + 6x_3 = 0$$
$$3x_1 + x_2 - 2x_3 = 0$$

Solution   This is the homogeneous version of the system in Example 1.3.1, page 7. Reducing successively, we obtain (after dividing the first equation by 2)

$$\begin{pmatrix} 1 & 2 & 3 & | & 0 \\ 4 & 5 & 6 & | & 0 \\ 3 & 1 & -2 & | & 0 \end{pmatrix} \xrightarrow[\substack{R_2 \to R_2 - 4R_1 \\ R_3 \to R_3 - 3R_1}]{} \begin{pmatrix} 1 & 2 & 3 & | & 0 \\ 0 & -3 & -6 & | & 0 \\ 0 & -5 & -11 & | & 0 \end{pmatrix} \xrightarrow[R_2 \to -\frac{1}{3}R_2]{} \begin{pmatrix} 1 & 2 & 3 & | & 0 \\ 0 & 1 & 2 & | & 0 \\ 0 & -5 & -11 & | & 0 \end{pmatrix}$$

$$\xrightarrow[\substack{R_1 \to R_1 - 2R_2 \\ R_3 \to R_3 + 5R_2}]{} \begin{pmatrix} 1 & 0 & -1 & | & 0 \\ 0 & 1 & 2 & | & 0 \\ 0 & 0 & -1 & | & 0 \end{pmatrix} \xrightarrow[R_3 \to -R_3]{} \begin{pmatrix} 1 & 0 & -1 & | & 0 \\ 0 & 1 & 2 & | & 0 \\ 0 & 0 & 1 & | & 0 \end{pmatrix} \xrightarrow[\substack{R_1 \to R_1 + R_3 \\ R_2 \to R_2 - 2R_3}]{} \begin{pmatrix} 1 & 0 & 0 & | & 0 \\ 0 & 1 & 0 & | & 0 \\ 0 & 0 & 1 & | & 0 \end{pmatrix}$$

Thus the system has the unique solution $(0, 0, 0)$. That is, the system has only the trivial solution.   ∎

EXAMPLE 2   **A Homogeneous System with an Infinite Number of Solutions**   Solve the homogeneous system

$$x_1 + 2x_2 - x_3 = 0$$
$$3x_1 - 3x_2 + 2x_3 = 0$$
$$-x_1 - 11x_2 + 6x_3 = 0$$

Solution   Using Gauss-Jordan elimination, we obtain, successively,

$$\begin{pmatrix} 1 & 2 & -1 & | & 0 \\ 3 & -3 & 2 & | & 0 \\ -1 & -11 & 6 & | & 0 \end{pmatrix} \xrightarrow[\substack{R_2 \to R_2 - 3R_1 \\ R_3 \to R_3 + R_1}]{} \begin{pmatrix} 1 & 2 & -1 & | & 0 \\ 0 & -9 & 5 & | & 0 \\ 0 & -9 & 5 & | & 0 \end{pmatrix}$$

$$\xrightarrow[R_2 \to -\frac{1}{9}R_2]{} \begin{pmatrix} 1 & 2 & -1 & | & 0 \\ 0 & 1 & -\frac{5}{9} & | & 0 \\ 0 & -9 & 5 & | & 0 \end{pmatrix} \xrightarrow[\substack{R_1 \to R_1 - 2R_2 \\ R_3 \to R_3 + 9R_2}]{} \begin{pmatrix} 1 & 0 & \frac{1}{9} & | & 0 \\ 0 & 1 & -\frac{5}{9} & | & 0 \\ 0 & 0 & 0 & | & 0 \end{pmatrix}$$

The augmented matrix is now in reduced row echelon form and, evidently, there are an infinite number of solutions given by $(-\frac{1}{9}x_3, \frac{5}{9}x_3, x_3)$. If $x_3 = 0$, for example, we obtain the trivial solution. If $x_3 = 1$ we obtain the solution $(-\frac{1}{9}, \frac{5}{9}, 1)$. ∎

EXAMPLE 3 **A Homogeneous System with More Unknowns Than Equations Has an Infinite Number of Solutions** Solve the system

$$
\begin{aligned}
x_1 + x_2 - x_3 &= 0 \\
4x_1 - 2x_2 + 7x_3 &= 0
\end{aligned}
\tag{2}
$$

Solution Row-reducing, we obtain

$$
\begin{pmatrix} 1 & 1 & -1 & \bigm| & 0 \\ 4 & -2 & 7 & \bigm| & 0 \end{pmatrix}
\xrightarrow{R_2 \to R_2 - 4R_1}
\begin{pmatrix} 1 & 1 & -1 & \bigm| & 0 \\ 0 & -6 & 11 & \bigm| & 0 \end{pmatrix}
$$

$$
\xrightarrow{R_2 \to -\frac{1}{6}R_2}
\begin{pmatrix} 1 & 1 & -1 & \bigm| & 0 \\ 0 & 1 & -\frac{11}{6} & \bigm| & 0 \end{pmatrix}
\xrightarrow{R_1 \to R_1 - R_2}
\begin{pmatrix} 1 & 0 & \frac{5}{6} & \bigm| & 0 \\ 0 & 1 & -\frac{11}{6} & \bigm| & 0 \end{pmatrix}
$$

Thus there are an infinite number of solutions given by $(-\frac{5}{6}x_3, \frac{11}{6}x_3, x_3)$. This is not surprising since system (2) contains three unknowns and only two equations. ∎

In fact, if there are more unknowns than equations, the homogeneous system (1) will always have an infinite number of solutions. To see this, note that if there were only the trivial solution, then row reduction would lead us to the system

$$
\begin{aligned}
x_1 \qquad\qquad &= 0 \\
x_2 \qquad\quad &= 0 \\
\vdots \\
x_n &= 0
\end{aligned}
$$

and, possibly, additional equations of the form $0 = 0$. But this system has at least as many equations as unknowns. Since row reduction does not change either the number of equations or the number of unknowns, we have a contradiction of our assumption that there were more unknowns than equations. Thus we have Theorem 1.

THEOREM 1 The homogeneous system (1) has an infinite number of solutions if $n > m$. ∎

PROBLEMS 1.4

In Problems 1–13 find all solutions to the homogeneous systems.

1. $\begin{aligned} 2x_1 - x_2 &= 0 \\ 3x_1 + 4x_2 &= 0 \end{aligned}$

2. $\begin{aligned} x_1 - 5x_2 &= 0 \\ -x_1 + 5x_2 &= 0 \end{aligned}$

3. $\begin{aligned} x_1 + x_2 - x_3 &= 0 \\ 2x_1 - 4x_2 + 3x_3 &= 0 \\ 3x_1 + 7x_2 - x_3 &= 0 \end{aligned}$

4. $\begin{aligned} x_1 + x_2 - x_3 &= 0 \\ 2x_1 - 4x_2 + 3x_3 &= 0 \\ -x_1 - 7x_2 + 6x_3 &= 0 \end{aligned}$

5. $x_1 + x_2 - x_3 = 0$
$2x_1 - 4x_2 + 3x_3 = 0$
$-5x_1 + 13x_2 - 10x_3 = 0$

6. $2x_1 + 3x_2 - x_3 = 0$
$6x_1 - 5x_2 + 7x_3 = 0$

7. $4x_1 - x_2 = 0$
$7x_1 + 3x_2 = 0$
$-8x_1 + 6x_2 = 0$

8. $x_1 - x_2 + 7x_3 - x_4 = 0$
$2x_1 + 3x_2 - 8x_3 + x_4 = 0$

9. $x_1 - 2x_2 + x_3 + x_4 = 0$
$3x_1 \quad\quad + 2x_3 - 2x_4 = 0$
$\quad\quad 4x_2 - x_3 - x_4 = 0$
$5x_1 \quad\quad + 3x_3 - x_4 = 0$

10. $-2x_1 \quad\quad\quad\quad + 7x_4 = 0$
$x_1 + 2x_2 - x_3 + 4x_4 = 0$
$3x_1 \quad\quad - x_3 + 5x_4 = 0$
$4x_1 + 2x_2 + 3x_3 \quad\quad = 0$

11. $2x_1 - x_2 = 0$
$3x_1 + 5x_2 = 0$
$7x_1 - 3x_2 = 0$
$-2x_1 + 3x_2 = 0$

12. $x_1 - 3x_2 = 0$
$-2x_1 + 6x_2 = 0$
$4x_1 - 12x_2 = 0$

13. $x_1 + x_2 - x_3 = 0$
$4x_1 - x_2 + 5x_3 = 0$
$-2x_1 + x_2 - 2x_3 = 0$
$3x_1 + 2x_2 - 6x_3 = 0$

14. Show that the homogeneous system

$$a_{11}x_1 + a_{12}x_2 = 0$$
$$a_{21}x_1 + a_{22}x_2 = 0$$

has an infinite number of solutions if and only if $a_{11}a_{22} - a_{12}a_{21} = 0$.

15. Consider the system

$$2x_1 - 3x_2 + 5x_3 = 0$$
$$-x_1 + 7x_2 - x_3 = 0$$
$$4x_1 - 11x_2 + kx_3 = 0$$

For what value of k will the system have nontrivial solutions?

*16. Consider the 3×3 homogeneous system

$$a_{11}x_1 + a_{12}x_2 + a_{13}x_3 = 0$$
$$a_{21}x_1 + a_{22}x_2 + a_{23}x_3 = 0$$
$$a_{31}x_1 + a_{32}x_2 + a_{33}x_3 = 0$$

Find conditions on the coefficients a_{ij} such that the zero solution is the only solution.

1.5 VECTORS

The study of vectors and matrices lies at the heart of linear algebra. The study of vectors began essentially with the work of the great Irish mathematician Sir William Rowan Hamilton (1805–1865).† His desire to find a way to represent certain ob-

---

† See the biographical sketch of Hamilton on page 33.

jects in the plane and in space led to the discovery of what he called *quaternions*. This notion led to the development of what we now call *vectors*. Throughout Hamilton's life, and for the remainder of the nineteenth century, there was considerable debate over the usefulness of quaternions and vectors. At the end of the century the great British physicist Lord Kelvin wrote that quaternions, "although beautifully ingenious, have been an unmixed evil to those who have touched them in any way [and] vectors . . . have never been of the slightest use to any creature."

But Kelvin was wrong. Today nearly all branches of classical and modern physics are represented by means of the language of vectors. Vectors are also used with increasing frequency in the social and biological sciences.†

On page 2 we described the solution to a system of two equations in two unknowns as a pair of numbers written (x, y). In Example 1.3.1 on page 7 we wrote the solution of a system of three equations in three unknowns as the triple of numbers $(4, -2, 3)$. Both (x, y) and $(4, -2, 3)$ are **vectors.**

n-Component Row Vector

We define an ***n*-component row vector** to be an **ordered** set of n numbers written as

$$(x_1, x_2, \ldots, x_n) \tag{1}$$

n-Component Column Vector

An ***n*-component column vector** is an **ordered** set of n numbers written as

$$\begin{pmatrix} x_1 \\ x_2 \\ \vdots \\ x_n \end{pmatrix} \tag{2}$$

Components of a vector

In (1) or (2), x_1 is called the **first component** of the vector, x_2 is the **second component,** and so on. In general, x_k is called the **kth component** of the vector.

For simplicity, we shall often refer to an n-component row vector as a **row vector** or an ***n*-vector.** Similarly, we shall use the term **column vector** (or n-vector) to denote an n-component column vector. Any vector whose entries are all zero is called a **zero vector.**

Zero vector

EXAMPLE 1   **Four Vectors**   The following are vectors:

i. $(3, 6)$ is a row vector (or a 2-vector).

---

† For interesting discussions of the development of modern vector analysis, consult the book by M. J. Crowe, *A History of Vector Analysis* (Notre Dame: University of Notre Dame Press, 1967) or Morris Kline's excellent book *Mathematical Thought from Ancient to Modern Times* (New York: Oxford University Press, 1972), Chapter 32.

ii. $\begin{pmatrix} 2 \\ -1 \\ 5 \end{pmatrix}$ is a column vector (or a 3-vector).

iii. $(2, -1, 0, 4)$ is a row vector (or a 4-vector).

iv. $\begin{pmatrix} 0 \\ 0 \\ 0 \\ 0 \\ 0 \end{pmatrix}$ is a column vector and a zero vector.

■

WARNING

The word "ordered" in the definition of a vector is essential. Two vectors with the same components written in different orders are *not* the same. Thus, for example, the row vectors $(1, 2)$ and $(2, 1)$ are not equal.   □

For the remainder of this text we shall denote vectors with boldface lowercase letters like **u**, **v**, **a**, **b**, **c**, and so on. A zero vector is denoted by **0**. Moreover, since it will usually be obvious whether a vector is a row or a column, we shall usually refer to row or column vectors simply as "vectors."

Vectors arise in a great number of ways. Suppose that the buyer for a manufacturing plant must order different quantities of steel, aluminum, oil, and paper. He can keep track of the quantities to be ordered with a single vector. The vector $\begin{pmatrix} 10 \\ 30 \\ 15 \\ 60 \end{pmatrix}$ indicates that he would order 10 units of steel, 30 units of aluminum, and so on.

Remark. We see here why the order in which the components of a vector are written is important. It is clear that the vectors $\begin{pmatrix} 30 \\ 15 \\ 60 \\ 10 \end{pmatrix}$ and $\begin{pmatrix} 10 \\ 30 \\ 15 \\ 60 \end{pmatrix}$ mean very different things to the buyer.

We now describe some properties of vectors. Since it would be repetitive to do so first for row vectors and then for column vectors, we shall give all definitions in terms of column vectors. Similar definitions hold for row vectors.

$\mathbb{R}$
$\mathbb{C}$
The components of all the vectors in this text are either real or complex numbers.† We denote the set of all real numbers by $\mathbb{R}$ and the set of all complex numbers by $\mathbb{C}$.

_____

† A complex number is a number of the form $a + ib$, where a and b are real numbers and $i = \sqrt{-1}$. A description of complex numbers is given in Appendix 2. We shall not encounter complex vectors again until Chapter 4; they will be especially useful in Chapter 6. Therefore, unless otherwise stated, we assume, for the time being, that all vectors have real components.

$\mathbb{R}^n$ We use the symbol $\mathbb{R}^n$ to denote the set of all n-vectors $\begin{pmatrix} a_1 \\ a_2 \\ \vdots \\ a_n \end{pmatrix}$, where each a_i is

$\mathbb{C}^n$ a real number. Similarly, we use the symbol $\mathbb{C}^n$ to denote the set of all n-vectors

$\begin{pmatrix} c_1 \\ c_2 \\ \vdots \\ c_n \end{pmatrix}$, where each c_i is a complex number. In Chapter 3 we shall discuss the sets $\mathbb{R}^2$

(vectors in the plane) and $\mathbb{R}^3$ (vectors in space). In Chapter 4 we shall examine arbitrary sets of vectors.

DEFINITION 1   **Equality of Vectors** Two column (or row) vectors **a** and **b** are **equal** if and only if†
they have the same number of components and their corresponding components are

equal. In symbols, the vectors $\mathbf{a} = \begin{pmatrix} a_1 \\ a_2 \\ \vdots \\ a_n \end{pmatrix}$ and $\mathbf{b} = \begin{pmatrix} b_1 \\ b_2 \\ \vdots \\ b_n \end{pmatrix}$ are equal if and only if

$a_1 = b_1, a_2 = b_2, \ldots, a_n = b_n.$

DEFINITION 2   **Addition of Vectors** Let $\mathbf{a} = \begin{pmatrix} a_1 \\ a_2 \\ \vdots \\ a_n \end{pmatrix}$ and $\mathbf{b} = \begin{pmatrix} b_1 \\ b_2 \\ \vdots \\ b_n \end{pmatrix}$ be n-vectors. Then the sum of **a**

and **b** is defined by

$$\mathbf{a} + \mathbf{b} = \begin{pmatrix} a_1 + b_1 \\ a_2 + b_2 \\ \vdots \\ a_n + b_n \end{pmatrix} \tag{3}$$

---

†The term "if and only if" applies to two statements. "Statement A if and only if statement B" means
that statements A and B are equivalent. That is, if statement A is true, then statement B is true, and if
statement B is true, then statement A is true. Put another way, it means that you cannot have one without
the other.

EXAMPLE 2   **The Sum of Two Vectors**

$$\begin{pmatrix} 1 \\ 2 \\ 4 \end{pmatrix} + \begin{pmatrix} -6 \\ 7 \\ 5 \end{pmatrix} = \begin{pmatrix} -5 \\ 9 \\ 9 \end{pmatrix}$$

∎

EXAMPLE 3   **The Sum of Two Vectors**

$$\begin{pmatrix} 2 \\ -1 \end{pmatrix} + \begin{pmatrix} -2 \\ 1 \end{pmatrix} = \begin{pmatrix} 0 \\ 0 \end{pmatrix}$$

∎

WARNING

It is essential that **a** and **b** have the same number of components. For example, the sum $\begin{pmatrix} 2 \\ 3 \end{pmatrix} + \begin{pmatrix} 1 \\ 2 \\ 3 \end{pmatrix}$ is not defined since 2-vectors and 3-vectors are different kinds of objects and cannot be added together. Moreover, it is not possible to add a row and a column vector together. For example, the sum $\begin{pmatrix} 1 \\ 2 \end{pmatrix} + (3, 5)$ is *not* defined.

□

Scalars

When dealing with vectors, we shall refer to numbers as **scalars** (which may be real or complex depending on whether the vectors in question are real or complex).

Historical Note. The term "scalar" originated with Hamilton. His definition of the quaternion included what he called a "real part" and an "imaginary part." In his paper "On Quaternions, or on a New System of Imaginaries in Algebra," in *Philosophical Magazine,* 3rd series, 25(1844):26–27, he wrote: "The algebraically *real* part may receive . . . all values contained on the one *scale* of progression of numbers from negative to positive infinity; we shall call it therefore the *scalar part,* or simply the *scalar* of the quaternion. . . ." In the same paper, Hamilton went on to define the imaginary part of his quaternion as the *vector* part. Although this was not the first usage of the word "vector," it was the first time it was used in the context of the definitions in this section. It is fair to say that the paper from which the preceding quotation was taken marks the beginning of modern vector analysis.

DEFINITION 3   **Multiplication of a Vector by a Scalar** Let $\mathbf{a} = \begin{pmatrix} a_1 \\ a_2 \\ \vdots \\ a_n \end{pmatrix}$ be a vector and α a scalar.

Then the product $\alpha\mathbf{a}$ is given by

$$\alpha \mathbf{a} = \begin{pmatrix} \alpha a_1 \\ \alpha a_2 \\ \vdots \\ \alpha a_n \end{pmatrix} \tag{4}$$

That is, to multiply a vector by a scalar, we simply multiply each component of the vector by the scalar.

EXAMPLE 4   **A Scalar Multiple of a Vector**

$$3 \begin{pmatrix} 2 \\ -1 \\ 4 \end{pmatrix} = \begin{pmatrix} 6 \\ -3 \\ 12 \end{pmatrix} \qquad \blacksquare$$

Note. Putting Definition 2 and Definition 3 together, we can define the difference of two vectors by

$$\mathbf{a} - \mathbf{b} = \mathbf{a} + (-1)\mathbf{b} \tag{5}$$

This means that if $\mathbf{a} = \begin{pmatrix} a_1 \\ a_2 \\ \vdots \\ a_n \end{pmatrix}$ and $\mathbf{b} = \begin{pmatrix} b_1 \\ b_2 \\ \vdots \\ b_n \end{pmatrix}$, then $\mathbf{a} - \mathbf{b} = \begin{pmatrix} a_1 - b_1 \\ a_2 - b_2 \\ \vdots \\ a_n - b_n \end{pmatrix}$

EXAMPLE 5   **The Sum of Scalar Multiples of Two Vectors**   Let $\mathbf{a} = \begin{pmatrix} 4 \\ 6 \\ 1 \\ 3 \end{pmatrix}$ and $\mathbf{b} = \begin{pmatrix} -2 \\ 4 \\ -3 \\ 0 \end{pmatrix}$.

Calculate $2\mathbf{a} - 3\mathbf{b}$.

Solution $2\mathbf{a} - 3\mathbf{b} = 2 \begin{pmatrix} 4 \\ 6 \\ 1 \\ 3 \end{pmatrix} + (-3) \begin{pmatrix} -2 \\ 4 \\ -3 \\ 0 \end{pmatrix} = \begin{pmatrix} 8 \\ 12 \\ 2 \\ 6 \end{pmatrix} + \begin{pmatrix} 6 \\ -12 \\ 9 \\ 0 \end{pmatrix} = \begin{pmatrix} 14 \\ 0 \\ 11 \\ 6 \end{pmatrix} \qquad \blacksquare$

Once we know how to add vectors and multiply them by scalars, we can prove a number of facts relating these operations. Several of these facts are given in Theorem 1 below. We prove parts (*ii*) and (*iii*) and leave the remaining parts as exercises (see Problems 21–23).

THEOREM 1   Let $\mathbf{a}$, $\mathbf{b}$, and $\mathbf{c}$ be n-vectors and let α and β be scalars. Then:

    i. $\mathbf{a} + \mathbf{0} = \mathbf{a}$

    ii. $0\mathbf{a} = \mathbf{0}$ (Note that the zero on the left is the number zero, whereas the zero on the right is a zero vector.)

    iii. $\mathbf{a} + \mathbf{b} = \mathbf{b} + \mathbf{a}$ (**commutative law**)

    iv. $(\mathbf{a} + \mathbf{b}) + \mathbf{c} = \mathbf{a} + (\mathbf{b} + \mathbf{c})$ (**associative law**)

    v. $\alpha(\mathbf{a} + \mathbf{b}) = \alpha\mathbf{a} + \alpha\mathbf{b}$ (**distributive law for scalar multiplication**)

    vi. $(\alpha + \beta)\mathbf{a} = \alpha\mathbf{a} + \beta\mathbf{a}$

    vii. $(\alpha\beta)\mathbf{a} = \alpha(\beta\mathbf{a})$

Proof of (ii) and (iii)

    ii. If $\mathbf{a} = \begin{pmatrix} a_1 \\ a_2 \\ \vdots \\ a_n \end{pmatrix}$, then $0\mathbf{a} = 0 \begin{pmatrix} a_1 \\ a_2 \\ \vdots \\ a_n \end{pmatrix} = \begin{pmatrix} 0 \cdot a_1 \\ 0 \cdot a_2 \\ \vdots \\ 0 \cdot a_n \end{pmatrix} = \begin{pmatrix} 0 \\ 0 \\ \vdots \\ 0 \end{pmatrix} = \mathbf{0}.$

    iii. Let $\mathbf{b} = \begin{pmatrix} b_1 \\ b_2 \\ \vdots \\ b_n \end{pmatrix}$. Then $\mathbf{a} + \mathbf{b} = \begin{pmatrix} a_1 + b_1 \\ a_2 + b_2 \\ \vdots \\ a_n + b_n \end{pmatrix} = \begin{pmatrix} b_1 + a_1 \\ b_2 + a_2 \\ \vdots \\ b_n + a_n \end{pmatrix} = \mathbf{b} + \mathbf{a}.$

Here we used the fact that for any two numbers x and y, $x + y = y + x$ and $0 \cdot x = 0$.   ∎

EXAMPLE 6   **Illustration of the Associative Law for Vector Addition**   To illustrate the associative law, we note that

$$\left[\begin{pmatrix} 3 \\ 1 \\ 2 \end{pmatrix} + \begin{pmatrix} -2 \\ 4 \\ -1 \end{pmatrix} \right] + \begin{pmatrix} 6 \\ -3 \\ 5 \end{pmatrix} = \begin{pmatrix} 1 \\ 5 \\ 1 \end{pmatrix} + \begin{pmatrix} 6 \\ -3 \\ 5 \end{pmatrix} = \begin{pmatrix} 7 \\ 2 \\ 6 \end{pmatrix}$$

while

$$\begin{pmatrix} 3 \\ 1 \\ 2 \end{pmatrix} + \left[\begin{pmatrix} -2 \\ 4 \\ -1 \end{pmatrix} + \begin{pmatrix} 6 \\ -3 \\ 5 \end{pmatrix} \right] = \begin{pmatrix} 3 \\ 1 \\ 2 \end{pmatrix} + \begin{pmatrix} 4 \\ 1 \\ 4 \end{pmatrix} = \begin{pmatrix} 7 \\ 2 \\ 6 \end{pmatrix}$$ ∎

    Example 6 illustrates the importance of the associative law of vector addition, since if we wish to add together three or more vectors, we can only do so by adding them together two at a time. The associative law tells us that we can do this in two different ways and still come up with the same answer. If this were not the case, the sum of three or more vectors would be more difficult to define since we would have to specify whether we wanted $(\mathbf{a} + \mathbf{b}) + \mathbf{c}$ or $\mathbf{a} + (\mathbf{b} + \mathbf{c})$ to be the definition of the sum $\mathbf{a} + \mathbf{b} + \mathbf{c}$.

Focus on . . .

Sir William Rowan Hamilton, 1805–1865

Sir William Rowan
Hamilton (*The
Granger Collection*)

Born in Dublin in 1805, where he spent most of his life, William Rowan Hamilton was without question Ireland's greatest mathematician. Hamilton's father (an attorney) and mother died when he was a small boy. His uncle, a linguist, took over the boy's education. By his fifth birthday Hamilton could read English, Hebrew, Latin, and Greek. By his thirteenth birthday he had mastered not only the languages of continental Europe, but also Sanscrit, Chinese, Persian, Arabic, Malay, Hindi, Bengali, and several others as well. Hamilton liked to write poetry, both as a child and as an adult, and his friends included the great English poets Samuel Taylor Coleridge and William Wordsworth. Hamilton's poetry was considered so bad, however, that it is fortunate that he developed other interests—especially in mathematics.

Although he enjoyed mathematics as a young boy, Hamilton's interest was greatly enhanced by a chance meeting at the age of 15 with Zerah Colburn, the American lightning calculator. Shortly afterward, Hamilton began to read important mathematical books of the time. In 1823, at the age of 18, he discovered an error in Simon Laplace's *Mécanique céleste* and wrote an impressive paper on the subject. A year later he entered Trinity College in Dublin.

Hamilton's university career was astonishing. At the age of 21, while still an undergraduate, he had so impressed the faculty that he was appointed Royal Astronomer of Ireland and Professor of Astronomy at the University. Shortly thereafter he wrote what is now considered a classic work on optics. Using only mathematical theory, he predicted conical refraction in certain types of crystals. Later this theory was confirmed by physicists. Largely because of this work, Hamilton was knighted in 1835.

Hamilton's first great purely mathematical paper appeared in 1833. In this work he described an algebraic way to manipulate pairs of real numbers. This work gives rules that are used today to add, subtract, multiply, and divide complex numbers. At first, however, Hamilton was unable to devise a multiplication for triples or n-tuples of numbers for $n > 2$. For 10 years he pondered this problem, and it is said that he solved it in an inspiration while walking on the Brougham Bridge in Dublin in 1843. The key was to discard the familiar commutative property of multiplication. The new objects he created were called *quaternions*, which were the precursors of what we now call *vectors*. Today a tablet embedded in the stone of the bridge tells the story.

For the rest of his life, Hamilton spent most of his time developing the algebra of quaternions. He felt that they would have revolutionary significance in mathematical physics. His monumental work on this subject, *Treatise on Quaternions*, was published in 1853. Thereafter, he worked on an enlarged

Here as he walked by

on the 16th of October 1843

Sir William Rowan Hamilton

in a flash of genius discovered

the fundamental formula for

quaternion multiplication

$$i^2 = j^2 = k^2 = ijk = -1$$

& cut it in a stone of this bridge

work, *Elements of Quaternions*. Although Hamilton died in 1865 before his *Elements* was completed, the work was published by his son in 1866.

Students of mathematics and physics know Hamilton in a variety of other contexts. In mathematical physics, for example, one encounters the Hamiltonian function, which often represents the total energy in a system, and the Hamilton-Jacobi differential equations of dynamics. In matrix theory, the Cayley-Hamilton theorem states that every matrix satisfies its own characteristic equation. We discuss this in Section 6.8.

Despite the great work he was doing, Hamilton's final years were a torment to him. His wife was a semi-invalid and he was plagued by alcoholism. It is therefore gratifying to point out that, during these last years, the newly formed American National Academy of Sciences elected Sir William Rowan Hamilton to be its first foreign associate.

PROBLEMS 1.5

In Problems 1–10 perform the indicated computation with $\mathbf{a} = \begin{pmatrix} -3 \\ 1 \\ 4 \end{pmatrix}$, $\mathbf{b} = \begin{pmatrix} 5 \\ -4 \\ 7 \end{pmatrix}$, and $\mathbf{c} = \begin{pmatrix} 2 \\ 0 \\ -2 \end{pmatrix}$.

1. $\mathbf{a} + \mathbf{b}$ 2. $3\mathbf{b}$ 3. $-2\mathbf{c}$

4. $\mathbf{b} + 3\mathbf{c}$ 5. $2\mathbf{a} - 5\mathbf{b}$ 6. $-3\mathbf{b} + 2\mathbf{c}$

7. $0\mathbf{c}$ 8. $\mathbf{a} + \mathbf{b} + \mathbf{c}$ 9. $3\mathbf{a} - 2\mathbf{b} + 4\mathbf{c}$

10. $3\mathbf{b} - 7\mathbf{c} + 2\mathbf{a}$

In Problems 11–20 perform the indicated computation with $\mathbf{a} = (3, -1, 4, 2)$, $\mathbf{b} = (6, 0, -1, 4)$, and $\mathbf{c} = (-2, 3, 1, 5)$. Of course, it is first necessary to extend the definitions in this section to row vectors.

11. $a + c$ **12.** $b - a$ **13.** $4c$

14. $-2b$ **15.** $2a - c$ **16.** $4b - 7a$

17. $a + b + c$ **18.** $c - b + 2a$ **19.** $3a - 2b + 4c$

20. $\alpha a + \beta b + \gamma c$

21. Let $a = \begin{pmatrix} a_1 \\ a_2 \\ \vdots \\ a_n \end{pmatrix}$ and let 0 denote the n-component zero column vector. Use Definitions 2

   and 3 to show that $a + 0 = a$ and $0a = 0$.

22. Let $a = \begin{pmatrix} a_1 \\ a_2 \\ \vdots \\ a_n \end{pmatrix}$, $b = \begin{pmatrix} b_1 \\ b_2 \\ \vdots \\ b_n \end{pmatrix}$, and $c = \begin{pmatrix} c_1 \\ c_2 \\ \vdots \\ c_n \end{pmatrix}$. Compute $(a + b) + c$ and $a + (b + c)$ and

   show that they are equal.

23. Let a and b be as in Problem 22 and let α and β be scalars. Compute $\alpha(a + b)$ and
   $\alpha a + \alpha b$ and show that they are equal. Similarly, compute $(\alpha + \beta)a$ and $\alpha a + \beta a$ and
   show that they are equal. Finally, show that $(\alpha\beta)a = \alpha(\beta a)$.

24. Find numbers α, β, and γ such that $(2, -1, 4) + (\alpha, \beta, \gamma) = 0$.

25. In the manufacture of a certain product, four raw materials are needed. The vector

   $d = \begin{pmatrix} d_1 \\ d_2 \\ d_3 \\ d_4 \end{pmatrix}$ represents a given factory's demand for each of the four raw materials to

   produce 1 unit of its product. If d is the demand vector for factory 1 and e is the demand
   vector for factory 2, what is represented by the vectors $d + e$ and $2d$?

26. Let $a = \begin{pmatrix} 1 \\ 3 \\ 2 \end{pmatrix}$, $b = \begin{pmatrix} -2 \\ 4 \\ 1 \end{pmatrix}$, and $c = \begin{pmatrix} 0 \\ 1 \\ 4 \end{pmatrix}$. Find a vector v such that

   $2a - b + 3v = 4c$.

27. With a, b, and c as in Problem 26, find a vector w such that $a - b + c - w = 0$.

1.6   MATRICES

Matrix

An **$m \times n$ matrix** A is a rectangular array of mn numbers arranged in m rows and n columns:

$$A = \begin{pmatrix} a_{11} & a_{12} & \cdots & a_{1j} & \cdots & a_{1n} \\ a_{21} & a_{22} & \cdots & a_{2j} & \cdots & a_{2n} \\ \vdots & \vdots & & \vdots & & \vdots \\ a_{i1} & a_{i2} & \cdots & a_{ij} & \cdots & a_{in} \\ \vdots & \vdots & & \vdots & & \vdots \\ a_{m1} & a_{m2} & \cdots & a_{mj} & \cdots & a_{mn} \end{pmatrix} \tag{1}$$

Rows and
columns of a
matrix

Component

Square matrix

Zero matrix

Size of a matrix

The symbol $m \times n$ is read "m by n." Unless stated otherwise, we shall always assume that the numbers in a matrix or vector are real. We call the row vector

$(a_{i1}, a_{i2} \cdots a_{in})$ **row i** and the column vector $\begin{pmatrix} a_{1j} \\ a_{2j} \\ \vdots \\ a_{mj} \end{pmatrix}$ **column j**. The **ijth compo-**

nent of A, denoted by a_{ij}, is the number appearing in the ith row and jth column of A. We shall sometimes write the matrix A as $A = (a_{ij})$. Usually, matrices will be denoted by capital letters.

If A is an $m \times n$ matrix with $m = n$, then A is called a **square matrix**. An $m \times n$ matrix with all components equal to zero is called the $m \times n$ **zero matrix**. An $m \times n$ matrix is said to have the **size $m \times n$**.

Historical Note. The term "matrix" was first used in 1850 by the British mathematician James Joseph Sylvester (1814–1897) to distinguish matrices from determinants (which we shall discuss in Chapter 2). In fact, the term "matrix" was intended to mean "mother of determinants."

EXAMPLE 1   **Five Matrices**   Five matrices of different sizes are given below:

$$\textbf{i. } A = \begin{pmatrix} 1 & 3 \\ 4 & 2 \end{pmatrix}, 2 \times 2 \text{ (square)} \qquad \textbf{ii. } A = \begin{pmatrix} -1 & 3 \\ 4 & 0 \\ 1 & -2 \end{pmatrix}, 3 \times 2$$

$$\textbf{iii. } \begin{pmatrix} -1 & 4 & 1 \\ 3 & 0 & 2 \end{pmatrix}, 2 \times 3 \qquad \textbf{iv. } \begin{pmatrix} 1 & 6 & -2 \\ 3 & 1 & 4 \\ 2 & -6 & 5 \end{pmatrix}, 3 \times 3 \text{ (square)}$$

$$\textbf{v. } \begin{pmatrix} 0 & 0 & 0 & 0 \\ 0 & 0 & 0 & 0 \end{pmatrix}, 2 \times 4 \text{ zero matrix}$$ ∎

Bracket Notation. In some books matrices are given in square brackets rather than parentheses. For example, the first two matrices in Example 1 can be written as

$$\textbf{i. } A = \begin{bmatrix} 1 & 3 \\ 4 & 2 \end{bmatrix} \qquad \textbf{ii. } A = \begin{bmatrix} -1 & 3 \\ 4 & 0 \\ 1 & -2 \end{bmatrix}$$

In this text we shall use the parentheses exclusively.

Throughout this book we shall refer to the ith row, the jth column, and the ijth component of a matrix for various numbers i and j. We illustrate these ideas in the next example.

EXAMPLE 2   **Finding Components of a Matrix**   Find the 1,2, 3,1 and 2,2 components of

$$A = \begin{pmatrix} 1 & 6 & 4 \\ 2 & -3 & 5 \\ 7 & 4 & 0 \end{pmatrix}$$

Solution The 1,2 component is the number in the first row and the second column. We have shaded the first row and the second column; the 1,2 component is 6:

2nd column
↓

$$1\text{st row} \rightarrow \begin{pmatrix} 1 & 6 & 4 \\ 2 & -3 & 5 \\ 7 & 4 & 0 \end{pmatrix}$$

From the shaded matrices below, we see that the 3,1 component is 7 and the 2,2 component is -3:

1st column
↓

$$\begin{pmatrix} 1 & 6 & 4 \\ 2 & -3 & 5 \\ 7 & 4 & 0 \end{pmatrix}$$

3rd row →

2nd column
↓

$$2\text{nd row} \rightarrow \begin{pmatrix} 1 & 6 & 4 \\ 2 & -3 & 5 \\ 7 & 4 & 0 \end{pmatrix}$$

■

Equality of matrices   Two matrices $A = (a_{ij})$ and $B = (b_{ij})$ are **equal** if (1) they have the same size, and (2) corresponding components are equal.

EXAMPLE 3   **Equal and Unequal Matrices**   Are the following matrices equal?

i. $\begin{pmatrix} 4 & 1 & 5 \\ 2 & -3 & 0 \end{pmatrix}$ and $\begin{pmatrix} 1+3 & 1 & 2+3 \\ 1+1 & 1-4 & 6-6 \end{pmatrix}$

ii. $\begin{pmatrix} -2 & 0 \\ 1 & 3 \end{pmatrix}$ and $\begin{pmatrix} 0 & -2 \\ 1 & 3 \end{pmatrix}$

iii. $\begin{pmatrix} 1 & 0 \\ 0 & 1 \end{pmatrix}$ and $\begin{pmatrix} 1 & 0 & 0 \\ 0 & 1 & 0 \end{pmatrix}$

Solution   i. Yes; both matrices are 2×3, and $1 + 3 = 4$, $2 + 3 = 5$, $1 + 1 = 2$, $1 - 4 = -3$, and $6 - 6 = 0$.

ii. No; $-2 \neq 0$, so the matrices are unequal because, for example, the 1,1 components are unequal. This is true even though the two matrices contain the same numbers. *Corresponding* components must be equal. This means that the 1,1 component in A must be equal to the 1,1 component in B, and so on.

iii. No; the first matrix is 2×2 and the second matrix is 2×3, so they do not have the same size.   ■

Each vector is a special kind of matrix. Thus, for example, the n-component row vector $(a_1, a_2, \ldots, a_n)$ is a $1 \times n$ matrix, whereas the n-component column

vector $\begin{pmatrix} a_1 \\ a_2 \\ \vdots \\ a_n \end{pmatrix}$ is an $n \times 1$ matrix.

Matrices, like vectors, arise in a great number of practical situations. For example, we saw in Section 1.5 how the vector $\begin{pmatrix} 10 \\ 30 \\ 15 \\ 60 \end{pmatrix}$ could represent order quantities for four different products used by one manufacturer. Suppose that there were five different plants. Then the 4×5 matrix

$$Q = \begin{pmatrix} 10 & 20 & 15 & 16 & 25 \\ 30 & 10 & 20 & 25 & 22 \\ 15 & 22 & 18 & 20 & 13 \\ 60 & 40 & 50 & 35 & 45 \end{pmatrix}$$

could represent the orders for the four products in each of the five plants. We can see, for example, that plant 4 orders 25 units of the second product while plant 2 orders 40 units of the fourth product.

Matrices, like vectors, can be added and multiplied by scalars.

DEFINITION 1   **Addition of Matrices** Let $A = (a_{ij})$ and $B = (b_{ij})$ be two $m \times n$ matrices. Then the sum of A and B is the $m \times n$ matrix $A + B$ given by

$$A + B = (a_{ij} + b_{ij}) = \begin{pmatrix} a_{11} + b_{11} & a_{12} + b_{12} & \cdots & a_{1n} + b_{1n} \\ a_{21} + b_{21} & a_{22} + b_{22} & \cdots & a_{2n} + b_{2n} \\ \vdots & \vdots & & \vdots \\ a_{m1} + b_{m1} & a_{m2} + b_{m2} & \cdots & a_{mn} + b_{mn} \end{pmatrix} \quad (2)$$

That is, $A + B$ is the $m \times n$ matrix obtained by adding the corresponding components of A and B.

WARNING   The sum of two matrices is defined only when both matrices have the same size. Thus, for example, it is not possible to add together the matrices $\begin{pmatrix} 1 & 2 & 3 \\ 4 & 5 & 6 \end{pmatrix}$ and $\begin{pmatrix} -1 & 0 \\ 2 & -5 \\ 4 & 7 \end{pmatrix}$. □

EXAMPLE 4   **The Sum of Two Matrices**

$$\begin{pmatrix} 2 & 4 & -6 & 7 \\ 1 & 3 & 2 & 1 \\ -4 & 3 & -5 & 5 \end{pmatrix} + \begin{pmatrix} 0 & 1 & 6 & -2 \\ 2 & 3 & 4 & 3 \\ -2 & 1 & 4 & 4 \end{pmatrix} = \begin{pmatrix} 2 & 5 & 0 & 5 \\ 3 & 6 & 6 & 4 \\ -6 & 4 & -1 & 9 \end{pmatrix}$$ ■

DEFINITION 2   **Multiplication of a Matrix by a Scalar** If $A = (a_{ij})$ is an $m \times n$ matrix and if α is a scalar, then the $m \times n$ matrix αA is given by

$$\alpha A = (\alpha a_{ij}) = \begin{pmatrix} \alpha a_{11} & \alpha a_{12} & \cdots & \alpha a_{1n} \\ \alpha a_{21} & \alpha a_{22} & \cdots & \alpha a_{2n} \\ \vdots & \vdots & & \vdots \\ \alpha a_{m1} & \alpha a_{m2} & \cdots & \alpha a_{mn} \end{pmatrix} \tag{3}$$

In other words, $\alpha A = (\alpha a_{ij})$ is the matrix obtained by multiplying each component of A by α. If $\alpha A = B = (b_{ij})$, then $b_{ij} = \alpha a_{ij}$ for $i = 1, 2, \ldots, m$ and $j = 1, 2, \ldots, n$.

EXAMPLE 5   **Scalar Multiples of Matrices**

Let $A = \begin{pmatrix} 1 & -3 & 4 & 2 \\ 3 & 1 & 4 & 6 \\ -2 & 3 & 5 & 7 \end{pmatrix}$. Then $2A = \begin{pmatrix} 2 & -6 & 8 & 4 \\ 6 & 2 & 8 & 12 \\ -4 & 6 & 10 & 14 \end{pmatrix}$,

$-3A = \begin{pmatrix} -3 & 9 & -12 & -6 \\ -9 & -3 & -12 & -18 \\ 6 & -9 & -15 & -21 \end{pmatrix}$, and $0A = \begin{pmatrix} 0 & 0 & 0 & 0 \\ 0 & 0 & 0 & 0 \\ 0 & 0 & 0 & 0 \end{pmatrix}$. ■

EXAMPLE 6   Let $A = \begin{pmatrix} 1 & 2 & 4 \\ -7 & 3 & -2 \end{pmatrix}$ and $B = \begin{pmatrix} 4 & 0 & 5 \\ 1 & -3 & 6 \end{pmatrix}$. Calculate $-2A + 3B$.

Solution $-2A + 3B = (-2)\begin{pmatrix} 1 & 2 & 4 \\ -7 & 3 & -2 \end{pmatrix} + (3)\begin{pmatrix} 4 & 0 & 5 \\ 1 & -3 & 6 \end{pmatrix}$

$= \begin{pmatrix} -2 & -4 & -8 \\ 14 & -6 & 4 \end{pmatrix} + \begin{pmatrix} 12 & 0 & 15 \\ 3 & -9 & 18 \end{pmatrix} = \begin{pmatrix} 10 & -4 & 7 \\ 17 & -15 & 22 \end{pmatrix}$ ■

The next theorem is similar to Theorem 1.5.1 on page 32. We prove part (*iii*) and leave the remaining parts of the proof as an exercise (see Problems 21–23).

THEOREM 1   Let A, B, and C be $m \times n$ matrices and let α be a scalar. Then:

   i. $A + 0 = A$

   ii. $0A = 0$

   iii. $A + B = B + A$ (**commutative law for matrix addition**)

   iv. $(A + B) + C = A + (B + C)$ (**associative law for matrix addition**)

v. $\alpha(A + B) = \alpha A + \alpha B$ (**distributive law for scalar multiplication**)

vi. $1A = A$

Note. The zero in part (*i*) of the theorem is the $m \times n$ zero matrix. In part (*ii*) the zero on the left is a scalar while the zero on the right is the $m \times n$ zero matrix.

Proof of (iii)   Let $A = \begin{pmatrix} a_{11} & a_{12} & \cdots & a_{1n} \\ a_{21} & a_{22} & \cdots & a_{2n} \\ \vdots & \vdots & & \vdots \\ a_{m1} & a_{m2} & \cdots & a_{mn} \end{pmatrix}$ and $B = \begin{pmatrix} b_{11} & b_{12} & \cdots & b_{1n} \\ b_{21} & b_{22} & \cdots & b_{2n} \\ \vdots & \vdots & & \vdots \\ b_{m1} & b_{m2} & \cdots & b_{mn} \end{pmatrix}$.

Then

$$A + B = \begin{pmatrix} a_{11} + b_{11} & a_{12} + b_{12} & \cdots & a_{1n} + b_{1n} \\ a_{21} + b_{21} & a_{22} + b_{22} & \cdots & a_{2n} + b_{2n} \\ \vdots & \vdots & & \vdots \\ a_{m1} + b_{m1} & a_{m2} + b_{m2} & \cdots & a_{mn} + b_{mn} \end{pmatrix}$$

$a + b = b + a$ for any real numbers a and b

$$\underset{=}{\downarrow} \begin{pmatrix} b_{11} + a_{11} & b_{12} + a_{12} & \cdots & b_{1n} + a_{1n} \\ b_{21} + a_{21} & b_{22} + a_{22} & \cdots & b_{2n} + a_{2n} \\ \vdots & \vdots & & \vdots \\ b_{m1} + a_{m1} & b_{m2} + a_{m2} & \cdots & b_{mn} + a_{mn} \end{pmatrix} = B + A$$
■

EXAMPLE 7   **Illustrating the Associative Law of Matrix Addition**   To illustrate the associative law we note that

$$\left[\begin{pmatrix} 1 & 4 & -2 \\ 3 & -1 & 0 \end{pmatrix} + \begin{pmatrix} 2 & -2 & 3 \\ 1 & -1 & 5 \end{pmatrix} \right] + \begin{pmatrix} 3 & -1 & 2 \\ 0 & 1 & 4 \end{pmatrix}$$

$$= \begin{pmatrix} 3 & 2 & 1 \\ 4 & -2 & 5 \end{pmatrix} + \begin{pmatrix} 3 & -1 & 2 \\ 0 & 1 & 4 \end{pmatrix} = \begin{pmatrix} 6 & 1 & 3 \\ 4 & -1 & 9 \end{pmatrix}$$

Similarly,

$$\begin{pmatrix} 1 & 4 & -2 \\ 3 & -1 & 0 \end{pmatrix} + \left[\begin{pmatrix} 2 & -2 & 3 \\ 1 & -1 & 5 \end{pmatrix} + \begin{pmatrix} 3 & -1 & 2 \\ 0 & 1 & 4 \end{pmatrix} \right]$$

$$= \begin{pmatrix} 1 & 4 & -2 \\ 3 & -1 & 0 \end{pmatrix} + \begin{pmatrix} 5 & -3 & 5 \\ 1 & 0 & 9 \end{pmatrix} = \begin{pmatrix} 6 & 1 & 3 \\ 4 & -1 & 9 \end{pmatrix}$$
■

As with vectors, the associative law for matrix addition enables us to define the sum of three or more matrices.

PROBLEMS 1.6

In Problems 1–12 perform the indicated computation with $A = \begin{pmatrix} 1 & 3 \\ 2 & 5 \\ -1 & 2 \end{pmatrix}$,

$B = \begin{pmatrix} -2 & 0 \\ 1 & 4 \\ -7 & 5 \end{pmatrix}$, and $C = \begin{pmatrix} -1 & 1 \\ 4 & 6 \\ -7 & 3 \end{pmatrix}$.

1. $3A$ **2.** $A + B$ **3.** $A - C$

4. $2C - 5A$ **5.** $0B$ (0 is the scalar zero)   **6.** $-7A + 3B$

7. $A + B + C$ **8.** $C - A - B$ **9.** $2A - 3B + 4C$

10. $7C - B + 2A$

11. Find a matrix D such that $2A + B - D$ is the 3×2 zero matrix.

12. Find a matrix E such that $A + 2B - 3C + E$ is the 3×2 zero matrix.

In Problems 13–20 perform the indicated computation with $A = \begin{pmatrix} 1 & -1 & 2 \\ 3 & 4 & 5 \\ 0 & 1 & -1 \end{pmatrix}$,

$B = \begin{pmatrix} 0 & 2 & 1 \\ 3 & 0 & 5 \\ 7 & -6 & 0 \end{pmatrix}$, and $C = \begin{pmatrix} 0 & 0 & 2 \\ 3 & 1 & 0 \\ 0 & -2 & 4 \end{pmatrix}$.

13. $A - 2B$ **14.** $3A - C$ **15.** $A + B + C$

16. $2A - B + 2C$ **17.** $C - A - B$ **18.** $4C - 2B + 3A$

19. Find a matrix D such that $A + B + C + D$ is the 3×3 zero matrix.

20. Find a matrix E such that $3C - 2B + 8A - 4E$ is the 3×3 zero matrix.

21. Let $A = (a_{ij})$ be an $m \times n$ matrix and let $\bar{0}$ denote the $m \times n$ zero matrix. Use Definitions 1 and 2 to show that $0A = \bar{0}$ and $\bar{0} + A = A$. Similarly, show that $1A = A$.

22. If $A = (a_{ij})$, $B = (b_{ij})$, and $C = (c_{ij})$ are $m \times n$ matrices, compute $(A + B) + C$ and $A + (B + C)$ and show that they are equal.

23. If α is a scalar and A and B are $m \times n$ matrices, compute $\alpha(A + B)$ and $\alpha A + \alpha B$ and show that they are equal.

24. Consider the "graph" joining the four points in Figure 1.2. Construct a 4×4 matrix having the property that $a_{ij} = 0$ if point i is not connected (joined by a line) to point j and $a_{ij} = 1$ if point i is connected to point j.

25. Do the same (this time constructing a 5×5 matrix) for the graph in Figure 1.3.

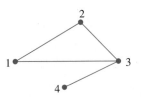

Figure 1.2

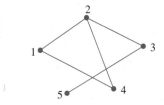

Figure 1.3

1.7 VECTOR AND MATRIX PRODUCTS

In this section we see how two matrices can be multiplied together. Quite obviously, we could define the product of two $m \times n$ matrices $A = (a_{ij})$ and $B = (b_{ij})$ to be the $m \times n$ matrix whose ijth component is $a_{ij}b_{ij}$. However, for just about all the important applications involving matrices, another kind of product is needed. Let us try to see why this is the case.

EXAMPLE 1 **The Product of a Demand Vector and a Price Vector**   Suppose that a manufacturer produces four items. The demand for the items is given by the **demand vector d** = (30   20   40   10) (a 1 × 4 matrix). The price per unit that the manufacturer receives for the items is given by the **price vector p** = $\begin{pmatrix} \$20 \\ \$15 \\ \$18 \\ \$40 \end{pmatrix}$ (a 4 × 1 matrix). If the demand is met, how much money will the manufacturer receive?

Solution   Demand for the first item is 30, and the manufacturer receives \$20 for each of the first item sold. Thus $(30)(20) = \$600$ is received from the sales of the first item. By continuing this reasoning, we see that the total amount of money received is

$$(30)(20) + (20)(15) + (40)(18) + (10)(40) = 600 + 300 + 720 + 400$$
$$= \$2020$$

We write this result as

$$(30 \quad 20 \quad 40 \quad 10)\begin{pmatrix} 20 \\ 15 \\ 18 \\ 40 \end{pmatrix} = 2020$$

That is, we multiplied a 4-component row vector and a 4-component column vector to obtain a scalar (real number).   ∎

In the last example we multiplied a row vector by a column vector and obtained a scalar. In general, we have the following definition.

DEFINITION 1   **Scalar Product**   Let $\mathbf{a} = \begin{pmatrix} a_1 \\ a_2 \\ \vdots \\ a_n \end{pmatrix}$ and $\mathbf{b} = \begin{pmatrix} b_1 \\ b_2 \\ \vdots \\ b_n \end{pmatrix}$ be two n-vectors. Then the **scalar product** of $\mathbf{a}$ and $\mathbf{b}$, denoted by $\mathbf{a} \cdot \mathbf{b}$, is given by

> ### Scalar Product
>
> $$\mathbf{a} \cdot \mathbf{b} = a_1 b_1 + a_2 b_2 + \cdots + a_n b_n \qquad (1)$$

Because of the notation in (1), the scalar product of two vectors is often called the **dot product** or **inner product** of the vectors. Note that the scalar product of two n-vectors is a scalar (that is, a number).

WARNING When taking the scalar product of $\mathbf{a}$ and $\mathbf{b}$, it is necessary that $\mathbf{a}$ and $\mathbf{b}$ have the same number of components. □

We shall often be taking the scalar product of a row vector and column vector. In this case we have

> ### Scalar Product
>
> $1 \times n$ row vector
> $$\downarrow$$
> $$(a_1, a_2, \ldots, a_n) \cdot \begin{pmatrix} b_1 \\ b_2 \\ \vdots \\ b_n \end{pmatrix} = a_1 b_1 + a_2 b_2 + \cdots + a_n b_n \qquad (2)$$
> This is a real number (a scalar)
> $$\uparrow$$
> $n \times 1$ column vector

EXAMPLE 2 **The Scalar Product of Two Vectors** Let $\mathbf{a} = \begin{pmatrix} 1 \\ -2 \\ 3 \end{pmatrix}$ and $\mathbf{b} = \begin{pmatrix} 3 \\ -2 \\ 4 \end{pmatrix}$. Calculate $\mathbf{a} \cdot \mathbf{b}$.

Solution $\mathbf{a} \cdot \mathbf{b} = (1)(3) + (-2)(-2) + (3)(4) = 3 + 4 + 12 = 19$ ■

EXAMPLE 3 **The Scalar Product of Two Vectors** Let $\mathbf{a} = (2, -3, 4, -6)$ and $\mathbf{b} = \begin{pmatrix} 1 \\ 2 \\ 0 \\ 3 \end{pmatrix}$. Compute $\mathbf{a} \cdot \mathbf{b}$.

Solution Here $\mathbf{a} \cdot \mathbf{b} = (2)(1) + (-3)(2) + (4)(0) + (-6)(3) = 2 - 6 + 0 - 18 = -22$. ■

The next theorem follows directly from the definition of the scalar product. We prove part **(ii)** and leave the remaining parts as an exercise.

THEOREM 1   Let **a**, **b**, and **c** be n-vectors and let α and β be scalars. Then:

  i. $\mathbf{a} \cdot \mathbf{0} = 0$

  ii. $\mathbf{a} \cdot \mathbf{b} = \mathbf{b} \cdot \mathbf{a}$ (commutative law for scalar product)

  iii. $\mathbf{a} \cdot (\mathbf{b} + \mathbf{c}) = \mathbf{a} \cdot \mathbf{b} + \mathbf{a} \cdot \mathbf{c}$ (distributive law for scalar product)

  iv. $(\alpha \mathbf{a}) \cdot \mathbf{b} = \alpha(\mathbf{a} \cdot \mathbf{b})$

Proof of (ii)   Let $\mathbf{a} = \begin{pmatrix} a_1 \\ a_2 \\ \vdots \\ a_n \end{pmatrix}$ and $\mathbf{b} = \begin{pmatrix} b_1 \\ b_2 \\ \vdots \\ b_n \end{pmatrix}$.

Then

$$ab = ba \text{ for}$$
$$\text{any two numbers } a \text{ and } b$$
$$\downarrow$$
$$\mathbf{a} \cdot \mathbf{b} = a_1 b_1 + a_2 b_2 + \cdots + a_n b_n = b_1 a_1 + b_2 a_2 + \cdots + b_n a_n = \mathbf{b} \cdot \mathbf{a} \quad \blacksquare$$

Note that there is *no* associative law for the scalar product. The expression $(\mathbf{a} \cdot \mathbf{b}) \cdot \mathbf{c} = \mathbf{a} \cdot (\mathbf{b} \cdot \mathbf{c})$ does not make sense because neither side of the equation is defined. For the left side, this follows from the fact that $\mathbf{a} \cdot \mathbf{b}$ is a scalar and the scalar product of the scalar $\mathbf{a} \cdot \mathbf{b}$ and the vector $\mathbf{c}$ is not defined.

We now define the product of two matrices.

DEFINITION 2   **Product of Two Matrices** Let $A = (a_{ij})$ be an $m \times n$ matrix, and let $B = (b_{ij})$ be an $n \times p$ matrix. Then the **product** of A and B is an $m \times p$ matrix $C = (c_{ij})$, where

$$c_{ij} = (i\text{th row of } A) \cdot (j\text{th column of } B) \tag{3}$$

That is, the ijth element of AB is the dot product of the ith row of A and the jth column of B. If we write this out, we obtain

$$c_{ij} = a_{i1}b_{1j} + a_{i2}b_{2j} + \cdots + a_{in}b_{nj} \tag{4}$$

WARNING   Two matrices can be multiplied together only if the number of columns of the first matrix is equal to the number of rows of the second. Otherwise the vectors that are the ith row of A and the jth column of B will not have the same number of components, and the dot product in equation (3) will not be defined. To illustrate this, we write the matrices A and B:

$$
i\text{th row of } A \rightarrow
\begin{pmatrix}
a_{11} & a_{12} & \cdots & a_{1n} \\
a_{21} & a_{22} & \cdots & a_{2n} \\
\vdots & \vdots & & \vdots \\
a_{i1} & a_{i2} & \cdots & a_{in} \\
\vdots & \vdots & & \vdots \\
a_{m1} & a_{m2} & \cdots & a_{mn}
\end{pmatrix}
\begin{pmatrix}
b_{11} & b_{12} & \cdots & b_{1j} & \cdots & b_{1p} \\
b_{21} & b_{22} & \cdots & b_{2j} & \cdots & b_{2p} \\
\vdots & \vdots & & \vdots & & \vdots \\
b_{n1} & b_{n2} & \cdots & b_{nj} & \cdots & b_{np}
\end{pmatrix}
$$

jth column of B

The shaded row and column vectors must have the same number of components. ☐

EXAMPLE 4   **The Product of Two 2 × 2 Matrices**   If $A = \begin{pmatrix} 1 & 3 \\ -2 & 4 \end{pmatrix}$ and $B = \begin{pmatrix} 3 & -2 \\ 5 & 6 \end{pmatrix}$, calculate AB and BA.

Solution A is a 2×2 matrix and B is a 2×2 matrix, so $C = AB = (2 \times 2) \times (2 \times 2)$ is also a 2×2 matrix. If $C = (c_{ij})$, what is c_{11}? We know that

$$c_{11} = (1\text{st row of } A) \cdot (1\text{st column of } B)$$

Rewriting the matrices, we have

1st column of B

1st row of $A \rightarrow \begin{pmatrix} \boxed{1 \quad 3} \\ -2 \quad 4 \end{pmatrix} \begin{pmatrix} \boxed{3} & -2 \\ \boxed{5} & 6 \end{pmatrix}$

Thus

$$c_{11} = (1 \quad 3)\begin{pmatrix} 3 \\ 5 \end{pmatrix} = 3 + 15 = 18$$

Similarly, to compute c_{12} we have

2nd column of B

1st row of $A \rightarrow \begin{pmatrix} \boxed{1 \quad 3} \\ -2 \quad 4 \end{pmatrix} \begin{pmatrix} 3 & \boxed{-2} \\ 5 & \boxed{6} \end{pmatrix}$

and

$$c_{12} = (1 \quad 3)\begin{pmatrix} -2 \\ 6 \end{pmatrix} = -2 + 18 = 16$$

Continuing, we find that

$$c_{21} = (-2 \quad 4) \begin{pmatrix} 3 \\ 5 \end{pmatrix} = -6 + 20 = 14$$

and

$$c_{22} = (-2 \quad 4) \begin{pmatrix} -2 \\ 6 \end{pmatrix} = 4 + 24 = 28$$

Thus

$$C = AB = \begin{pmatrix} 18 & 16 \\ 14 & 28 \end{pmatrix}$$

Similarly, leaving out the intermediate steps, we see that

$$C' = BA = \begin{pmatrix} 3 & -2 \\ 5 & 6 \end{pmatrix} \begin{pmatrix} 1 & 3 \\ -2 & 4 \end{pmatrix} = \begin{pmatrix} 3+4 & 9-8 \\ 5-12 & 15+24 \end{pmatrix} = \begin{pmatrix} 7 & 1 \\ -7 & 39 \end{pmatrix} \quad \blacksquare$$

Remark. Example 4 illustrates an important fact: *Matrix products do not, in general, commute.* That is, $AB \neq BA$ in general. It sometimes happens that $AB = BA$, but this will be the exception, not the rule. In fact, as the next example illustrates, it may occur that AB is defined while BA is not. Thus we must be careful of *order* when multiplying two matrices together.

EXAMPLE 5   **The Product of a 2 × 3 and a 3 × 3 Matrix Is Defined While the Product of a 3 × 3 and a 2 × 3 Matrix Is Not Defined**

Let $A = \begin{pmatrix} 2 & 0 & -3 \\ 4 & 1 & 5 \end{pmatrix}$ and $B = \begin{pmatrix} 7 & -1 & 4 & 7 \\ 2 & 5 & 0 & -4 \\ -3 & 1 & 2 & 3 \end{pmatrix}$. Calculate AB.

Solution   We first note that A is a 2 × 3 matrix and B is a 3 × 4 matrix. Hence the number of columns of A equals the number of rows of B. The product AB is therefore defined and is a 2 × 4 matrix. Let $AB = C = (c_{ij})$. Then

$$c_{11} = (2 \quad 0 \quad -3) \cdot \begin{pmatrix} 7 \\ 2 \\ -3 \end{pmatrix} = 23 \qquad c_{12} = (2 \quad 0 \quad -3) \cdot \begin{pmatrix} -1 \\ 5 \\ 1 \end{pmatrix} = -5$$

$$c_{13} = (2 \quad 0 \quad -3) \cdot \begin{pmatrix} 4 \\ 0 \\ 2 \end{pmatrix} = 2 \qquad c_{14} = (2 \quad 0 \quad -3) \cdot \begin{pmatrix} 7 \\ -4 \\ 3 \end{pmatrix} = 5$$

$$c_{21} = (4 \quad 1 \quad 5) \cdot \begin{pmatrix} 7 \\ 2 \\ -3 \end{pmatrix} = 15 \qquad c_{22} = (4 \quad 1 \quad 5) \cdot \begin{pmatrix} -1 \\ 5 \\ 1 \end{pmatrix} = 6$$

$$c_{23} = (4 \quad 1 \quad 5) \cdot \begin{pmatrix} 4 \\ 0 \\ 2 \end{pmatrix} = 26 \qquad c_{24} = (4 \quad 1 \quad 5) \cdot \begin{pmatrix} 7 \\ -4 \\ 3 \end{pmatrix} = 39$$

Hence $AB = \begin{pmatrix} 23 & -5 & 2 & 5 \\ 15 & 6 & 26 & 39 \end{pmatrix}$. This completes the problem. Note that the product BA is *not* defined since the number of columns of B (four) is not equal to the number of rows of A (two). ■

EXAMPLE 6
Direct and Indirect Contact with a Contagious Disease In this example we show how matrix multiplication can be used to model the spread of a contagious disease. Suppose that four individuals have contracted such a disease. This group has contacts with six people in a second group. We can represent these contacts, called *direct contacts,* by a 4×6 matrix. An example of such a matrix is given below.

Direct Contact Matrix: First and second groups

$$A = \begin{pmatrix} 0 & 1 & 0 & 0 & 1 & 0 \\ 1 & 0 & 0 & 1 & 0 & 1 \\ 0 & 0 & 0 & 1 & 1 & 0 \\ 1 & 0 & 0 & 0 & 0 & 1 \end{pmatrix}$$

Here we set $a_{ij} = 1$ if the ith person in the first group has made contact with the jth person in the second group. For example, the 1 in the 2,4 position means that the second person in the first (infected) group has been in contact with the fourth person in the second group. Now suppose that a third group of five people has had a variety of direct contact with individuals of the second group. We can also represent this by a matrix.

Direct Contact Matrix: Second and third groups

$$B = \begin{pmatrix} 0 & 0 & 1 & 0 & 1 \\ 0 & 0 & 0 & 1 & 0 \\ 0 & 1 & 0 & 0 & 0 \\ 1 & 0 & 0 & 0 & 1 \\ 0 & 0 & 0 & 1 & 0 \\ 0 & 0 & 1 & 0 & 0 \end{pmatrix}$$

Note that $b_{64} = 0$, which means that the sixth person in the second group has had no contact with the fourth person in the third group.

The *indirect* or *second-order* contacts between the individuals in the first and third groups is represented by the 4×5 matrix $C = AB$. To see this, observe that a person in group 3 can be infected from someone in group 2, who in turn has been infected by someone in group 1. For example, since $a_{24} = 1$ and $b_{45} = 1$, we see that, indirectly, the fifth person in group 3 has contact (through the fourth person in group 2) with the second person in group 1. The total number of indirect contacts between the second person in group 1 and the fifth person in group 3 is given by

$$c_{25} = a_{21}b_{15} + a_{22}b_{25} + a_{23}b_{35} + a_{24}b_{45} + a_{25}b_{55} + a_{26}b_{65}$$
$$= 1 \cdot 1 + 0 \cdot 0 + 0 \cdot 0 + 1 \cdot 1 + 0 \cdot 0 + 1 \cdot 0 = 2$$

We now compute.

Indirect Contact Matrix. First and third groups

$$C = AB = \begin{pmatrix} 0 & 0 & 0 & 2 & 0 \\ 1 & 0 & 2 & 0 & 2 \\ 1 & 0 & 0 & 1 & 1 \\ 0 & 0 & 2 & 0 & 1 \end{pmatrix}$$

We observe that only the second person in group 3 has no indirect contacts with the disease. The fifth person in this group has $2 + 1 + 1 = 4$ indirect contacts. ■

We have seen that for matrix multiplication the commutative law does not hold. The next theorem shows that the associative law does hold.

THEOREM 2   **Associative Law for Matrix Multiplication** Let $A = (a_{ij})$ be an $n \times m$ matrix, $B = (b_{ij})$ an $m \times p$ matrix, and $C = (c_{ij})$ a $p \times q$ matrix. Then the **associative law**

$$\boxed{A(BC) = (AB)C} \tag{5}$$

holds and ABC, defined by either side of (5), is an $n \times q$ matrix. ■

The proof of this theorem is not difficult, but it is somewhat tedious. It is best given using the summation notation. For that reason let us defer it until the end of the section.

EXAMPLE 7   **Illustrating the Associative Law of Matrix Multiplication**   Verify the associative law for $A = \begin{pmatrix} 1 & -3 \\ 0 & 2 \end{pmatrix}$, $B = \begin{pmatrix} 2 & -1 & 4 \\ 3 & 1 & 5 \end{pmatrix}$, and $C = \begin{pmatrix} 0 & -2 & 1 \\ 4 & 3 & 2 \\ -5 & 0 & 6 \end{pmatrix}$.

Solution   We first note that A is 2×2, B is 2×3, and C is 3×3. Hence all products used in the statement of the associative law are defined and the resulting product will be a

2×3 matrix. We then calculate

$$AB = \begin{pmatrix} 1 & -3 \\ 0 & 2 \end{pmatrix} \begin{pmatrix} 2 & -1 & 4 \\ 3 & 1 & 5 \end{pmatrix} = \begin{pmatrix} -7 & -4 & -11 \\ 6 & 2 & 10 \end{pmatrix}$$

$$(AB)C = \begin{pmatrix} -7 & -4 & -11 \\ 6 & 2 & 10 \end{pmatrix} \begin{pmatrix} 0 & -2 & 1 \\ 4 & 3 & 2 \\ -5 & 0 & 6 \end{pmatrix} = \begin{pmatrix} 39 & 2 & -81 \\ -42 & -6 & 70 \end{pmatrix}$$

Similarly,

$$BC = \begin{pmatrix} 2 & -1 & 4 \\ 3 & 1 & 5 \end{pmatrix} \begin{pmatrix} 0 & -2 & 1 \\ 4 & 3 & 2 \\ -5 & 0 & 6 \end{pmatrix} = \begin{pmatrix} -24 & -7 & 24 \\ -21 & -3 & 35 \end{pmatrix}$$

$$A(BC) = \begin{pmatrix} 1 & -3 \\ 0 & 2 \end{pmatrix} \begin{pmatrix} -24 & -7 & 24 \\ -21 & -3 & 35 \end{pmatrix} = \begin{pmatrix} 39 & 2 & -81 \\ -42 & -6 & 70 \end{pmatrix}$$

Thus $(AB)C = A(BC)$. ∎

From now on we shall write the product of three matrices simply as ABC. We can do this because $(AB)C = A(BC)$; thus we get the same answer no matter how the multiplication is carried out (provided that we do not commute any of the matrices).

The associative law can be extended to longer products. For example, suppose that AB, BC, and CD are defined. Then

$$\boxed{ABCD = A(B(CD)) = ((AB)C)D = A(BC)D = (AB)(CD)} \tag{6}$$

There are two distributive laws for matrix multiplication.

THEOREM 3   **Distributive Laws for Matrix Multiplication** If all the following sums and products are defined, then

$$\boxed{A(B + C) = AB + AC} \tag{7}$$

and

$$\boxed{(A + B)C = AC + BC} \tag{8}$$
∎

In order to prove Theorems 2 and 3 and to discuss many other things in this book, we need to use the *summation notation*. If this is not familiar to you, then continue reading. Otherwise skip to the proofs of Theorems 2 and 3.

The Σ Notation

A sum can be written†, with $N \geq M$,

$$a_M + a_{M+1} + a_{M+2} + \cdots + a_N = \sum_{k=M}^{N} a_k \tag{9}$$

Summation sign

Index of summation

which is read "the sum of the terms a_k as k goes from M to N." In this context, Σ is called the **summation sign** and k is called the **index of summation.**

EXAMPLE 8   **Interpreting the Summation Notation**   Write out the sum $\Sigma_{k=1}^{5} b_k$.

Solution   Starting at $k = 1$ and ending at $k = 5$, we obtain

$$\sum_{k=1}^{5} b_k = b_1 + b_2 + b_3 + b_4 + b_5$$

■

EXAMPLE 9   **Interpreting the Summation Notation**   Write out the sum $\Sigma_{k=3}^{6} c_k$.

Solution   Starting at $k = 3$ and ending at $k = 6$, we obtain

$$\sum_{k=3}^{6} c_k = c_3 + c_4 + c_5 + c_6$$

■

EXAMPLE 10   **Interpreting the Summation Notation**   Calculate $\Sigma_{-2}^{3} k^2$.

Solution   Here $a_k = k^2$, and k ranges from -2 to 3.

$$\sum_{-2}^{3} k^2 = (-2)^2 + (-1)^2 + (0)^2 + 1^2 + 2^2 + 3^2$$

$$= 4 + 1 + 0 + 1 + 4 + 9 = 19$$

■

Note. As in Example 10, the index of summation can take on negative integer values or zero.

EXAMPLE 11   **Writing a Sum Using the Summation Notation**   Write the sum $S_8 = 1 - 2 + 3 - 4 + 5 - 6 + 7 - 8$ by using the summation sign.

Solution   Since $1 = (-1)^2$, $-2 = (-1)^3 \cdot 2$, $3 = (-1)^4 \cdot 3$, . . . , we have

$$S_8 = \sum_{k=1}^{8} (-1)^{k+1} k$$

■

---

† The Greek letter Σ (sigma) was first used to denote a sum by the Swiss mathematician Leonhard Euler (1707–1783).

EXAMPLE 12   **Writing the Scalar Product Using the Summation Notation**   Equation (1) for the scalar product can be written compactly using the summation notation:

$$\mathbf{a} \cdot \mathbf{b} = a_1b_1 + a_2b_2 + \cdots + a_nb_n = \sum_{i=1}^{n} a_ib_i$$

Formula (4) for the ijth component of the product AB can be written

$$c_{ij} = a_{i1}b_{1j} + a_{i2}b_{2j} + \cdots + a_{in}b_{nj} = \sum_{k=1}^{n} a_{ik}b_{kj} \qquad (10)$$

■

The sigma notation has a number of useful properties. For example,

$$\sum_{k=1}^{n} ca_k = ca_1 + ca_2 + ca_3 + \cdots + ca_n$$

$$= c(a_1 + a_2 + a_3 + \cdots + a_n) = c \sum_{k=1}^{n} a_k$$

This and other facts are summarized below.

Facts About the Sigma Notation

Let $\{a_n\}$ and $\{b_n\}$ be real sequences, and let c be a real number. Then

$$\sum_{k=M}^{N} ca_k = c \sum_{k=M}^{N} a_k \qquad (11)$$

$$\sum_{k=M}^{N} (a_k + b_k) = \sum_{k=M}^{N} a_k + \sum_{k=M}^{N} b_k \qquad (12)$$

$$\sum_{k=M}^{N} (a_k - b_k) = \sum_{k=M}^{N} a_k - \sum_{k=M}^{N} b_k \qquad (13)$$

$$\sum_{k=M}^{N} a_k = \sum_{k=M}^{m} a_k + \sum_{k=m+1}^{N} a_k \qquad \text{if } M < m < N \qquad (14)$$

The proofs of these facts are left as exercises (see Problems 81–83).

We now use the summation notation to prove the associative and distributive laws.

**Proofs of
Theorems 2
and 3**

Associative Law Since A is $n \times m$ and B is $m \times p$, AB is $n \times p$. Thus $(AB)C = (n \times p) \times (p \times q)$ is an $n \times q$ matrix. Similarly, BC is $m \times q$ and $A(BC)$ is $n \times q$ so that $(AB)C$ and $A(BC)$ are both of the same size. We must show that the ijth component of $(AB)C$ equals the ijth component of $A(BC)$. Define $D = (d_{ij}) = AB$. Then

from (10)
$$\downarrow$$
$$d_{ij} = \sum_{k=1}^{m} a_{ik} b_{kj}$$

The ijth component of $(AB)C = DC$ is

$$\sum_{l=1}^{p} d_{il} c_{lj} = \sum_{l=1}^{p} \left(\sum_{k=1}^{m} a_{ik} b_{kl} \right) c_{lj} = \sum_{k=1}^{m} \sum_{l=1}^{p} a_{ik} b_{kl} c_{lj}$$

Next we define $E = (e_{ij}) = BC$. Then

$$e_{kj} = \sum_{l=1}^{p} b_{kl} c_{lj}$$

and the ijth component of $A(BC) = AE$ is

$$\sum_{k=1}^{m} a_{ik} e_{kj} = \sum_{k=1}^{m} \sum_{l=1}^{p} a_{ik} b_{kl} c_{lj}$$

Thus the ijth component of $(AB)C$ is equal to the ijth component of $A(BC)$. This proves the associative law. ∎

Distributive Laws We prove the first distributive law [equation (7)]. The proof of the second one [equation (8)] is virtually identical and is therefore omitted. Let A be $n \times m$ and let B and C be $m \times p$. The the kjth component of $B + C$ is $b_{kj} + c_{kj}$ and the ijth component of $A(B + C)$ is

From (12)
$$\downarrow$$
$$\sum_{k=1}^{m} a_{ik}(b_{kj} + c_{kj}) = \sum_{k=1}^{m} a_{ik} b_{kj} + \sum_{k=1}^{m} a_{ik} c_{kj} = ij\text{th component of } AB \text{ plus}$$

the ijth component of AC and this proves equation (7). ∎

Focus on . . .

Arthur Cayley and the Algebra of Matrices

Arthur Cayley
(*Library of
Congress*)

The algebra of matrices, that is, the rules by which matrices can be added and multiplied, was developed by the English mathematician Arthur Cayley (1821–1895) in 1857. Cayley was born at Richmond, in Surrey (near London), and was educated at Trinity College, Cambridge, graduating in 1842. In that year he placed first in the very difficult test for the Smith's prize. For a period of several years he studied and practiced law, always being careful not to let his legal practice prevent him from working on mathematics. While a student of the bar he went to Dublin and attended Hamilton's lectures on quaternions. When the Sadlerian professorship was established at Cambridge in 1863, Cayley was offered the chair, which he accepted, thus giving up a lucrative future in the legal profession for the modest provision of an academic life. But then he could devote *all* of his time to mathematics.

Cayley ranks as the third most prolific writer of mathematics in the history of the subject, being surpassed only by Euler and Cauchy. He began publishing while still an undergraduate student at Cambridge, put out between 200 and 300 papers during his years of legal practice, and continued his prolific publication the rest of his long life. The massive *Collected Mathematical Papers* of Cayley contains 966 papers and fills 13 large quarto volumes averaging about 600 pages per volume. There is scarcely an area in pure mathematics that has not been touched and enriched by the genius of Cayley.

Besides developing matrix theory, Cayley made pioneering contributions to analytic geometry, the theory of determinants, higher-dimensional geometry, the theory of curves and surfaces, the study of binary forms, the theory of elliptic functions, and the development of invariant theory.

Cayley's mathematical style reflects his legal training, for his papers are severe, direct, methodical, and clear. He possessed a phenomenal memory and seemed never to forget anything he had once seen or read. He also possessed a singularly serene, even, and gentle temperament. He has been called "the mathematicians' mathematician."

Cayley developed an unusual avidity for novel reading. He read novels while traveling, while waiting for meetings to start, and at any odd moments that presented themselves. During his life he read thousands of novels, not only in English, but also in Greek, French, German, and Italian. He took great delight in painting, especially in water colors, and he exhibited a marked talent as a water colorist. He was also an ardent student of botany and nature in general.

Cayley was, in the true British tradition, an amateur mountain climber, and he made frequent trips to the Continent for long walks and mountain scaling. A story is told that he claimed the reason he undertook mountain climbing was that, though he found the ascent arduous and tiring, the grand feeling of exhilaration he attained when he conquered the peak was like that he experienced when he solved a difficult mathematics problem or completed an intricate mathematical theory, and it was easier for him to attain the desired feeling by climbing the mountain.

Matrices arose with Cayley in connection with linear transformations of the type

$$x' = ax + by$$
$$y' = cx + dy$$
(15)

where a, b, c, d are real numbers, and which may be thought of as functions that take the vector (x, y) into the vector (x', y'). We shall discuss linear transformations in great detail in Chapter 5. Here we observe that the transformation (15) is completely determined by the four coefficients a, b, c, d, and so they can be symbolized by the square array

$$\begin{pmatrix} a & b \\ c & d \end{pmatrix}$$

which we have called a 2×2 matrix. Since two transformations like (15) are identical if and only if they possess the same coefficients, Cayley defined two matrices

$$\begin{pmatrix} a & b \\ c & d \end{pmatrix} \quad \text{and} \quad \begin{pmatrix} e & f \\ g & h \end{pmatrix}$$

to be equal if and only if $a = e$, $b = f$, $c = g$, and $d = h$.

Now suppose that the transformation (15) is followed by a second transformation

$$x'' = ex' + fy'$$
$$y'' = gx' + hy'$$
(16)

Then

$$x'' = e(ax + by) + f(cx + dy) = (ea + fc)x + (eb + fd)y$$

and

$$y'' = g(ax + by) + h(cx + dy) = (ga + hc)x + (gb + hd)y$$

This led Cayley to the following definition for the product of two matrices:

$$\begin{pmatrix} e & f \\ g & h \end{pmatrix}\begin{pmatrix} a & b \\ c & d \end{pmatrix} = \begin{pmatrix} ea + fc & eb + fd \\ ga + hc & gb + hd \end{pmatrix}$$

which is, of course, a special case of the general definition of the matrix product we gave on page 44.

It is interesting to observe how, in mathematics, very simple observations can sometimes lead to important, and far reaching, definitions and theorems.

PROBLEMS   1.7

In Problems 1–7 calculate the scalar product of the two vectors.

1. $\begin{pmatrix} 2 \\ 3 \\ -5 \end{pmatrix}; \begin{pmatrix} 3 \\ 0 \\ 4 \end{pmatrix}$

2. $(1, 2, -1, 0); (3, -7, 4, -2)$

3. $\begin{pmatrix} 5 \\ 7 \end{pmatrix}; \begin{pmatrix} 3 \\ -2 \end{pmatrix}$

4. $(8, 3, 1); (7, -4, 3)$

5. $(a, b); (c, d)$

6. $\begin{pmatrix} x \\ y \\ z \end{pmatrix}; \begin{pmatrix} y \\ z \\ x \end{pmatrix}$

7. $(-1, -3, 4, 5); (-1, -3, 4, 5)$

8. Let **a** be an n-vector. Show that $\mathbf{a} \cdot \mathbf{a} \geq 0$.

9. Find conditions on a vector **a** such that $\mathbf{a} \cdot \mathbf{a} = 0$.

In Problems 10–14 perform the indicated computation with $\mathbf{a} = \begin{pmatrix} 1 \\ -2 \\ 4 \end{pmatrix}$, $\mathbf{b} = \begin{pmatrix} 0 \\ -3 \\ -7 \end{pmatrix}$, and $\mathbf{c} = \begin{pmatrix} 4 \\ -1 \\ 5 \end{pmatrix}$.

10. $(2\mathbf{a}) \cdot (3\mathbf{b})$

11. $\mathbf{a} \cdot (\mathbf{b} + \mathbf{c})$

12. $\mathbf{c} \cdot (\mathbf{a} - \mathbf{b})$

13. $(2\mathbf{b}) \cdot (3\mathbf{c} - 5\mathbf{a})$

14. $(\mathbf{a} - \mathbf{c}) \cdot (3\mathbf{b} - 4\mathbf{a})$

In Problems 15–29 perform the indicated computation.

15. $\begin{pmatrix} 2 & 3 \\ -1 & 2 \end{pmatrix}\begin{pmatrix} 4 & 1 \\ 0 & 6 \end{pmatrix}$

16. $\begin{pmatrix} 3 & -2 \\ 1 & 4 \end{pmatrix}\begin{pmatrix} -5 & 6 \\ 1 & 3 \end{pmatrix}$

17. $\begin{pmatrix} 1 & -1 \\ 1 & 1 \end{pmatrix}\begin{pmatrix} -1 & 0 \\ 2 & 3 \end{pmatrix}$

18. $\begin{pmatrix} -5 & 6 \\ 1 & 3 \end{pmatrix}\begin{pmatrix} 3 & -2 \\ 1 & 4 \end{pmatrix}$

19. $\begin{pmatrix} -4 & 5 & 1 \\ 0 & 4 & 2 \end{pmatrix}\begin{pmatrix} 3 & -1 & 1 \\ 5 & 6 & 4 \\ 0 & 1 & 2 \end{pmatrix}$

20. $\begin{pmatrix} 7 & 1 & 4 \\ 2 & -3 & 5 \end{pmatrix}\begin{pmatrix} 1 & 6 \\ 0 & 4 \\ -2 & 3 \end{pmatrix}$

21. $\begin{pmatrix} 1 & 6 \\ 0 & 4 \\ -2 & 3 \end{pmatrix}\begin{pmatrix} 7 & 1 & 4 \\ 2 & -3 & 5 \end{pmatrix}$

22. $\begin{pmatrix} 1 & 4 & -2 \\ 3 & 0 & 4 \end{pmatrix}\begin{pmatrix} 0 & 1 \\ 2 & 3 \end{pmatrix}$

23. $\begin{pmatrix} 1 & 4 & 6 \\ -2 & 3 & 5 \\ 1 & 0 & 4 \end{pmatrix}\begin{pmatrix} 2 & -3 & 5 \\ 1 & 0 & 6 \\ 2 & 3 & 1 \end{pmatrix}$

24. $\begin{pmatrix} 2 & -3 & 5 \\ 1 & 0 & 6 \\ 2 & 3 & 1 \end{pmatrix}\begin{pmatrix} 1 & 4 & 6 \\ -2 & 3 & 5 \\ 1 & 0 & 4 \end{pmatrix}$

25. $(1 \quad 4 \quad 0 \quad 2)\begin{pmatrix} 3 & -6 \\ 2 & 4 \\ 1 & 0 \\ -2 & 3 \end{pmatrix}$

26. $\begin{pmatrix} 3 & 2 & 1 & -2 \\ -6 & 4 & 0 & 3 \end{pmatrix}\begin{pmatrix} 1 \\ 4 \\ 0 \\ 2 \end{pmatrix}$

27. $\begin{pmatrix} 3 & -2 & 1 \\ 4 & 0 & 6 \\ 5 & 1 & 9 \end{pmatrix}\begin{pmatrix} 1 & 0 & 0 \\ 0 & 1 & 0 \\ 0 & 0 & 1 \end{pmatrix}$

28. $\begin{pmatrix} 1 & 0 & 0 \\ 0 & 1 & 0 \\ 0 & 0 & 1 \end{pmatrix}\begin{pmatrix} 3 & -2 & 1 \\ 4 & 0 & 6 \\ 5 & 1 & 9 \end{pmatrix}$

29. $\begin{pmatrix} a & b & c \\ d & e & f \\ g & h & j \end{pmatrix}\begin{pmatrix} 1 & 0 & 0 \\ 0 & 1 & 0 \\ 0 & 0 & 1 \end{pmatrix}$, where $a, b, c, d, e, f, g, h, j$ are real numbers.

30. Find a matrix $A = \begin{pmatrix} a & b \\ c & d \end{pmatrix}$ such that $A\begin{pmatrix} 2 & 3 \\ 1 & 2 \end{pmatrix} = \begin{pmatrix} 1 & 0 \\ 0 & 1 \end{pmatrix}$.

*31. Let a_{11}, a_{12}, a_{21}, and a_{22} be given real numbers such that $a_{11}a_{22} - a_{12}a_{21} \neq 0$.
Find numbers b_{11}, b_{12}, b_{21}, and b_{22} such that $\begin{pmatrix} a_{11} & a_{12} \\ a_{21} & a_{22} \end{pmatrix}\begin{pmatrix} b_{11} & b_{12} \\ b_{21} & b_{22} \end{pmatrix} = \begin{pmatrix} 1 & 0 \\ 0 & 1 \end{pmatrix}$.

32. Verify the associative law for multiplication for the matrices $A = \begin{pmatrix} 2 & -1 & 4 \\ 1 & 0 & 6 \end{pmatrix}$,

$$B = \begin{pmatrix} 1 & 0 & 1 \\ 2 & -1 & 2 \\ 3 & -2 & 0 \end{pmatrix}, \text{ and } C = \begin{pmatrix} 1 & 6 \\ -2 & 4 \\ 0 & 5 \end{pmatrix}.$$

33. As in Example 6, suppose that a group of people have contracted a contagious disease.
These persons have contacts with a second group who in turn have contacts with a third
group. Let $A = \begin{pmatrix} 1 & 0 & 1 & 0 \\ 0 & 1 & 1 & 0 \\ 1 & 0 & 0 & 1 \end{pmatrix}$ represent the contacts between the contagious group
and the members of group 2, and let

$$B = \begin{pmatrix} 1 & 0 & 1 & 0 & 0 \\ 0 & 0 & 0 & 1 & 0 \\ 1 & 1 & 0 & 0 & 0 \\ 0 & 0 & 1 & 0 & 1 \end{pmatrix}$$

represent the contacts between groups 2 and 3. (a) How many people are in each group?
(b) Find the matrix of indirect contacts between groups 1 and 3.

34. Answer the questions of Problem 33 for $A = \begin{pmatrix} 1 & 0 & 1 & 1 & 0 \\ 0 & 1 & 0 & 1 & 1 \end{pmatrix}$ and

$$B = \begin{pmatrix} 1 & 0 & 0 & 0 & 0 & 0 & 1 \\ 0 & 1 & 0 & 1 & 0 & 0 & 0 \\ 1 & 1 & 0 & 0 & 1 & 1 & 1 \\ 0 & 0 & 0 & 1 & 1 & 0 & 1 \\ 0 & 1 & 0 & 0 & 0 & 0 & 0 \end{pmatrix}$$

ORTHOGONAL VECTORS

Two vectors **a** and **b** are said to be **orthogonal** if $\mathbf{a} \cdot \mathbf{b} = 0$. In Problems 35–39 determine
which pairs of vectors are orthogonal.†

35. $\begin{pmatrix} 2 \\ -3 \end{pmatrix}; \begin{pmatrix} 3 \\ 2 \end{pmatrix}$ **36.** $\begin{pmatrix} 2 \\ -3 \end{pmatrix}; \begin{pmatrix} -3 \\ 2 \end{pmatrix}$ **37.** $\begin{pmatrix} 1 \\ 4 \\ -7 \end{pmatrix}; \begin{pmatrix} 2 \\ 3 \\ 2 \end{pmatrix}$

38. (1, 0, 1, 0); (0, 1, 0, 1)   **39.** $\begin{pmatrix} a \\ 0 \\ b \\ 0 \\ c \end{pmatrix}; \begin{pmatrix} 0 \\ d \\ 0 \\ e \\ 0 \end{pmatrix}$

---

†We shall be dealing extensively with orthogonal vectors in Chapters 3 and 4.

40. Determine a number α such that $(1, -2, 3, 5)$ is orthogonal to $(-4, \alpha, 6, -1)$.

41. Determine all numbers α and β such that the vectors $\begin{pmatrix} 1 \\ -\alpha \\ 2 \\ 3 \end{pmatrix}$ and $\begin{pmatrix} 4 \\ 5 \\ -2\beta \\ 7 \end{pmatrix}$ are orthogonal.

42. Using the definition of the scalar product, prove Theorem 1.

43. A manufacturer of custom-designed jewelry has orders for two rings, three pairs of earrings, five pins, and one necklace. The manufacturer estimates that it takes 1 hour of labor to make a ring, $1\frac{1}{2}$ hours to make a pair of earrings, $\frac{1}{2}$ hour for each pin, and 2 hours to make a necklace.
   a. Express the manufacturer's orders as a row vector.
   b. Express the hourly requirements for the various types of jewelry as a column vector.
   c. Use the scalar product to calculate the total number of hours it will require to complete all the orders.

44. A tourist returned from a European trip with the following foreign currency: 1000 Austrian schillings, 20 British pounds, 100 French francs, 5000 Italian lire, and 50 German marks. In American money, a schilling was worth $0.055, the pound $1.80, the franc $0.20, the lira $0.001, and the mark $0.40.
   a. Express the quantity of each currency by means of a row vector.
   b. Express the value of each currency in American money by means of a column vector.
   c. Use the scalar product to compute how much the tourist's foreign currency was worth in American money.

45. A company pays its executives a salary and gives them shares of its stock as an annual bonus. Last year the president of the company received $80,000 and 50 shares of stock, each of the three vice-presidents was paid $45,000 and 20 shares of stock, and the treasurer was paid $40,000 and 10 shares of stock.
   a. Express the payments to the executives in money and stock by means of a 2×3 matrix.
   b. Express the number of executives of each rank by means of a column vector.
   c. Use matrix multiplication to calculate the total amount of money and the total number of shares of stock the company paid these executives last year.

46. Sales, unit gross profits, and unit taxes for sales of a large corporation are given in the following table:

Month	Product Sales of Item			Item	Unit Profit (in hundreds of dollars)	Unit Taxes (in hundreds of dollars)
	I	II	III			
January	4	2	20	I	3.5	1.5
February	6	1	9	II	2.75	2
March	5	3	12	III	1.5	0.6
April	8	2.5	20			

Find a matrix that shows total profits and taxes in each of the 4 months.

47. Let A be a square matrix. Then A^2 is defined simply as AA. Calculate $\begin{pmatrix} 2 & -1 \\ 4 & 6 \end{pmatrix}^2$.

48. Calculate A^2, where $A = \begin{pmatrix} 1 & -2 & 4 \\ 2 & 0 & 3 \\ 1 & 1 & 5 \end{pmatrix}$.

49. Calculate A^3, where $A = \begin{pmatrix} -1 & 2 \\ 3 & 4 \end{pmatrix}$.

50. Calculate A^2, A^3, A^4, and A^5, where

$$A = \begin{pmatrix} 0 & 1 & 0 & 0 \\ 0 & 0 & 1 & 0 \\ 0 & 0 & 0 & 1 \\ 0 & 0 & 0 & 0 \end{pmatrix}$$

51. Calculate A^2, A^3, A^4, and A^5, where

$$A = \begin{pmatrix} 0 & 1 & 0 & 0 & 0 \\ 0 & 0 & 1 & 0 & 0 \\ 0 & 0 & 0 & 1 & 0 \\ 0 & 0 & 0 & 0 & 1 \\ 0 & 0 & 0 & 0 & 0 \end{pmatrix}$$

52. An $n \times n$ matrix A has the property that AB is the zero matrix for any $n \times n$ matrix B. Prove that A is the zero matrix.

53. A **probability matrix** is a square matrix having two properties: (*i*) every component is nonnegative (≥ 0) and (*ii*) the sum of the elements in each row is 1. The following are probability matrices:

$$P = \begin{pmatrix} \frac{1}{3} & \frac{1}{3} & \frac{1}{3} \\ \frac{1}{4} & \frac{1}{2} & \frac{1}{4} \\ 0 & 0 & 1 \end{pmatrix} \quad \text{and} \quad Q = \begin{pmatrix} \frac{1}{6} & \frac{1}{6} & \frac{2}{3} \\ 0 & 1 & 0 \\ \frac{1}{5} & \frac{1}{5} & \frac{3}{5} \end{pmatrix}$$

Show that PQ is a probability matrix.

*54. Let P be a probability matrix. Show that P^2 is a probability matrix.

**55. Let P and Q be probability matrices of the same size. Prove that PQ is a probability matrix.

56. Prove formula (6) by using the associative law [equation (5)].

*57. A round robin tennis tournament can be organized in the following way. Each of the n players plays all the others, and the results are recorded in an $n \times n$ matrix R as follows:

$$R_{ij} = \begin{cases} 1 & \text{if the } i\text{th player beats the } j\text{th player} \\ 0 & \text{if the } i\text{th player loses to the } j\text{th player} \\ 0 & \text{if } i = j \end{cases}$$

The ith player is then assigned the score

$$S_i = \sum_{j=1}^{n} R_{ij} + \frac{1}{2} \sum_{j=1}^{n} (R^2)_{ij}\dagger$$

$\dagger (R^2)_{ij}$ is the ijth component of the matrix R^2.

a.   In a tournament between four players

$$R = \begin{pmatrix} 0 & 1 & 0 & 0 \\ 0 & 0 & 1 & 1 \\ 1 & 0 & 0 & 0 \\ 1 & 0 & 1 & 0 \end{pmatrix}$$

Rank the players according to their scores.

b.   Interpret the meaning of the score.

58. Let O be the $m \times n$ zero matrix and let A be an $n \times p$ matrix. Show that $OA = O_1$, where O_1 is the $m \times p$ zero matrix.

59. Verify the distributive law [equation (7)] for the matrices

$$A = \begin{pmatrix} 1 & 2 & 4 \\ 3 & -1 & 0 \end{pmatrix} \qquad B = \begin{pmatrix} 2 & 7 \\ -1 & 4 \\ 6 & 0 \end{pmatrix} \qquad C = \begin{pmatrix} -1 & 2 \\ 3 & 7 \\ 4 & 1 \end{pmatrix}$$

In Problems 60–67 evaluate the given sums.

60. $\displaystyle\sum_{k=1}^{4} 2k$

61. $\displaystyle\sum_{i=1}^{3} i^3$

62. $\displaystyle\sum_{k=0}^{6} 1$

63. $\displaystyle\sum_{k=1}^{8} 3^k$

64. $\displaystyle\sum_{i=2}^{5} \frac{1}{1+i}$

65. $\displaystyle\sum_{j=5}^{7} \frac{2j+3}{j-2}$

66. $\displaystyle\sum_{i=1}^{3}\sum_{j=1}^{4} ij$

67. $\displaystyle\sum_{k=1}^{3}\sum_{j=2}^{4} k^2 j^3$

In Problems 68–80 write each sum using the Σ notation.

68. $1 + 2 + 4 + 8 + 16$

69. $1 - 3 + 9 - 27 + 81 - 243$

70. $\dfrac{2}{3} + \dfrac{3}{4} + \dfrac{4}{5} + \dfrac{5}{6} + \dfrac{6}{7} + \dfrac{7}{8} + \cdots + \dfrac{n}{n+1}$

71. $1 + 2^{1/2} + 3^{1/3} + 4^{1/4} + 5^{1/5} + \cdots + n^{1/n}$

72. $1 + x^3 + x^6 + x^9 + x^{12} + x^{15} + x^{18} + x^{21}$

73. $-1 + \dfrac{1}{a} - \dfrac{1}{a^2} + \dfrac{1}{a^3} - \dfrac{1}{a^4} + \dfrac{1}{a^5} - \dfrac{1}{a^6} + \dfrac{1}{a^7} - \dfrac{1}{a^8} + \dfrac{1}{a^9}$

74. $1 \cdot 3 + 3 \cdot 5 + 5 \cdot 7 + 7 \cdot 9 + 9 \cdot 11 + 11 \cdot 13 + 13 \cdot 15 + 15 \cdot 17$

75. $2^2 \cdot 4 + 3^2 \cdot 6 + 4^2 \cdot 8 + 5^2 \cdot 10 + 6^2 \cdot 12 + 7^2 \cdot 14$

76. $a_{11} + a_{12} + a_{13} + a_{21} + a_{22} + a_{23}$

77. $a_{11} + a_{12} + a_{21} + a_{22} + a_{31} + a_{32}$

78. $a_{21} + a_{22} + a_{23} + a_{24} + a_{31} + a_{32} + a_{33} + a_{34} + a_{41} + a_{42} + a_{43} + a_{44}$

79. $a_{31}b_{12} + a_{32}b_{22} + a_{33}b_{32} + a_{34}b_{42} + a_{35}b_{52}$

80. $a_{21}b_{11}c_{15} + a_{21}b_{12}c_{25} + a_{21}b_{13}c_{35} + a_{21}b_{14}c_{45}$
$\quad + a_{22}b_{21}c_{15} + a_{22}b_{22}c_{25} + a_{22}b_{23}c_{35} + a_{22}b_{24}c_{45}$
$\quad + a_{23}b_{31}c_{15} + a_{23}b_{32}c_{25} + a_{23}b_{33}c_{35} + a_{23}b_{34}c_{45}$

81. Prove formula (12) by writing out the terms in

$$\sum_{k=M}^{N} (a_k + b_k)$$

82. Prove formula (13).

[*Hint:* Use (11) to show that $\sum\limits_{k=M}^{N} (-a_k) = -\sum\limits_{k=M}^{N} a_n$. Then use (12).]

83. Prove formula (14).

1.8 MATRICES AND LINEAR SYSTEMS OF EQUATIONS

In Section 1.3, page 15, we discussed the following systems of m equations in n unknowns:

$$\begin{array}{cccccc}
a_{11}x_1 & + & a_{12}x_2 & + \cdots + & a_{1n}x_n & = b_1 \\
a_{21}x_1 & + & a_{22}x_2 & + \cdots + & a_{2n}x_n & = b_2 \\
\vdots & & \vdots & & \vdots & \vdots \\
a_{m1}x_1 & + & a_{m2}x_2 & + \cdots + & a_{mn}x_n & = b_m
\end{array} \tag{1}$$

Let

$$A = \begin{pmatrix}
a_{11} & a_{12} & \cdots & a_{1n} \\
a_{21} & a_{22} & \cdots & a_{2n} \\
\vdots & \vdots & & \vdots \\
a_{m1} & a_{m2} & \cdots & a_{mn}
\end{pmatrix}$$

be the coefficient matrix, $\mathbf{x}$ the vector $\begin{pmatrix} x_1 \\ x_2 \\ \vdots \\ x_n \end{pmatrix}$, and $\mathbf{b}$ the vector $\begin{pmatrix} b_1 \\ b_2 \\ \vdots \\ b_m \end{pmatrix}$. Since A is an $m \times n$ matrix and $\mathbf{x}$ is an $n \times 1$ matrix, the matrix product $A\mathbf{x}$ is defined as an $m \times 1$ matrix. It is not difficult to see that system (1) can be written as

---

Matrix Form of a Linear System of Equations

$$A\mathbf{x} = \mathbf{b}$$

$$(2)$$

EXAMPLE 1   **Writing a System in Matrix Form**   Consider the system

$$\begin{array}{rrrr}
2x_1 & + 4x_2 & + 6x_3 & = 18 \\
4x_1 & + 5x_2 & + 6x_3 & = 24 \\
3x_1 & + \ x_2 & - 2x_3 & = 4
\end{array} \tag{3}$$

(See Example 1.3.1 on page 7.) This can be written in form $A\mathbf{x} = \mathbf{b}$ with

$$A = \begin{pmatrix} 2 & 4 & 6 \\ 4 & 5 & 6 \\ 3 & 1 & -2 \end{pmatrix}, \ \mathbf{x} = \begin{pmatrix} x_1 \\ x_2 \\ x_3 \end{pmatrix}, \text{ and } \mathbf{b} = \begin{pmatrix} 18 \\ 24 \\ 4 \end{pmatrix}.$$

■

It is obviously easier to write out system (1) in the form $A\mathbf{x} = \mathbf{b}$. There are many other advantages, too. In Section 1.9 we shall see how a square system can be solved almost at once if we know a matrix called the *inverse* of A. Even without that, as we saw in Section 1.3, computations are much easier to write down by using an augmented matrix.

If $\mathbf{b} = \begin{pmatrix} 0 \\ 0 \\ \vdots \\ 0 \end{pmatrix}$ is the $m \times 1$ zero vector, then system (1) is homogeneous (see Section 1.4) and can be written

$$A\mathbf{x} = \mathbf{0} \qquad \text{Matrix form of a homogeneous system of equations}$$

There is a fundamental relationship between homogeneous and nonhomogeneous systems. Let A be an $m \times n$ matrix

$$\mathbf{x} = \begin{pmatrix} x_1 \\ x_2 \\ \vdots \\ x_n \end{pmatrix}, \qquad \mathbf{b} = \begin{pmatrix} b_1 \\ b_2 \\ \vdots \\ b_m \end{pmatrix}, \quad \text{and} \quad \mathbf{0} \overset{m \text{ zeros}}{=} \begin{pmatrix} 0 \\ 0 \\ \vdots \\ 0 \end{pmatrix}$$

The general nonhomogeneous system can be written as

$$A\mathbf{x} = \mathbf{b} \tag{4}$$

Associated homogeneous system

With A and $\mathbf{x}$ as in (4), we define the **associated homogeneous system** by

$$A\mathbf{x} = \mathbf{0} \tag{5}$$

THEOREM 1   Let $\mathbf{x}_1$ and $\mathbf{x}_2$ be solutions of the nonhomogeneous system (4). Then their difference, $\mathbf{x}_1 - \mathbf{x}_2$, is a solution of the related homogeneous system (5).

By the distributive
law (7) on page 49
↓

Proof $A(\mathbf{x}_1 - \mathbf{x}_2) = A\mathbf{x}_1 - A\mathbf{x}_2 = \mathbf{b} - \mathbf{b} = \mathbf{0}.$ ∎

COROLLARY   Let $\mathbf{x}$ be a particular solution to the nonhomogeneous system (4) and let $\mathbf{y}$ be another solution to (4). Then there exists a solution $\mathbf{h}$ to the homogeneous system (5) such that

$$\mathbf{y} = \mathbf{x} + \mathbf{h} \tag{6}$$

Proof   If $\mathbf{h}$ is defined by $\mathbf{h} = \mathbf{y} - \mathbf{x}$, then $\mathbf{h}$ solves (5) by Theorem 1 and $\mathbf{y} = \mathbf{x} + \mathbf{h}$. ∎

Theorem 1 and its corollary are very useful. They tell us that

---

In order to find all solutions to the nonhomogeneous system (4), it is sufficient to find *one* solution to (4) and all solutions to the associated homogeneous system (5).

Remark. A very similar result holds for solutions of homogeneous and nonhomogeneous linear differential equations (See Problems 23 and 24). One of the many nice things about mathematics is that seemingly very different topics are closely interrelated.

EXAMPLE 2   **Writing an Infinite Number of Solutions as a Particular Solution to a Nonhomogeneous System Plus Solutions to the Homogeneous System**
Find all solutions to the nonhomogeneous system

$$x + 2x_2 - x_3 = 2$$
$$2x_1 + 3x_2 + 5x_3 = 5$$
$$-x_1 - 3x_2 + 8x_3 = -1$$

by using the result given above.

Solution   First we find one solution by row reduction:

$$\begin{pmatrix} 1 & 2 & -1 & | & 2 \\ 2 & 3 & 5 & | & 5 \\ -1 & -3 & 8 & | & -1 \end{pmatrix} \xrightarrow[R_3 \to R_3 + R_1]{R_2 \to R_2 - 2R_1} \begin{pmatrix} 1 & 2 & -1 & | & 2 \\ 0 & -1 & 7 & | & 1 \\ 0 & -1 & 7 & | & 1 \end{pmatrix}$$

$$\xrightarrow[R_3 \to R_3 - R_2]{R_1 \to R_1 + 2R_2} \begin{pmatrix} 1 & 0 & 13 & | & 4 \\ 0 & -1 & 7 & | & 1 \\ 0 & 0 & 0 & | & 0 \end{pmatrix}$$

We see that there are an infinite number of solutions. Setting $x_3 = 0$ (any other number would do), we obtain $x_1 = 4$ and $x_2 = -1$. So one particular solution is $\mathbf{x_p} = (4, -1, 0)$.

Row reduction of the associated homogeneous system leads to

$$\begin{pmatrix} 1 & 0 & 13 & | & 0 \\ 0 & -1 & 7 & | & 0 \\ 0 & 0 & 0 & | & 0 \end{pmatrix}$$

Therefore all solutions to the homogeneous system satisfy

$$x_1 = -13x_3, \qquad x_2 = 7x_3$$

or

$$\mathbf{x_h} = (x_1, x_2, x_3) = (-13x_3, 7x_3, x_3) = x_3(-13, 7, 1)$$

Thus each solution to (6) can be written

$$\mathbf{x} = \mathbf{x_p} + \mathbf{x_h} = (4, -1, 0) + x_3(-13, 7, 1)$$

for an appropriate value of x_3. For example, $x_3 = 0$ yields the solution $(4, -1, 0)$, whereas $x_3 = 2$ gives the solution $(-22, 13, 2)$.   ∎

PROBLEMS   1.8

In Problems 1–6 write the given system in the form $A\mathbf{x} = \mathbf{b}$.

1. $2x_1 - x_2 = 3$
$\quad 4x_1 + 5x_2 = 7$

2. $x_1 - x_2 + 3x_3 = 11$
$\quad 4x_1 + x_2 - x_3 = -4$
$\quad 2x_1 - x_2 + 3x_3 = 10$

3. $3x_1 + 6x_2 - 7x_3 = 0$
$\quad 2x_1 - x_2 + 3x_3 = 1$

4. $4x_1 - x_2 + x_3 - x_4 = -7$
$\quad 3x_1 + x_2 - 5x_3 + 6x_4 = 8$
$\quad 2x_1 - x_2 + x_3 \quad\quad = 9$

5. $\quad\quad x_2 - x_3 = 7$
$\quad x_1 \quad\quad + x_3 = 2$
$\quad 3x_1 + 2x_2 \quad\quad = -5$

6. $2x_1 + 3x_2 - x_3 = 0$
$\quad -4x_1 + 2x_2 + x_3 = 0$
$\quad 7x_1 + 3x_2 - 9x_3 = 0$

In Problems 7–15 write out the system of equations represented by the given augmented matrix.

7. $\begin{pmatrix} 1 & 1 & -1 & | & 7 \\ 4 & -1 & 5 & | & 4 \\ 6 & 1 & 3 & | & 20 \end{pmatrix}$

8. $\begin{pmatrix} 0 & 1 & | & 2 \\ 1 & 0 & | & 3 \end{pmatrix}$

9. $\begin{pmatrix} 2 & 0 & 1 & | & 2 \\ -3 & 4 & 0 & | & 3 \\ 0 & 5 & 6 & | & 5 \end{pmatrix}$

10. $\begin{pmatrix} 2 & 3 & 1 & | & 2 \\ 0 & 4 & 1 & | & 3 \\ 0 & 0 & 0 & | & 0 \end{pmatrix}$

11. $\begin{pmatrix} 1 & 0 & 0 & 0 & | & 2 \\ 0 & 1 & 0 & 0 & | & 3 \\ 0 & 0 & 1 & 0 & | & -5 \\ 0 & 0 & 0 & 1 & | & 6 \end{pmatrix}$

12. $\begin{pmatrix} 2 & 3 & 1 & | & 0 \\ 4 & -1 & 5 & | & 0 \\ 3 & 6 & -7 & | & 0 \end{pmatrix}$

13. $\begin{pmatrix} 6 & 2 & 1 & | & 2 \\ -2 & 3 & 1 & | & 4 \\ 0 & 0 & 0 & | & 2 \end{pmatrix}$

14. $\begin{pmatrix} 3 & 1 & 5 & | & 6 \\ 2 & 3 & 2 & | & 4 \end{pmatrix}$

15. $\begin{pmatrix} 7 & 2 & | & 1 \\ 3 & 1 & | & 2 \\ 6 & 9 & | & 3 \end{pmatrix}$

16. Find a matrix A and vectors $\mathbf{x}$ and $\mathbf{b}$ such that the system represented by the following augmented matrix can be written in the form $A\mathbf{x} = \mathbf{b}$ and solve the system.

$$\begin{pmatrix} 2 & 0 & 0 & | & 3 \\ 0 & 4 & 0 & | & 5 \\ 0 & 0 & -5 & | & 2 \end{pmatrix}$$

In Problems 17–22 find all solutions to the given nonhomogeneous system by first finding one solution (if possible) and then finding all solutions to the associated homogeneous system.

17. $\quad x_1 - 3x_2 = 2$
$\quad -2x_1 + 6x_2 = -4$

18. $\quad x_1 - x_2 + x_3 = 6$
$\quad 3x_1 - 3x_2 + 3x_3 = 18$

19. $\begin{aligned} x_1 - x_2 - x_3 &= 2 \\ 2x_1 + x_2 + 2x_3 &= 4 \\ x_1 - 4x_2 - 5x_3 &= 2 \end{aligned}$

20. $\begin{aligned} x_1 - x_2 - x_3 &= 2 \\ 2x_1 + x_2 + 2x_3 &= 4 \\ x_1 - 4x_2 - 5x_3 &= 2 \end{aligned}$

21. $\begin{aligned} x_1 + x_2 - x_3 + 2x_4 &= 3 \\ 3x_1 + 2x_2 + x_3 - x_4 &= 5 \end{aligned}$

22. $\begin{aligned} x_1 - x_2 + x_3 - x_4 &= -2 \\ -2x_1 + 3x_2 - x_3 + 2x_4 &= 5 \\ 4x_1 - 2x_2 + 2x_3 - 3x_4 &= 6 \end{aligned}$

Calculus †**23.** Consider the linear, homogeneous second-order differential equation

$$y''(x) + a(x)y'(x) + b(x)y(x) = 0 \tag{7}$$

where $a(x)$ and $b(x)$ are continuous and the unknown function y is assumed to have a second derivative. Show that if y_1 and y_2 are solutions to (7), then $c_1y_1 + c_2y_2$ is a solution for any constants y_1 and y_2.

Calculus **24.** Suppose that y_p and y_q are solutions to the nonhomogeneous equation

$$y''(x) + a(x)y'(x) + b(x)y(x) = f(x) \tag{8}$$

Show that $y_p - y_q$ is a solution to (7).

1.9 THE INVERSE OF A SQUARE MATRIX

In this section we define two kinds of matrices that are central to matrix theory. We begin with a simple example. Let $A = \begin{pmatrix} 2 & 5 \\ 1 & 3 \end{pmatrix}$ and $B = \begin{pmatrix} 3 & -5 \\ -1 & 2 \end{pmatrix}$. Then an easy computation shows that $AB = BA = I_2$, where $I_2 = \begin{pmatrix} 1 & 0 \\ 0 & 1 \end{pmatrix}$. The matrix I_2 is called the 2×2 *identity matrix*. The matrix B is called the *inverse* of A and is written A^{-1}.

DEFINITION 1   **Identity Matrix** The $n \times n$ **identity matrix** I_n is the $n \times n$ matrix with 1's down the **main diagonal‡** and 0's everywhere else. That is,

$$I_n = (b_{ij}) \quad \text{where} \quad b_{ij} = \begin{cases} 1 & \text{if } i = j \\ 0 & \text{if } i \neq j \end{cases} \tag{1}$$

---

† The symbol Calculus indicates that calculus is needed to solve the problem.

‡ The main diagonal of $A = (a_{ij})$ consists of the components a_{11}, a_{22}, a_{33}, and so on. Unless otherwise stated, we shall refer to the main diagonal simply as the **diagonal.**

EXAMPLE 1   **Two Identity Matrices**

$$I_3 = \begin{pmatrix} 1 & 0 & 0 \\ 0 & 1 & 0 \\ 0 & 0 & 1 \end{pmatrix} \quad \text{and} \quad I_5 = \begin{pmatrix} 1 & 0 & 0 & 0 & 0 \\ 0 & 1 & 0 & 0 & 0 \\ 0 & 0 & 1 & 0 & 0 \\ 0 & 0 & 0 & 1 & 0 \\ 0 & 0 & 0 & 0 & 1 \end{pmatrix}$$ ∎

THEOREM 1   Let A be a square $n \times n$ matrix. Then

$$AI_n = I_nA = A$$

That is, I_n commutes with every $n \times n$ matrix and leaves it unchanged after multiplication on the left or right.

Note. I_n functions for $n \times n$ matrices the way the number 1 functions for real numbers (since $1 \cdot a = a \cdot 1 = a$ for every real number a).

Proof   Let c_{ij} be the ijth element of AI_n. Then

$$c_{ij} = a_{i1}b_{1j} + a_{i2}b_{2j} + \cdots + a_{ij}b_{jj} + \cdots + a_{in}b_{nj}$$

But, from (1), this sum is equal to a_{ij}. Thus $AI_n = A$. In a similar fashion we can show that $I_nA = A$, and this proves the theorem.   ∎

Notation. From now on we shall write the identity matrix simply as I, since if A is $n \times n$, the products IA and AI are defined only if I is also $n \times n$.

DEFINITION 2   **The Inverse of a Matrix** Let A and B be $n \times n$ matrices. Suppose that

$$AB = BA = I$$

Then B is called the **inverse** of A and is written as A^{-1}. We then have

$$AA^{-1} = A^{-1}A = I$$

If A has an inverse, then A is said to be **invertible**.

Remark 1. From this definition it immediately follows that $(A^{-1})^{-1} = A$ if A is invertible.

Remark 2. This definition does *not* state that every square matrix has an inverse. In fact there are many square matrices that have no inverse. (See, for instance, Example 3 below.)

In Definition 2 we defined *the* inverse of a matrix. This statement suggests that inverses are unique. This is indeed the case, as the following theorem shows.

THEOREM 2   If a square matrix A is invertible, then its inverse is unique.

Proof   Suppose B and C are two inverses for A. We can show that $B = C$. By definition, we have $AB = BA = I$ and $AC = CA = I$. Then $B(AC) = BI = B$ and $(BA)C = IC = C$. But $B(AC) = (BA)C$ by the associative law of matrix multiplication. Hence $B = C$ and the theorem is proved.   ∎

Another important fact about inverses is given below.

THEOREM 3   Let A and B be invertible $n \times n$ matrices. Then AB is invertible and

$$(AB)^{-1} = B^{-1}A^{-1}$$

Proof   To prove this result, we refer to Definition 2. That is, $B^{-1}A^{-1} = (AB)^{-1}$ if and only if $B^{-1}A^{-1}(AB) = (AB)(B^{-1}A^{-1}) = I$. But this follows since

Equation (6) on page 49

$$(B^{-1}A^{-1})(AB) = B^{-1}(A^{-1}A)B = B^{-1}IB = B^{-1}B = I$$

and

$$(AB)(B^{-1}A^{-1}) = A(BB^{-1})A^{-1} = AIA^{-1} = AA^{-1} = I.$$ ∎

Consider the system of n equations in n unknowns

$$A\mathbf{x} = \mathbf{b}$$

and suppose that A is invertible. Then

$$A^{-1}A\mathbf{x} = A^{-1}\mathbf{b} \qquad \text{we multiplied on the left by } A^{-1}$$
$$I\mathbf{x} = A^{-1}\mathbf{b} \qquad A^{-1}A = I$$
$$\mathbf{x} = A^{-1}\mathbf{b} \qquad I\mathbf{x} = \mathbf{x}$$

This is a solution to the system because

$$A\mathbf{x} = A(A^{-1}\mathbf{b}) = (AA^{-1})\mathbf{b} = I\mathbf{b} = \mathbf{b}$$

If $\mathbf{y}$ is a vector with $A\mathbf{y} = \mathbf{b}$, then the computation above shows that $\mathbf{y} = A^{-1}\mathbf{b}$. That is, $\mathbf{y} = \mathbf{x}$. We have shown the following:

> If A is invertible, the system $A\mathbf{x} = \mathbf{b}$ has the unique solution $\mathbf{x} = A^{-1}\mathbf{b}$.

(2)

This is one of the reasons we study matrix inverses.

There are two basic questions that come to mind once we have defined the inverse of a matrix.

Question 1.   What matrices have inverses?

Question 2.   If a matrix has an inverse, how can we compute it?

We answer both questions in this section. Rather than starting by giving you what seems to be a set of arbitrary rules, we look first at what happens in the 2×2 case.

EXAMPLE 2   **Finding the Inverse of a 2 × 2 Matrix**   Let $A = \begin{pmatrix} 2 & -3 \\ -4 & 5 \end{pmatrix}$. Compute A^{-1} if it exists.

Solution   Suppose that A^{-1} exists. We write $A^{-1} = \begin{pmatrix} x & y \\ z & w \end{pmatrix}$ and use the fact that $AA^{-1} = I$. Then

$$AA^{-1} = \begin{pmatrix} 2 & -3 \\ -4 & 5 \end{pmatrix}\begin{pmatrix} x & y \\ z & w \end{pmatrix} = \begin{pmatrix} 2x - 3z & 2y - 3w \\ -4x + 5z & -4y + 5w \end{pmatrix} = \begin{pmatrix} 1 & 0 \\ 0 & 1 \end{pmatrix}$$

The last two matrices can be equal only if each of their corresponding components are equal. This means that

$$2x \quad\;\; - 3z \quad\quad\;\; = 1 \tag{3}$$
$$2y \quad\quad - 3w = 0 \tag{4}$$
$$-4x \quad\quad + 5z \quad\quad = 0 \tag{5}$$
$$-4y \quad\quad + 5w = 1 \tag{6}$$

This is a system of four equations in four unknowns. Note that there are two equations involving x and z only [equations (3) and (5)] and two equations involving y and w only [equations (4) and (6)]. We write these two systems in augmented-matrix form:

$$\left(\begin{array}{cc|c} 2 & -3 & 1 \\ -4 & 5 & 0 \end{array}\right) \tag{7}$$

$$\left(\begin{array}{cc|c} 2 & -3 & 0 \\ -4 & 5 & 1 \end{array}\right) \tag{8}$$

Now, we know from Section 1.3 that if system (7) (in the variables x and z) has a unique solution, then Gauss-Jordan elimination of (7) will result in

$$\begin{pmatrix} 1 & 0 & \big| & x \\ 0 & 1 & \big| & z \end{pmatrix}$$

where (x, z) is the unique pair of numbers that satisfies $2x - 3z = 1$ and $-4x + 5z = 0$. Similarly, row reduction of (8) will result in

$$\begin{pmatrix} 1 & 0 & \big| & y \\ 0 & 1 & \big| & w \end{pmatrix}$$

where (y, w) is the unique pair of numbers that satisfies $2y - 3w = 0$ and $-4y + 5w = 1$.

Since the coefficient matrices in (7) and (8) are the same, we can perform the row reductions on the two augmented matrices simultaneously by considering the new augmented matrix

$$\begin{pmatrix} 2 & -3 & \big| & 1 & 0 \\ -4 & 5 & \big| & 0 & 1 \end{pmatrix} \tag{9}$$

If A is invertible, then the system defined by (3), (4), (5), and (6) has a unique solution and, by what we said above, row reduction will result in

$$\begin{pmatrix} 1 & 0 & \big| & x & y \\ 0 & 1 & \big| & z & w \end{pmatrix}$$

We now carry out the computation, noting that the matrix on the left in (9) is A and the matrix on the right in (9) is I:

$$\begin{pmatrix} 2 & -3 & \big| & 1 & 0 \\ -4 & 5 & \big| & 0 & 1 \end{pmatrix} \xrightarrow{R_1 \to \frac{1}{2}R_1} \begin{pmatrix} 1 & -\frac{3}{2} & \big| & \frac{1}{2} & 0 \\ -4 & 5 & \big| & 0 & 1 \end{pmatrix}$$

$$\xrightarrow{R_2 \to R_2 + 4R_1} \begin{pmatrix} 1 & -\frac{3}{2} & \big| & \frac{1}{2} & 0 \\ 0 & -1 & \big| & 2 & 1 \end{pmatrix}$$

$$\xrightarrow{R_2 \to -R_2} \begin{pmatrix} 1 & -\frac{3}{2} & \big| & \frac{1}{2} & 0 \\ 0 & 1 & \big| & -2 & -1 \end{pmatrix}$$

$$\xrightarrow{R_1 \to R_1 + \frac{3}{2}R_2} \begin{pmatrix} 1 & 0 & \big| & -\frac{5}{2} & -\frac{3}{2} \\ 0 & 1 & \big| & -2 & -1 \end{pmatrix}$$

Thus $x = -\frac{5}{2}$, $y = -\frac{3}{2}$, $z = -2$, $w = -1$ and $\begin{pmatrix} x & y \\ z & w \end{pmatrix} = \begin{pmatrix} -\frac{5}{2} & -\frac{3}{2} \\ -2 & -1 \end{pmatrix}$. We compute

$$\begin{pmatrix} 2 & -3 \\ -4 & 5 \end{pmatrix}\begin{pmatrix} -\frac{5}{2} & -\frac{3}{2} \\ -2 & -1 \end{pmatrix} = \begin{pmatrix} 1 & 0 \\ 0 & 1 \end{pmatrix}$$

and

$$\begin{pmatrix} -\frac{5}{2} & -\frac{3}{2} \\ -2 & -1 \end{pmatrix}\begin{pmatrix} 2 & -3 \\ -4 & 5 \end{pmatrix} = \begin{pmatrix} 1 & 0 \\ 0 & 1 \end{pmatrix}$$

Thus A is invertible and $A^{-1} = \begin{pmatrix} -\frac{5}{2} & -\frac{3}{2} \\ -2 & -1 \end{pmatrix}$.   ∎

EXAMPLE 3   **A 2 × 2 Matrix That Is Not Invertible**   Let $A = \begin{pmatrix} 1 & 2 \\ -2 & -4 \end{pmatrix}$. Calculate A^{-1} if it exists.

Solution   If $A^{-1} = \begin{pmatrix} x & y \\ z & w \end{pmatrix}$ exists, then

$$AA^{-1} = \begin{pmatrix} 1 & 2 \\ -2 & -4 \end{pmatrix}\begin{pmatrix} x & y \\ z & w \end{pmatrix} = \begin{pmatrix} x + 2z & y + 2w \\ -2x - 4z & -2y - 4w \end{pmatrix} = \begin{pmatrix} 1 & 0 \\ 0 & 1 \end{pmatrix}$$

This leads to the system

$$
\begin{aligned}
x \quad\;\; + 2z \qquad\;\; &= 1 \\
y \qquad + 2w &= 0 \\
-2x \qquad - 4z \qquad\;\; &= 0 \\
-2y \qquad - 4w &= 1
\end{aligned}
\tag{10}
$$

Using the same reasoning as in Example 1, we can write this system in the augmented-matrix form $(A|I)$ and row-reduce:

$$\begin{pmatrix} 1 & 2 & | & 1 & 0 \\ -2 & -4 & | & 0 & 1 \end{pmatrix} \xrightarrow{R_2 \to R_2 + 2R_1} \begin{pmatrix} 1 & 2 & | & 1 & 0 \\ 0 & 0 & | & 2 & 1 \end{pmatrix}$$

This is as far as we can go. The last line reads $0 = 2$ or $0 = 1$, depending on which of the two systems of equations (in x and z or in y and w) is being solved. Thus system (10) is inconsistent and A is not invertible.   ∎

The last two examples illustrate a procedure that always works when you are trying to find the inverse of a matrix.

Procedure for Computing the Inverse of a Square Matrix A

Step 1.   Write the augmented matrix $(A|I)$.

Step 2.   Use row reduction to reduce the matrix A to its reduced row echelon form.

Step 3.   Decide if A is invertible.
    a.   If A can be reduced to the identity matrix I, then A^{-1} will be the matrix to the right of the vertical bar.
    b.   If the row reduction of A leads to a row of zeros to the left of the vertical bar, then A is not invertible.

Remark. We can rephrase (a) and (b) as follows:

> *A square matrix A is invertible if and only if its reduced row echelon form is the identity matrix.*

Let $A = \begin{pmatrix} a_{11} & a_{12} \\ a_{21} & a_{22} \end{pmatrix}$. Then, as in equation (1.2.7), page 4, we define

Determinant of a
2×2 matrix

$$\boxed{\text{Determinant of } A = a_{11}a_{22} - a_{12}a_{21}} \tag{11}$$

We abbreviate the determinant of A by det A.

THEOREM 4   Let A be a 2×2 matrix. Then:

i. A is invertible if and only if det $A \neq 0$.

ii. If det $A \neq 0$, then

$$\boxed{A^{-1} = \frac{1}{\det A}\begin{pmatrix} a_{22} & -a_{12} \\ -a_{21} & a_{11} \end{pmatrix}^\dagger} \tag{12}$$

Proof   First suppose that det $A \neq 0$ and let $B = (1/\det A)\begin{pmatrix} a_{22} & -a_{12} \\ -a_{21} & a_{11} \end{pmatrix}$. Then

$$BA = \frac{1}{\det A}\begin{pmatrix} a_{22} & -a_{12} \\ -a_{21} & a_{11} \end{pmatrix}\begin{pmatrix} a_{11} & a_{12} \\ a_{21} & a_{22} \end{pmatrix}$$

$$= \frac{1}{a_{11}a_{22} - a_{12}a_{21}}\begin{pmatrix} a_{22}a_{11} - a_{12}a_{21} & 0 \\ 0 & -a_{21}a_{12} + a_{11}a_{22} \end{pmatrix} = \begin{pmatrix} 1 & 0 \\ 0 & 1 \end{pmatrix} = I$$

Similarly, $AB = I$, which shows that A is invertible and that $B = A^{-1}$. We still must show that if A is invertible, then det $A \neq 0$. To do so we consider the system

$$\begin{aligned} a_{11}x_1 + a_{12}x_2 &= b_1 \\ a_{21}x_1 + a_{22}x_2 &= b_2 \end{aligned} \tag{13}$$

We do this because we know from our Summing Up Theorem (Theorem 1.2.1, page 5) that if this system has a unique solution, then its determinant is nonzero. The system can be written in the form

$$A\mathbf{x} = \mathbf{b} \tag{14}$$

---

†This formula can be obtained directly by applying our procedure for computing an inverse (see Problem 46).

with $\mathbf{x} = \begin{pmatrix} x_1 \\ x_2 \end{pmatrix}$ and $\mathbf{b} = \begin{pmatrix} b_1 \\ b_2 \end{pmatrix}$. Then, since A is invertible, we see from (2) that system (14) has a unique solution given by

$$\mathbf{x} = A^{-1}\mathbf{b}$$

But by Theorem 1.2.1 the fact that system (13) has a unique solution implies that $\det A \neq 0$. This completes the proof. ∎

EXAMPLE 4   **Calculating the Inverse of a 2 × 2 Matrix**   Let $A = \begin{pmatrix} 2 & -4 \\ 1 & 3 \end{pmatrix}$. Calculate A^{-1} if it exists.

Solution   We find that $\det A = (2)(3) - (-4)(1) = 10$; hence A^{-1} exists. From equation (12) we get

$$A^{-1} = \frac{1}{10}\begin{pmatrix} 3 & 4 \\ -1 & 2 \end{pmatrix} = \begin{pmatrix} \frac{3}{10} & \frac{4}{10} \\ -\frac{1}{10} & \frac{2}{10} \end{pmatrix}$$

Check

$$A^{-1}A = \frac{1}{10}\begin{pmatrix} 3 & 4 \\ -1 & 2 \end{pmatrix}\begin{pmatrix} 2 & -4 \\ 1 & 3 \end{pmatrix} = \frac{1}{10}\begin{pmatrix} 10 & 0 \\ 0 & 10 \end{pmatrix} = \begin{pmatrix} 1 & 0 \\ 0 & 1 \end{pmatrix}$$

and

$$AA^{-1} = \begin{pmatrix} 2 & -4 \\ 1 & 3 \end{pmatrix}\begin{pmatrix} \frac{3}{10} & \frac{4}{10} \\ -\frac{1}{10} & \frac{2}{10} \end{pmatrix} = \begin{pmatrix} 1 & 0 \\ 0 & 1 \end{pmatrix}$$ ∎

EXAMPLE 5   **A 2 × 2 Matrix That Is Not Invertible**   Let $A = \begin{pmatrix} 1 & 2 \\ -2 & -4 \end{pmatrix}$. Calculate A^{-1} if it exists.

Solution   We find that $\det A = (1)(-4) - (2)(-2) = -4 + 4 = 0$, so that A^{-1} does not exist, as we saw in Example 3. ∎

The procedure described above works for $n \times n$ matrices where $n > 2$. We illustrate this with a number of examples.

EXAMPLE 6   **Calculating the Inverse of a 3 × 3 Matrix**   Let $A = \begin{pmatrix} 2 & 4 & 6 \\ 4 & 5 & 6 \\ 3 & 1 & -2 \end{pmatrix}$ (see Example 1.3.1 on page 7). Calculate A^{-1} if it exists.

Solution   We first put I next to A in an augmented-matrix form

$$\left(\begin{array}{ccc|ccc} 2 & 4 & 6 & 1 & 0 & 0 \\ 4 & 5 & 6 & 0 & 1 & 0 \\ 3 & 1 & -2 & 0 & 0 & 1 \end{array}\right)$$

and then carry out the row reduction.

$$\xrightarrow{R_1 \to \frac{1}{2}R_1} \begin{pmatrix} 1 & 2 & 3 & \frac{1}{2} & 0 & 0 \\ 4 & 5 & 6 & 0 & 1 & 0 \\ 3 & 1 & -2 & 0 & 0 & 1 \end{pmatrix} \xrightarrow[R_3 \to R_3 - 3R_1]{R_2 \to R_2 - 4R_1} \begin{pmatrix} 1 & 2 & 3 & \frac{1}{2} & 0 & 0 \\ 0 & -3 & -6 & -2 & 1 & 0 \\ 0 & -5 & -11 & -\frac{3}{2} & 0 & 1 \end{pmatrix}$$

$$\xrightarrow{R_2 \to -\frac{1}{3}R_2} \begin{pmatrix} 1 & 2 & 3 & \frac{1}{2} & 0 & 0 \\ 0 & 1 & 2 & \frac{2}{3} & -\frac{1}{3} & 0 \\ 0 & -5 & -11 & -\frac{3}{2} & 0 & 1 \end{pmatrix} \xrightarrow[R_3 \to R_3 + 5R_2]{R_1 \to R_1 - 2R_2} \begin{pmatrix} 1 & 0 & -1 & -\frac{5}{6} & \frac{2}{3} & 0 \\ 0 & 1 & 2 & \frac{2}{3} & -\frac{1}{3} & 0 \\ 0 & 0 & -1 & \frac{11}{6} & -\frac{5}{3} & 1 \end{pmatrix}$$

$$\xrightarrow{R_3 \to -R_3} \begin{pmatrix} 1 & 0 & -1 & -\frac{5}{6} & \frac{2}{3} & 0 \\ 0 & 1 & 2 & \frac{2}{3} & -\frac{1}{3} & 0 \\ 0 & 0 & 1 & -\frac{11}{6} & \frac{5}{3} & -1 \end{pmatrix} \xrightarrow[R_2 \to R_2 - 2R_3]{R_1 \to R_1 + R_3} \begin{pmatrix} 1 & 0 & 0 & -\frac{8}{3} & \frac{7}{3} & -1 \\ 0 & 1 & 0 & \frac{13}{3} & -\frac{11}{3} & 2 \\ 0 & 0 & 1 & -\frac{11}{6} & \frac{5}{3} & -1 \end{pmatrix}$$

Since A has now been reduced to I, we have

$$A^{-1} = \begin{pmatrix} -\frac{8}{3} & \frac{7}{3} & -1 \\ \frac{13}{3} & -\frac{11}{3} & 2 \\ -\frac{11}{6} & \frac{5}{3} & -1 \end{pmatrix} = \frac{1}{6} \begin{pmatrix} -16 & 14 & -6 \\ 26 & -22 & 12 \\ -11 & 10 & -6 \end{pmatrix}$$

We factor out $\frac{1}{6}$ to make computations easier.

Check

$$A^{-1}A = \frac{1}{6} \begin{pmatrix} -16 & 14 & -6 \\ 26 & -22 & 12 \\ -11 & 10 & -6 \end{pmatrix} \begin{pmatrix} 2 & 4 & 6 \\ 4 & 5 & 6 \\ 3 & 1 & -2 \end{pmatrix} = \frac{1}{6} \begin{pmatrix} 6 & 0 & 0 \\ 0 & 6 & 0 \\ 0 & 0 & 6 \end{pmatrix} = I.$$

We can also verify that $AA^{-1} = I$.   ∎

WARNING   It is easy to make numerical errors in computing A^{-1}. Therefore it is important to check the computations by verifying that $A^{-1}A = I$.   □

EXAMPLE 7   **Calculating the Inverse of a 3×3 Matrix**   Let $A = \begin{pmatrix} 2 & 4 & 3 \\ 0 & 1 & -1 \\ 3 & 5 & 7 \end{pmatrix}$. Calculate A^{-1} if it exists.

Solution   Proceeding as in Example 6, we obtain, successively, the following augmented matrices:

$$\begin{pmatrix} 2 & 4 & 3 & 1 & 0 & 0 \\ 0 & 1 & -1 & 0 & 1 & 0 \\ 3 & 5 & 7 & 0 & 0 & 1 \end{pmatrix} \xrightarrow{R_1 \to \frac{1}{2}R_1} \begin{pmatrix} 1 & 2 & \frac{3}{2} & \frac{1}{2} & 0 & 0 \\ 0 & 1 & -1 & 0 & 1 & 0 \\ 3 & 5 & 7 & 0 & 0 & 1 \end{pmatrix}$$

$$\xrightarrow{R_3 \to R_3 - 3R_1} \begin{pmatrix} 1 & 2 & \frac{3}{2} & \frac{1}{2} & 0 & 0 \\ 0 & 1 & -1 & 0 & 1 & 0 \\ 0 & -1 & \frac{5}{2} & -\frac{3}{2} & 0 & 1 \end{pmatrix} \xrightarrow[R_3 \to R_3 + R_2]{R_1 \to R_1 - 2R_2} \begin{pmatrix} 1 & 0 & \frac{7}{2} & \frac{1}{2} & -2 & 0 \\ 0 & 1 & -1 & 0 & 1 & 0 \\ 0 & 0 & \frac{3}{2} & -\frac{3}{2} & 1 & 1 \end{pmatrix}$$

$$\xrightarrow{R_3 \to \frac{2}{3}R_3} \begin{pmatrix} 1 & 0 & \frac{7}{2} & \frac{1}{2} & -2 & 0 \\ 0 & 1 & -1 & 0 & 1 & 0 \\ 0 & 0 & 1 & -1 & \frac{2}{3} & \frac{2}{3} \end{pmatrix} \xrightarrow[R_2 \to R_2 + R_3]{R_1 \to R_1 - \frac{7}{2}R_3} \begin{pmatrix} 1 & 0 & 0 & 4 & -\frac{13}{3} & -\frac{7}{3} \\ 0 & 1 & 0 & -1 & \frac{5}{3} & \frac{2}{3} \\ 0 & 0 & 1 & -1 & \frac{2}{3} & \frac{2}{3} \end{pmatrix}$$

Thus

$$A^{-1} = \begin{pmatrix} 4 & -\frac{13}{3} & -\frac{7}{3} \\ -1 & \frac{5}{3} & \frac{2}{3} \\ -1 & \frac{2}{3} & \frac{2}{3} \end{pmatrix}$$

Check

$$A^{-1}A = \begin{pmatrix} 4 & -\frac{13}{3} & -\frac{7}{3} \\ -1 & \frac{5}{3} & \frac{2}{3} \\ -1 & \frac{2}{3} & \frac{2}{3} \end{pmatrix} \begin{pmatrix} 2 & 4 & 3 \\ 0 & 1 & -1 \\ 3 & 5 & 7 \end{pmatrix} = \begin{pmatrix} 1 & 0 & 0 \\ 0 & 1 & 0 \\ 0 & 0 & 1 \end{pmatrix}$$
■

EXAMPLE 8   **A 3 × 3 Matrix That Is Not Invertible**   Let $A = \begin{pmatrix} 1 & -3 & 4 \\ 2 & -5 & 7 \\ 0 & -1 & 1 \end{pmatrix}$. Calculate A^{-1} if it exists.

Solution   Proceeding as before, we obtain, successively,

$$\left(\begin{array}{ccc|ccc} 1 & -3 & 4 & 1 & 0 & 0 \\ 2 & -5 & 7 & 0 & 1 & 0 \\ 0 & -1 & 1 & 0 & 0 & 1 \end{array}\right) \xrightarrow{R_2 \to R_2 - 2R_1} \left(\begin{array}{ccc|ccc} 1 & -3 & 4 & 1 & 0 & 0 \\ 0 & 1 & -1 & -2 & 1 & 0 \\ 0 & -1 & 1 & 0 & 0 & 1 \end{array}\right)$$

$$\xrightarrow[R_3 \to R_3 + R_2]{R_1 \to R_1 + 3R_2} \left(\begin{array}{ccc|ccc} 1 & 0 & 1 & -5 & 3 & 0 \\ 0 & 1 & -1 & -2 & 1 & 0 \\ 0 & 0 & 0 & -2 & 1 & 1 \end{array}\right)$$

This is as far as we can go. The matrix A *cannot* be reduced to the identity matrix, and we can conclude that A is *not* invertible.
■

There is another way to see the result of the last example.. Let **b** be any 3-vector and consider the system $A\mathbf{x} = \mathbf{b}$. If we tried to solve this by Gaussian elimination, we would end up with an equation that reads $0 = c \neq 0$ as in Example 3, or $0 = 0$. This is case (*ii*) or (*iii*) of Section 1.3 (see page 15). That is, the system either has no solution or it has an infinite number of solutions. The one possibility ruled out is the case in which the system has a unique solution. But if A^{-1} existed, then there would be a unique solution given by $\mathbf{x} = A^{-1}\mathbf{b}$. We are left to conclude that

---

If row reduction of A produces a row of zeros, then A is *not* invertible.

DEFINITION 3   **Row Equivalent Matrices** Suppose that by elementary row operations we can transform the matrix A into the matrix B. Then A and B are said to be **row equivalent.**

The reasoning used above can be used to prove the following theorem (see Problem 47).

THEOREM 5   Let A be an $n \times n$ matrix.

   i. A is invertible if and only if A is row equivalent to the identity matrix I_n; that is, the reduced row echelon form of A is I_n.

   ii. A is invertible if and only if the system $A\mathbf{x} = \mathbf{b}$ has a unique solution for every n-vector $\mathbf{b}$.

   iii. If A is invertible, then this unique solution is given by $\mathbf{x} = A^{-1}\mathbf{b}$.   ■

EXAMPLE 9   **Using the Inverse to Solve a System of Equations**   Solve the system

$$
\begin{aligned}
2x_1 + 4x_2 + 3x_3 &= 6 \\
x_2 - x_3 &= -4 \\
3x_1 + 5x_2 + 7x_3 &= 7
\end{aligned}
$$

Solution   This system can be written as $A\mathbf{x} = \mathbf{b}$, where $A = \begin{pmatrix} 2 & 4 & 3 \\ 0 & 1 & -1 \\ 3 & 5 & 7 \end{pmatrix}$ and $\mathbf{b} = \begin{pmatrix} 6 \\ -4 \\ 7 \end{pmatrix}$. In Example 7 we found that A^{-1} exists and

$$
A^{-1} = \begin{pmatrix} 4 & -\frac{13}{3} & -\frac{7}{3} \\ -1 & \frac{5}{3} & \frac{2}{3} \\ -1 & \frac{2}{3} & \frac{2}{3} \end{pmatrix}
$$

Thus the unique solution is given by

$$
\mathbf{x} = \begin{pmatrix} x_1 \\ x_2 \\ x_3 \end{pmatrix} = A^{-1}\mathbf{b} = \begin{pmatrix} 4 & -\frac{13}{3} & -\frac{7}{3} \\ -1 & \frac{5}{3} & \frac{2}{3} \\ -1 & \frac{2}{3} & \frac{2}{3} \end{pmatrix} \begin{pmatrix} 6 \\ -4 \\ 7 \end{pmatrix} = \begin{pmatrix} 25 \\ -8 \\ -4 \end{pmatrix}
$$

  ■

EXAMPLE 10   **The Technology and Leontief Matrices: Modeling the 1958 American Economy**
In the Leontief input-output model described in Example 1.3.8 on page 16, we obtained the system

$$
\begin{aligned}
a_{11}x_1 + a_{12}x_2 + \cdots + a_{1n}x_n + e_1 &= x_1 \\
a_{21}x_1 + a_{22}x_2 + \cdots + a_{2n}x_n + e_2 &= x_2 \\
\vdots \qquad \vdots \qquad\qquad \vdots \qquad \vdots \quad &\; \vdots \\
a_{n1}x_1 + a_{n2}x_2 + \cdots + a_{nn}x_n + e_n &= x_n
\end{aligned}
\tag{15}
$$

which can be written as

$$
A\mathbf{x} + \mathbf{e} = \mathbf{x} = I\mathbf{x}
$$

or

$$
(I - A)\mathbf{x} = \mathbf{e} \tag{16}
$$

The matrix A of internal demands is called the **technology matrix,** and the matrix

$I - A$ is called the **Leontief matrix.** If the Leontief matrix is invertible, then systems (15) and (16) have unique solutions.

Leontief used his model to analyze the 1958 U.S. economy.† He divided the economy into 81 sectors and grouped them into six families of related sectors. For simplicity, we treat each family of sectors as a single sector so that we can treat the U.S. economy as an economy with six industries. These industries are listed in Table 1.1.

Table 1.1

Sector	Examples
Final nonmetal (FN)	Furniture, processed food
Final metal (FM)	Household appliances, motor vehicles
Basic metal (BM)	Machine-shop products, mining
Basic nonmetal (BN)	Agriculture, printing
Energy (E)	Petroleum, coal
Services (S)	Amusements, real estate

The input-output table, Table 1.2, gives internal demands in 1958 based on Leontief's figures. The units in the table are millions of dollars. Thus, for example, the number 0.173 in the 6,5 position means that in order to produce $1 million worth of energy, it is necessary to provide $0.173 million = $173,000 worth of services. Similarly, the 0.037 in the 4,2 position means that in order to produce $1 million worth of final metal, it is necessary to expend $0.037 million = $37,000 on basic nonmetal products.

Table 1.2   Internal Demands in 1958 U.S. Economy

	FN	FM	BM	BN	E	S
FN	0.170	0.004	0	0.029	0	0.008
FM	0.003	0.295	0.018	0.002	0.004	0.016
BM	0.025	0.173	0.460	0.007	0.011	0.007
BN	0.348	0.037	0.021	0.403	0.011	0.048
E	0.007	0.001	0.039	0.025	0.358	0.025
S	0.120	0.074	0.104	0.123	0.173	0.234

Finally, Leontief estimated the external demands on the 1958 U.S. economy (in millions of dollars) as listed in Table 1.3.

---

†*Scientific American* (April 1965): 26–27.

Table 1.3 External Demands on 1958 U.S. Economy (Millions of Dollars)

FN	$99,640
FM	$75,548
BM	$14,444
BN	$33,501
E	$23,527
S	$263,985

In order to run the U.S. economy in 1958 and to meet all external demands, how many units in each of the six sectors had to be produced?

Solution   The technology matrix is given by

$$A = \begin{pmatrix} 0.170 & 0.004 & 0 & 0.029 & 0 & 0.008 \\ 0.003 & 0.295 & 0.018 & 0.002 & 0.004 & 0.016 \\ 0.025 & 0.173 & 0.460 & 0.007 & 0.011 & 0.007 \\ 0.348 & 0.037 & 0.021 & 0.403 & 0.011 & 0.048 \\ 0.007 & 0.001 & 0.039 & 0.025 & 0.358 & 0.025 \\ 0.120 & 0.074 & 0.104 & 0.123 & 0.173 & 0.234 \end{pmatrix}$$

and

$$\mathbf{e} = \begin{pmatrix} 99,640 \\ 75,548 \\ 14,444 \\ 33,501 \\ 23,527 \\ 263,985 \end{pmatrix}$$

To obtain the Leontief matrix, we subtract to obtain

$$I - A = \begin{pmatrix} 1 & 0 & 0 & 0 & 0 & 0 \\ 0 & 1 & 0 & 0 & 0 & 0 \\ 0 & 0 & 1 & 0 & 0 & 0 \\ 0 & 0 & 0 & 1 & 0 & 0 \\ 0 & 0 & 0 & 0 & 1 & 0 \\ 0 & 0 & 0 & 0 & 0 & 1 \end{pmatrix}$$

$$- \begin{pmatrix} 0.170 & 0.004 & 0 & 0.029 & 0 & 0.008 \\ 0.003 & 0.295 & 0.018 & 0.002 & 0.004 & 0.016 \\ 0.025 & 0.173 & 0.460 & 0.007 & 0.011 & 0.007 \\ 0.348 & 0.037 & 0.021 & 0.403 & 0.011 & 0.048 \\ 0.007 & 0.001 & 0.039 & 0.025 & 0.358 & 0.025 \\ 0.120 & 0.074 & 0.104 & 0.123 & 0.173 & 0.234 \end{pmatrix}$$

$$
= \begin{pmatrix}
0.830 & -0.004 & 0 & -0.029 & 0 & -0.008 \\
-0.003 & 0.705 & -0.018 & -0.002 & -0.004 & -0.016 \\
-0.025 & -0.173 & 0.540 & -0.007 & -0.011 & -0.007 \\
-0.348 & -0.037 & -0.021 & 0.597 & -0.011 & -0.048 \\
-0.007 & -0.001 & -0.039 & -0.025 & 0.642 & -0.025 \\
-0.120 & -0.074 & -0.104 & -0.123 & -0.173 & 0.766
\end{pmatrix}
$$

The computation of the inverse of a 6×6 matrix is a tedious affair. Carrying three decimal places on a calculator, we obtain the matrix below. Intermediate steps are omitted:

$$
(I - A)^{-1} = \begin{pmatrix}
1.234 & 0.014 & 0.006 & 0.064 & 0.007 & 0.018 \\
0.017 & 1.436 & 0.057 & 0.012 & 0.020 & 0.032 \\
0.071 & 0.465 & 1.877 & 0.036 & 0.045 & 0.031 \\
0.751 & 0.134 & 0.100 & 1.740 & 0.066 & 0.124 \\
0.060 & 0.045 & 0.130 & 0.082 & 1.578 & 0.059 \\
0.339 & 0.236 & 0.307 & 0.312 & 0.376 & 1.349
\end{pmatrix}
$$

Therefore the "ideal" output vector is given by

$$
\mathbf{x} = (I - A)^{-1}\mathbf{e} = \begin{pmatrix}
1.234 & 0.014 & 0.006 & 0.064 & 0.007 & 0.018 \\
0.017 & 1.436 & 0.057 & 0.012 & 0.020 & 0.032 \\
0.071 & 0.465 & 1.877 & 0.036 & 0.045 & 0.031 \\
0.751 & 0.134 & 0.100 & 1.740 & 0.066 & 0.124 \\
0.060 & 0.045 & 0.130 & 0.082 & 1.578 & 0.059 \\
0.339 & 0.236 & 0.307 & 0.312 & 0.376 & 1.349
\end{pmatrix} \begin{pmatrix}
99,640 \\
75,548 \\
14,444 \\
33,501 \\
23,527 \\
263,985
\end{pmatrix}
$$

$$
= \begin{pmatrix}
131,161 \\
120,324 \\
79,764 \\
178,936 \\
66,703 \\
431,456
\end{pmatrix}
$$

This means that it would require approximately 131,161 units ($131,161 million worth) of final nonmetal products, 120,324 units of final metal products, 79,764 units of basic metal products, 178,936 units of basic nonmetal products, 66,703 units of energy and 431,456 service units to run the U.S. economy and to meet the external demands in 1958. ∎

   In Section 1.2 we encountered the first form of our Summing Up Theorem (Theorem 1.2.1, page 5). We are now ready to improve upon it. The next theorem states that several statements involving inverse, uniqueness of solutions, row equivalence, and determinants are equivalent. At this point we can prove the equivalence of parts (i), (ii), (iii), and (iv). We shall finish the proof after we have developed some basic theory about determinants (see Theorem 2.4.4 on page 146).

THEOREM 6   **Summing Up Theorem—View 2** Let A be an $n \times n$ matrix. Then the following five statements are equivalent. That is, each statement implies the other four (so that if one is true, all are true and if one is false, all are false).

    i. A is invertible.

    ii. The only solution to the homogeneous system $A\mathbf{x} = \mathbf{0}$ is the trivial solution $(\mathbf{x} = \mathbf{0})$.

    iii. The system $A\mathbf{x} = \mathbf{b}$ has a unique solution for every n-vector $\mathbf{b}$.

    iv. A is row equivalent to the $n \times n$ identity matrix I_n; that is, the reduced row echelon form of A is I_n.

    v. $\det A \neq 0$. (So far, $\det A$ is only defined if A is a 2×2 matrix.)

Proof   We have already seen that statements (i) and (iii) are equivalent [Theorem 5, part (ii)] and that (i) and (iv) are equivalent [Theorem 5, part (i)]. We shall show that (ii) and (iv) are equivalent. Suppose that (ii) holds. That is, suppose that $A\mathbf{x} = \mathbf{0}$ has only the trivial solution $\mathbf{x} = \mathbf{0}$. If we write out this system, we obtain

$$
\begin{aligned}
a_{11}x_1 + a_{12}x_2 + \cdots + a_{1n}x_n &= 0 \\
a_{21}x_1 + a_{22}x_2 + \cdots + a_{2n}x_n &= 0 \\
\vdots \qquad \vdots \qquad\qquad \vdots \\
a_{n1}x_1 + a_{n2}x_2 + \cdots + a_{nn}x_n &= 0
\end{aligned}
\tag{17}
$$

If A were not equivalent to I_n, then row reduction of the augmented matrix associated with (17) would leave us with a row of zeros. But if, say, the last row is zero, then the last equation reads $0 = 0$. Then the homogeneous system reduces to one with $n - 1$ equations in n unknowns, which by Theorem 1.4.1 on page 25 has an infinite number of solutions. But we assumed that $\mathbf{x} = \mathbf{0}$ was the only solution to system (17). This contradiction shows that A is row equivalent to I_n. Conversely, suppose that (iv) holds; that is, suppose that A is row equivalent to I_n. Then by Theorem 5, part (i), A is invertible and by Theorem 5, part (iii), the unique solution to $A\mathbf{x} = \mathbf{0}$ is $\mathbf{x} = A^{-1}\mathbf{0} = \mathbf{0}$. Thus (ii) and (iv) are equivalent. In Theorem 1.2.1 we showed that (i) and (v) are equivalent in the 2×2 case. We shall prove the equivalence of (i) and (v) in Section 2.4. ∎

Remark. We could add another statement to the theorem. Suppose the system $A\mathbf{x} = \mathbf{b}$ has a unique solution. Let R be a matrix in row echelon form that is row equivalent to A. Then R cannot have a row of zeros because if it did, it could not be reduced to the identity matrix.† Thus the row echelon form of A must look like this:

$$\begin{pmatrix} 1 & r_{12} & r_{13} & \cdots & r_{1n} \\ 0 & 1 & r_{23} & \cdots & r_{2n} \\ 0 & 0 & 1 & \cdots & r_{3n} \\ \vdots & \vdots & \vdots & & \vdots \\ 0 & 0 & 0 & \cdots & 1 \end{pmatrix} \qquad (18)$$

That is, R is a matrix with 1's down the diagonal and 0's below it. We thus have Theorem 7.

THEOREM 7   If any of the statements in Theorem 6 holds, then the row echelon form of A has the form of matrix (18). ∎

We have seen that in order to verify that $B = A^{-1}$, we have to check that $AB = BA = I$. It turns out that only half this work has to be done.

THEOREM 8   Let A and B be $n \times n$ matrices. Then A is invertible and $B = A^{-1}$ if (i) $BA = I$ or (ii) $AB = I$.

Remark. This theorem simplifies the work in checking that one matrix is the inverse of another.

Proof   i.   We assume that $BA = I$. Consider the homogeneous system $A\mathbf{x} = \mathbf{0}$. Multiplying both sides of this equation on the left by B, we obtain

$$BA\mathbf{x} = B\mathbf{0} \qquad (19)$$

But $BA = I$ and $B\mathbf{0} = \mathbf{0}$, so (19) becomes $I\mathbf{x} = \mathbf{0}$ or $\mathbf{x} = \mathbf{0}$. This shows that $\mathbf{x} = \mathbf{0}$ is the only solution to $A\mathbf{x} = \mathbf{0}$ and, by Theorem 6, parts (i) and (ii), this means that A is invertible. We still have to show that $B = A^{-1}$. Let $A^{-1} = C$. Then $AC = I$. Thus

$$BAC = B(AC) = BI = B \quad \text{and} \quad BAC = (BA)C = IC = C$$

Hence $B = C$ and part (i) is proved.

ii.   Let $AB = I$. Then, from part (i) $A = B^{-1}$. From Definition 2 this means that $AB = BA = I$, which proves that A is invertible and that $B = A^{-1}$. This completes the proof. ∎

---

†Note that if the ith row of R contains only zeros, then the homogeneous system $R\mathbf{x} = \mathbf{0}$ contains more unknowns than equations (since the ith equation is the zero equation) and the system has an infinite number of solutions. But then $A\mathbf{x} = \mathbf{0}$ has an infinite number of solutions, which is a contradiction of our assumption.

PROBLEMS   1.9

In Problems 1–15 determine whether the given matrix is invertible. If it is, calculate the inverse.

1. $\begin{pmatrix} 2 & 1 \\ 3 & 2 \end{pmatrix}$

2. $\begin{pmatrix} -1 & 6 \\ 2 & -12 \end{pmatrix}$

3. $\begin{pmatrix} 0 & 1 \\ 1 & 0 \end{pmatrix}$

4. $\begin{pmatrix} 1 & 1 \\ 3 & 3 \end{pmatrix}$

5. $\begin{pmatrix} a & a \\ b & b \end{pmatrix}$

6. $\begin{pmatrix} 1 & 1 & 1 \\ 0 & 2 & 3 \\ 5 & 5 & 1 \end{pmatrix}$

7. $\begin{pmatrix} 3 & 2 & 1 \\ 0 & 2 & 2 \\ 0 & 0 & -1 \end{pmatrix}$

8. $\begin{pmatrix} 1 & 1 & 1 \\ 0 & 1 & 1 \\ 0 & 0 & 1 \end{pmatrix}$

9. $\begin{pmatrix} 1 & 6 & 2 \\ -2 & 3 & 5 \\ 7 & 12 & -4 \end{pmatrix}$

10. $\begin{pmatrix} 3 & 1 & 0 \\ 1 & -1 & 2 \\ 1 & 1 & 1 \end{pmatrix}$

11. $\begin{pmatrix} 2 & -1 & 4 \\ -1 & 0 & 5 \\ 19 & -7 & 3 \end{pmatrix}$

12. $\begin{pmatrix} 1 & 2 & 3 \\ 1 & 1 & 2 \\ 0 & 1 & 2 \end{pmatrix}$

13. $\begin{pmatrix} 1 & 1 & 1 & 1 \\ 1 & 2 & -1 & 2 \\ 1 & -1 & 2 & 1 \\ 1 & 3 & 3 & 2 \end{pmatrix}$

14. $\begin{pmatrix} 1 & 0 & 2 & 3 \\ -1 & 1 & 0 & 4 \\ 2 & 1 & -1 & 3 \\ -1 & 0 & 5 & 7 \end{pmatrix}$

15. $\begin{pmatrix} 1 & -3 & 0 & -2 \\ 3 & -12 & -2 & -6 \\ -2 & 10 & 2 & 5 \\ -1 & 6 & 1 & 3 \end{pmatrix}$

16. Show that if A, B, and C are invertible matrices, then ABC is invertible and $(ABC)^{-1} = C^{-1}B^{-1}A^{-1}$.

17. If $A_1, A_2, \ldots, A_m$ are invertible $n \times n$ matrices, show that $A_1 A_2 \ldots A_m$ is invertible and calculate its inverse.

18. Show that the matrix $\begin{pmatrix} 3 & 4 \\ -2 & -3 \end{pmatrix}$ is equal to its own inverse.

19. Show that the matrix $\begin{pmatrix} a_{11} & a_{12} \\ a_{21} & a_{22} \end{pmatrix}$ is equal to its own inverse if $A = \pm I$ or if $a_{11} = -a_{22}$ and $a_{21}a_{12} = 1 - a_{11}^2$.

20. Find the output vector $\mathbf{x}$ in the Leontief input–output model if $n = 3$, $\mathbf{e} = \begin{pmatrix} 30 \\ 20 \\ 40 \end{pmatrix}$, and

$$A = \begin{pmatrix} \frac{1}{5} & \frac{1}{5} & 0 \\ \frac{2}{5} & \frac{2}{5} & \frac{3}{5} \\ \frac{1}{5} & \frac{1}{10} & \frac{2}{5} \end{pmatrix}.$$

*21. Suppose that A is $n \times m$ and B is $m \times n$ so that AB is $n \times n$. Show that AB is not invertible if $n > m$. [*Hint*: Show that there is a nonzero vector $\mathbf{x}$ such that $AB\mathbf{x} = \mathbf{0}$ and then apply Theorem 6.]

*22. Use the methods of this section to find the inverses of the following matrices with complex entries:

a. $\begin{pmatrix} i & 2 \\ 1 & -i \end{pmatrix}$

b. $\begin{pmatrix} 1-i & 0 \\ 0 & 1+i \end{pmatrix}$

c. $\begin{pmatrix} 1 & i & 0 \\ -i & 0 & 1 \\ 0 & 1+i & 1-i \end{pmatrix}$

23. Show that for every real number θ the matrix $\begin{pmatrix} \sin\theta & \cos\theta & 0 \\ \cos\theta & -\sin\theta & 0 \\ 0 & 0 & 1 \end{pmatrix}$ is invertible and find its inverse.

24. Calculate the inverse of $A = \begin{pmatrix} 2 & 0 & 0 \\ 0 & 3 & 0 \\ 0 & 0 & 4 \end{pmatrix}$.

25. A square matrix $A = (a_{ij})$ is called **diagonal** if all its elements off the main diagonal are zero. That is, $a_{ij} = 0$ if $i \neq j$. (The matrix of Problem 24 is diagonal.) Show that a diagonal matrix is invertible if and only if each of its diagonal components is nonzero.

26. Let

$$A = \begin{pmatrix} a_{11} & 0 & \cdots & 0 \\ 0 & a_{22} & \cdots & 0 \\ \vdots & \vdots & \ddots & \vdots \\ 0 & 0 & \cdots & a_{nn} \end{pmatrix}$$

be a diagonal matrix such that each of its diagonal components is nonzero. Calculate A^{-1}.

27. Calculate the inverse of $A = \begin{pmatrix} 2 & 1 & -1 \\ 0 & 3 & 4 \\ 0 & 0 & 5 \end{pmatrix}$.

28. Show that the matrix $A = \begin{pmatrix} 1 & 0 & 0 \\ -2 & 0 & 0 \\ 4 & 6 & 1 \end{pmatrix}$ is not invertible.

*29. A square matrix is called **upper (lower) triangular** if all its elements below (above) the main diagonal are zero. (The matrix of Problem 27 is upper triangular and the matrix of Problem 28 is lower triangular.) Show that an upper or lower triangular matrix is invertible if and only if each of its diagonal elements is nonzero.

*30. Show that the inverse of an invertible upper triangular matrix is upper triangular. [*Hint:* First prove the result for a 3×3 matrix.]

In Problems 31 and 32 a matrix is given. In each case show that the matrix is not invertible by finding a nonzero vector $\mathbf{x}$ such that $A\mathbf{x} = \mathbf{0}$.

31. $\begin{pmatrix} 2 & -1 \\ -4 & 2 \end{pmatrix}$

32. $\begin{pmatrix} 1 & -1 & 3 \\ 0 & 4 & -2 \\ 2 & -6 & 8 \end{pmatrix}$

33. A factory for the construction of quality furniture has two divisions: a machine shop where the parts of the furniture are fabricated, and an assembly and finishing division where the parts are put together into the finished product. Suppose that there are 12 employees in the machine shop and 20 in the assembly and finishing division and that each employee works an 8-hour day. Suppose further that the factory produces only two products: chairs and tables. A chair requires $\frac{384}{17}$ hours of machine shop time and $\frac{480}{17}$ hours of assembly and finishing time. A table requires $\frac{240}{17}$ hours of machine shop time and $\frac{640}{17}$ hours of assembly and finishing time. Assuming that there is an unlimited demand for these products and that the manufacturer wishes to keep all employees busy, how many chairs and how many tables can this factory produce each day?

34. A witch's magic cupboard contains 10 ounces of ground four-leaf clovers and 14 ounces of powdered mandrake root. The cupboard will replenish itself automatically provided she uses up exactly all her supplies. A batch of love potion requires $3\frac{1}{13}$ ounces of ground four-leaf clovers and $2\frac{2}{13}$ ounces of powdered mandrake root. One recipe of a well-known (to witches) cure for the common cold requires $5\frac{5}{13}$ ounces of four-leaf clovers and $10\frac{10}{13}$ ounces of mandrake root. How much of the love potion and the cold remedy should the witch make in order to use up the supply in the cupboard exactly?

35. A farmer feeds his cattle a mixture of two types of feed. One standard unit of type A feed supplies a steer with 10% of its minimum daily requirement of protein and 15% of its requirement of carbohydrates. Type B feed contains 12% of the requirement of protein and 8% of the requirement of carbohydrates in a standard unit. If the farmer wishes to feed his cattle exactly 100% of their minimum daily requirement of protein and carbohydrates, how many units of each type of feed should he give a steer each day?

36. A much simplified version of an input-output table for the 1958 Israeli economy divides that economy into three sectors—agriculture, manufacturing, and energy—with the following result.†

	Agriculture	Manufacturing	Energy
Agriculture	0.293	0	0
Manufacturing	0.014	0.207	0.017
Energy	0.044	0.010	0.216

   a. How many units of agricultural production are required to produce one unit of agricultural output?
   b. How many units of agricultural production are required to produce 200,000 units of agricultural output?
   c. How many units of agricultural product go into the production of 50,000 units of energy?
   d. How many units of energy go into the production of 50,000 units of agricultural products?

37. Continuing Problem 36, exports (in thousands of Israeli pounds) in 1958 were

Agriculture	13,213
Manufacturing	17,597
Energy	1,786

   a. Compute the technology and Leontief matrices.
   b. Determine the number of Israeli pounds worth of agricultural products, manufactured goods, and energy required to run this model of the Israeli economy and export the stated value of products.

---

† Wassily Leontief, *Input-Output Economics* (New York: Oxford University Press, 1966), 54–57.

In Problems 38–45 compute the row echelon form of the given matrix and use it to determine directly whether the given matrix is invertible.

38. The matrix of Problem 3.   **39.** The matrix of Problem 1.

40. The matrix of Problem 4.   **41.** The matrix of Problem 7.

42. The matrix of Problem 9.   **43.** The matrix of Problem 11.

44. The matrix of Problem 13.   **45.** The matrix of Problem 14.

46. Let $A = \begin{pmatrix} a_{11} & a_{12} \\ a_{21} & a_{22} \end{pmatrix}$ and assume that $a_{11}a_{22} - a_{12}a_{21} \neq 0$. Derive formula (12) by row-reducing the augmented matrix $\begin{pmatrix} a_{11} & a_{12} & | & 1 & 0 \\ a_{21} & a_{22} & | & 0 & 1 \end{pmatrix}$.

47. Prove parts (*i*) and (*ii*) of Theorem 5.

1.10   THE TRANSPOSE OF A MATRIX

Corresponding to every matrix is another matrix, which, as we shall see in Chapter 2, has properties very similar to those of the original matrix.

DEFINITION 1   **Transpose**   Let $A = (a_{ij})$ be an $m \times n$ matrix. Then the **transpose** of A, written A^t, is the $n \times m$ matrix obtained by interchanging the rows and columns of A. Succinctly, we may write $A^t = (a_{ji})$. In other words,

$$\text{if } A = \begin{pmatrix} a_{11} & a_{12} & \cdots & a_{1n} \\ a_{21} & a_{22} & \cdots & a_{2n} \\ \vdots & \vdots & & \vdots \\ a_{m1} & a_{m2} & \cdots & a_{mn} \end{pmatrix}, \quad \text{then} \quad A^t = \begin{pmatrix} a_{11} & a_{21} & \cdots & a_{m1} \\ a_{12} & a_{22} & \cdots & a_{m2} \\ \vdots & \vdots & & \vdots \\ a_{1n} & a_{2n} & \cdots & a_{mn} \end{pmatrix} \quad (1)$$

Simply put, the *i*th row of A is the *i*th column of A^t and the *j*th column of A is the *j*th row of A^t.

EXAMPLE 1   **Finding the Transposes of Three Matrices**   Find the transposes of the matrices

$$A = \begin{pmatrix} 2 & 3 \\ 1 & 4 \end{pmatrix} \qquad B = \begin{pmatrix} 2 & 3 & 1 \\ -1 & 4 & 6 \end{pmatrix} \qquad C = \begin{pmatrix} 1 & 2 & -6 \\ 2 & -3 & 4 \\ 0 & 1 & 2 \\ 2 & -1 & 5 \end{pmatrix}$$

Solution   Interchanging the rows and columns of each matrix, we obtain

$$A^t = \begin{pmatrix} 2 & 1 \\ 3 & 4 \end{pmatrix} \qquad B^t = \begin{pmatrix} 2 & -1 \\ 3 & 4 \\ 1 & 6 \end{pmatrix} \qquad C^t = \begin{pmatrix} 1 & 2 & 0 & 2 \\ 2 & -3 & 1 & -1 \\ -6 & 4 & 2 & 5 \end{pmatrix}$$

Note, for example, that 4 is the component in row 2 and column 3 of C while 4 is the component in row 3 and column 2 of C^t. That is, the 23 element of C is the 32 element of C^t. ∎

THEOREM 1   Suppose $A = (a_{ij})$ is an $n \times m$ matrix and $B = (b_{ij})$ is an $m \times p$ matrix. Then:

  i. $(A^t)^t = A$.   (2)

  ii. $(AB)^t = B^t A^t$ (3)

  iii. If A and B are $n \times m$, then $(A + B)^t = A^t + B^t$.   (4)

  iv. If A is invertible, then A^t is invertible and $(A^t)^{-1} = (A^{-1})^t$.   (5)

Proof   **i.** This follows directly from the definition of the transpose.

  ii. First we note that AB is an $n \times p$ matrix, so $(AB)^t$ is $p \times n$. Also, B^t is $p \times m$ and A^t is $m \times n$, so $B^t A^t$ is $p \times n$. Thus both matrices in equation (3) have the same size. Now the ijth element of AB is $\sum_{k=1}^{m} a_{ik} b_{kj}$ and this is the jith element of $(AB)^t$. Let $C = B^t$ and $D = A^t$. Then the ijth element c_{ij} of C is b_{ji} and the ijth element d_{ij} of D is a_{ji}. Thus the jith element of $CD =$ the jith element of

$$B^t A^t = \sum_{k=1}^{m} c_{jk} d_{ki} = \sum_{k=1}^{m} b_{kj} a_{ik} = \sum_{k=1}^{m} a_{ik} b_{kj} = \text{the } ji\text{th element of } (AB)^t. \text{ This}$$

completes the proof of part (ii).

  iii. This part is left as an exercise (see Problem 11).

  iv. Let $A^{-1} = B$. Then $AB = BA = I$ so that, from part (ii), $(AB)^t = B^t A^t = I^t = I$ and $(BA)^t = A^t B^t = I$. Thus A^t is invertible and B^t is the inverse of A^t; that is, $(A^t)^{-1} = B^t = (A^{-1})^t$. ∎

The transpose plays an important role in matrix theory. We shall see in succeeding chapters that A and A^t have many properties in common. Since columns of A^t are rows of A, we shall be able to use facts about the transpose to conclude that just about anything which is true about the rows of a matrix is true about its columns.

We conclude this section with an important definition.

DEFINITION 2   **Symmetric Matrix** The $n \times m$ (square) matrix A is called **symmetric** if $A^t = A$.

EXAMPLE 2   **Four Symmetric Matrices**   The following four matrices are symmetric:

$$I \quad A = \begin{pmatrix} 1 & 2 \\ 2 & 3 \end{pmatrix} \quad B = \begin{pmatrix} 1 & -4 & 2 \\ -4 & 7 & 5 \\ 2 & 5 & 0 \end{pmatrix} \quad C = \begin{pmatrix} -1 & 2 & 4 & 6 \\ 2 & 7 & 3 & 5 \\ 4 & 3 & 8 & 0 \\ 6 & 5 & 0 & -4 \end{pmatrix} \quad \blacksquare$$

We shall see the importance of symmetric matrices in Chapters 5 and 6.

Another Way to Write the Scalar Product

Let $\mathbf{a} = \begin{pmatrix} a_1 \\ a_2 \\ \vdots \\ a_n \end{pmatrix}$ and $\mathbf{b} = \begin{pmatrix} b_1 \\ b_2 \\ \vdots \\ b_n \end{pmatrix}$ be two n-component column vectors. Then, from equation (1) on page 43,

$$\mathbf{a} \cdot \mathbf{b} = a_1 b_1 + a_2 b_2 + \cdots + a_n b_n$$

Now $\mathbf{a}$ is an $n \times 1$ matrix so $\mathbf{a}^t$ is a $1 \times n$ matrix and

$$\mathbf{a}^t = (a_1 \ a_2 \cdots a_n)$$

Then $\mathbf{a}^t \mathbf{b}$ is a 1×1 matrix (or scalar), and by the definition of matrix multiplication,

$$\mathbf{a}^t \mathbf{b} = (a_1 \ a_2 \cdots a_n) \begin{pmatrix} b_1 \\ b_2 \\ \vdots \\ b_n \end{pmatrix} = a_1 b_1 + a_2 b_2 + \cdots + a_n b_n$$

Thus, if $\mathbf{a}$ and $\mathbf{b}$ are n-component column vectors, then

$$\boxed{\mathbf{a} \cdot \mathbf{b} = \mathbf{a}^t \mathbf{b}} \tag{6}$$

Formula (6) will be useful to us later in the book.

PROBLEMS   1.10

In Problems 1–10 find the transpose of the given matrix.

1. $\begin{pmatrix} -1 & 4 \\ 6 & 5 \end{pmatrix}$ **2.** $\begin{pmatrix} 3 & 0 \\ 1 & 2 \end{pmatrix}$ **3.** $\begin{pmatrix} 2 & 3 \\ -1 & 2 \\ 1 & 4 \end{pmatrix}$ **4.** $\begin{pmatrix} 2 & -1 & 0 \\ 1 & 5 & 6 \end{pmatrix}$

5. $\begin{pmatrix} 1 & 2 & 3 \\ -1 & 0 & 4 \\ 1 & 5 & 5 \end{pmatrix}$ **6.** $\begin{pmatrix} 1 & 2 & 3 \\ 2 & 4 & -5 \\ 3 & -5 & 7 \end{pmatrix}$ **7.** $\begin{pmatrix} 1 & 0 & 1 & 0 \\ 0 & 1 & 0 & 1 \end{pmatrix}$ **8.** $\begin{pmatrix} 2 & -1 \\ 2 & 4 \\ 1 & 6 \\ 1 & 5 \end{pmatrix}$

9. $\begin{pmatrix} a & b & c \\ d & e & f \\ g & h & j \end{pmatrix}$ **10.** $\begin{pmatrix} 0 & 0 & 0 \\ 0 & 0 & 0 \end{pmatrix}$

11. Let A and B be $n \times m$ matrices. Show, using Definition 1, that $(A + B)^t = A^t + B^t$.

12. Find numbers α and β such that $\begin{pmatrix} 2 & \alpha & 3 \\ 5 & -6 & 2 \\ \beta & 2 & 4 \end{pmatrix}$ is symmetric.

13. If A and B are symmetric $n \times n$ matrices, prove that $A + B$ is symmetric.

14. If A and B are symmetric $n \times n$ matrices, show that $(AB)^t = BA$.

15. Show that for any matrix A, the product matrix AA^t is defined and is a symmetric matrix.

16. Show that every diagonal matrix (see Problem 1.9.25, page 25) is symmetric.

17. Show that the transpose of every upper triangular matrix (see Problem 1.9.29 on page 25) is lower triangular.

18. A square matrix is called **skew-symmetric** if $A^t = -A$ (that is, $a_{ij} = -a_{ji}$). Which of the following matrices are skew-symmetric?

a. $\begin{pmatrix} 1 & -6 \\ 6 & 0 \end{pmatrix}$ **b.** $\begin{pmatrix} 0 & -6 \\ 6 & 0 \end{pmatrix}$ **c.** $\begin{pmatrix} 2 & -2 & -2 \\ 2 & 2 & -2 \\ 2 & 2 & 2 \end{pmatrix}$ **d.** $\begin{pmatrix} 0 & 1 & -1 \\ -1 & 0 & 2 \\ 1 & -2 & 0 \end{pmatrix}$

19. Let A and B be $n \times n$ skew-symmetric matrices. Show that $A + B$ is skew-symmetric.

20. If A is skew-symmetric, show that every component on the main diagonal of A is zero.

21. If A and B are skew-symmetric $n \times n$ matrices, show that $(AB)^t = BA$, so that AB is symmetric if and only if A and B commute.

22. Let A be an $n \times n$ matrix. Show that the matrix $\frac{1}{2}(A + A^t)$ is symmetric.

23. Let A be an $n \times n$ matrix. Show that the matrix $\frac{1}{2}(A - A^t)$ is skew-symmetric.

*24. Show that any square matrix can be written in a unique way as the sum of a symmetric matrix and a skew-symmetric matrix.

*25. Let $A = \begin{pmatrix} a_{11} & a_{12} \\ a_{21} & a_{22} \end{pmatrix}$ be a matrix with nonnegative entries having the properties that (i) $a_{11}^2 + a_{12}^2 = 1$ and $a_{21}^2 + a_{22}^2 = 1$ and (ii) $\begin{pmatrix} a_{11} \\ a_{12} \end{pmatrix} \cdot \begin{pmatrix} a_{21} \\ a_{22} \end{pmatrix} = 0$. Show that A is invertible and that $A^{-1} = A^t$.

In Problems 26–29 compute $(A^t)^{-1}$ and $(A^{-1})^t$ and show that they are equal.

26. $A = \begin{pmatrix} 1 & 2 \\ 3 & 4 \end{pmatrix}$ **27.** $A = \begin{pmatrix} 2 & 1 \\ 3 & 2 \end{pmatrix}$

28. $A = \begin{pmatrix} 3 & 2 & 1 \\ 0 & 2 & 2 \\ 0 & 0 & -1 \end{pmatrix}$ **29.** $A = \begin{pmatrix} 1 & 1 & 1 \\ 0 & 2 & 3 \\ 5 & 5 & 1 \end{pmatrix}$

1.11   ELEMENTARY MATRICES AND MATRIX INVERSES

Let A be an $m \times n$ matrix. Then, as we shall soon see, we can perform elementary row operations on A by multiplying A on the left by an appropriate matrix. The elementary row operations are:

i. Multiply the ith row by a nonzero number c. $R_i \rightarrow cR_i$

ii. Add a multiple of the ith row to the jth row. $R_j \rightarrow R_j + cR_i$

iii. Permute (interchange) the ith and jth rows. $R_i \rightleftarrows R_j$

DEFINITION 1   **Elementary Matrix**   An $n \times n$ (square) matrix E is called an **elementary matrix** if it can be obtained from the $n \times n$ identity matrix I_n by a *single* elementary row operation.

Notation. We denote an elementary matrix by E or by cR_i, $R_j + cR_i$, or $R_i \rightleftarrows R_j$, depending on how the matrix is obtained from I.

EXAMPLE 1   **Three Elementary Matrices**   We obtain three elementary matrices.

i. $\begin{pmatrix} 1 & 0 & 0 \\ 0 & 1 & 0 \\ 0 & 0 & 1 \end{pmatrix} \xrightarrow{R_2 \rightarrow 5R_2} \begin{pmatrix} 1 & 0 & 0 \\ 0 & 5 & 0 \\ 0 & 0 & 1 \end{pmatrix} = 5R_2$ Matrix obtained by multiplying second row of I by 5

ii. $\begin{pmatrix} 1 & 0 & 0 \\ 0 & 1 & 0 \\ 0 & 0 & 1 \end{pmatrix} \xrightarrow{R_3 \rightarrow R_3 - 3R_1} \begin{pmatrix} 1 & 0 & 0 \\ 0 & 1 & 0 \\ -3 & 0 & 1 \end{pmatrix} = R_3 - 3R_1$ Matrix obtained by multiplying first row of I by -3 and adding it to the third row

iii. $\begin{pmatrix} 1 & 0 & 0 \\ 0 & 1 & 0 \\ 0 & 0 & 1 \end{pmatrix} \xrightarrow{R_2 \rightleftarrows R_3} \begin{pmatrix} 1 & 0 & 0 \\ 0 & 0 & 1 \\ 0 & 1 & 0 \end{pmatrix} = R_2 \rightleftarrows R_3$ Matrix obtained by permuting the second and third rows of I ∎

The proof of the following theorem is left as an exercise (See Problems 54–56).

THEOREM 1   To perform an elementary row operation on the $m \times n$ matrix A, multiply A on the left by the appropriate elementary matrix.   ∎

EXAMPLE 2   **Performing Elementary Row Operations by Multiplying by Elementary Matrices**   Let $A = \begin{pmatrix} 1 & 3 & 2 & 1 \\ 4 & 2 & 3 & -5 \\ 3 & 1 & -2 & 4 \end{pmatrix}$. Perform the following elementary row operations on A by multiplying A on the left by an appropriate elementary matrix.

i. Multiply the second row by 5.

ii. Multiply the first row by -3 and add it to the third row.

iii. Permute the second and third rows.

Solution   Since A is a 3×4 matrix, each elementary matrix E must be 3×3 since E must be square and E is multiplying A on the left. We use the results of Example 1.

$$\textbf{i. } (5R_2)\, A = \begin{pmatrix} 1 & 0 & 0 \\ 0 & 5 & 0 \\ 0 & 0 & 1 \end{pmatrix} \begin{pmatrix} 1 & 3 & 2 & 1 \\ 4 & 2 & 3 & -5 \\ 3 & 1 & -2 & 4 \end{pmatrix} = \begin{pmatrix} 1 & 3 & 2 & 1 \\ 20 & 10 & 15 & -25 \\ 3 & 1 & -2 & 4 \end{pmatrix}$$

$$\textbf{ii. } (R_3 - 3R_1)\, A = \begin{pmatrix} 1 & 0 & 0 \\ 0 & 1 & 0 \\ -3 & 0 & 1 \end{pmatrix} \begin{pmatrix} 1 & 3 & 2 & 1 \\ 4 & 2 & 3 & -5 \\ 3 & 1 & -2 & 4 \end{pmatrix} = \begin{pmatrix} 1 & 3 & 2 & 1 \\ 4 & 2 & 3 & -5 \\ 0 & -8 & -8 & 1 \end{pmatrix}$$

$$\textbf{iii. } (R_2 \rightleftarrows R_3)\, A = \begin{pmatrix} 1 & 0 & 0 \\ 0 & 0 & 1 \\ 0 & 1 & 0 \end{pmatrix} \begin{pmatrix} 1 & 3 & 2 & 1 \\ 4 & 2 & 3 & -5 \\ 3 & 1 & -2 & 4 \end{pmatrix} = \begin{pmatrix} 1 & 3 & 2 & 1 \\ 3 & 1 & -2 & 4 \\ 4 & 2 & 3 & -5 \end{pmatrix} \quad\blacksquare$$

Consider the following three products:

$$\begin{pmatrix} 1 & 0 & 0 \\ 0 & c & 0 \\ 0 & 0 & 1 \end{pmatrix} \begin{pmatrix} 1 & 0 & 0 \\ 0 & 1/c & 0 \\ 0 & 0 & 1 \end{pmatrix} = \begin{pmatrix} 1 & 0 & 0 \\ 0 & 1 & 0 \\ 0 & 0 & 1 \end{pmatrix} \tag{1}$$

$$\begin{pmatrix} 1 & 0 & 0 \\ 0 & 1 & 0 \\ c & 0 & 1 \end{pmatrix} \begin{pmatrix} 1 & 0 & 0 \\ 0 & 1 & 0 \\ -c & 0 & 1 \end{pmatrix} = \begin{pmatrix} 1 & 0 & 0 \\ 0 & 1 & 0 \\ 0 & 0 & 1 \end{pmatrix} \tag{2}$$

$$\begin{pmatrix} 1 & 0 & 0 \\ 0 & 0 & 1 \\ 0 & 1 & 0 \end{pmatrix} \begin{pmatrix} 1 & 0 & 0 \\ 0 & 0 & 1 \\ 0 & 1 & 0 \end{pmatrix} = \begin{pmatrix} 1 & 0 & 0 \\ 0 & 1 & 0 \\ 0 & 0 & 1 \end{pmatrix} \tag{3}$$

Equations (1), (2), and (3) suggest that each elementary matrix is invertible and that its inverse is of the same type (Table 1.4). These facts follow from Theorem 1. Evidently, if the operations $R_j \rightarrow R_j + cR_i$ followed by $R_j \rightarrow R_j - cR_i$ are performed on the matrix A, the matrix A is unchanged. Also, $R_i \rightarrow cR_i$ followed by $R_i \rightarrow \dfrac{1}{c}R_i$ and permuting the same two rows twice leave the matrix A unchanged. We have

$$(cR_i)^{-1} = \frac{1}{c}R_i \tag{4}$$

$$(R_j + cR_i)^{-1} = R_j - cR_i \tag{5}$$

$$(R_i \rightleftarrows R_j)^{-1} = R_i \rightleftarrows R_j \tag{6}$$

Table 1.4

Elementary Matrix Type E	Effect of Multiplying A on the Left by E	Symbolic Representation of Elementary Row Operation	When Multiplied on the Left, E^{-1} Does the Following	Symbolic Representation of Inverse Operation
Multiplication	Multiplies ith row of A by $c \neq 0$	cR_i	Multiplies ith row of A by $\dfrac{1}{c}$	$\dfrac{1}{c}R_i$
Addition	Multiplies ith row of A by c and adds it to jth row	$R_j + cR_i$	Multiplies ith row of A by $-c$ and adds it to jth row	$R_j - cR_i$
Permutation	Permutes the ith and jth rows of A	$R_i \rightleftarrows R_j$	Permutes the ith and jth rows of A	$R_i \rightleftarrows R_j$

Equation (6) indicates that

Every elementary permutation matrix is its own inverse.

We summarize our results.

THEOREM 2 Each elementary matrix is invertible. The inverse of an elementary matrix is a matrix of the same type. ∎

THEOREM 3 A square matrix is invertible if and only if it is the product of elementary matrices.

Proof Let $A = E_1 E_2 \cdots E_m$ where each E_i is elementary. By Theorem 2, each E_i is invertible. Moreover, by Theorem 1.9.3 on page 66. A is invertible† and

$$A^{-1} = E_m^{-1} E_{m-1}^{-1} \cdots E_2^{-1} E_1^{-1}$$

Conversely, suppose that A is invertible. According to Theorem 1.9.6 (the Summing Up Theorem), A is row equivalent to the identity matrix. This means that

---

† Here we have used the generalization of Theorem 1.9.3 to more than two matrices. See, for example, Problem 1.9.16 on page 80.

A can be reduced to I by a finite number, say, m, of elementary row operations. By Theorem 1, each such operation is accomplished by multiplying A on the left by an elementary matrix. This means that there are elementary matrices $E_1, E_2, \ldots, E_m$ such that

$$E_m E_{m-1} \cdots E_2 E_1 A = I,$$

Thus from Theorem 1.9.8 on page 79,

$$E_m E_{m-1} \cdots E_2 E_1 = A^{-1}$$

and since each E_i is invertible by Theorem 2,

$$A = (A^{-1})^{-1} = (E_m E_{m-1} \cdots E_2 E_1)^{-1} = E_1^{-1} E_2^{-1} \cdots E_{m-1}^{-1} E_m^{-1} \qquad (7)$$

Since the inverse of an elementary matrix is an elementary matrix, we have written A as a product of elementary matrices and the proof is complete. ∎

EXAMPLE 3   **Writing an Invertible Matrix as the Product of Elementary Matrices**   Show that the matrix $A = \begin{pmatrix} 2 & 4 & 6 \\ 4 & 5 & 6 \\ 3 & 1 & -2 \end{pmatrix}$ is invertible and write it as the product of elementary matrices.

Solution   We have encountered this matrix before, first in Example 1.3.1 on page 7. To solve the problem, we reduce A to I and keep track of the elementary row operations. In Example 1.9.6 on page 71 we did reduce A to I by using the following operations:

① $\frac{1}{2}R_1$ ② $R_2 - 4R_1$ ③ $R_3 - 3R_1$ ④ $-\frac{1}{3}R_2$
⑤ $R_1 - 2R_2$ ⑥ $R_3 + 5R_2$ ⑦ $-R_3$ ⑧ $R_1 + R_3$
⑨ $R_2 - 2R_3$

A^{-1} was obtained by starting with I and applying these nine elementary row operations. Thus A^{-1} is the product of nine elementary matrices:

$$A^{-1} = \underbrace{\begin{pmatrix} 1 & 0 & 0 \\ 0 & 1 & -2 \\ 0 & 0 & 1 \end{pmatrix}}_{R_2 - 2R_3} \underbrace{\begin{pmatrix} 1 & 0 & 1 \\ 0 & 1 & 0 \\ 0 & 0 & 1 \end{pmatrix}}_{R_1 + R_3} \underbrace{\begin{pmatrix} 1 & 0 & 0 \\ 0 & 1 & 0 \\ 0 & 0 & -1 \end{pmatrix}}_{-R_3} \underbrace{\begin{pmatrix} 1 & 0 & 0 \\ 0 & 1 & 0 \\ 0 & 5 & 1 \end{pmatrix}}_{R_3 + 5R_2} \underbrace{\begin{pmatrix} 1 & -2 & 0 \\ 0 & 1 & 0 \\ 0 & 0 & 1 \end{pmatrix}}_{R_1 - 2R_2}$$

$$\times \underbrace{\begin{pmatrix} 1 & 0 & 0 \\ 0 & -\frac{1}{3} & 0 \\ 0 & 0 & 1 \end{pmatrix}}_{-\frac{1}{3}R_2} \underbrace{\begin{pmatrix} 1 & 0 & 0 \\ 0 & 1 & 0 \\ -3 & 0 & 1 \end{pmatrix}}_{R_3 - 3R_1} \underbrace{\begin{pmatrix} 1 & 0 & 0 \\ -4 & 1 & 0 \\ 0 & 0 & 1 \end{pmatrix}}_{R_2 - 4R_1} \underbrace{\begin{pmatrix} \frac{1}{2} & 0 & 0 \\ 0 & 1 & 0 \\ 0 & 0 & 1 \end{pmatrix}}_{\frac{1}{2}R_1}$$

Then $A = (A^{-1})^{-1} =$ the product of the inverses of the nine matrices in the opposite order:

$$\begin{pmatrix} 2 & 4 & 6 \\ 4 & 5 & 6 \\ 3 & 1 & -2 \end{pmatrix} = \begin{pmatrix} 2 & 0 & 0 \\ 0 & 1 & 0 \\ 0 & 0 & 1 \end{pmatrix} \begin{pmatrix} 1 & 0 & 0 \\ 4 & 1 & 0 \\ 0 & 0 & 1 \end{pmatrix} \begin{pmatrix} 1 & 0 & 0 \\ 0 & 1 & 0 \\ 3 & 0 & 1 \end{pmatrix} \begin{pmatrix} 1 & 0 & 0 \\ 0 & -3 & 0 \\ 0 & 0 & 0 \end{pmatrix} \begin{pmatrix} 1 & 2 & 0 \\ 0 & 1 & 0 \\ 0 & 0 & 1 \end{pmatrix}$$

$$\qquad\qquad 2R_1 \qquad R_2 + 4R_i \quad R_3 + 3R_1 \qquad -3R_2 \qquad R_1 + 2R_2$$

$$\times \begin{pmatrix} 1 & 0 & 0 \\ 0 & 1 & 0 \\ 0 & -5 & 1 \end{pmatrix} \begin{pmatrix} 1 & 0 & 0 \\ 0 & 1 & 0 \\ 0 & 0 & -1 \end{pmatrix} \begin{pmatrix} 1 & 0 & -1 \\ 0 & 1 & 0 \\ 0 & 0 & 1 \end{pmatrix} \begin{pmatrix} 1 & 0 & 0 \\ 0 & 1 & 2 \\ 0 & 0 & 1 \end{pmatrix} \quad \blacksquare$$

$$\qquad\quad R_3 - 5R_2 \qquad -R_3 \qquad R_1 - R_3 \qquad R_2 + 2R_3$$

We can use Theorem 3 to extend our Summing Up Theorem, last seen on page 78.

THEOREM 4   **Summing Up Theorem—View 3** Let A be an $n \times n$ matrix. Then the following six statements are equivalent. That is, each one implies the other five (so that if one statement is true, all are true, and if one is false, all are false).

    i. A is invertible.

    ii. The only solution to the homogeneous system $A\mathbf{x} = \mathbf{0}$ is the trivial solution $(\mathbf{x} = \mathbf{0})$.

    iii. The system $A\mathbf{x} = \mathbf{b}$ has a unique solution for every n-vector $\mathbf{b}$.

    iv. A is row equivalent to the $n \times n$ identity matrix I_n; that is, the reduced row echelon form of A is I_n.

    v. A can be written as a product of elementary matrices.

    vi. $\det A \neq 0$ (so far, $\det A$ is defined only if A is a 2×2 matrix). $\blacksquare$

There is one further result that will prove very useful in Section 2.3. First we need a definition.

DEFINITION 2   **Upper Triangular Matrix and Lower Triangular Matrix** A square matrix is called **upper (lower) triangular** if all its components below (above) the main diagonal are zero.

EXAMPLE 4   **Two Upper Triangular and Two Lower Triangular Matrices**   Matrices U and V are upper triangular while matrices L and M are lower triangular:

$$U = \begin{pmatrix} 2 & -3 & 5 \\ 0 & 1 & 6 \\ 0 & 0 & 2 \end{pmatrix} \qquad V = \begin{pmatrix} 1 & 5 \\ 0 & -2 \end{pmatrix}$$

$$L = \begin{pmatrix} 0 & 0 \\ 5 & 1 \end{pmatrix} \qquad M = \begin{pmatrix} 2 & 0 & 0 & 0 \\ -5 & 4 & 0 & 0 \\ 6 & 1 & 2 & 0 \\ 3 & 0 & 1 & 5 \end{pmatrix}$$ ∎

THEOREM 5   Let A be an $n \times n$ matrix. Then A can be written as a product of elementary matrices and an upper triangular matrix U. In the product the elementary matrices are on the left and the upper triangular matrix is on the right.

Proof   Gaussian elimination to solve the system $A\mathbf{x} = \mathbf{b}$ will result in an upper triangular matrix. To see this, observe that Gaussian elimination will terminate when the matrix is in row echelon form—and the row echelon form of an $n \times n$ matrix is upper triangular. We denote the row echelon form of A by U. Then A is reduced to U by a sequence of elementary row operations each of which can be obtained by multiplication by an elementary matrix. Thus

$$U = E_m E_{m-1} \cdots E_2 E_1 A$$

and

$$A = E_1^{-1} E_2^{-1} \cdots E_{m-1}^{-1} E_m^{-1} U$$

Since the inverse of an elementary matrix is an elementary matrix, we have written A as the product of elementary matrices and U.   ∎

EXAMPLE 5   **Writing a Matrix as the Product of Elementary Matrices and an Upper Triangular Matrix**   Write the matrix

$$A = \begin{pmatrix} 3 & 6 & 9 \\ 2 & 5 & 1 \\ 1 & 1 & 8 \end{pmatrix}$$

as the product of elementary matrices and an upper triangular matrix.

Solution   We row-reduce A to obtain its row echelon form:

$$\begin{pmatrix} 3 & 6 & 9 \\ 2 & 5 & 1 \\ 1 & 1 & 8 \end{pmatrix} \xrightarrow{R_1 \to \frac{1}{3}R_1} \begin{pmatrix} 1 & 2 & 3 \\ 2 & 5 & 1 \\ 1 & 1 & 8 \end{pmatrix}$$

$$\xrightarrow[R_3 \to R_3 - R_1]{R_2 \to R_2 - 2R_1} \begin{pmatrix} 1 & 2 & 3 \\ 0 & 1 & -5 \\ 0 & -1 & 5 \end{pmatrix} \xrightarrow{R_3 \to R_3 + R_2} \begin{pmatrix} 1 & 2 & 3 \\ 0 & 1 & -5 \\ 0 & 0 & 0 \end{pmatrix} = U$$

Then working backward, we see that

$$U = \begin{pmatrix} 1 & 2 & 3 \\ 0 & 1 & -5 \\ 0 & 0 & 0 \end{pmatrix} = \begin{pmatrix} 1 & 0 & 0 \\ 0 & 1 & 0 \\ 0 & 1 & 1 \end{pmatrix} \begin{pmatrix} 1 & 0 & 0 \\ 0 & 1 & 0 \\ -1 & 0 & 1 \end{pmatrix}$$
$$\underset{R_3 + R_2}{} \qquad \underset{R_3 - R_1}{}$$

$$\times \begin{pmatrix} 1 & 0 & 0 \\ -2 & 1 & 0 \\ 0 & 0 & 1 \end{pmatrix} \begin{pmatrix} \frac{1}{3} & 0 & 0 \\ 0 & 1 & 0 \\ 0 & 0 & 1 \end{pmatrix} \begin{pmatrix} 3 & 6 & 9 \\ 2 & 5 & 1 \\ 1 & 1 & 8 \end{pmatrix}$$
$$\underset{R_2 - 2R_1}{} \qquad \underset{\frac{1}{3}R_1}{} \qquad \underset{A}{}$$

and, taking inverses of the four elementary matrices, we obtain

$$A = \begin{pmatrix} 3 & 6 & 9 \\ 2 & 5 & 1 \\ 1 & 1 & 8 \end{pmatrix} = \begin{pmatrix} 3 & 0 & 0 \\ 0 & 1 & 0 \\ 0 & 0 & 1 \end{pmatrix} \begin{pmatrix} 1 & 0 & 0 \\ 2 & 1 & 0 \\ 0 & 0 & 1 \end{pmatrix}$$
$$\underset{3R_1}{} \qquad \underset{R_2 + 2R_1}{}$$

$$\times \begin{pmatrix} 1 & 0 & 0 \\ 0 & 1 & 0 \\ 1 & 0 & 1 \end{pmatrix} \begin{pmatrix} 1 & 0 & 0 \\ 0 & 1 & 0 \\ 0 & -1 & 1 \end{pmatrix} \begin{pmatrix} 1 & 2 & 3 \\ 0 & 1 & -5 \\ 0 & 0 & 0 \end{pmatrix}$$
$$\underset{R_3 + R_1}{} \qquad \underset{R_3 - R_2}{} \qquad \underset{U}{} \qquad \blacksquare$$

PROBLEMS   1.11

In Problems 1–12 determine which matrices are elementary matrices.

1. $\begin{pmatrix} 0 & 1 \\ 1 & 0 \end{pmatrix}$ 2. $\begin{pmatrix} 1 & 0 \\ 1 & 1 \end{pmatrix}$ 3. $\begin{pmatrix} 0 & 1 \\ 1 & 1 \end{pmatrix}$

4. $\begin{pmatrix} 1 & 0 \\ 0 & 2 \end{pmatrix}$ 5. $\begin{pmatrix} 3 & 0 \\ 0 & 3 \end{pmatrix}$ 6. $\begin{pmatrix} 0 & 1 & 0 \\ 1 & 0 & 0 \\ 0 & 0 & 1 \end{pmatrix}$

7. $\begin{pmatrix} 0 & 1 & 0 \\ 0 & 0 & 1 \\ 1 & 0 & 0 \end{pmatrix}$ 8. $\begin{pmatrix} 1 & 0 & 0 \\ 2 & 1 & 0 \\ 3 & 0 & 1 \end{pmatrix}$ 9. $\begin{pmatrix} 1 & 0 & 0 \\ 2 & 1 & 0 \\ 0 & 0 & 1 \end{pmatrix}$

10. $\begin{pmatrix} 1 & 0 & 0 & 0 \\ 0 & 1 & 0 & 0 \\ 0 & 0 & 1 & 0 \\ 0 & 1 & 0 & 1 \end{pmatrix}$ 11. $\begin{pmatrix} 1 & 0 & 0 & 0 \\ 1 & 1 & 0 & 0 \\ 0 & 0 & 1 & 0 \\ 0 & 0 & 1 & 1 \end{pmatrix}$ 12. $\begin{pmatrix} 1 & -1 & 0 & 0 \\ 0 & 1 & 0 & 0 \\ 0 & 0 & 1 & 0 \\ 0 & 0 & 0 & 1 \end{pmatrix}$

In Problems 13–20 write the 3×3 elementary matrix that will carry out the given row operation on a 3×5 matrix A by left multiplication.

13. $R_2 \to 4R_2$ **14.** $R_2 \to R_2 + 2R_1$ **15.** $R_1 \to R_1 - 3R_2$

16. $R_1 \to R_1 + 4R_3$ **17.** $R_1 \rightleftarrows R_3$ **18.** $R_2 \rightleftarrows R_3$

19. $R_2 \to R_2 + R_3$ **20.** $R_3 \to -R_3$

In Problems 21–30 find the elementary matrix E such that $EA = B$.

21. $A = \begin{pmatrix} 2 & 3 \\ -1 & 4 \end{pmatrix}$, $B = \begin{pmatrix} 2 & 3 \\ 2 & -8 \end{pmatrix}$ **22.** $A = \begin{pmatrix} 2 & 3 \\ -1 & 4 \end{pmatrix}$, $B = \begin{pmatrix} 2 & 3 \\ -5 & -2 \end{pmatrix}$

23. $A = \begin{pmatrix} 2 & 3 \\ -1 & 4 \end{pmatrix}$, $B = \begin{pmatrix} 0 & 11 \\ -1 & 4 \end{pmatrix}$ **24.** $A = \begin{pmatrix} 2 & 3 \\ -1 & 4 \end{pmatrix}$, $B = \begin{pmatrix} -1 & 4 \\ 2 & 3 \end{pmatrix}$

25. $A = \begin{pmatrix} 1 & 2 \\ 3 & 4 \\ 5 & 6 \end{pmatrix}$, $B = \begin{pmatrix} 5 & 6 \\ 3 & 4 \\ 1 & 2 \end{pmatrix}$ **26.** $A = \begin{pmatrix} 1 & 2 \\ 3 & 4 \\ 5 & 6 \end{pmatrix}$, $B = \begin{pmatrix} 1 & 2 \\ 0 & -2 \\ 5 & 6 \end{pmatrix}$

27. $A = \begin{pmatrix} 1 & 2 \\ 3 & 4 \\ 5 & 6 \end{pmatrix}$, $B = \begin{pmatrix} -1 & -2 \\ 3 & 4 \\ 5 & 6 \end{pmatrix}$ **28.** $A = \begin{pmatrix} 1 & 2 \\ 3 & 4 \\ 5 & 6 \end{pmatrix}$, $B = \begin{pmatrix} -5 & -6 \\ 3 & 4 \\ 5 & 6 \end{pmatrix}$

29. $A = \begin{pmatrix} 1 & 2 & 5 & 2 \\ 0 & -1 & 3 & 4 \\ 5 & 0 & -2 & 7 \end{pmatrix}$, $B = \begin{pmatrix} 1 & 2 & 5 & 2 \\ 0 & -1 & 3 & 4 \\ 0 & -10 & -27 & -3 \end{pmatrix}$

30. $A = \begin{pmatrix} 1 & 2 & 5 & 2 \\ 0 & -1 & 3 & 4 \\ 5 & 0 & -2 & 7 \end{pmatrix}$, $B = \begin{pmatrix} 1 & 0 & 11 & 10 \\ 0 & -1 & 3 & 4 \\ 5 & 0 & -2 & 7 \end{pmatrix}$

In Problems 31–40 find the inverse of the given elementary matrix.

31. $\begin{pmatrix} 0 & 1 \\ 1 & 0 \end{pmatrix}$ **32.** $\begin{pmatrix} 1 & 3 \\ 0 & 1 \end{pmatrix}$ **33.** $\begin{pmatrix} 1 & 0 \\ 0 & 4 \end{pmatrix}$

34. $\begin{pmatrix} 0 & 1 & 0 \\ 1 & 0 & 0 \\ 0 & 0 & 1 \end{pmatrix}$ **35.** $\begin{pmatrix} 1 & -2 & 0 \\ 0 & 1 & 0 \\ 0 & 0 & 1 \end{pmatrix}$ **36.** $\begin{pmatrix} 1 & 0 & 0 \\ 0 & 1 & 0 \\ -2 & 0 & 1 \end{pmatrix}$

37. $\begin{pmatrix} 1 & 0 & 0 \\ 0 & -\frac{1}{2} & 0 \\ 0 & 0 & 1 \end{pmatrix}$ **38.** $\begin{pmatrix} 1 & 0 & 1 & 0 \\ 0 & 1 & 0 & 0 \\ 0 & 0 & 1 & 0 \\ 0 & 0 & 0 & 1 \end{pmatrix}$ **39.** $\begin{pmatrix} 1 & 0 & 0 & 5 \\ 0 & 1 & 0 & 0 \\ 0 & 0 & 1 & 0 \\ 0 & 0 & 0 & 1 \end{pmatrix}$

40. $\begin{pmatrix} 1 & 0 & 0 & 0 \\ 0 & 1 & 0 & 0 \\ 0 & -3 & 1 & 0 \\ 0 & 0 & 0 & 1 \end{pmatrix}$

In Problems 41–48 show that each matrix is invertible and write the matrix as the product of elementary matrices.

41. $\begin{pmatrix} 2 & 1 \\ 3 & 2 \end{pmatrix}$ **42.** $\begin{pmatrix} 1 & 2 \\ 3 & 4 \end{pmatrix}$ **43.** $\begin{pmatrix} 1 & 1 & 1 \\ 0 & 2 & 3 \\ 5 & 5 & 1 \end{pmatrix}$

44. $\begin{pmatrix} 3 & 2 & 1 \\ 0 & 2 & 2 \\ 0 & 0 & -1 \end{pmatrix}$ **45.** $\begin{pmatrix} 0 & -1 & 0 \\ 0 & 1 & -1 \\ 1 & 0 & 1 \end{pmatrix}$ **46.** $\begin{pmatrix} 2 & 0 & 4 \\ 0 & 1 & 1 \\ 3 & -1 & 1 \end{pmatrix}$

47. $\begin{pmatrix} 2 & 0 & 0 & 0 \\ 0 & 3 & 0 & 0 \\ 0 & 0 & -4 & 0 \\ 0 & 0 & 0 & 5 \end{pmatrix}$ **48.** $\begin{pmatrix} 2 & 1 & 0 & 0 \\ 0 & 2 & 1 & 0 \\ 0 & 0 & 2 & 1 \\ 0 & 0 & 0 & 2 \end{pmatrix}$

49. Let $A = \begin{pmatrix} a & b \\ 0 & c \end{pmatrix}$ where $ac \neq 0$. Write A as the product of three elementary matrices and conclude that A is invertible.

50. Let $A = \begin{pmatrix} a & b & c \\ 0 & d & e \\ 0 & 0 & f \end{pmatrix}$ where $adf \neq 0$. Write A as the product of six elementary matrices and conclude that A is invertible.

*51. Let A be an $n \times n$ upper triangular matrix. Prove that if each diagonal component of A is nonzero, then A is invertible [*Hint:* Look at Problems 49 and 50].

*52. Show that if A is an $n \times n$ upper triangular matrix with nonzero diagonal components, then A^{-1} is upper triangular.

*53. Use Theorem 1.10.1(iv) on page 84 and the result of Problem 52 to show that if A is an $n \times n$ lower triangular matrix with nonzero diagonal components, then A is invertible and A^{-1} is lower triangular.

54. Show that if P_{ij} is the $n \times n$ matrix obtained by permuting the ith and jth rows of I_n, then $P_{ij}A$ is the matrix obtained from A by permuting its ith and jth rows.

55. Let A_{ij} be the matrix with c in the jith position, 1's down the diagonal, and 0's everywhere else. Show that $A_{ij}A$ is the matrix obtained from A by multiplying the ith row of A by c and adding it to the jth row.

56. Let M_i be the matrix with c in the ii position, 1's in the other diagonal positions, and 0's everywhere else. Show that M_iA is the matrix obtained by A by multiplying the ith row of A by c.

In Problems 57–62 write each square matrix as the product of elementary matrices and an upper triangular matrix.

57. $A = \begin{pmatrix} 1 & 2 \\ 2 & 4 \end{pmatrix}$ **58.** $A = \begin{pmatrix} 2 & -3 \\ -4 & 6 \end{pmatrix}$ **59.** $A = \begin{pmatrix} 0 & 0 \\ 1 & 0 \end{pmatrix}$

60. $A = \begin{pmatrix} 1 & -1 & 2 \\ 2 & 1 & 4 \\ 4 & -1 & 8 \end{pmatrix}$ **61.** $A = \begin{pmatrix} 1 & -3 & 3 \\ 0 & -3 & 1 \\ 1 & 0 & 2 \end{pmatrix}$ **62.** $A = \begin{pmatrix} 1 & 0 & 0 \\ 2 & 3 & 0 \\ -1 & 4 & 0 \end{pmatrix}$

1.12   GRAPH THEORY: AN APPLICATION OF MATRICES

In recent years much attention has been focused on a relatively new area of mathematical research called **graph theory.** Graphs, which we shall define shortly, are useful in studying the interrelationships among components in networks that arise in commerce, the social sciences, medicine, and many other areas. For example, graphs are useful in studying family ties in a tribal society, the spread of a communicable disease, or a network of commercial flights connecting a given number of major cities. Graph theory is a vast subject. In this section we merely give some definitions and show how close the relationship is between graph theory and matrix theory.

We now illustrate how a graph can arise in practice.

EXAMPLE 1   **Representing a Communication System by a Graph**   Suppose we are studying a communication system linked by telephone wires. In this system there are five stations. In the following table we indicate available lines to and from the stations

Station	1	2	3	4	5
1		✔			
2	✔				✔
3				✔	
4		✔	✔		
5	✔			✔	

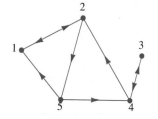

Figure 1.4
The graph showing lines from one station to another

For example, the check in the 1, 2 box indicates that there is a line from station 1 to station 2. The information in the table can be represented by a directed graph, as illustrated in Figure 1.4.   ■

Directed graph
Vertices
Edges

In general, a **directed graph** is a collection of n points called **vertices,** denoted by $V_1, V_2, \ldots, V_n$, together with a finite number of **edges** joining various pairs of vertices. Any directed graph can be represented by an $n \times n$ matrix where the number in the ij position is the number of edges joining the ith vertex to the jth vertex.

EXAMPLE 2   **The Matrix Representation of a Directed Graph**   The matrix representation of the graph in Figure 1.4 is

$$A = \begin{pmatrix} 0 & 1 & 0 & 0 & 0 \\ 1 & 0 & 0 & 0 & 1 \\ 0 & 0 & 0 & 1 & 0 \\ 0 & 1 & 1 & 0 & 0 \\ 1 & 0 & 0 & 1 & 0 \end{pmatrix} \tag{1}$$

■

EXAMPLE 3   **The Matrix Representations of Two Directed Graphs**   Find the matrix representations of the directed graphs in Figure 1.5.

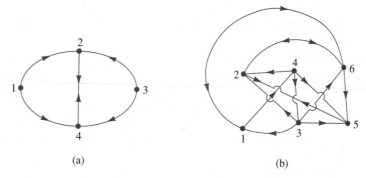

(a)   (b)

Figure 1.5   Two directed graphs

Solution

(a) $A = \begin{pmatrix} 0 & 1 & 0 & 1 \\ 0 & 0 & 0 & 1 \\ 0 & 1 & 0 & 1 \\ 0 & 1 & 0 & 0 \end{pmatrix}$ (b) $A = \begin{pmatrix} 0 & 0 & 0 & 1 & 0 & 1 \\ 0 & 0 & 0 & 0 & 1 & 1 \\ 1 & 1 & 0 & 0 & 1 & 1 \\ 0 & 1 & 1 & 0 & 1 & 0 \\ 0 & 1 & 0 & 0 & 0 & 0 \\ 1 & 1 & 0 & 0 & 1 & 0 \end{pmatrix}$ ∎

EXAMPLE 4   **Obtaining a Graph from Its Matrix Representation**   Sketch a graph represented by the matrix

$$A = \begin{pmatrix} 0 & 1 & 1 & 0 & 1 \\ 1 & 0 & 0 & 1 & 0 \\ 0 & 1 & 0 & 0 & 0 \\ 1 & 0 & 1 & 0 & 1 \\ 0 & 1 & 1 & 1 & 0 \end{pmatrix}$$

Solution   Since A is a 5×5 matrix, the graph has five vertices. See Figure 1.6.

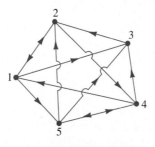

Figure 1.6   The directed graph that is represented by A ∎

Remark. In the examples considered, we have had directed graphs that satisfy the following two conditions:

i. No vertex is connected to itself.

ii. At most, one edge leads from one vertex to another.

Incidence matrix The matrix representing a directed graph satisfying these conditions is called an **incidence matrix.** In general, however, it is possible to have either a 1 on the main diagonal of a representing matrix (indicating an edge from a vertex to itself) or an integer bigger than 1 in the matrix (indicating more than one path from one vertex to another). To avoid more complicated (but treatable) situations, we have assumed, and shall continue to assume, that (*i*) and (*ii*) are satisfied.

EXAMPLE 5   **A Directed Graph That Depicts Domination in a Group**   Directed graphs are often used by sociologists to study group interactions. In many group situations certain individuals dominate others. This domination may be physical, intellectual, or emotional. To be more specific, we assume that in a certain setting involving six persons, a sociologist has been able to determine who dominates whom. (This may have been done by psychological testing, by filling out questionnaires, or simply by observation.) The directed graph in Figure 1.7 indicates the sociologist's findings.

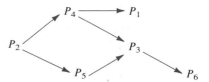

Figure 1.7   The graph shows who dominates whom in the group

The matrix representation of this graph is

$$A = \begin{pmatrix} 0 & 0 & 0 & 0 & 0 & 0 \\ 0 & 0 & 0 & 1 & 1 & 0 \\ 0 & 0 & 0 & 0 & 0 & 1 \\ 1 & 0 & 1 & 0 & 0 & 0 \\ 0 & 0 & 1 & 0 & 0 & 0 \\ 0 & 0 & 0 & 0 & 0 & 0 \end{pmatrix}$$

∎

There would be little point in introducing matrix representations of graphs if all we could do with them was to write them down. There are several facts that can be asked about graphs that may not be apparent. To illustrate this point, consider the graph in Figure 1.8.

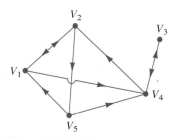

Figure 1.8   There are paths from V_1 to V_5 even though there is no edge from V_1 to V_5. One such path is $V_1 \rightarrow V_2 \rightarrow V_5$

We can see that although there is no edge from V_1 to V_5, it is possible to send a message between these vertices. In fact, there are at least two ways to do this:

$$V_1 \rightarrow V_2 \rightarrow V_5 \tag{2}$$

and

$$V_1 \rightarrow V_4 \rightarrow V_2 \rightarrow V_5 \tag{3}$$

Path
Chain

A route from one vertex to another is called a **path** or **chain.** The path from V_1 to V_5 in (2) is called a **2-chain** because we are traversing two edges. Path (3) is called a **3-chain.** In general, a path traversing n edges (and therefore passing through $n + 1$ vertices) is called an ***n*-chain.** Now, returning to our graph, we see that we can also go from V_1 to V_5 along the 5-chain

$$V_1 \rightarrow V_4 \rightarrow V_3 \rightarrow V_4 \rightarrow V_2 \rightarrow V_5 \tag{4}$$

However, it would be somewhat foolish to do so, since part of the path gains us nothing. A path in which some vertex is encountered more than once is called **redundant.** The 5-chain (4) is redundant because vertex 4 is encountered twice.

It is of great interest to be able to determine the shortest path (if any) joining two vertices in a directed graph. There is a theorem that shows us how this can be done, but first we make an interesting observation. As we have seen, the matrix representation of the graph in Figure 1.4 is given by

$$A = \begin{pmatrix} 0 & 1 & 0 & 0 & 0 \\ 1 & 0 & 0 & 0 & 1 \\ 0 & 0 & 0 & 1 & 0 \\ 0 & 1 & 1 & 0 & 0 \\ 1 & 0 & 0 & 1 & 0 \end{pmatrix}$$

We compute

$$A^2 = \begin{pmatrix} 0 & 1 & 0 & 0 & 0 \\ 1 & 0 & 0 & 0 & 1 \\ 0 & 0 & 0 & 1 & 0 \\ 0 & 1 & 1 & 0 & 0 \\ 1 & 0 & 0 & 1 & 0 \end{pmatrix} \begin{pmatrix} 0 & 1 & 0 & 0 & 0 \\ 1 & 0 & 0 & 0 & 1 \\ 0 & 0 & 0 & 1 & 0 \\ 0 & 1 & 1 & 0 & 0 \\ 1 & 0 & 0 & 1 & 0 \end{pmatrix} = \begin{pmatrix} 1 & 0 & 0 & 0 & 1 \\ 1 & 1 & 0 & 1 & 0 \\ 0 & 1 & 1 & 0 & 0 \\ 1 & 0 & 0 & 1 & 1 \\ 0 & 2 & 1 & 0 & 0 \end{pmatrix}$$

Let us look more closely at the components of A^2. For example, the 1 in the 2,4 position is the dot product of the second row of A and the fourth column of A:

$$(1 \quad 0 \quad 0 \quad 0 \quad 1) \begin{pmatrix} 0 \\ 0 \\ 1 \\ 0 \\ 1 \end{pmatrix} = 1$$

In the second row the last 1 represents the link

$$V_2 \twoheadrightarrow V_5$$

In the fourth column the last 1 represents the link

$$V_5 \twoheadrightarrow V_4$$

Multiplying these 1's together represents the 2-chain

$$V_2 \twoheadrightarrow V_5 \twoheadrightarrow V_4$$

Similarly, the 2 in the 5,2 position of A^2 is the dot product of the fifth row and second column of A:

$$(1 \quad 0 \quad 0 \quad 1 \quad 0) \begin{pmatrix} 1 \\ 0 \\ 0 \\ 1 \\ 0 \end{pmatrix} = 2$$

Reasoning as above, we see that this indicates the pair of 2-chains

$$V_5 \twoheadrightarrow V_1 \twoheadrightarrow V_2$$

and

$$V_5 \twoheadrightarrow V_4 \twoheadrightarrow V_2$$

If we generalize these facts, we can prove the following results:

THEOREM 1   If A is the incidence matrix of a directed graph, then the ijth component of A^2 gives the number of 2-chains from vertex i to vertex j.   ∎

Using this theorem, we can show that the number of 3-chains joining vertex i to vertex j is the ijth component of A^3. In Example 2

$$A^3 = \begin{pmatrix} 1 & 1 & 0 & 1 & 0 \\ 1 & 2 & 1 & 0 & 1 \\ 1 & 0 & 0 & 1 & 1 \\ 1 & 2 & 1 & 1 & 0 \\ 2 & 0 & 0 & 1 & 2 \end{pmatrix}$$

For example, the two 3-chains from vertex 4 to vertex 2 are

$$V_4 \rightarrow V_3 \rightarrow V_4 \rightarrow V_2$$

and

$$V_4 \rightarrow V_2 \rightarrow V_1 \rightarrow V_2$$

Both of these are redundant. The two 3-chains from vertex 5 to vertex 1 are

$$V_5 \rightarrow V_4 \rightarrow V_2 \rightarrow V_1$$

and

$$V_5 \rightarrow V_1 \rightarrow V_2 \rightarrow V_1$$

The following theorem answers the question posed earlier about finding a smallest chain linking two vertices.

THEOREM 2   Let A be an incidence matrix of a directed graph. Let $a_{ij}^{(n)}$ denote the ijth component of A^n.

   i.   If $a_{ij}^{(n)} = k$, then there are exactly k n-chains from vertex i to vertex j.

   ii.   Moreover, if $a_{ij}^{(m)} = 0$ for all $m < n$ and $a_{ij}^{(n)} \neq 0$, then the shortest chain from vertex i to vertex j is an n-chain.   ∎

EXAMPLE 6   **Computing Chains by Taking Powers of the Incidence Matrix**   In Example 2 we have

$$A = \begin{pmatrix} 0 & 1 & 0 & 0 & 0 \\ 1 & 0 & 0 & 0 & 1 \\ 0 & 0 & 0 & 1 & 0 \\ 0 & 1 & 1 & 0 & 0 \\ 1 & 0 & 0 & 1 & 0 \end{pmatrix}, \quad A^2 = \begin{pmatrix} 1 & 0 & 0 & 0 & 1 \\ 1 & 1 & 0 & 1 & 0 \\ 0 & 1 & 1 & 0 & 0 \\ 1 & 0 & 0 & 1 & 1 \\ 0 & 2 & 1 & 0 & 0 \end{pmatrix}, \quad A^3 = \begin{pmatrix} 1 & 1 & 0 & 1 & 0 \\ 1 & 2 & 1 & 0 & 1 \\ 1 & 0 & 0 & 1 & 1 \\ 1 & 2 & 1 & 1 & 0 \\ 2 & 0 & 0 & 1 & 2 \end{pmatrix}$$

$$A^4 = \begin{pmatrix} 1 & 2 & 1 & 0 & 1 \\ 3 & 1 & 0 & 2 & 2 \\ 1 & 2 & 1 & 1 & 0 \\ 2 & 2 & 1 & 1 & 2 \\ 2 & 3 & 1 & 2 & 0 \end{pmatrix}, \quad \text{and} \quad A^5 = \begin{pmatrix} 3 & 1 & 0 & 2 & 2 \\ 3 & 5 & 2 & 2 & 1 \\ 2 & 2 & 1 & 1 & 2 \\ 4 & 3 & 1 & 3 & 2 \\ 3 & 4 & 2 & 1 & 3 \end{pmatrix}$$

Since $a_{13}^{(1)} = a_{13}^{(2)} = a_{13}^{(3)} = 0$ and $a_{13}^{(4)} = 1$, we see that the shortest path from vertex 1 to vertex 3 is a 4-chain. It is given by

$$V_1 \rightarrow V_2 \rightarrow V_5 \rightarrow V_4 \rightarrow V_3$$

Note also that there are five 5-chains (all of which are redundant) joining vertex 2 to itself.   ∎

EXAMPLE 7   **Indirect Domination in a Group**   In our example from sociology (Example 5) a chain (which is not an edge) represents indirect control of one person over another. That is, if Peter dominates Paul, who dominates Mary, then we can see that Peter exercises some control (albeit indirect) over Mary. To determine who has direct or indirect control over whom, we need only compute powers of the incidence matrix A. We have

$$A = \begin{pmatrix} 0 & 0 & 0 & 0 & 0 & 0 \\ 0 & 0 & 0 & 1 & 1 & 0 \\ 0 & 0 & 0 & 0 & 0 & 1 \\ 1 & 0 & 1 & 0 & 0 & 0 \\ 0 & 0 & 1 & 0 & 0 & 0 \\ 0 & 0 & 0 & 0 & 0 & 0 \end{pmatrix}, \quad A^2 = \begin{pmatrix} 0 & 0 & 0 & 0 & 0 & 0 \\ 1 & 0 & 2 & 0 & 0 & 0 \\ 0 & 0 & 0 & 0 & 0 & 0 \\ 0 & 0 & 0 & 0 & 0 & 1 \\ 0 & 0 & 0 & 0 & 0 & 1 \\ 0 & 0 & 0 & 0 & 0 & 0 \end{pmatrix}$$

and

$$A^3 = \begin{pmatrix} 0 & 0 & 0 & 0 & 0 & 0 \\ 0 & 0 & 0 & 0 & 0 & 2 \\ 0 & 0 & 0 & 0 & 0 & 0 \\ 0 & 0 & 0 & 0 & 0 & 0 \\ 0 & 0 & 0 & 0 & 0 & 0 \\ 0 & 0 & 0 & 0 & 0 & 0 \end{pmatrix}$$

As was apparent from the graph on page 98, these matrices show that person P_2 has direct or indirect control over everyone else. He or she has direct control over P_4 and P_5, second-order control over P_1 and P_3, and third-order control over P_6.

Note. In real situations things are much more complex. There may be hundreds of stations in a communications network or hundreds of individuals in a dominant-passive sociological study. In these cases matrices are essential for dealing with the huge amount of data that has to be analyzed.   ∎

PROBLEMS   1.12

In Problems 1–4 find the matrix representation of the given directed graph.

1.

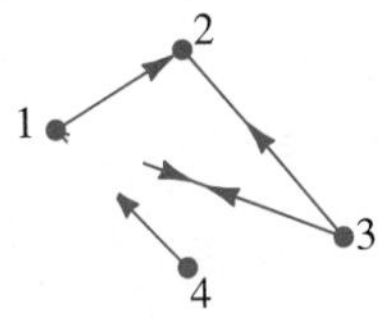

Figure 1.9

2.

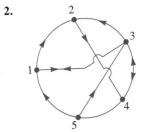

Figure 1.10

3.

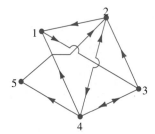

Figure 1.11

4.

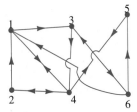

Figure 1.12

In Problems 5–7 draw graphs that are represented by the given matrices.

5.
$$\begin{pmatrix} 0 & 1 & 0 & 1 \\ 1 & 0 & 0 & 0 \\ 1 & 1 & 0 & 1 \\ 1 & 0 & 1 & 0 \end{pmatrix}$$

6.
$$\begin{pmatrix} 0 & 1 & 0 & 1 & 0 \\ 0 & 0 & 1 & 1 & 1 \\ 1 & 1 & 0 & 0 & 0 \\ 1 & 0 & 0 & 0 & 0 \\ 0 & 1 & 1 & 1 & 0 \end{pmatrix}$$

7.
$$\begin{pmatrix} 0 & 1 & 1 & 1 & 0 & 0 \\ 1 & 0 & 0 & 0 & 1 & 0 \\ 1 & 1 & 0 & 1 & 0 & 1 \\ 0 & 0 & 1 & 0 & 0 & 0 \\ 0 & 0 & 0 & 1 & 0 & 1 \\ 1 & 1 & 0 & 0 & 1 & 0 \end{pmatrix}$$

8. Determine the number of 2-, 3-, and 4-chains linking the vertices in the graph of Problem 2.

9. Do the same for the graph of Problem 3.

10. Prove that the shortest path linking two vertices in a directed graph is not redundant.

11. If A is the incidence matrix of a directed graph, show that $A + A^2$ represents the total number of 1-step or 2-step links between vertices.

12. Describe the direct and indirect dominance given by the following graph:

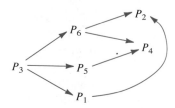

Figure 1.13

SUMMARY

- An ***n*-component row vector** is an ordered set of n numbers written as $(x_1, x_2, \ldots, x_n)$. (p. 27)
- An ***n*-component column vector** is an ordered set of n numbers written as (p. 27)

$$\begin{pmatrix} x_1 \\ x_2 \\ \vdots \\ x_n \end{pmatrix}$$

- A vector all of whose components are zero is called a **zero vector.** (p. 27)
- **Vector addition** and **multiplication by scalars** are defined by (pp. 29, 30)

$$\mathbf{a} + \mathbf{b} = \begin{pmatrix} a_1 + b_1 \\ a_2 + b_2 \\ \vdots \\ a_n + b_n \end{pmatrix} \quad \text{and} \quad \alpha\mathbf{a} = \begin{pmatrix} \alpha a_1 \\ \alpha a_2 \\ \vdots \\ \alpha a_n \end{pmatrix}$$

- An $m \times n$ **matrix** is a rectangular array of mn numbers arranged in m rows and n columns: (p. 35)

$$A = \begin{pmatrix} a_{11} & a_{12} \cdots a_{1n} \\ a_{21} & a_{22} \cdots a_{2n} \\ \vdots & \vdots \qquad \vdots \\ a_{m1} & a_{m2} \cdots a_{mn} \end{pmatrix}$$

The matrix A is also written as $A = (a_{ij})$.

- A matrix all of whose components are zero is called a **zero matrix.** (p. 36)
- If A and B are $m \times n$ matrices, then $A + B$ and αA (α a scalar) are $m \times n$ matrices: (pp. 38, 39)

The ijth component of $A + B$ is $a_{ij} + b_{ij}$.

The ijth component of αA is αa_{ij}.

- The **Scalar product** of two n-component vectors:

$$\mathbf{a} \cdot \mathbf{b} = (a_1 \ a_2 \cdots a_n) \cdot \begin{pmatrix} b_1 \\ b_2 \\ \vdots \\ b_n \end{pmatrix} = a_1 b_1 + a_2 b_2 + \cdots + a_n b_n = \sum_{i=1}^{n} a_i b_i$$

(pp. 42, 43, 51)

- **Matrix Products·**Let A be an $m \times n$ matrix and let B be an $n \times p$ matrix. Then AB is an $m \times p$ matrix and (p. 44)

$$ij\text{th component of } AB = (i\text{th row of } A) \cdot (j\text{th column of } B)$$

$$= a_{i1} b_{1j} + a_{i2} b_{2j} + \cdots + a_{in} b_{nj} = \sum_{k=1}^{n} a_{ik} b_{kj}$$

- Matrix products do not in general commute; that is, it is usually the case that $AB \neq BA$. (p. 46)
- **Associative Law for Matrix Multiplication·**If A is an $n \times m$ matrix, B is $m \times p$ and C is $p \times q$, then

$$A(BC) = (AB)C$$

and both $A(BC)$ and $(AB)C$ are $n \times q$ matrices. (p. 48)
- **Distributive Laws for Matrix Multiplication·**If all the products are defined, then (p. 49)

$$A(B + C) = AB + AC \quad \text{and} \quad (A + B)C = AC + BC$$

- The **coefficient matrix** of the system

$$\begin{array}{c} a_{11}x_1 + a_{12}x_2 + \cdots + a_{1n}x_n = b_1 \\ a_{21}x_1 + a_{22}x_2 + \cdots + a_{2n}x_n = b_2 \\ \vdots \qquad \vdots \qquad \quad \vdots \qquad \vdots \\ a_{m1}x_1 + a_{m2}x_2 + \cdots + a_{mn}x_n = b_m \end{array}$$

is the matrix
(pp. 9, 60)

$$A = \begin{pmatrix} a_{11} & a_{12} & \cdots & a_{1n} \\ a_{21} & a_{22} & \cdots & a_{2n} \\ \vdots & \vdots & & \vdots \\ a_{m1} & a_{m2} & \cdots & a_{mn} \end{pmatrix}$$

- The system above can be written using the **augmented matrix** (p. 9)

$$\begin{pmatrix} a_{11} & a_{12} & \cdots & a_{1n} & \bigm| & b_1 \\ a_{21} & a_{22} & \cdots & a_{2n} & \bigm| & b_2 \\ \vdots & \vdots & & \vdots & & \vdots \\ a_{m1} & a_{m2} & \cdots & a_{mn} & \bigm| & b_m \end{pmatrix}$$

It can also be written as $A\mathbf{x} = \mathbf{b}$, where
(p. 60)

$$\mathbf{x} = \begin{pmatrix} x_1 \\ x_2 \\ \vdots \\ x_n \end{pmatrix} \quad \text{and} \quad \mathbf{b} = \begin{pmatrix} b_1 \\ b_2 \\ \vdots \\ b_m \end{pmatrix}$$

- A matrix is in **reduced row echelon form** if the four conditions on page 12 hold. (p. 12)
- A matrix is in **row echelon form** if the first three conditions on page 12 hold. (p. 13)
- The three **elementary row operations** are: (pp. 9, 10)

Multiply the ith row of a matrix by c: $R_i \rightarrow cR_i$, where $c \neq 0$.

Multiply the ith row by c and add it to the jth row: $R_j \rightarrow R_j + cR_i$, where $c \neq 0$.

Permute the ith and jth rows: $R_i \rightleftarrows R_j$.

- The process of applying elementary row operations to a matrix is called **row reduction.** (p. 9)
- **Gauss-Jordan elimination** is the process of solving a system of equations by row-reducing its augmented matrix to its reduced row echelon form. (pp. 8, 14)
- **Gaussian elimination** is the process of solving a system of equations by row-reducing its augmented matrix to row echelon form and using **back substitution.** (p. 14)
- A system that has one or more solutions is called **consistent.** (p. 15)
- A system that has no solution is called **inconsistent.** (pp. 11, 15)
- A system having solutions has either a unique solution or an infinite number of solutions. (p. 15)
- A **homogeneous** system of m equations in n unknowns is a system of the form (p. 23)

$$\begin{aligned} a_{11}x_1 + a_{12}x_2 + \cdots + a_{1n}x_n &= 0 \\ a_{21}x_1 + a_{22}x_2 + \cdots + a_{2n}x_n &= 0 \\ \vdots \qquad\qquad \vdots \qquad\qquad \vdots \qquad\quad \vdots \\ a_{m1}x_1 + a_{m2}x_2 + \cdots + a_{mn}x_n &= 0 \end{aligned}$$

- A homogeneous system always has the **trivial solution** (or **zero solution**) (p. 24)

$$x_1 = x_2 = \cdots = x_n = 0$$

- Solutions other than the zero solution to a homogeneous system are called **nontrivial solutions.** (p. 24)
- The homogeneous system above has an infinite number of solutions if there are more unknowns than equations ($n > m$). (p. 25)
- The $n \times n$ **identity matrix**, I_n, is the $n \times n$ matrix with 1's down the **main diagonal** and 0's everywhere else. I_n is usually denoted by I. (p. 64)

- If A is a square matrix, then $AI = IA = A$.   (p. 65)
- The $n \times n$ matrix A is **invertible** if there is an $n \times n$ matrix A^{-1} such that   (p. 65)

$$AA^{-1} = A^{-1}A = I$$

In this case the A^{-1} is called the **inverse** of A.

- If A is invertible, the inverse is unique.   (p. 66)
- If A and B are invertible $n \times n$ matrices, then AB is invertible and   (p. 66)

$$(AB)^{-1} = B^{-1}A^{-1}$$

- To determine whether an $n \times n$ matrix A is invertible:

   i. Write the augmented matrix $(A|I)$.   (p. 69)

   ii. Use row reduction to reduce A to its reduced row echelon form.

   iii. a. If the reduced row echelon form of A is I, then A^{-1} will be the matrix to the right of the vertical bar.

   b. If the reduced row echelon form of A contains a row of zeros, then A is not invertible.

- If $A = \begin{pmatrix} a_{11} & a_{12} \\ a_{21} & a_{22} \end{pmatrix}$, then A is invertible if and only if   (p. 70)

$$\text{determinant of } A = \det A = a_{11}a_{22} - a_{12}a_{21} \neq 0$$

In that case

$$A^{-1} = \frac{1}{\det A}\begin{pmatrix} a_{22} & -a_{12} \\ -a_{21} & a_{11} \end{pmatrix}$$

- Two matrices A and B are **row equivalent** if A can be transformed into B by row reduction.   (p. 73)
- Let A be an $n \times n$ matrix. If $AB = I$ or $BA = I$, then A is invertible and $B = A^{-1}$.   (p. 79)
- If $A = (a_{ij})$, then the **transpose of A,** written A^t, is given by $A^t = (a_{ji})$.   (p. 83)
  That is, A^t is obtained by interchanging the rows and columns of A.
- **Facts About Transpose** · If all sums and products are defined and if A is invertible, then   (p. 84)

$$(A^t)^t = A \qquad (AB)^t = B^tA^t \qquad (A + B)^t = A^t + B^t \qquad (A^t)^{-1} = (A^{-1})^t$$

- A square matrix A is **symmetric** if $A^t = A$.   (p. 84)
- An **elementary matrix** is a square matrix obtained by performing exactly one of the elementary row operations on the identity matrix. The three types of elementary matrices are: (p. 87)

cR_i multiply the ith row of I by c, $c \neq 0$

$R_j + cR_i$ multiply the ith row of I by c and add it to the jth row, $c \neq 0$

$R_i \rightleftarrows R_j$ permute the ith and jth rows

- A square matrix is invertible if and only if it is the product of elementary matrices.   (p. 89)
- Any square matrix can be written as the product of elementary matrices and one upper triangular matrix.   (p. 92)
- **Summing Up Theorem** · Let A be an $n \times n$ matrix. Then the following are equivalent: (p. 91)

   i. A is invertible.

   ii. The only solution to the homogeneous system $A\mathbf{x} = \mathbf{0}$ is the trivial solution ($\mathbf{x} = \mathbf{0}$).

   iii. The system $A\mathbf{x} = \mathbf{b}$ has a unique solution for every n-vector $\mathbf{b}$.

iv. A is row equivalent to the $n \times n$ identity matrix I_n.

v. A can be written as a product of elementary matrices.

vi. $\det A \neq 0$ (so far, $\det A$ is defined only if A is a 2×2 matrix).

REVIEW EXERCISES

In Exercises 1–14 find all solutions (if any) to the given systems.

1.
$$3x_1 + 6x_2 = 9$$
$$-2x_1 + 3x_2 = 4$$

2.
$$3x_1 + 6x_2 = 9$$
$$2x_1 + 4x_2 = 6$$

3.
$$3x_1 - 6x_2 = 9$$
$$-2x_1 + 4x_2 = 6$$

4.
$$x_1 + x_2 + x_3 = 2$$
$$2x_1 - x_2 + 2x_3 = 4$$
$$-3x_1 + 2x_2 + 3x_3 = 8$$

5.
$$x_1 + x_2 + x_3 = 0$$
$$2x_1 - x_2 + 2x_3 = 0$$
$$-3x_1 + 2x_2 + 3x_3 = 0$$

6.
$$x_1 + x_2 + x_3 = 2$$
$$2x_1 - x_2 + 2x_3 = 4$$
$$-x_1 + 4x_2 + x_3 = 2$$

7.
$$x_1 + x_2 + x_3 = 2$$
$$2x_1 - x_2 + 2x_3 = 4$$
$$-x_1 + 4x_2 + x_3 = 3$$

8.
$$x_1 + x_2 + x_3 = 0$$
$$2x_1 - x_2 + 2x_3 = 0$$
$$-x_1 + 4x_2 + x_3 = 0$$

9.
$$2x_1 + x_2 - 3x_3 = 0$$
$$4x_1 - x_2 + x_3 = 0$$

10.
$$x_1 + x_2 = 0$$
$$2x_1 + x_2 = 0$$
$$3x_1 + x_2 = 0$$

11.
$$x_1 + x_2 = 1$$
$$2x_1 + x_2 = 3$$
$$3x_1 + x_2 = 4$$

12.
$$x_1 + x_2 + x_3 + x_4 = 4$$
$$2x_1 - 3x_2 - x_3 + 4x_4 = 7$$
$$-2x_1 + 4x_2 + x_3 - 2x_4 = 1$$
$$5x_1 - x_2 + 2x_3 + x_4 = -1$$

13.
$$x_1 + x_2 + x_3 + x_4 = 0$$
$$2x_1 - 3x_2 - x_3 + 4x_4 = 0$$
$$-2x_1 + 4x_2 + x_3 - 2x_4 = 0$$
$$5x_1 - x_2 + 2x_3 + x_4 = 0$$

14.
$$x_1 + x_2 + x_3 + x_4 = 0$$
$$2x_1 - 3x_2 - x_3 + 4x_4 = 0$$
$$-2x_1 + 4x_2 + x_3 - 2x_4 = 0$$

In Exercises 15–22 perform the indicated computations.

15. $3\begin{pmatrix} -2 & 1 \\ 0 & 4 \\ 2 & 3 \end{pmatrix}$

16. $\begin{pmatrix} 1 & 0 & 3 \\ 2 & -1 & 6 \end{pmatrix} + \begin{pmatrix} 2 & 0 & 4 \\ -2 & 5 & 8 \end{pmatrix}$

17. $5\begin{pmatrix} 2 & 1 & 3 \\ -1 & 2 & 4 \\ -6 & 1 & 5 \end{pmatrix} - 3\begin{pmatrix} -2 & 1 & 4 \\ 5 & 0 & 7 \\ 2 & -1 & 3 \end{pmatrix}$

18. $\begin{pmatrix} 2 & 3 \\ -1 & 4 \end{pmatrix}\begin{pmatrix} 5 & -1 \\ 2 & 7 \end{pmatrix}$

19. $\begin{pmatrix} 2 & 3 & 1 & 5 \\ 0 & 6 & 2 & 4 \end{pmatrix}\begin{pmatrix} 5 & 7 & 1 \\ 2 & 0 & 3 \\ 1 & 0 & 0 \\ 0 & 5 & 6 \end{pmatrix}$

20. $\begin{pmatrix} 2 & 3 & 5 \\ -1 & 6 & 4 \\ 1 & 0 & 6 \end{pmatrix}\begin{pmatrix} 0 & -1 & 2 \\ 3 & 1 & 2 \\ -7 & 3 & 5 \end{pmatrix}$

21. $\begin{pmatrix} 1 & 0 & 3 & -1 & 5 \\ 2 & 1 & 6 & 2 & 5 \end{pmatrix} \begin{pmatrix} 7 & 1 \\ 2 & 3 \\ -1 & 0 \\ 5 & 6 \\ 2 & 3 \end{pmatrix}$ **22.** $\begin{pmatrix} 1 & -1 & 2 \\ 3 & 5 & 6 \\ 2 & 4 & -1 \end{pmatrix} \begin{pmatrix} 2 \\ 1 \\ 3 \end{pmatrix}$

In Exercises 23–26 determine whether the given matrix is in row echelon form (but not reduced row echelon form), reduced row echelon form, or neither.

23. $\begin{pmatrix} 1 & 0 & 0 & 0 \\ 0 & 1 & 0 & 2 \\ 0 & 0 & 1 & 3 \end{pmatrix}$ **24.** $\begin{pmatrix} 1 & 8 & 1 & 0 \\ 0 & 1 & 5 & -7 \\ 0 & 0 & 1 & 4 \end{pmatrix}$ **25.** $\begin{pmatrix} 1 & 0 \\ 0 & 3 \\ 0 & 0 \end{pmatrix}$

26. $\begin{pmatrix} 1 & 0 & 2 & 0 \\ 0 & 1 & 3 & 0 \end{pmatrix}$

In Exercises 27 and 28 reduce the matrix to row echelon form and reduced row echelon form.

27. $\begin{pmatrix} 2 & 8 & -2 \\ 1 & 0 & -6 \end{pmatrix}$ **28.** $\begin{pmatrix} 1 & -1 & 2 & 4 \\ -1 & 2 & 0 & 3 \\ 2 & 3 & -1 & 1 \end{pmatrix}$

In Exercises 29–33 calculate the row echelon form and the inverse of the given matrix (if the inverse exists).

29. $\begin{pmatrix} 2 & 3 \\ -1 & 4 \end{pmatrix}$ **30.** $\begin{pmatrix} -1 & 2 \\ 2 & -4 \end{pmatrix}$ **31.** $\begin{pmatrix} 1 & 2 & 0 \\ 2 & 1 & -1 \\ 3 & 1 & 1 \end{pmatrix}$

32. $\begin{pmatrix} -1 & 2 & 0 \\ 4 & 1 & -3 \\ 2 & 5 & -3 \end{pmatrix}$ **33.** $\begin{pmatrix} 2 & 0 & 4 \\ -1 & 3 & 1 \\ 0 & 1 & 2 \end{pmatrix}$

In Exercises 34–36 first write the system in the form $A\mathbf{x} = \mathbf{b}$, then calculate A^{-1}, and, finally, use matrix multiplication to obtain the solution vector.

34. $\begin{aligned} x_1 - 3x_2 &= 4 \\ 2x_1 + 5x_2 &= 7 \end{aligned}$ **35.** $\begin{aligned} x_1 + 2x_2 \phantom{{}+ x_3} &= 3 \\ 2x_1 + x_2 - x_3 &= -1 \\ 3x_1 + x_2 + x_3 &= 7 \end{aligned}$ **36.** $\begin{aligned} 2x_1 \phantom{{}+ 3x_2} + 4x_3 &= 7 \\ -x_1 + 3x_2 + x_3 &= -4 \\ x_2 + 2x_3 &= 5 \end{aligned}$

In Exercises 37–42 calculate the transpose of the given matrix and determine whether the matrix is symmetric or skew-symmetric.†

37. $\begin{pmatrix} 2 & 3 & 1 \\ -1 & 0 & 2 \end{pmatrix}$ **38.** $\begin{pmatrix} 4 & 6 \\ 6 & 4 \end{pmatrix}$ **39.** $\begin{pmatrix} 2 & 3 & 1 \\ 3 & -6 & -5 \\ 1 & -5 & 9 \end{pmatrix}$

40. $\begin{pmatrix} 0 & 5 & 6 \\ -5 & 0 & 4 \\ -6 & -4 & 0 \end{pmatrix}$ **41.** $\begin{pmatrix} 1 & -1 & 4 & 6 \\ -1 & 2 & 5 & 7 \\ 4 & 5 & 3 & -8 \\ 6 & 7 & -8 & 9 \end{pmatrix}$ **42.** $\begin{pmatrix} 0 & 1 & -1 & 1 \\ -1 & 0 & 1 & -2 \\ 1 & 1 & 0 & 1 \\ 1 & -2 & -1 & 0 \end{pmatrix}$

---

†From Problem 1.10.18 on page 86 we have: A is skew-symmetric if $A^t = -A$.

In Exercises 43–47 find a 3 × 3 elementary matrix that will carry out the given row operation.

43. $R_2 \rightarrow -2R_2$ **44.** $R_1 \rightarrow R_1 + 2R_2$ **45.** $R_3 \rightarrow R_3 - 5R_1$

46. $R_3 \rightleftarrows R_1$ **47.** $R_2 \rightarrow R_2 + \frac{1}{5}R_3$

In Exercises 48–50 find the inverse of the elementary matrix.

48. $\begin{pmatrix} 1 & 3 \\ 0 & 1 \end{pmatrix}$ **49.** $\begin{pmatrix} 0 & 1 & 0 \\ 1 & 0 & 0 \\ 0 & 0 & 1 \end{pmatrix}$ **50.** $\begin{pmatrix} 1 & 0 & 0 \\ 0 & 1 & 0 \\ 0 & 0 & -\frac{1}{3} \end{pmatrix}$

In Exercises 51 and 52 write the matrix as the product of elementary matrices.

51. $\begin{pmatrix} 2 & -1 \\ -1 & 1 \end{pmatrix}$ **52.** $\begin{pmatrix} 1 & 0 & 3 \\ 2 & 1 & -5 \\ 3 & 2 & 4 \end{pmatrix}$

In Exercises 53 and 54 write each matrix as the product of elementary matrices and one upper triangular matrix.

53. $\begin{pmatrix} 2 & -1 \\ -4 & 2 \end{pmatrix}$ **54.** $\begin{pmatrix} 1 & -2 & 3 \\ 2 & 0 & 4 \\ 1 & 2 & 1 \end{pmatrix}$

In Exercises 55 and 56 find the matrix that represents each graph.

55.

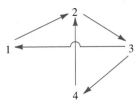

Figure 1.14

56.

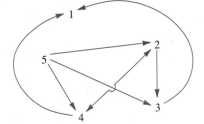

Figure 1.15

57. Draw the graph represented by the following matrix. $\begin{pmatrix} 0 & 0 & 1 & 1 & 0 \\ 0 & 0 & 0 & 1 & 1 \\ 1 & 0 & 0 & 0 & 0 \\ 0 & 1 & 1 & 0 & 1 \\ 1 & 0 & 1 & 0 & 0 \end{pmatrix}$

2

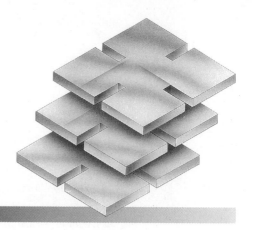

Determinants

2.1 DEFINITIONS

Let $A = \begin{pmatrix} a_{11} & a_{12} \\ a_{21} & a_{22} \end{pmatrix}$ be a 2×2 matrix. In Section 1.9 on page 70 we defined the determinant of A by

$$\det A = a_{11}a_{22} - a_{12}a_{21} \qquad (1)$$

We shall often denote $\det A$ by

$$|A| = \begin{vmatrix} a_{11} & a_{12} \\ a_{21} & a_{22} \end{vmatrix} \qquad (2)$$

We showed that A is invertible if and only if $\det A \neq 0$. As we shall see, this important theorem is valid for $n \times n$ matrices.

In this chapter we develop some of the basic properties of determinants and see how they can be used to calculate inverses and solve systems of n linear equations in n unknowns.

We define the determinant of an $n \times n$ matrix *inductively*. In other words, we use our knowledge of a 2×2 determinant to define a 3×3 determinant, use this to define a 4×4 determinant, and so on. We start by defining a 3×3 determinant.†

---

† There are several ways to define a determinant and this is one of them. It is important to realize that "det" is a function which assigns a *number* to a *square* matrix.

DEFINITION 1 **3 × 3 Determinant** Let $A = \begin{pmatrix} a_{11} & a_{12} & a_{13} \\ a_{21} & a_{22} & a_{23} \\ a_{31} & a_{32} & a_{33} \end{pmatrix}$. Then

$$\det A = |A| = a_{11}\begin{vmatrix} a_{22} & a_{23} \\ a_{32} & a_{33} \end{vmatrix} - a_{12}\begin{vmatrix} a_{21} & a_{23} \\ a_{31} & a_{33} \end{vmatrix} + a_{13}\begin{vmatrix} a_{21} & a_{22} \\ a_{31} & a_{32} \end{vmatrix} \qquad (3)$$

Note the minus sign before the second term on the right side of (3).

EXAMPLE 1 **Calculating a 3 × 3 Determinant**   Let $A = \begin{pmatrix} 3 & 5 & 2 \\ 4 & 2 & 3 \\ -1 & 2 & 4 \end{pmatrix}$. Calculate $|A|$.

Solution

$$|A| = \begin{vmatrix} 3 & 5 & 2 \\ 4 & 2 & 3 \\ -1 & 2 & 4 \end{vmatrix} = 3\begin{vmatrix} 2 & 3 \\ 2 & 4 \end{vmatrix} - 5\begin{vmatrix} 4 & 3 \\ -1 & 4 \end{vmatrix} + 2\begin{vmatrix} 4 & 2 \\ -1 & 2 \end{vmatrix}$$

$$= 3 \cdot 2 - 5 \cdot 19 + 2 \cdot 10 = -69 \qquad \blacksquare$$

EXAMPLE 2 **Calculating a 3 × 3 Determinant**   Calculate $\begin{vmatrix} 2 & -3 & 5 \\ 1 & 0 & 4 \\ 3 & -3 & 9 \end{vmatrix}$.

Solution

$$\begin{vmatrix} 2 & -3 & 5 \\ 1 & 0 & 4 \\ 3 & -3 & 9 \end{vmatrix} = 2\begin{vmatrix} 0 & 4 \\ -3 & 9 \end{vmatrix} - (-3)\begin{vmatrix} 1 & 4 \\ 3 & 9 \end{vmatrix} + 5\begin{vmatrix} 1 & 0 \\ 3 & -3 \end{vmatrix}$$

$$= 2 \cdot 12 + 3(-3) + 5(-3) = 0 \qquad \blacksquare$$

There is another method for calculalting 3×3 determinants. From equation (3) we have

$$\begin{vmatrix} a_{11} & a_{12} & a_{13} \\ a_{21} & a_{22} & a_{23} \\ a_{31} & a_{32} & a_{33} \end{vmatrix} = a_{11}(a_{22}a_{33} - a_{23}a_{32}) - a_{12}(a_{21}a_{33} - a_{23}a_{31}) + a_{13}(a_{21}a_{32} - a_{22}a_{31})$$

or

$$|A| = a_{11}a_{22}a_{33} + a_{12}a_{23}a_{31} + a_{13}a_{21}a_{32} - a_{13}a_{22}a_{31} \qquad (4)$$

$$- a_{12}a_{21}a_{33} - a_{11}a_{32}a_{23}$$

We write A and adjoin its first two columns to it:

We then calculate the six products, put minus signs before the products with arrows pointing upward, and add. This gives the sum in equation (4).

EXAMPLE 3   **Calculating a 3 × 3 Determinant by Using the New Method**

Calculate $\begin{vmatrix} 3 & 5 & 2 \\ 4 & 2 & 3 \\ -1 & 2 & 4 \end{vmatrix}$ by using this new method.

Solution   Writing $\begin{vmatrix} 3 & 5 & 2 \\ 4 & 2 & 3 \\ -1 & 2 & 4 \end{vmatrix} \begin{matrix} 3 & 5 \\ 4 & 2 \\ -1 & 2 \end{matrix}$ and multiplying as indicated, we obtain

$$|A| = (3)(2)(4) + (5)(3)(-1) + (2)(4)(2) - (-1)(2)(2) - 2(3)(3) - (4)(4)(5)$$
$$= 24 - 15 + 16 + 4 - 18 - 80 = -69 \qquad \blacksquare$$

WARNING   The method given above will *not* work for $n \times n$ determinants if $n > 3$. If you try something analogous for 4×4 or higher-order determinants, you will get the wrong answer. $\square$

Before defining $n \times n$ determinants, we first note that in equation (3) $\begin{pmatrix} a_{22} & a_{23} \\ a_{32} & a_{33} \end{pmatrix}$ is the matrix obtained by deleting the first row and first column of A; $\begin{pmatrix} a_{21} & a_{23} \\ a_{31} & a_{33} \end{pmatrix}$ is the matrix obtained by deleting the first row and second column of A; and $\begin{pmatrix} a_{21} & a_{22} \\ a_{31} & a_{32} \end{pmatrix}$ is the matrix obtained by deleting the first row and third column of A. If we denote these three matrices by M_{11}, M_{12}, and M_{13}, respectively, and if $A_{11} = \det M_{11}$, $A_{12} = -\det M_{12}$, and $A_{13} = \det M_{13}$, then equation (3) can be written

$$\boxed{\det A = |A| = a_{11}A_{11} + a_{12}A_{12} + a_{13}A_{13}} \qquad (5)$$

DEFINITION 2   **Minor** Let A be an $n \times n$ matrix and let M_{ij} be the $(n-1) \times (n-1)$ matrix obtained from A by deleting the ith row and jth column of A. M_{ij} is called the **ijth minor** of A.

EXAMPLE 4   **Finding Two Minors of a 3 × 3 Matrix**   Let $A = \begin{pmatrix} 2 & -1 & 4 \\ 0 & 1 & 5 \\ 6 & 3 & -4 \end{pmatrix}$.
Find M_{13} and M_{32}.

Solution   Deleting the first row and third column of A, we obtain $M_{13} = \begin{pmatrix} 0 & 1 \\ 6 & 3 \end{pmatrix}$. Similarly,

by eliminating the third row and second column, we obtain $M_{32} = \begin{pmatrix} 2 & 4 \\ 0 & 5 \end{pmatrix}$. ∎

EXAMPLE 5   **Finding Two Minors of a 4×4 Matrix**   Let $A = \begin{pmatrix} 1 & -3 & 5 & 6 \\ 2 & 4 & 0 & 3 \\ 1 & 5 & 9 & -2 \\ 4 & 0 & 2 & 7 \end{pmatrix}$.
Find M_{32} and M_{24}.

Solution   Deleting the third row and second column of A, we find that $M_{32} = \begin{pmatrix} 1 & 5 & 6 \\ 2 & 0 & 3 \\ 4 & 2 & 7 \end{pmatrix}$;
similarly, $M_{24} = \begin{pmatrix} 1 & -3 & 5 \\ 1 & 5 & 9 \\ 4 & 0 & 2 \end{pmatrix}$. ∎

DEFINITION 3   **Cofactor**   Let A be an $n \times n$ matrix. The **ijth cofactor** of A, denoted by A_{ij}, is given by

$$A_{ij} = (-1)^{i+j}|M_{ij}| \tag{6}$$

That is, the ijth cofactor of A is obtained by taking the determinant of the ijth minor and multiplying it by $(-1)^{i+j}$. Note that

$$(-1)^{i+j} = \begin{cases} 1 & \text{if } i+j \text{ is even} \\ -1 & \text{if } i+j \text{ is odd} \end{cases}$$

Remark.   Definition 3 makes sense because we are going to define an $n \times n$ determinant with the assumption that we already know what an $(n-1) \times (n-1)$ determinant is.

EXAMPLE 6   **Finding Two Cofactors of a 4×4 Matrix**   In Example 5 we have

$$A_{32} = (-1)^{3+2}|M_{32}| = - \begin{vmatrix} 1 & 5 & 6 \\ 2 & 0 & 3 \\ 4 & 2 & 7 \end{vmatrix} = -8$$

$$A_{24} = (-1)^{2+4} \begin{vmatrix} 1 & -3 & 5 \\ 1 & 5 & 9 \\ 4 & 0 & 2 \end{vmatrix} = -192$$

∎

We now consider the general $n \times n$ matrix. Here

$$
A = \begin{pmatrix}
a_{11} & a_{12} & \cdots & a_{1n} \\
a_{21} & a_{22} & \cdots & a_{2n} \\
\vdots & \vdots & & \vdots \\
a_{n1} & a_{n2} & \cdots & a_{nn}
\end{pmatrix}
\tag{7}
$$

DEFINITION 4   ***n* × *n* Determinant** Let A be an $n \times n$ matrix. Then the determinant of A, written $\det A$ or $|A|$, is given by

$$
\det A = |A| = a_{11}A_{11} + a_{12}A_{12} + a_{13}A_{13} + \cdots + a_{1n}A_{1n}
\tag{8}
$$

$$
= \sum_{k=1}^{n} a_{1k}A_{1k}
$$

The expression on the right side of (8) is called an **expansion of cofactors.**

Remark. In equation (8) we defined the determinant by expanding by cofactors using components of A in the first row. We shall see in the next section (Theorem 2.2.1) that we get the same answer if we expand by cofactors in any row or column.

EXAMPLE 7   **Calculating the Determinant of a 4 × 4 Matrix**   Calculate $\det A$, where

$$
A = \begin{pmatrix}
1 & 3 & 5 & 2 \\
0 & -1 & 3 & 4 \\
2 & 1 & 9 & 6 \\
3 & 2 & 4 & 8
\end{pmatrix}
$$

Solution

$$
\begin{vmatrix}
1 & 3 & 5 & 2 \\
0 & -1 & 3 & 4 \\
2 & 1 & 9 & 6 \\
3 & 2 & 4 & 8
\end{vmatrix} = a_{11}A_{11} + a_{12}A_{12} + a_{13}A_{13} + a_{14}A_{14}
$$

$$
= 1\begin{vmatrix} -1 & 3 & 4 \\ 1 & 9 & 6 \\ 2 & 4 & 8 \end{vmatrix} - 3\begin{vmatrix} 0 & 3 & 4 \\ 2 & 9 & 6 \\ 3 & 4 & 8 \end{vmatrix} + 5\begin{vmatrix} 0 & -1 & 4 \\ 2 & 1 & 6 \\ 3 & 2 & 8 \end{vmatrix} - 2\begin{vmatrix} 0 & -1 & 3 \\ 2 & 1 & 9 \\ 3 & 2 & 4 \end{vmatrix}
$$

$$
= 1(-92) - 3(-70) + 5(2) - 2(-16) = 160 \qquad\blacksquare
$$

It is clear that calculating the determinant of an $n \times n$ matrix can be tedious. To calculate a 4 × 4 determinant, we must calculate four 3 × 3 determinants. To calculate a 5 × 5 determinant, we must calculate five 4 × 4 determinants—which is the same as calculating 20 3 × 3 determinants. Fortunately, techniques exist for greatly simplifying these computations. Some of these methods are discussed in the next section. There are, however, some matrices whose determinants can easily be calculated. We begin by repeating a definition given on page 91.

DEFINITION 5   **Triangular Matrix** A square matrix is called **upper triangular** if all its components below the diagonal are zero. It is **lower triangular** if all its components above the diagonal are zero. A matrix is called **diagonal** if all its elements not on the diagonal are zero; that is, $A = (a_{ij})$ is upper triangular if $a_{ij} = 0$ for $i > j$, lower triangular if $a_{ij} = 0$ for $i < j$, and diagonal if $a_{ij} = 0$ for $i \neq j$. Note that a diagonal matrix is both upper and lower triangular.

EXAMPLE 8   **Six Triangular Matrices**

The matrices $A = \begin{pmatrix} 2 & 1 & 7 \\ 0 & 2 & -5 \\ 0 & 0 & 1 \end{pmatrix}$ and $B = \begin{pmatrix} -2 & 3 & 0 & 1 \\ 0 & 0 & 2 & 4 \\ 0 & 0 & 1 & 3 \\ 0 & 0 & 0 & -2 \end{pmatrix}$ are upper triangular;

$C = \begin{pmatrix} 5 & 0 & 0 \\ 2 & 3 & 0 \\ -1 & 2 & 4 \end{pmatrix}$ and $D = \begin{pmatrix} 0 & 0 \\ 1 & 0 \end{pmatrix}$ are lower triangular; I and $E =$

$\begin{pmatrix} 2 & 0 & 0 \\ 0 & -7 & 0 \\ 0 & 0 & -4 \end{pmatrix}$ are diagonal. Note that the matrix E is both upper and lower triangular.

■

EXAMPLE 9   **The Determinant of a Lower Triangular Matrix**   The matrix

$$A = \begin{pmatrix} a_{11} & 0 & 0 & 0 \\ a_{21} & a_{22} & 0 & 0 \\ a_{31} & a_{32} & a_{33} & 0 \\ a_{41} & a_{42} & a_{43} & a_{44} \end{pmatrix}$$

is lower triangular. Compute det A.

Solution

$$\det A = a_{11}A_{11} + 0A_{12} + 0A_{13} + 0A_{14} = a_{11}A_{11}$$

$$= a_{11} \begin{vmatrix} a_{22} & 0 & 0 \\ a_{32} & a_{33} & 0 \\ a_{42} & a_{43} & a_{44} \end{vmatrix}$$

$$= a_{11}a_{22} \begin{vmatrix} a_{33} & 0 \\ a_{43} & a_{44} \end{vmatrix}$$

$$= a_{11}a_{22}a_{33}a_{44}$$

■

Example 9 can be generalized to prove the following theorem.

THEOREM 1   Let $A = (a_{ij})$ be an upper† or lower triangular $n \times n$ matrix. Then

$$\det A = a_{11}a_{22}a_{33} \cdots a_{nn} \tag{9}$$

That is: The determinant of a triangular matrix equals the product of its diagonal components. ∎

EXAMPLE 10   **The Determinants of Six Triangular Matrices**   The determinants of the six matrices in Example 8 are $|A| = 2 \cdot 2 \cdot 1 = 4$; $|B| = (-2)(0)(1)(-2) = 0$; $|C| = 5 \cdot 3 \cdot 4 = 60$; $|D| = 0$; $|I| = 1$; $|E| = (2)(-7)(-4) = 56$. ∎

Geometric Interpretation of the 2 × 2 Determinant

Let $A = \begin{vmatrix} a & b \\ c & d \end{vmatrix}$. In Figure 2.1 we plot the points (a, c) and (b, d) in the xy-plane

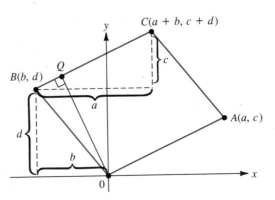

Figure 2.1 Q is on the line segment BC and is also on a line perpendicular to BC that passes through the origin. The area of the parallelogram is $\overline{OQ} \times \overline{OA}$

and draw lines from $(0, 0)$ to each of these points. We assume that the two lines are not collinear. This is the same as assuming that (b, d) is not a multiple of (a, c).

The **area generated by** A is defined as the area of the parallelogram with three vertices at $(0, 0)$, (a, c), and (b, d).

---

†The proof for the upper triangular case is more difficult at this stage, but it will be just the same once we know that det A can be evaluated by expanding in any column (Theorem 2.2.1).

THEOREM 2   The area generated by $A = |\det A|$.

Proof   We assume that neither a nor c is zero. The proof for $a = 0$ or $c = 0$ is left as an exercise (see Problem 17).

The Area of a Parallelogram = Base × Height. The base of the parallelogram in Figure 2.1 has length $\overline{OA} = \sqrt{a^2 + c^2}$. The height of the parallelogram is $\overline{0Q}$, where $0Q$ is a line segment perpendicular to BC. From the figure we see that the coordinates of C, the fourth vertex of the parallelogram are $x = a + b$ and $y = c + d$. Thus

$$\text{Slope of } BC = \frac{\Delta y}{\Delta x} = \frac{(c + d) - d}{(a + b) - b} = \frac{c}{a}$$

Then the equation of the line passing through B and C is

$$\frac{y - d}{x - b} = \frac{c}{a} \quad \text{or} \quad y = \frac{c}{a}x + d - \frac{bc}{a}$$

Fact (*vi*) on page 2
$$\downarrow$$
$$\text{Slope of } 0Q = -\frac{1}{\text{slope of } BC} = -\frac{a}{c}$$

The equation of the line passing through $(0, 0)$ and Q is

$$\frac{y - 0}{x - 0} = -\frac{a}{c} \quad \text{or} \quad y = -\frac{a}{c}x$$

Q is the intersection of BC and $0Q$ so it satisfies both equations. At the point of intersection, we have

$$\frac{c}{a}x + d - \frac{bc}{a} = -\frac{a}{c}x$$

$$\left(\frac{c}{a} + \frac{a}{c}\right)x = \frac{bc}{a} - d$$

$$\frac{a^2 + c^2}{ac}x = \frac{bc - ad}{a}$$

$$x = \frac{ac(bc - ad)}{a(a^2 + c^2)} = \frac{c(bc - ad)}{a^2 + c^2} = -\frac{c(ad - bc)}{a^2 + c^2} = -\frac{c \det A}{a^2 + c^2}$$

and

$$y = -\frac{a}{c}x = -\frac{a}{c} \cdot -\frac{c \det A}{a^2 + c^2} = \frac{a \det A}{a^2 + c^2}$$

Thus Q has coordinates $\left(\dfrac{-c \det A}{a^2 + c^2}, \dfrac{a \det A}{a^2 + c^2}\right)$

and

$$\overline{0Q} = \text{distance from } (0, 0) \text{ to } Q = \sqrt{\frac{c^2(\det A)^2}{(a^2 + c^2)^2} + \frac{a^2(\det A)^2}{(a^2 + c^2)^2}}$$

$$= \sqrt{\frac{(c^2 + a^2)(\det A)^2}{(c^2 + a^2)^2}} = \sqrt{\frac{(\det A)^2}{a^2 + c^2}} = \frac{|\det A|}{\sqrt{a^2 + c^2}}$$

Finally,

$$\text{Area of parallelogram} = \overline{0A} \times \overline{0Q} = \sqrt{a^2 + c^2} \times \frac{|\det A|}{\sqrt{a^2 + c^2}} = |\det A|. \qquad \blacksquare$$

We shall be able to give a much simpler proof of this theorem when we discuss the cross product of two vectors in Section 4.4.

PROBLEMS   2.1

In Problems 1–10 calculate the determinant.

1. $\begin{vmatrix} 1 & 0 & 3 \\ 0 & 1 & 4 \\ 2 & 1 & 0 \end{vmatrix}$
   2. $\begin{vmatrix} -1 & 1 & 0 \\ 2 & 1 & 4 \\ 1 & 5 & 6 \end{vmatrix}$
   3. $\begin{vmatrix} 3 & -1 & 4 \\ 6 & 3 & 5 \\ 2 & -1 & 6 \end{vmatrix}$

4. $\begin{vmatrix} -1 & 0 & 6 \\ 0 & 2 & 4 \\ 1 & 2 & -3 \end{vmatrix}$
   5. $\begin{vmatrix} -2 & 3 & 1 \\ 4 & 6 & 5 \\ 0 & 2 & 1 \end{vmatrix}$
   6. $\begin{vmatrix} 5 & -2 & 1 \\ 6 & 0 & 3 \\ -2 & 1 & 4 \end{vmatrix}$

7. $\begin{vmatrix} 2 & 0 & 3 & 1 \\ 0 & 1 & 4 & 2 \\ 0 & 0 & 1 & 5 \\ 1 & 2 & 3 & 0 \end{vmatrix}$
   8. $\begin{vmatrix} -3 & 0 & 0 & 0 \\ -4 & 7 & 0 & 0 \\ 5 & 8 & -1 & 0 \\ 2 & 3 & 0 & 6 \end{vmatrix}$
   9. $\begin{vmatrix} -2 & 0 & 0 & 7 \\ 1 & 2 & -1 & 4 \\ 3 & 0 & -1 & 5 \\ 4 & 2 & 3 & 0 \end{vmatrix}$

10. $\begin{vmatrix} 2 & 3 & -1 & 4 & 5 \\ 0 & 1 & 7 & 8 & 2 \\ 0 & 0 & 4 & -1 & 5 \\ 0 & 0 & 0 & -2 & 8 \\ 0 & 0 & 0 & 0 & 6 \end{vmatrix}$

11. Show that if A and B are diagonal $n \times n$ matrices, then $\det AB = \det A \det B$.

*12. Show that if A and B are lower triangular matrices, then $\det AB = \det A \det B$.

13. Show that, in general, it is not true that $\det (A + B) = \det A + \det B$.

14. Show that if A is triangular, then $\det A \neq 0$ if and only if all the diagonal components of A are nonzero.

15. Prove Theorem 1 for a lower triangular matrix.

**16. More on the geometric interpretation of the determinant: Let $\mathbf{u}_1$ and $\mathbf{u}_2$ be two

2-vectors and let $\mathbf{v}_1 = A\mathbf{u}_1$ and $\mathbf{v}_2 = A\mathbf{u}_2$. Show that (area generated by $\mathbf{v}_1$ and $\mathbf{v}_2$) = (area generated by $\mathbf{u}_1$ and $\mathbf{u}_2$) $|\det A|$

17. Prove Theorem 2 when A has coordinates $(0, c)$ or $(a, 0)$.

2.2 PROPERTIES OF DETERMINANTS

Determinants have many properties that can make computations easier. We begin to describe these properties by stating a theorem from which everything else follows. The proof of this theorem is difficult and is deferred to the next section.

THEOREM 1 **Basic Theorem.** Let

$$A = \begin{pmatrix} a_{11} & a_{12} & \cdots & a_{1n} \\ a_{21} & a_{22} & \cdots & a_{2n} \\ \vdots & \vdots & & \vdots \\ a_{n1} & a_{n2} & \cdots & a_{nn} \end{pmatrix}$$

be an $n \times n$ matrix. Then

$$\det A = a_{i1}A_{i1} + a_{i2}A_{i2} + \cdots + a_{in}A_{in} = \sum_{k=1}^{n} a_{ik}A_{ik} \qquad (1)$$

for $i = 1, 2, \ldots, n$. That is, we can calculate $\det A$ by expanding by cofactors in *any* row of A. Furthermore,

$$\det A = a_{1j}A_{1j} + a_{2j}A_{2j} + \cdots + a_{nj}A_{nj} = \sum_{k=1}^{n} a_{kj}A_{kj} \qquad (2)$$

Since the jth column of A is $\begin{pmatrix} a_{1j} \\ a_{2j} \\ \vdots \\ a_{nj} \end{pmatrix}$, equation (2) indicates that we can calculate

$\det A$ by expanding by cofactors in any column of A. ∎

EXAMPLE 1 **Obtaining the Determinant by Expanding in the Second Row or the Third Column** For $A = \begin{pmatrix} 3 & 5 & 2 \\ 4 & 2 & 3 \\ -1 & 2 & 4 \end{pmatrix}$, we saw in Example 2.1.1 on page 111 that

$\det A = -69$. Expanding in the second row, we obtain

$$\det A = 4A_{21} + 2A_{22} + 3A_{23}$$

$$= 4(-1)^{2+1}\begin{vmatrix} 5 & 2 \\ 2 & 4 \end{vmatrix} + 2(-1)^{2+2}\begin{vmatrix} 3 & 2 \\ -1 & 4 \end{vmatrix} + 3(-1)^{2+3}\begin{vmatrix} 3 & 5 \\ -1 & 2 \end{vmatrix}$$

$$= -4(16) + 2(14) - 3(11) = -69$$

Similarly, if we expand in the third column, say, we obtain

$$\det A = 2A_{13} + 3A_{23} + 4A_{33}$$

$$= 2(-1)^{1+3}\begin{vmatrix} 4 & 2 \\ -1 & 2 \end{vmatrix} + 3(-1)^{2+3}\begin{vmatrix} 3 & 5 \\ -1 & 2 \end{vmatrix} + 4(-1)^{3+3}\begin{vmatrix} 3 & 5 \\ 4 & 2 \end{vmatrix}$$

$$= 2(10) - 3(11) + 4(-14) = -69$$

You should verify that we get the same answer if we expand in the third row or the first or second column ∎

We now list and prove some additional properties of determinants. In each case we assume that A is an $n \times n$ matrix.† We shall see that these properties can be used to reduce greatly the work involved in evaluating a determinant.

Property 1   If any row or column of A is the zero vector, then $\det A = 0$.

Proof   Suppose the ith row of A contains all zeros. That is, $a_{ij} = 0$ for $j = 1, 2, \ldots, n$. Then $\det A = a_{i1}A_{i1} + a_{i2}A_{i2} + \cdots + a_{in}A_{in} = 0 + 0 + \cdots + 0 = 0$. The same proof works if the jth column is the zero vector. ∎

EXAMPLE 2   **If A Has a Row or Column of Zeros, Then $\det A = 0$**   It is easy to verify that

$$\begin{vmatrix} 2 & 3 & 5 \\ 0 & 0 & 0 \\ 1 & -2 & 4 \end{vmatrix} = 0 \quad \text{and} \quad \begin{vmatrix} -1 & 3 & 0 & 1 \\ 4 & 2 & 0 & 5 \\ -1 & 6 & 0 & 4 \\ 2 & 1 & 0 & 1 \end{vmatrix} = 0$$

∎

Property 2   If the ith row or the jth column of A is multiplied by the constant c, then $\det A$ is multiplied by c. That is, if we call this new matrix B, then

$$|B| = \begin{vmatrix} a_{11} & a_{12} & \cdots & a_{1n} \\ a_{21} & a_{22} & \cdots & a_{2n} \\ \vdots & \vdots & & \vdots \\ ca_{i1} & ca_{i2} & \cdots & ca_{in} \\ \vdots & \vdots & & \vdots \\ a_{n1} & a_{n2} & \cdots & a_{nn} \end{vmatrix} = c\begin{vmatrix} a_{11} & a_{12} & \cdots & a_{1n} \\ a_{21} & a_{22} & \cdots & a_{2n} \\ \vdots & \vdots & & \vdots \\ a_{i1} & a_{i2} & \cdots & a_{in} \\ \vdots & \vdots & & \vdots \\ a_{n1} & a_{n2} & \cdots & a_{nn} \end{vmatrix} = c|A| \qquad (3)$$

---

† The proofs of these properties are given in terms of the rows of a matrix. Using Theorem 1 the same properties can be proved for columns.

Proof   To prove (3) we expand in the ith row of A to obtain

$$\det B = ca_{i1}A_{i1} + ca_{i2}A_{i2} + \cdots ca_{in}A_{in}$$
$$= c(a_{i1}A_{i1} + a_{i2}A_{i2} + \cdots + a_{in}A_{in}) = c \det A$$

A similar proof works for columns.   ■

EXAMPLE 3   **Illustration of Property 2**   Let $A = \begin{pmatrix} 1 & -1 & 2 \\ 3 & 1 & 4 \\ 0 & -2 & 5 \end{pmatrix}$. Then $\det A = 16$. If we

multiply the second row by 4, we have $B = \begin{pmatrix} 1 & -1 & 2 \\ 12 & 4 & 16 \\ 0 & -2 & 5 \end{pmatrix}$ and $\det B = 64 =$

$4 \det A$. If the third column is multiplied by -3, we obtain $C = \begin{pmatrix} 1 & -1 & -6 \\ 3 & 1 & -12 \\ 0 & -2 & -15 \end{pmatrix}$

and $\det C = -48 = -3 \det A$.   ■

Remark. Using Property 2 we can prove (see Problem 28) the following interesting fact: For any scalar α and $n \times n$ matrix A, $\det \alpha A = \alpha^n \det A$.

Property 3   Let

$$A = \begin{pmatrix} a_{11} & a_{12} & \cdots & a_{1j} & \cdots & a_{1n} \\ a_{21} & a_{22} & \cdots & a_{2j} & \cdots & a_{2n} \\ \vdots & \vdots & & \vdots & & \vdots \\ a_{n1} & a_{n2} & \cdots & a_{nj} & \cdots & a_{nn} \end{pmatrix}, \quad B = \begin{pmatrix} a_{11} & a_{12} & \cdots & \alpha_{1j} & \cdots & a_{1n} \\ a_{21} & a_{22} & \cdots & \alpha_{2j} & \cdots & a_{2n} \\ \vdots & \vdots & & \vdots & & \vdots \\ a_{n1} & a_{n2} & \cdots & \alpha_{nj} & \cdots & a_{nn} \end{pmatrix},$$

and $\quad C = \begin{pmatrix} a_{11} & a_{12} & \cdots & a_{1j} + \alpha_{1j} & \cdots & a_{1n} \\ a_{21} & a_{22} & \cdots & a_{2j} + \alpha_{2j} & \cdots & a_{2n} \\ \vdots & \vdots & & \vdots & & \vdots \\ a_{n1} & a_{n2} & \cdots & a_{nj} + \alpha_{nj} & \cdots & a_{nn} \end{pmatrix}$

Then

$$\boxed{\det C = \det A + \det B} \qquad (4)$$

In other words, suppose that A, B, and C are identical except for the jth column and that the jth column of C is the sum of the jth columns of A and B. Then $\det C = \det A + \det B$. The same statement is true for rows.

Proof   We expand $\det C$ in the jth column to obtain

$$\det C = (a_{1j} + \alpha_{1j})A_{1j} + (a_{2j} + \alpha_{2j})A_{2j} + \cdots + (a_{nj} + \alpha_{nj})A_{nj}$$
$$= (a_{1j}A_{1j} + a_{2j}A_{2j} + \cdots + a_{nj}A_{nj})$$
$$+ (\alpha_{1j}A_{1j} + \alpha_{2j}A_{2j} + \cdots + \alpha_{nj}A_{nj}) = \det A + \det B \quad \blacksquare$$

EXAMPLE 4   **Illustration of Property 3**   Let $A = \begin{pmatrix} 1 & -1 & 2 \\ 3 & 1 & 4 \\ 0 & -2 & 5 \end{pmatrix}$, $B = \begin{pmatrix} 1 & -6 & 2 \\ 3 & 2 & 4 \\ 0 & 4 & 5 \end{pmatrix}$,

and $C = \begin{pmatrix} 1 & -1-6 & 2 \\ 3 & 1+2 & 4 \\ 0 & -2+4 & 5 \end{pmatrix} = \begin{pmatrix} 1 & -7 & 2 \\ 3 & 3 & 4 \\ 0 & 2 & 5 \end{pmatrix}$. Then $\det A = 16$, $\det B = 108$,

and $\det C = 124 = \det A + \det B$. $\quad \blacksquare$

Property 4   Interchanging any two rows (or columns) of A has the effect of multiplying $\det A$ by -1.

Proof   We prove the statement for rows and assume first that two adjacent rows are interchanged. That is, we assume that the ith and $(i+1)$st rows are interchanged. Let

$$A = \begin{pmatrix} a_{11} & a_{12} & \cdots & a_{1n} \\ a_{21} & a_{22} & \cdots & a_{2n} \\ \vdots & \vdots & & \vdots \\ a_{i1} & a_{i2} & \cdots & a_{in} \\ a_{i+1,1} & a_{i+1,2} & \cdots & a_{i+1,n} \\ \vdots & \vdots & & \vdots \\ a_{n1} & a_{n2} & & a_{nn} \end{pmatrix} \quad \text{and} \quad B = \begin{pmatrix} a_{11} & a_{12} & \cdots & a_{1n} \\ a_{21} & a_{22} & \cdots & a_{2n} \\ \vdots & \vdots & & \vdots \\ a_{i+1,1} & a_{i+1,2} & \cdots & a_{i+1,n} \\ a_{i1} & a_{i2} & \cdots & a_{in} \\ \vdots & \vdots & & \vdots \\ a_{n1} & a_{n2} & \cdots & a_{nn} \end{pmatrix}$$

Then, expanding $\det A$ in its ith row and $\det B$ in its $(i+1)$st row, we obtain

$$\det A = a_{i1}A_{i1} + a_{i2}A_{i2} + \cdots + a_{in}A_{in} \tag{5}$$
$$\det B = a_{i1}B_{i+1,1} + a_{i2}B_{i+1,2} + \cdots + a_{in}B_{i+1,n}$$

Here $A_{ij} = (-1)^{i+j}|M_{ij}|$, where M_{ij} is obtained by crossing off the ith row and jth column of A. Notice now that if we cross off the $(i+1)$st row and jth column of B, we obtain the same M_{ij}. Thus

$$B_{i+1,j} = (-1)^{i+1+j}|M_{ij}| = -(-1)^{i+j}|M_{ij}| = -A_{ij}$$

so that, from equations (5), $\det B = -\det A$.

Now suppose that $i < j$ and that the ith and jth rows are to be interchanged. We can do this by interchanging adjacent rows several times. It will take $j - i$ interchanges to move row j into the ith row. Then row i will be in the $(i+1)$st row and it will take an additional $j - i - 1$ interchanges to move row i into the jth row. To illustrate, we interchange rows 2 and 6:†

$$
\begin{matrix}
1 & & 1 & & 1 & & 1 & & 1 & & 1 & & 1 & & 1 \\
2 & & 2 & & 2 & & 2 & & 6 & & 6 & & 6 & & 6 \\
3 & & 3 & & 3 & & 6 & & 2 & & 3 & & 3 & & 3 \\
4 & \rightarrow & 4 & \rightarrow & 6 & \rightarrow & 3 & \rightarrow & 3 & \rightarrow & 2 & \rightarrow & 4 & \rightarrow & 4 \\
5 & & 6 & & 4 & & 4 & & 4 & & 4 & & 2 & & 5 \\
6 & & 5 & & 5 & & 5 & & 5 & & 5 & & 5 & & 2 \\
7 & & 7 & & 7 & & 7 & & 7 & & 7 & & 7 & & 7
\end{matrix}
$$

$\underbrace{\qquad\qquad\qquad\qquad}$ $\underbrace{\qquad\qquad\qquad}$

$6 - 2 = 4$ interchanges to $6 - 2 - 1 = 3$ interchanges to get
move the 6 into the 2 position   the 2 into the 6 position

Finally, the total number of interchanges of adjacent rows is $(j - i) + (j - i - 1) = 2j - 2i - 1$, which is odd. Thus det A is multiplied by -1 an odd number of times, which is what we needed to show. ∎

EXAMPLE 5   **Illustration of Property 4**   Let $A = \begin{pmatrix} 1 & -1 & 2 \\ 3 & 1 & 4 \\ 0 & -2 & 5 \end{pmatrix}$. By interchanging the first

and third rows, we obtain $B = \begin{pmatrix} 0 & -2 & 5 \\ 3 & 1 & 4 \\ 1 & -1 & 2 \end{pmatrix}$. By interchanging the first and second

columns of A, we obtain $C = \begin{pmatrix} -1 & 1 & 2 \\ 1 & 3 & 4 \\ -2 & 0 & 5 \end{pmatrix}$. Then, by direct calculation, we find

that det $A = 16$ and det $B =$ det $C = -16$. ∎

Property 5   If A has two equal rows or columns, then det $A = 0$.

Proof   Suppose the ith and jth rows of A are equal. By interchanging these rows we get a matrix B having the property that det $B = -\det A$ (from Property 4). But since row $i =$ row j, interchanging them gives us the same matrix. Thus $A = B$ and det $A =$ det $B = -\det A$. Thus $2 \det A = 0$, which can happen only if det $A = 0$. ∎

EXAMPLE 6   **Illustration of Property 5**   By direct calculation we can verify that for $A = \begin{pmatrix} 1 & -1 & 2 \\ 5 & 7 & 3 \\ 1 & -1 & 2 \end{pmatrix}$ [two equal rows] and $B = \begin{pmatrix} 5 & 2 & 2 \\ 3 & -1 & -1 \\ -2 & 4 & 4 \end{pmatrix}$ [two equal columns], det $A =$ det $B = 0$ ∎

---

† Note that all the numbers here refer to rows.

Property 6   If one row (column) of A is a constant multiple of another row (column), then $\det A = 0$.

Proof   Let $(a_{j1}, a_{j2}, \ldots, a_{jn}) = c(a_{i1}, a_{i2}, \ldots, a_{in})$. Then, from Property 2,

$$
\det A = c
\begin{vmatrix}
a_{11} & a_{12} & \cdots & a_{1n} \\
a_{21} & a_{22} & \cdots & a_{2n} \\
\vdots & \vdots & & \vdots \\
a_{i1} & a_{i2} & \cdots & a_{in} \\
\vdots & \vdots & & \vdots \\
a_{i1} & a_{i2} & \cdots & a_{in} \\
\vdots & \vdots & & \vdots \\
a_{n1} & a_{n2} & \cdots & a_{nn}
\end{vmatrix}
= 0 \qquad \text{(from Property 5)}
$$

with jth row$\rightarrow$ pointing to the $(a_{i1}, a_{i2}, \cdots, a_{in})$ row. ■

EXAMPLE 7   **Illustration of Property 6**
$$
\begin{vmatrix}
2 & -3 & 5 \\
1 & 7 & 2 \\
-4 & 6 & -10
\end{vmatrix}
= 0 \text{ since the third row is } -2
$$
times the first row. ■

EXAMPLE 8   **Another Illustration of Property 6**
$$
\begin{vmatrix}
2 & 4 & 1 & 12 \\
-1 & 1 & 0 & 3 \\
0 & -1 & 9 & -3 \\
7 & 3 & 6 & 9
\end{vmatrix}
= 0 \text{ since the fourth}
$$
column is three times the second column. ■

Property 7   If a multiple of one row (column) of A is added to another row (column) of A, then the determinant is unchanged.

Proof   Let B be the matrix obtained by adding c times the ith row of A to the jth row of A. Then

$$
\det B =
\begin{vmatrix}
a_{11} & a_{12} & \cdots & a_{1n} \\
a_{21} & a_{22} & \cdots & a_{2n} \\
\vdots & \vdots & & \vdots \\
a_{i1} & a_{i2} & \cdots & a_{in} \\
\vdots & \vdots & & \vdots \\
a_{j1} + ca_{i1} & a_{j2} + ca_{i2} & \cdots & a_{jn} + ca_{in} \\
\vdots & \vdots & & \vdots \\
a_{n1} & a_{n2} & \cdots & a_{nn}
\end{vmatrix}
$$

$$= \begin{vmatrix} a_{11} & a_{12} & \cdots & a_{1n} \\ a_{21} & a_{22} & \cdots & a_{2n} \\ \vdots & \vdots & & \vdots \\ a_{i1} & a_{i2} & \cdots & a_{in} \\ \vdots & \vdots & & \vdots \\ a_{j1} & a_{j2} & \cdots & a_{jn} \\ \vdots & \vdots & & \vdots \\ a_{n1} & a_{n2} & \cdots & a_{nn} \end{vmatrix} + \begin{vmatrix} a_{11} & a_{12} & \cdots & a_{1n} \\ a_{21} & a_{22} & \cdots & a_{2n} \\ \vdots & \vdots & & \vdots \\ a_{i1} & a_{i2} & \cdots & a_{in} \\ \vdots & \vdots & & \vdots \\ ca_{i1} & ca_{i2} & \cdots & ca_{in} \\ \vdots & \vdots & & \vdots \\ a_{n1} & a_{n2} & \cdots & a_{nn} \end{vmatrix}$$

(from Property 3)

$$= \det A + 0 = \det A \qquad \text{(the zero comes from Property 6)} \qquad \blacksquare$$

EXAMPLE 9   Illustration of Property 7   Let $A = \begin{pmatrix} 1 & -1 & 2 \\ 3 & 1 & 4 \\ 0 & -2 & 5 \end{pmatrix}$. Then $\det A = 16$. If we

multiply the third row by 4 and add it to the second row, we obtain a new matrix B given by

$$B = \begin{pmatrix} 1 & -1 & 2 \\ 3+4(0) & 1+4(-2) & 4+5(4) \\ 0 & -2 & 5 \end{pmatrix} = \begin{pmatrix} 1 & -1 & 2 \\ 3 & -7 & 24 \\ 0 & -2 & 5 \end{pmatrix}$$

and $\det B = 16 = \det A$. $\blacksquare$

The properties discussed above make it much easier to evaluate high-order determinants. We simply "row-reduce" the determinant, using Property 7, until the determinant is in an easily evaluated form. The most common goal will be to use Property 7 repeatedly until either (1) the new determinant has a row (column) of zeros or one row (column) is a multiple of another row (column)—in which case the determinant is zero—or (2) the new matrix is triangular so that its determinant is the product of its diagonal elements.

EXAMPLE 10   Using the Properties of a Determinant to Calculate a 4 × 4 Determinant
Calculate

$$|A| = \begin{vmatrix} 1 & 3 & 5 & 2 \\ 0 & -1 & 3 & 4 \\ 2 & 1 & 9 & 6 \\ 3 & 2 & 4 & 8 \end{vmatrix}$$

Solution   (See Example 2.1.7, page 114)

There is already a zero in the first column, so it is simplest to reduce other elements in the first column to zero. We then continue to reduce, aiming for a triangular matrix.

Multiply the first row by -2 and add it to the third row and multiply the first row by -3 and add it to the fourth row.

$$|A| = \begin{vmatrix} 1 & 3 & 5 & 2 \\ 0 & -1 & 3 & 4 \\ 0 & -5 & -1 & 2 \\ 0 & -7 & -11 & 2 \end{vmatrix}$$

Multiply the second row by -5 and -7 and add it to the third and fourth rows, respectively.

$$= \begin{vmatrix} 1 & 3 & 5 & 2 \\ 0 & -1 & 3 & 4 \\ 0 & 0 & -16 & -18 \\ 0 & 0 & -32 & -26 \end{vmatrix}$$

Factor out -16 from the third row (using Property 2).

$$= -16 \begin{vmatrix} 1 & 3 & 5 & 2 \\ 0 & -1 & 3 & 4 \\ 0 & 0 & 1 & \frac{9}{8} \\ 0 & 0 & -32 & -26 \end{vmatrix}$$

Multiply the third row by 32 and add it to the fourth row.

$$= -16 \begin{vmatrix} 1 & 3 & 5 & 2 \\ 0 & -1 & 3 & 4 \\ 0 & 0 & 1 & \frac{9}{8} \\ 0 & 0 & 0 & 10 \end{vmatrix}$$

Now we have an upper triangular matrix and $|A| = -16(1)(-1)(1)(10) = (-16)(-10) = 160$. ∎

EXAMPLE 11   **Using the Properties to Calculate a 4 × 4 Determinant**   Calculate

$$|A| = \begin{vmatrix} -2 & 1 & 0 & 4 \\ 3 & -1 & 5 & 2 \\ -2 & 7 & 3 & 1 \\ 3 & -7 & 2 & 5 \end{vmatrix}$$

Solution   There are a number of ways to proceed here and it is not apparent which way will get us the answer most quickly. However, since there is already one zero in the first row, we begin our reduction in that row.

Multiply the second column by 2 and -4 and add it to the first and fourth columns, respectively.

$$|A| = \begin{vmatrix} 0 & 1 & 0 & 0 \\ 1 & -1 & 5 & 6 \\ 12 & 7 & 3 & -27 \\ -11 & -7 & 2 & 33 \end{vmatrix}$$

Interchange the first two columns.

$$= -\begin{vmatrix} 1 & 0 & 0 & 0 \\ -1 & 1 & 5 & 6 \\ 7 & 12 & 3 & -27 \\ -7 & -11 & 2 & 33 \end{vmatrix}$$

Multiply the second column by -5 and -6 and add it to the third and fourth columns, respectively.

$$= -\begin{vmatrix} 1 & 0 & 0 & 0 \\ -1 & 1 & 0 & 0 \\ 7 & 12 & -57 & -99 \\ -7 & -11 & 57 & 99 \end{vmatrix}$$

Since the fourth column is now a multiple of the third column (column 4 = $\frac{99}{57}$ × column 3), we see that $|A| = 0$. ∎

EXAMPLE 12 **Using the Properties to Calculate a 5 × 5 Determinant** Calculate

$$|A| = \begin{vmatrix} 1 & -2 & 3 & -5 & 7 \\ 2 & 0 & -1 & -5 & 6 \\ 4 & 7 & 3 & -9 & 4 \\ 3 & 1 & -2 & -2 & 3 \\ -5 & -1 & 3 & 7 & -9 \end{vmatrix}$$

Solution Adding first row 2 and then row 4 to row 5, we obtain

$$|A| = \begin{vmatrix} 1 & -2 & 3 & -5 & 7 \\ 2 & 0 & -1 & -5 & 6 \\ 4 & 7 & 3 & -9 & 4 \\ 3 & 1 & -2 & -2 & 3 \\ 0 & 0 & 0 & 0 & 0 \end{vmatrix} = 0 \qquad \text{(from Property 1)}$$

This example illustrates the fact that a little looking before beginning the computations can simplify matters considerably. ∎

There are three additional facts about determinants that will be very useful to us.

THEOREM 2 Let A be an $n \times n$ matrix. Then

$$\boxed{a_{i1}A_{j1} + a_{i2}A_{j2} + \cdots + a_{in}A_{jn} = 0 \qquad \text{if } i \neq j} \tag{6}$$

Note. From Theorem 1 the sum in equation (6) equals det A if $i = j$.

Proof   Let

$$B = \begin{pmatrix} a_{11} & a_{12} & \cdots & a_{1n} \\ a_{21} & a_{22} & \cdots & a_{2n} \\ \vdots & \vdots & & \vdots \\ a_{i1} & a_{i2} & \cdots & a_{in} \\ \vdots & \vdots & & \vdots \\ a_{i1} & a_{i2} & \cdots & a_{in} \\ \vdots & \vdots & & \vdots \\ a_{n1} & a_{n2} & \cdots & a_{nn} \end{pmatrix} \quad j\text{th row} \rightarrow$$

Then since two rows of B are equal, $\det B = 0$. But $B = A$ except in the jth row. Thus if we calculate $\det B$ by expanding in the jth row of B, we obtain the sum in (6) and the theorem is proved. Note that when we expand in the jth row, the jth row is deleted in computing the cofactors of B. Thus $B_{jk} = A_{jk}$ for $k = 1, 2, \ldots, n$. ∎

THEOREM 3   Let A be an $n \times n$ matrix. Then

$$\det A = \det A^t \tag{7}$$

Proof   This proof uses mathematical induction. If you are unfamiliar with this important method of proof, refer to Appendix 1. We first prove the theorem in the case $n = 2$. If

$$|A| = \begin{vmatrix} a_{11} & a_{12} \\ a_{21} & a_{22} \end{vmatrix} = a_{11}a_{22} - a_{12}a_{21}$$

then

$$|A^t| = \begin{vmatrix} a_{11} & a_{21} \\ a_{12} & a_{22} \end{vmatrix} = a_{11}a_{22} - a_{21}a_{12} = |A|$$

so the theorem is true for $n = 2$. Next we assume the theorem to be true for $(n-1) \times (n-1)$ matrices and prove it for $n \times n$ matrices. This will prove the theorem. Let $B = A^t$. Then

$$|A| = \begin{vmatrix} a_{11} & a_{12} & \cdots & a_{1n} \\ a_{21} & a_{22} & \cdots & a_{2n} \\ \vdots & \vdots & & \vdots \\ a_{n1} & a_{n2} & \cdots & a_{nn} \end{vmatrix} \quad \text{and} \quad |A^t| = |B| = \begin{vmatrix} a_{11} & a_{21} & \cdots & a_{n1} \\ a_{12} & a_{22} & \cdots & a_{n2} \\ \vdots & \vdots & & \vdots \\ a_{1n} & a_{2n} & \cdots & a_{nn} \end{vmatrix}$$

We expand $|A|$ in the first row and expand $|B|$ in the first column. This gives us

$$|A| = a_{11}A_{11} + a_{12}A_{12} + \cdots + a_{1n}A_{1n}$$

$$|B| = a_{11}B_{11} + a_{12}B_{21} + \cdots + a_{1n}B_{n1}$$

We need to show that $A_{1k} = B_{k1}$ for $k = 1, 2, \ldots, n$. But $A_{1k} = (-1)^{1+k}|M_{1k}|$ and $B_{k1} = (-1)^{k+1}|N_{k1}|$, where M_{1k} is the $1k$th minor of A and N_{k1} is the $k1$st minor of B. Then

$$|M_{1k}| = \begin{vmatrix} a_{21} & a_{22} & \cdots & a_{2,k-1} & a_{2,k+1} & \cdots & a_{2n} \\ a_{31} & a_{32} & \cdots & a_{3,k-1} & a_{3,k+1} & \cdots & a_{3n} \\ \vdots & \vdots & & \vdots & \vdots & & \vdots \\ a_{n1} & a_{n2} & \cdots & a_{n,k-1} & a_{n,k+1} & \cdots & a_{nn} \end{vmatrix}$$

and

$$|N_{k1}| = \begin{vmatrix} a_{21} & a_{31} & \cdots & a_{n1} \\ a_{22} & a_{32} & \cdots & a_{n2} \\ \vdots & \vdots & & \vdots \\ a_{2,k-1} & a_{3,k-1} & \cdots & a_{n,k-1} \\ a_{2,k+1} & a_{3,k+1} & \cdots & a_{n,k+1} \\ \vdots & \vdots & & \vdots \\ a_{2n} & a_{3n} & \cdots & a_{nn} \end{vmatrix}$$

Clearly $M_{1k} = N^t_{k1}$, and since both are $(n-1) \times (n-1)$ matrices, the induction hypothesis tells us that $|M_{1k}| = |N_{k1}|$. Thus $A_{1k} = B_{k1}$ and the proof is complete. ∎

EXAMPLE 13   **A Matrix and Its Transpose Have the Same Determinant**

Let $A = \begin{pmatrix} 1 & -1 & 2 \\ 3 & 1 & 4 \\ 0 & -2 & 5 \end{pmatrix}$. Then $A^t = \begin{pmatrix} 1 & 3 & 0 \\ -1 & 1 & -2 \\ 2 & 4 & 5 \end{pmatrix}$ and it is easy to verify that $|A| = |A^t| = 16$. ∎

THEOREM 4   Let A and B be $n \times n$ matrices. Then

$$\boxed{\det AB = \det A \det B} \tag{8}$$

That is: *The determinant of the product is the product of the determinants.*

Proof   The proof, using elementary matrices, is given in Section 2.3. You are asked in Problem 38 to verify the result in the 2×2 case. ∎

EXAMPLE 14   **Illustration of Fact That det AB = det A det B**   Verify equation (8) for

$$A = \begin{pmatrix} 1 & -1 & 2 \\ 3 & 1 & 4 \\ 0 & -2 & 5 \end{pmatrix} \text{ and } B = \begin{pmatrix} 1 & -2 & 3 \\ 0 & -1 & 4 \\ 2 & 0 & -2 \end{pmatrix}.$$

Solution   det A = 16 and det B = −8. We calculate

$$AB = \begin{pmatrix} 1 & -1 & 2 \\ 3 & 1 & 4 \\ 0 & -2 & 5 \end{pmatrix}\begin{pmatrix} 1 & -2 & 3 \\ 0 & -1 & 4 \\ 2 & 0 & -2 \end{pmatrix} = \begin{pmatrix} 5 & -1 & -5 \\ 11 & -7 & 5 \\ 10 & 2 & -18 \end{pmatrix}$$

and det AB = −128 = (16)(−8) = det A det B.   ∎

WARNING   The determinant of the sum is *not* equal to the sum of the determinants. That is,

$$\det (A + B) \neq \det A + \det B$$

For example, let $A = \begin{pmatrix} 1 & 2 \\ 3 & 4 \end{pmatrix}$ and $B = \begin{pmatrix} 3 & 0 \\ -2 & 2 \end{pmatrix}$. Then $A + B = \begin{pmatrix} 4 & 2 \\ 1 & 6 \end{pmatrix}$:

$$\det A = -2, \quad \det B = 6, \quad \text{and}$$
$$\det (A + B) = 22 \neq \det A + \det B = -2 + 6 = 4 \qquad \square$$

PROBLEMS   2.2

In Problems 1–20 evaluate the determinant by using the methods of this section.

1. $\begin{vmatrix} 3 & -5 \\ 2 & 6 \end{vmatrix}$

2. $\begin{vmatrix} 4 & 1 \\ 0 & -3 \end{vmatrix}$

3. $\begin{vmatrix} -1 & 0 & 2 \\ 3 & 1 & 4 \\ 2 & 0 & -6 \end{vmatrix}$

4. $\begin{vmatrix} 2 & 1 & -1 \\ 3 & -2 & 0 \\ 5 & 1 & 6 \end{vmatrix}$

5. $\begin{vmatrix} -3 & 2 & 4 \\ 1 & -1 & 2 \\ -1 & 4 & 0 \end{vmatrix}$

6. $\begin{vmatrix} 0 & -2 & 3 \\ 1 & 2 & -3 \\ 4 & 0 & 5 \end{vmatrix}$

7. $\begin{vmatrix} -2 & 3 & 6 \\ 4 & 1 & 8 \\ -2 & 0 & 0 \end{vmatrix}$

8. $\begin{vmatrix} 2 & -1 & 3 \\ 4 & 0 & 6 \\ 5 & -2 & 3 \end{vmatrix}$

9. $\begin{vmatrix} 1 & -1 & 2 & 4 \\ 0 & -3 & 5 & 6 \\ 1 & 4 & 0 & 3 \\ 0 & 5 & -6 & 7 \end{vmatrix}$

10. $\begin{vmatrix} 2 & -3 & 1 & 4 \\ 0 & -2 & 0 & 0 \\ 3 & 7 & -1 & 2 \\ 4 & 1 & -3 & 8 \end{vmatrix}$

11. $\begin{vmatrix} 1 & 1 & -1 & 0 \\ -3 & 4 & 6 & 0 \\ 2 & 5 & -1 & 3 \\ 4 & 0 & 3 & 0 \end{vmatrix}$

12.
$$
\begin{vmatrix}
3 & -1 & 2 & 1 \\
4 & 3 & 1 & -2 \\
-1 & 0 & 2 & 3 \\
6 & 2 & 5 & 2
\end{vmatrix}
$$

13.
$$
\begin{vmatrix}
2 & 0 & 0 & 0 \\
0 & 0 & 3 & 0 \\
0 & -1 & 0 & 0 \\
0 & 0 & 0 & 4
\end{vmatrix}
$$

14.
$$
\begin{vmatrix}
0 & a & 0 & 0 \\
b & 0 & 0 & 0 \\
0 & 0 & 0 & c \\
0 & 0 & d & 0
\end{vmatrix}
$$

15.
$$
\begin{vmatrix}
1 & 2 & 0 & 0 \\
3 & -2 & 0 & 0 \\
0 & 0 & 1 & -5 \\
0 & 0 & 7 & 2
\end{vmatrix}
$$

16.
$$
\begin{vmatrix}
a & b & 0 & 0 \\
c & d & 0 & 0 \\
0 & 0 & a & -b \\
0 & 0 & c & d
\end{vmatrix}
$$

17.
$$
\begin{vmatrix}
2 & -1 & 0 & 4 & 1 \\
3 & 1 & -1 & 2 & 0 \\
3 & 2 & -2 & 5 & 1 \\
0 & 0 & 4 & -1 & 6 \\
3 & 2 & 1 & -1 & 1
\end{vmatrix}
$$

18.
$$
\begin{vmatrix}
1 & -1 & 2 & 0 & 0 \\
3 & 1 & 4 & 0 & 0 \\
2 & -1 & 5 & 0 & 0 \\
0 & 0 & 0 & 2 & 3 \\
0 & 0 & 0 & -1 & 4
\end{vmatrix}
$$

19.
$$
\begin{vmatrix}
a & 0 & 0 & 0 & 0 \\
0 & 0 & b & 0 & 0 \\
0 & 0 & 0 & 0 & c \\
0 & 0 & 0 & d & 0 \\
0 & e & 0 & 0 & 0
\end{vmatrix}
$$

20.
$$
\begin{vmatrix}
2 & 5 & -6 & 8 & 0 \\
0 & 1 & -7 & 6 & 0 \\
0 & 0 & 0 & 4 & 0 \\
0 & 2 & 1 & 5 & 1 \\
4 & -1 & 5 & 3 & 0
\end{vmatrix}
$$

In Problems 21–27 compute the determinant assuming that

$$
\begin{vmatrix}
a_{11} & a_{12} & a_{13} \\
a_{21} & a_{22} & a_{23} \\
a_{31} & a_{32} & a_{33}
\end{vmatrix} = 8
$$

21.
$$
\begin{vmatrix}
a_{31} & a_{32} & a_{33} \\
a_{21} & a_{22} & a_{23} \\
a_{11} & a_{12} & a_{13}
\end{vmatrix}
$$

22.
$$
\begin{vmatrix}
a_{31} & a_{32} & a_{33} \\
a_{11} & a_{12} & a_{13} \\
a_{21} & a_{22} & a_{23}
\end{vmatrix}
$$

23.
$$
\begin{vmatrix}
a_{11} & a_{12} & a_{13} \\
2a_{21} & 2a_{22} & 2a_{23} \\
a_{31} & a_{32} & a_{33}
\end{vmatrix}
$$

24.
$$
\begin{vmatrix}
-3a_{11} & -3a_{12} & -3a_{13} \\
2a_{21} & 2a_{22} & 2a_{23} \\
5a_{31} & 5a_{32} & 5a_{33}
\end{vmatrix}
$$

25.
$$
\begin{vmatrix}
a_{11} & 2a_{13} & a_{12} \\
a_{21} & 2a_{23} & a_{22} \\
a_{31} & 2a_{33} & a_{32}
\end{vmatrix}
$$

26.
$$
\begin{vmatrix}
a_{11} - a_{12} & a_{12} & a_{13} \\
a_{21} - a_{22} & a_{22} & a_{23} \\
a_{31} - a_{32} & a_{32} & a_{33}
\end{vmatrix}
$$

27.
$$
\begin{vmatrix}
2a_{11} - 3a_{21} & 2a_{12} - 3a_{22} & 2a_{13} - 3a_{23} \\
a_{31} & a_{32} & a_{33} \\
a_{21} & a_{22} & a_{23}
\end{vmatrix}
$$

28. Using Property 2, show that if α is a number and A is an $n \times n$ matrix, then $\det \alpha A = \alpha^n \det A$.

*29. Show that

$$
\begin{vmatrix}
1+x_1 & x_2 & x_3 & \cdots & x_n \\
x_1 & 1+x_2 & x_3 & \cdots & x_n \\
x_1 & x_2 & 1+x_3 & \cdots & x_n \\
\vdots & \vdots & \vdots & & \vdots \\
x_1 & x_2 & x_3 & \cdots & 1+x_n
\end{vmatrix} = 1 + x_1 + x_2 + \cdots + x_n
$$

*30. A matrix is **skew-symmetric** if $A^t = -A$. If A is an $n \times n$ skew-symmetric matrix, show that $\det A = (-1)^n \det A$.

31. Using the result of Problem 30, show that if A is a skew-symmetric $n \times n$ matrix and n is odd, then $\det A = 0$.

32. A matrix A is called **orthogonal** if A is invertible and $A^{-1} = A^t$. Show that if A is orthogonal, then $\det A = \pm 1$.

**33. Let Δ denote the triangle in the plane with vertices at (x_1, y_1), (x_2, y_2), and (x_3, y_3). Show that the area of the triangle is given by

$$
\text{Area of } \Delta = \pm \frac{1}{2} \begin{vmatrix} 1 & x_1 & y_1 \\ 1 & x_2 & y_2 \\ 1 & x_3 & y_3 \end{vmatrix}
$$

Under what circumstances will this determinant equal zero?

**34. Three lines, no two of which are parallel, determine a triangle in the plane. Suppose that the lines are given by

$$
\begin{aligned}
a_{11}x + a_{12}y + a_{13} &= 0 \\
a_{21}x + a_{22}y + a_{23} &= 0 \\
a_{31}x + a_{32}y + a_{33} &= 0
\end{aligned}
$$

Show that the area determined by the lines is

$$
\frac{\pm 1}{2A_{13}A_{23}A_{33}} \begin{vmatrix} A_{11} & A_{12} & A_{13} \\ A_{21} & A_{22} & A_{23} \\ A_{31} & A_{32} & A_{33} \end{vmatrix}
$$

35. The 3×3 **Vandermonde†** **determinant** is given by

$$
D_3 = \begin{vmatrix} 1 & 1 & 1 \\ a_1 & a_2 & a_3 \\ a_1^2 & a_2^2 & a_3^2 \end{vmatrix}
$$

Show that $D_3 = (a_2 - a_1)(a_3 - a_1)(a_3 - a_2)$.

36. $D_4 = \begin{vmatrix} 1 & 1 & 1 & 1 \\ a_1 & a_2 & a_3 & a_4 \\ a_1^2 & a_2^2 & a_3^2 & a_4^2 \\ a_1^3 & a_2^3 & a_3^3 & a_4^3 \end{vmatrix}$ is the 4×4 Vandermonde determinant.

Show that $D_4 = (a_2 - a_1)(a_3 - a_1)(a_4 - a_1)(a_3 - a_2)(a_4 - a_2)(a_4 - a_3)$.

---

†A. T. Vandermonde (1735–1796) was a French mathematician.

**37. a.   Define the $n \times n$ Vandermonde determinant D_n.

   b.   Show that $D_n = \prod\limits_{\substack{i=1 \\ j>i}}^{n} (a_j - a_i)$, where $\prod$ stands for the word "product." Note that

   the product in Problem 36 can be written $\prod\limits_{\substack{i=1 \\ j>i}}^{4} (a_j - a_i)$.

38.   Let $A = \begin{pmatrix} a_{11} & a_{12} \\ a_{21} & a_{22} \end{pmatrix}$ and $B = \begin{pmatrix} b_{11} & b_{12} \\ b_{21} & b_{22} \end{pmatrix}$.

   a.   Write out the product AB.

   b.   Compute det A, det B, and det AB.

   c.   Show that det $AB = ($det $A)($det $B)$.

39.   The $n \times n$ matrix A is called **nilpotent** if $A^k = 0$, the $n \times n$ zero matrix, for some integer $k \geq 1$. Show that the following matrices are nilpotent and find the smallest k such that $A^k = 0$.

   a. $\begin{pmatrix} 0 & 2 \\ 0 & 0 \end{pmatrix}$ **b.** $\begin{pmatrix} 0 & 1 & 3 \\ 0 & 0 & 4 \\ 0 & 0 & 0 \end{pmatrix}$

40.   Show that if A is nilpotent, then det $A = 0$.

41.   The matrix A is called **idempotent** if $A^2 = A$. What are the possible values for det A if A is idempotent?

2.3   PROOFS OF THREE IMPORTANT THEOREMS AND SOME HISTORY (If time permits)

Earlier in this book we cited three theorems that are central in the theory of matrices and determinants. The proofs of these theorems are more difficult than those proofs we have already given. Work through these proofs slowly; the reward will be a deeper understanding of some of the important ideas in linear algebra.

THEOREM 1   **Basic Theorem**   Let $A = (a_{ij})$ be an $n \times n$ matrix. Then

$$\det A = a_{11}A_{11} + a_{12}A_{12} + \cdots + a_{1n}A_{1n}$$
$$= a_{i1}A_{i1} + a_{i2}A_{i2} + \cdots + a_{in}A_{in} \tag{1}$$
$$= a_{1j}A_{1j} + a_{2j}A_{2j} + \cdots + a_{nj}A_{nj} \tag{2}$$

for $i = 1, 2, \ldots, n$ and $j = 1, 2, \ldots, n$.

Note. The first equality is Definition 2.1.4 of the determinant by cofactor expansion in the first row; the second equality says that the expansion by cofactors in any other row yields the determinant; the third equality says that expansion by cofactors in any column gives the determinant.

Proof   We prove equality (1) by mathematical induction. For the 2×2 matrix $A = \begin{pmatrix} a_{11} & a_{12} \\ a_{21} & a_{22} \end{pmatrix}$, we first expand the first row by cofactors: $\det A = a_{11}A_{11} + a_{12}A_{12} = a_{11}(a_{22}) + a_{12}(-a_{21}) = a_{11}a_{22} - a_{12}a_{21}$. Similarly, expanding in the second row, we obtain $a_{21}A_{21} + a_{22}A_{22} = a_{21}(-a_{12}) + a_{22}(a_{11}) = a_{11}a_{22} - a_{12}a_{21}$. Thus we get the same result by expanding in any row of a 2×2 matrix, and this proves equality (1) in the 2×2 case.

We now assume that equality (1) holds for all $(n-1) \times (n-1)$ matrices. We must show that it holds for $n \times n$ matrices. Our procedure will be to expand by cofactors in the first and ith rows and show that the expansions are identical. If we expand in the first row, then a typical term in the cofactor expansion is

$$a_{1k}A_{1k} = (-1)^{1+k}a_{1k}|M_{1k}| \tag{3}$$

Note that this is the only place in the expansion of $|A|$ that the term a_{1k} occurs since another typical term is $a_{1m}A_{1m} = (-1)^{1+m}|M_{1m}|$, $k \neq m$, and M_{1m} is obtained by deleting the first row and mth column of A (and a_{1k} is in the first row of A). Since M_{1k} is an $(n-1) \times (n-1)$ matrix, we can by the induction hypothesis calculate $|M_{1k}|$ by expanding in the ith row of A (which is the $(i-1)$st row of M_{1k}). A typical term in this expansion is

$$a_{il} \text{ (cofactor of } a_{il} \text{ in } M_{1k}) \qquad (k \neq l) \tag{4}$$

For the reasons outlined above, this is the only term in the expansion of $|M_{1k}|$ in the ith row of A that contains the term a_{il}. Substituting (4) into (3), we find that

$$(-1)^{1+k}a_{1k}a_{il} \text{ (cofactor of } a_{il} \text{ in } M_{1k}) \qquad (k \neq l) \tag{5}$$

is the only occurrence of the term $a_{1k}a_{il}$ in the cofactor expansion of $\det A$ in the first row.

Now if we expand by cofactors in the ith row of A (where $i \neq 1$), a typical term is

$$(-1)^{i+l}a_{il}|M_{il}| \tag{6}$$

and a typical term in the expansion of $|M_{il}|$ in the first row of M_{il} is

$$a_{1k} \text{ (cofactor of } a_{1k} \text{ in } M_{il}) \qquad (k \neq l) \tag{7}$$

and, inserting (7) in (6), we find that the only occurrence of the term $a_{il}a_{1k}$ in the expansion of $\det A$ along its ith row is

$$(-1)^{i+l}a_{1k}a_{il}(\text{cofactor of } a_{1k} \text{ in } M_{il}) \qquad (k \neq l) \tag{8}$$

If we can show that the expressions in (5) and (8) are the same, then (1) will be proved, for the term in (5) is the only occurrence of $a_{1k}a_{il}$ in the first row expansion, the term in (8) is the only occurrence of $a_{1k}a_{il}$ in the ith row expansion, and k, i, and l are arbitrary. This will show that the sums of the terms in the first and ith row expansions are the same.

Now let $M_{1i,kl}$ denote the $(n-2) \times (n-2)$ matrix obtained by deleting the first and ith rows and kth and lth columns of A. (This is called a **second-order minor** of A.) We first suppose that $k < l$. Then

$$M_{1k} = \begin{pmatrix} a_{21} & \cdots & a_{2,k-1} & a_{2,k+1} & \cdots & a_{2l} & \cdots & a_{2n} \\ \vdots & & \vdots & \vdots & & \vdots & & \vdots \\ a_{i1} & \cdots & a_{i,k-1} & a_{i,k+1} & \cdots & a_{il} & \cdots & a_{in} \\ \vdots & & \vdots & \vdots & & \vdots & & \vdots \\ a_{n1} & \cdots & a_{n,k-1} & a_{n,k+1} & \cdots & a_{nl} & \cdots & a_{nn} \end{pmatrix} \tag{9}$$

$$M_{il} = \begin{pmatrix} a_{11} & \cdots & a_{1k} & \cdots & a_{1,l-1} & a_{1,l+1} & \cdots & a_{1n} \\ \vdots & & \vdots & & \vdots & \vdots & & \vdots \\ a_{i-1,1} & \cdots & a_{i-1,k} & \cdots & a_{i-1,l-1} & a_{i-1,l+1} & \cdots & a_{i-1,n} \\ a_{i+1,1} & \cdots & a_{i+1,k} & \cdots & a_{i+1,l-1} & a_{i+1,l+1} & \cdots & a_{i+1,n} \\ \vdots & & \vdots & & \vdots & \vdots & & \vdots \\ a_{n1} & \cdots & a_{nk} & \cdots & a_{n,l-1} & a_{n,l+1} & \cdots & a_{nn} \end{pmatrix} \tag{10}$$

From (9) and (10) we see that

$$\text{Cofactor of } a_{il} \text{ in } M_{1k} = (-1)^{(i-1)+(l-1)}|M_{1i,kl}| \tag{11}$$

$$\text{Cofactor of } a_{1k} \text{ in } M_{il} = (-1)^{1+k}|M_{1i,kl}| \tag{12}$$

Thus (5) becomes

$$(-1)^{1+k}a_{1k}a_{il}(-1)^{(i-1)+(l-1)}|M_{1i,kl}| = (-1)^{i+k+l-1}a_{1k}a_{il}|M_{1i,kl}| \tag{13}$$

and (8) becomes

$$(-1)^{i+l}a_{1k}a_{il}(-1)^{1+k}|M_{1i,kl}| = (-1)^{i+k+l+1}a_{1k}a_{il}|M_{1i,kl}| \tag{14}$$

But $(-1)^{i+k+l-1} = (-1)^{i+k+l+1}$, so that the right sides of equations (13) and (14) are equal. Hence expressions (5) and (8) are equal and (1) is proved in the case $k < l$. If $k > l$, then, by similar reasoning, we find that

$$\text{Cofactor of } a_{il} \text{ in } M_{1k} = (-1)^{(i-1)+l}|M_{1i,kl}|$$

$$\text{Cofactor of } a_{1k} \text{ in } M_{il} = (-1)^{1+(k-1)}|M_{1i,kl}|$$

so that (5) becomes

$$(-1)^{1+k}a_{1k}a_{il}(-1)^{(i-1)+l}|M_{1i,kl}| = (-1)^{i+k+l}a_{1k}a_{il}|M_{1i,kl}|$$

and (8) becomes

$$(-1)^{i+l}a_{1k}a_{il}(-1)^{1+k-1}|M_{1i,kl}| = (-1)^{i+k+l}a_{1k}a_{il}|M_{1i,kl}|$$

This completes the proof of equation (1).

To prove equation (2) we go through a similar process. If we expand in the kth and lth columns, we find that the only occurrences of the term $a_{1k}a_{il}$ will be given by (5) and (8). (See Problems 1 and 2.) This shows that the expansion by cofactors in any two columns is the same and that each is equal to the expansion along any row. This completes the proof. ∎

We now wish to prove that for any two $n \times n$ matrices A and B, det $AB =$ det A det B. The proof is difficult and involves a number of steps. We shall make use of a number of facts about elementary matrices proved in Section 1.10.

We begin by computing the determinants of elementary matrices.

LEMMA 1   Let E be an elementary matrix:

   i. If E is represented by $R_i \rightleftarrows R_j$, then det $E = -1$.   **(15)**

   ii. If E is represented by $R_j \rightarrow R_j + cR_i$, then det $E = 1$.   **(16)**

   iii. If E is represented by $R_i \rightarrow cR_i$, then det $E = c$.   **(17)**

Proof   **i.** det $I = 1$. E is obtained from I by interchanging the ith and jth rows of I. From Property 4 on page 121, det $E = (-1)$ det $I = -1$.

   ii. E is obtained from I by multiplying the ith row of I by c and adding it to the jth row. Thus by Property 7 on page 124 det $E =$ det $I = 1$.

   iii. E is obtained from I by multiplying the ith row of I by c. Thus, from Property 2 on page 120, det $E = c$ det $I = c$.   ■

LEMMA 2   Let B be an $n \times n$ matrix and let E be an elementary matrix. Then

$$\det EB = \det E \det B \qquad\qquad \textbf{(18)}$$

■

The proof of this lemma follows from Lemma 1 and the results relating elementary row operations to determinants discussed in Section 2.2. The steps in the proof are indicated in Problems 6 to 8.

LEMMA 3   Let T be an upper triangular matrix. Then T is invertible if and only if det $T \neq 0$.
Proof   Let

$$T = \begin{pmatrix} a_{11} & a_{12} & a_{13} & \cdots & a_{1n} \\ 0 & a_{22} & a_{23} & \cdots & a_{2n} \\ 0 & 0 & a_{33} & \cdots & a_{3n} \\ \vdots & \vdots & \vdots & & \vdots \\ 0 & 0 & 0 & \cdots & a_{nn} \end{pmatrix} \qquad\qquad \textbf{(19)}$$

From Theorem 2.1.1 on page 116,

$$\det T = a_{11}a_{22}\cdots a_{nn} \qquad\qquad \textbf{(20)}$$

Thus det $T \neq 0$ if and only if each of its diagonal components is nonzero.

If det $T \neq 0$, then T can be row reduced to I in the following way:

i. For $i = 1, 2, \ldots, n$, divide the ith row of T by $a_{ii} \neq 0$ to obtain

$$\begin{pmatrix} 1 & a'_{12} & \cdots & a'_{1n} \\ 0 & 1 & \cdots & a'_{2n} \\ \vdots & \vdots & & \vdots \\ 0 & 0 & \cdots & 1 \end{pmatrix}$$

ii. Use the 1 in the jth diagonal component to make zero each component above it in the jth column.

Thus T is row-equivalent to I and so is invertible by Theorem 1.9.6 on page 78 (the Summing Up Theorem).

Suppose that $\det T = 0$. Then at least one of the diagonal components of T is zero. Let a_{ii} be the first such component. Then T can be written

$$T = \begin{pmatrix} a_{11} & a_{12} & \cdots & a_{1,i-1} & a_{1i} & a_{1,i+1} & \cdots & a_{1n} \\ 0 & a_{22} & \cdots & a_{2,i-1} & a_{2i} & a_{2,i+1} & \cdots & a_{2n} \\ \vdots & \vdots & & \vdots & \vdots & \vdots & & \vdots \\ 0 & 0 & \cdots & a_{i-1,i-1} & a_{i-1,i} & a_{i-1,i+1} & \cdots & a_{i-1,n} \\ 0 & 0 & \cdots & 0 & 0 & a_{i,i+1} & \cdots & a_{in} \\ 0 & 0 & \cdots & 0 & 0 & a_{i+1,i+1} & \cdots & a_{i+1,n} \\ \vdots & \vdots & & \vdots & \vdots & \vdots & \ddots & \vdots \\ 0 & 0 & \cdots & 0 & 0 & 0 & \cdots & a_{nn} \end{pmatrix} \qquad (21)$$

Consider the homogeneous system

$$\begin{aligned} a_{11}x_1 + a_{12}x_2 + \cdots + a_{1,i-1}x_{i-1} + \quad a_{1i}x_i &= 0 \\ a_{22}x_2 + \cdots + a_{2,i-1}x_{i-1} + \quad a_{2i}x_i &= 0 \\ \vdots \qquad\qquad \vdots \quad\quad \\ a_{i-1,i-1}x_{i-1} + a_{i-1,i}x_i &= 0 \end{aligned}$$

This is a system of $i - 1$ equations in i unknowns. By Theorem 1.4.1 on page 25, the system has nontrivial solutions.

Let $\begin{pmatrix} x_1 \\ x_2 \\ \vdots \\ x_i \end{pmatrix}$ be such a solution where not all of $x_1, x_2, \ldots, x_i$ are zero. Let

$$\mathbf{x}^* = \begin{pmatrix} x_1 \\ x_2 \\ \vdots \\ x_i \\ 0 \\ \vdots \\ 0 \end{pmatrix} \Big\} \; n - i \text{ 0's}$$

From (21) and the way the x's were chosen, we see that $T\mathbf{x}^* = \mathbf{0}$; that is, the equation $T\mathbf{x} = \mathbf{0}$ has a nontrivial solution. Using Theorem 1.9.6 again [part (ii)] we conclude that T is not invertible. ∎

The following theorem is very important.

THEOREM 2   Let A be an $n \times n$ matrix. Then A is invertible if and only if $\det A \neq 0$.

Proof   From Theorem 1.11.5 on page 92 we know that there are elementary matrices $E_1, E_2, \ldots, E_m$ and a triangular matrix T such that

$$A = E_1 E_2 \cdots E_m T \tag{22}$$

Using Lemma 2 m times, we see that

$$\det A = \det E_1 \det (E_2 E_3 \cdots E_m T)$$
$$= \det E_1 \det E_2 \det(E_3 \cdots E_m T)$$
$$\vdots$$
$$= \det E_1 \det E_2 \cdots \det E_{m-1} \det(E_m T)$$

or

$$\det A = \det E_1 \det E_2 \cdots \det E_{m-1} \det E_m \det T \tag{23}$$

By Lemma 1, $\det E_i \neq 0$. We conclude that $\det A \neq 0$ if and only if $\det T \neq 0$.

Now suppose that A is invertible. Then, using (22) and the fact that every elementary matrix is invertible, we can write T as the product of invertible matrices. Thus T is invertible and, by Lemma 3, $\det T \neq 0$. Thus $\det A \neq 0$.

If $\det A \neq 0$, then, by (23), $\det T \neq 0$ so T is invertible (by Lemma 3). Then the right side of (22) is the product of invertible matrices, and so A is invertible. This completes the proof. ∎

We can now, finally, prove the main result.

THEOREM 3   Let A and B be $n \times n$ matrices. Then

$$\det AB = \det A \det B \tag{24}$$

Proof   *Case 1:* $\det A = \det B = 0$. Then, by Theorem 2, B is not invertible so, by Theorem 1.9.6, there is an n-vector $\mathbf{x} \neq \mathbf{0}$ such that $B\mathbf{x} = \mathbf{0}$. Then $(AB)\mathbf{x} = A(B\mathbf{x}) = A\mathbf{0} = \mathbf{0}$. Therefore, again by Theorem 1.9.6, AB is not invertible. By Theorem 2,

$$0 = \det AB = 0 \cdot 0 = \det A \det B$$

Case 2: $\det A = 0$ and $\det B \neq 0$. A is not invertible so there is an n-vector $\mathbf{y} \neq \mathbf{0}$ such that $A\mathbf{y} = \mathbf{0}$. Since $\det B \neq 0$, B is invertible and there is a unique vector $\mathbf{x} \neq \mathbf{0}$

such that $B\mathbf{x} = \mathbf{y}$. Then $AB\mathbf{x} = A(B\mathbf{x}) = A\mathbf{y} = \mathbf{0}$. Thus AB is not invertible, so

$$\det AB = 0 = 0 \det B = \det A \det B$$

Case 3: $\det A \neq 0$. A is invertible and can be written as the product of elementary matrices:

$$A = E_1 E_2 \cdots E_m$$

Then

$$AB = E_1 E_2 \cdots E_m B$$

Using the result of Lemma 2 repeatedly, we see that

$$\begin{aligned}
\det AB &= \det(E_1 E_2 \cdots E_m B) \\
&= \det E_1 \det E_2 \cdots \det E_m \det B \\
&= \det(E_1 E_2 \cdots E_m) \det B \\
&= \det A \det B
\end{aligned}$$

$\blacksquare$

Focus on . . .

A Short History of Determinants

Gottfried Wilhelm
Leibniz
*(David Eugene Smith
Collection, Rare
Book and Manuscript
Library, Columbia
University)*

Determinants appeared in mathematical literature over a century before matrices. As pointed out in the note on page 36, the term *matrix* was coined by James Joseph Sylvester and was intended to mean "mother of determinants."

Some of the greatest mathematicians of the eighteenth and nineteenth centuries helped to develop properties of determinants. Most historians believe that the theory of determinants originated with the German mathematician Gottfried Wilhelm Leibniz (1646–1716), who, with Newton, was the co-inventor of calculus. Leibniz used determinants in 1693 in reference to systems of simultaneous linear equations. Some believe, however, that a Japanese mathematician, Seki Kōwa, did the same thing about 10 years earlier.

The most prolific contributor to the theory of determinants was the French mathematician Augustin-Louis Cauchy (1789–1857). Cauchy wrote an 84-page memoir in 1812 that contained the first proof of the theorem $\det AB = \det A \det B$. In 1840 Cauchy defined the characteristic equation of the matrix A to be the polynomial equation $\det(A - \lambda I) = 0$. We shall discuss this equation in great detail in Chapter 6.

Cauchy made many other contributions to mathematics. In his 1829 calculus textbook *Leçons sur le calcul différential*, he gave the first reasonably clear definition of a limit.

Augustin-Louis
Cauchy
*(David Eugene Smith
Collection, Rare
Book and Manuscript
Library, Columbia
University)*

Cauchy wrote extensively in both pure and applied mathematics. Only Euler wrote more. Cauchy contributed to many areas including real and complex function theory, probability theory, geometry, wave propagation theory, and infinite series.

Cauchy is credited with setting a new standard of rigor in mathematical publication. After Cauchy, it was much more difficult to publish a paper based on intuition; a strict adherence to formal proof was demanded.

The sheer volume of Cauchy's publication was overwhelming. When the French Academy of Sciences began publishing its journal *Comptes Rendus* in 1835, Cauchy sent his work there to be published. Soon the printing bill for Cauchy's work alone became so large that the Academy placed a limit of four pages on each published paper. This rule is still in force today.

Some other mathematicians are worthy of mention here. The expansion of a determinant by cofactors was first used by the French mathematician Pierre-Simon Laplace (1749–1827). Laplace is best known for the Laplace transform studied in applied mathematics courses.

A major contributor to determinant theory (second only to Cauchy) was the German mathematician Carl Gustav Jacobi (1804–1851). It was with him that the word "determinant" gained final acceptance. Jacobi first used the determinant applied to functions in the setting of the theory of functions of several variables. This determinant was later named the *Jacobian* by Sylvester. Students today study Jacobians in second year calculus classes.

Finally, no history of determinants would be complete without citing the text book *An Elementary Theory of Determinants,* written in 1867 by Charles Dodgson (1832–1898). In this book Dodgson gives conditions such that systems of equations have nontrivial solutions. These conditions are written in terms of the determinants of the minors of coefficient matrices. Charles Dodgson is better known by his pen name Lewis Carroll. Under this pen name he wrote his much better known book *Alice in Wonderland.*

PROBLEMS   2.3

1. Show that if A is expanded along its kth column, then the only occurrence of the term $a_{1k}a_{il}$ is given by equation (5).

2. Show that if A is expanded along its lth column, then the only occurrence of the term $a_{1k}a_{il}$ is given by equation (8).

3. Show that if A is expanded along its kth column, then the only occurrence of the term $a_{ik}a_{jl}$ is $(-1)^{i+k}a_{ik}a_{jl}$ (cofactor of a_{jl} in M_{ik}) for $l \neq k$.

4. Let $A = \begin{pmatrix} 1 & 5 & 7 \\ 2 & -1 & 3 \\ 4 & 5 & -2 \end{pmatrix}$. Compute det A by expanding in each of the rows and columns.

5. Do the same for the matrix $A = \begin{pmatrix} 1 & -1 & 4 \\ 0 & 1 & 5 \\ -3 & 7 & 2 \end{pmatrix}$.

6. Let E be the representation of $R_i \rightleftarrows R_j$ and let B be an $n \times n$ matrix. Show that $\det EB = \det E \det B$. [*Hint:* Describe the matrix EB and then use equation (15) and Property 4.]

7. Let E be the representation of $R_j \rightarrow R_j + cR_i$ and let B be an $n \times n$ matrix. Show that $\det EB = \det E \det B$. [Hint: Describe the matrix EB and then use equation (16) and Property 7.]

8. Let E be the representation of $R_i \rightarrow cR_i$ and let B be an $n \times n$ matrix. Show that $\det EB = \det E \det B$. [*Hint:* Describe the matrix EB and then use equation (7) and Property 2.]

2.4 DETERMINANTS AND INVERSES

In this section we see how matrix inverses can be calculated by using determinants. Moreover, we complete the task, begun in Chapter 1, of proving the important Summing Up Theorem (see Theorems 1.9.6 on page 78 and 1.11.4 on page 91), which shows the equivalence of various properties of matrices. We begin with a simple result.

THEOREM 1 If A is invertible, then $\det A \neq 0$ and

$$\det A^{-1} = \frac{1}{\det A} \tag{1}$$

Proof According to Theorem 2.3.2 on page 138, $\det A \neq 0$. From Theorem 2.2.4, page 129

$$1 = \det I = \det AA^{-1} = \det A \det A^{-1} \tag{2}$$

which implies that

$$\det A^{-1} = 1/\det A. \qquad \blacksquare$$

Before using determinants to calculate inverses, we need to define the *adjoint* of a matrix $A = (a_{ij})$. Let $B = (A_{ij})$ be the matrix of cofactors of A. (Remember that a cofactor, defined on page 113, is a number.) Then

$$B = \begin{pmatrix} A_{11} & A_{12} & \cdots & A_{1n} \\ A_{21} & A_{22} & \cdots & A_{2n} \\ \vdots & \vdots & & \vdots \\ A_{n1} & A_{n2} & \cdots & A_{nn} \end{pmatrix} \tag{3}$$

DEFINITION 1   **The Adjoint** Let A be an $n \times n$ matrix and let B, given by (3), denote the matrix of its cofactors. Then the **adjoint** of A, written adj A, is the transpose of the $n \times n$ matrix B; that is,

$$\text{adj } A = B^t = \begin{pmatrix} A_{11} & A_{21} & \cdots & A_{n1} \\ A_{12} & A_{22} & \cdots & A_{n2} \\ \vdots & \vdots & & \vdots \\ A_{1n} & A_{2n} & \cdots & A_{nn} \end{pmatrix} \tag{4}$$

EXAMPLE 1   **Computing the Adjoint of a 3 × 3 Matrix**   Let $A = \begin{pmatrix} 2 & 4 & 3 \\ 0 & 1 & -1 \\ 3 & 5 & 7 \end{pmatrix}$. Compute adj A.

Solution   We have $A_{11} = \begin{vmatrix} 1 & -1 \\ 5 & 7 \end{vmatrix} = 12$, $A_{12} = -\begin{vmatrix} 0 & -1 \\ 3 & 7 \end{vmatrix} = -3$, $A_{13} = -3$, $A_{21} = -13$, $A_{22} = 5$, $A_{23} = 2$, $A_{31} = -7$, $A_{32} = 2$, and $A_{33} = 2$. Thus $B = \begin{pmatrix} 12 & -3 & -3 \\ -13 & 5 & 2 \\ -7 & 2 & 2 \end{pmatrix}$ and adj $A = B^t = \begin{pmatrix} 12 & -13 & -7 \\ -3 & 5 & 2 \\ -3 & 2 & 2 \end{pmatrix}$. ∎

EXAMPLE 2   **Computing the Adjoint of a 4 × 4 Matrix**   Let

$$A = \begin{pmatrix} 1 & -3 & 0 & -2 \\ 3 & -12 & -2 & -6 \\ -2 & 10 & 2 & 5 \\ -1 & 6 & 1 & 3 \end{pmatrix}$$

Calculate adj A.

Solution   This is more tedious since we have to compute sixteen 3×3 determinants. For example, we have $A_{12} = -\begin{vmatrix} 3 & -2 & -6 \\ -2 & 2 & 5 \\ -1 & 1 & 3 \end{vmatrix} = -1$, $A_{24} = \begin{vmatrix} 1 & -3 & 0 \\ -2 & 10 & 2 \\ -1 & 6 & 1 \end{vmatrix} = -2$, and $A_{43} = -\begin{vmatrix} 1 & -3 & -2 \\ 3 & -12 & -6 \\ -2 & 10 & 5 \end{vmatrix} = 3$. Completing these calculations, we find that

$$B = \begin{pmatrix} 0 & -1 & 0 & 2 \\ -1 & 1 & -1 & -2 \\ 0 & 2 & -3 & -3 \\ -2 & -2 & 3 & 2 \end{pmatrix}$$

$$\operatorname{adj} A = B^t = \begin{pmatrix} 0 & -1 & 0 & -2 \\ -1 & 1 & 2 & -2 \\ 0 & -1 & -3 & 3 \\ 2 & -2 & -3 & 2 \end{pmatrix}$$ ∎

EXAMPLE 3 **The Adjoint of a 2 × 2 Matrix** Let $A = \begin{pmatrix} a_{11} & a_{12} \\ a_{21} & a_{22} \end{pmatrix}$. Then $\operatorname{adj} A = \begin{pmatrix} A_{11} & A_{21} \\ A_{12} & A_{22} \end{pmatrix} = \begin{pmatrix} a_{22} & -a_{12} \\ -a_{21} & a_{11} \end{pmatrix}$. ∎

WARNING In taking the adjoint of a matrix, do not forget to transpose the matrix of cofactors. □

THEOREM 2 Let A be an $n \times n$ matrix. Then

$$(A)(\operatorname{adj} A) = \begin{pmatrix} \det A & 0 & 0 & \cdots & 0 \\ 0 & \det A & 0 & \cdots & 0 \\ 0 & 0 & \det A & \cdots & 0 \\ \vdots & \vdots & \vdots & & \vdots \\ 0 & 0 & 0 & \cdots & \det A \end{pmatrix} = (\det A)I \quad (5)$$

Proof Let $C = (c_{ij}) = (A)(\operatorname{adj} A)$. Then

$$C = \begin{pmatrix} a_{11} & a_{12} & \cdots & a_{1n} \\ a_{21} & a_{22} & \cdots & a_{2n} \\ \vdots & \vdots & & \vdots \\ a_{n1} & a_{n2} & \cdots & a_{nn} \end{pmatrix} \begin{pmatrix} A_{11} & A_{21} & \cdots & A_{n1} \\ A_{12} & A_{22} & \cdots & A_{n2} \\ \vdots & \vdots & & \vdots \\ A_{1n} & A_{2n} & \cdots & A_{nn} \end{pmatrix} \quad (6)$$

We have

$$c_{ij} = (i\text{th row of } A) \cdot (j\text{th column of adj } A)$$
$$= (a_{i1} \ \ a_{i2} \cdots a_{in}) \cdot \begin{pmatrix} A_{j1} \\ A_{j2} \\ \vdots \\ A_{jn} \end{pmatrix}$$

Thus $$c_{ij} = a_{i1}A_{j1} + a_{i2}A_{j2} + \cdots + a_{in}A_{jn} \quad (7)$$

Now if $i = j$, the sum in (7) equals $a_{i1}A_{i1} + a_{i2}A_{i2} + \cdots + a_{in}A_{in}$, which is the expansion of $\det A$ in the ith row of A. On the other hand, if $i \neq j$, then from Theorem 2.2.2 on page 127, the sum in (7) equals zero. Thus

$$c_{ij} = \begin{cases} \det A & \text{if } i = j \\ 0 & \text{if } i \neq j \end{cases}$$

This proves the theorem. ∎

We can now state the main result.

THEOREM 3   Let A be an $n \times n$ matrix. Then A is invertible if and only if det $A \neq 0$. If det $A \neq 0$, then

$$A^{-1} = \frac{1}{\det A} \text{ adj } A \tag{8}$$

Note that Theorem 1.9.4 on page 70 for 2×2 matrices is a special case of this theorem.

Proof   The first part of the theorem is Theorem 2.3.2. If det $A \neq 0$, then

$$(A)\left(\frac{1}{\det A} \text{ adj } A\right) = \frac{1}{\det A}\left[A(\text{adj } A)\right] \overset{\text{Theorem 2}}{=} \frac{1}{\det A}(\det A)I = I$$

But, by Theorem 1.9.8 on page 79, if $AB = I$, then $B = A^{-1}$. Thus

$$(1/\det A) \text{ adj } A = A^{-1} \qquad \blacksquare$$

EXAMPLE 4   **Using the Determinant and the Adjoint to Calculate an Inverse**

Let $A = \begin{pmatrix} 2 & 4 & 3 \\ 0 & 1 & -1 \\ 3 & 5 & 7 \end{pmatrix}$. Determine whether A is invertible and calculate A^{-1} if it is.

Solution   Since det $A = 3 \neq 0$, we see that A is invertible. From Example 1,

$$\text{adj } A = \begin{pmatrix} 12 & -13 & -7 \\ -3 & 5 & 2 \\ -3 & 2 & 2 \end{pmatrix}$$

Thus

$$A^{-1} = \frac{1}{3}\begin{pmatrix} 12 & -13 & -7 \\ -3 & 5 & 2 \\ -3 & 2 & 2 \end{pmatrix} = \begin{pmatrix} 4 & -\frac{13}{3} & -\frac{7}{3} \\ -1 & \frac{5}{3} & \frac{2}{3} \\ -1 & \frac{2}{3} & \frac{2}{3} \end{pmatrix}$$

Check.

$$A^{-1}A = \frac{1}{3}\begin{pmatrix} 12 & -13 & -7 \\ -3 & 5 & 2 \\ -3 & 2 & 2 \end{pmatrix}\begin{pmatrix} 2 & 4 & 3 \\ 0 & 1 & -1 \\ 3 & 5 & 7 \end{pmatrix} = \frac{1}{3}\begin{pmatrix} 3 & 0 & 0 \\ 0 & 3 & 0 \\ 0 & 0 & 3 \end{pmatrix} = I \qquad \blacksquare$$

EXAMPLE 5   **Calculating the Inverse of a 4 × 4 Matrix Using the Determinant and the Adjoint**   Let

$$A = \begin{pmatrix} 1 & -3 & 0 & -2 \\ 3 & -12 & -2 & -6 \\ -2 & 10 & 2 & 5 \\ -1 & 6 & 1 & 3 \end{pmatrix}$$

Determine whether A is invertible and, if so, calculate A^{-1}.

Solution   Using properties of determinants, we compute

$$\begin{vmatrix} 1 & -3 & 0 & -2 \\ 3 & -12 & -2 & -6 \\ -2 & 10 & 2 & 5 \\ -1 & 6 & 1 & 3 \end{vmatrix}$$

Multiply the first column by 3 and 2 and add it to the second and fourth columns, respectively.

$$= \begin{vmatrix} 1 & 0 & 0 & 0 \\ 3 & -3 & -2 & 0 \\ -2 & 4 & 2 & 1 \\ -1 & 3 & 1 & 1 \end{vmatrix}$$

Expand in the first row.

$$= \begin{vmatrix} -3 & -2 & 0 \\ 4 & 2 & 1 \\ 3 & 1 & 1 \end{vmatrix} = -1$$

Thus $\det A = -1 \neq 0$ and A^{-1} exists. By Example 2, we have

$$\text{adj } A = \begin{pmatrix} 0 & -1 & 0 & -2 \\ -1 & 1 & 2 & -2 \\ 0 & -1 & -3 & 3 \\ 2 & -2 & -3 & 2 \end{pmatrix}$$

Thus

$$A^{-1} = \frac{1}{-1} \begin{pmatrix} 0 & -1 & 0 & -2 \\ -1 & 1 & 2 & -2 \\ 0 & -1 & -3 & 3 \\ 2 & -2 & -3 & 2 \end{pmatrix} = \begin{pmatrix} 0 & 1 & 0 & 2 \\ 1 & -1 & -2 & 2 \\ 0 & 1 & 3 & -3 \\ -2 & 2 & 3 & -2 \end{pmatrix}$$

Check.

$$AA^{-1} = \begin{pmatrix} 1 & -3 & 0 & -2 \\ 3 & -12 & -2 & -6 \\ -2 & 10 & 2 & 5 \\ -1 & 6 & 1 & 3 \end{pmatrix} \begin{pmatrix} 0 & 1 & 0 & 2 \\ 1 & -1 & -2 & 2 \\ 0 & 1 & 3 & -3 \\ -2 & 2 & 3 & -2 \end{pmatrix}$$

$$= \begin{pmatrix} 1 & 0 & 0 & 0 \\ 0 & 1 & 0 & 0 \\ 0 & 0 & 1 & 0 \\ 0 & 0 & 0 & 1 \end{pmatrix}$$

■

Note. As you have noticed, if $n > 3$ it is generally easier to compute A^{-1} by row reduction than by using adj A since, even for the 4×4 case, it is necessary to calculate 17 determinants (16 for the adjoint plus det A). Nevertheless, Theorem 3 is very important since, before you do any row reduction, the calculation of det A (if it can be done easily) will tell you whether or not A^{-1} exists.

We last saw our Summing Up Theorem (Theorems 1.2.1, 1.9.6, and 1.11.4) in Section 1.11. This is the theorem that ties together many of the concepts developed in the first chapters of this book.

THEOREM 4   **Summing Up Theorem—View 4** Let A be an $n \times n$ matrix. Then the following six statements are equivalent. That is, each one implies the other five (so that, if one is true, all are true):

i. A is invertible.

ii. The only solution to the homogeneous system $A\mathbf{x} = \mathbf{0}$ is the trivial solution ($\mathbf{x} = \mathbf{0}$).

iii. The system $A\mathbf{x} = \mathbf{b}$ has a unique solution for every n-vector $\mathbf{b}$.

iv. A is row equivalent to the $n \times n$ identity matrix I_n.

v. A is the product of elementary matrices.

vi. det $A \neq 0$.

Proof   In Theorem 1.9.6 we proved the equivalence of parts (i), (ii), (iii), and (iv). In Theorem 1.11.3 we proved the equivalence of parts (i) and (v). Theorem 1 (or Theorem 2.3.2) proves the equivalence of (i) and (vi).   ∎

PROBLEMS 2.4

In Problems 1–12 use the methods of this section to determine whether the given matrix is invertible. If so, compute the inverse.

1. $\begin{pmatrix} 3 & 2 \\ 1 & 2 \end{pmatrix}$

2. $\begin{pmatrix} 3 & 6 \\ -4 & -8 \end{pmatrix}$

3. $\begin{pmatrix} 0 & 1 \\ 1 & 0 \end{pmatrix}$

4. $\begin{pmatrix} 1 & 1 & 1 \\ 0 & 2 & 3 \\ 5 & 5 & 1 \end{pmatrix}$

5. $\begin{pmatrix} 3 & 2 & 1 \\ 0 & 2 & 2 \\ 0 & 1 & -1 \end{pmatrix}$

6. $\begin{pmatrix} 1 & 1 & 1 \\ 0 & 1 & 1 \\ 0 & 0 & 1 \end{pmatrix}$

7. $\begin{pmatrix} 1 & 2 & 3 \\ 1 & 1 & 2 \\ 0 & 1 & 2 \end{pmatrix}$

8. $\begin{pmatrix} 3 & 1 & 0 \\ 1 & -1 & 2 \\ 1 & 1 & 1 \end{pmatrix}$

9. $\begin{pmatrix} 2 & -1 & 4 \\ -1 & 0 & 5 \\ 19 & -7 & 3 \end{pmatrix}$

10. $\begin{pmatrix} 1 & 6 & 2 \\ -2 & 3 & 5 \\ 7 & 12 & -4 \end{pmatrix}$

11. $\begin{pmatrix} 1 & 1 & 1 & 1 \\ 1 & 2 & -1 & 2 \\ 1 & -1 & 2 & 1 \\ 1 & 3 & 3 & 2 \end{pmatrix}$

12. $\begin{pmatrix} 1 & -3 & 0 & -2 \\ 3 & -12 & -2 & -6 \\ -2 & 10 & 2 & 5 \\ -1 & 6 & 1 & 3 \end{pmatrix}$

13. Use determinants to show that an $n \times n$ matrix A is invertible if and only if A^t is invertible.

14. For $A = \begin{pmatrix} 1 & 1 \\ 2 & 5 \end{pmatrix}$, verify that $\det A^{-1} = 1/\det A$.

15. For $A = \begin{pmatrix} 1 & -1 & 3 \\ 4 & 1 & 6 \\ 2 & 0 & -2 \end{pmatrix}$, verify that $\det A^{-1} = 1/\det A$.

16. For what values of α is the matrix $\begin{pmatrix} \alpha & -3 \\ 4 & 1 - \alpha \end{pmatrix}$ not invertible?

17. For what values of α does the matrix $\begin{pmatrix} -\alpha & \alpha - 1 & \alpha + 1 \\ 1 & 2 & 3 \\ 2 - \alpha & \alpha + 3 & \alpha + 7 \end{pmatrix}$ not have an inverse?

18. Suppose that the $n \times n$ matrix A is not invertible. Show that $(A)(\text{adj } A)$ is the zero matrix.

2.5 CRAMER'S RULE

In this section we examine an old method for solving systems with the same number of unknowns as equations. Consider the system of n equations in n unknowns.

$$
\begin{aligned}
a_{11}x_1 + a_{12}x_2 + \cdots + a_{1n}x_n &= b_1 \\
a_{21}x_1 + a_{22}x_2 + \cdots + a_{2n}x_n &= b_2 \\
&\vdots \\
a_{n1}x_1 + a_{n2}x_2 + \cdots + a_{nn}x_n &= b_n
\end{aligned}
\tag{1}
$$

which can be written in the form

$$
A\mathbf{x} = \mathbf{b}
\tag{2}
$$

We suppose that $\det A \neq 0$. Then system (2) has a unique solution given by $\mathbf{x} = A^{-1}\mathbf{b}$. We can develop a method for finding that solution without row reduction and without computing A^{-1}.

Let $D = \det A$. We define n new matrices:

$$
A_1 = \begin{pmatrix} b_1 & a_{12} & \cdots & a_{1n} \\ b_2 & a_{22} & \cdots & a_{2n} \\ \vdots & \vdots & & \vdots \\ b_n & a_{n2} & \cdots & a_{nn} \end{pmatrix}, \quad A_2 = \begin{pmatrix} a_{11} & b_1 & \cdots & a_{1n} \\ a_{21} & b_2 & \cdots & a_{2n} \\ \vdots & \vdots & & \vdots \\ a_{n1} & b_n & \cdots & a_{nn} \end{pmatrix}, \ldots,
$$

$$
A_n = \begin{pmatrix} a_{11} & a_{12} & \cdots & b_1 \\ a_{21} & a_{22} & \cdots & b_2 \\ \vdots & \vdots & & \vdots \\ a_{n1} & a_{n2} & \cdots & b_n \end{pmatrix}
$$

That is, A_i is the matrix obtained by replacing the ith column of A with $\mathbf{b}$. Finally, let $D_1 = \det A_1$, $D_2 = \det A_2, \ldots, D_n = \det A_n$.

THEOREM 1   **Cramer's Rule** Let A be an $n \times n$ matrix and suppose that $\det A \neq 0$. Then the unique solution to the system $A\mathbf{x} = \mathbf{b}$ is given by

$$x_1 = \frac{D_1}{D}, \; x_2 = \frac{D_2}{D}, \; \ldots, \; x_i = \frac{D_i}{D}, \; \ldots, \; x_n = \frac{D_n}{D} \tag{3}$$

Proof   The solution to $A\mathbf{x} = \mathbf{b}$ is $\mathbf{x} = A^{-1}\mathbf{b}$. But

$$A^{-1}\mathbf{b} = \frac{1}{D}(\text{adj } A)\mathbf{b} = \frac{1}{D}\begin{pmatrix} A_{11} & A_{21} & \cdots & A_{n1} \\ A_{12} & A_{22} & \cdots & A_{n2} \\ \vdots & \vdots & & \vdots \\ A_{1n} & A_{2n} & \cdots & A_{nn} \end{pmatrix}\begin{pmatrix} b_1 \\ b_2 \\ \vdots \\ b_n \end{pmatrix} \tag{4}$$

Now $(\text{adj } A)\mathbf{b}$ is an n-vector, the jth component of which is

$$(A_{1j} \; A_{2j} \; \ldots \; A_{nj}) \cdot \begin{pmatrix} b_1 \\ b_2 \\ \vdots \\ b_n \end{pmatrix} = b_1 A_{1j} + b_2 A_{2j} + \cdots + b_n A_{nj} \tag{5}$$

Consider the matrix A_j:

$$A_j = \begin{pmatrix} a_{11} & a_{12} & \cdots & b_1 & \cdots & a_{1n} \\ a_{21} & a_{22} & \cdots & b_2 & \cdots & a_{2n} \\ \vdots & \vdots & & \vdots & & \vdots \\ a_{n1} & a_{n2} & \cdots & b_n & \cdots & a_{nn} \end{pmatrix} \tag{6}$$
$$\uparrow$$
$$j\text{th column}$$

If we expand the determinant of A_j in its jth column, we obtain

$$D_j = b_1 (\text{cofactor of } b_1) + b_2 (\text{cofactor of } b_2) + \cdots \tag{7}$$
$$+ b_n (\text{cofactor of } b_n)$$

But to find the cofactor of b_i, say, we delete the ith row and jth column of A_j (since b_i is in the jth column of A_j). But the jth column of A_j is $\mathbf{b}$ and, with this deleted, we simply have the ij minor, M_{ij}, of A. Thus

$$\text{Cofactor of } b_i \text{ in } A_j = A_{ij}$$

so that (7) becomes

$$D_j = b_1 A_{1j} + b_2 A_{2j} + \cdots + b_n A_{nj} \tag{8}$$

But this is the same as the right side of (5). Thus the ith component of $(\text{adj } A)\mathbf{b}$ is D_i, and we have

$$\mathbf{x} = \begin{pmatrix} x_1 \\ x_2 \\ \vdots \\ x_n \end{pmatrix} = A^{-1}\mathbf{b} = \frac{1}{D}(\text{adj } A)\mathbf{b} = \frac{1}{D}\begin{pmatrix} D_1 \\ D_2 \\ \vdots \\ D_n \end{pmatrix} = \begin{pmatrix} D_1/D \\ D_2/D \\ \vdots \\ D_n/D \end{pmatrix}$$

and the proof is complete. ∎

Historical note: Cramer's rule is named for the Swiss mathematician Gabriel Cramer (1704–1752). Cramer published the rule in 1750 in his *Introduction to the Analysis of Lines of Algebraic Curves*. Actually, there is much evidence to suggest that the rule was known as early as 1729 to Colin Maclaurin (1698–1746), who was probably the most outstanding British mathematician in the years following the death of Newton. Cramer's rule is one of the most famous results in the history of mathematics. For almost 200 years it was central in the teaching of algebra and the theory of equations. Because of the great number of computations involved, the rule is used today less frequently. However, the result was very important in its time.

EXAMPLE 1   **Solving a 3 × 3 System Using Cramer's Rule**   Solve, using Cramer's rule, the system

$$\begin{aligned} 2x_1 + 4x_2 + 6x_3 &= 18 \\ 4x_1 + 5x_2 + 6x_3 &= 24 \\ 3x_1 + x_2 - 2x_3 &= 4 \end{aligned} \qquad (9)$$

Solution   We have solved this before—using row reduction in Example 1.3.1 on page 7. We could also solve it by calculating A^{-1} (Example 1.9.6, page 71) and then finding $A^{-1}\mathbf{b}$. We now solve it by using Cramer's rule. First we have

$$D = \begin{vmatrix} 2 & 4 & 6 \\ 4 & 5 & 6 \\ 3 & 1 & -2 \end{vmatrix} = 6 \neq 0$$

so that system (9) has a unique solution. Then $D_1 = \begin{vmatrix} 18 & 4 & 6 \\ 24 & 5 & 6 \\ 4 & 1 & -2 \end{vmatrix} = 24,$

$$D_2 = \begin{vmatrix} 2 & 18 & 6 \\ 4 & 24 & 6 \\ 3 & 4 & -2 \end{vmatrix} = -12 \quad \text{and} \quad D_3 = \begin{vmatrix} 2 & 4 & 18 \\ 4 & 5 & 24 \\ 3 & 1 & 4 \end{vmatrix} = 18. \quad \text{Hence } x_1 = \frac{D_1}{D} =$$

$\frac{24}{6} = 4$, $x_2 = \dfrac{D_2}{D} = -\dfrac{12}{6} = -2$ and $x_3 = \dfrac{D_3}{D} = \dfrac{18}{6} = 3.$ ∎

EXAMPLE 2 **Solving a 4 × 4 System Using Cramer's Rule** Show that the system

$$x_1 + 3x_2 + 5x_3 + 2x_4 = 2$$
$$-x_2 + 3x_3 + 4x_4 = 0$$
$$2x_1 + x_2 + 9x_3 + 6x_4 = -3$$
$$3x_1 + 2x_2 + 4x_3 + 8x_4 = -1$$

(10)

has a unique solution and find it by using Cramer's rule.

Solution We saw in Example 2.2.10 on page 125 that

$$|A| = \begin{vmatrix} 1 & 3 & 5 & 2 \\ 0 & -1 & 3 & 4 \\ 2 & 1 & 9 & 6 \\ 3 & 2 & 4 & 8 \end{vmatrix} = 160 \neq 0$$

Thus the system has a unique solution. To find it we compute $D_1 = -464$; $D_2 = 280$; $D_3 = -56$; $D_4 = 112$. Thus $x_1 = D_1/D = -464/160$, $x_2 = D_2/D = 280/160$, $x_3 = D_3/D = -56/160$, and $x_4 = D_4/D = 112/160$. These solutions can be verified by direct substitution into system (10). ∎

PROBLEMS 2.5

In Problems 1–9 solve the given system by using Cramer's rule.

1. $2x_1 + 3x_2 = -1$
$-7x_1 + 4x_2 = 47$

2. $3x_1 - x_2 = 0$
$4x_1 + 2x_2 = 5$

3. $2x_1 + x_2 + x_3 = 6$
$3x_1 - 2x_2 - 3x_3 = 5$
$8x_1 + 2x_2 + 5x_3 = 11$

4. $x_1 + x_2 + x_3 = 8$
$4x_2 - x_3 = -2$
$3x_1 - x_2 + 2x_3 = 0$

5. $2x_1 + 2x_2 + x_3 = 7$
$x_1 + 2x_2 - x_3 = 0$
$-x_1 + x_2 + 3x_3 = 1$

6. $2x_1 + 5x_2 - x_3 = -1$
$4x_1 + x_2 + 3x_3 = 3$
$-2x_1 + 2x_2 = 0$

7. $2x_1 + x_2 - x_3 = 4$
$x_1 + x_3 = 2$
$-x_2 + 5x_3 = 1$

8. $x_1 + x_2 + x_3 + x_4 = 6$
$2x_1 - x_3 - x_4 = 4$
$3x_3 + 6x_4 = 3$
$x_1 - x_4 = 5$

9. $x_1 - x_4 = 7$
$2x_2 + x_3 = 2$
$4x_1 - x_2 = -3$
$3x_3 - 5x_4 = 2$

*10. Consider the triangle in Figure 2.2.

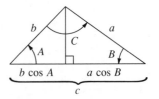

Figure 2.2

a. Show, using elementary trigonometry, that

$$c \cos A \qquad\qquad + a \cos C = b$$
$$b \cos A + a \cos B \qquad\qquad = c$$
$$c \cos B + b \cos C = a$$

b. If the system of part (a) is thought of as a system of three equations in the three unknowns cos A, cos B, and cos C, show that the determinant of the system is nonzero.

c. Use Cramer's rule to solve for cos C.

d. Use part (c) to prove the **law of cosines;** $c^2 = a^2 + b^2 - 2ab \cos C$.

SUMMARY

- The **determinant** of a 2×2 matrix $A = \begin{pmatrix} a_{11} & a_{12} \\ a_{21} & a_{22} \end{pmatrix}$ is given by

$$\text{Determinant of } A = \det A = |A| = a_{11}a_{22} - a_{12}a_{21} \qquad \text{(p. 110)}$$

- 3×3 determinant

$$\det \begin{pmatrix} a_{11} & a_{12} & a_{13} \\ a_{21} & a_{22} & a_{23} \\ a_{31} & a_{32} & a_{33} \end{pmatrix} = a_{11}\begin{vmatrix} a_{22} & a_{23} \\ a_{32} & a_{33} \end{vmatrix} - a_{12}\begin{vmatrix} a_{21} & a_{23} \\ a_{31} & a_{33} \end{vmatrix} + a_{13}\begin{vmatrix} a_{21} & a_{22} \\ a_{31} & a_{32} \end{vmatrix} \qquad \text{(p. 111)}$$

- The **ijth minor** of the $n \times n$ matrix A, denoted by M_{ij}, is the $(n-1) \times (n-1)$ matrix obtained by crossing off the ith row and jth column of A. (p. 112)
- The **ijth cofactor** of A, denoted by A_{ij}, is given by (p. 113)

$$A_{ij} = (-i)^{i+j} \det M_{ij}$$

- **$n \times n$ Determinant**·Let A be a $n \times n$ matrix. Then (p. 114)

$$\det A = a_{11}A_{11} + a_{12}A_{12} + \cdots + A_{1n}A_{1n} = \sum_{k=1}^{n} a_{1k}A_{1k}$$

The sum above is called the **expansion of det A by cofactors.**

- If A is an upper triangular, lower triangular, or diagonal $n \times n$ matrix with diagonal components (p. 116) $a_{11}, a_{22}, \ldots, a_{nn}$, then

$$\det A = a_{11} a_{22} \cdots a_{nn}$$

- **Basic Theorem**·If A is an $n \times n$ matrix, then

$$\det A = a_{i1}A_{i1} + a_{i2}A_{i2} + \cdots + a_{in}A_{in} = \sum_{k=1}^{n} a_{ik}A_{ik}$$

and (pp. 119, 133)

$$\det A = a_{1j}A_{1j} + a_{2j}A_{2j} + \cdots + a_{nj}A_{nj} = \sum_{k=1}^{n} a_{kj}A_{kj}$$

for $i = 1, 2, \ldots, n$ and $j = 1, 2, \ldots, n$.

That is, the determinant of A can be obtained by expanding in any row or column of A.
- If any row or column A is the zero vector, then det $A = 0$. (p. 120)
- If any row (column) of A is multiplied by the constant c, then det A is multiplied by c. (p. 120)
- If A and B are two $n \times n$ matrices that are equal except in the jth column (ith row) and C is the matrix that is identical to A and B except that the jth column (ith row) of C is the sum of the jth column of A and the jth column of B (ith row of A and ith row of B), then det $C =$ det $A +$ det B. (p. 121)
- Interchanging any two rows or columns of A has the effect of multiplying det A by -1. (p. 122)
- If any row (column) of A is multiplied by a constant and added to any other row (column) of A, then det A is unchanged. (p. 124)
- If one row (column) of A is a multiple of another row (column) of A, then det $A = 0$. (p. 124)
- det $A =$ det A^t. (p. 128)
- det $AB =$ det A det B (pp. 129, 138)
- If A is invertible, then det $A \neq 0$ and (p. 141)

$$\det A^{-1} = \frac{1}{\det A}$$

- Let A be an $n \times n$ matrix. The **adjoint** of A, denoted by adj A, is the $n \times n$ matrix whose ijth component is A_{ji}, the jith cofactor of A. (p. 142)
- If det $A \neq 0$, then A is invertible and (p. 144)

$$A^{-1} = \frac{1}{\det A} \text{ adj } A$$

- **Summing Up Theorem** · Let A be an $n \times n$ matrix. Then the following six statements are equivalent: (p. 146)

  i. A is invertible.

  ii. The only solution to the homogeneous system $A\mathbf{x} = \mathbf{0}$ is the trivial solution ($\mathbf{x} = \mathbf{0}$).

  iii. The system $A\mathbf{x} = \mathbf{b}$ has a unique solution for every n-vector $\mathbf{b}$.

  iv. A is row equivalent to the $n \times n$ identity matrix I_n.

  v. A is the product of elementary matrices.

  vi. det $A \neq 0$.

- **Cramer's Rule** · Let A be an $n \times n$ matrix with det $A \neq 0$. Then the unique solution to the system $A\mathbf{x} = \mathbf{b}$ is given by (p. 148)

$$x_1 = \frac{D_1}{\det A}, \ x_2 = \frac{D_2}{\det A}, \ \ldots, \ x_n = \frac{D_n}{\det A}$$

where D_j is the determinant of the matrix obtained by replacing the jth column of A by the column vector $\mathbf{b}$.

REVIEW EXERCISES

In Exercises 1–8 calculate the determinant.

1. $\begin{vmatrix} -1 & 2 \\ 0 & 4 \end{vmatrix}$

2. $\begin{vmatrix} -3 & 5 \\ -7 & 4 \end{vmatrix}$

3. $\begin{vmatrix} 1 & -2 & 3 \\ 0 & 4 & 5 \\ 0 & 0 & 6 \end{vmatrix}$

4. $\begin{vmatrix} 5 & 0 & 0 \\ 6 & 2 & 0 \\ 10 & 100 & 6 \end{vmatrix}$ **5.** $\begin{vmatrix} 1 & -1 & 2 \\ 3 & 4 & 2 \\ -2 & 3 & 4 \end{vmatrix}$ **6.** $\begin{vmatrix} 3 & 1 & -2 \\ 4 & 0 & 5 \\ -6 & 1 & 3 \end{vmatrix}$

7. $\begin{vmatrix} 1 & -1 & 2 & 3 \\ 4 & 0 & 2 & 5 \\ -1 & 2 & 3 & 7 \\ 5 & 1 & 0 & 4 \end{vmatrix}$ **8.** $\begin{vmatrix} 3 & 15 & 17 & 19 \\ 0 & 2 & 21 & 60 \\ 0 & 0 & 1 & 50 \\ 0 & 0 & 0 & -1 \end{vmatrix}$

In Exercises 9–14 use determinants to calculate the inverse (if one exists).

9. $\begin{pmatrix} -3 & 4 \\ 2 & 1 \end{pmatrix}$ **10.** $\begin{pmatrix} 3 & -5 & 7 \\ 0 & 2 & 4 \\ 0 & 0 & -3 \end{pmatrix}$ **11.** $\begin{pmatrix} 1 & -1 & 2 \\ 3 & 1 & 4 \\ 5 & -1 & 8 \end{pmatrix}$

12. $\begin{pmatrix} 1 & 1 & 1 \\ 1 & 0 & 1 \\ 0 & 1 & 1 \end{pmatrix}$ **13.** $\begin{pmatrix} 2 & 1 & 0 & 0 \\ 0 & -1 & 3 & 0 \\ 1 & 0 & 0 & -2 \\ 3 & 0 & -1 & 0 \end{pmatrix}$ **14.** $\begin{pmatrix} 3 & -1 & 2 & 4 \\ 1 & 1 & 0 & 3 \\ -2 & 4 & 1 & 5 \\ 6 & -4 & 1 & 2 \end{pmatrix}$

In Exercises 15–18 solve the system by using Cramer's rule.

15. $2x_1 - x_2 = 3$
$3x_1 + 2x_2 = 5$

16. $x_1 - x_2 + x_3 = 7$
$2x_1 \qquad - 5x_3 = 4$
$\qquad 3x_2 - x_3 = 2$

17. $2x_1 + 3x_2 - x_3 = 5$
$-x_1 + 2x_2 + 3x_3 = 0$
$4x_1 - x_2 + x_3 = -1$

18. $x_1 \qquad - x_3 + x_4 = 7$
$\qquad 2x_2 + 2x_3 - 3x_4 = -1$
$4x_1 - x_2 - x_3 \qquad = 0$
$-2x_1 + x_2 + 4x_3 \qquad = 2$

3

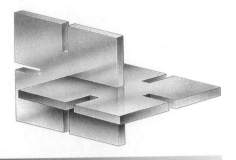

Vectors in $\mathbb{R}^2$ and $\mathbb{R}^3$

In Section 1.5 we defined column and row vectors as ordered sets of n real numbers. In the next chapter we shall define other kinds of sets of vectors, called *vector spaces*.

The study of arbitrary vector spaces is, initially, quite abstract. For that reason it is helpful to have a "store" of easily visualized vectors that we can use as examples.

In this chapter we discuss basic properties of vectors in the xy-plane and in real, three-dimensional space. Students who have studied multivariable calculus will have seen this material before. In that case it should be covered briefly, as a review. For others, coverage of this chapter will provide the examples that can make the material in Chapters 4 and 5 a great deal more comprehensible.

3.1 VECTORS IN THE PLANE

As we defined it in Section 1.5, $\mathbb{R}^2$ is the set of vectors (x_1, x_2) with x_1 and x_2 real numbers. Since any point in the plane can be written in the form (x, y), it is apparent that any point in the plane can be thought of as a vector in $\mathbb{R}^2$, and vice versa. Thus the terms "the plane" and "$\mathbb{R}^2$" are often used interchangeably. However, for a variety of physical applications (including the notions of force, velocity, acceleration, and momentum), it is important to think of a vector not as a point but as an entity having "length" and "direction." Now we shall see how this is done.

Directed line
segment

Let P and Q be two points in the plane. Then the **directed line segment** from P to Q, denoted by $\overrightarrow{PQ}$, is the straight-line segment that extends from P to Q (see

Figure 3.1*a*). Note that the directed line segments $\overrightarrow{PQ}$ and $\overrightarrow{QP}$ are different since they point in opposite directions (Figure 3.1*b*).

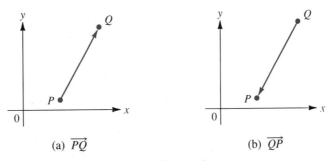

(a) $\overrightarrow{PQ}$ (b) $\overrightarrow{QP}$

Figure 3.1   The directed line segments $\overrightarrow{PQ}$ and $\overrightarrow{QP}$ point in opposite directions

Initial point

Terminal point

Equivalent
directed line
segment

The point P in the directed line segment $\overrightarrow{PQ}$ is called the **initial point** of the segment and the point Q is called the **terminal point.** The two major properties of a directed line segment are its magnitude (length) and its direction. If two directed line segments $\overrightarrow{PQ}$ and $\overrightarrow{RS}$ have the same magnitude and direction, we say that they are **equivalent** no matter where they are located with respect to the origin. The directed line segments in Figure 3.2 are all equivalent.

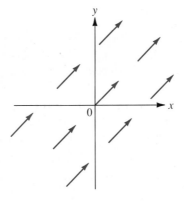

Figure 3.2   A set of equivalent directed line segments

DEFINITION 1   **Geometric Definition of a Vector**   The set of all directed line segments equivalent to a given directed line segment is called a **vector.** Any directed line segment in that set is called a **representation** of the vector.

Remark. The directed line segments in Figure 3.2 are all representations of the same vector.

From Definition 1 we see that a given vector **v** can be represented in many different ways. Let $\overrightarrow{PQ}$ be a representation of **v**. Then, without changing magnitude or direction, we can move $\overrightarrow{PQ}$ in a parallel way so that its initial point is shifted to the origin. We then obtain the directed line segment $\overrightarrow{OR}$, which is another representation of the vector **v** (see Figure 3.3). Now suppose that R has the Cartesian

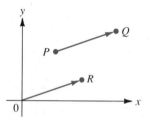

Figure 3.3 We can move $\overrightarrow{PQ}$ to obtain an equivalent directed line segment with its initial point at the origin. Note that $\overrightarrow{OR}$ and $\overrightarrow{PQ}$ are parallel and have the same length

coordinates (a, b). Then we can describe the directed line segment $\overrightarrow{OR}$ by the coordinates (a, b). That is, $\overrightarrow{OR}$ is the directed line segment with initial point $(0, 0)$ and terminal point (a, b). Since one representation of a vector is as good as another, we can write the vector **v** as (a, b).

DEFINITION 2   **Algebraic Definition of a Vector**   A **vector v** in the xy-plane is an ordered pair of real numbers (a, b). The numbers a and b are called the **components** of the vector **v**. The **zero vector** is the vector $(0, 0)$.

Remark 1. With this definition, a point in the xy-plane can be thought of as a vector originating at the origin and terminating at that point.

Remark 2. The zero vector has a magnitude of zero. Therefore, since the initial and terminal points coincide, we say that the zero vector has *no direction*.

Remark 3. We emphasize that Definitions 1 and 2 describe precisely the same objects. Each point of view (geometric and algebraic) has its advantages. Definition 2 is the definition of a 2-vector that we have been using all along.

Magnitude or length of a vector

Since a vector is really a set of equivalent line segments, we define the **magnitude** or **length** of a vector as the length of any one of its representations and its **direction** as the direction of any one of its representations. Using the representation $\overrightarrow{OR}$ and writing the vector $\mathbf{v} = (a, b)$, we find that

$$\boxed{|\mathbf{v}| = \text{magnitude of } \mathbf{v} = \sqrt{a^2 + b^2}} \tag{1}$$

This follows from the Pythagorean theorem (see Figure 3.4). We have used the notation $|\mathbf{v}|$ to denote the magnitude of $\mathbf{v}$. Note that $|\mathbf{v}|$ is a *scalar*.

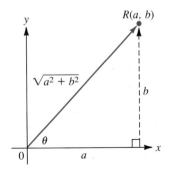

Figure 3.4 The magnitude of a vector with x-coordinate a and y-coordinate b is $\sqrt{a^2 + b^2}$

EXAMPLE 1 **Calculating the Magnitudes of Six Vectors**   Calculate the magnitudes of the vectors **(i)** $(2, 2)$; **(ii)** $(2, 2\sqrt{3})$; **(iii)** $(-2\sqrt{3}, 2)$; **(iv)** $(-3, -3)$; **(v)** $(6, -6)$; **(vi)** $(0, 3)$.

Solution
   i. $|\mathbf{v}| = \sqrt{2^2 + 2^2} = \sqrt{8} = 2\sqrt{2}$
  ii. $|\mathbf{v}| = \sqrt{2^2 + (2\sqrt{3})^2} = 4$
 iii. $|\mathbf{v}| = \sqrt{(-2\sqrt{3})^2 + 2^2} = 4$
  iv. $|\mathbf{v}| = \sqrt{(-3)^2 + (-3)^2} = \sqrt{18} = 3\sqrt{2}$
   v. $|\mathbf{v}| = \sqrt{6^2 + (-6)^2} = \sqrt{72} = 6\sqrt{2}$
  vi. $|\mathbf{v}| = \sqrt{0^2 + 3^2} = \sqrt{9} = 3$ ∎

Direction of a vector

We now define the **direction** of the vector $\mathbf{v} = (a, b)$ to be the angle θ, measured in radians, that the vector makes with the positive x-axis. By convention, we choose θ such that $0 \le \theta < 2\pi$. It follows from Figure 3.4 that if $a \ne 0$, then

$$\tan \theta = \frac{b}{a} \tag{2}$$

Note: Tan θ is periodic of period π, so if $a \ne 0$ there are always *two* numbers in $[0, 2\pi)$ such that $\tan \theta = b/a$. For example, $\tan \pi/4 = \tan 5\pi/4 = 1$. In order to determine θ uniquely, we need to determine the quadrant of $\mathbf{v}$ as we shall see in the next example.

EXAMPLE 2 **Calculating the Directions of Six Vectors**   Calculate the directions of the vectors in Example 1.

Solution   These six vectors are depicted in Figure 3.5.

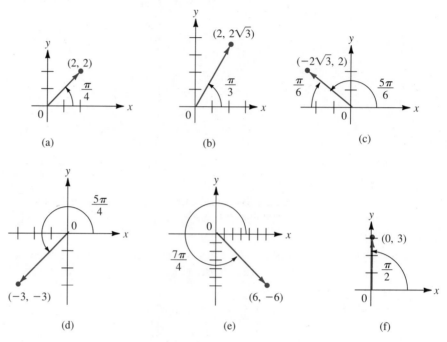

Figure 3.5   The directions of six vectors

i. Here **v** is in the first quadrant and since $\tan \theta = 2/2 = 1$, $\theta = \pi/4$.

ii. Here $\theta = \tan^{-1} 2\sqrt{3}/2 = \tan^{-1}\sqrt{3} = \pi/3$ (since **v** is in the first quadrant).

iii. We see that **v** is in the second quadrant and, since $\tan^{-1} 2/2\sqrt{3} = \tan^{-1}1/\sqrt{3} = \pi/6$, we see from Figure 3.5c that $\theta = \pi - (\pi/6) = 5\pi/6$.

iv. Here **v** is in the third quadrant and, since $\tan^{-1} 1 = \pi/4$, we find that $\theta = \pi + (\pi/4) = 5\pi/4$.

v. Since **v** is in the fourth quadrant and $\tan^{-1}(-1) = -\pi/4$, we get $\theta = 2\pi - (\pi/4) = 7\pi/4$.

vi. We cannot use equation (2) because b/a is undefined. However, we see in Figure 5f that $\theta = \pi/2$. ∎

In general, if $b > 0$

$$\text{Direction of } (0, b) = \frac{\pi}{2} \quad \text{and} \quad \text{direction of } (0, -b) = \frac{3\pi}{2} \qquad b > 0$$

In Section 1.5 we defined vector addition and scalar multiplication. What do these concepts mean geometrically? We start with scalar multiplication. If $\mathbf{v} = (a, b)$, then $\alpha\mathbf{v} = (\alpha a, \alpha b)$. We find that

$$|\alpha\mathbf{v}| = \sqrt{\alpha^2 a^2 + \alpha^2 b^2} = |\alpha|\sqrt{a^2 + b^2} = |\alpha|\,|\mathbf{v}| \tag{3}$$

That is,

> Multiplying a vector by a nonzero scalar has the effect of multiplying the length of the vector by the absolute value of that scalar.

Moreover, if $\alpha > 0$, then $\alpha\mathbf{v}$ is in the same quadrant as $\mathbf{v}$, and therefore the direction of $\alpha\mathbf{v}$ is the *same* as the direction of $\mathbf{v}$ since $\tan^{-1}(\alpha b/\alpha a) = \tan^{-1}(b/a)$. If $\alpha < 0$, then $\alpha\mathbf{v}$ points in the direction opposite to that of $\mathbf{v}$. In other words,

> Direction of $\alpha\mathbf{v}$ = direction of $\mathbf{v}$, if $\alpha > 0$
> Direction of $\alpha\mathbf{v}$ = direction of $\mathbf{v} + \pi$, if $\alpha < 0$ **(4)**

EXAMPLE 3

Multiplying a Vector by a Scalar   Let $\mathbf{v} = (1, 1)$. Then $|\mathbf{v}| = \sqrt{1 + 1} = \sqrt{2}$ and $|2\mathbf{v}| = |(2, 2)| = \sqrt{2^2 + 2^2} = \sqrt{8} = 2\sqrt{2} = 2|\mathbf{v}|$. Further, $|-2\mathbf{v}| = \sqrt{(-2)^2 + (-2)^2} = 2\sqrt{2} = 2|\mathbf{v}|$. Moreover, the direction of $2\mathbf{v}$ is $\pi/4$, whereas

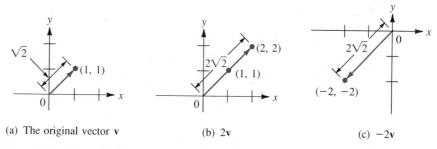

(a) The original vector $\mathbf{v}$ (b) $2\mathbf{v}$ (c) $-2\mathbf{v}$

Figure 3.6 The vector $2\mathbf{v}$ has the same direction as $\mathbf{v}$ and twice its magnitude. The vector $-2\mathbf{v}$ has the opposite direction of $\mathbf{v}$ and twice its magnitude

the direction of $-2\mathbf{v}$ is $5\pi/4$ (see Figure 3.6).   ∎

Now suppose we add the vectors $\mathbf{u} = (a_1, b_1)$ and $\mathbf{v} = (a_2, b_2)$ as in Figure 3.7. From the figure we see that the vector $\mathbf{u} + \mathbf{v} = (a_1 + a_2, b_1 + b_2)$ can be

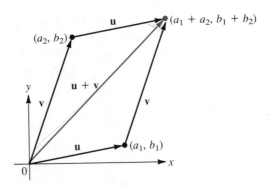

Figure 3.7   The parallelogram rule for adding vectors

obtained by shifting the representation of the vector **v** so that its initial point coincides with the terminal point (a_1, b_1) of the vector **u**. We can therefore obtain the vector **u** + **v** by drawing a parallelogram with one vertex at the origin and sides **u** and **v**. Then **u** + **v** is the vector that points from the origin along the diagonal of the parallelogram.

Note. Since a straight line is the shortest distance between two points, it immediately follows from Figure 3.7 that

$$
\boxed{\begin{array}{c} \textbf{Triangle Inequality} \\[4pt] |\mathbf{u} + \mathbf{v}| \le |\mathbf{u}| + |\mathbf{v}| \end{array}}
\tag{5}
$$

For reasons obvious from Figure 3.7, inequality (5) is called the **triangle inequality.**

We can also use Figure 3.7 to obtain a geometric representation of the vector **u** − **v**. Since **u** = **u** − **v** + **v**, the vector **u** − **v** is the vector that must be added to **v** to obtain **u**. This fact is illustrated in Figure 3.8*a*. A similar fact is illustrated in Figure 3.8*b*.

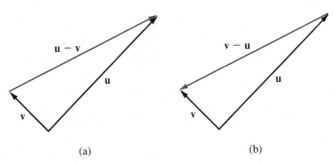

(a)   (b)

Figure 3.8   The vectors **u** − **v** and **v** − **u** have the same magnitude but point in opposite directions

There are two special vectors in $\mathbb{R}^2$ that allow us to represent other vectors in $\mathbb{R}^2$ in a convenient way. We denote the vector $(1, 0)$ by the symbol **i** and the vector $(0, 1)$ by the vector **j**. (See Figure 3.9.) If $\mathbf{v} = (a, b)$ is any vector in the plane, then

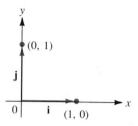

Figure 3.9 The vectors **i** and **j**

since $(a, b) = a(1, 0) + b(0, 1)$, we may write

$$\mathbf{v} = (a, b) = a\mathbf{i} + b\mathbf{j} \tag{6}$$

With this representation we say that **v** is *resolved into its horizontal and vertical components*. The vectors **i** and **j** have two properties:

i. Neither one is a multiple of the other. (In the terminology of Chapter 4, they are *linearly independent*.)

ii. Any vector **v** can be written in terms of **i** and **j** as in equation (6).†

Historical Note. The symbols **i** and **j** were first used by Hamilton. He defined his quaternion as a quantity of the form $a + b\mathbf{i} + c\mathbf{j} + d\mathbf{k}$, where a is the "scalar part" and $b\mathbf{i} + c\mathbf{j} + d\mathbf{k}$ the "vector part." In Section 3.3 we shall write vectors in space in the form $b\mathbf{i} + c\mathbf{j} + d\mathbf{k}$.

Under these two conditions **i** and **j** are said to form a **basis** in $\mathbb{R}^2$. We shall discuss bases in arbitrary vector spaces in Chapter 4.

We now define a kind of a vector that is very useful in certain applications.

DEFINITION 3   **Unit Vector** A **unit vector u** is a vector that has length 1.

EXAMPLE 4   **A Unit Vector**   The vector $\mathbf{u} = (1/2)\mathbf{i} + (\sqrt{3}/2)\mathbf{j}$ is a unit vector since

$$|\mathbf{u}| = \sqrt{\left(\frac{1}{2}\right)^2 + \left(\frac{\sqrt{3}}{2}\right)^2} = \sqrt{\frac{1}{4} + \frac{3}{4}} = 1$$

∎

---

† In equation (6) we say that **v** can be written as a *linear combination* of **i** and **j**. We shall discuss the notion of linear combination in Section 4.5.

Let $\mathbf{u} = a\mathbf{i} + b\mathbf{j}$ be a unit vector. Then $|\mathbf{u}| = \sqrt{a^2 + b^2} = 1$, so that $a^2 + b^2 = 1$ and $\mathbf{u}$ can be represented by a point on the unit circle (see Figure 3.10).

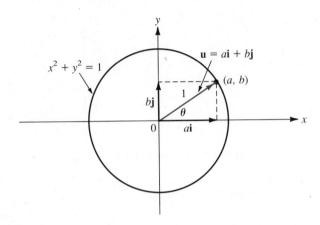

Figure 3.10   The terminal point of a unit vector with initial point at the origin lies on the unit circle

If θ is the direction of $\mathbf{u}$, then we immediately see that $a = \cos \theta$ and $b = \sin \theta$. Thus any unit vector $\mathbf{u}$ can be written in the form

$$\mathbf{u} = (\cos \theta)\mathbf{i} + (\sin \theta)\mathbf{j} \tag{7}$$

where θ is the direction of $\mathbf{u}$.

EXAMPLE 5   **Writing a Unit Vector as $(\cos \theta)\mathbf{i} + (\sin \theta)\mathbf{j}$**   The unit vector $\mathbf{u} = (1/2)\mathbf{i} + (\sqrt{3}/2)\mathbf{j}$ of Example 4 can be written in the form of (7) with $\theta = \cos^{-1}(1/2) = \pi/3$.   ■

We also have (see Problem 17).

Let $\mathbf{v}$ be any nonzero vector. Then $\mathbf{u} = \mathbf{v}/|\mathbf{v}|$ is a unit vector having the same direction as $\mathbf{v}$.

EXAMPLE 6   **Finding a Unit Vector Having the Same Direction as a Given Vector**   Find a unit vector having the same direction as $\mathbf{v} = 2\mathbf{i} - 3\mathbf{j}$.

Solution   Here $|\mathbf{v}| = \sqrt{4 + 9} = \sqrt{13}$, so $\mathbf{u} = \mathbf{v}/|\mathbf{v}| = (2/\sqrt{13})\mathbf{i} - (3/\sqrt{13})\mathbf{j}$ is the required unit vector.   ■

We conclude this section with a summary of the properties of vectors (Table 3.1).

Table 3.1

Object	Intuitive Definition	Expression in terms of components if $\mathbf{u} = u_1\mathbf{i} + u_2\mathbf{j}$, $\mathbf{v} = v_1\mathbf{i} + v_2\mathbf{j}$, and $\mathbf{u} = (u_1, u_2)$, $\mathbf{v} = (v_1, v_2)$
Vector $\mathbf{v}$	An object having magnitude and direction	$v_1\mathbf{i} + v_2\mathbf{j}$ or (v_1, v_2)
$|\mathbf{v}|$	Magnitude (or length) of $\mathbf{v}$	$\sqrt{v_1^2 + v_2^2}$
$\alpha\mathbf{v}$	$\nearrow\mathbf{v}$ $\nearrow\alpha\mathbf{v}$ (In this sketch $\alpha = 2$)	$\alpha v_1\mathbf{i} + \alpha v_2\mathbf{j}$ or $(\alpha v_1, \alpha v_2)$
$-\mathbf{v}$	$\nearrow\mathbf{v}$ $\swarrow -\mathbf{v}$	$-v_1\mathbf{i} - v_2\mathbf{j}$ or $(-v_1 - v_2)$ or $-(v_1, v_2)$
$\mathbf{u} + \mathbf{v}$		$(u_1 + v_1)\mathbf{i} + (u_2 + v_2)\mathbf{j}$ or $(u_1 + v_1, u_2 + v_2)$
$\mathbf{u} - \mathbf{v}$		$(u_1 - v_1)\mathbf{i} + (u_2 - v_2)\mathbf{j}$ or $(u_1 - v_1, u_2 - v_2)$

PROBLEMS 3.1

In Problems 1–12 find the magnitude and direction of the given vector.

1. $\mathbf{v} = (4, 4)$ **2.** $\mathbf{v} = (-4, 4)$ **3.** $\mathbf{v} = (4, -4)$

4. $\mathbf{v} = (-4, -4)$ **5.** $\mathbf{v} = (\sqrt{3}, 1)$ **6.** $\mathbf{v} = (1, \sqrt{3})$

7. $\mathbf{v} = (-1, \sqrt{3})$ **8.** $\mathbf{v} = (1, -\sqrt{3})$ **9.** $\mathbf{v} = (-1, -\sqrt{3})$

10. $\mathbf{v} = (1, 2)$ **11.** $\mathbf{v} = (-5, 8)$ **12.** $\mathbf{v} = (11, -14)$

13. Let $\mathbf{u} = (2, 3)$ and $\mathbf{v} = (-5, 4)$. Find: **(a)** $3\mathbf{u}$; **(b)** $\mathbf{u} + \mathbf{v}$; **(c)** $\mathbf{v} - \mathbf{u}$; **(d)** $2\mathbf{u} - 7\mathbf{v}$. Sketch these vectors.

14. Let $\mathbf{u} = 2\mathbf{i} - 3\mathbf{j}$ and $\mathbf{v} = -4\mathbf{i} + 6\mathbf{j}$. Find: **(a)** $\mathbf{u} + \mathbf{v}$; **(b)** $\mathbf{u} - \mathbf{v}$; **(c)** $3\mathbf{u}$; **(d)** $-7\mathbf{v}$; **(e)** $8\mathbf{u} - 3\mathbf{v}$; **(f)** $4\mathbf{v} - 6\mathbf{u}$. Sketch these vectors.

15. Show that the vectors $\mathbf{i}$ and $\mathbf{j}$ are unit vectors.

16. Show that the vector $(1/\sqrt{2})\mathbf{i} + (1/\sqrt{2})\mathbf{j}$ is a unit vector.

17. Show that if $\mathbf{v} = a\mathbf{i} + b\mathbf{j} \neq \mathbf{0}$, then $\mathbf{u} = (a/\sqrt{a^2 + b^2})\mathbf{i} + (b/\sqrt{a^2 + b^2})\mathbf{j}$ is a unit vector having the same direction as $\mathbf{v}$.

In Problems 18–21 find a unit vector having the same direction as the given vector.

18. $\mathbf{v} = 2\mathbf{i} + 3\mathbf{j}$ **19.** $\mathbf{v} = \mathbf{i} - \mathbf{j}$

20. $\mathbf{v} = -3\mathbf{i} + 4\mathbf{j}$ **21.** $\mathbf{v} = a\mathbf{i} + a\mathbf{j}; \; a \neq 0.$

22. If $\mathbf{v} = a\mathbf{i} + b\mathbf{j}$, show that $a/\sqrt{a^2 + b^2} = \cos\theta$ and $b/\sqrt{a^2 + b^2} = \sin\theta$, where θ is the direction of $\mathbf{v}$.

23. If $\mathbf{v} = 2\mathbf{i} - 3\mathbf{j}$, find $\sin\theta$ and $\cos\theta$.

24. If $\mathbf{v} = -3\mathbf{i} + 8\mathbf{j}$, find $\sin\theta$ and $\cos\theta$.

A vector $\mathbf{v}$ has a direction opposite to that of a vector $\mathbf{u}$ if direction $\mathbf{v} =$ direction $\mathbf{u} + \pi$. In Problems 25–28 find a unit vector $\mathbf{v}$ that has a direction opposite the direction of the given vector $\mathbf{u}$.

25. $\mathbf{u} = \mathbf{i} + \mathbf{j}$ **26.** $\mathbf{u} = 2\mathbf{i} - 3\mathbf{j}$

27. $\mathbf{u} = -3\mathbf{i} + 4\mathbf{j}$ **28.** $\mathbf{u} = -2\mathbf{i} + 3\mathbf{j}$

29. Let $\mathbf{u} = 2\mathbf{i} - 3\mathbf{j}$ and $\mathbf{v} = -\mathbf{i} + 2\mathbf{j}$. Find a unit vector having the same direction as:
(a) $\mathbf{u} + \mathbf{v}$; (b) $2\mathbf{u} - 3\mathbf{v}$; (c) $3\mathbf{u} + 8\mathbf{v}$.

30. Let $P = (c, d)$ and $Q = (c + a, d + b)$. Show that the magnitude of $\overrightarrow{PQ}$ is $\sqrt{a^2 + b^2}$.

31. Show that the direction of $\overrightarrow{PQ}$ in Problem 30 is the same as the direction of the vector (a, b). [*Hint:* If $R = (a, b)$, show that the line passing through the points P and Q is parallel to the line passing through the points 0 and R.]

In Problems 32–35 find a vector $\mathbf{v}$ having the given magnitude and direction.

32. $|\mathbf{v}| = 3; \; \theta = \pi/6$ **33.** $|\mathbf{v}| = 8; \; \theta = \pi/3$

34. $|\mathbf{v}| = 1; \; \theta = \pi/4$ **35.** $|\mathbf{v}| = 6; \; \theta = 2\pi/3$.

*36. Show algebraically (that is, strictly from the definitions of vector addition and magnitude) that for any two vectors $\mathbf{u}$ and $\mathbf{v}$, $|\mathbf{u} + \mathbf{v}| \leq |\mathbf{u}| + |\mathbf{v}|$.

37. Show that if neither $\mathbf{u}$ nor $\mathbf{v}$ is the zero vector, then $|\mathbf{u} + \mathbf{v}| = |\mathbf{u}| + |\mathbf{v}|$ if and only if $\mathbf{u}$ is a positive scalar multiple of $\mathbf{v}$.

3.2 THE SCALAR PRODUCT AND PROJECTIONS IN $\mathbb{R}^2$

In Section 1.7 we defined the scalar product of two vectors. If $\mathbf{u} = (a_1, b_1)$ and $\mathbf{v} = (a_2, b_2)$, then

$$\boxed{\mathbf{u} \cdot \mathbf{v} = a_1 a_2 + b_1 b_2} \tag{1}$$

We now see how the scalar product can be interpreted geometrically.

DEFINITION 1   **Angle Between Vectors** Let $\mathbf{u}$ and $\mathbf{v}$ be two nonzero vectors. Then the **angle** φ **between u and v** is defined to be the smallest nonnegative angle† between the

---

† This angle will be in the interval $[0, \pi]$.

representations of **u** and **v** that have the origin as their initial points. If $\mathbf{u} = \alpha\mathbf{v}$ for some scalar α, then we define $\varphi = 0$ if $\alpha > 0$ and $\varphi = \pi$ if $\alpha < 0$.

This definition is illustrated in Figure 3.11. Note that φ can always be chosen to be a nonnegative angle in the interval $[0, \pi]$.

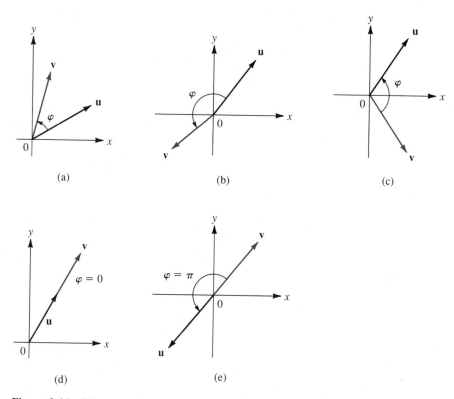

Figure 3.11   The angle φ between two vectors

THEOREM 1   Let **v** be a vector. Then

$$|\mathbf{v}|^2 = \mathbf{v} \cdot \mathbf{v} \qquad (2)$$

Proof   Let $\mathbf{v} = (a, b)$. Then

$$|\mathbf{v}|^2 = a^2 + b^2$$

and

$$\mathbf{v} \cdot \mathbf{v} = (a, b) \cdot (a, b) = a \cdot a + b \cdot b = a^2 + b^2 = |\mathbf{v}|^2 \qquad \blacksquare$$

THEOREM 2   Let **u** and **v** be two nonzero vectors. If φ is the angle between them, then

$$\cos \varphi = \frac{\mathbf{u} \cdot \mathbf{v}}{|\mathbf{u}|\,|\mathbf{v}|} \tag{3}$$

Proof   The law of cosines (see Problem 2.5.10, page 150) states that in the triangle of Figure 3.12

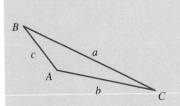

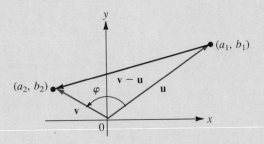

Figure 3.12   A triangle with sides a, b, and c

Figure 3.13   A triangle with sides $|\mathbf{u}|$, $|\mathbf{v}|$, and $|\mathbf{v} - \mathbf{u}|$

$$c^2 = a^2 + b^2 - 2ab \cos C$$

We now place the representations of **u** and **v** with initial points at the origin so that $\mathbf{u} = (a_1, b_1)$ and $\mathbf{v} = (a_2, b_2)$ (see Figure 3.13). Then, from the law of cosines, $|\mathbf{v} - \mathbf{u}|^2 = |\mathbf{v}|^2 + |\mathbf{u}|^2 - 2|\mathbf{u}|\,|\mathbf{v}| \cos \varphi$. But

$$\underset{\underset{\text{from (2)}}{\downarrow}}{}\qquad\qquad \underset{\underset{\text{Theorem 1 }(iii)\text{ on page 44}}{\downarrow}}{}$$

$$|\mathbf{v} - \mathbf{u}|^2 = (\mathbf{v} - \mathbf{u}) \cdot (\mathbf{v} - \mathbf{u}) = \mathbf{v} \cdot \mathbf{v} - 2\mathbf{u} \cdot \mathbf{v} + \mathbf{u} \cdot \mathbf{u}$$

$$= |\mathbf{v}|^2 - 2\mathbf{u} \cdot \mathbf{v} + |\mathbf{u}|^2$$

Thus, after simplification, we obtain $-2\mathbf{u} \cdot \mathbf{v} = -2|\mathbf{u}|\,|\mathbf{v}| \cos \varphi$, from which the theorem follows.   ∎

Remark. Using Theorem 1 we could define the scalar product $\mathbf{u} \cdot \mathbf{v}$ by

$$\mathbf{u} \cdot \mathbf{v} = |\mathbf{u}|\,|\mathbf{v}| \cos \varphi$$

EXAMPLE 1   **Computing the Angle Between Two Vectors**   Find the angle between the vectors $\mathbf{u} = 2\mathbf{i} + 3\mathbf{j}$ and $\mathbf{v} = -7\mathbf{i} + \mathbf{j}$.

Solution $\mathbf{u} \cdot \mathbf{v} = -14 + 3 = -11$, $|\mathbf{u}| = \sqrt{2^2 + 3^2} = \sqrt{13}$, and $|\mathbf{v}| = \sqrt{(-7)^2 + 1^2} = \sqrt{50}$. Hence

$$\cos \varphi = \frac{\mathbf{u} \cdot \mathbf{v}}{|\mathbf{u}|\,|\mathbf{v}|} = \frac{-11}{\sqrt{13}\,\sqrt{50}} = \frac{-11}{\sqrt{650}} \approx -0.431455497\dagger$$

so

$$\varphi = \cos^{-1}(-0.431455497) \approx 2.0169\ddagger \ (\approx 115.6°) \qquad \blacksquare$$

DEFINITION 2 **Parallel Vectors** Two nonzero vectors $\mathbf{u}$ and $\mathbf{v}$ are **parallel** if the angle between them is zero or π. Note that parallel vectors have the same or opposite directions.

EXAMPLE 2 **Two Parallel Vectors** Show that the vectors $\mathbf{u} = (2, -3)$ and $\mathbf{v} = (-4, 6)$ are parallel.

Solution $$\cos \varphi = \frac{\mathbf{u} \cdot \mathbf{v}}{|\mathbf{u}|\,|\mathbf{v}|} = \frac{-8 - 18}{\sqrt{13}\,\sqrt{52}} = \frac{-26}{\sqrt{13}(2\sqrt{13})} = \frac{-26}{2(13)} = -1$$

Hence $\varphi = \pi$ (so that $\mathbf{u}$ and $\mathbf{v}$ have opposite directions). $\qquad \blacksquare$

THEOREM 3 If $\mathbf{u} \neq \mathbf{0}$, then $\mathbf{v} = \alpha \mathbf{u}$ for some nonzero constant α if and only if $\mathbf{u}$ and $\mathbf{v}$ are parallel.

Proof The proof is left as an exercise (see Problem 35). $\qquad \blacksquare$

DEFINITION 3 **Orthogonal Vectors** The nonzero vectors $\mathbf{u}$ and $\mathbf{v}$ are called **orthogonal** (or **perpendicular**) if the angle between them is $\pi/2$.

EXAMPLE 3 **Two Orthogonal Vectors** Show that the vectors $\mathbf{u} = 3\mathbf{i} - 4\mathbf{j}$ and $\mathbf{v} = 4\mathbf{i} + 3\mathbf{j}$ are orthogonal.

Solution $\mathbf{u} \cdot \mathbf{v} = 3 \cdot 4 - 4 \cdot 3 = 0$. This implies that $\cos \varphi = (\mathbf{u} \cdot \mathbf{v})/(|\mathbf{u}|\,|\mathbf{v}|) = 0$. Since φ is in the interval $[0, \pi]$, $\varphi = \pi/2$. $\qquad \blacksquare$

THEOREM 4 The nonzero vectors $\mathbf{u}$ and $\mathbf{v}$ are orthogonal if and only if $\mathbf{u} \cdot \mathbf{v} = 0$.

Proof This proof is also left as an exercise (see Problem 36). $\qquad \blacksquare$

---

† These numbers, like others in the text, were obtained with a hand calculator.

‡ When doing this computation yourself, make certain that your calculator is set to radian mode.

EXAMPLE 4   **Determining a Way to Make Two Vectors Orthogonal or Parallel**
Let $\mathbf{u} = \mathbf{i} + 4\mathbf{j}$ and $\mathbf{v} = 3\mathbf{i} + \alpha\mathbf{j}$. Determine α such that **(i)** $\mathbf{u}$ and $\mathbf{v}$ are orthogonal; **(ii)** $\mathbf{u}$ and $\mathbf{v}$ are parallel.

Solution   **i.**   We have $\mathbf{u} \cdot \mathbf{v} = 3 + 4\alpha$. For $\mathbf{u}$ and $\mathbf{v}$ to be orthogonal, we must have $\mathbf{u} \cdot \mathbf{v} = 0$. This implies that $3 + 4\alpha = 0$ or $\alpha = -\frac{3}{4}$.

ii.   Here we must have $\varphi = 0$ or π so that $\cos \varphi = \pm 1$. Then

$$\cos \varphi = \frac{\mathbf{u} \cdot \mathbf{v}}{|\mathbf{u}|\,|\mathbf{v}|} = \frac{3 + 4\alpha}{\sqrt{17}\,\sqrt{9 + \alpha^2}} = \pm 1$$

Squaring both sides of this last equation, we obtain

$$9 + 24\alpha + 16\alpha^2 = 17(9 + \alpha^2) = 153 + 17\alpha^2.$$

This leads to the quadratic equation

$$\alpha^2 - 24\alpha + 144 = 0 = (\alpha - 12)^2,$$

with the single solution $\alpha = 12$.

Check. $3\mathbf{i} + 12\mathbf{j} = 3(\mathbf{i} + 4\mathbf{j})$, so that $\mathbf{v} = 3\mathbf{u}$ and $\mathbf{u}$ and $\mathbf{v}$ are parallel by Theorem 3.   ∎

A number of interesting problems involve the notion of the projection of one vector along another. Before defining this, we prove the following theorem.

THEOREM 5   Let $\mathbf{v}$ be a nonzero vector. Then for any other vector $\mathbf{u}$, the vector

$$\mathbf{w} = \mathbf{u} - \frac{\mathbf{u} \cdot \mathbf{v}}{|\mathbf{v}|^2}\mathbf{v}$$

is orthogonal to $\mathbf{v}$.

Proof
$$\mathbf{w} \cdot \mathbf{v} = \left[\mathbf{u} - \frac{(\mathbf{u} \cdot \mathbf{v})\mathbf{v}}{|\mathbf{v}|^2}\right] \cdot \mathbf{v} = \mathbf{u} \cdot \mathbf{v} - \frac{(\mathbf{u} \cdot \mathbf{v})(\mathbf{v} \cdot \mathbf{v})}{|\mathbf{v}|^2}$$

$$= \mathbf{u} \cdot \mathbf{v} - \frac{(\mathbf{u} \cdot \mathbf{v})|\mathbf{v}|^2}{|\mathbf{v}|^2} = \mathbf{u} \cdot \mathbf{v} - \mathbf{u} \cdot \mathbf{v} = 0 \qquad ∎$$

The vectors $\mathbf{u}$, $\mathbf{v}$, $\mathbf{w}$ are illustrated in Figure 3.14.

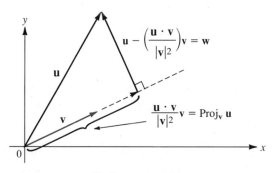

Figure 3.14   The vector $\mathbf{w} = \mathbf{u} - \dfrac{\mathbf{u} \cdot \mathbf{v}}{|\mathbf{v}|^2}\mathbf{v}$ is orthogonal to $\mathbf{v}$

DEFINITION 4 **Projection** Let **u** and **v** be nonzero vectors. Then the **projection** of **u** on **v** is a vector, denoted by proj$_\mathbf{v}$ **u**, which is defined by

$$\text{proj}_\mathbf{v} \, \mathbf{u} = \frac{\mathbf{u} \cdot \mathbf{v}}{|\mathbf{v}|^2}\mathbf{v} \qquad (4)$$

The **component** of **u** in the direction **v** is $\dfrac{\mathbf{u} \cdot \mathbf{v}}{|\mathbf{v}|}$ (5)

Note that **v**/|**v**| is a unit vector in the direction of **v**.

Remark 1. From Figures 3.14 and 3.15 and the fact that $\cos \varphi = (\mathbf{u} \cdot \mathbf{v})/(|\mathbf{u}| \, |\mathbf{v}|)$, we find that

v and proj$_\mathbf{v}$ **u** have (*i*) the same direction if $\mathbf{u} \cdot \mathbf{v} > 0$ and (*ii*) opposite directions if $\mathbf{u} \cdot \mathbf{v} < 0$.

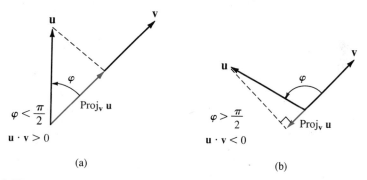

Figure 3.15 (a) **v** and Proj$_\mathbf{v}$ **u** have the same direction if $\mathbf{u} \cdot \mathbf{v} > 0$. (b) **v** and proj$_\mathbf{v}$ **u** have opposite directions if $\mathbf{u} \cdot \mathbf{v} < 0$

Remark 2. Proj$_\mathbf{v}$ **u** can be thought of as the "**v**-component" of the vector **u**.

Remark 3. If **u** and **v** are orthogonal, then $\mathbf{u} \cdot \mathbf{v} = 0$ so that proj$_\mathbf{v}$ **u** = **0**.

Remark 4. An alternative definition of projection is: If **u** and **v** are nonzero vectors, then proj$_\mathbf{v}$ **u** is the unique vector having these properties:

i. Proj$_\mathbf{v}$ **u** is parallel to **v**.

ii. $\mathbf{u} - \text{proj}_\mathbf{v} \, \mathbf{u}$ is orthogonal to **v**.

EXAMPLE 5   **Calculating a Projection**   Let $\mathbf{u} = 2\mathbf{i} + 3\mathbf{j}$ and $\mathbf{v} = \mathbf{i} + \mathbf{j}$. Calculate $\text{proj}_{\mathbf{v}} \mathbf{u}$.

Solution $\text{Proj}_{\mathbf{v}} \mathbf{u} = (\mathbf{u} \cdot \mathbf{v})\mathbf{v}/|\mathbf{v}|^2 = [5/(\sqrt{2})^2]\mathbf{v} = (5/2)\mathbf{i} + (5/2)\mathbf{j}$ (see Figure 3.16).

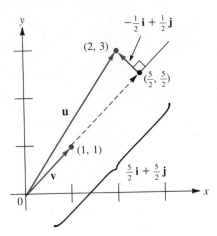

Figure 3.16   The projection of $(2, 3)$ on $(1, 1)$ is $(\frac{5}{2}, \frac{5}{2})$ ∎

EXAMPLE 6   **Calculating a Projection**   Let $\mathbf{u} = 2\mathbf{i} - 3\mathbf{j}$ and $\mathbf{v} = \mathbf{i} + \mathbf{j}$. Find $\text{proj}_{\mathbf{v}} \mathbf{u}$.

Solution   Here $(\mathbf{u} \cdot \mathbf{v})/|\mathbf{v}|^2 = -\frac{1}{2}$; hence $\text{proj}_{\mathbf{v}} \mathbf{u} = -\frac{1}{2}\mathbf{i} - \frac{1}{2}\mathbf{j}$ (see Figure 3.17).

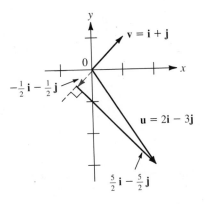

Figure 3.17   The projection of $2\mathbf{i} - 3\mathbf{j}$ on $\mathbf{i} + \mathbf{j}$ is $-\frac{1}{2}\mathbf{i} - \frac{1}{2}\mathbf{j}$ ∎

PROBLEMS   3.2

In Problems 1–8 calculate both the scalar product of the two vectors and the cosine of the angle between them.

1. $\mathbf{u} = \mathbf{i} + \mathbf{j}; \mathbf{v} = \mathbf{i} - \mathbf{j}$

2. $\mathbf{u} = 3\mathbf{i}; \mathbf{v} = -7\mathbf{j}$

3. $\mathbf{u} = -5\mathbf{i}; \mathbf{v} = 18\mathbf{j}$

4. $\mathbf{u} = \alpha\mathbf{i}; \mathbf{v} = \beta\mathbf{j}; \alpha, \beta$ real

5. $\mathbf{u} = 2\mathbf{i} + 5\mathbf{j}$; $\mathbf{v} = 5\mathbf{i} + 2\mathbf{j}$ 6. $\mathbf{u} = 2\mathbf{i} + 5\mathbf{j}$; $\mathbf{v} = 5\mathbf{i} - 2\mathbf{j}$

7. $\mathbf{u} = -3\mathbf{i} + 4\mathbf{j}$; $\mathbf{v} = -2\mathbf{i} - 7\mathbf{j}$ 8. $\mathbf{u} = 4\mathbf{i} + 5\mathbf{j}$; $\mathbf{v} = 5\mathbf{i} - 4\mathbf{j}$

9. Show that for any real numbers α and β, the vectors $\mathbf{u} = \alpha\mathbf{i} + \beta\mathbf{j}$ and $\mathbf{v} = \beta\mathbf{i} - \alpha\mathbf{j}$ are orthogonal.

10. Let $\mathbf{u}$, $\mathbf{v}$, and $\mathbf{w}$ denote three arbitrary vectors. Explain why the product $\mathbf{u} \cdot \mathbf{v} \cdot \mathbf{w}$ is *not defined*.

In Problems 11–16 determine whether the given vectors are orthogonal, parallel, or neither. Then sketch each pair.

11. $\mathbf{u} = 3\mathbf{i} + 5\mathbf{j}$; $\mathbf{v} = -6\mathbf{i} - 10\mathbf{j}$ \\   12. $\mathbf{u} = 2\mathbf{i} + 3\mathbf{j}$; $\mathbf{v} = 6\mathbf{i} - 4\mathbf{j}$

13. $\mathbf{u} = 2\mathbf{i} + 3\mathbf{j}$; $\mathbf{v} = 6\mathbf{i} + 4\mathbf{j}$ *neither*   14. $\mathbf{u} = 2\mathbf{i} + 3\mathbf{j}$; $\mathbf{v} = -6\mathbf{i} + 4\mathbf{j}$

15. $\mathbf{u} = 7\mathbf{i}$; $\mathbf{v} = -23\mathbf{j}$ *orthog*   16. $\mathbf{u} = 2\mathbf{i} - 6\mathbf{j}$; $\mathbf{v} = -\mathbf{i} + 3\mathbf{j}$

17. Let $\mathbf{u} = 3\mathbf{i} + 4\mathbf{j}$ and $\mathbf{v} = \mathbf{i} + \alpha\mathbf{j}$. Determine α such that:
    a. $\mathbf{u}$ and $\mathbf{v}$ are orthogonal.   b. $\mathbf{u}$ and $\mathbf{v}$ are parallel.
    c. The angle between $\mathbf{u}$ and $\mathbf{v}$ is $\pi/4$.   d. The angle between $\mathbf{u}$ and $\mathbf{v}$ is $\pi/3$.

18. Let $\mathbf{u} = -2\mathbf{i} + 5\mathbf{j}$ and $\mathbf{v} = \alpha\mathbf{i} - 2\mathbf{j}$. Determine α such that
    a. $\mathbf{u}$ and $\mathbf{v}$ are orthogonal.   b. $\mathbf{u}$ and $\mathbf{v}$ are parallel.
    c. The angle between $\mathbf{u}$ and $\mathbf{v}$ is $2\pi/3$.   d. The angle between $\mathbf{u}$ and $\mathbf{v}$ is $\pi/3$.

19. In Problem 17 show that there is no value of α for which $\mathbf{u}$ and $\mathbf{v}$ have opposite directions.

20. In Problem 18 show that there is no value of α for which $\mathbf{u}$ and $\mathbf{v}$ have the same direction.

In Problems 21–30 calculate $\text{proj}_\mathbf{v}\ \mathbf{u}$.

21. $\mathbf{u} = 3\mathbf{i}$; $\mathbf{v} = \mathbf{i} + \mathbf{j}$ 22. $\mathbf{u} = -5\mathbf{j}$; $\mathbf{v} = \mathbf{i} + \mathbf{j}$

23. $\mathbf{u} = 2\mathbf{i} + \mathbf{j}$; $\mathbf{v} = \mathbf{i} - 2\mathbf{j}$ 24. $\mathbf{u} = 2\mathbf{i} + 3\mathbf{j}$; $\mathbf{v} = 4\mathbf{i} + \mathbf{j}$

25. $\mathbf{u} = \mathbf{i} + \mathbf{j}$; $\mathbf{v} = 2\mathbf{i} - 3\mathbf{j}$ 26. $\mathbf{u} = \mathbf{i} + \mathbf{j}$; $\mathbf{v} = 2\mathbf{i} + 3\mathbf{j}$

27. $\mathbf{u} = \alpha\mathbf{i} + \beta\mathbf{j}$; $\mathbf{v} = \mathbf{i} + \mathbf{j}$; α, β real and positive

28. $\mathbf{u} = \mathbf{i} + \mathbf{j}$; $\mathbf{v} = \alpha\mathbf{i} + \beta\mathbf{j}$, α, β real and positive

29. $\mathbf{u} = \alpha\mathbf{i} - \beta\mathbf{j}$; $\mathbf{v} = \mathbf{i} + \mathbf{j}$; α, β real and positive with $\alpha > \beta$

30. $\mathbf{u} = \alpha\mathbf{i} - \beta\mathbf{j}$; $\mathbf{v} = \mathbf{i} + \mathbf{j}$; α, β real and positive with $\alpha < \beta$.

31. Let $\mathbf{u} = a_1\mathbf{i} + b_1\mathbf{j}$ and $\mathbf{v} = a_2\mathbf{i} + b_2\mathbf{j}$. Give a condition on a_1, b_1, a_2, and b_2 which will ensure that $\mathbf{v}$ and $\text{proj}_\mathbf{v}\ \mathbf{u}$ have the same direction.

32. In Problem 31 give a condition which will ensure that $\mathbf{v}$ and $\text{proj}_\mathbf{v}\ \mathbf{u}$ have opposite directions.

33. Let $P = (2, 3)$, $Q = (5, 7)$, $R = (2, -3)$, and $S = (1, 2)$. Calculate $\text{proj}_{\overrightarrow{PQ}}\overrightarrow{RS}$ and $\text{proj}_{\overrightarrow{RS}}\overrightarrow{PQ}$.

34. Let $P = (-1, 3)$, $Q = (2, 4)$, $R = (-6, -2)$, and $S = (3, 0)$. Calculate $\text{proj}_{\overrightarrow{PQ}}\overrightarrow{RS}$ and $\text{proj}_{\overrightarrow{RS}}\overrightarrow{PQ}$.

35. Prove that the nonzero vectors $\mathbf{u}$ and $\mathbf{v}$ are parallel if and only if $\mathbf{v} = \alpha\mathbf{u}$ for some constant α. [*Hint:* Show that $\cos \varphi = \pm 1$ if and only if $\mathbf{v} = \alpha\mathbf{u}$.]

36. Prove that $\mathbf{u}$ and $\mathbf{v}$ are orthogonal if and only if $\mathbf{u} \cdot \mathbf{v} = 0$.

37. Show that the vector $\mathbf{v} = a\mathbf{i} + b\mathbf{j}$ is orthogonal to the line $ax + by + c = 0$.

38. Show that the vector $\mathbf{u} = b\mathbf{i} - a\mathbf{j}$ is parallel to the line $ax + by + c = 0$.

39. A triangle has vertices $(1, 3)$, $(4, -2)$, and $(-3, 6)$. Find the cosine of each of its angles.

40. A triangle has vertices (a_1, b_1), (a_2, b_2), and (a_3, b_3). Find a formula for the cosines of each of its angles.

*41. The **Cauchy-Schwarz inequality** states that for any real numbers a_1, a_2, b_1, and b_2,

$$\left| \sum_{k=1}^{2} a_k b_k \right| \le \left(\sum_{k=1}^{2} a_k^2 \right)^{1/2} \left(\sum_{k=1}^{2} b_k^2 \right)^{1/2}$$

Use the scalar product to prove this formula. Under what circumstances can the inequality be replaced by an equality?

*42. Prove that the shortest distance between a point and a line is measured along a line through the point and perpendicular to the line.

43. Find the distance between $P = (2, 3)$ and the line through the points $Q = (-1, 7)$ and $R = (3, 5)$.

44. Find the distance between $(3, 7)$ and the line along the vector $\mathbf{v} = 2\mathbf{i} - 3\mathbf{j}$ passing through the origin.

45. Let A be a 2×2 matrix such that each column is a unit vector and the two columns are orthogonal. Show that A is invertible and that $A^{-1} = A^t$. (A is called an **orthogonal matrix.**)

3.3 VECTORS IN SPACE

We have seen how any point in a plane can be represented as an ordered pair of real numbers. Analogously, any point in space can be represented by an **ordered triple** of real numbers

$$(a, b, c) \tag{1}$$

$\mathbb{R}^3$
Origin
x-axis
y-axis
z-axis

Vectors of the form (1) constitute $\mathbb{R}^3$. To represent a point in space, we begin by choosing a point in $\mathbb{R}^3$. We call this point the **origin,** denoted by 0. Then we draw three mutually perpendicular axes, which we label the **x-axis,** the **y-axis,** and the **z-axis.** These axes can be selected in a variety of ways, but the most common selection has the x- and y-axes drawn horizontally with the z-axis vertical. On each axis, we choose a positive direction and measure distance along each axis as the number of units in this positive direction measured from the origin.

The two basic systems of drawing these axes are depicted in Figure 3.18. If the axes are placed as in Figure 3.18a, then the system is called a **right-handed system;** if they are placed as in Figure 3.18b, the system is a **left-handed system.** In the figures the arrows indicate the positive directions on the axes. The reason for this choice of terms is as follows: In a right-handed system, if you place your right

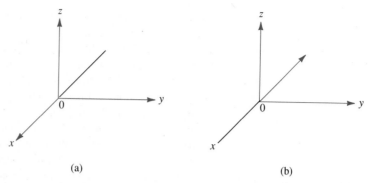

(a)   (b)

Figure 3.18   (a) A right-handed system.   (b) A left-handed system

hand so that your index finger points in the positive direction of the x-axis while your middle finger points in the positive direction of the y-axis, then your thumb will point in the positive direction of the z-axis. This concept is illustrated in Figure 3.19. For a left-handed system, the same rule works for your left hand. For the

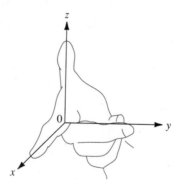

Figure 3.19   A right hand indicates directions in a right-handed system

remainder of this text, we shall follow common practice and depict the coordinate axes using a right-handed system.

Coordinate planes

The three axes in our system determine three **coordinate planes,** which we call the xy-plane, the xz-plane, and the yz-plane. The xy-plane contains the x- and y-axes and is simply the plane with which we have been dealing to this point in most of this book. The xz- and yz-planes can be thought of in a similar way.

Having built our structure of coordinate axes and planes, we can describe any point P in $\mathbb{R}^3$ in a unique way:

$$P = (x, y, z)$$ **(2)**

where the first coordinate x is the directed distance from the yz-plane to P (measured in the positive direction of the x-axis and along a line parallel to the x-axis), the second coordinate y is the directed distance from the xz-plane to P (measured in the positive direction of the y-axis and along a line parallel to the y-axis), and the third coordinate z is the directed distance from the xy-plane to P (measured in the positive direction of the z-axis and along a line parallel to the z-axis).

In this system the three coordinate planes divide $\mathbb{R}^3$ into eight **octants,** just as in $\mathbb{R}^2$ the two coordinate axes divide the plane into four quadrants. The first octant is always chosen to be the one in which the three coordinates are positive.

The coordinate system we have just established is often referred to as the **rectangular coordinate system** or the **Cartesian coordinate system.** Once we are comfortable with the notion of depicting a point in this system, we can generalize many of our ideas from the plane.

Cartesian coordinate system in $\mathbb{R}^3$

THEOREM 1   Let $P = (x_1, y_1, z_1)$ and $Q = (x_2, y_2, z_2)$ be two points in space. Then the distance $\overline{PQ}$ between P and Q is given by

$$\overline{PQ} = \sqrt{(x_1 - x_2)^2 + (y_1 - y_2)^2 + (z_1 - z_2)^2} \qquad (3)$$

You are asked to prove this result in Problem 39.   ∎

EXAMPLE 1   **Calculating the Distance Between Two Points in $\mathbb{R}^3$**   Calculate the distance between the points $(3, -1, 6)$ and $(-2, 3, 5)$.

Solution $$\overline{PQ} = \sqrt{[3 - (-2)]^2 + (-1 - 3)^2 + (6 - 5)^2} = \sqrt{42} \qquad ∎$$

In Sections 3.1 and 3.2 we developed geometric properties of vectors in the plane. Given the similarity between the coordinate systems in $\mathbb{R}^2$ and $\mathbb{R}^3$, it should come as no surprise that vectors in $\mathbb{R}^2$ and $\mathbb{R}^3$ have very similar structures. We now develop the notion of a vector in space. The development will closely follow the development in the last two sections, and therefore some of the details will be omitted.

Let P and Q be two distinct points in $\mathbb{R}^3$. Then the **directed line segment $\overrightarrow{PQ}$** is the straight-line segment that extends from P to Q. Two directed line segments are **equivalent** if they have the same magnitude and direction. A **vector** in $\mathbb{R}^3$ is the set of all directed line segments equivalent to a given directed line segment, and any directed line segment $\overrightarrow{PQ}$ in that set is called a **representation** of the vector.

Directed line segment

Vector in $\mathbb{R}^3$

So far, our definitions are identical. For convenience, we choose P to be the origin so that the vector $\mathbf{v} = \overrightarrow{0Q}$ can be described by the coordinates (x, y, z) of the point Q. Then the magnitude of $\mathbf{v} = |\mathbf{v}| = \sqrt{x^2 + y^2 + z^2}$ (from Theorem 1).

EXAMPLE 2 **Calculating the Magnitude of a Vector in $\mathbb{R}^3$**   Let $\mathbf{v} = (1, 3, -2)$. Find $|\mathbf{v}|$.

Solution $$|\mathbf{v}| = \sqrt{1^2 + 3^2 + (-2)^2} = \sqrt{14}.$$ ■

Let $\mathbf{u} = (x_1, y_1, z_1)$ and $\mathbf{v} = (x_2, y_2, z_2)$ be two vectors and let α be a real number (scalar). Then we define

Vector Addition and Scalar Multiplication in $\mathbb{R}^3$

$$\mathbf{u} + \mathbf{v} = (x_1 + x_2, y_1 + y_2, z_1 + z_2)$$

and

$$\alpha\mathbf{u} = (\alpha x_1, \alpha y_1, \alpha z_1)$$

This is the same definition of vector addition and scalar multiplication we had before; it is illustrated in Figure 3.20.

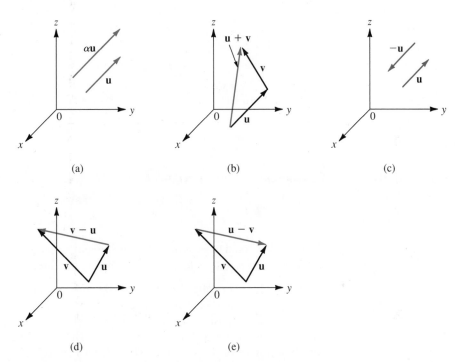

(a)   (b)   (c)

(d)   (e)

Figure 3.20   Illustration of vector addition and scalar multiplication in $\mathbb{R}^3$

Unit vector   A **unit vector** $\mathbf{u}$ is a vector with magnitude 1. If $\mathbf{v}$ is any nonzero vector, then $\mathbf{u} = \mathbf{v}/|\mathbf{v}|$ is a unit vector having the same direction as $\mathbf{v}$.

EXAMPLE 3   **Finding a Unit Vector in $\mathbb{R}^3$**   Find a unit vector having the same direction as $\mathbf{v} = (2, 4, -3)$.

Solution   Since $\mathbf{v} = \sqrt{2^2 + 4^2 + (-3)^2} = \sqrt{29}$, we have

$$\mathbf{u} = (2/\sqrt{29}, 4/\sqrt{29}, -3/\sqrt{29}).$$

■

We can now formally define the direction of a vector in $\mathbb{R}^3$. We cannot define it to be the angle θ the vector makes with the positive x-axis, since, for example, if $0 < \theta < \pi/2$, then there are an *infinite number* of vectors making the angle θ with the positive x-axis, and these together form a cone (see Figure 3.21).

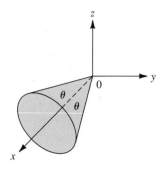

Figure 3.21   All the vectors which lie on this cone make the angle θ with the positive x-axis

DEFINITION 1   **Direction in $\mathbb{R}^3$** The **direction** of a nonzero vector $\mathbf{v}$ in $\mathbb{R}^3$ is defined as the unit vector $\mathbf{u} = \mathbf{v}/|\mathbf{v}|$.

Remark. We could have defined the direction of a vector $\mathbf{v}$ in $\mathbb{R}^2$ in this way. For if $\mathbf{u} = \mathbf{v}/|\mathbf{v}|$, then $\mathbf{u} = (\cos\theta, \sin\theta)$, where θ is the direction of $\mathbf{v}$.

It would still be satisfying to define the direction of a vector in terms of some angles. Let $\mathbf{v}$ be the vector $\overrightarrow{OP}$ depicted in Figure 3.22. We define α to be the angle between $\mathbf{v}$ and the positive x-axis, β the angle between $\mathbf{v}$ and the positive y-axis, and γ the angle between $\mathbf{v}$ and the positive z-axis. The angles α, β, and γ are called the

Direction angles   **direction angles** of the vector $\mathbf{v}$. Then, from Figure 3.22,

$$\cos\alpha = \frac{x_0}{|\mathbf{v}|} \qquad \cos\beta = \frac{y_0}{|\mathbf{v}|} \qquad \cos\gamma = \frac{z_0}{|\mathbf{v}|} \tag{4}$$

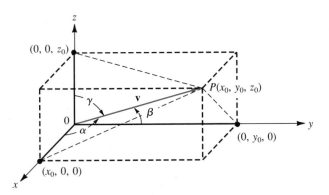

Figure 3.22 The vector **v** makes the angle α with the positive x-axis, β with the positive y-axis, and γ with the positive z-axis

If **v** is a unit vector, then $|\mathbf{v}| = 1$ and

$$\cos \alpha = x_0 \qquad \cos \beta = y_0 \qquad \cos \gamma = z_0 \qquad (5)$$

Direction cosines

By definition, each of these three angles lies in the interval $[0, \pi]$. The cosines of these angles are called the **direction cosines** of the vector **v**. Note, from equations (4), that

$$\cos^2 \alpha + \cos^2 \beta + \cos^2 \gamma = \frac{x_0^2 + y_0^2 + z_0^2}{|\mathbf{v}|^2} = \frac{x_0^2 + y_0^2 + z_0^2}{x_0^2 + y_0^2 + z_0^2} = 1 \qquad (6)$$

If α, β, and γ are any three numbers between zero and π such that condition (6) is satisfied, then they uniquely determine a unit vector given by $\mathbf{u} = (\cos \alpha, \cos \beta, \cos \gamma)$.

Direction numbers

Remark. If $\mathbf{v} = (a, b, c)$ and $|\mathbf{v}| \neq 1$, then the numbers a, b, and c are called **direction numbers** of the vector **v**.

EXAMPLE 4

Finding the Direction Cosines of a Vector in $\mathbb{R}^3$ Find the direction cosines of the vector $\mathbf{v} = (4, -1, 6)$.

Solution

The direction of **v** is $\mathbf{v}/|\mathbf{v}| = \mathbf{v}/\sqrt{53} = (4/\sqrt{53}, -1/\sqrt{53}, 6/\sqrt{53})$. Then $\cos \alpha = 4/\sqrt{53} \approx 0.5494$, $\cos \beta = -1/\sqrt{53} \approx -0.1374$, and $\cos \gamma = 6/\sqrt{53} \approx 0.8242$. From these we use a table of cosines or a hand calculator to obtain $\alpha \approx 56.7° \approx 0.989$ rad, $\beta \approx 97.9° \approx 1.71$ rad, and $\gamma = 34.5° \approx 0.602$ rad. The vector, along with the angles α, β, and γ, is sketched in Figure 3.23.

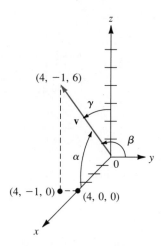

Figure 3.23   The direction cosines of $(4, -1, 6)$ are $\cos \alpha$, $\cos \beta$, and $\cos \gamma$

EXAMPLE 5   **Finding a Vector in $\mathbb{R}^3$ Given Its Magnitude and Direction Cosines**   Find a vector **v** of magnitude 7 whose direction cosines are $1/\sqrt{6}$, $1/\sqrt{3}$, and $1/\sqrt{2}$.

Solution   Let $\mathbf{u} = (1/\sqrt{6}, 1/\sqrt{3}, 1/\sqrt{2})$. Then **u** is a unit vector since $|\mathbf{u}| = 1$. Thus the direction of **v** is given by **u** and $\mathbf{v} = |\mathbf{v}|\,\mathbf{u} = 7\mathbf{u} = (7/\sqrt{6}, 7/\sqrt{3}, 7/\sqrt{2})$.

Note. We can solve this problem because $(1/\sqrt{6})^2 + (1/\sqrt{3})^2 + (1/\sqrt{2})^2 = 1$.

It is interesting to note that if **v** in $\mathbb{R}^2$ is a unit vector and we write $\mathbf{v} = (\cos \theta)\mathbf{i} + (\sin \theta)\mathbf{j}$, where θ is the direction of **v**, then $\cos \theta$ and $\sin \theta$ are the direction cosines of **v**. Here $\alpha = \theta$ and we define β to be the angle that **v** makes with the y-axis (see Figure 3.24). Then $\beta = (\pi/2) - \alpha$ so that $\cos \beta = \cos (\pi/2 - \alpha) =$

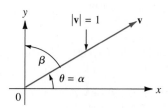

Figure 3.24   If $\beta = \dfrac{\pi}{2} - \theta = \dfrac{\pi}{2} - \alpha$ and **v** is a unit vector, then

$\mathbf{v} = \cos \theta\,\mathbf{i} + \sin \theta\,\mathbf{j} = \cos \alpha\mathbf{i} + \cos \beta\mathbf{j}$

$\sin \alpha$ and $\mathbf{v}$ can be written in the "direction cosine" form

$$\mathbf{v} = \cos \alpha \, \mathbf{i} + \cos \beta \, \mathbf{j}$$

In Section 3.1 we saw how any vector in the plane can be written in terms of the basis vectors $\mathbf{i}$ and $\mathbf{j}$. To extend this idea to $\mathbb{R}^3$, we define

$$\boxed{\mathbf{i} = (1, 0, 0) \qquad \mathbf{j} = (0, 1, 0) \qquad \mathbf{k} = (0, 0, 1)} \qquad (7)$$

Here $\mathbf{i}$, $\mathbf{j}$, and $\mathbf{k}$ are unit vectors. The vector $\mathbf{i}$ lies along the x-axis, $\mathbf{j}$ along the y-axis, and $\mathbf{k}$ along the z-axis. These are sketched in Figure 3.25. If $\mathbf{v} = (x, y, z)$ is

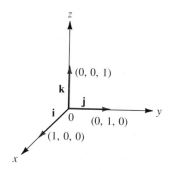

Figure 3.25   The basis vectors $\mathbf{i}$, $\mathbf{j}$, and $\mathbf{k}$ in $\mathbb{R}^3$

any vector in $\mathbb{R}^3$, then

$$\mathbf{v} = (x, y, z) = (x, 0, 0) + (0, y, 0) + (0, 0, z) = x\mathbf{i} + y\mathbf{j} + z\mathbf{k}$$

That is: *Any vector $\mathbf{v}$ in $\mathbb{R}^3$ can be written in a unique way in terms of the vectors $\mathbf{i}$, $\mathbf{j}$, and $\mathbf{k}$.*

The definition of the scalar product in $\mathbb{R}^3$ is, of course, the definition we have already seen in Section 1.7. Note that $\mathbf{i} \cdot \mathbf{i} = 1$, $\mathbf{j} \cdot \mathbf{j} = 1$, $\mathbf{k} \cdot \mathbf{k} = 1$, $\mathbf{i} \cdot \mathbf{j} = 0$, $\mathbf{j} \cdot \mathbf{k} = 0$, and $\mathbf{i} \cdot \mathbf{k} = 0$.

THEOREM 2   If φ denotes the smallest positive angle between two nonzero vectors $\mathbf{u}$ and $\mathbf{v}$, we have

$$\boxed{\cos \varphi = \frac{\mathbf{u} \cdot \mathbf{v}}{|\mathbf{u}| \, |\mathbf{v}|}} \qquad (8)$$

Proof   The proof is almost identical to the proof of Theorem 3.2.2 on page 166 and is left as an exercise (see Problem 40).   ∎

EXAMPLE 6   **Computing the Cosine of the Angle Between Two Vectors in $\mathbb{R}^3$**   Calculate the cosine of the angle between $\mathbf{u} = 3\mathbf{i} - \mathbf{j} + 2\mathbf{k}$ and $\mathbf{v} = 4\mathbf{i} + 3\mathbf{j} - \mathbf{k}$.

Solution $\mathbf{u} \cdot \mathbf{v} = 7$, $|\mathbf{u}| = \sqrt{14}$, and $|\mathbf{v}| = \sqrt{26}$ so that $\cos \varphi = 7/\sqrt{(14)(26)} = 7/\sqrt{364} \approx 0.3669$ and $\varphi \approx 68.5° \approx 1.2$ rad. ∎

DEFINITION 2   **Parallel and Orthogonal Vectors** Two nonzero vectors $\mathbf{u}$ and $\mathbf{v}$ are:

  i. **Parallel** if the angle between them is zero or π

  ii. **Orthogonal** (or **perpendicular**) if the angle between them is $\pi/2$

THEOREM 3   **i.** If $\mathbf{u} \neq \mathbf{0}$, then $\mathbf{u}$ and $\mathbf{v}$ are parallel if and only if $\mathbf{v} = \alpha\mathbf{u}$ for some constant $\alpha \neq 0$.

  ii. If $\mathbf{u}$ and $\mathbf{v}$ are nonzero, then $\mathbf{u}$ and $\mathbf{v}$ are orthogonal if and only if $\mathbf{u} \cdot \mathbf{v} = 0$.

Proof   Again the proof is easy and is left as an exercise (see Problem 41). ∎

EXAMPLE 7   **Two Parallel Vectors in $\mathbb{R}^3$**   Show that the vectors $\mathbf{u} = \mathbf{i} + 3\mathbf{j} - 4\mathbf{k}$ and $\mathbf{v} = -2\mathbf{i} - 6\mathbf{j} + 8\mathbf{k}$ are parallel.

Solution $\mathbf{u} \cdot \mathbf{v} = -52$, $|\mathbf{u}| = \sqrt{26}$, and $|\mathbf{v}| = \sqrt{104} = 2\sqrt{26}$. Hence $\mathbf{u} \cdot \mathbf{v}/|\mathbf{u}| \, |\mathbf{v}| = -52/(\sqrt{26} \cdot 2\sqrt{26}) = -1$ so that $\cos \theta = -1$, $\theta = \pi$, and $\mathbf{u}$ and $\mathbf{v}$ are parallel (but have opposite directions). Another way to see this is to note that $\mathbf{v} = -2\mathbf{u}$ so that, by Theorem 3, $\mathbf{u}$ and $\mathbf{v}$ are parallel (see Figure 3.26).

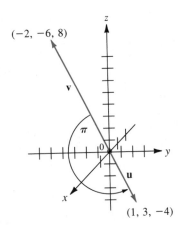

Figure 3.26   The angle between $\mathbf{u}$ and $\mathbf{v}$ is π, so that $\mathbf{u}$ and $\mathbf{v}$ are parallel. Note that $\mathbf{v} = -2\mathbf{u}$

EXAMPLE 8   **Determining When Two Vectors in $\mathbb{R}^3$ are Orthogonal**   Find a number α such that $\mathbf{u} = 8\mathbf{i} - 2\mathbf{j} + 4\mathbf{k}$ and $\mathbf{v} = 2\mathbf{i} + 3\mathbf{j} + \alpha\mathbf{k}$ are orthogonal.

Solution   We must have $0 = \mathbf{u} \cdot \mathbf{v} = 10 + 4\alpha$ so that $\alpha = -\frac{5}{2}$. The vectors $\mathbf{u}$ and $\mathbf{v}$ are sketched in Figure 3.27.

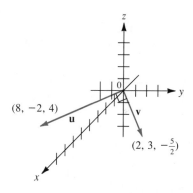

Figure 3.27   The vectors $(8, -2, 4)$ and $(2, 3, -\frac{5}{2})$ are orthogonal

We now turn to the definition of the projection of one vector on another. First we state the theorem which is the analog of Theorem 3.2.5 (and which has an identical proof).

THEOREM 4   Let $\mathbf{v}$ be a nonzero vector. Then for any other vector $\mathbf{u}$,

$$\mathbf{w} = \mathbf{u} - \frac{\mathbf{u} \cdot \mathbf{v}}{|\mathbf{v}|^2}\mathbf{v}$$

is orthogonal to $\mathbf{v}$.

DEFINITION 3   **Projection**   Let $\mathbf{u}$ and $\mathbf{v}$ be nonzero vectors. Then the **projection** of $\mathbf{u}$ on $\mathbf{v}$, denoted by $\text{proj}_\mathbf{v}\, \mathbf{u}$, is defined by

$$\text{proj}_\mathbf{v}\, \mathbf{u} = \frac{\mathbf{u} \cdot \mathbf{v}}{|\mathbf{v}|^2}\mathbf{v} \tag{9}$$

The **component** of $\mathbf{u}$ in the direction $\mathbf{v}$ is given by $\dfrac{\mathbf{u} \cdot \mathbf{v}}{|\mathbf{v}|}$.   **(10)**

EXAMPLE 9   **Calculating a Projection in $\mathbb{R}^3$**   Let $\mathbf{u} = 2\mathbf{i} + 3\mathbf{j} + \mathbf{k}$ and $\mathbf{v} = \mathbf{i} + 2\mathbf{j} - 6\mathbf{k}$. Find $\text{proj}_\mathbf{v}\,\mathbf{u}$.

Solution   Here $(\mathbf{u} \cdot \mathbf{v})/|\mathbf{v}|^2 = 2/41$ and $\text{proj}_\mathbf{v}\,\mathbf{u} = \frac{2}{41}\mathbf{i} + \frac{4}{41}\mathbf{j} - \frac{12}{41}\mathbf{k}$. The component of $\mathbf{u}$ in the direction $\mathbf{v}$ is $(\mathbf{u} \cdot \mathbf{v})/|\mathbf{v}| = 2/\sqrt{41}$. ∎

Note that, as in the planar case, $\text{proj}_\mathbf{v}\,\mathbf{u}$ is a vector that has the same direction as $\mathbf{v}$ if $\mathbf{u} \cdot \mathbf{v} > 0$ and the direction opposite to that of $\mathbf{v}$ if $\mathbf{u} \cdot \mathbf{v} < 0$.

PROBLEMS   3.3

In Problems 1–3 find the distance between the two points.

1. $(3, -4, 3)$; $(3, 2, 5)$ **2.** $(3, -4, 7)$; $(3, -4, 9)$

3. $(-2, 1, 3)$; $(4, 1, 3)$

In Problems 4–17 find the magnitude and the direction cosines of the given vector.

4. $\mathbf{v} = 3\mathbf{j}$ **5.** $\mathbf{v} = -3\mathbf{i}$ **6.** $\mathbf{v} = 4\mathbf{i} - \mathbf{j}$

7. $\mathbf{v} = \mathbf{i} + 2\mathbf{k}$ **8.** $\mathbf{v} = \mathbf{i} - \mathbf{j} + \mathbf{k}$ **9.** $\mathbf{v} = \mathbf{i} + \mathbf{j} - \mathbf{k}$

10. $\mathbf{v} = -\mathbf{i} + \mathbf{j} + \mathbf{k}$ **11.** $\mathbf{v} = \mathbf{i} - \mathbf{j} - \mathbf{k}$ **12.** $\mathbf{v} = -\mathbf{i} + \mathbf{j} - \mathbf{k}$

13. $\mathbf{v} = -\mathbf{i} - \mathbf{j} + \mathbf{k}$ **14.** $\mathbf{v} = -\mathbf{i} - \mathbf{j} - \mathbf{k}$ **15.** $\mathbf{v} = 2\mathbf{i} + 5\mathbf{j} - 7\mathbf{k}$

16. $\mathbf{v} = -3\mathbf{i} - 3\mathbf{j} + 8\mathbf{k}$ **17.** $\mathbf{v} = -2\mathbf{i} - 3\mathbf{j} - 4\mathbf{k}$

18. The three direction angles of a certain unit vector are the same and are between zero and $\pi/2$. What is the vector?

19. Find a vector of magnitude 12 that has the same direction as the vector in Problem 18.

20. Show that there is no unit vector whose direction angles are $\pi/6$, $\pi/3$, and $\pi/4$.

21. Let $P = (2, 1, 4)$ and $Q = (3, -2, 8)$. Find a unit vector in the direction $\overrightarrow{PQ}$.

22. Let $P = (-3, 1, 7)$ and $Q = (8, 1, 7)$. Find a unit vector whose direction is opposite that of $\overrightarrow{PQ}$.

23. In Problem 22 find all points R such that $\overrightarrow{PR} \perp \overrightarrow{PQ}$.

*24. Show that the set of points which satisfy the condition of Problem 23 and the condition $|\overrightarrow{PR}| = 1$ forms a circle.

25. Triangle Inequality If $\mathbf{u}$ and $\mathbf{v}$ are in $\mathbb{R}^3$, show that $|\mathbf{u} + \mathbf{v}| \leq |\mathbf{u}| + |\mathbf{v}|$.

26. Under what circumstances can the inequality in Problem 25 be replaced by an equals sign?

In Problems 27–38 let $\mathbf{u} = 2\mathbf{i} - 3\mathbf{j} + 4\mathbf{k}$, $\mathbf{v} = -2\mathbf{i} - 3\mathbf{j} + 5\mathbf{k}$, $\mathbf{w} = \mathbf{i} - 7\mathbf{j} + 3\mathbf{k}$, and $\mathbf{t} = 3\mathbf{i} + 4\mathbf{j} + 5\mathbf{k}$.

27. Calculate $\mathbf{u} + \mathbf{v}$.   **28.** Calculate $2\mathbf{u} - 3\mathbf{v}$.

29. Calculate $\mathbf{t} + 3\mathbf{w} - \mathbf{v}$.   **30.** Calculate $2\mathbf{u} - 7\mathbf{w} + 5\mathbf{v}$.

31. Calculate $2\mathbf{v} + 7\mathbf{t} - \mathbf{w}$.   **32.** Calculate $\mathbf{u} \cdot \mathbf{v}$.

33. Calculate $|\mathbf{w}|$.   **34.** Calculate $\mathbf{u} \cdot \mathbf{w} - \mathbf{w} \cdot \mathbf{t}$.

35. Calculate the angle between **u** and **w**.

36. Calculate the angle between **t** and **w**.

37. Calculate $\text{proj}_{\mathbf{u}}\,\mathbf{v}$.   **38.** Calculate $\text{proj}_{\mathbf{t}}\,\mathbf{w}$.

39. Prove Theorem 1. [*Hint:* Use the Pythagorean theorem twice in Figure 3.28.]

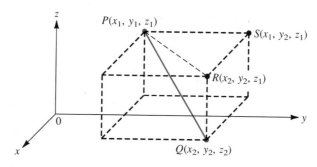

Figure 3.28

40. Prove Theorem 2.

41. Prove Theorem 3.

42. Prove Theorem 4.

3.4 THE CROSS PRODUCT OF TWO VECTORS

To this point the only product of vectors that we have considered has been the scalar or dot product. We now define a new product, called the *cross product* (or *vector product*), which is defined only in $\mathbb{R}^3$.

Historical Note. The cross product was defined by Hamilton in one of a series of papers published in *Philosophical Magazine* between the years 1844 and 1850.

DEFINITION 1   **Cross Product** Let $\mathbf{u} = a_1\mathbf{i} + b_1\mathbf{j} + c_1\mathbf{k}$ and $\mathbf{v} = a_2\mathbf{i} + b_2\mathbf{j} + c_2\mathbf{k}$. Then the **cross product (vector product)** of **u** and **v**, denoted by $\mathbf{u} \times \mathbf{v}$, is a new vector defined by

$$\mathbf{u} \times \mathbf{v} = (b_1c_2 - c_1b_2)\mathbf{i} + (c_1a_2 - a_1c_2)\mathbf{j} + (a_1b_2 - b_1a_2)\mathbf{k} \qquad \textbf{(1)}$$

Note that the result of the cross product is a vector, whereas the result of the scalar product is a scalar.

Here the cross product seems to have been defined somewhat arbitrarily. There are obviously many ways to define a vector product. Why was this definition chosen? We answer that question in this section by demonstrating some of the properties of the cross product and illustrating some of its uses.

EXAMPLE 1   **Calculating the Cross Product of Two Vectors**   Let $\mathbf{u} = \mathbf{i} - \mathbf{j} + 2\mathbf{k}$ and $\mathbf{v} = 2\mathbf{i} + 3\mathbf{j} - 4\mathbf{k}$. Calculate $\mathbf{w} = \mathbf{u} \times \mathbf{v}$.

Solution   Using formula (1), we obtain

$$\mathbf{w} = [(-1)(-4) - (2)(3)]\mathbf{i} + [(2)(2) - (1)(-4)]\mathbf{j} + [(1)(3) - (-1)(2)]\mathbf{k}$$
$$= -2\mathbf{i} + 8\mathbf{j} + 5\mathbf{k}$$

Note. In this example $\mathbf{u} \cdot \mathbf{w} = (\mathbf{i} - \mathbf{j} + 2\mathbf{k}) \cdot (-2\mathbf{i} + 8\mathbf{j} + 5\mathbf{k}) = -2 - 8 + 10 = 0$. Similarly, $\mathbf{v} \cdot \mathbf{w} = 0$. That is, $\mathbf{u} \times \mathbf{v}$ is orthogonal to both $\mathbf{u}$ and $\mathbf{v}$. As we shall shortly see, the cross product of $\mathbf{u}$ and $\mathbf{v}$ is always orthogonal to $\mathbf{u}$ and $\mathbf{v}$. ■

Before continuing our discussion of the uses of the cross product, we observe that there is an easy way to calculate $\mathbf{u} \times \mathbf{v}$ by using determinants.

THEOREM 1

$$\mathbf{u} \times \mathbf{v} = \begin{vmatrix} \mathbf{i} & \mathbf{j} & \mathbf{k} \\ a_1 & b_1 & c_1 \\ a_2 & b_2 & c_2 \end{vmatrix}^\dagger$$

Proof

$$\begin{vmatrix} \mathbf{i} & \mathbf{j} & \mathbf{k} \\ a_1 & b_1 & c_1 \\ a_2 & b_2 & c_2 \end{vmatrix} = \mathbf{i} \begin{vmatrix} b_1 & c_1 \\ b_2 & c_2 \end{vmatrix} - \mathbf{j} \begin{vmatrix} a_1 & c_1 \\ a_2 & c_2 \end{vmatrix} + \mathbf{k} \begin{vmatrix} a_1 & b_1 \\ a_2 & b_2 \end{vmatrix}$$

$$= (b_1 c_2 - c_1 b_2)\mathbf{i} + (c_1 a_2 - a_1 c_2)\mathbf{j} + (a_1 b_2 - b_1 a_2)\mathbf{k}$$

which is equal to $\mathbf{u} \times \mathbf{v}$ according to Definition 1. ■

EXAMPLE 2   **Using Theorem 1 to Calculate a Cross Product**   Calculate $\mathbf{u} \times \mathbf{v}$, where $\mathbf{u} = 2\mathbf{i} + 4\mathbf{j} - 5\mathbf{k}$ and $\mathbf{v} = -3\mathbf{i} - 2\mathbf{j} + \mathbf{k}$.

Solution $$\mathbf{u} \times \mathbf{v} = \begin{vmatrix} \mathbf{i} & \mathbf{j} & \mathbf{k} \\ 2 & 4 & -5 \\ -3 & -2 & 1 \end{vmatrix} = (4 - 10)\mathbf{i} - (2 - 15)\mathbf{j} + (-4 + 12)\mathbf{k}$$

$$= -6\mathbf{i} + 13\mathbf{j} + 8\mathbf{k}$$ ■

---

†This is not really a determinant, because $\mathbf{i}$, $\mathbf{j}$, and $\mathbf{k}$ are not numbers. However, using determinant notation, Theorem 1 helps us remember how to calculate a cross product.

The following theorem summarizes some properties of the cross product. Its proof is left as an exercise (see Problems 32–35).

THEOREM 2   Let **u**, **v**, and **w** be vectors in $\mathbb{R}^3$ and let α be a scalar. Then:

  i. $\mathbf{u} \times \mathbf{0} = \mathbf{0} \times \mathbf{u} = \mathbf{0}$.

  ii. $\mathbf{u} \times \mathbf{v} = -(\mathbf{v} \times \mathbf{u})$ **(anticommutative property for the vector product).**

  iii. $(\alpha\mathbf{u}) \times \mathbf{v} = \alpha(\mathbf{u} \times \mathbf{v})$.

  iv. $\mathbf{u} \times (\mathbf{v} + \mathbf{w}) = (\mathbf{u} \times \mathbf{v}) + (\mathbf{u} \times \mathbf{w})$ **(distributive property for the vector product).**

  v. $(\mathbf{u} \times \mathbf{v}) \cdot \mathbf{w} = \mathbf{u} \cdot (\mathbf{v} \times \mathbf{w})$. (This is called the **scalar triple product** of **u**, **v**, and **w**.)

  vi. $\mathbf{u} \cdot (\mathbf{u} \times \mathbf{v}) = \mathbf{v} \cdot (\mathbf{u} \times \mathbf{v}) = 0$. (That is, $\mathbf{u} \times \mathbf{v}$ is orthogonal to both **u** and **v**.)

  vii. If **u** and **v** are parallel, then $\mathbf{u} \times \mathbf{v} = \mathbf{0}$. ∎

Part (*vi*) is the most commonly used part of this theorem. We restate it below:

> The cross product $\mathbf{u} \times \mathbf{v}$ is orthogonal to both **u** and **v**.

We now know that $\mathbf{u} \times \mathbf{v}$ is a vector orthogonal to **u** and **v**. But there are always *two* unit vectors orthogonal to **u** and **v** (see Figure 3.29). The vectors **n** and $-\mathbf{n}$ (**n** stands for **normal**) are both orthogonal to **u** and **v**. Which one is in the direction of $\mathbf{u} \times \mathbf{v}$? The answer is given by the **right-hand rule.** If the right hand is placed so that the index finger points in the direction of **u** while the middle finger points in the direction of **v**, then the thumb points in the direction of $\mathbf{u} \times \mathbf{v}$ (see Figure 3.30).

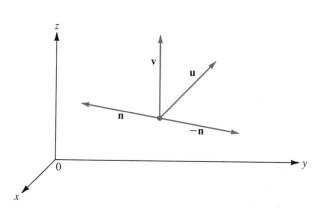

Figure 3.29 There are exactly two vectors **n** and $-\mathbf{n}$ that are orthogonal to two nonparallel vectors **u** and **v** in $\mathbb{R}^3$

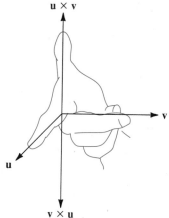

Figure 3.30 The direction of $\mathbf{u} \times \mathbf{v}$ can be determined using the right-hand rule

Having discussed the direction of the vector $\mathbf{u} \times \mathbf{v}$, we now turn to a discussion of its magnitude.

THEOREM 3   If φ is the angle between $\mathbf{u}$ and $\mathbf{v}$, then

$$|\mathbf{u} \times \mathbf{v}| = |\mathbf{u}|\,|\mathbf{v}|\, \sin \varphi \qquad (2)$$

Proof   It is not difficult to show (by comparing components) that $|\mathbf{u} \times \mathbf{v}|^2 = |\mathbf{u}|^2|\mathbf{v}|^2 - (\mathbf{u} \cdot \mathbf{v})^2$ (see Problem 31). Then since $(\mathbf{u} \cdot \mathbf{v})^2 = |\mathbf{u}|^2|\mathbf{v}|^2 \cos^2 \varphi$ (from Theorem 3.3.2, page 179),

$$|\mathbf{u} \times \mathbf{v}|^2 = |\mathbf{u}|^2|\mathbf{v}|^2 - |\mathbf{u}|^2|\mathbf{v}|^2 \cos^2 \varphi = |\mathbf{u}|^2|\mathbf{v}|^2 (1 - \cos^2 \theta)$$
$$= |\mathbf{u}|^2|\mathbf{v}|^2 \sin^2 \varphi$$

and the theorem follows after taking the square root of both sides. Note that $\sin \varphi \geq 0$ because $0 \leq \varphi \leq \pi$.   ∎

There is an interesting geometric interpretation of Theorem 3. The vectors $\mathbf{u}$ and $\mathbf{v}$ are sketched in Figure 3.31 and can be thought of as two adjacent sides of a

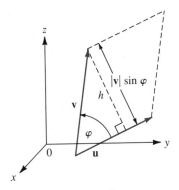

Figure 3.31 φ is the angle between $\mathbf{u}$ and $\mathbf{v}$. $\dfrac{h}{|\mathbf{v}|} = \sin \varphi$, so that $h = |\mathbf{v}| \sin \varphi$

parallelogram. Then, from elementary geometry, we see that

$$\text{Area of the parallelogram} = |\mathbf{u}|\,|\mathbf{v}|\, \sin \varphi = |\mathbf{u} \times \mathbf{v}| \qquad (3)$$

EXAMPLE 3   **Finding the Area of a Parallelogram in $\mathbb{R}^3$**   Find the area of the parallelogram with consecutive vertices at $P = (1, 3, -2)$, $Q = (2, 1, 4)$, and $R = (-3, 1, 6)$.

Solution   The parallelogram is sketched in Figure 3.32. We have

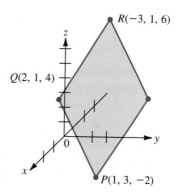

Figure 3.32   A parallelogram in $\mathbb{R}^3$

$$\text{Area} = |\overrightarrow{PQ} \times \overrightarrow{QR}| = |(\mathbf{i} - 2\mathbf{j} + 6\mathbf{k}) \times (-5\mathbf{i} + 2\mathbf{k})|$$

$$= \begin{vmatrix} \mathbf{i} & \mathbf{j} & \mathbf{k} \\ 1 & -2 & 6 \\ -5 & 0 & 2 \end{vmatrix} = |-4\mathbf{i} - 32\mathbf{j} - 10\mathbf{k}| = \sqrt{1140} \text{ square units}$$

∎

Geometric Interpretation of 2 × 2 Determinants (Revisited)

In Section 2.1 (page 116) we discussed the geometric meaning of a 2×2 determinant. We look at the same problem now. Using the cross product, we obtain the result of Section 2.1 much more easily. Let A be a 2×2 matrix and let $\mathbf{u}$ and $\mathbf{v}$ be two 2-vectors. Let $\mathbf{u} = \begin{pmatrix} u_1 \\ u_2 \end{pmatrix}$ and $\mathbf{v} = \begin{pmatrix} v_1 \\ v_2 \end{pmatrix}$. These vectors are given in Figure 3.33.

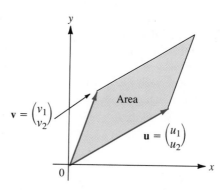

Figure 3.33   The area of the shaded region is the area generated by $\mathbf{u}$ and $\mathbf{v}$

The **area generated** by **u** *and* **v** is defined to be the area of the parallelogram given in the figure. We can think of **u** and **v** as vectors in $\mathbb{R}^3$ lying in the *xy*-plane. Then

$$\mathbf{u} = \begin{pmatrix} u_1 \\ u_2 \\ 0 \end{pmatrix}, \quad \mathbf{v} = \begin{pmatrix} v_1 \\ v_2 \\ 0 \end{pmatrix}, \text{ and}$$

$$\text{Area generated by } \mathbf{u} \text{ and } \mathbf{v} = |\mathbf{u} \times \mathbf{v}| = \begin{vmatrix} \mathbf{i} & \mathbf{j} & \mathbf{k} \\ u_1 & u_2 & 0 \\ v_1 & v_2 & 0 \end{vmatrix}$$

$$= |(u_1 v_2 - u_2 v_1)\mathbf{k}| = |u_1 v_2 - u_2 v_1|^\dagger$$

Now let $A = \begin{pmatrix} a_{11} & a_{12} \\ a_{21} & a_{22} \end{pmatrix}$, $\mathbf{u}' = A\mathbf{u}$, and $\mathbf{v}' = A\mathbf{v}$. Then $\mathbf{u}' = \begin{pmatrix} a_{11}u_1 + a_{12}u_2 \\ a_{21}u_1 + a_{22}u_2 \end{pmatrix}$

and $\mathbf{v}' = \begin{pmatrix} a_{11}v_1 + a_{12}v_2 \\ a_{21}v_1 + a_{22}v_2 \end{pmatrix}$. What is the area generated by $\mathbf{u}'$ and $\mathbf{v}'$? Following the preceding steps, we calculate

Area generated by $\mathbf{u}'$ and $\mathbf{v}' =$

$$|\mathbf{u}' \times \mathbf{v}'| = \left| \begin{vmatrix} \mathbf{i} & \mathbf{j} & \mathbf{k} \\ a_{11}u_1 + a_{12}u_2 & a_{21}u_1 + a_{22}u_2 & 0 \\ a_{11}v_1 + a_{12}v_2 & a_{21}v_1 + a_{22}v_2 & 0 \end{vmatrix} \right|$$

$$= |(a_{11}u_1 + a_{12}u_2)(a_{21}v_1 + a_{22}v_2) - (a_{21}u_1 + a_{22}u_2)(a_{11}v_1 + a_{12}v_2)|$$

It is simple algebra to verify that the last expression is equal to

$$|(a_{11}a_{22} - a_{12}a_{21})(u_1 v_2 - u_2 v_1)| = \pm\det A \text{ (area generated by } \mathbf{u} \text{ and } \mathbf{v})$$

Thus (in this context): *The determinant has the effect of multiplying area.* In Problem 39 you are asked to show that in a certain sense a 3×3 determinant has the effect of multiplying volume.

Geometric Interpretation of the Scalar Triple Product

Let **u**, **v**, and **w** be the three vectors that are not in the same plane. Then they form the sides of a **parallelepiped** in space (see Figure 3.34). Let us compute its volume. The base of the parallelepiped is a parallelogram. Its area, from (3), is equal to $|\mathbf{u} \times \mathbf{v}|$.

The vector $\mathbf{u} \times \mathbf{v}$ is orthogonal to both **u** and **v**, and is therefore orthogonal to the parallelogram determined by **u** and **v**. The height of the parallelepiped, h, is measured along a vector orthogonal to the parallelogram.

$\dagger$ Note that this is the absolute value of $\det \begin{pmatrix} u_1 & v_1 \\ u_2 & v_2 \end{pmatrix}$.

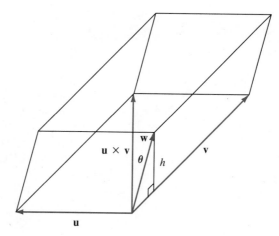

Figure 3.34 Three vectors **u**, **v**, and **w** that are not in the same plane determine a parallelepiped in $\mathbb{R}^3$

From our discussion of projections on page 169, we see that h is the absolute value of the component of **w** in the (orthogonal) direction **u** ⨉ **v**. Thus from equation (10) on page 181

$$h = \text{component of } \mathbf{w} \text{ in the direction } \mathbf{u} \times \mathbf{v} = \left| \frac{\mathbf{w} \cdot (\mathbf{u} \times \mathbf{v})}{|\mathbf{u} \times \mathbf{v}|} \right|$$

Thus

Volume of parallelepiped = area of base ⨉ height

$$= |\mathbf{u} \times \mathbf{v}| \left[\frac{|\mathbf{w} \cdot (\mathbf{u} \times \mathbf{v})|}{|\mathbf{u} \times \mathbf{v}|} \right] = |\mathbf{w} \cdot (\mathbf{u} \times \mathbf{v})|$$

That is,

> The volume of the parallelepiped determined by the three vectors **u**, **v**, and **w** is equal to $|(\mathbf{u} \times \mathbf{v}) \cdot \mathbf{w}|$ = absolute value of the scalar triple product of **u**, **v**, and **w**.
>
> **(4)**

We can derive an interesting and useful fact from (4). If **w** is in the plane of **u** and **v**, then **w** is perpendicular to **u** ⨉ **v**, which means that $\mathbf{w} \cdot (\mathbf{u} \times \mathbf{v}) = 0$. Conversely, if $(\mathbf{u} \times \mathbf{v}) \cdot \mathbf{w} = 0$, then **w** is perpendicular to $(\mathbf{u} \times \mathbf{v})$, so that **w** is in the plane determined by **u** and **v**. We conclude that

> Three vectors **u**, **v**, and **w** are coplanar if and only if their scalar triple product is zero.

Focus on . . .

Josiah Willard Gibbs and the Origins of Vector Analysis

Josiah Willard Gibbs
*(The Granger
Collection,
New York)*

As we have already noted, the study of vectors originated with Hamilton's invention of quaternions. Quaternions were developed by Hamilton and others as mathematical tools for the exploration of physical space. But the results were disappointing because quaternions proved to be too complicated for quick mastery and easy application. Fortunately, there was a solution. Quaternions contained a scalar part and a vector part and difficulties arose when these parts were treated simultaneously. Scientists soon learned that many problems could be dealt with by considering the vector part separately, and the study of vector analysis began.

This work was due principally to the American physicist Josiah Willard Gibbs (1839–1903). As a native of New Haven, Connecticut, Gibbs studied mathematics and physics at Yale University, receiving a doctorate in physics in 1863. He then studied mathematics and physics further in Paris, Berlin, and Heidelberg. In 1871 he was appointed professor of mathematical physics at Yale. He was a highly original physicist who published widely in mathematical physics. Gibbs' book *Vector Analysis* appeared in 1881 and again in 1884. In 1902 he published his *Elementary Principles of Statistical Mechanics*. Students of applied mathematics encounter the curious **Gibbs' phenomenon** of Fourier series.

Gibbs' pioneering book *Vector Analysis* was actually a small pamphlet, printed for private distribution—primarily for the use of his students. Nevertheless, it created great excitement among those who were looking for an alternative to quaternions, and the book soon became widely known. The material was finally turned into a standard book by E. B. Wilson. The book, *Vector Analysis* by Gibbs and Wilson was based on Gibbs' lectures. It was published in 1901.

Gibbs' work is encountered by every student of elementary physics. In introductory physics a vector in space is regarded as a directed line segment, or arrow. Gibbs gave definitions of equality, addition, and multiplication of vectors; these are essentially the definitions given in this chapter. In particular, the vector part of a quaternion was written as $a\mathbf{i} + b\mathbf{j} + c\mathbf{k}$, and this is one way we now depict vectors in $\mathbb{R}^3$.

Gibbs defined the scalar product initially only for the vectors $\mathbf{i}$, $\mathbf{j}$, $\mathbf{k}$:

$$\mathbf{i} \cdot \mathbf{i} = \mathbf{j} \cdot \mathbf{j} = \mathbf{k} \cdot \mathbf{k} = 1$$

$$\mathbf{i} \cdot \mathbf{j} = \mathbf{j} \cdot \mathbf{i} = \mathbf{i} \cdot \mathbf{k} = \mathbf{k} \cdot \mathbf{i} = \mathbf{j} \cdot \mathbf{k} = \mathbf{k} \cdot \mathbf{j} = 0$$

The more general definition followed soon thereafter. Gibbs applied the scalar product in a problem involving force. (Remember, he was first a physicist.) If $\mathbf{F}$ is a force vector of magnitude $|F|$ acting in the direction of the segment $\overrightarrow{0Q}$ (see Figure 3.35), then the effectiveness of this force in pushing an object along the segment $\overrightarrow{0P}$ (i.e., along the vector $\mathbf{u}$) is given by $\mathbf{F} \cdot \mathbf{u}$. If $|\mathbf{u}| = 1$, then $\mathbf{F} \cdot \mathbf{u}$ is

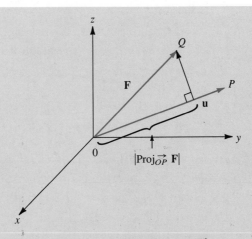

Figure 3.35 The effectiveness of $\mathbf{F}$ in the direction of $\overrightarrow{0P}$ is the component of $\mathbf{F}$ in the direction $\overrightarrow{0P}(=\mathbf{u})$ if $\mathbf{u} = 1$

the component of $\mathbf{F}$ in the direction $\mathbf{u}$. The cross product, too, has physical significance. Suppose that a force vector $\mathbf{F}$ acts at a point P in space in the direction $\overrightarrow{PQ}$. If $\mathbf{u}$ denotes the vector represented by $\overrightarrow{0P}$, then the moment of force exerted by $\mathbf{F}$ around the origin is the vector $\mathbf{u} \times \mathbf{F}$ (see Figure 3.36).

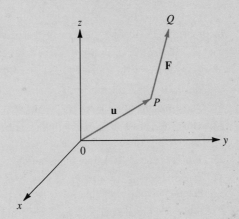

Figure 3.36 The vector $\mathbf{u} \times \mathbf{F}$ is the moment of force around the origin

Both the scalar and cross products of vectors appear prominently in physical applications involving multivariable calculus. These include the famous Maxwell equations of electromagnetism.

In studying mathematics at the end of the twentieth century, we must not lose sight of the fact that much of modern mathematics was developed to solve real-world problems. Vectors were developed by Gibbs and others to make it easier to analyze physical phenomena. In that role they have been hugely successful.

PROBLEMS 3.4

In Problems 1–20 find the cross product $\mathbf{u} \times \mathbf{v}$.

1. $\mathbf{u} = \mathbf{i} - 2\mathbf{j}; \mathbf{v} = 3\mathbf{k}$

2. $\mathbf{u} = 3\mathbf{i} - 7\mathbf{j}; \mathbf{v} = \mathbf{i} + \mathbf{k}$

3. $\mathbf{u} = \mathbf{i} - \mathbf{j}; \mathbf{v} = \mathbf{j} + \mathbf{k}$

4. $\mathbf{u} = -7\mathbf{k}; \mathbf{v} = \mathbf{j} + 2\mathbf{k}$

5. $\mathbf{u} = -2\mathbf{i} + 3\mathbf{j}; \mathbf{v} = 7\mathbf{i} + 4\mathbf{k}$

6. $\mathbf{u} = a\mathbf{i} + b\mathbf{j}; \mathbf{v} = c\mathbf{i} + d\mathbf{j}$

7. $\mathbf{u} = a\mathbf{i} + b\mathbf{k}; \mathbf{v} = c\mathbf{i} + d\mathbf{k}$

8. $\mathbf{u} = a\mathbf{j} + b\mathbf{k}; \mathbf{v} = c\mathbf{i} + d\mathbf{k}$

9. $\mathbf{u} = 2\mathbf{i} - 3\mathbf{j} + \mathbf{k}; \mathbf{v} = \mathbf{i} + 2\mathbf{j} + \mathbf{k}$

10. $\mathbf{u} = 3\mathbf{i} - 4\mathbf{j} + 2\mathbf{k}; \mathbf{v} = 6\mathbf{i} - 3\mathbf{j} + 5\mathbf{k}$

11. $\mathbf{u} = -3\mathbf{i} - 2\mathbf{j} + \mathbf{k}; \mathbf{v} = 6\mathbf{i} + 4\mathbf{j} - 2\mathbf{k}$

12. $\mathbf{u} = \mathbf{i} + 7\mathbf{j} - 3\mathbf{k}; \mathbf{v} = -\mathbf{i} - 7\mathbf{j} + 3\mathbf{k}$

13. $\mathbf{u} = \mathbf{i} - 7\mathbf{j} - 3\mathbf{k}; \mathbf{v} = -\mathbf{i} + 7\mathbf{j} - 3\mathbf{k}$

14. $\mathbf{u} = 2\mathbf{i} - 3\mathbf{j} + 5\mathbf{k}; \mathbf{v} = 3\mathbf{i} - \mathbf{j} - \mathbf{k}$

15. $\mathbf{u} = 10\mathbf{i} + 7\mathbf{j} - 3\mathbf{k}; \mathbf{v} = -3\mathbf{i} + 4\mathbf{j} - 3\mathbf{k}$

16. $\mathbf{u} = 2\mathbf{i} + 4\mathbf{j} - 6\mathbf{k}; \mathbf{v} = -\mathbf{i} - \mathbf{j} + 3\mathbf{k}$

17. $\mathbf{u} = 2\mathbf{i} - \mathbf{j} + \mathbf{k}; \mathbf{v} = 4\mathbf{i} + 2\mathbf{j} + 2\mathbf{k}$

18. $\mathbf{u} = 3\mathbf{i} - \mathbf{j} + 8\mathbf{k}; \mathbf{v} = \mathbf{i} + \mathbf{j} - 4\mathbf{k}$

19. $\mathbf{u} = a\mathbf{i} + a\mathbf{j} + a\mathbf{k}; \mathbf{v} = b\mathbf{i} + b\mathbf{j} + b\mathbf{k}$

20. $\mathbf{u} = a\mathbf{i} + b\mathbf{j} + c\mathbf{k}; \mathbf{v} = a\mathbf{i} + b\mathbf{j} - c\mathbf{k}$

21. Find two unit vectors orthogonal to both $\mathbf{u} = 2\mathbf{i} - 3\mathbf{j}$ and $\mathbf{v} = 4\mathbf{j} + 3\mathbf{k}$.

22. Find two unit vectors orthogonal to both $\mathbf{u} = \mathbf{i} + \mathbf{j} + \mathbf{k}$ and $\mathbf{v} = \mathbf{i} - \mathbf{j} - \mathbf{k}$.

23. Use the cross product to find the sine of the angle φ between the vectors $\mathbf{u} = 2\mathbf{i} + \mathbf{j} - \mathbf{k}$ and $\mathbf{v} = -3\mathbf{i} - 2\mathbf{j} + 4\mathbf{k}$.

24. Use the scalar product to calculate the cosine of the angle φ between the vectors of Problem 23. Then show that for the values you have calculated, $\sin^2 \varphi + \cos^2 \varphi = 1$.

In Problems 25–30 find the area of the parallelogram with the given adjacent vertices.

25. $(1, -2, 3); (2, 0, 1); (0, 4, 0)$

26. $(-2, 1, 1); (2, 2, 3); (-1, -2, 4)$

27. $(-2, 1, 0); (1, 4, 2); (-3, 1, 5)$

28. $(7, -2, -3); (-4, 1, 6); (5, -2, 3)$

29. $(a, 0, 0); (0, b, 0); (0, 0, c)$

30. $(a, b, 0); (a, 0, b); (0, a, b)$

31. Show that $|\mathbf{u} \times \mathbf{v}|^2 = |\mathbf{u}|^2|\mathbf{v}|^2 - (\mathbf{u} \cdot \mathbf{v})^2$. [*Hint:* Write out in terms of components.]

32. Use Properties 1, 4, 2, and 3 (in that order) in Section 2.2 to prove parts (*i*), (*ii*), (*iii*), and (*iv*) of Theorem 2.

33. Prove Theorem 2(*v*) by writing out the components of each side of the equality.

34. Prove Theorem 2(*vi*). [*Hint:* Use parts (*ii*) and (*v*) and the fact that the scalar product is commutative to show that $\mathbf{u} \cdot (\mathbf{u} \times \mathbf{v}) = -\mathbf{u} \cdot (\mathbf{u} \times \mathbf{v})$.]

35. Prove Theorem 2(*vii*). [*Hint:* Use Theorem 3.3.3 on page 180 and Property 6 on page 124.]

36. Show that if $\mathbf{u} = (a_1, b_1, c_1)$, $\mathbf{v} = (a_2, b_2, c_2)$, and $\mathbf{w} = (a_3, b_3, c_3)$, then

$$\mathbf{u} \cdot (\mathbf{v} \times \mathbf{w}) = \begin{vmatrix} a_1 & b_1 & c_1 \\ a_2 & b_2 & c_2 \\ a_3 & b_3 & c_3 \end{vmatrix}$$

37. Calculate the volume of the parallelepiped determined by the vectors $\mathbf{i} - \mathbf{j}$, $3\mathbf{i} + 2\mathbf{k}$, $-7\mathbf{j} + 3\mathbf{k}$.

38. Calculate the volume of the parallelepiped determined by the vectors $\overrightarrow{PQ}$, $\overrightarrow{PR}$, and $\overrightarrow{PS}$, where $P = (2, 1, -1)$, $Q = (-3, 1, 4)$, $R = (-1, 0, 2)$, and $S = (-3, -1, 5)$.

39. The **volume generated by three vectors $\mathbf{u}$, $\mathbf{v}$, and $\mathbf{w}$ in $\mathbb{R}^3$ is defined to be the volume of the parallelepiped whose sides are $\mathbf{u}$, $\mathbf{v}$, and $\mathbf{w}$ (as in Figure 3.34). Let A be a 3×3 matrix and let $\mathbf{u}_1 = A\mathbf{u}$, $\mathbf{v}_1 = A\mathbf{v}$, and $\mathbf{w}_1 = A\mathbf{w}$. Show that

> Volume generated by $\mathbf{u}_1$, $\mathbf{v}_1$, $\mathbf{w}_1 = (\pm\det A)$(volume generated by $\mathbf{u}$, $\mathbf{v}$, $\mathbf{w}$)

This shows that just as the determinant of a 2×2 matrix multiplies area, the determinant of a 3×3 matrix multiplies volume.

40. Let $A = \begin{pmatrix} 2 & 3 & 1 \\ 4 & -1 & 5 \\ 1 & 0 & 6 \end{pmatrix}$, $\mathbf{u} = \begin{pmatrix} 2 \\ -1 \\ 0 \end{pmatrix}$, $\mathbf{v} = \begin{pmatrix} 1 \\ 0 \\ 4 \end{pmatrix}$, and $\mathbf{w} = \begin{pmatrix} -1 \\ 3 \\ 2 \end{pmatrix}$.

    a. Calculate the volume generated by $\mathbf{u}$, $\mathbf{v}$, and $\mathbf{w}$.
    b. Calculate the volume generated by $A\mathbf{u}$, $A\mathbf{v}$, and $A\mathbf{w}$.
    c. Calculate $\det A$.
    d. Show that [volume in part (b)] $= (\pm\det A) \times$ [volume in part (a)].

41. The **triple cross product** of three vectors in $\mathbb{R}^3$ is defined to be the vector $\mathbf{u} \times (\mathbf{v} \times \mathbf{w})$. Show that

$$\mathbf{u} \times (\mathbf{v} \times \mathbf{w}) = (\mathbf{u} \cdot \mathbf{w})\mathbf{v} - (\mathbf{u} \cdot \mathbf{v})\mathbf{w}$$

3.5 LINES AND PLANES IN SPACE

In the plane $\mathbb{R}^2$ we can find the equation of a line if we know either two points on the line or one point and the slope of the line. In $\mathbb{R}^3$ our intuition tells us that the basic ideas are the same. Since two points determine a line, we should be able to calculate the equation of a line in space if we know two points on it. Alternatively, if we know one point and the direction of a line, we should also be able to find its equation.

We begin with two points $P = (x_1, y_1, z_1)$ and $Q = (x_2, y_2, z_2)$ on a line L. A vector parallel to L is a vector with representation $\overrightarrow{PQ}$. Thus

$$\mathbf{v} = (x_2 - x_1)\mathbf{i} + (y_2 - y_1)\mathbf{j} + (z_2 - z_1)\mathbf{k} \tag{1}$$

is a vector parallel to L. Now let $R = (x, y, z)$ be another point on the line. Then $\overrightarrow{PR}$ is parallel to $\overrightarrow{PQ}$, which is parallel to $\mathbf{v}$, so that, by Theorem 3.3.3 on page 180,

$$\overrightarrow{PR} = t\mathbf{v} \tag{2}$$

for some real number t. Now look at Figure 3.37. From this figure we have (in each of the three possible cases)

$$\overrightarrow{OR} = \overrightarrow{OP} + \overrightarrow{PR} \tag{3}$$

And, combining (2) and (3), we get

$$\overrightarrow{PR} = \overrightarrow{OR} - \overrightarrow{OP} = t\mathbf{v}$$

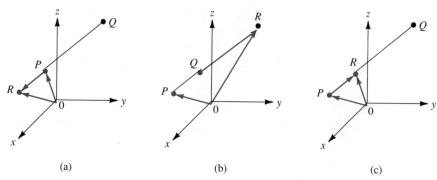

(a)   (b)   (c)

Figure 3.37   In all three cases $\overrightarrow{OR} = \overrightarrow{OP} + \overrightarrow{PR}$

or

$$\overrightarrow{OR} = \overrightarrow{OP} + t\mathbf{v} \tag{4}$$

Vector equation of a line   Equation (4) is called the **vector equation** of the line L. For if R is on L, then (4) is satisfied for some real number t. Conversely, if (4) is satisfied, then, reversing our steps, we see that $\overrightarrow{PR}$ is parallel to $\mathbf{v}$, which means that R is on L.

If we write out the components of equation (4), we obtain

$$x\mathbf{i} + y\mathbf{j} + z\mathbf{k} = x_1\mathbf{i} + y_1\mathbf{j} + z_1\mathbf{k} + t(x_2 - x_1)\mathbf{i} + t(y_2 - y_1)\mathbf{j} + t(z_2 - z_1)\mathbf{k}$$

or

$$\begin{aligned} x &= x_1 + t(x_2 - x_1) \\ y &= y_1 + t(y_2 - y_1) \\ z &= z_1 + t(z_2 - z_1) \end{aligned} \tag{5}$$

Parametric equations of a line   Equations (5) are called **parametric equations** of a line.

Finally, solving for t in (5), and defining $x_2 - x_1 = a$, $y_2 - y_1 = b$, and $z_2 - z_1 = c$, we find that

$$\frac{x - x_1}{a} = \frac{y - y_1}{b} = \frac{z - z_1}{c} \tag{6}$$

Symmetric equations of a line   Equations (6) are called **symmetric equations** of the line. Here a, b, and c are direction numbers of the vector $\mathbf{v}$. Of course, equations (6) are valid only if a, b, and c are nonzero.

EXAMPLE 1   **Determining Equations of a Line**   Find a vector equation, parametric equations, and symmetric equations of the line L passing through the points $P = (2, -1, 6)$ and $Q = (3, 1, -2)$.

Solution   First we calculate $v = (3 - 2)i + [1 - (-1)]j + (-2 - 6)k = i + 2j - 8k$. Then, from (4), if $R = (x, y, z)$ is on the line, we obtain $\overrightarrow{OR} = xi + yj + zk = \overrightarrow{OP} + tv = 2i - j + 6k + t(i + 2j - 8k)$ or

$$x = 2 + t \qquad y = -1 + 2t \qquad z = 6 - 8t \qquad \text{parametric equations}$$

Finally, since $a = 1$, $b = 2$, and $c = -8$, we find the symmetric equations

$$\frac{x - 2}{1} = \frac{y + 1}{2} = \frac{z - 6}{-8} \qquad \text{symmetric equations} \qquad (7)$$

To check this, we verify that $(2, -1, 6)$ and $(3, 1, -2)$ are indeed on the line. We have [after plugging these points into (7)]

$$\frac{2 - 2}{1} = \frac{-1 + 1}{2} = \frac{6 - 6}{-8} = 0$$

$$\frac{3 - 2}{1} = \frac{1 + 1}{2} = \frac{-2 - 6}{-8} = 1$$

Other points on the line can be found. If $t = 3$, for example, we obtain

$$3 = \frac{x - 2}{1} = \frac{y + 1}{2} = \frac{2 - 6}{-8}$$

which yield the point $(5, 5, -18)$.   ■

EXAMPLE 2   **Finding Symmetric Equations of a Line**   Find symmetric equations of the line passing through the point $(1, -2, 4)$ and parallel to the vector $v = i + j - k$.

Solution   We use formula (6) with $P = (x_1, y_1, z_1) = (1, -2, 4)$ and v as above so that $a = 1$, $b = 1$ and $c = -1$. This gives us

$$\frac{x - 1}{1} = \frac{y + 2}{1} = \frac{z - 4}{-1}$$ ■

What happens if one of the direction numbers a, b, or c is zero?

EXAMPLE 3   **Finding Symmetric Equations of a Line When One Direction Number Is Zero**
Find symmetric equations of the line containing the points $P = (3, 4, -1)$ and $Q = (-2, 4, 6)$.

Solution   Here $v = -5i + 7k$ and $a = -5$, $b = 0$, $c = 7$. Then a parametric representation of the line is $x = 3 - 5t$, $y = 4$, and $z = -1 + 7t$. Solving for t, we find that

$$\frac{x - 3}{-5} = \frac{z + 1}{7} \qquad \text{and} \quad y = 4$$

The equation $y = 4$ is the equation of a plane parallel to the xz-plane, so we have obtained an equation of a line in that plane. ∎

EXAMPLE 4   **Finding Symmetric Equations of a Line in the xy-plane**   Find symmetric equations of the line in the xy-plane that passes through the points $(x_1, y_1, 0)$ and $(x_2, y_2, 0)$, where $x_1 \neq x_2$ and $y_1 \neq y_2$.

Solution   Here $\mathbf{v} = (x_2 - x_1)\mathbf{i} + (y_2 - y_1)\mathbf{j}$, and we obtain

$$\frac{x - x_1}{x_2 - x_1} = \frac{y - y_1}{y_2 - y_1} \quad \text{and} \quad z = 0$$

We can rewrite this as

$$y - y_1 = \left(\frac{y_2 - y_1}{x_2 - x_1} \right)(x - x_1)$$

Here $(y_2 - y_1)/(x_2 - x_1) = m$, the slope of the line. When $x = 0$,

$$y = y_1 - [(y_2 - y_1)/(x_2 - x_1)]x_1 = b,$$

the y-intercept of the line. That is, $y = mx + b$, which is the slope-intercept form of a line in the xy-plane. Thus we see that the symmetric equations of a line in space are really a generalization of the slope-intercept equation of a line in the plane. ∎

What happens if two of the direction numbers are zero?

EXAMPLE 5   **Finding Symmetric Equations of a Line When Two Direction Numbers Are Zero**   Find symmetric equations of the line passing through the points $P = (2, 3, -2)$ and $Q = (2, -1, -2)$.

Solution   Here $\mathbf{v} = -4\mathbf{j}$, so that $a = 0$, $b = -4$, and $c = 0$. A parametric representation of the line is, by equation (5), given by $x = 2$, $y = 3 - 4t$, $z = -2$. Now $x = 2$ is the equation of a plane parallel to the yz-plane, whereas $z = -2$ is the equation of a plane parallel to the xy-plane. Their intersection is the line $x = 2$, $z = -2$, which is parallel to the y-axis. In fact, the equation $y = 3 - 4t$ says, essentially, that y can take on any value (while x and z remain fixed). ∎

WARNING   The parametric or symmetric equations of a line are *not* unique. To see this, simply start with two other points on the line. □

EXAMPLE 6   **Illustration That Symmetric Equations of a Line Are Not Unique**   In Example 1 the line contains the point $(5, 5, -18)$. Choose $P = (5, 5, -18)$ and $Q = (3, 1, -2)$. We find that $\mathbf{v} = -2\mathbf{i} - 4\mathbf{j} + 16\mathbf{k}$, so that $x = 5 - 2t$, $y = 5 - 4t$, and $z = -18 + 16t$. (Note that if $t = \frac{3}{2}$, we obtain $(x, y, z) = (2, -1, 6)$.) The symmetric equations are now

$$\frac{x-5}{-2} = \frac{y-5}{-4} = \frac{z+18}{16}$$

∎

The equation of a line in space is obtained by specifying a point on the line and a vector *parallel* to this line. We can derive the equation of a plane in space by specifying a point in the plane and a vector orthogonal to every vector in the plane. This orthogonal vector is called a **normal vector** and is denoted by **n** (see Figure 3.38).

Normal vector

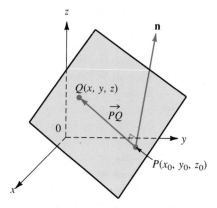

Figure 3.38   The vector **n** is orthogonal to every vector in the plane

DEFINITION 1   **Plane** Let P be a point in space and let **n** be a given nonzero vector. Then the set of all points Q for which $\overrightarrow{PQ} \cdot \mathbf{n} = 0$ constitutes a **plane** in $\mathbb{R}^3$.

Notation. We usually denote a plane by the symbol π.

Let $P = (x_0, y_0, z_0)$ be a fixed point on a plane with normal vector $\mathbf{n} = a\mathbf{i} + b\mathbf{j} + c\mathbf{k}$. If $Q = (x, y, z)$ is any other point on the plane, then $\overrightarrow{PQ} = (x - x_0)\mathbf{i} + (y - y_0)\mathbf{j} + (z - z_0)\mathbf{k}$. Since $\overrightarrow{PQ} \perp \mathbf{n}$, we have $\overrightarrow{PQ} \cdot \mathbf{n} = 0$. But this implies that

$$a(x - x_0) + b(y - y_0) + c(z - z_0) = 0 \qquad \textbf{(8)}$$

A more common way to write the equation of a plane is easily derived from (8):

Standard Equation of a Plane

$$ax + by + cz = d \qquad\qquad \textbf{(9)}$$

where

$$d = ax_0 + by_0 + cz_0 = \overrightarrow{0P} \cdot \mathbf{n}$$

EXAMPLE 7   **Finding an Equation of the Plane Passing Through a Given Point with Given Normal Vector**   Find the plane π passing through the point $(2, 5, 1)$ having the normal vector $\mathbf{n} = \mathbf{i} - 2\mathbf{j} + 3\mathbf{k}$.

Solution   From (8), we immediately obtain $(x - 2) - 2(y - 5) + 3(z - 1) = 0$ or

$$x - 2y + 3z = -5 \tag{10}$$

This plane is sketched in Figure 3.39.

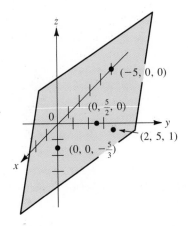

Figure 3.39   The plane $x - 2y + 3z = -5$

Remark. The plane is easily sketched by setting $x = y = 0$ in equation (10) to obtain $(0, 0, -\frac{5}{3})$, $x = z = 0$ to obtain $(0, \frac{5}{2}, 0)$, and $y = z = 0$ to obtain $(-5, 0, 0)$. These three points all lie on the plane.   ∎

The three coordinate planes are represented as follows:

i. The *xy-plane* passes through the origin $(0, 0, 0)$, and any vector lying along the z-axis is normal to it. The simplest such vector is $\mathbf{k}$. Thus, from (8), we obtain $0(x - 0) + 0(y - 0) + 1(z - 0) = 0$, which yields

$$z = 0 \tag{11}$$

as the equation of the xy-plane. (This result should not be very surprising.)

ii. The *xz-plane* has the equation

$$y = 0 \tag{12}$$

iii. The *yz-plane* has the equation

$$x = 0 \tag{13}$$

Three points which are not collinear determine a plane since they determine two nonparallel vectors that intersect at a point (see Figure 3.40).

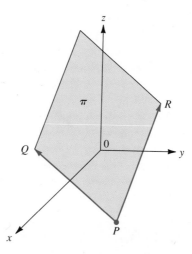

Figure 3.40   The points P, Q, and R determine a plane as long as they are not collinear

EXAMPLE 8

Finding the Equation of a Plane Passing Through Three Given Points   Find the equation of the plane passing through the points $P = (1, 2, 1)$, $Q = (-2, 3, -1)$, and $R = (1, 0, 4)$.

Solution   The vectors $\overrightarrow{PQ} = -3\mathbf{i} + \mathbf{j} - 2\mathbf{k}$ and $\overrightarrow{QR} = 3\mathbf{i} - 3\mathbf{j} + 5\mathbf{k}$ lie on the plane and are therefore orthogonal to the normal vector so that

$$\mathbf{n} = \overrightarrow{PQ} \times \overrightarrow{QR} = \begin{vmatrix} \mathbf{i} & \mathbf{j} & \mathbf{k} \\ -3 & 1 & -2 \\ 3 & -3 & 5 \end{vmatrix} = -\mathbf{i} + 9\mathbf{j} + 6\mathbf{k}$$

and we obtain

$$\pi: \quad -(x - 1) + 9(y - 2) + 6(z - 1) = 0$$

or

$$-x + 9y + 6z = 23$$

Note that if we choose another point, say, Q, we get the equation $-(x + 2) +$

$9(y - 3) + 6(z + 1) = 0$, which reduces to $-x + 9y + 6z = 23$. This plane is sketched in Figure 3.41.

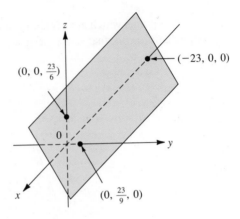

Figure 3.41   The plane $-x + 9y + 6z = 23$

DEFINITION 2   **Parallel Planes**   Two planes are **parallel†** if their normal vectors are parallel, that is, if the cross product of their normal vectors is zero.

Two parallel planes are drawn in Figure 3.42.

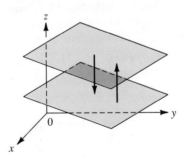

Figure 3.42   Two parallel planes

EXAMPLE 9   **Two Parallel Planes**   The planes π_1: $2x + 3y - z = 3$ and π_2: $-4x - 6y + 2z = 8$ are parallel since $\mathbf{n}_1 = 2\mathbf{i} + 3\mathbf{j} - \mathbf{k}$, $\mathbf{n}_2 = -4\mathbf{i} - 6\mathbf{j} + 2\mathbf{k} = -2\mathbf{n}_1$ (and $\mathbf{n}_1 \times \mathbf{n}_2 = \mathbf{0}$). ∎

---

†Note that two parallel planes could be coincident. For example, the planes $x + y + z = 1$ and $2x + 2y + 2z = 2$ are coincident (the same).

If two planes are not parallel, then they intersect in a straight line.

EXAMPLE 10

Finding Points of Intersection of Planes   Find all points of intersection of the planes $2x - y - z = 3$ and $x + 2y + 3z = 7$.

Solution

When the planes intersect, we have $x + 2y + 3z = 7$ and $2x - y - z = 3$. Solving this system of two equations in three unknowns by row reduction, we obtain, successively,

$$\begin{pmatrix} 1 & 2 & 3 & | & 7 \\ 2 & -1 & -1 & | & 3 \end{pmatrix} \xrightarrow{R_2 \to R_2 - 2R_1} \begin{pmatrix} 1 & 2 & 3 & | & 7 \\ 0 & -5 & -7 & | & -11 \end{pmatrix}$$

$$\xrightarrow{R_2 \to -\frac{1}{5}R_2} \begin{pmatrix} 1 & 2 & 3 & | & 7 \\ 0 & 1 & \frac{7}{5} & | & \frac{11}{5} \end{pmatrix} \xrightarrow{R_1 \to R_1 - 2R_2} \begin{pmatrix} 1 & 0 & \frac{1}{5} & | & \frac{13}{5} \\ 0 & 1 & \frac{7}{5} & | & \frac{11}{5} \end{pmatrix}$$

Thus $y = \frac{11}{5} - (\frac{7}{5})z$ and $x = \frac{13}{5} - (\frac{1}{5})z$. Finally, setting $z = t$, we obtain a parametric representation of the line of intersection: $x = \frac{13}{5} - \frac{1}{5}t$, $y = \frac{11}{5} - \frac{7}{5}t$, and $z = t$. ∎

Another Geometric Way to Look at a Plane

Consider equation (9)

$$ax + by + cz = d$$

where at least one of the numbers a, b, c is nonzero. Let $\mathbf{a} = (a, b, c)$ and $\mathbf{x} = (x, y, z)$. Then (9) can be written

$$\mathbf{a} \cdot \mathbf{x} = d \qquad (14)$$

or, since $\mathbf{a} \neq \mathbf{0}$, we can divide both sides of (14) by $|\mathbf{a}|$ to obtain

$$\frac{\mathbf{a}}{|\mathbf{a}|} \cdot \mathbf{x} = \frac{d}{|\mathbf{a}|} \qquad (15)$$

Let $\mathbf{v} = \mathbf{a}/|\mathbf{a}|$, a unit vector. Then using (15) and equations (9) and (10) on page 181, we obtain

$$\text{Proj}_{\mathbf{v}}\, \mathbf{x} = \frac{\mathbf{v} \cdot \mathbf{x}}{|\mathbf{v}|^2} \overset{\overset{|v|=1}{\downarrow}}{\mathbf{v}} = \frac{d}{|\mathbf{a}|}\mathbf{v} \qquad (16)$$

and the component of $\mathbf{x}$ in the direction $\mathbf{a}$ (= the direction of $\mathbf{v}$) is $d/|\mathbf{a}|$. That is,

> The plane is the set of vectors $\mathbf{x}$ such that the component of $\mathbf{x}$ in the direction $\mathbf{a}$ is $\dfrac{d}{|\mathbf{a}|}$.

This is illustrated in Figure 3.43. Vectors $\mathbf{x}$ such that $\dfrac{\mathbf{a}}{|\mathbf{a}|} \cdot \mathbf{x} > \dfrac{d}{|\mathbf{a}|}$ and $\dfrac{\mathbf{a}}{|\mathbf{a}|} \cdot \mathbf{x} < \dfrac{d}{|\mathbf{a}|}$ line on opposite sides of the plane.

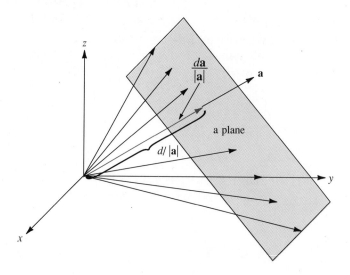

Figure 3.43   The plane consists of all vectors $\mathbf{x}$ such that $|\text{Proj}_{\mathbf{a}}\,\mathbf{x}| = \dfrac{d}{|\mathbf{a}|}$

PROBLEMS   3.5

In Problems 1–14 find a vector equation, parametric equations, and symmetric equations of the indicated line.

1. Containing $(2, 1, 3)$ and $(1, 2, -1)$

2. Containing $(1, -1, 1)$ and $(-1, 1, -1)$

3. Containing $(-4, 1, 3)$ and $(-4, 0, 1)$

4. Containing $(2, 3, -4)$ and $(2, 0, -4)$

5. Containing $(1, 2, 3)$ and $(3, 2, 1)$

6. Containing $(7, 1, 3)$ and $(-1, -2, 3)$

7. Containing $(2, 2, 1)$ and parallel to $2\mathbf{i} - \mathbf{j} - \mathbf{k}$

8. Containing $(-1, -6, 2)$ and parallel to $4\mathbf{i} + \mathbf{j} - 3\mathbf{k}$

9. Containing $(-1, -2, 5)$ and parallel to $-3\mathbf{j} + 7\mathbf{k}$

10. Containing $(-2, 3, -2)$ and parallel to $4\mathbf{k}$

11. Containing (a, b, c) and parallel to $d\mathbf{i} + e\mathbf{j}$

12. Containing (a, b, c) and parallel to $d\mathbf{k}$

13. Containing $(4, 1, -6)$ and parallel to $(x - 2)/3 = (y + 1)/6 = (z - 5)/2$

14. Containing $(3, 1, -2)$ and parallel to $(x + 1)/3 = (y + 3)/2 = (z - 2)/(-4)$

15. Let L_1 be given by

$$\frac{x - x_1}{a_1} = \frac{y - y_1}{b_1} = \frac{z - z_1}{c_1}$$

and L_2 be given by

$$\frac{x - x_1}{a_2} = \frac{y - y_1}{b_2} = \frac{z - z_1}{c_2}$$

Show that L_1 is orthogonal to L_2 if and only if $a_1 a_2 + b_1 b_2 + c_1 c_2 = 0$.

16. Show that the lines

$$L_1: \quad \frac{x - 3}{2} = \frac{y + 1}{4} = \frac{z - 2}{-1} \quad \text{and} \quad L_2: \quad \frac{x - 3}{5} = \frac{y + 1}{-2} = \frac{z - 3}{2}$$

are orthogonal.

17. Show that the lines

$$L_1: \quad \frac{x - 1}{1} = \frac{y + 3}{2} = \frac{z + 3}{3} \quad \text{and} \quad L_2: \quad \frac{x - 3}{3} = \frac{y - 1}{6} = \frac{z - 8}{9}$$

are parallel.

Lines in $\mathbb{R}^3$ that do not have the same direction need not have a point in common.

18. Show that the lines L_1: $\quad x = 1 + t, y = -3 + 2t, z = -2 - t$ and L_2: $\quad x = 17 + 3s$, $y = 4 + s, z = -8 - s$ have the point $(2, -1, -3)$ in common.

19. Show that the lines L_1: $\quad x = 2 - t, y = 1 + t, z = -2t$ and L_2: $\quad x = 1 + s, y = -2s$, $z = 3 + 2s$ do *not* have a point in common.

20. Let L be given in its vector form $\overrightarrow{OR} = \overrightarrow{OP} + t\mathbf{v}$. Find a number t such that $\overrightarrow{OR}$ is perpendicular to $\mathbf{v}$.

21. Use the result of Problem 20 to find the distance between the line L (containing P and parallel to $\mathbf{v}$) and the origin when
   a. $P = (2, 1, -4)$; $\mathbf{v} = \mathbf{i} + \mathbf{j} + \mathbf{k}$
   b. $P = (1, 2, -3)$; $\mathbf{v} = 3\mathbf{i} - \mathbf{j} - \mathbf{k}$
   c. $P = (-1, 4, 2)$; $\mathbf{v} = -\mathbf{i} + \mathbf{j} + 2\mathbf{k}$

In Problems 22–25 find a line L orthogonal to the two given lines and passing through the given point.

22. $\dfrac{x + 2}{-3} = \dfrac{y - 1}{4} = \dfrac{z}{-5}; \dfrac{x - 3}{7} = \dfrac{y + 2}{-2} = \dfrac{z - 8}{3}; (1, -3, 2)$

23. $\dfrac{x - 2}{-4} = \dfrac{y + 3}{-7} = \dfrac{z + 1}{3}; \dfrac{x + 2}{3} = \dfrac{y - 5}{-4} = \dfrac{z + 3}{-2}; (-4, 7, 3)$

24. $x = 3 - 2t; \quad y = 4 + 3t; \quad z = -7 + 5t; \quad x = -2 + 4s, \quad y = 3 - 2s, \quad z = 3 + s;$ $(-2, 3, 4)$

25. $x = 4 + 10t, y = -4 - 8t, z = 3 + 7t; x = -2t, y = 1 + 4t, z = -7 - 3t; (4, 6, 0)$

*26. Calculate the distance between the lines

$$L_1: \quad \frac{x - 2}{3} = \frac{y - 5}{2} = \frac{z - 1}{-1} \quad \text{and} \quad L_2: \quad \frac{x - 4}{-4} = \frac{y - 5}{4} = \frac{z + 2}{1}$$

[*Hint:* The distance is measured along a vector **v** that is perpendicular to both L_1 and L_2. Let P be a point on L_1 and Q a point on L_2. Then the length of the projection of $\overrightarrow{PQ}$ on **v** is the distance between the lines, measured along a vector that is perpendicular to them both.]

*27. Find the distance between the lines

$$L_1: \quad \frac{x+2}{3} = \frac{y-7}{-4} = \frac{z-2}{4} \quad \text{and} \quad L_2: \quad \frac{x-1}{-3} = \frac{y+2}{4} = \frac{z+1}{1}$$

In Problems 28–41 find the equation of the plane.

28. $P = (0, 0, 0)$; $\mathbf{n} = \mathbf{i}$ **29.** $P = (0, 0, 0)$; $\mathbf{n} = \mathbf{j}$

30. $P = (0, 0, 0)$; $\mathbf{n} = \mathbf{k}$ **31.** $P = (1, 2, 3)$; $\mathbf{n} = \mathbf{i} + \mathbf{j}$

32. $P = (1, 2, 3)$; $\mathbf{n} = \mathbf{i} + \mathbf{k}$ **33.** $P = (1, 2, 3)$; $\mathbf{n} = \mathbf{j} + \mathbf{k}$

34. $P = (2, -1, 6)$; $\mathbf{n} = 3\mathbf{i} - \mathbf{j} + 2\mathbf{k}$ **35.** $P = (-4, -7, 5)$; $\mathbf{n} = -3\mathbf{i} - 4\mathbf{j} + \mathbf{k}$

36. $P = (-3, 11, 2)$; $\mathbf{n} = 4\mathbf{i} + \mathbf{j} - 7\mathbf{k}$ **37.** $P = (3, -2, 5)$; $\mathbf{n} = 2\mathbf{i} - 7\mathbf{j} - 8\mathbf{k}$

38. Containing $(1, 2, -4)$, $(2, 3, 7)$, and $(4, -1, 3)$

39. Containing $(-7, 1, 0)$, $(2, -1, 3)$, and $(4, 1, 6)$

40. Containing $(1, 0, 0)$, $(0, 1, 0)$, and $(0, 0, 1)$

41. Containing $(2, 3, -2)$, $(4, -1, -1)$, and $(3, 1, 2)$

Two planes are **orthogonal** if their normal vectors are orthogonal. In Problems 42–46 determine whether the given planes are parallel, orthogonal, coincident (that is, the same), or none of these.

42. π_1: $x + y + z = 2$; π_2: $2x + 2y + 2z = 4$

43. π_1: $x - y + z = 3$; π_2: $-3x + 3y - 3z = -9$

44. π_1: $2x - y + z = 3$; π_2: $x + y - z = 7$

45. π_1: $2x - y + z = 3$; π_2: $x + y + z = 3$

46. π_1: $3x - 2y + 7z = 4$; π_2: $-2x + 4y + 2z = 16$

In Problems 47–49 find the equation of the set of all points of intersection of the two planes.

47. π_1: $x - y + z = 2$; π_2: $2x - 3y + 4z = 7$

48. π_1: $3x - y + 4z = 3$; π_2: $-4x - 2y + 7z = 8$

49. π_1: $-2x - y + 17z = 4$; π_2: $2x - y - z = -7$

*50. Let π be a plane, P a point on the plane, $\mathbf{n}$ a vector normal to the plane, and Q a point not on the plane (see Figure 3.44). Show that the perpendicular distance D from Q to the

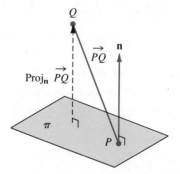

Figure 3.44

plane is given by

$$D = |\text{proj}_{\mathbf{n}} \overrightarrow{PQ}| = \frac{|\overrightarrow{PQ} \cdot \mathbf{n}|}{|\mathbf{n}|}$$

In Problems 51–53 find the distance from the given point to the given plane.

51. $(4, 0, 1)$; $2x - y + 8z = 3$ **52.** $(-7, -2, -1)$; $-2x + 8z = -5$

53. $(-3, 0, 2)$; $-3x + y + 5z = 0$

54. Prove that the distance between the plane $ax + by + cz = d$ and the point (x_0, y_0, z_0) is given by

$$D = \frac{|ax_0 + by_0 + cz_0 - d|}{\sqrt{a^2 + b^2 + c^2}}$$

The **angle between two planes** is defined to be the acute† angle between their normal vectors. In Problems 55–57 find the angle between the two planes.

55. The two planes of Problem 47   **56.** The two planes of Problem 48

57. The two planes of Problem 49

*58. Let $\mathbf{u}$ and $\mathbf{v}$ be two nonparallel, nonzero vectors in a plane π. Show that if $\mathbf{w}$ is any other vector in π, then there exist scalars α and β such that $\mathbf{w} = \alpha\mathbf{u} + \beta\mathbf{v}$. This is called the **parametric representation** of the plane π. [*Hint:* Draw a parallelogram in which $\alpha\mathbf{u}$ and $\beta\mathbf{v}$ form adjacent sides and the diagonal vector is $\mathbf{w}$.]

*59. Three vectors $\mathbf{u}$, $\mathbf{v}$, and $\mathbf{w}$ are called **coplanar** if they all lie in the same plane π. Show that if $\mathbf{u}$, $\mathbf{v}$ and $\mathbf{w}$ all pass through the origin, then they are coplanar if and only if the scalar triple product equals zero: $\mathbf{u} \cdot (\mathbf{v} \times \mathbf{w}) = 0$.

In Problems 60–64 determine whether the three given position vectors (that is, one endpoint at the origin) are coplanar. If they are coplanar, find the equation of the plane containing them.

60. $\mathbf{u} = 2\mathbf{i} - 3\mathbf{j} + 4\mathbf{k}$; $\mathbf{v} = 7\mathbf{i} - 2\mathbf{j} + 3\mathbf{k}$; $\mathbf{w} = 9\mathbf{i} - 5\mathbf{j} + 7\mathbf{k}$

61. $\mathbf{u} = -3\mathbf{i} + \mathbf{j} + 8\mathbf{k}$; $\mathbf{v} = -2\mathbf{i} - 3\mathbf{j} + 5\mathbf{k}$; $\mathbf{w} = 2\mathbf{i} + 14\mathbf{j} - 4\mathbf{k}$

62. $\mathbf{u} = 2\mathbf{i} + \mathbf{j} - 2\mathbf{k}$; $\mathbf{v} = 2\mathbf{i} - \mathbf{j} - 2\mathbf{k}$; $\mathbf{w} = 2\mathbf{i} - \mathbf{j} + 2\mathbf{k}$

63. $\mathbf{u} = 3\mathbf{i} - 2\mathbf{j} + \mathbf{k}$; $\mathbf{v} = \mathbf{i} + \mathbf{j} - 5\mathbf{k}$; $\mathbf{w} = -\mathbf{i} + 5\mathbf{j} - 16\mathbf{k}$

64. $\mathbf{u} = 2\mathbf{i} - \mathbf{j} - \mathbf{k}$; $\mathbf{v} = 4\mathbf{i} + 3\mathbf{j} + 2\mathbf{k}$; $\mathbf{w} = 6\mathbf{i} + 7\mathbf{j} + 5\mathbf{k}$

SUMMARY

- The **directed line segment** extending from P to Q in $\mathbb{R}^2$ or $\mathbb{R}^3$, denoted by $\overrightarrow{PQ}$, is the straight-line segment that extends from P to Q.   (pp. 154, 174)
- Two directed line segments in $\mathbb{R}^2$ or $\mathbb{R}^3$ are **equivalent** if they have the same magnitude (length) and direction.   (pp. 155, 174)
- **Geometric Definition of a Vector** · A **vector** in $\mathbb{R}^2(\mathbb{R}^3)$ is the set of all directed line segments in $\mathbb{R}^3(\mathbb{R}^3)$ equivalent to a given directed line segment. One representation of a vector has its initial point at the origin as is denoted by $\overrightarrow{0P}$.   (pp. 155, 174)

---

†Recall that an acute angle α is an angle between $0°$ and $90°$; that is, $\alpha \in (0, \pi/2)$.

- **Algebraic Definition of a Vector.** A **vector v** in the *xy*-plane ($\mathbb{R}^2$) is an ordered pair of real numbers (a, b). The numbers a and b are called the **components** of the vector **v**. The **zero vector** is the vector $(0, 0)$. In $\mathbb{R}^3$, a vector **v** is an **ordered triple** of real numbers (a, b, c). The **zero vector** in $\mathbb{R}^3$ is the vector $(0, 0, 0)$.   (pp. 156, 174)
- The geometric and algebraic definitions of a vector in $\mathbb{R}^2[\mathbb{R}^3]$ are related in the following way: If $\mathbf{v} = (a, b)[(a, b, c)]$, then one representation of **v** is $\overrightarrow{OR}$ where $R = (a, b)[R = (a, b, c)]$.   (p. 156)
- If $\mathbf{v} = (a, b)$, then the **magnitude of v**, denoted by $|\mathbf{v}|$ is given by $|\mathbf{v}| = \sqrt{a^2 + b^2}$. If $\mathbf{v} = (a, b, c)$, then $|\mathbf{v}| = \sqrt{a^2 + b^2 + c^2}$.   (pp. 156, 174)
- If **v** is a vector in $\mathbb{R}^2$, then the **direction of v** is the angle in $[0, 2\pi)$ that any representation of **v** makes with the positive *x*-axis.   (p. 157)
- **Triangle Inequality.** In $\mathbb{R}^2$ or $\mathbb{R}^3$

$$|\mathbf{u} + \mathbf{v}| \le |\mathbf{u}| + |\mathbf{v}|$$ (p. 160)

- In $\mathbb{R}^2$ let $\mathbf{i} = (1, 0)$ and $\mathbf{j} = (0, 1)$. Then $\mathbf{v} = (a, b)$ can be written $\mathbf{v} = a\mathbf{i} + b\mathbf{j}$ (p. 161)
- In $\mathbb{R}^3$ let $\mathbf{i} = (1, 0, 0)$, $\mathbf{j} = (0, 1, 0)$, and $\mathbf{k} = (0, 0, 1)$. Then $\mathbf{v} = (a, b, c)$ can be written   (p. 179)

$$\mathbf{v} = a\mathbf{i} + b\mathbf{j} + c\mathbf{k}$$

- A **unit vector u** in $\mathbb{R}^2$ or $\mathbb{R}^3$ is a vector which satisfies $|\mathbf{u}| = 1$. In $\mathbb{R}^2$ a unit vector can be written   (pp. 161, 175)

$$\mathbf{u} = (\cos\,\theta)\mathbf{i} + (\sin\,\theta)\mathbf{j}$$

where θ is the direction of **u**.
- Let $\mathbf{u} = (a_1, b_1)$ and $\mathbf{v} = (a_2, b_2)$. Then the **scalar product** or **dot product** of **u** and **v**, written $\mathbf{u} \cdot \mathbf{v}$, is given by   (p. 164)

$$\mathbf{u} \cdot \mathbf{v} = a_1a_2 + b_1b_2$$

If $\mathbf{u} = (a_1, b_1, c_1)$ and $\mathbf{v} = (a_2, b_2, c_2)$, then

$$\mathbf{u} \cdot \mathbf{v} = a_1a_2 + b_1b_2 + c_1c_2$$

- The **angle** φ between two vectors **u** and **v** in $\mathbb{R}^2$ or $\mathbb{R}^3$ is the unique number in $[0, \pi]$ that satisfies   (pp. 164, 179)

$$\cos\,\varphi = \frac{\mathbf{u} \cdot \mathbf{v}}{|\mathbf{u}|\,|\mathbf{v}|}$$

- Two vectors in $\mathbb{R}^2$ or $\mathbb{R}^3$ are **parallel** if the angle between them is 0 or π. They are parallel if and only if one is a scalar multiple of the other.   (pp. 167, 180)
- Two vectors in $\mathbb{R}^2$ or $\mathbb{R}^3$ are **orthogonal** if the angle between them is $\pi/2$. They are orthogonal if and only if their scalar product is zero.   (pp. 167, 180)
- Let **u** and **v** be two nonzero vectors in $\mathbb{R}^2$ or $\mathbb{R}^3$. Then the **projection** of **u** on **v** is a vector, denoted by $\text{Proj}_{\mathbf{v}}\,\mathbf{u}$, which is defined by   (pp. 169, 181)

$$\text{Proj}_{\mathbf{v}}\,\mathbf{u} = \frac{\mathbf{u} \cdot \mathbf{v}}{|\mathbf{v}|^2}\mathbf{v}$$

The vector $\dfrac{\mathbf{u} \cdot \mathbf{v}}{|\mathbf{v}|}$ is called the **component** of **u** in the direction **v**.

- $\text{Proj}_{\mathbf{v}}\,\mathbf{u}$ is parallel to **v** and $\mathbf{u} - \text{proj}_{\mathbf{v}}\,\mathbf{u}$ is orthogonal to **v**.   (pp. 169, 181)
- The **direction** of a vector **v** in $\mathbb{R}^3$ is the unit vector   (p. 176)

$$u = \frac{v}{|v|}$$

- If $v = (a, b, c)$, then $\cos \alpha = \dfrac{a}{|v|}$, $\cos \beta \dfrac{b}{|v|}$ and $\cos \gamma = \dfrac{c}{|v|}$ are called the **direction cosines** of v.

(p. 177)

- Let $u = a_1 i + b_1 j + c_1 k$ and $v = a_2 i + b_2 j + c_2 k$. Then the **cross product** or **vector product** of u and v, denoted by $u \times v$, is given by

(pp. 183, 184)

$$u \times v = \begin{vmatrix} i & j & k \\ a_1 & b_1 & c_1 \\ a_2 & b_2 & c_2 \end{vmatrix}$$

- **Properties of the Cross Product**

(p. 185)

  i. $u \times 0 = 0 \times u = 0$.

  ii. $u \times v = -v \times u$.

  iii. $(\alpha u) \times v = \alpha(u \times v)$.

  iv. $u \times (v + w) = (u \times v) + (u \times w)$.

  v. $(u \times v) \cdot w = u \cdot (v \times w)$ (the **scalar triple product**).

  vi. $u \times v$ is orthogonal to both u and v.

  vii. If u and v are parallel, then $u \times v = 0$.

- If φ is the angle between u and v, then

(p. 186)

$|u \times v| = |u|\,|v| \sin \varphi = $ area of the parallelogram with sides u and v

- Let $P = (x_1, y_1, z_1)$ and $Q = (x_2, y_2, z_2)$ be two points on a line L in $\mathbb{R}^3$. Let $v = (x_2 - x_1)i + (y_2 - y_1)j + (z_2 - z_1)k$ and let $a = x_2 - x_1$, $b = y_2 - y_1$, and $c = z_2 - z_1$.

Vector equation of the line: $\overrightarrow{OR} = \overrightarrow{OP} + tv$

(p. 194)

Parametric equations of the line: $x = x_1 + at$

$$y = y_1 + bt$$

(p. 194)

$$z = z_1 + ct$$

Symmetric equations of the line: $\dfrac{x - x_1}{a} = \dfrac{y - y_1}{b} = \dfrac{z - z_1}{c}$, if a, b, and c are nonzero.

(p. 194)

- Let P be a point in $\mathbb{R}^3$ and let n be a given nonzero vector. Then the set of all points Q for which $\overrightarrow{PQ} \cdot n = 0$ constitutes a plane in $\mathbb{R}^3$. The vector n is called the **normal vector** of the plane. (p. 197)
- If $n = ai + bj + ck$ and $P = (x_0, y_0, z_0)$, then the equation of the plane can be written (p. 197)

$$ax + by + cz = d$$

where

$$d = ax_0 + by_0 + cz_0 = \overrightarrow{OP} \cdot n$$

- The **xy-plane** has the equation $z = 0$; the **xz-plane** has the equation $y = 0$; the **yz-plane** has the equation $x = 0$.

(pp. 198, 199)

- Two planes are **parallel** if their normal vectors are parallel. If two planes are not parallel, then they intersect in a straight line.

(p. 200)

REVIEW EXERCISES

In Exercises 1–6 find the magnitude and direction of the given vector.

1. $\mathbf{v} = (3, 3)$ **2.** $\mathbf{v} = -3\mathbf{i} + 3\mathbf{j}$ **3.** $\mathbf{v} = (2, -2\sqrt{3})$

4. $\mathbf{v} = (\sqrt{3}, 1)$ **5.** $\mathbf{v} = -12\mathbf{i} - 12\mathbf{j}$ **6.** $\mathbf{v} = \mathbf{i} + 4\mathbf{j}$

In Exercises 7–10 write the vector $\mathbf{v}$ that is represented by $\overrightarrow{PQ}$ in the form $a\mathbf{i} + b\mathbf{j}$. Sketch $\overrightarrow{PQ}$ and $\mathbf{v}$.

7. $P = (2, 3); Q = (4, 5)$ **8.** $P = (1, -2); Q = (7, 12)$

9. $P = (-1, -6); Q = (3, -4)$ **10.** $P = (-1, 3); Q = (3, -1)$

11. Let $\mathbf{u} = (2, 1)$ and $\mathbf{v} = (-3, 4)$. Find **(a)** $5\mathbf{u}$; **(b)** $\mathbf{u} - \mathbf{v}$; **(c)** $-8\mathbf{u} + 5\mathbf{v}$.

12. Let $\mathbf{u} = -4\mathbf{i} + \mathbf{j}$ and $\mathbf{v} = -3\mathbf{i} - 4\mathbf{j}$. Find **(a)** $-3\mathbf{v}$; **(b)** $\mathbf{u} + \mathbf{v}$; **(c)** $3\mathbf{u} - 6\mathbf{v}$.

In Exercises 13–19 find a unit vector having the same direction as the given vector.

13. $\mathbf{v} = \mathbf{i} + \mathbf{j}$ **14.** $\mathbf{v} = -\mathbf{i} + \mathbf{j}$ **15.** $\mathbf{v} = 2\mathbf{i} + 5\mathbf{j}$

16. $\mathbf{v} = -7\mathbf{i} + 3\mathbf{j}$ **17.** $\mathbf{v} = 3\mathbf{i} + 4\mathbf{j}$ **18.** $\mathbf{v} = -2\mathbf{i} - 2\mathbf{j}$

19. $\mathbf{v} = a\mathbf{i} - a\mathbf{j}$

20. If $\mathbf{v} = 4\mathbf{i} - 7\mathbf{j}$, find $\sin \theta$ and $\cos \theta$, where θ is the direction of $\mathbf{v}$.

21. Find a unit vector with direction opposite to that of $\mathbf{v} = 5\mathbf{i} + 2\mathbf{j}$.

22. Find two unit vectors orthogonal to $\mathbf{v} = \mathbf{i} - \mathbf{j}$.

23. Find a unit vector with direction opposite to that of $\mathbf{v} = 10\mathbf{i} - 7\mathbf{j}$.

In Exercises 24–27 find a vector $\mathbf{v}$ having the given magnitude and direction.

24. $|\mathbf{v}| = 2; \theta = \pi/3$ **25.** $|\mathbf{v}| = 1; \theta = \pi/2$

26. $|\mathbf{v}| = 4; \theta = \pi$ **27.** $|\mathbf{v}| = 7; \theta = 5\pi/6$

In Exercises 28–31 calculate the scalar product of the two vectors and the cosine of the angle between them.

28. $\mathbf{u} = \mathbf{i} - \mathbf{j}; \mathbf{v} = \mathbf{i} + 2\mathbf{j}$ **29.** $\mathbf{u} = -4\mathbf{i}; \mathbf{v} = 11\mathbf{j}$

30. $\mathbf{u} = 4\mathbf{i} - 7\mathbf{j}; \mathbf{v} = 5\mathbf{i} + 6\mathbf{j}$ **31.** $\mathbf{u} = -\mathbf{i} - 2\mathbf{j}; \mathbf{v} = 4\mathbf{i} + 5\mathbf{j}$

In Exercises 32–37 determine whether the given vectors are orthogonal, parallel, or neither. Then sketch each pair.

32. $\mathbf{u} = 2\mathbf{i} - 6\mathbf{j}; \mathbf{v} = -\mathbf{i} + 3\mathbf{j}$ **33.** $\mathbf{u} = 4\mathbf{i} - 5\mathbf{j}; \mathbf{v} = 5\mathbf{i} - 4\mathbf{j}$

34. $\mathbf{u} = 4\mathbf{i} - 5\mathbf{j}; \mathbf{v} = -5\mathbf{i} + 4\mathbf{j}$ **35.** $\mathbf{u} = -7\mathbf{i} - 7\mathbf{j}; \mathbf{v} = \mathbf{i} + \mathbf{j}$

36. $\mathbf{u} = -7\mathbf{i} - 7\mathbf{j}; \mathbf{v} = -\mathbf{i} + \mathbf{j}$ **37.** $\mathbf{u} = -7\mathbf{i} - 7\mathbf{j}; \mathbf{v} = -\mathbf{i} - \mathbf{j}$

38. Let $\mathbf{u} = 2\mathbf{i} + 3\mathbf{j}$ and $\mathbf{v} = 4\mathbf{i} + \alpha\mathbf{j}$. Determine α such that
   a. $\mathbf{u}$ and $\mathbf{v}$ are orthogonal.
   b. $\mathbf{u}$ and $\mathbf{v}$ are parallel.
   c. The angle between $\mathbf{u}$ and $\mathbf{v}$ is $\pi/4$.
   d. The angle between $\mathbf{u}$ and $\mathbf{v}$ is $\pi/6$.

In Exercises 39–44 calculate $\text{proj}_v \, \mathbf{u}$.

39. $\mathbf{u} = 14\mathbf{i}; \mathbf{v} = \mathbf{i} + \mathbf{j}$

40. $\mathbf{u} = 14\mathbf{i}, \mathbf{v} = \mathbf{i} - \mathbf{j}$

41. $\mathbf{u} = 3\mathbf{i} - 2\mathbf{j}; \mathbf{v} = 3\mathbf{i} + 2\mathbf{j}$

42. $\mathbf{u} = 3\mathbf{i} + 2\mathbf{j}; \mathbf{v} = \mathbf{i} - 3\mathbf{j}$

43. $\mathbf{u} = 2\mathbf{i} - 5\mathbf{j}; \mathbf{v} = -3\mathbf{i} - 7\mathbf{j}$

44. $\mathbf{u} = 4\mathbf{i} - 5\mathbf{j}; \mathbf{v} = -3\mathbf{i} - \mathbf{j}$

45. Let $P = (3, -2)$, $Q = (4, 7)$, $R = (-1, 3)$, and $S = (2, -1)$. Calculate $\text{proj}_{\overrightarrow{PQ}} \, \overrightarrow{RS}$ and $\text{proj}_{\overrightarrow{RS}} \, \overrightarrow{PQ}$.

In Exercises 46–48 find the distance between the two given points.

46. $(4, -1, 7); (-5, 1, 3)$

47. $(-2, 4, -8); (0, 0, 6)$

48. $(2, -7, 0); (0, 5, -8)$

In Exercises 49–51 find the magnitude and the direction cosines of the given vector.

49. $\mathbf{v} = 3\mathbf{j} + 11\mathbf{k}$

50. $\mathbf{v} = \mathbf{i} - 2\mathbf{j} - 3\mathbf{k}$

51. $\mathbf{v} = -4\mathbf{i} + \mathbf{j} + 6\mathbf{k}$

52. Find a unit vector in the direction of $\overrightarrow{PQ}$, where $P = (3, -1, 2)$ and $Q = (-4, 1, 7)$.

53. Find a unit vector whose direction is opposite that of $\overrightarrow{PQ}$, where $P = (1, -3, 0)$ and $Q = (-7, 1, -4)$.

In Exercises 54–61 let $\mathbf{u} = \mathbf{i} - 2\mathbf{j} + 3\mathbf{k}$, $\mathbf{v} = -3\mathbf{i} + 2\mathbf{j} + 5\mathbf{k}$, and $\mathbf{w} = 2\mathbf{i} - 4\mathbf{j} + \mathbf{k}$. Calculate

54. $\mathbf{u} - \mathbf{v}$

55. $3\mathbf{v} + 5\mathbf{w}$

56. $\text{proj}_v \, \mathbf{w}$

57. $\text{proj}_w \, \mathbf{u}$

58. $2\mathbf{u} - 4\mathbf{v} + 7\mathbf{w}$

59. $\mathbf{u} \cdot \mathbf{w} - \mathbf{w} \cdot \mathbf{v}$

60. The angle between $\mathbf{u}$ and $\mathbf{v}$

61. The angle between $\mathbf{v}$ and $\mathbf{w}$

In Exercises 62–65 find the cross product $\mathbf{u} \times \mathbf{v}$.

62. $\mathbf{u} = 3\mathbf{i} - \mathbf{j}; \mathbf{v} = 2\mathbf{i} + 4\mathbf{k}$

63. $\mathbf{u} = 7\mathbf{j}; \mathbf{v} = \mathbf{i} - \mathbf{k}$

64. $\mathbf{u} = 4\mathbf{i} - \mathbf{j} + 7\mathbf{k}; \mathbf{v} = -7\mathbf{i} + \mathbf{j} - 2\mathbf{k}$

65. $\mathbf{u} = -2\mathbf{i} + 3\mathbf{j} - 4\mathbf{k};$ $\mathbf{v} = -3\mathbf{i} + \mathbf{j} - 10\mathbf{k}$

66. Find two unit vectors orthogonal to both $\mathbf{u} = \mathbf{i} - \mathbf{j} + 3\mathbf{k}$ and $\mathbf{v} = -2\mathbf{i} - 3\mathbf{j} + 4\mathbf{k}$.

67. Calculate the area of the parallelogram with the adjacent vertices $(1, 4, -2)$, $(-3, 1, 6)$, and $(1, -2, 3)$.

In Exercises 68–71 find a vector equation, parametric equations, and symmetric equations of the given line.

68. Containing $(3, -1, 4)$ and $(-1, 6, 2)$

69. Containing $(-4, 1, 0)$ and $(3, 0, 7)$

70. Containing $(3, 1, 2)$ and parallel to $3\mathbf{i} - \mathbf{j} - \mathbf{k}$

71. Containing $(1, -2, -3)$ and parallel to $(x + 1)/5 = (y - 2)/(-3) = (z - 4)/2$

72. Show that the lines L_1: $x = 3 - 2t$, $y = 4 + t$, $z = -2 + 7t$ and L_2: $x = -3 + s$, $y = 2 - 4s$, $z = 1 + 6s$ have no points of intersection.

73. Find the distance from the origin to the line passing through the point $(3, 1, 5)$ and having the direction $\mathbf{v} = 2\mathbf{i} - \mathbf{j} + \mathbf{k}$.

74. Find the equation of the line passing through $(-1, 2, 4)$ and orthogonal to L_1: $(x - 1)/4 = (y + 6)/3 = z/(-2)$ and L_2: $(x + 3)/5 = (y - 1)/1 = (z + 3)/4$.

In Exercises 75–77 find the equation of the plane containing the given point and orthogonal to the given normal vector.

75. $P = (1, 3, -2)$; $\mathbf{n} = \mathbf{i} + \mathbf{k}$

76. $P = (1, -4, 6)$; $\mathbf{n} = 2\mathbf{j} - 3\mathbf{k}$

77. $P = (-4, 1, 6)$; $\mathbf{n} = 2\mathbf{i} - 3\mathbf{j} + 5\mathbf{k}$

78. Find the equation of the plane containing the points $(-2, 4, 1)$, $(3, -7, 5)$, and $(-1, -2, -1)$.

79. Find all points of intersection of the planes π_1: $-x + y + z = 3$ and π_2: $-4x + 2y - 7z = 5$.

80. Find all points of intersection of the planes π_1: $-4x + 6y + 8z = 12$ and π_2: $2x - 3y - 4z = 5$.

81. Find all points of intersection of the planes π_1: $3x - y + 4z = 8$ and π_2: $-3x - y - 11z = 0$.

82. Find the distance from $(1, -2, 3)$ to the plane $2x - y - z = 6$.

83. Find the angle between the planes of Exercise 79.

84. Show that the position vectors $\mathbf{u} = \mathbf{i} - 2\mathbf{j} + \mathbf{k}$, $\mathbf{v} = 3\mathbf{i} + 2\mathbf{j} - 3\mathbf{k}$, and $\mathbf{w} = 9\mathbf{i} - 2\mathbf{j} - 3\mathbf{k}$ are coplanar and find the equation of the plane containing them.

4

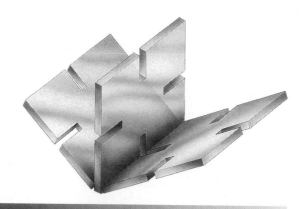

Vector Spaces

4.1 INTRODUCTION

As we saw in the last chapter, the sets $\mathbb{R}^2$ (vectors in the plane) and $\mathbb{R}^3$ (vectors in space) have a number of nice properties. We can add two vectors in $\mathbb{R}^2$ and obtain another vector in $\mathbb{R}^2$. Under addition, vectors in $\mathbb{R}^2$ commute and obey the associative law. If $\mathbf{x} \in \mathbb{R}^2$, then $\mathbf{x} + \mathbf{0} = \mathbf{x}$ and $\mathbf{x} + (-\mathbf{x}) = \mathbf{0}$. We can multiply vectors in $\mathbb{R}^2$ by scalars and obtain a number of distributive laws. The same properties also hold in $\mathbb{R}^3$.

The sets $\mathbb{R}^2$ and $\mathbb{R}^3$ are called *vector spaces*. Intuitively, we can say that a vector space is a set of objects that obey the rules described in the previous paragraph.

In this chapter we make a seemingly great leap from the concrete world of solving equations and dealing with easily visualized vectors to the abstract world of arbitrary vector spaces. There is a great advantage in doing so. Once we have established a fact about vector spaces in general, we can apply that fact to *every* vector space. Otherwise, we would have to prove that fact again and again, once for each new vector space we encounter (and there is an endless supply of them). But, as you will see, many of the abstract theorems we shall prove are really no more difficult than the ones already encountered.

4.2 DEFINITION AND BASIC PROPERTIES

DEFINITION 1   **Real Vector Space** A **real vector space** V is a set of objects, called **vectors,** together with two operations called **addition** and **scalar multiplication** that satisfy the ten axioms listed below.

Notation. If **x** and **y** are in V and if α is a real number, then we write $\mathbf{x} + \mathbf{y}$ for the sum of **x** and **y** and $\alpha\mathbf{x}$ for the scalar product of α and **x**.

Before we list the properties satisfied by vectors in a vector space, two things should be mentioned. First, while it might be helpful to think of $\mathbb{R}^2$ or $\mathbb{R}^3$ when dealing with a vector space, it often occurs that a vector space may appear to be very different from these comfortable spaces. (We shall see this shortly.) Second, Definition 1 gives a definition of a *real* vector space. The word "real" means that the scalars we use are real numbers. It would be just as easy to define a *complex* vector space by using complex numbers instead of real ones. This book deals primarily with real vector spaces, but generalizations to other sets of scalars present little difficulty.

AXIOMS OF A VECTOR SPACE

i. If $\mathbf{x} \in V$ and $\mathbf{y} \in V$, then $\mathbf{x} + \mathbf{y} \in V$ (**closure under addition**).

ii. For all **x**, **y**, and **z** in V, $(\mathbf{x} + \mathbf{y}) + \mathbf{z} = \mathbf{x} + (\mathbf{y} + \mathbf{z})$ (**associative law of vector addition**).

iii. There is a vector $\mathbf{0} \in V$ such that for all $\mathbf{x} \in V$, $\mathbf{x} + \mathbf{0} = \mathbf{0} + \mathbf{x} = \mathbf{x}$ (**0** is called the **zero vector** or **additive identity**).

iv. If $\mathbf{x} \in V$, there is a vector $-\mathbf{x}$ in V such that $\mathbf{x} + (-\mathbf{x}) = \mathbf{0}$ ($-\mathbf{x}$ is called the **additive inverse** of **x**).

v. If **x** and **y** are in V, then $\mathbf{x} + \mathbf{y} = \mathbf{y} + \mathbf{x}$ (**commutative law of vector addition**).

vi. If $\mathbf{x} \in V$ and α is a scalar, then $\alpha\mathbf{x} \in V$ (**closure under scalar multiplication**).

vii. If **x** and **y** are in V and α is a scalar, then $\alpha(\mathbf{x} + \mathbf{y}) = \alpha\mathbf{x} + \alpha\mathbf{y}$ (**first distributive law**).

viii. If $\mathbf{x} \in V$ and α and β are scalars, then $(\alpha + \beta)\mathbf{x} = \alpha\mathbf{x} + \beta\mathbf{x}$ (**second distributive law**).

ix. If $\mathbf{x} \in V$ and α and β are scalars, then $\alpha(\beta\mathbf{x}) = (\alpha\beta)\mathbf{x}$ (**associative law of scalar multiplication**).

x. For every vector $\mathbf{x} \in V$, $1\mathbf{x} = \mathbf{x}$ (the scalar 1 is called a **multiplicative identity**).

EXAMPLE 1   **The Space $\mathbb{R}^n$**

Let $V = \mathbb{R}^n = \{(x_1, x_2, \ldots, x_n): x_i \in \mathbb{R} \text{ for } i = 1, 2, \ldots, n\}$.

From Section 1.5 (see Theorem 1.5.1, page 32) we see that V satisfies all the axioms of a vector space if we take the set of scalars to be $\mathbb{R}$.   ∎

EXAMPLE 2   **A Trivial Vector Space**   Let $V = \{0\}$. That is, V consists of the single number 0. Since $0 + 0 = 1 \cdot 0 = 0 + (0 + 0) = (0 + 0) + 0 = 0$, we see that V is a vector space. It is often referred to as a **trivial** vector space.   ∎

EXAMPLE 3   **A Set That Is Not a Vector Space**   Let $V = \{1\}$. That is, V consists of the single number 1. This is *not* a vector space since it violates axiom (i)—the closure axiom. To see this we simply note that $1 + 1 = 2 \notin V$. It also violates other axioms as well. However, all we need to show is that it violates one axiom in order to prove that V is not a vector space.   ∎

EXAMPLE 4   **The Set of Points in $\mathbb{R}^2$ That Lie on a Line Passing Through the Origin Constitutes a Vector Space**   Let

$$V = \{(x, y): \ y = mx, \text{ where } m \text{ is a fixed real number and}$$
$$x \text{ is an arbitrary real number}\}.$$

That is, V consists of all points lying on the line $y = mx$ passing through the origin with slope m. To show that V is a vector space, we check each of the axioms.

i. Suppose that $\mathbf{x} = (x_1, y_1)$ and $\mathbf{y} = (x_2, y_2)$ are in V. Then $y_1 = mx_1$, $y_2 = mx_2$, and

$$\mathbf{x} + \mathbf{y} = (x_1, y_1) + (x_2, y_2) = (x_1, mx_1) + (x_2, mx_2) = (x_1 + x_2, mx_1 + mx_2)$$
$$= (x_1 + x_2, m(x_1 + x_2)) \in V$$

Thus axiom (i) is satisfied.

ii. Let $\mathbf{z} = (x_3, y_3) = (x_3, mx_3)$ be in V. Then

$$(\mathbf{x} + \mathbf{y}) + \mathbf{z} = [(x_1, y_1) + (x_2, y_2)] + (x_3, y_3)$$
$$= (x_1 + x_2, mx_1 + mx_2) + (x_3, mx_3)$$
$$= (x_1 + x_2 + x_3, mx_1 + mx_2 + mx_3)$$
$$= (x_1, mx_1) + (x_2 + x_3, mx_2 + mx_3)$$
$$= (x_1, y_1) + [(x_2, y_2) + (x_3, y_3)] = \mathbf{x} + (\mathbf{y} + \mathbf{z})$$

iii. Let $\mathbf{0} = (0, 0) = (0, m \cdot 0) \in V$. Then

$$\mathbf{x} + \mathbf{0} = (x_1, mx_1) + (0, 0) = (x_1 + 0, mx_1 + 0) = (x_1, mx_1) = \mathbf{x}$$

iv. If $\mathbf{x} = (x_1, mx_1)$, let $-\mathbf{x} = (-x_1, -mx_1)$. Then

$$\mathbf{x} + (-\mathbf{x}) = (x_1 - x_1, mx_1 - mx_1) = (0, 0) = \mathbf{0}$$

v. $\mathbf{x} + \mathbf{y} = (x_1, mx_1) + (x_2, mx_2) = (x_1 + x_2, mx_1 + mx_2)$
$$= (x_2 + x_1, mx_2 + mx_1)$$
$$= (x_2, mx_2) + (x_1, mx_1) = \mathbf{y} + \mathbf{x}$$

vi. $\alpha\mathbf{x} = \alpha(x_1, mx_1) = (\alpha x_1, \alpha mx_1) = (\alpha x_1, m(\alpha x_1)) \in V$

vii. $\alpha(\mathbf{x} + \mathbf{y}) = \alpha(x_1 + x_2, mx_1 + mx_2) = (\alpha x_1 + \alpha x_2, \alpha mx_1 + \alpha mx_2)$

$\qquad\qquad = (\alpha x_1, \alpha mx_1) + (\alpha x_2, \alpha mx_2)$

$\qquad\qquad = \alpha(x_1, mx_1) + \alpha(x_2, mx_2) = \alpha\mathbf{x} + \alpha\mathbf{y}$

viii. $(\alpha + \beta)\mathbf{x} = (\alpha + \beta)(x_1, mx_1) = ((\alpha + \beta)x_1, (\alpha + \beta)mx_1)$

$\qquad\qquad = (\alpha x_1 + \beta x_1, \alpha mx_1 + \beta mx_1) = (\alpha x_1, \alpha mx_1) + (\beta x_1, \beta mx_1)$

$\qquad\qquad = \alpha(x_1, mx_1) + \beta(x_1, mx_1) = \alpha\mathbf{x} + \beta\mathbf{x}$

ix. $\alpha(\beta\mathbf{x}) = \alpha[\beta(x_1, mx_1)] = \alpha(\beta x_1, \beta mx_1) = (\alpha\beta x_1, \alpha\beta mx_1)$

$\qquad\qquad = \alpha\beta(x_1, mx_1) = \alpha\beta\mathbf{x}$

x. $1\mathbf{x} = 1(x_1, mx_1) = (1 \cdot x_1, 1 \cdot mx_1) = (x_1, mx_1) = \mathbf{x}$

Thus all ten axioms are satisfied, and we see that the set of points in the plane lying on a straight line passing through the origin constitutes a vector space. ∎

Note. Checking all ten axioms can be tedious. From now on we shall check only those axioms that are not immediately obvious.

EXAMPLE 5   **The Set of Points in $\mathbb{R}^2$ Lying on a Line Not Passing Through the Origin Does Not Constitute a Vector Space**   Let $V = \{(x, y): y = 2x + 1, x \in \mathbb{R}\}$. That is, V is the set of points lying on the line $y = 2x + 1$. V is *not* a vector space because closure is violated, as in Example 3. To see this, let us suppose that (x_1, y_1) and (x_2, y_2) are in V. Then

$$(x_1, y_1) + (x_2, y_2) = (x_1 + x_2, y_1 + y_2)$$

If this last vector were in V, we would have

$$y_1 + y_2 = 2(x_1 + x_2) + 1 = 2x_1 + 2x_2 + 1$$

But $y_1 = 2x_1 + 1$ and $y_2 = 2x_2 + 1$ so that

$$y_1 + y_2 = (2x_1 + 1) + (2x_2 + 1) = 2x_1 + 2x_2 + 2$$

Hence we conclude that

$$(x_1 + x_2, y_1 + y_2) \notin V \qquad \text{if } (x_1, y_1) \in V \quad \text{and} \quad (x_2, y_2) \in V.$$

An easier way to see that V is not a vector space is to observe that $\mathbf{0} = (0, 0)$ is not in V because $0 \neq 2 \cdot 0 + 1$. ∎

EXAMPLE 6   **The Set of Points in $\mathbb{R}^3$ Lying on a Plane Passing Through the Origin Constitutes a Vector Space**   Let $V = \{(x, y, z); ax + by + cz = 0\}$. That is, V is the set of points in $\mathbb{R}^3$ lying on the plane with normal vector (a, b, c) and passing through the origin.

Suppose (x_1, y_1, z_1) and (x_2, y_2, z_2) are in V. Then $(x_1, y_1, z_1) + (x_2, y_2, z_2) = (x_1 + x_2, y_1 + y_2, z_1 + z_2) \in V$ because

$$a(x_1 + x_2) + b(y_1 + y_2) + c(z_1 + z_2)$$
$$= (ax_1 + by_1 + cz_1) + (ax_2 + by_2 + cz_2) = 0 + 0 = 0;$$

hence axiom (i) is satisfied. The other axioms are easily verified. Thus the set of points lying on a plane in $\mathbb{R}^3$ that passes through the origin constitutes a vector space. ∎

EXAMPLE 7   **The Vector Space P_n**   Let $V = P_n$, the set of polynomials with real coefficients of degree less than or equal to n.† If $p \in P_n$, then

$$p(x) = a_n x^n + a_{n-1} x^{n-1} + \cdots + a_1 x + a_0$$

where each a_i is real. The sum $p(x) + q(x)$ is defined in the obvious way: If $q(x) = b_n x^n + b_{n-1} x^{n-1} + \cdots + b_1 x + b_0$, then

$$p(x) + q(x) = (a_n + b_n)x^n + (a_{n-1} + b_{n-1})x^{n-1} + \cdots + (a_1 + b_1)x + (a_0 + b_0)$$

Clearly the sum of two polynomials of degree less than or equal to n is another polynomial with degree less than or equal to n, so axiom (i) is satisfied. Properties (ii) and (v) to (x) are obvious. If we define the zero polynomial $\mathbf{0} = 0x^n + 0x^{n-1} + \cdots + 0x + 0$, then clearly $\mathbf{0} \in P_n$ and axiom (iii) is satisfied. Finally, letting $-p(x) = -a_n x^n - a_{n-1} x^{n-1} - \cdots - a_1 x - a_0$, we see that axiom ($iv$) holds, so P_n is a real vector space. ∎

EXAMPLE 8   **The Vector Spaces $C[0, 1]$ and $C[a, b]$**   Let $V = C[0, 1] =$ the set of real-valued continuous functions defined on the interval $[0, 1]$. We define

$$(f + g)x = f(x) + g(x) \text{ and } (\alpha f)(x) = \alpha[f(x)].$$

Since the sum of continuous functions is continuous, axiom (i) is satisfied and the other axioms are easily verified with $\mathbf{0} =$ the zero function and $(-f)(x) = -f(x)$. Similarly, $C[a, b]$, the set of real-valued functions defined and continuous on $[a, b]$, constitutes a vector space. ∎

EXAMPLE 9   **The Vector Space M_{34}**   Let $V = M_{34}$ denote the set of 3×4 matrices with real components. Then with the usual sum and scalar multiplication of matrices, it is again easy to verify that M_{34} is a vector space with $\mathbf{0}$ being the 3×4 zero matrix. If $A = (a_{ij})$ is in M_{34}, then $-A = (-a_{ij})$ is also in M_{34}. ∎

EXAMPLE 10   **The Vector Space M_{mn}**   In an identical manner we see that M_{mn}, the set of $m \times n$ matrices with real components, forms a vector space for any positive integers m and n. ∎

---

† Constant functions (including the function $f(x) = 0$) are said to be polynomials of **degree zero.**

EXAMPLE 11   **A Set of Invertible Matrices Might Not Form a Vector Space**   Let S_3 denote the set of invertible 3×3 matrices. Define the "sum" $A + B$ by $A + B = AB$. If A and B are invertible, then AB is invertible (by Theorem 1.9.3, page 66) so that axiom (*i*) is satisfied. Axiom (*ii*) is simply the associative law for matrix multiplication (Theorem 1.7.2, page 48); axioms (*iii*) and (*iv*) are satisfied with $\mathbf{0} = I_3$ and $-A = A^{-1}$. Axiom (*v*) fails, however, since in general $AB \neq BA$ so that S_3 is not a vector space. ■

EXAMPLE 12   **A Set of Points in a Half Plane Might Not Form a Vector Space**   Let $V = \{(x, y): y \geq 0\}$. V consists of the points in $\mathbb{R}^2$ in the upper half plane (the first two quadrants). If $y_1 \geq 0$ and $y_2 \geq 0$, then $y_1 + y_2 \geq 0$; hence if $(x_1, y_1) \in V$ and $(x_2, y_2) \in V$, then $(x_1 + x_2, y_1 + y_2) \in V$. V is not a vector space, however, since the vector $(1, 1)$, for example, does not have an inverse in V because $(-1, -1) \notin V$. Moreover, axiom (*vi*) fails since if $(x, y) \in V$, then $\alpha(x, y) \notin V$ if $\alpha < 0$. ■

EXAMPLE 13   **The Space $\mathbb{C}^n$**   Let $V = \mathbb{C}^n = \{(c_1, c_2, \ldots, c_n): c_i$ is a complex number for $i = 1, 2, \ldots, n\}$ and the set of scalars is the set of complex numbers. It is not difficult to verify that $\mathbb{C}^n$, too, is a vector space. ■

As these examples suggest, there are many different kinds of vector spaces and many kinds of sets that are *not* vector spaces. Before leaving this section, let us prove some elementary results about vector spaces.

THEOREM 1   Let V be a vector space. Then

    i. $\alpha \mathbf{0} = \mathbf{0}$ for every real number α.

    ii. $0 \cdot \mathbf{x} = \mathbf{0}$ for every $\mathbf{x} \in V$.

    iii. If $\alpha \mathbf{x} = \mathbf{0}$, then $\alpha = 0$ or $\mathbf{x} = \mathbf{0}$ (or both).

    iv. $(-1)\mathbf{x} = -\mathbf{x}$ for every $\mathbf{x} \in V$.

Proof   **i.** By axiom (*iii*), $\mathbf{0} + \mathbf{0} = \mathbf{0}$; and from axiom (*vii*),

$$\alpha(\mathbf{0} + \mathbf{0}) = \alpha\mathbf{0} + \alpha\mathbf{0} = \alpha\mathbf{0} \tag{1}$$

Adding $-\alpha\mathbf{0}$ to both sides of the last equation in (1) and using the associative law (axiom *ii*), we obtain

$$[\alpha\mathbf{0} + \alpha\mathbf{0}] + (-\alpha\mathbf{0}) = \alpha\mathbf{0} + (-\alpha\mathbf{0})$$
$$\alpha\mathbf{0} + [\alpha\mathbf{0} + (-\alpha\mathbf{0})] = \mathbf{0}$$
$$\alpha\mathbf{0} + \mathbf{0} = \mathbf{0}$$
$$\alpha\mathbf{0} = \mathbf{0}$$

ii. Essentially the same proof as used in part (i) works. We start with $0 + 0 = 0$ and use axiom $(viii)$ to see that $0\mathbf{x} = (0 + 0)\mathbf{x} = 0\mathbf{x} + 0\mathbf{x}$ or $0\mathbf{x} + (-0\mathbf{x}) = 0\mathbf{x} + [0\mathbf{x} + (-0\mathbf{x})]$ or $\mathbf{0} = 0\mathbf{x} + \mathbf{0} = 0\mathbf{x}$.

iii. Let $\alpha\mathbf{x} = \mathbf{0}$. If $\alpha \neq 0$, we multiply both sides of the equation by $1/\alpha$ to obtain $(1/\alpha)(\alpha\mathbf{x}) = (1/\alpha)\mathbf{0} = \mathbf{0}$ [by part (i)]. But $(1/\alpha)(\alpha\mathbf{x}) = 1\mathbf{x} = \mathbf{x}$ (by axiom ix), so $\mathbf{x} = \mathbf{0}$.

iv. We start with the fact that $1 + (-1) = 0$. Then, using part (ii), we obtain

$$\mathbf{0} = 0\mathbf{x} = [1 + (-1)]\mathbf{x} = 1\mathbf{x} + (-1)\mathbf{x} = \mathbf{x} + (-1)\mathbf{x} \qquad (2)$$

We add $-\mathbf{x}$ to both sides of (2) to obtain

$$\mathbf{0} + (-\mathbf{x}) = \mathbf{x} + (-1)\mathbf{x} + (-\mathbf{x}) = \mathbf{x} + (-\mathbf{x}) + (-1)\mathbf{x}$$
$$= \mathbf{0} + (-1)\mathbf{x} = (-1)\mathbf{x}$$

Thus $-\mathbf{x} = (-1)\mathbf{x}$. Note that we were able to reverse the order of addition in the preceding equation by using the commutative law (axiom v).   ■

Remark. Part (iii) of Theorem 1 is not so obvious as it seems. There are objects which have the property that $xy = 0$ does not imply that either x or y is zero. As an example, we look at the multiplication of 2×2 matrices. If $A = \begin{pmatrix} 0 & 1 \\ 0 & 0 \end{pmatrix}$ and $B = \begin{pmatrix} 0 & -2 \\ 0 & 0 \end{pmatrix}$, then neither A nor B is zero, although, as is easily verified, the product $AB = 0$, the zero matrix.

PROBLEMS   4.2

In Problems 1–20 determine whether the given set is a vector space. If it is not, list the axioms that do not hold.

1. The set of diagonal $n \times n$ matrices under the usual matrix addition and the usual scalar multiplication.

2. The set of diagonal $n \times n$ matrices under multiplication (that is, $A + B = AB$).

3. $\{(x, y): y \leq 0;\ x, y \text{ real}\}$ with the usual addition and scalar multiplication of vectors.

4. The vectors in the plane lying in the first quadrant.

5. The set of vectors in $\mathbb{R}^3$ in the form (x, x, x).

6. The set of polynomials of degree 4 under the operations of Example 7.

7. The set of $n \times n$ symmetric matrices (see Section 1.10) under the usual addition and scalar multiplication.

8. The set of 2×2 matrices having the form $\begin{pmatrix} 0 & a \\ b & 0 \end{pmatrix}$ under the usual addition and scalar multiplication.

9. The set of matrices of the form $\begin{pmatrix} 1 & \alpha \\ \beta & 1 \end{pmatrix}$ with the matrix operations of addition and scalar multiplication.

10. The set consisting of the single vector $(0, 0)$ under the usual operations in $\mathbb{R}^2$.

11. The set of polynomials of degree $\leq n$ with zero constant term.

12. The set of polynomials of degree $\leq n$ with positive constant term a_0.

13. The set of continuous functions in $[0, 1]$ with $f(0) = 0$ and $f(1) = 0$ under the operations of Example 8.

14. The set of points in $\mathbb{R}^3$ lying on a line passing through the origin.

15. The set of points in $\mathbb{R}^3$ lying on the line $x = t + 1$, $y = 2t$, $z = t - 1$.

16. $\mathbb{R}^2$ with addition defined by $(x_1, y_1) + (x_2, y_2) = (x_1 + x_2 + 1, y_1 + y_2 + 1)$ and ordinary scalar multiplication.

17. The set of Problem 16 with scalar multiplication defined by $\alpha(x, y) = (\alpha + \alpha x - 1, \alpha + \alpha y - 1)$.

18. The set consisting of one object with addition defined by *object* + *object* = *object* and scalar multiplication defined by $\alpha(\textit{object}) = \textit{object}$.

† Calculus 19. The set of differentiable functions defined on $[0, 1]$ with the operations of Example 8.

*20. The set of real numbers of the form $a + b\sqrt{2}$, where a and b are rational numbers, under the usual addition of real numbers and with scalar multiplication defined only for rational scalars.

21. Show that in a vector space the additive identity element is unique.

22. Show that in a vector space each vector has a unique additive inverse.

23. If $\mathbf{x}$ and $\mathbf{y}$ are vectors in a vector space V, show that there is a unique vector $\mathbf{z} \in V$ such that $\mathbf{x} + \mathbf{z} = \mathbf{y}$.

24. Show that the set of positive real numbers forms a vector space under the operations $x + y = xy$ and $\alpha x = x^\alpha$.

* Calculus 25. Consider the homogeneous second-order differential equation

$$y''(x) + a(x)y'(x) + b(x)y(x) = 0$$

where $a(x)$ and $b(x)$ are continuous functions. Show that the set of solutions to the equation is a vector space under the usual rules for adding functions and multiplying them by real numbers.

4.3   SUBSPACES

From Example 4.2.1, page 212, we know that $\mathbb{R}^2 = \{(x, y): x \in \mathbb{R} \text{ and } y \in \mathbb{R}\}$ is a vector space. In Example 4.2.4, page 213, we saw that $V = \{(x, y): y = mx\}$ is also a vector space. Moreover, it is clear that $V \subset \mathbb{R}^2$. That is, $\mathbb{R}^2$ has a subset that is also a vector space. In fact, all vector spaces have subsets that are also vector spaces. We examine these important subsets in this section.

---

† Calculus This symbol is used throughout the book to indicate that the problem or example uses calculus.

DEFINITION 1   **Subspace** Let H be a nonempty subset of a vector space V and suppose that H is itself a vector space under the operations of addition and scalar multiplication defined on V. Then H is said to be a **subspace** of V.

We can say that the subspace H **inherits** the operations from the "parent" vector space V.

We encounter many examples of subspaces in this chapter. But first we prove a result that makes it relatively easy to determine whether a subset of V is indeed a subspace of V.

THEOREM 1   A nonempty subset H of the vector space V is a subspace of V if the two closure rules hold:

Rules for Checking Whether a Nonempty Subset is a Subspace

i.   If $\mathbf{x} \in H$ and $\mathbf{y} \in H$, then $\mathbf{x} + \mathbf{y} \in H$.

ii. If $\mathbf{x} \in H$, then $\alpha \mathbf{x} \in H$ for every scalar α.

Proof   Evidently, if H is a vector space, then the two closure rules must hold. Conversely, to show that H is a vector space, we must show that axioms (i) to (x) on page 212 hold under the operations of vector addition and scalar multiplication defined in V. The two closure operations [axioms (i) and (vi)] hold by hypothesis. Since vectors in H are also in V, the associative, commutative, distributive, and multiplicative identity laws [axioms (ii), (v), (vii), $(viii)$, (ix), and (x)] hold. Let $\mathbf{x} \in H$. Then $0\mathbf{x} \in H$ by hypothesis (ii). But by Theorem 4.2.1, page 216, (part ii), $0\mathbf{x} = \mathbf{0}$. Thus $\mathbf{0} \in H$ and axiom (iii) holds. Finally, by part (ii), $(-1)\mathbf{x} \in H$ for every $\mathbf{x} \in H$. By Theorem 4.2.1 (part iv), $-\mathbf{x} = (-1)\mathbf{x} \in H$ so that axiom (iv) also holds and the proof is complete.   ∎

This theorem shows that to test whether H is a subspace of V, it is sufficient to verify that

---

$\mathbf{x} + \mathbf{y}$ and $\alpha \mathbf{x}$ are in H when $\mathbf{x}$ and $\mathbf{y}$ are H and α is a scalar.

The preceding proof contains a fact that is important enough to mention explicitly

---

Every nonempty subspace of a vector space V contains $\mathbf{0}$.   **(1)**

This fact will often make it easy to see that a particular subset of V is *not* a vector space. That is, if a subset does not contain **0**, then it is not a subspace. Note that the zero vector in H, a subspace of V, is the same as the zero vector in V.

We now give some examples of subspaces.

EXAMPLE 1   **The Trivial Subspace**   For any vector space V, the subset $\{\mathbf{0}\}$ consisting of the zero vector alone is a subspace since $\mathbf{0} + \mathbf{0} = \mathbf{0}$ and $\alpha\mathbf{0} = \mathbf{0}$ for every real number α [part (i) of Theorem 4.2.1]. It is called the **trivial subspace.**   ∎

EXAMPLE 2   **A Vector Space is a Subspace of Itself**   For every vector space V, V is a subspace of itself.   ∎

The first two examples show that every vector space V contains two subspaces $\{\mathbf{0}\}$ and V (which coincide if $V = \{\mathbf{0}\}$). It is more interesting to find other subspaces.

Proper subspaces   Subspaces other than $\{\mathbf{0}\}$ and $\mathbf{V}$ are called **proper subspaces.**

EXAMPLE 3   **A Proper Subspace of $\mathbb{R}^2$**   Let $H = \{(x, y): y = mx\}$ (see Example 4.2.4, page 213). Then, as we have already mentioned, H is a subspace of $\mathbb{R}^2$. As we shall see in Section 4.6 (Problem 15 on page 252), if H is a proper subspace of $\mathbb{R}^2$, then H consists of the set of points lying on a straight line through the origin; that is, a set of points lying on a straight line passing through the origin is the only kind of proper subspace of $\mathbb{R}^2$.   ∎

EXAMPLE 4   **A Proper Subspace of $\mathbb{R}^3$**
Let $H = \{(x, y, z): x = at, y = bt, \text{ and } z = ct; a, b, c, t \text{ real}\}$. Then H consists of the vectors in $\mathbb{R}^3$ lying on a straight line passing through the origin. To see that H is a subspace of $\mathbb{R}^3$, let $\mathbf{x} = (at_1, bt_1, ct_1) \in H$ and $\mathbf{y} = (at_2, bt_2, ct_2) \in H$. Then

$$\mathbf{x} + \mathbf{y} = (a(t_1 + t_2), b(t_1 + t_2), c(t_1 + t_2)) \in H$$

and

$$\alpha\mathbf{x} = (a(\alpha t_1), b(\alpha t_2), c(\alpha t_3)) \in H.$$

Thus H is a subspace of $\mathbb{R}^3$.   ∎

EXAMPLE 5   **Another Proper Subspace of $\mathbb{R}^3$**   Let $\pi = \{(x, y, z): ax + by + cz = 0; a, b, c \text{ real}\}$. Then, as we saw in Example 4.2.6, page 214, π is a vector space; thus π is a subspace of $\mathbb{R}^3$.   ∎

We shall prove in Section 4.6 that sets of vectors lying on lines and planes through the origin are the only proper subspaces of $\mathbb{R}^3$.

Before studying more examples, we note that *not every vector space has proper subspaces*.

EXAMPLE 6 $\mathbb{R}$ **Has No Proper Subspace**   Let H be a subspace of $\mathbb{R}$.† If $H \neq \{0\}$, then H contains a nonzero real number α. Then, by axiom (vi), $1 = (1/\alpha)\alpha \in H$ and $\beta 1 = \beta \in H$ for every real number β. Thus if H is not the trivial subspace then $H = \mathbb{R}$. That is, $\mathbb{R}$ has *no* proper subspace.   ∎

EXAMPLE 7   **Some Proper Subspaces of P_n**   If P_n denotes the vector space of polynomials of degree $\leq n$ (Example 4.2.7, page 215), and if $0 \leq m < n$, then P_m is a proper subspace of P_n, as is easily verified.   ∎

EXAMPLE 8   **A Proper Subspace of M_{mn}**   Let M_{mn} (Example 4.2.10, page 215) denote the vector space of $m \times n$ matrices with real components and let $H = \{A \in M_{mn}: a_{11} = 0\}$. By the definition of matrix addition and scalar multiplication it is clear that the two closure axioms hold, so that H is a subspace.   ∎

EXAMPLE 9   **A Subset That Is Not a Subspace of M_{nn}**   Let $V = M_{nn}$ (the $n \times n$ matrices) and let $H = \{A \in M_{nn}: A$ is invertible$\}$. Then H is not a subspace since the $n \times n$ zero matrix is not in H.   ∎

EXAMPLE 10   **A Proper Subspace of $C[0, 1]$** $P_n[0, 1]$‡ $\subset C[0, 1]$ (see Example 4.2.8, page 215) because every polynomial is continuous and P_n is a vector space for every integer n, so that each $P_n[0, 1]$ is a subspace of $C[0, 1]$.   ∎

EXAMPLE 11
Calculus
   $C^1[0, 1]$ **Is a Proper Subspace of $C[0, 1]$**   Let $C^1[0, 1]$ denote the set of functions with continuous first derivatives defined on $[0, 1]$. Since every differentiable function is continuous, we have $C^1[0, 1] \subset C[0, 1]$. Since the sum and scalar multiple of two differentiable functions are differentiable, we see that $C^1[0, 1]$ is a subspace of $C[0, 1]$. It is a proper subspace because not every continuous function is differentiable.   ∎

EXAMPLE 12
Calculus
   Another Proper Subspace of $C[0, 1]$ If $f \in C[0, 1]$, then $\int_0^1 f(x)\, dx$ exists. Let $H = \{f \in C[0, 1]: \int_0^1 f(x)\, dx = 0\}$. If $f \in H$ and $g \in H$, then $\int_0^1 [f(x) + g(x)]\, dx = \int_0^1 f(x)\, dx + \int_0^1 g(x)\, dx = 0 + 0 = 0$ and $\int_0^1 \alpha f(x)\, dx = \alpha \int_0^1 f(x)\, dx = 0$. Thus $f + g$ and αf are in H for every real number α. This shows that H is a proper subspace of $C[0, 1]$.   ∎

As the last three examples illustrate, a vector space can have a great number and variety of proper subspaces. Before leaving this section, we prove an interesting fact about subspaces.

---

† Note that $\mathbb{R}$ is a vector space over itself; that is, $\mathbb{R}$ is a vector space where the scalars are taken to be the reals. This is Example 4.2.1, page 212, with $n = 1$.

‡ $P_n[0, 1]$ denotes the set of polynomials defined on the interval $[0, 1]$ of degree $\leq n$.

THEOREM 2   Let H_1 and H_2 be subspaces of a vector space V. Then $H_1 \cap H_2$ is a subspace of V.

Proof   Note that $H_1 \cap H_2$ is nonempty because it contains $\mathbf{0}$. Let $\mathbf{x}_1 \in H_1 \cap H_2$ and $\mathbf{x}_2 \in H_1 \cap H_2$. Then, since H_1 and H_2 are subspaces, $\mathbf{x}_1 + \mathbf{x}_2 \in H_1$ and $\mathbf{x}_1 + \mathbf{x}_2 \in H_2$. This means that $\mathbf{x}_1 + \mathbf{x}_2 \in H_1 \cap H_2$. Similarly, $\alpha \mathbf{x}_1 \in H_1 \cap H_2$. Thus the two closure axioms are satisfied and $H_1 \cap H_2$ is a subspace.† ∎

EXAMPLE 13   **The Intersection of Two Subspaces of $\mathbb{R}^3$ Is a Subspace**   In $\mathbb{R}^3$ let $H_1 = \{(x, y, z): 2x - y - z = 0\}$ and $H_2 = \{(x, y, z): x + 2y + 3z = 0\}$. Then H_1 and H_2 consist of vectors lying on planes through the origin and are, by Example 5, subspaces of $\mathbb{R}^3$. $H_1 \cap H_2$ is the intersection of the two planes which we compute as in Example 10 in Section 3.5:

$$x + 2y + 3z = 0$$
$$2x - y - z = 0$$

or, row-reducing:

$$\begin{pmatrix} 1 & 2 & 3 & | & 0 \\ 2 & -1 & -1 & | & 0 \end{pmatrix} \xrightarrow{R_2 \to R_2 - 2R_1} \begin{pmatrix} 1 & 2 & 3 & | & 0 \\ 0 & -5 & -7 & | & 0 \end{pmatrix} \xrightarrow{R_2 \to -\frac{1}{5}R_2}$$

$$\begin{pmatrix} 1 & 2 & 3 & | & 0 \\ 0 & 1 & \frac{7}{5} & | & 0 \end{pmatrix} \xrightarrow{R_1 \to R_1 - 2R_2} \begin{pmatrix} 1 & 0 & \frac{1}{5} & | & 0 \\ 0 & 1 & \frac{7}{5} & | & 0 \end{pmatrix}$$

Thus all solutions to the homogeneous system are given by $(-\frac{1}{5}z, -\frac{7}{5}z, z)$. Setting $z = t$, we obtain the parametric equations of a line L in $\mathbb{R}^3$: $x = -\frac{1}{5}t$, $y = -\frac{7}{5}t$, $z = t$. As we saw in Example 4, the set of vectors on L constitutes a subspace of $\mathbb{R}^3$. ∎

Remark. It is not true that if H_1 and H_2 is a subspace of V, then $H_1 \cup H_2$ is necessarily a subspace of V (it may or may not be). For example, $H_1 = \{(x, y): y = 2x\}$ and $\{(x, y): y = 3x\}$ are subspaces of $\mathbb{R}^2$, but $H_1 \cup H_2$ is not a subspace. To see this, observe that $(1, 2) \in H_1$ and $(1, 3) \in H_2$, so that both $(1, 2)$ and $(1, 3)$ are in $H_1 \cup H_2$. But $(1, 2) + (1, 3) = (2, 5) \notin H_1 \cup H_2$ because $(2, 5) \notin H_1$ and $(2, 5) \notin H_2$. Thus $H_1 \cup H_2$ is not closed under addition, and is therefore not a subspace.

PROBLEMS   4.3

In Problems 1–20 determine whether the given subset H of the vector space V is a subspace of V.

1. $V = \mathbb{R}^2$; $H = \{(x, y): y \geq 0\}$ **2.** $V = \mathbb{R}^2$; $H = \{(x, y): x = y\}$

---

†Note, in particular, that as $\mathbf{0} \in H_1$ and $\mathbf{0} \in H_2$, we have $\mathbf{0} \in H_1 \cap H_2$.

3. $V = \mathbb{R}^3$; $H = $ the xy-plane

4. $V = \mathbb{R}^2$; $H = \{(x, y): x^2 + y^2 \le 1\}$

5. $V = M_{nn}$; $H = \{D \in M_{nn}: D \text{ is diagonal}\}$

6. $V = M_{nn}$; $H = \{T \in M_{nn}: T \text{ is upper triangular}\}$

7. $V = M_{nn}$; $H = \{S \in M_{nn}: S \text{ is symmetric}\}$

8. $V = M_{mn}$; $H = \{A \in M_{mn}: a_{ij} = 0\}$

9. $V = M_{22}$; $H = \left\{ A \in M_{22}: A = \begin{pmatrix} a & b \\ -b & c \end{pmatrix} \right\}$

10. $V = M_{22}$; $H = \left\{ A \in M_{22}: A = \begin{pmatrix} a & 1+a \\ 0 & 0 \end{pmatrix} \right\}$

11. $V = M_{22}$; $H = \left\{ A \in M_{22}: A = \begin{pmatrix} 0 & a \\ b & 0 \end{pmatrix} \right\}$

12. $V = P_4$; $H = \{p \in P_4: \deg p = 4\}$

13. $V = P_4$; $H = \{p \in P_4: p(0) = 0\}$

14. $V = P_n$; $H = \{p \in P_n: p(0) = 0\}$

15. $V = P_n$; $H = \{p \in P_n: p(0) = 1\}$

16. $V = C[0, 1]$; $H = \{f \in C[0, 1]: f(0) = f(1) = 0\}$

17. $V = C[0, 1]$; $H = \{f \in C[0, 1]: f(0) = 2\}$

Calculus 18. $V = C^1[0, 1]$; $H = \{f \in C^1[0, 1]: f'(0) = 0\}$

Calculus 19. $V = C[a, b]$, where a and b are real numbers and $a < b$; $H = \{f \in C[a, b]: \int_a^b f(x)\, dx = 0\}$

Calculus 20. $V = C[a, b]$; $H = \{f \in C[a, b]: \int_a^b f(x)\, dx = 1\}$

21. Let $V = M_{22}$; let $H_1 = \{A \in M_{22}: a_{11} = 0\}$ and $H_2 = \left\{ A \in M_{22}: A = \begin{pmatrix} -b & a \\ a & b \end{pmatrix} \right\}$.

    a. Show that H_1 and H_2 are subspaces.

    b. Describe the subset $H = H_1 \cap H_2$ and show that it is a subspace.

Calculus 22. If $V = C[0, 1]$, let H_1 denote the subspace of Example 10 and H_2 denote the subspace of Example 11. Describe the set $H_1 \cap H_2$ and show that it is a subspace.

23. Let A be an $n \times m$ matrix and let $H = \{\mathbf{x} \in \mathbb{R}^m: A\mathbf{x} = \mathbf{0}\}$. Show that H is a subspace of $\mathbb{R}^m$. H is called the **kernel** of the matrix A.

24. In Problem 23 let $H = \{\mathbf{x} \in \mathbb{R}^m: A\mathbf{x} \ne \mathbf{0}\}$. Show that H is not a subspace of $\mathbb{R}^m$.

25. Let $H = \{(x, y, z, w): ax + by + cz + dw = 0\}$, where a, b, c, and d are real numbers not all zero. Show that H is a proper subspace of $\mathbb{R}^4$. H is called a **hyperplane** in $\mathbb{R}^4$.

26. Let $H = \{(x_1, x_2, \ldots, x_n): a_1x_1 + a_2x_2 + \cdots + a_nx_n = 0\}$, where $a_1, a_2, \ldots, a_n$ are real numbers not all zero. Show that H is a proper subspace of $\mathbb{R}^n$. H, as in Problem 25, is called a **hyperplane** in $\mathbb{R}^n$.

27. Let H_1 and H_2 be subspaces of a vector space V. Let $H_1 + H_2 = \{\mathbf{v}: \mathbf{v} = \mathbf{v}_1 + \mathbf{v}_2 \text{ with } \mathbf{v}_1 \in H_1 \text{ and } \mathbf{v}_2 \in H_2\}$. Show that $H_1 + H_2$ is a subspace of V.

28. Let $\mathbf{v}_1$ and $\mathbf{v}_2$ be two vectors in $\mathbb{R}^2$. Show that $H = \{\mathbf{v}: \mathbf{v} = a\mathbf{v}_1 + b\mathbf{v}_2; a, b \text{ real}\}$ is a subspace of $\mathbb{R}^2$.

*29. In Problem 28 show that if $\mathbf{v}_1$ and $\mathbf{v}_2$ are not collinear, then $H = \mathbb{R}^2$.

*30. Let $\mathbf{v}_1, \mathbf{v}_2, \ldots, \mathbf{v}_n$ be arbitrary vectors in a vector space V. Let $H = \{\mathbf{v} \in V: \mathbf{v} = a_1\mathbf{v}_1 + a_2\mathbf{v}_2 + \cdots + a_n\mathbf{v}_n, \text{ where } a_1, a_2, \ldots, a_n \text{ are scalars}\}$. Show that H is a subspace of V. H is called the subspace *spanned* by the vectors $\mathbf{v}_1, \mathbf{v}_2, \ldots, \mathbf{v}_n$.

4.4   LINEAR COMBINATION AND SPAN

We have seen that every vector $\mathbf{v} = (a, b, c)$ in $\mathbb{R}^3$ can be written in the form

$$\mathbf{v} = a\mathbf{i} + b\mathbf{j} + c\mathbf{k}$$

In this case we say that $\mathbf{v}$ is a *linear combination* of the three vectors $\mathbf{i}$, $\mathbf{j}$, and $\mathbf{k}$. More generally, we have the following definition.

DEFINITION 1   **Linear Combination** Let $\mathbf{v}_1$, $\mathbf{v}_2$, . . . , $\mathbf{v}_n$ be vectors in a vector space V. Then any expression of the form

$$a_1\mathbf{v}_1 + a_2\mathbf{v}_2 + \cdots + a_n\mathbf{v}_n \tag{1}$$

where a_1, a_2, . . . , a_n are scalars is called a **linear combination** of $\mathbf{v}_1$, $\mathbf{v}_2$, . . . , $\mathbf{v}_n$.

EXAMPLE 1   **A Linear Combination in $\mathbb{R}^3$**   In $\mathbb{R}^3$, $\begin{pmatrix} -7 \\ 7 \\ 7 \end{pmatrix}$ is a linear combination of $\begin{pmatrix} -1 \\ 2 \\ 4 \end{pmatrix}$ and $\begin{pmatrix} 5 \\ -3 \\ 1 \end{pmatrix}$ since $\begin{pmatrix} -7 \\ 7 \\ 7 \end{pmatrix} = 2\begin{pmatrix} -1 \\ 2 \\ 4 \end{pmatrix} - \begin{pmatrix} 5 \\ -3 \\ 1 \end{pmatrix}$ ∎

EXAMPLE 2   **A Linear Combination in M_{23}**

In M_{23}, $\begin{pmatrix} -3 & 2 & 8 \\ -1 & 9 & 3 \end{pmatrix} = 3\begin{pmatrix} -1 & 0 & 4 \\ 1 & 1 & 5 \end{pmatrix} + 2\begin{pmatrix} 0 & 1 & -2 \\ -2 & 3 & -6 \end{pmatrix}$,

which shows that $\begin{pmatrix} -3 & 2 & 8 \\ -1 & 9 & 3 \end{pmatrix}$ is a linear combination of

$\begin{pmatrix} -1 & 0 & 4 \\ 1 & 1 & 5 \end{pmatrix}$ and $\begin{pmatrix} 0 & 1 & -2 \\ -2 & 3 & -6 \end{pmatrix}$.   ∎

EXAMPLE 3   **Linear Combinations in P_n**   In P_n every polynomial can be written as a linear combination of the "monomials" $1, x, x^2, \ldots x^n$.   ∎

DEFINITION 2   **Span**   The vectors $\mathbf{v}_1, \mathbf{v}_2, \ldots, \mathbf{v}_n$ in a vector space V are said to **span** V if every vector in V can be written as a linear combination of them. That is, for every $\mathbf{v} \in V$ there are scalars $a_1, a_2, \ldots, a_n$ such that

$$\mathbf{v} = a_1\mathbf{v}_1 + a_2\mathbf{v}_2 + \cdots + a_n\mathbf{v}_n \tag{2}$$

EXAMPLE 4   **Sets of Vectors That Span $\mathbb{R}^2$ and $\mathbb{R}^3$**   We saw in Section 3.1 that the vectors

$$\mathbf{i} = \begin{pmatrix} 1 \\ 0 \end{pmatrix} \text{ and } \mathbf{j} = \begin{pmatrix} 0 \\ 1 \end{pmatrix} \text{ span } \mathbb{R}^2. \text{ In Section 3.3 we saw that } \mathbf{i} = \begin{pmatrix} 1 \\ 0 \\ 0 \end{pmatrix}, \mathbf{j} = \begin{pmatrix} 0 \\ 1 \\ 0 \end{pmatrix}, \text{ and}$$

$$\mathbf{k} = \begin{pmatrix} 0 \\ 0 \\ 1 \end{pmatrix} \text{ span } \mathbb{R}^3.$$

∎

We now look briefly at spanning sets of some other vector spaces.

EXAMPLE 5   **$n + 1$ Vectors That Span P_n**   From Example 3 it follows that the monomials $1, x, x^2, \ldots, x^n$ span P_n.   ∎

EXAMPLE 6   **Four Vectors That Span M_{22}**   Since $\begin{pmatrix} a & b \\ c & d \end{pmatrix} = a\begin{pmatrix} 1 & 0 \\ 0 & 0 \end{pmatrix} + b\begin{pmatrix} 0 & 1 \\ 0 & 0 \end{pmatrix} +$

$c\begin{pmatrix} 0 & 0 \\ 1 & 0 \end{pmatrix} + d\begin{pmatrix} 0 & 0 \\ 0 & 1 \end{pmatrix}$, we see that $\begin{pmatrix} 1 & 0 \\ 0 & 0 \end{pmatrix}, \begin{pmatrix} 0 & 1 \\ 0 & 0 \end{pmatrix}, \begin{pmatrix} 0 & 0 \\ 1 & 0 \end{pmatrix}$, and $\begin{pmatrix} 0 & 0 \\ 0 & 1 \end{pmatrix}$ span M_{22}.   ∎

EXAMPLE 7   **No Finite Set of Polynomials Spans P**   Let P denote the vector space of polynomials. Then no *finite* set of polynomials spans P. To see this suppose that $p_1, p_2, \ldots, p_m$ are polynomials. Let p_k be the polynomial of largest degree in this set and let $N = \deg p_k$. Then the polynomial $p(x) = x^{N+1}$ cannot be written as a linear combination of $p_1, p_2, \ldots, p_m$. For example if $N = 3$, then $x^4 \neq c_0 + c_1 x + c_2 x^2 + c_3 x^3$ for any scalars c_0, c_1, c_2, and c_3.   ∎

We now turn to another way of finding subspaces of a vector space V.

DEFINITION 3   **Span of a Set of Vectors** Let $v_1, v_2, \ldots, v_k$ be k vectors in a vector space V. The **span** of $\{v_1, v_2, \ldots, v_k\}$ is the set of linear combinations of $v_1, v_2, \ldots, v_k$. That is,

$$\text{span }\{v_1, v_2, \ldots, v_k\} = \{v\colon v = a_1 v_1 + a_2 v_2 + \cdots + a_k v_k\} \tag{3}$$

where $a_1, a_2, \ldots, a_k$ are scalars.

THEOREM 1   Span $\{v_1, v_2, \ldots, v_k\}$ is a subspace of V.

Proof   The proof is easy and is left as an exercise (see Problem 16).   ∎

EXAMPLE 8   **The Span of Two Vectors in $\mathbb{R}^3$**   Let $v_1 = (2, -1, 4)$ and $v_2 = (4, 1, 6)$. Then $H = \text{span }\{v_1, v_2\} = \{v\colon v = a_1(2, -1, 4) + a_2(4, 1, 6)\}$. What does H look like? If $v = (x, y, z) \in H$, then we have $x = 2a_1 + 4a_2$, $y = -a_1 + a_2$, and $z = 4a_1 + 6a_2$. If we think of (x, y, z) as being fixed, then we can view these equations as a system of three equations in the two unknowns a_1, a_2. We solve this system in the usual way:

$$\begin{pmatrix} -1 & 1 & | & y \\ 2 & 4 & | & x \\ 4 & 6 & | & z \end{pmatrix} \xrightarrow{R_1 \to -R_1} \begin{pmatrix} 1 & -1 & | & -y \\ 2 & 4 & | & x \\ 4 & 6 & | & z \end{pmatrix} \xrightarrow[R_3 \to R_3 - 4R_1]{R_2 \to R_2 - 2R_1} \begin{pmatrix} 1 & -1 & | & -y \\ 0 & 6 & | & x + 2y \\ 0 & 10 & | & z + 4y \end{pmatrix}$$

$$\xrightarrow{R_2 \to \frac{1}{6}R_2} \begin{pmatrix} 1 & -1 & | & -y \\ 0 & 1 & | & (x+2y)/6 \\ 0 & 10 & | & z + 4y \end{pmatrix} \xrightarrow[R_3 \to R_3 - 10R_2]{R_1 \to R_1 + R_2} \begin{pmatrix} 1 & 0 & | & x/6 - 2y/3 \\ 0 & 1 & | & x/6 + y/3 \\ 0 & 0 & | & -5x/3 + 2y/3 + z \end{pmatrix}$$

From Chapter 1 we see that the system has a solution only if $-5x/3 + 2y/3 + z = 0$; or, multiplying through by -3, if

$$5x - 2y - 3z = 0 \tag{4}$$

Equation (4) is the equation of a plane in $\mathbb{R}^3$ passing through the origin.   ∎

The last example can be generalized to prove the following interesting fact:

> *The span of two nonzero vectors in $\mathbb{R}^3$ that are not parallel is a plane passing through the origin.*

For a suggested proof see Problems 19 and 20.

We can give a geometric interpretation of this result. Look at the vectors in Figure 4.1. We know (from Section 3.1) the geometric interpretation of the vectors

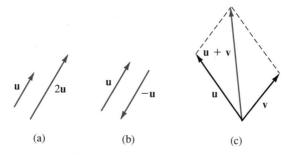

Figure 4.1 **u** + **v** is obtained from the parallelogram rule

$2\mathbf{u}$, $-\mathbf{u}$, and $\mathbf{u} + \mathbf{v}$, for example. Using these, we see that any other vector in the plane of **u** and **v** can be obtained as a linear combination of **u** and **v**. Figure 4.2 shows how in four different situations a third vector **w** in the plane of **u** and **v** can be written as $\alpha\mathbf{u} + \beta\mathbf{v}$ for appropriate choices of the numbers α and β.

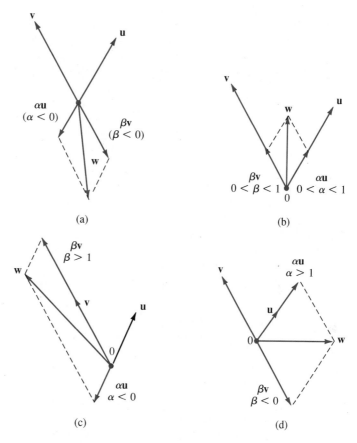

Figure 4.2 In each case $\mathbf{w} = \alpha\mathbf{u} + \beta\mathbf{v}$ for appropriate choices of α and β

Remark. In Definitions 2 and 3 we used the word "span" in two different ways: as a verb and as a noun. We emphasize that

$$\overset{\text{verb}}{\downarrow}$$

A set of vectors $v_1, v_2, \ldots, v_n$ *span* V if every vector in V can be written as a linear combination of $v_1, v_2, \ldots, v_n$.

but

$$\overset{\text{noun}}{\downarrow}$$

The *span* of the n vectors $v_1, v_2, \ldots, v_k$ is the set of linear combinations of these vectors.

These two concepts are different—even though they use the same word.

We close this section by citing a useful result. Its proof is not difficult and is left as an exercise (see Problem 21).

THEOREM 2   Let $v_1, v_2, \ldots, v_n, v_{n+1}$ be $n + 1$ vectors that are in a vector space V. If $v_1, v_2, \ldots, v_n$ span V, then $v_1, v_2, \ldots, v_n, v_{n+1}$ also span V. That is, the addition of one (or more) vectors to a spanning set yields another spanning set.   ∎

PROBLEMS   4.4

In Problems 1–13 determine whether the given set of vectors spans the given vector space.

1. In $\mathbb{R}^2$: $\begin{pmatrix} 1 \\ 2 \end{pmatrix}, \begin{pmatrix} 3 \\ 4 \end{pmatrix}$

2. In $\mathbb{R}^2$: $\begin{pmatrix} 1 \\ 1 \end{pmatrix}, \begin{pmatrix} 2 \\ 1 \end{pmatrix}, \begin{pmatrix} 2 \\ 2 \end{pmatrix}$

3. In $\mathbb{R}^2$: $\begin{pmatrix} 1 \\ 1 \end{pmatrix}, \begin{pmatrix} 2 \\ 2 \end{pmatrix}, \begin{pmatrix} 5 \\ 5 \end{pmatrix}$

4. In $\mathbb{R}^3$: $\begin{pmatrix} 1 \\ 2 \\ 3 \end{pmatrix}, \begin{pmatrix} -1 \\ 2 \\ 3 \end{pmatrix}, \begin{pmatrix} 5 \\ 2 \\ 3 \end{pmatrix}$

5. In $\mathbb{R}^3$: $\begin{pmatrix} 1 \\ 1 \\ 1 \end{pmatrix}, \begin{pmatrix} 0 \\ 1 \\ 1 \end{pmatrix}, \begin{pmatrix} 0 \\ 0 \\ 1 \end{pmatrix}$

6. In $\mathbb{R}^3$: $\begin{pmatrix} 2 \\ 0 \\ 1 \end{pmatrix}, \begin{pmatrix} 3 \\ 1 \\ 2 \end{pmatrix}, \begin{pmatrix} 1 \\ 1 \\ 1 \end{pmatrix}, \begin{pmatrix} 7 \\ 3 \\ 5 \end{pmatrix}$

7. In $\mathbb{R}^3$: $(1, -1, 2), (1, 1, 2), (0, 0, 1)$

8. In $\mathbb{R}^3$: $(1, -1, 2), (-1, 1, 2), (0, 0, 1)$

9. In P_2: $1 - x$, $3 - x^2$

10. In P_2: $1 - x$, $3 - x^2$, x

11. In M_{22}: $\begin{pmatrix} 2 & 1 \\ 0 & 0 \end{pmatrix}, \begin{pmatrix} 0 & 0 \\ 2 & 1 \end{pmatrix}, \begin{pmatrix} 3 & -1 \\ 0 & 0 \end{pmatrix}, \begin{pmatrix} 0 & 0 \\ 3 & 1 \end{pmatrix}$

12. In M_{22}: $\begin{pmatrix} 1 & 0 \\ 1 & 0 \end{pmatrix}, \begin{pmatrix} 1 & 2 \\ 0 & 0 \end{pmatrix}, \begin{pmatrix} 4 & -1 \\ 3 & 0 \end{pmatrix}, \begin{pmatrix} -2 & 5 \\ 6 & 0 \end{pmatrix}$

13. In M_{23}: $\begin{pmatrix} 1 & 0 & 0 \\ 0 & 0 & 0 \end{pmatrix}, \begin{pmatrix} 0 & 1 & 0 \\ 0 & 0 & 0 \end{pmatrix}, \begin{pmatrix} 0 & 0 & 1 \\ 0 & 0 & 0 \end{pmatrix},$

$\begin{pmatrix} 0 & 0 & 0 \\ 1 & 0 & 0 \end{pmatrix}, \begin{pmatrix} 0 & 0 & 0 \\ 0 & 1 & 0 \end{pmatrix}, \begin{pmatrix} 0 & 0 & 0 \\ 0 & 0 & 1 \end{pmatrix}$

14. Show that two polynomials cannot span P_2.

*15. If $p_1, p_2, \ldots, p_m$ span P_n, show that $m \geq n + 1$.

16. Show that if $\mathbf{u}$ and $\mathbf{v}$ are in span $\{\mathbf{v}_1, \mathbf{v}_2, \ldots, \mathbf{v}_k\}$, then $\mathbf{u} + \mathbf{v}$ and $\alpha\mathbf{u}$ are in span $\{\mathbf{v}_1, \mathbf{v}_2, \ldots, \mathbf{v}_k\}$ [*Hint:* Using the definition of span write $\mathbf{u} + \mathbf{v}$ and $\alpha\mathbf{u}$ as linear combinations of $\mathbf{v}_1, \mathbf{v}_2, \ldots, \mathbf{v}_k$.]

17. Show that the infinite set $\{1, x, x^2, x^3, \ldots\}$ spans P, the vector space of polynomials.

18. Let H be a subspace of V containing $\mathbf{v}_1, \mathbf{v}_2, \ldots, \mathbf{v}_n$. Show that span $\{\mathbf{v}_1, \mathbf{v}_2, \ldots, \mathbf{v}_n\} \subseteq H$. That is, span $\{\mathbf{v}_1, \mathbf{v}_2, \ldots, \mathbf{v}_n\}$ is the *smallest* subspace of V containing $\mathbf{v}_1, \mathbf{v}_2, \ldots, \mathbf{v}_n$.

19. Let $\mathbf{v}_1 = (x_1, y_1, z_1)$ and $\mathbf{v}_2 = (x_2, y_2, z_2)$ be in $\mathbb{R}^3$. Show that if $\mathbf{v}_2 = c\mathbf{v}_1$, then span $\{\mathbf{v}_1, \mathbf{v}_2\}$ is a line passing through the origin.

**20. In Problem 19 assume that $\mathbf{v}_1$ and $\mathbf{v}_2$ are not parallel. Show that $H = $ span $\{\mathbf{v}_1, \mathbf{v}_2\}$ is a plane passing through the origin. What is the equation of that plane? [*Hint:* If $(x, y, z) \in H$, write $\mathbf{v} = a_1\mathbf{v}_1 + a_2\mathbf{v}_2$ and find a condition relating x, y, and z such that the resulting 3×2 system has a solution.]

21. Prove Theorem 2. [*Hint:* If $\mathbf{v} \in V$, write $\mathbf{v}$ as a linear combination of $\mathbf{v}_1, \mathbf{v}_2, \ldots, \mathbf{v}_n, \mathbf{v}_{n+1}$ with the coefficient of $\mathbf{v}_{n+1}$ equal to zero.]

22. Show that M_{22} can be spanned by invertible matrices.

*23. Let $\{\mathbf{u}_1, \mathbf{u}_2, \ldots, \mathbf{u}_n\}$ and $\{\mathbf{v}_1, \mathbf{v}_2, \ldots, \mathbf{v}_n\}$ be $2n$ vectors in a vector space V. Suppose that

$$\mathbf{v}_1 = a_{11}\mathbf{u}_1 + a_{12}\mathbf{u}_2 + \cdots + a_{1n}\mathbf{u}_n$$
$$\mathbf{v}_2 = a_{21}\mathbf{u}_1 + a_{22}\mathbf{u}_2 + \cdots + a_{2n}\mathbf{u}_n$$
$$\vdots \qquad \vdots \qquad \vdots \qquad \qquad \vdots$$
$$\mathbf{v}_n = a_{n1}\mathbf{u}_1 + a_{n2}\mathbf{u}_2 + \cdots + a_{nn}\mathbf{u}_n$$

Show that if

$$\begin{vmatrix} a_{11} & a_{12} & \cdots & a_{1n} \\ a_{21} & a_{22} & \cdots & a_{2n} \\ \vdots & \vdots & & \vdots \\ a_{n1} & a_{n2} & \cdots & a_{nn} \end{vmatrix} \neq 0$$

then span $\{\mathbf{u}_1, \mathbf{u}_2, \ldots, \mathbf{u}_n\} = $ span $\{\mathbf{v}_1, \mathbf{v}_2, \ldots, \mathbf{v}_n\}$.

4.5   LINEAR INDEPENDENCE

In the study of linear algebra, one of the central ideas is that of the linear dependence or independence of vectors. In this section we define what we mean by linear independence and show how it is related to the theory of homogeneous systems of equations and determinants.

Is there a special relationship between the vectors $\mathbf{v}_1 = \begin{pmatrix} 1 \\ 2 \end{pmatrix}$ and $\mathbf{v}_2 = \begin{pmatrix} 2 \\ 4 \end{pmatrix}$? Of course, we see that $\mathbf{v}_2 = 2\mathbf{v}_1$ or, writing this equation in another way,

$$2\mathbf{v}_1 - \mathbf{v}_2 = \mathbf{0} \tag{1}$$

That is, the zero vector can be written as a linear combination of $\mathbf{v}_1$ and $\mathbf{v}_2$. What is special about the vectors $\mathbf{v}_1 = \begin{pmatrix} 1 \\ 2 \\ 3 \end{pmatrix}$, $\mathbf{v}_2 = \begin{pmatrix} -4 \\ 1 \\ 5 \end{pmatrix}$, and $\mathbf{v}_3 = \begin{pmatrix} -5 \\ 8 \\ 19 \end{pmatrix}$? This question is more difficult to answer at first glance. It is easy to verify, however, that $\mathbf{v}_3 = 3\mathbf{v}_1 + 2\mathbf{v}_2$, or, rewriting,

$$3\mathbf{v}_1 + 2\mathbf{v}_2 - \mathbf{v}_3 = \mathbf{0} \tag{2}$$

Now we have written the zero vector as a linear combination of $\mathbf{v}_1$, $\mathbf{v}_2$, and $\mathbf{v}_3$. It appears that the two vectors in equation (1) and the three vectors in (2) are more closely related than an arbitrary pair of 2-vectors or an arbitrary triple of 3-vectors. In each case we say that the vectors are *linearly dependent*. In general, we have the following important definition.

DEFINITION 1   **Linear Dependence and Independence** Let $\mathbf{v}_1, \mathbf{v}_2, \ldots, \mathbf{v}_n$ be n vectors in a vector space V. Then the vectors are said to be **linearly dependent** if there exist n scalars $c_1, c_2, \ldots, c_n$ *not all zero* such that

$$c_1\mathbf{v}_1 + c_2\mathbf{v}_2 + \cdots + c_n\mathbf{v}_n = \mathbf{0} \tag{3}$$

If the vectors are not linearly dependent, they are said to be **linearly independent.**

Putting this another way, $\mathbf{v}_1, \mathbf{v}_2, \ldots, \mathbf{v}_n$ are linearly independent if the equation $c_1\mathbf{v}_1 + c_2\mathbf{v}_2 + \cdots + c_n\mathbf{v}_n = \mathbf{0}$ holds only for $c_1 = c_2 = \cdots = c_n = 0$. They are linearly dependent if the zero vector in V can be written as a linear combination of $\mathbf{v}_1, \mathbf{v}_2, \ldots, \mathbf{v}_n$ with not all the coefficients equal to zero.

How do we determine whether a set of vectors is linearly dependent or independent? The case for 2-vectors is easy.

THEOREM 1   Two vectors in a vector space V are linearly dependent if and only if one is a scalar multiple of the other.

Proof   First suppose that $\mathbf{v}_2 = c\mathbf{v}_1$ for some scalar $c \neq 0$. Then $c\mathbf{v}_1 - \mathbf{v}_2 = \mathbf{0}$ and $\mathbf{v}_1$ and $\mathbf{v}_2$ are linearly dependent. On the other hand, suppose that $\mathbf{v}_1$ and $\mathbf{v}_2$ are linearly dependent. Then there are constants c_1 and c_2, not both zero, such that $c_1\mathbf{v}_1 + c_2\mathbf{v}_2 = \mathbf{0}$. If $c_1 \neq 0$, then, dividing by c_1, we obtain $\mathbf{v}_1 + (c_2/c_1)\mathbf{v}_2 = \mathbf{0}$ or

$$\mathbf{v}_1 = \left(-\frac{c_2}{c_1} \right)\mathbf{v}_2$$

That is, $\mathbf{v}_1$ is a scalar multiple of $\mathbf{v}_2$. If $c_1 = 0$, then $c_2 \neq 0$, and hence $\mathbf{v}_2 = \mathbf{0} = 0\mathbf{v}_1$. ∎

EXAMPLE 1   **Two Linearly Dependent Vectors in $\mathbb{R}^4$**   The vectors $\mathbf{v}_1 = \begin{pmatrix} 2 \\ -1 \\ 0 \\ 3 \end{pmatrix}$ and

$\mathbf{v}_2 = \begin{pmatrix} -6 \\ 3 \\ 0 \\ -9 \end{pmatrix}$ are linearly dependent since $\mathbf{v}_2 = -3\mathbf{v}_1$. ∎

EXAMPLE 2   **Two Linearly Independent Vectors in $\mathbb{R}^3$**   The vectors $\begin{pmatrix} 1 \\ 2 \\ 4 \end{pmatrix}$ and $\begin{pmatrix} 2 \\ 5 \\ -3 \end{pmatrix}$ are

linearly independent; if they were not, we would have $\begin{pmatrix} 2 \\ 5 \\ -3 \end{pmatrix} = c\begin{pmatrix} 1 \\ 2 \\ 4 \end{pmatrix} = \begin{pmatrix} c \\ 2c \\ 4c \end{pmatrix}$.

Then $2 = c$, $5 = 2c$, and $-3 = 4c$, which is clearly impossible for any number c. ∎

EXAMPLE 3   **Determining Whether Three Vectors in $\mathbb{R}^3$ Are Linearly Dependent or Independent**   Determine whether the vectors $\begin{pmatrix} 1 \\ -2 \\ 3 \end{pmatrix}$, $\begin{pmatrix} 2 \\ -2 \\ 0 \end{pmatrix}$, and $\begin{pmatrix} 0 \\ 1 \\ 7 \end{pmatrix}$ are linearly dependent or independent.

Solution   Suppose that $c_1\begin{pmatrix} 1 \\ -2 \\ 3 \end{pmatrix} + c_2\begin{pmatrix} 2 \\ -2 \\ 0 \end{pmatrix} + c_3\begin{pmatrix} 0 \\ 1 \\ 7 \end{pmatrix} = \mathbf{0} = \begin{pmatrix} 0 \\ 0 \\ 0 \end{pmatrix}$. Then, multiplying through and adding, we have $\begin{pmatrix} c_1 + 2c_2 \\ -2c_1 - 2c_2 + c_3 \\ 3c_1 \qquad + 7c_3 \end{pmatrix} = \begin{pmatrix} 0 \\ 0 \\ 0 \end{pmatrix}$. This yields a homogeneous system of three equations in the three unknowns c_1, c_2, and c_3:

$$\begin{aligned} c_1 + 2c_2 \qquad\quad &= 0 \\ -2c_1 - 2c_2 + c_3 &= 0 \\ 3c_1 \qquad\quad + 7c_3 &= 0 \end{aligned} \qquad\qquad \textbf{(4)}$$

Thus the vectors will be linearly dependent if and only if system (4) has nontrivial solutions. We write system (4) using an augmented matrix and then row-reduce:

$$\begin{pmatrix} 1 & 2 & 0 & | & 0 \\ -2 & -2 & 1 & | & 0 \\ 3 & 0 & 7 & | & 0 \end{pmatrix} \xrightarrow[R_3 \to R_3 - 3R_1]{R_2 \to R_2 + 2R_1} \begin{pmatrix} 1 & 2 & 0 & | & 0 \\ 0 & 2 & 1 & | & 0 \\ 0 & -6 & 7 & | & 0 \end{pmatrix}$$

$$\xrightarrow{R_2 \to \frac{1}{2}R_2} \begin{pmatrix} 1 & 2 & 0 & | & 0 \\ 0 & 1 & \frac{1}{2} & | & 0 \\ 0 & -6 & 7 & | & 0 \end{pmatrix} \xrightarrow[R_3 \to R_3 + 6R_2]{R_1 \to R_1 - 2R_2} \begin{pmatrix} 1 & 0 & -1 & | & 0 \\ 0 & 1 & \frac{1}{2} & | & 0 \\ 0 & 0 & 10 & | & 0 \end{pmatrix}$$

$$\xrightarrow{R_3 \to \frac{1}{10}R_3} \begin{pmatrix} 1 & 0 & -1 & | & 0 \\ 0 & 1 & \frac{1}{2} & | & 0 \\ 0 & 0 & 1 & | & 0 \end{pmatrix} \xrightarrow[R_2 \to R_2 - \frac{1}{2}R_3]{R_1 \to R_1 + R_3} \begin{pmatrix} 1 & 0 & 0 & | & 0 \\ 0 & 1 & 0 & | & 0 \\ 0 & 0 & 1 & | & 0 \end{pmatrix}$$

The last system of equations reads $c_1 = 0$, $c_2 = 0$, $c_3 = 0$. Hence (4) has no non-trivial solutions and the given vectors are linearly independent.   ∎

EXAMPLE 4   **Determining Whether Three Vectors in $\mathbb{R}^3$ Are Linearly Dependent or Independent**   Determine whether the vectors $\begin{pmatrix} 1 \\ -3 \\ 0 \end{pmatrix}$, $\begin{pmatrix} 3 \\ 0 \\ 4 \end{pmatrix}$, and $\begin{pmatrix} 11 \\ -6 \\ 12 \end{pmatrix}$ are linearly dependent or independent.

Solution   The equation $c_1 \begin{pmatrix} 1 \\ -3 \\ 0 \end{pmatrix} + c_2 \begin{pmatrix} 3 \\ 0 \\ 4 \end{pmatrix} + c_3 \begin{pmatrix} 11 \\ -6 \\ 12 \end{pmatrix} = \begin{pmatrix} 0 \\ 0 \\ 0 \end{pmatrix}$ leads to the homogeneous system

$$\begin{aligned} c_1 + 3c_2 + 11c_3 &= 0 \\ -3c_1 \qquad\quad - 6c_3 &= 0 \\ 4c_2 + 12c_3 &= 0 \end{aligned} \qquad\qquad \textbf{(5)}$$

Writing system (5) in augmented-matrix form and row-reducing, we obtain, successively,

$$\begin{pmatrix} 1 & 3 & 11 & | & 0 \\ -3 & 0 & -6 & | & 0 \\ 0 & 4 & 12 & | & 0 \end{pmatrix} \xrightarrow{R_2 \to R_2 + 3R_1} \begin{pmatrix} 1 & 3 & 11 & | & 0 \\ 0 & 9 & 27 & | & 0 \\ 0 & 4 & 12 & | & 0 \end{pmatrix}$$

$$\xrightarrow{R_2 \to \frac{1}{9}R_2} \begin{pmatrix} 1 & 3 & 11 & | & 0 \\ 0 & 1 & 3 & | & 0 \\ 0 & 4 & 12 & | & 0 \end{pmatrix} \xrightarrow[R_3 \to R_3 - 4R_2]{R_1 \to R_1 - 3R_2} \begin{pmatrix} 1 & 0 & 2 & | & 0 \\ 0 & 1 & 3 & | & 0 \\ 0 & 0 & 0 & | & 0 \end{pmatrix}$$

We can stop here since the theory of Section 1.4 shows us that system (5) has an infinite number of solutions. For example, the last augmented matrix reads

$$c_1 \qquad + 2c_3 = 0$$
$$c_2 + 3c_3 = 0$$

If we choose $c_3 = 1$, we have $c_2 = -3$ and $c_1 = -2$ so that, as is easily verified,

$$-2\begin{pmatrix} 1 \\ -3 \\ 0 \end{pmatrix} - 3\begin{pmatrix} 3 \\ 0 \\ 4 \end{pmatrix} + \begin{pmatrix} 11 \\ -6 \\ 12 \end{pmatrix} = \begin{pmatrix} 0 \\ 0 \\ 0 \end{pmatrix}$$ and the vectors are linearly dependent. ∎

Geometric Interpretation of Linear Dependence in $\mathbb{R}^3$

In Example 3 we found three vectors in $\mathbb{R}^3$ that were linearly independent. In Example 4 we found three vectors that were dependent. What does this mean geometrically?

Suppose that $\mathbf{u}$, $\mathbf{v}$, and $\mathbf{w}$ are three linearly dependent vectors in $\mathbb{R}^3$. Then there are constants c_1, c_2, and c_3, not all zero, such that

$$c_1\mathbf{u} + c_2\mathbf{v} + c_3\mathbf{w} = \mathbf{0} \qquad \text{(6)}$$

Suppose that $c_3 \neq 0$ (a similar result holds if $c_1 \neq 0$ or $c_2 \neq 0$). Then we may divide both sides of (6) by c_3 and rearrange terms to obtain

$$\mathbf{w} = -\frac{c_1}{c_3}\mathbf{u} - \frac{c_2}{c_3}\mathbf{v} = A\mathbf{u} + B\mathbf{v}$$

where $A = -c_1/c_3$ and $B = -c_2/c_3$. We now show that $\mathbf{u}$, $\mathbf{v}$, and $\mathbf{w}$ are coplanar. We compute

$$\mathbf{w} \cdot (\mathbf{u} \times \mathbf{v}) = (A\mathbf{u} + B\mathbf{v}) \cdot (\mathbf{u} \times \mathbf{v}) = A[\mathbf{u} \cdot (\mathbf{u} \times \mathbf{v})] + B[\mathbf{v} \cdot (\mathbf{u} \times \mathbf{v})]$$
$$= A \cdot 0 + B \cdot 0 = 0$$

because $\mathbf{u}$ and $\mathbf{v}$ are both orthogonal to $\mathbf{u} \times \mathbf{v}$ (see page 185). Let $\mathbf{n} = \mathbf{u} \times \mathbf{v}$. Then $\mathbf{u}$ and $\mathbf{v}$ lie in the plane consisting of those vectors passing through the origin that are orthogonal to $\mathbf{n}$. But $\mathbf{w}$ is in the same plane because $\mathbf{w} \cdot \mathbf{n} = \mathbf{w} \cdot (\mathbf{u} \times \mathbf{v}) = 0$. This shows that $\mathbf{u}$, $\mathbf{v}$, and $\mathbf{w}$ are coplanar.

In Problem 59 you are asked to show that if $\mathbf{u}$, $\mathbf{v}$, and $\mathbf{w}$ are coplanar, then they are linearly dependent. We conclude that

> Three vectors in $\mathbb{R}^3$ are linearly dependent if and only if they are coplanar.

Figure 4.3 illustrates this fact using the vectors in Examples 3 and 4.

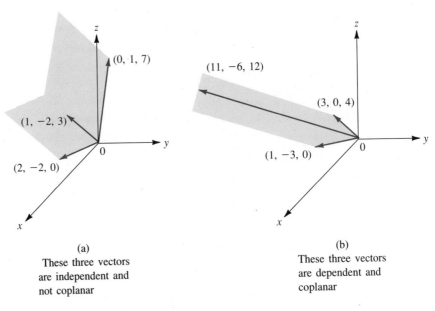

(a)
These three vectors
are independent and
not coplanar

(b)
These three vectors
are dependent and
coplanar

Figure 4.3   Two sets of three vectors

The theory of homogeneous systems can tell us something about the linear dependence or independence of vectors.

THEOREM 2   A set of n vectors in $\mathbb{R}^m$ is always linearly dependent if $n > m$.

Proof   Let $\mathbf{v}_1, \mathbf{v}_2, \ldots, \mathbf{v}_n$ be n vectors in $\mathbb{R}^m$ and let us try to find constants $c_1, c_2, \ldots, c_n$ not all zero such that

$$c_1\mathbf{v}_1 + c_2\mathbf{v}_2 + \cdots + c_n\mathbf{v}_n = \mathbf{0} \tag{7}$$

Let $\mathbf{v}_1 = \begin{pmatrix} a_{11} \\ a_{21} \\ \vdots \\ a_{m1} \end{pmatrix}$, $\mathbf{v}_2 = \begin{pmatrix} a_{12} \\ a_{22} \\ \vdots \\ a_{m2} \end{pmatrix}$, $\ldots$, $\mathbf{v}_n = \begin{pmatrix} a_{1n} \\ a_{2n} \\ \vdots \\ a_{mn} \end{pmatrix}$. Then equation (7) becomes

$$
\begin{aligned}
a_{11}c_1 + a_{12}c_2 + \cdots + a_{1n}c_n &= 0 \\
a_{21}c_1 + a_{22}c_2 + \cdots + a_{2n}c_n &= 0 \\
\vdots \qquad\quad \vdots \qquad\qquad\quad \vdots \quad & \ \ \vdots \\
a_{m1}c_1 + a_{m2}c_2 + \cdots + a_{mn}c_n &= 0
\end{aligned}
\tag{8}
$$

But system (8) is system (1.4.1) on page 23 and, according to Theorem 1.4.1, this system has an infinite number of solutions if $n > m$. Thus there are scalars $c_1, c_2, \ldots, c_n$ not all zero that satisfy (8) and the vectors $\mathbf{v}_1, \mathbf{v}_2, \ldots, \mathbf{v}_n$ are therefore linearly dependent.   ∎

EXAMPLE 5   **Four Vectors in $\mathbb{R}^3$ Are Linearly Dependent**   The vectors $\begin{pmatrix} 2 \\ -3 \\ 4 \end{pmatrix}$, $\begin{pmatrix} 4 \\ 7 \\ -6 \end{pmatrix}$, $\begin{pmatrix} 18 \\ -11 \\ 4 \end{pmatrix}$, and $\begin{pmatrix} 2 \\ -7 \\ 3 \end{pmatrix}$ are linearly dependent since they comprise a set of four 3-vectors.

∎

There is a very important (and obvious) corollary to Theorem 2.

COROLLARY   A set of linearly independent vectors in $\mathbb{R}^n$ contains at most n vectors.   ∎

Note. We can rephrase the corollary as follows: If we have n linearly independent n-vectors, then we cannot add any more vectors without making the set linearly dependent.

From system (8) we can make another important observation whose proof is left as an exercise (see Problem 27).

THEOREM 3   Let

$$A = \begin{pmatrix} a_{11} & a_{12} & \cdots & a_{1n} \\ a_{21} & a_{22} & \cdots & a_{2n} \\ \vdots & \vdots & & \vdots \\ a_{m1} & a_{m2} & \cdots & a_{mn} \end{pmatrix}$$

Then the columns of A, considered as vectors, are linearly dependently if and only if system (8), which can be written $A\mathbf{c} = \mathbf{0}$, has nontrivial solutions. Here $\mathbf{c} = \begin{pmatrix} c_1 \\ c_2 \\ \vdots \\ c_n \end{pmatrix}$.   ∎

EXAMPLE 6   **Writing Solutions to a Homogeneous System as Linear Combinations of Linearly Independent Solution Vectors**   Consider the homogeneous system

$$\begin{aligned} x_1 + 2x_2 - x_3 + 2x_4 &= 0 \\ 3x_1 + 7x_2 + x_3 + 4x_4 &= 0 \end{aligned} \qquad (9)$$

We solve this by row reduction:

$$\begin{pmatrix} 1 & 2 & -1 & 2 & | & 0 \\ 3 & 7 & 1 & 4 & | & 0 \end{pmatrix} \xrightarrow{R_2 \to R_2 - 3R_1} \begin{pmatrix} 1 & 2 & -1 & 2 & | & 0 \\ 0 & 1 & 4 & -2 & | & 0 \end{pmatrix}$$

$$\xrightarrow{R_1 \to R_1 - 2R_2} \begin{pmatrix} 1 & 0 & -9 & 6 & | & 0 \\ 0 & 1 & 4 & -2 & | & 0 \end{pmatrix}$$

The last system is

$$x_1 \qquad - 9x_3 + 6x_4 = 0$$

$$x_2 + 4x_3 - 2x_4 = 0$$

We see that this system has an infinite number of solutions, which we write as a linear combination of column vectors:

$$\begin{pmatrix} x_1 \\ x_2 \\ x_3 \\ x_4 \end{pmatrix} = \begin{pmatrix} 9x_3 - 6x_4 \\ -4x_3 + 2x_4 \\ x_3 \\ x_4 \end{pmatrix} = x_3 \begin{pmatrix} 9 \\ -4 \\ 1 \\ 0 \end{pmatrix} + x_4 \begin{pmatrix} -6 \\ 2 \\ 0 \\ 1 \end{pmatrix} \tag{10}$$

Note that $\begin{pmatrix} 9 \\ -4 \\ 1 \\ 0 \end{pmatrix}$ and $\begin{pmatrix} -6 \\ 2 \\ 0 \\ 1 \end{pmatrix}$ are linearly independent solutions to (9) because neither one is a multiple of the other. (You should verify that they are solutions.) Since x_3 and x_4 are arbitrary real numbers, we see, from (10), that we can express all solutions to the system (9) in terms of two linearly independent solution vectors. ∎

The next two theorems follow directly from Theorem 3.

THEOREM 4   Let $\mathbf{v}_1, \mathbf{v}_2, \ldots, \mathbf{v}_n$ be n vectors in $\mathbb{R}^n$ and let A be the $n \times n$ matrix whose columns are $\mathbf{v}_1, \mathbf{v}_2, \ldots, \mathbf{v}_n$. Then $\mathbf{v}_1, \mathbf{v}_2, \ldots, \mathbf{v}_n$ are linearly independent if and only if the only solution to the homogeneous system $A\mathbf{x} = \mathbf{0}$ is the trivial solution $\mathbf{x} = \mathbf{0}$.

Proof   This is Theorem 3 in the case $m = n$. ∎

THEOREM 5   Let A be an $n \times n$ matrix. Then $\det A \neq 0$ if and only if the columns of A are linearly independent.

Proof   From Theorem 4 and the Summing Up Theorem (see page 146). Columns of A are linearly independent $\Leftrightarrow \mathbf{0}$ is the only solution to $A\mathbf{x} = \mathbf{0} \Leftrightarrow \det A \neq 0$. Here, $\Leftrightarrow$ stands for the words "if and only if." ∎

Theorem 5 enables us to extend our Summing Up Theorem.

THEOREM 6   **Summing Up Theorem—View 5** Let A be an $n \times n$ matrix. Then each of the following seven statements are equivalent; that is, each one implies the other six (so that if one is true, all are true).

  i. A is invertible.

  ii. The only solution to the homogeneous system $A\mathbf{x} = \mathbf{0}$ is the trivial solution ($\mathbf{x} = \mathbf{0}$).

  iii. The system $A\mathbf{x} = \mathbf{b}$ has a unique solution for every n-vector $\mathbf{b}$.

  iv. A is row equivalent to the $n \times n$ identity matrix I_n.

  v. A can be written as the product of elementary matrices.

  vi. $\det A \neq 0$.

  vii. The columns (and rows) of A are linearly independent.

Proof   The only part not proved is that the rows of A are linearly independent if $\det A \neq 0$. If the columns are independent, then $\det A \neq 0$. Then $\det A^t = \det A \neq 0$ (see Theorem 2.2.3 on page 128). Thus the columns of A^t are linearly independent. But the columns of A^t are the rows of A, so the rows of A are independent. ■

The following theorem ties together the ideas of linear independence and spanning sets in $\mathbb{R}^n$.

THEOREM 7   Any set of n linearly independent vectors in $\mathbb{R}^n$ spans $\mathbb{R}^n$.

Proof   Let $\mathbf{v}_1 = \begin{pmatrix} a_{11} \\ a_{21} \\ \vdots \\ a_{n1} \end{pmatrix}$, $\mathbf{v}_2 = \begin{pmatrix} a_{12} \\ a_{22} \\ \vdots \\ a_{n2} \end{pmatrix}$, $\ldots$, $\mathbf{v}_n = \begin{pmatrix} a_{1n} \\ a_{2n} \\ \vdots \\ a_{nn} \end{pmatrix}$ be linearly independent and let

$\mathbf{v} = \begin{pmatrix} x_1 \\ x_2 \\ \vdots \\ x_n \end{pmatrix}$ be a vector in $\mathbb{R}^n$. We must show that there exist scalars $c_1, c_2, \ldots, c_n$

such that

$$\mathbf{v} = c_1 \mathbf{v}_1 + c_2 \mathbf{v}_2 + \cdots + c_n \mathbf{v}_n$$

That is,

$$\begin{pmatrix} x_1 \\ x_2 \\ \vdots \\ x_n \end{pmatrix} = c_1 \begin{pmatrix} a_{11} \\ a_{21} \\ \vdots \\ a_{n1} \end{pmatrix} + c_2 \begin{pmatrix} a_{12} \\ a_{22} \\ \vdots \\ a_{n2} \end{pmatrix} + \cdots + c_n \begin{pmatrix} a_{1n} \\ a_{2n} \\ \vdots \\ a_{nn} \end{pmatrix} \tag{11}$$

In (11) we multiply through, add, and equate components to obtain a system of n equations in the n unknowns $c_1, c_2, \ldots, c_n$:

$$a_{11}c_1 + a_{12}c_2 + \cdots + a_{1n}c_n = x_1$$
$$a_{21}c_1 + a_{22}c_2 + \cdots + a_{2n}c_n = x_2 \tag{12}$$
$$\vdots \qquad \vdots \qquad \qquad \vdots \qquad \vdots$$
$$a_{n1}c_1 + a_{n2}c_2 + \cdots + a_{nn}c_n = x_n$$

We can write (12) as $A\mathbf{c} = \mathbf{v}$, where

$$A = \begin{pmatrix} a_{11} & a_{12} & \cdots & a_{1n} \\ a_{21} & a_{22} & \cdots & a_{2n} \\ \vdots & \vdots & & \vdots \\ a_{n1} & a_{n2} & \cdots & a_{nn} \end{pmatrix} \quad \text{and} \quad \mathbf{c} = \begin{pmatrix} c_1 \\ c_2 \\ \vdots \\ c_n \end{pmatrix}$$

But $\det A \neq 0$ since the columns of A are linearly independent. So system (12) has a unique solution $\mathbf{c}$ by Theorem 6 and the theorem is provided. ∎

Remark. This theorem not only shows that $\mathbf{v}$ can be written as a linear combination of the independent vectors $\mathbf{v}_1, \mathbf{v}_2, \ldots, \mathbf{v}_n$, but also that this can be done in *only one way* (since the solution vector $\mathbf{c}$ is unique).

EXAMPLE 7   **Three Vectors in $\mathbb{R}^3$ Span $\mathbb{R}^3$ If Their Determinant is Nonzero**   The vectors $(2, -1, 4)$, $(1, 0, 2)$, and $(3, -1, 5)$ span $\mathbb{R}^3$ because $\begin{vmatrix} 2 & 1 & 3 \\ -1 & 0 & -1 \\ 4 & 2 & 5 \end{vmatrix} = -1 \neq 0$, so that they are independent. ∎

Every example we have done so far has been in the space $\mathbb{R}^n$. This is not so much of a restriction as it seems. In Section 5.4 (Theorem 6) we shall show that many different looking vector spaces have essentially the same properties. For example, we shall see that the space P_n is essentially the same as the space $\mathbb{R}^{n+1}$. We shall say that two such vector spaces are *isomorphic*.

This very powerful result will have to wait until Chapter 5. In the meantime, we shall do some examples in spaces other than $\mathbb{R}^n$.

EXAMPLE 8   **Three Linearly Independent Matrices in M_{23}**   In M_{23} let $A_1 = \begin{pmatrix} 1 & 0 & 2 \\ 3 & 1 & -1 \end{pmatrix}$, $A_2 = \begin{pmatrix} -1 & 1 & 4 \\ 2 & 3 & 0 \end{pmatrix}$, and $A_3 = \begin{pmatrix} -1 & 0 & 1 \\ 1 & 2 & 1 \end{pmatrix}$. Determine whether A_1, A_2, and A_3 are linearly dependent or independent.

Solution   Suppose that $c_1 A_1 + c_2 A_2 + c_3 A_3 = 0$. Then

$$\begin{pmatrix} 0 & 0 & 0 \\ 0 & 0 & 0 \end{pmatrix} = c_1 \begin{pmatrix} 1 & 0 & 2 \\ 3 & 1 & -1 \end{pmatrix} + c_2 \begin{pmatrix} -1 & 1 & 4 \\ 2 & 3 & 0 \end{pmatrix} + c_3 \begin{pmatrix} -1 & 0 & 1 \\ 1 & 2 & 1 \end{pmatrix}$$

$$= \begin{pmatrix} c_1 - c_2 - c_3 & c_2 & 2c_1 + 4c_2 + c_3 \\ 3c_1 + 2c_2 + c_3 & c_1 + 3c_2 + 2c_3 & -c_1 + c_3 \end{pmatrix}$$

This gives us a homogeneous system of six equations in the three unknowns c_1, c_2, and c_3, and it is quite easy to verify that the only solution is $c_1 = c_2 = c_3 = 0$. Thus the three matrices are linearly independent. ∎

EXAMPLE 9   **Four Linearly Independent Polynomials in P_3**   In P_3 determine whether the polynomials 1, x, x^2, and x^3 are linearly dependent or independent.

Solution   Suppose that $c_1 + c_2 x + c_3 x^2 + c_4 x^3 = 0$. This must hold for every real number x. In particular, if $x = 0$, we obtain $c_1 = 0$. Then, setting $x = 1, -1, 2$, we obtain, successively,

$$c_2 + c_3 + c_4 = 0$$
$$-c_2 + c_3 - c_4 = 0$$
$$2c_2 + 4c_3 + 8c_4 = 0$$

The determinant of this homogeneous system is

$$\begin{vmatrix} 1 & 1 & 1 \\ -1 & 1 & -1 \\ 2 & 4 & 8 \end{vmatrix} = 12 \neq 0$$

so that the system has the unique solution $c_2 = c_3 = c_4 = 0$ and the four polynomials are linearly independent. We can see this in another way. We know that any polynomial of degree 3 has at most three real roots. But if $c_1 + c_2 x + c_3 x^2 + c_4 x^3 = 0$ for some nonzero constants c_1, c_2, c_3, and c_4 and for every real number x, then we have constructed a cubic polynomial for which every real number is a root. This clearly is impossible. ∎

EXAMPLE 10   **Three Linearly Dependent Polynomials in P_2**   In P_2 determine whether the polynomials $x - 2x^2$, $x^2 - 4x$, and $-7x + 8x^2$ are linearly dependent or independent.

Solution   Let $c_1(x - 2x^2) + c_2(x^2 - 4x) + c_3(-7x + 8x^2) = 0$. Then, rearranging terms, we obtain

$$(c_1 - 4c_2 - 7c_3)x = 0$$
$$(-2c_1 + c_2 + 8c_3)x^2 = 0$$

These equations hold for every x if and only if

$$c_1 - 4c_2 - 7c_3 = 0$$

and

$$-2c_1 + c_2 + 8c_3 = 0$$

But by Theorem 1.4.1, page 25 this system of two equations in three unknowns has an infinite number of solutions. This shows that the polynomials are linearly dependent.

If we solve this homogeneous system, we obtain, successively,

$$\begin{pmatrix} 1 & -4 & -7 & | & 0 \\ -2 & 1 & 8 & | & 0 \end{pmatrix} \xrightarrow{R_2 \rightarrow R_2 + 2R_1} \begin{pmatrix} 1 & -4 & -7 & | & 0 \\ 0 & -7 & -6 & | & 0 \end{pmatrix}$$

$$\xrightarrow{R_2 \rightarrow -\frac{1}{7}R_2} \begin{pmatrix} 1 & -4 & -7 & | & 0 \\ 0 & 1 & \frac{6}{7} & | & 0 \end{pmatrix} \xrightarrow{R_1 \rightarrow R_1 + 4R_2} \begin{pmatrix} 1 & 0 & -\frac{25}{7} & | & 0 \\ 0 & 1 & \frac{6}{7} & | & 0 \end{pmatrix}$$

Thus c_3 can be chosen arbitrarily, $c_1 = \frac{25}{7}c_3$ and $c_2 = -\frac{6}{7}c_3$. If $c_3 = 7$, for example, then $c_1 = 25$, $c_2 = -6$, and we have

$$25(x - 2x^2) - 6(x^2 - 4x) + 7(-7x + 8x^2) = 0. \qquad \blacksquare$$

PROBLEMS 4.5

In Problems 1–22 determine whether the given set of vectors is linearly dependent or independent.

1. $\begin{pmatrix} 1 \\ 2 \end{pmatrix}; \begin{pmatrix} -1 \\ -3 \end{pmatrix}$

2. $\begin{pmatrix} 2 \\ -1 \\ 4 \end{pmatrix}; \begin{pmatrix} 4 \\ -2 \\ 7 \end{pmatrix}$

3. $\begin{pmatrix} 2 \\ -1 \\ 4 \end{pmatrix}; \begin{pmatrix} 4 \\ -2 \\ 8 \end{pmatrix}$

4. $\begin{pmatrix} -2 \\ 3 \end{pmatrix}; \begin{pmatrix} 4 \\ 7 \end{pmatrix}$

5. $\begin{pmatrix} -3 \\ 2 \end{pmatrix}; \begin{pmatrix} 1 \\ 10 \end{pmatrix}; \begin{pmatrix} 4 \\ -5 \end{pmatrix}$

6. $\begin{pmatrix} 1 \\ 0 \\ 1 \end{pmatrix}; \begin{pmatrix} 0 \\ 1 \\ 1 \end{pmatrix}; \begin{pmatrix} 1 \\ 1 \\ 0 \end{pmatrix}$

7. $\begin{pmatrix} 1 \\ 0 \\ 0 \end{pmatrix}; \begin{pmatrix} 0 \\ 1 \\ 0 \end{pmatrix}; \begin{pmatrix} 0 \\ 0 \\ 1 \end{pmatrix}$

8. $\begin{pmatrix} -3 \\ 4 \\ 2 \end{pmatrix}; \begin{pmatrix} 7 \\ -1 \\ 3 \end{pmatrix}; \begin{pmatrix} 1 \\ 2 \\ 8 \end{pmatrix}$

9. $\begin{pmatrix} -3 \\ 4 \\ 2 \end{pmatrix}; \begin{pmatrix} 7 \\ -1 \\ 3 \end{pmatrix}; \begin{pmatrix} 1 \\ 1 \\ 8 \end{pmatrix}$

10. $\begin{pmatrix} 1 \\ -2 \\ 1 \\ 1 \end{pmatrix}; \begin{pmatrix} 3 \\ 0 \\ 2 \\ -2 \end{pmatrix}; \begin{pmatrix} 0 \\ 4 \\ -1 \\ -1 \end{pmatrix}; \begin{pmatrix} 5 \\ 0 \\ 3 \\ -1 \end{pmatrix}$

11. $\begin{pmatrix} 1 \\ -2 \\ 1 \\ 1 \end{pmatrix}; \begin{pmatrix} 3 \\ 0 \\ 2 \\ -2 \end{pmatrix}; \begin{pmatrix} 0 \\ 4 \\ -1 \\ 1 \end{pmatrix}; \begin{pmatrix} 5 \\ 0 \\ 3 \\ -1 \end{pmatrix}$

12. $\begin{pmatrix} 1 \\ -1 \\ 2 \end{pmatrix}; \begin{pmatrix} 4 \\ 0 \\ 0 \end{pmatrix}; \begin{pmatrix} -2 \\ 3 \\ 5 \end{pmatrix}; \begin{pmatrix} 7 \\ 1 \\ 2 \end{pmatrix}$

13. In P_2: $1 - x$, x

14. In P_2: $-x$, $x^2 - 2x$, $3x + 5x^2$

15. In P_2: $1 - x$, $1 + x$, x^2

16. In P_3: x, $x^2 - x$, $x^3 - x$

17. In P_3: $2x$, $x^3 - 3$, $1 + x - 4x^3$, $x^3 + 18x - 9$

18. In M_{22}: $\begin{pmatrix} 2 & -1 \\ 4 & 0 \end{pmatrix}$, $\begin{pmatrix} 0 & -3 \\ 1 & 5 \end{pmatrix}$, $\begin{pmatrix} 4 & 1 \\ 7 & -5 \end{pmatrix}$

19. In M_{22}: $\begin{pmatrix} 1 & -1 \\ 0 & 6 \end{pmatrix}$, $\begin{pmatrix} -1 & 0 \\ 3 & 1 \end{pmatrix}$, $\begin{pmatrix} 1 & 1 \\ -1 & 2 \end{pmatrix}$, $\begin{pmatrix} 0 & 1 \\ 1 & 0 \end{pmatrix}$

20. In M_{22}: $\begin{pmatrix} -1 & 0 \\ 1 & 2 \end{pmatrix}$, $\begin{pmatrix} 2 & 3 \\ 7 & -4 \end{pmatrix}$, $\begin{pmatrix} 8 & -5 \\ 7 & 6 \end{pmatrix}$, $\begin{pmatrix} 4 & -1 \\ 2 & 3 \end{pmatrix}$, $\begin{pmatrix} 2 & 3 \\ -1 & 4 \end{pmatrix}$

*21. In $C[0, 1]$: $\sin x$, $\cos x$

*22. In $C[0, 1]$: x, $\sqrt{x}$, $\sqrt[3]{x}$

23. Determine a condition on the numbers a, b, c, and d such that the vectors $\begin{pmatrix} a \\ b \end{pmatrix}$ and $\begin{pmatrix} c \\ d \end{pmatrix}$ are linearly dependent.

*24. Find a condition on the numbers a_{ij} such that the vectors $\begin{pmatrix} a_{11} \\ a_{21} \\ a_{31} \end{pmatrix}$, $\begin{pmatrix} a_{12} \\ a_{22} \\ a_{32} \end{pmatrix}$, and $\begin{pmatrix} a_{13} \\ a_{23} \\ a_{33} \end{pmatrix}$ are linearly dependent.

25. For what value(s) of α will the vectors $\begin{pmatrix} 1 \\ 2 \\ 3 \end{pmatrix}$, $\begin{pmatrix} 2 \\ -1 \\ 4 \end{pmatrix}$, $\begin{pmatrix} 3 \\ \alpha \\ 4 \end{pmatrix}$ be linearly dependent?

26. For what value(s) of α are the vectors $\begin{pmatrix} 2 \\ -3 \\ 1 \end{pmatrix}$, $\begin{pmatrix} -4 \\ 6 \\ -2 \end{pmatrix}$, $\begin{pmatrix} \alpha \\ 1 \\ 2 \end{pmatrix}$ linearly dependent? [*Hint:* Look carefully.]

27. Prove Theorem 3. [*Hint:* Look closely at system (8).]

28. Prove that if the vectors v_1, v_2, . . . , v_n are linearly dependent vectors in $\mathbb{R}^m$ and if v_{n+1} is any other vector in $\mathbb{R}^m$, then the set v_1, v_2, . . . , v_n, v_{n+1} is linearly dependent.

29. Show that if v_1, v_2, . . . , v_n $(n \geq 2)$ are linearly independent, then so too are v_1, v_2, . . . , v_k, where $k < n$.

30. Show that if the nonzero vectors v_1 and v_2 in $\mathbb{R}^n$ are orthogonal (see page 56), then the set $\{v_1, v_2\}$ is linearly independent.

*31. Suppose that v_1 is orthogonal to v_2 and v_3 and that v_2 is orthogonal to v_3. If v_1, v_2, and v_3 are nonzero, show that the set $\{v_1, v_2, v_3\}$ is linearly independent.

32. Let A be a square $(n \times n)$ matrix whose columns are the vectors v_1, v_2, . . . , v_n. Show that v_1, v_2, . . . , v_n are linearly independent if and only if the row echelon form of A does not contain a row of zeros.

In Problems 33–37 write the solutions to the given homogeneous systems in terms of one or more linearly independent vectors.

33. $x_1 + x_2 + x_3 = 0$

34. $\begin{aligned} x_1 - x_2 + 7x_3 - x_4 &= 0 \\ 2x_1 + 3x_2 - 8x_3 + x_4 &= 0 \end{aligned}$

35. $\begin{aligned} x_1 + 2x_2 - x_3 &= 0 \\ 2x_1 + 5x_2 + 4x_3 &= 0 \end{aligned}$

36. $\begin{aligned} x_1 + x_2 + x_3 - x_4 - x_5 &= 0 \\ -2x_1 + 3x_2 + x_3 + 4x_4 - 6x_5 &= 0 \end{aligned}$

37. $x_1 + 2x_2 - 3x_3 + 5x_4 = 0$

38. Let $\mathbf{u} = (1, 2, 3)$.
    a. Let $H = \{\mathbf{v} \in \mathbb{R}^3 : \mathbf{u} \cdot \mathbf{v} = 0\}$. Show that H is a subspace of $\mathbb{R}^3$.
    b. Find two linearly independent vectors in H. Call them $\mathbf{x}$ and $\mathbf{y}$.
    c. Compute $\mathbf{w} = \mathbf{x} \times \mathbf{y}$.
    d. Show that $\mathbf{u}$ and $\mathbf{w}$ are linearly dependent.
    e. Give a geometric interpretation of parts (*a*) and (*c*) and explain why (*d*) must be true.

Remark. If $V = \{\mathbf{v} \in \mathbb{R}^3 : \mathbf{v} = \alpha\mathbf{u}$ for some real number $\alpha\}$, then V is a subspace of $\mathbb{R}^3$ and H is called the **orthogonal complement** of **V**.

39. Choose a vector $\mathbf{u} \neq \mathbf{0}$ in $\mathbb{R}^3$. Repeat the steps of Problem 38, starting with the vector you have chosen.

40. Show that any four polynomials in P_2 are linearly dependent.

41. Show that two polynomials cannot span P_2.

*42. Show that any $n + 2$ polynomials in P_n are linearly dependent.

43. Show that any subset of a set of linearly independent vectors is linearly independent. [*Note:* This generalizes Problem 29.]

44. Show that any seven matrices in M_{32} are linearly dependent.

*45. Prove that any $mn + 1$ matrices in M_{mn} are linearly dependent.

46. Let S_1 and S_2 be two finite, linearly independent sets in a vector space V. Show that $S_1 \cap S_2$ is a linearly independent set.

*47. Show that in P_n the polynomials $1, x, x^2, \ldots, x^n$ are linearly independent. [*Hint:* This is certainly true if $n = 1$. Assume that $1, x, x^2, \ldots, x^{n-1}$ are linearly independent and show how this implies that $1, x, x^2, \ldots, x^n$ are also independent. This will complete the proof by mathematical induction.]

48. Let $\{\mathbf{v}_1, \mathbf{v}_2, \ldots, \mathbf{v}_n\}$ be a linearly independent set. Show that the vectors $\mathbf{v}_1, \mathbf{v}_1 + \mathbf{v}_2, \mathbf{v}_1 + \mathbf{v}_2 + \mathbf{v}_3, \ldots, \mathbf{v}_1 + \mathbf{v}_2 + \cdots + \mathbf{v}_n$ are linearly independent.

49. Let $S = \{\mathbf{v}_1, \mathbf{v}_2, \ldots, \mathbf{v}_n\}$ be a linearly dependent set of nonzero vectors in a vector space V. Show that at least one of the vectors in S can be written as a linear combination of the vectors that precede it. That is, show that there is an integer $k \leq n$ and scalars $a_1, a_2, \ldots, a_{k-1}$ such that $\mathbf{v}_k = a_1\mathbf{v}_1 + a_2\mathbf{v}_2 + \cdots + a_{k-1}\mathbf{v}_{k-1}$.

50. Let $\{\mathbf{v}_1, \mathbf{v}_2, \ldots, \mathbf{v}_n\}$ be a set of vectors having the property that the set $\{\mathbf{v}_i, \mathbf{v}_j\}$ is linearly dependent when $i \neq j$. Show that each vector in the set is a multiple of a single vector in the set.

$\boxed{\text{Calculus}}$ **51.** Let f and g be in $C^1[0 \quad 1]$. Then the **Wronskian**† of f and g is defined by

$$W(f, g)(x) = \begin{vmatrix} f(x) & g(x) \\ f'(x) & g'(x) \end{vmatrix}$$

Show that if f and g are linearly dependent, then $W(f, g)(x) = 0$ for every $x \in [0, 1]$.

---

†Named after the Polish mathematician Józef Maria Hoene-Wroński (1778–1853). Hoene-Wroński spent most of his adult life in France. He worked on the theory of determinants and was also known for his critical writings in the philosophy of mathematics.

52. Determine a suitable definition for the Wronskian of the functions $f_1, f_2, \ldots, f_n \in$ $C^{(n-1)} [0, 1]$.† [Calculus]

53. Suppose that $\mathbf{u}$, $\mathbf{v}$, and $\mathbf{w}$ are linearly independent. Prove or disprove: $\mathbf{u} + \mathbf{v}$, $\mathbf{u} + \mathbf{w}$, and $\mathbf{v} + \mathbf{w}$ are linearly independent.

54. For what real values of c are the vectors $(1 - c, 1 + c)$ and $(1 + c, 1 - c)$ linearly independent?

55. Show that the vectors $(1, a, a^2)$, $(1, b, b^2)$, and $(1, c, c^2)$ are linearly independent if $a \neq b$, $a \neq c$, and $b \neq c$.

56. Let $\{\mathbf{v}_1, \mathbf{v}_2, \ldots, \mathbf{v}_n\}$ be a linearly independent set and suppose that $\mathbf{v} \notin$ span $\{\mathbf{v}_1, \mathbf{v}_2, \ldots, \mathbf{v}_n\}$. Show that $\{\mathbf{v}_1, \mathbf{v}_2, \ldots, \mathbf{v}_n, \mathbf{v}\}$ is a linearly independent set.

57. Find a set of three linearly independent vectors in $\mathbb{R}^3$ that contains the vectors $\begin{pmatrix} 2 \\ 1 \\ 2 \end{pmatrix}$ and $\begin{pmatrix} -1 \\ 3 \\ 4 \end{pmatrix}$. $\left[\textit{Hint:} \text{ Find a vector } \mathbf{v} \notin \text{span} \left\{ \begin{pmatrix} 2 \\ 1 \\ 2 \end{pmatrix}, \begin{pmatrix} -1 \\ 3 \\ 4 \end{pmatrix} \right\}. \right]$

58. Find a set of three linearly independent vectors in P_2 that contains the polynomials $1 - x^2$ and $1 + x^2$.

59. Suppose that $\mathbf{u} = \begin{pmatrix} u_1 \\ u_2 \\ u_3 \end{pmatrix}$, $\mathbf{v} = \begin{pmatrix} v_1 \\ v_2 \\ v_3 \end{pmatrix}$, and $\mathbf{w} = \begin{pmatrix} w_1 \\ w_2 \\ w_3 \end{pmatrix}$ are coplanar.

   a. Show that there exist constants a, b, and c not all zero such that

$$au_1 + bu_2 + cu_3 = 0$$
$$av_1 + bv_2 + cv_3 = 0$$
$$aw_1 + bw_2 + cw_3 = 0$$

   b. Explain why

$$\det \begin{pmatrix} u_1 & u_2 & u_3 \\ v_1 & v_2 & v_3 \\ w_1 & w_2 & w_3 \end{pmatrix} = 0$$

   c. Use Theorem 3 to show that $\mathbf{u}$, $\mathbf{v}$, and $\mathbf{w}$ are linearly dependent.

4.6 BASIS AND DIMENSION

We have seen that in $\mathbb{R}^2$ it is convenient to write vectors in terms of the vectors $\mathbf{i} = \begin{pmatrix} 1 \\ 0 \end{pmatrix}$ and $\mathbf{j} = \begin{pmatrix} 0 \\ 1 \end{pmatrix}$. In $\mathbb{R}^3$ we wrote vectors in terms of $\begin{pmatrix} 1 \\ 0 \\ 0 \end{pmatrix}$, $\begin{pmatrix} 0 \\ 1 \\ 0 \end{pmatrix}$, and $\begin{pmatrix} 0 \\ 0 \\ 1 \end{pmatrix}$. We now generalize this idea.

---

† $C^{(n-1)} [0, 1]$ is the set of functions whose $(n-1)$st derivatives are defined and continuous on $[0, 1]$.

DEFINITION 1   **Basis**   A set of vectors $\{v_1, v_2, \ldots, v_n\}$ is a **basis** for a vector space V if

i. $\{v_1, v_2, \ldots, v_n\}$ is linearly independent.

ii. $\{v_1, v_2, \ldots, v_n\}$ spans V.

We have already seen quite a few examples of bases. In Theorem 4.5.7, for instance, we saw that any set of n linearly independent vectors in $\mathbb{R}^n$ spans $\mathbb{R}^n$. Thus

> *Every set of n linearly independent vectors in $\mathbb{R}^n$ is a basis in $\mathbb{R}^n$.*

In $\mathbb{R}^n$ we define

$$\mathbf{e}_1 = \begin{pmatrix} 1 \\ 0 \\ 0 \\ \vdots \\ 0 \end{pmatrix}, \; \mathbf{e}_2 = \begin{pmatrix} 0 \\ 1 \\ 0 \\ \vdots \\ 0 \end{pmatrix}, \; \mathbf{e}_3 = \begin{pmatrix} 0 \\ 0 \\ 1 \\ \vdots \\ 0 \end{pmatrix}, \; \ldots, \; \mathbf{e}_n = \begin{pmatrix} 0 \\ 0 \\ 0 \\ \vdots \\ 1 \end{pmatrix}$$

Standard basis

Then since the $\mathbf{e}_i$'s are the columns of the identity matrix (which has determinant 1), $\{\mathbf{e}_1, \mathbf{e}_2, \ldots, \mathbf{e}_n\}$ is linearly independent and therefore constitutes a basis in $\mathbb{R}^n$. This special basis is called the **standard basis** in $\mathbb{R}^n$. We now find bases for some other spaces.

EXAMPLE 1   **Standard Basis for P_n**   By Example 4.5.9, page 239, the polynomials $1, x, x^2, x^3$ are linearly independent in P_3. By Example 4.4.3, page 224, these polynomials span P_3. Thus $\{1, x, x^2, x^3\}$ is a basis for P_3. In general, the monomials $\{1, x, x^2, x^3, \ldots, x^n\}$ constitute a basis for P_n. This is called the **standard basis** for P_n.   ∎

EXAMPLE 2   **Standard Basis for M_{22}**   We saw in Example 4.4.6, page 225, that $\begin{pmatrix} 1 & 0 \\ 0 & 0 \end{pmatrix}$, $\begin{pmatrix} 0 & 1 \\ 0 & 0 \end{pmatrix}$, $\begin{pmatrix} 0 & 0 \\ 1 & 0 \end{pmatrix}$, and $\begin{pmatrix} 0 & 0 \\ 0 & 1 \end{pmatrix}$ span M_{22}. If $\begin{pmatrix} c_1 & c_2 \\ c_3 & c_4 \end{pmatrix} = c_1 \begin{pmatrix} 1 & 0 \\ 0 & 0 \end{pmatrix} + c_2 \begin{pmatrix} 0 & 1 \\ 0 & 0 \end{pmatrix} + c_3 \begin{pmatrix} 0 & 0 \\ 1 & 0 \end{pmatrix} + c_4 \begin{pmatrix} 0 & 0 \\ 0 & 1 \end{pmatrix} = \begin{pmatrix} 0 & 0 \\ 0 & 0 \end{pmatrix}$, then, obviously, $c_1 = c_2 = c_3 = c_4 = 0$. Thus these four matrices are linearly independent and form a basis for M_{22}. This is called the **standard basis** for M_{22}.   ∎

EXAMPLE 3   **A Basis for a Subspace of $\mathbb{R}^3$**   Find a basis for the set of vectors lying on the plane

$$\pi = \left\{ \begin{pmatrix} x \\ y \\ z \end{pmatrix} : 2x - y + 3z = 0 \right\}$$

Solution We saw in Example 4.2.6 that π is a vector space. To find a basis, we first note that if x and z are chosen arbitrarily and if $\begin{pmatrix} x \\ y \\ z \end{pmatrix} \in \pi$, then $y = 2x + 3z$. Thus vectors in

π have the form $\begin{pmatrix} x \\ 2x + 3z \\ z \end{pmatrix}$. Since x and z are arbitrary, we choose some simple

values for them. Choosing $x = 1$, $z = 0$, we obtain $\mathbf{v}_1 = \begin{pmatrix} 1 \\ 2 \\ 0 \end{pmatrix}$; choosing $x = 0$,

$z = 1$, we get $\mathbf{v}_2 = \begin{pmatrix} 0 \\ 3 \\ 1 \end{pmatrix}$. Then $\begin{pmatrix} x \\ 2x + 3z \\ z \end{pmatrix} = x \begin{pmatrix} 1 \\ 2 \\ 0 \end{pmatrix} + z \begin{pmatrix} 0 \\ 3 \\ 1 \end{pmatrix}$. Thus $\{\mathbf{v}_1, \mathbf{v}_2\}$ span

π, and since they are obviously linearly independent (because one is not a multiple of the other), they form a basis for π. We can see this more directly by writing

$$\begin{pmatrix} x \\ 2x + 3z \\ z \end{pmatrix} = \begin{pmatrix} x \\ 2x \\ 0 \end{pmatrix} + \begin{pmatrix} 0 \\ 3z \\ z \end{pmatrix} = x \begin{pmatrix} 1 \\ 2 \\ 0 \end{pmatrix} + z \begin{pmatrix} 0 \\ 3 \\ 1 \end{pmatrix}$$

which shows that $\begin{pmatrix} 1 \\ 2 \\ 0 \end{pmatrix}$ and $\begin{pmatrix} 0 \\ 3 \\ 1 \end{pmatrix}$ span π. ∎

If $\mathbf{v}_1, \mathbf{v}_2, \ldots, \mathbf{v}_n$ is a basis for V, then any other vector $\mathbf{v} \in V$ can be written $\mathbf{v} = c_1\mathbf{v}_1 + c_2\mathbf{v}_2 + \cdots + c_n\mathbf{v}_n$. Can it be written in another way as a linear combination of the $\mathbf{v}_i$'s? The answer is *no*. (See the remark following the proof of Theorem 4.5.7, page 238, in the case $V = \mathbb{R}^n$.)

THEOREM 1 If $\{\mathbf{v}_1, \mathbf{v}_2, \ldots, \mathbf{v}_n\}$ is a basis for V and if $\mathbf{v} \in V$, then there exists a *unique* set of scalars $c_1, c_2, \ldots, c_n$ such that $\mathbf{v} = c_1\mathbf{v}_1 + c_2\mathbf{v}_2 + \cdots + c_n\mathbf{v}_n$.

Proof At least one such set of scalars exists because $\{\mathbf{v}_1, \mathbf{v}_2, \ldots, \mathbf{v}_n\}$ spans V. Suppose then that $\mathbf{v}$ can be written in two ways as a linear combination of the basis vectors. That is, suppose that

$$\mathbf{v} = c_1\mathbf{v}_1 + c_2\mathbf{v}_2 + \cdots + c_n\mathbf{v}_n = d_1\mathbf{v}_1 + d_2\mathbf{v}_2 + \cdots + d_n\mathbf{v}_n$$

Then, subtracting, we obtain the equation

$$(c_1 - d_1)\mathbf{v}_1 + (c_2 - d_2)\mathbf{v}_2 + \cdots + (c_n - d_n)\mathbf{v}_n = \mathbf{0}$$

But since the $\mathbf{v}_i$'s are linearly independent, this equation can hold only if $c_1 - d_1 = c_2 - d_2 = \cdots = c_n - d_n = 0$. Thus $c_1 = d_1, c_2 = d_2, \ldots, c_n = d_n$ and the theorem is proved.   ∎

We have seen that vector spaces may have many bases. A question naturally arises: Do all bases contain the same number of vectors? In $\mathbb{R}^3$ the answer is certainly yes. To see this we note that any three linearly independent vectors in $\mathbb{R}^3$ form a basis. But fewer than three vectors cannot form a basis since, as we saw in Section 4.4, the span of two linearly independent vectors in $\mathbb{R}^3$ is a plane in $\mathbb{R}^3$—and a plane is not all of $\mathbb{R}^3$. Similarly, a set of four or more vectors in $\mathbb{R}^3$ cannot be linearly independent; for if the first three vectors in the set are linearly independent, then they form a basis, and therefore all other vectors in the set can be written as a linear combination of the first three. Thus all bases in $\mathbb{R}^3$ contain three vectors. The next theorem tells us that the answer to the question posed above is *yes* for all vector spaces.

THEOREM 2   If $\{\mathbf{u}_1, \mathbf{u}_2, \ldots, \mathbf{u}_m\}$ and $\{\mathbf{v}_1, \mathbf{v}_2, \ldots, \mathbf{v}_n\}$ are bases for the vector space V, then $m = n$; that is, any two bases in a vector space V have the same number of vectors.

Proof†   Let $S_1 = \{\mathbf{u}_1, \ldots, \mathbf{u}_m\}$ and $S_2 = \{\mathbf{v}_1, \ldots, \mathbf{v}_n\}$ be two bases for V. We must show that $m = n$. We prove this by showing that if $m > n$, then S_1 is a linearly dependent set, which contradicts the hypothesis that S_1 is a basis. This will show that $m \leq n$. The same proof will then show that $n \leq m$, and this will prove the theorem. Hence all we must show is that if $m > n$, then S_1 is dependent. Since S_2 constitutes a basis, we can write each $\mathbf{u}_i$ as a linear combination of the $\mathbf{v}_i$'s. We have

$$
\begin{aligned}
\mathbf{u}_1 &= a_{11}\mathbf{v}_1 + a_{12}\mathbf{v}_2 + \cdots + a_{1n}\mathbf{v}_n \\
\mathbf{u}_2 &= a_{21}\mathbf{v}_1 + a_{22}\mathbf{v}_2 + \cdots + a_{2n}\mathbf{v}_n \\
&\;\;\vdots \qquad\quad \vdots \qquad\quad \vdots \qquad\qquad \vdots \\
\mathbf{u}_m &= a_{m1}\mathbf{v}_1 + a_{m2}\mathbf{v}_2 + \cdots + a_{mn}\mathbf{v}_n
\end{aligned}
\tag{1}
$$

To show that S_1 is dependent, we must find scalars $c_1, c_2, \ldots, c_m$, not all zero, such that

$$c_1\mathbf{u}_1 + c_2\mathbf{u}_2 + \cdots + c_m\mathbf{u}_m = \mathbf{0} \tag{2}$$

Inserting (1) into (2), we obtain

$$
c_1(a_{11}\mathbf{v}_1 + a_{12}\mathbf{v}_2 + \cdots + a_{1n}\mathbf{v}_n) + c_2(a_{21}\mathbf{v}_1 + a_{22}\mathbf{v}_2 + \cdots + a_{2n}\mathbf{v}_n) \\
+ \cdots + c_m(a_{m1}\mathbf{v}_1 + a_{m2}\mathbf{v}_2 + \cdots + a_{mn}\mathbf{v}_n) = \mathbf{0} \tag{3}
$$

---

† This proof is given for vector spaces with bases containing a finite number of vectors. We also treat the scalars as though they were real numbers. However, the proof works in the complex case as well.

Equation (3) can be rewritten as

$$(a_{11}c_1 + a_{21}c_2 + \cdots + a_{m1}c_m)\mathbf{v}_1 + (a_{12}c_1 + a_{22}c_2 + \cdots + a_{m2}c_m)\mathbf{v}_2$$
$$+ \cdots + (a_{1n}c_1 + a_{2n}c_2 + \cdots + a_{mn}c_m)\mathbf{v}_n = \mathbf{0} \quad (4)$$

But since $\mathbf{v}_1, \mathbf{v}_2, \ldots, \mathbf{v}_n$ are linearly independent, we must have

$$\begin{aligned}
a_{11}c_1 + a_{21}c_2 + \cdots + a_{m1}c_m &= 0 \\
a_{12}c_1 + a_{22}c_2 + \cdots + a_{m2}c_m &= 0 \\
\vdots \qquad \vdots \qquad \qquad \vdots \qquad &\ \ \vdots \\
a_{1n}c_1 + a_{2n}c_2 + \cdots + a_{mn}c_m &= 0
\end{aligned} \quad (5)$$

System (5) is a homogeneous system of n equations in the m unknowns $c_1, c_2, \ldots, c_m$, and since $m > n$, Theorem 1.4.1, page 25, tells us that the system has an infinite number of solutions. Thus there are scalars $c_1, c_2, \ldots, c_m$, not all zero, such that (2) is satisfied, and therefore S_1 is a linearly dependent set. This contradiction proves that $m \leq n$, and, by exchanging the roles of S_1 and S_2, we can show that $n \leq m$ and the proof is complete. ∎

With this theorem we can define one of the central concepts in linear algebra.

DEFINITION 2 **Dimension** If the vector space V has a finite basis, then the **dimension** of V is the number of vectors in every basis and V is called a **finite dimensional vector space.** Otherwise V is called an **infinite dimensional vector space.** If $V = \{0\}$, then V is said to be **zero dimensional.**

Notation. We write the dimension of V as dim V.

Remark. We have not proved that every vector space has a basis. This very difficult proof appears in Section 4.13. But we do not need this fact for Definition 2 to make sense; for *if* V has a finite basis, then V is finite dimensional. Otherwise V is infinite dimensional. Thus, in order to show that V is infinite dimensional, it is only necessary to show that V does not have a finite basis. We can do this by showing that V contains an infinite number of linearly independent vectors (see Example 7 below). It is not necessary to construct an infinite basis for V.

EXAMPLE 4 **The Dimension of $\mathbb{R}^n$** Since n linearly independent vectors in $\mathbb{R}^n$ constitute a basis, we see that

$$\dim \mathbb{R}^n = n \qquad ∎$$

EXAMPLE 5 **The Dimension of P_n** By Example 1 and Problem 4.5.47, page 242, the polynomials $\{1, x, x^2, \ldots, x^n\}$ constitute a basis in P_n. Thus dim $P_n = n + 1$. ∎

EXAMPLE 6   **The Dimension of M_{mn}**   In M_{mn} let A_{ij} be the $m \times n$ matrix with a 1 in the ijth position and a zero everywhere else. It is easy to show that the A_{ij} for $i = 1, 2, \ldots, m$ and $j = 1, 2, \ldots, n$ form a basis for M_{mn}. Thus dim $M_{mn} = mn$.   ∎

EXAMPLE 7   **P Is Infinite Dimensional**   In Example 4.4.7, page 225, we saw that no finite set of polynomials spans P. Thus P has no finite basis, and is therefore an infinite dimensional vector space.   ∎

There are a number of theorems that tell us something about the dimension of a vector space.

THEOREM 3   Suppose that dim $V = n$. If $\mathbf{u}_1, \mathbf{u}_2, \ldots, \mathbf{u}_m$ is a set of m linearly independent vectors in V, then $m \leq n$.

Proof   Let $\mathbf{v}_1, \mathbf{v}_2, \ldots, \mathbf{v}_n$ be a basis for V. If $m > n$, then, as in the proof of Theorem 2, we can find constants $c_1, c_2, \ldots, c_m$ not all zero such that equation (2) is satisfied. This would contradict the linear independence of the $\mathbf{u}_i$'s. Thus $m \leq n$.   ∎

THEOREM 4   Let H be a subspace of the finite dimensional vector space V. Then H is finite dimensional and

$$\boxed{\dim H \leq \dim V} \qquad (6)$$

Proof   Let dim $V = n$. Any set of linearly independent vectors in H is also a linearly independent set in V. By Theorem 3, any linearly independent set in H can contain at most n vectors. Hence H is finite dimensional. Moreover, since any basis in H is a linearly independent set, we see that dim $H \leq n$.   ∎

Theorem 4 has some interesting consequences. We give two of them here.

EXAMPLE 8   **$C[0, 1]$ and $C^1[0, 1]$ Are Infinite Dimensional**   Let $P[0, 1]$ denote the set of polynomials defined on the interval $[0, 1]$. Then $P[0, 1] \subset C[0, 1]$. If $C[0, 1]$ were finite dimensional, then $P[0, 1]$ would be finite dimensional also. But, by Example 7, this is not the case. Hence $C[0, 1]$ is infinite dimensional. Similarly, since $P[0, 1] \subset C^1[0, 1]$ (since every polynomial is differentiable), we also see that $C^1[0, 1]$ is infinite dimensional.   ∎

In general,

Any vector space containing an infinite dimensional subspace is infinite dimensional.

EXAMPLE 9   **The Subspaces of $\mathbb{R}^3$**   We can use Theorem 4 to find *all* subspaces of $\mathbb{R}^3$. Let H be a subspace of $\mathbb{R}^3$. Then there are four possibilities: $H = \{\mathbf{0}\}$; dim $H = 1$, dim $H = 2$, and dim $H = 3$. If dim $H = 3$, then H contains a basis of three linearly independent vectors $\mathbf{v}_1$, $\mathbf{v}_2$, $\mathbf{v}_3$ in $\mathbb{R}^3$. But then $\mathbf{v}_1$, $\mathbf{v}_2$, $\mathbf{v}_3$ also form a basis for $\mathbb{R}^3$. Thus $H = $ span $\{\mathbf{v}_1, \mathbf{v}_2, \mathbf{v}_3\} = \mathbb{R}^3$. Hence the only way to get a *proper* subspace of $\mathbb{R}^3$ is to have dim $H = 1$ or dim $H = 2$. If dim $H = 1$, then H has a basis consisting of the one vector $\mathbf{v} = (a, b, c)$. Let $\mathbf{x}$ be in H. Then $\mathbf{x} = t(a, b, c)$ for some real number t [since (a, b, c) spans H]. If $\mathbf{x} = (x, y, z)$, this means that $x = at$, $y = bt$, $z = ct$. But this is the equation of a line in $\mathbb{R}^3$ passing through the origin with direction vector (a, b, c).

Now suppose dim $H = 2$ and let $\mathbf{v}_1 = (a_1, b_1, c_1)$ and $\mathbf{v}_2 = (a_2, b_2, c_2)$ be a basis for H. If $\mathbf{x} = (x, y, z) \in H$, then there exist real numbers s and t such that $\mathbf{x} = s\mathbf{v}_1 + t\mathbf{v}_2$ or $(x, y, z) = s(a_1, b_1, c_1) + t(a_2, b_2, c_2)$. Then

$$
\begin{aligned}
x &= sa_1 + ta_2 \\
y &= sb_1 + tb_2 \\
z &= sc_1 + tc_2
\end{aligned}
\tag{7}
$$

Let $\mathbf{v}_3 = (\alpha, \beta, \gamma) = \mathbf{v}_1 \times \mathbf{v}_2$. Then, from Theorem 3.4.2 on page 185, part (vi), we have $\mathbf{v}_3 \cdot \mathbf{v}_1 = 0$ and $\mathbf{v}_3 \cdot \mathbf{v}_2 = 0$. Now we calculate

$$
\begin{aligned}
\alpha x + \beta y + \gamma z &= \alpha(sa_1 + ta_2) + \beta(sb_1 + tb_2) + \gamma(sc_1 + tc_2) \\
&= (\alpha a_1 + \beta b_1 + \gamma c_1)s + (\alpha a_2 + \beta b_2 + \gamma c_2)t \\
&= (\mathbf{v}_3 \cdot \mathbf{v}_1)s + (\mathbf{v}_3 \cdot \mathbf{v}_2)t = 0
\end{aligned}
$$

Thus, if $(x, y, z) \in H$, then $\alpha x + \beta y + \gamma z = 0$, which shows that H is a plane passing through the origin with normal vector $\mathbf{v}_3 = \mathbf{v}_1 \times \mathbf{v}_2$. Therefore we have proved that

> *The only proper subspaces of $\mathbb{R}^3$ are sets of vectors lying on a single line or a single plane passing through the origin.*

■

EXAMPLE 10   **Solution Space and Kernel**   Let A be an $m \times n$ matrix and let $S = \{\mathbf{x} \in \mathbb{R}^n;$ $A\mathbf{x} = \mathbf{0}\}$. Let $\mathbf{x}_1 \in S$ and $\mathbf{x}_2 \in S$; then $A(\mathbf{x}_1 + \mathbf{x}_2) = A\mathbf{x}_1 + A\mathbf{x}_2 = \mathbf{0} + \mathbf{0} = \mathbf{0}$ and $A(\alpha\mathbf{x}_1) = \alpha(A\mathbf{x}_1) = \alpha\mathbf{0} = \mathbf{0}$, so that S is a subspace of $\mathbb{R}^n$ and dim $S \leq n$. S is called the **solution space** of the homogeneous system $A\mathbf{x} = \mathbf{0}$. It is also called the **kernel** of the matrix A. ■

EXAMPLE 11   **Finding a Basis for the Solution Space of a Homogeneous System**   Find a basis for (and the dimension of) the solution space S of the homogeneous system

$$
\begin{aligned}
x + 2y - z &= 0 \\
2x - y + 3z &= 0
\end{aligned}
$$

Solution   Here $A = \begin{pmatrix} 1 & 2 & -1 \\ 2 & -1 & 3 \end{pmatrix}$. Since A is a 2×3 matrix, S is a subspace of $\mathbb{R}^3$. Row-reducing, we find, successively,

$$\begin{pmatrix} 1 & 2 & -1 & | & 0 \\ 2 & -1 & 3 & | & 0 \end{pmatrix} \xrightarrow{R_2 \to R_2 - 2R_1} \begin{pmatrix} 1 & 2 & -1 & | & 0 \\ 0 & -5 & 5 & | & 0 \end{pmatrix}$$

$$\xrightarrow{R_2 \to -\frac{1}{5}R_2} \begin{pmatrix} 1 & 2 & -1 & | & 0 \\ 0 & 1 & -1 & | & 0 \end{pmatrix} \xrightarrow{R_1 \to R_1 - 2R_2} \begin{pmatrix} 1 & 0 & 1 & | & 0 \\ 0 & 1 & -1 & | & 0 \end{pmatrix}$$

Then $y = z$ and $x = -z$, so that all solutions are of the form $\begin{pmatrix} -z \\ z \\ z \end{pmatrix}$. Thus $\begin{pmatrix} -1 \\ 1 \\ 1 \end{pmatrix}$ is a basis for S and dim $S = 1$. Note that S is the set of vectors lying on the straight line

$$x = -t, \; y = t, \; z = t.$$ ■

EXAMPLE 12   **Finding a Basis for the Solution Space of a Homogeneous System**   Find a basis for the solution space S of the system

$$2x - y + 3z = 0$$
$$4x - 2y + 6z = 0$$
$$-6x + 3y - 9z = 0$$

Solution   Row-reducing as above, we obtain

$$\begin{pmatrix} 2 & -1 & 3 & | & 0 \\ 4 & -2 & 6 & | & 0 \\ -6 & 3 & -9 & | & 0 \end{pmatrix} \xrightarrow[R_3 \to R_3 + 3R_2]{R_2 \to R_2 - 2R_1} \begin{pmatrix} 2 & -1 & 3 & | & 0 \\ 0 & 0 & 0 & | & 0 \\ 0 & 0 & 0 & | & 0 \end{pmatrix}$$

giving the single equation $2x - y + 3z = 0$. S is a plane and, by Example 3, a basis is given by $\begin{pmatrix} 1 \\ 2 \\ 0 \end{pmatrix}$ and $\begin{pmatrix} 0 \\ 3 \\ 1 \end{pmatrix}$ and dim $S = 2$. Note that we have shown that any solution to the homogeneous equation can be written as

$$c_1 \begin{pmatrix} 1 \\ 2 \\ 0 \end{pmatrix} + c_2 \begin{pmatrix} 0 \\ 3 \\ 1 \end{pmatrix}$$

For example, if $c_1 = 2$ and $c_2 = -3$, we obtain the solution

$$\mathbf{x} = 2 \begin{pmatrix} 1 \\ 2 \\ 0 \end{pmatrix} - 3 \begin{pmatrix} 0 \\ 3 \\ 1 \end{pmatrix} = \begin{pmatrix} 2 \\ 4 \\ 0 \end{pmatrix} + \begin{pmatrix} 0 \\ -9 \\ -3 \end{pmatrix} = \begin{pmatrix} 2 \\ -5 \\ -3 \end{pmatrix}$$ ■

Before leaving this section, we prove a result that is very useful in finding bases in an arbitrary vector space. We have seen that n linearly independent vectors in $\mathbb{R}^n$ constitute a basis for $\mathbb{R}^n$. This fact holds in *any* finite dimensional vector space.

THEOREM 5 Any n linearly independent vectors in a vector space V of dimension n constitute a basis for V.

Proof Let $\mathbf{v}_1, \mathbf{v}_2, \ldots, \mathbf{v}_n$ be the n vectors. If they span V, then they constitute a basis. If they do not, then there is a vector $\mathbf{u} \in V$ such that $\mathbf{u} \notin \text{span} \{\mathbf{v}_1, \mathbf{v}_2, \ldots, \mathbf{v}_n\}$. This means that the $n + 1$ vectors $\mathbf{v}_1, \mathbf{v}_2, \ldots, \mathbf{v}_n, \mathbf{u}$ are linearly independent. To see this, note that if

$$c_1\mathbf{v}_1 + c_2\mathbf{v}_2 + \cdots + c_n\mathbf{v}_n + c_{n+1}\mathbf{u} = 0 \tag{8}$$

then $c_{n+1} = 0$, for if not we could write $\mathbf{u}$ as a linear combination of $\mathbf{v}_1, \mathbf{v}_2, \ldots, \mathbf{v}_n$ by dividing equation (8) by c_{n+1} and putting all terms except $\mathbf{u}$ on the right-hand side. But if $c_{n+1} = 0$, then (8) reads

$$c_1\mathbf{v}_1 + c_2\mathbf{v}_2 + \cdots + c_n\mathbf{v}_n = 0$$

which means that $c_1 = c_2 = \cdots = c_n = 0$ since the $\mathbf{v}_i$'s are linearly independent. Now let $W = \text{span} \{\mathbf{v}_1, \mathbf{v}_2, \ldots, \mathbf{v}_n, \mathbf{u}\}$. Then as all the vectors in brackets are in V, W is a subspace of V. Since $\mathbf{v}_1, \mathbf{v}_2, \ldots, \mathbf{v}_n, \mathbf{u}$ are linearly independent, they form a basis for W. Thus $\dim W = n + 1$. But from Theorem 4, $\dim W \leq n$. This contradiction shows that there is *no* vector $\mathbf{u} \in V$ such that $\mathbf{u} \notin \text{span} \{\mathbf{v}_1, \mathbf{v}_2, \ldots, \mathbf{v}_n\}$. Thus $\mathbf{v}_1, \mathbf{v}_2, \ldots, \mathbf{v}_n$ span V and therefore constitute a basis for V. ∎

PROBLEMS 4.6

In Problems 1–10 determine whether the given set of vectors is a basis for the given vector space.

1. In P_2: $1 - x^2, x$

2. In P_2: $-3x, 1 + x^2, x^2 - 5$

3. In P_2: $x^2 - 1, x^2 - 2, x^2 - 3$

4. In P_3: $1, 1 + x, 1 + x^2, 1 + x^3$

5. In P_3: $3, x^3 - 4x + 6, x^2$

6. In M_{22}: $\begin{pmatrix} 3 & 1 \\ 0 & 0 \end{pmatrix}, \begin{pmatrix} 3 & 2 \\ 0 & 0 \end{pmatrix}, \begin{pmatrix} -5 & 1 \\ 0 & 6 \end{pmatrix}, \begin{pmatrix} 0 & 1 \\ 0 & -7 \end{pmatrix}$

7. In M_{22}: $\begin{pmatrix} a & 0 \\ 0 & 0 \end{pmatrix}, \begin{pmatrix} 0 & b \\ 0 & 0 \end{pmatrix}, \begin{pmatrix} 0 & 0 \\ c & 0 \end{pmatrix}, \begin{pmatrix} 0 & 0 \\ 0 & d \end{pmatrix}$, where $abcd \neq 0$

8. In M_{22}: $\begin{pmatrix} -1 & 0 \\ 3 & 1 \end{pmatrix}, \begin{pmatrix} 2 & 1 \\ 1 & 4 \end{pmatrix}, \begin{pmatrix} -6 & 1 \\ 5 & 8 \end{pmatrix}, \begin{pmatrix} 7 & -2 \\ 1 & 0 \end{pmatrix}, \begin{pmatrix} 0 & 1 \\ 0 & 0 \end{pmatrix}$

9. $H = \{(x, y) \in \mathbb{R}^2 : x + y = 0\}$; $(1, -1)$

10. $H = \{(x, y) \in \mathbb{R}^2 : x + y = 0\}$; $(1, -1), (-3, 3)$

11. Find a basis in $\mathbb{R}^3$ for the set of vectors in the plane $2x - y - z = 0$.

12. Find a basis in $\mathbb{R}^3$ for the set of vectors in the plane $3x - 2y + 6z = 0$.

13. Find a basis in $\mathbb{R}^3$ for the set of vectors on the line $x/2 = y/3 = z/4$.

14. Find a basis in $\mathbb{R}^3$ for the set of vectors on the line $x = 3t$, $y = -2t$, $z = t$.

15. Show that the only proper subspaces of $\mathbb{R}^2$ are straight lines passing through the origin.

16. In $\mathbb{R}^4$ let $H = \{(x, y, z, w): ax + by + cz + dw = 0\}$, where $abcd \neq 0$.
   a.   Show that H is a subspace of $\mathbb{R}^4$.
   b.   Find a basis for H.
   c.   What is dim H?

*17. In $\mathbb{R}^n$ a **hyperplane** is a subspace of dimension $n - 1$. If H is a hyperplane in $\mathbb{R}^n$ show that

$$H = \{(x_1, x_2, \ldots, x_n): a_1x_1 + a_2x_2 + \cdots + a_nx_n = 0\}$$

where $a_1, a_2, \ldots, a_n$ are fixed real numbers, not all of which are zero.

18. In $\mathbb{R}^5$ find a basis for the hyperplane

$$H = \{(x_1, x_2, x_3, x_4, x_5): 2x_1 - 3x_2 + x_3 + 4x_4 - x_5 = 0\}$$

In Problems 19–23 find a basis for the solution space of the given homogeneous system.

19. $x - y = 0$
        $-2x + 2y = 0$

20. $x - 2y = 0$
        $3x + y = 0$

21. $x - y - z = 0$
        $2x - y + z = 0$

22. $x - 3y + z = 0$
        $-2x + 2y - 3z = 0$
        $4x - 8y + 5z = 0$

23. $2x - 6y + 4z = 0$
        $-x + 3y - 2z = 0$
        $-3x + 9y - 6z = 0$

24. Find a basis for D_3, the vector space of diagonal 3×3 matrices. What is the dimension of D_3?

25. What is the dimension of D_n, the space of diagonal $n \times n$ matrices?

26. Let S_{nn} denote the vector space of symmetric $n \times n$ matrices. Show that S_{nn} is a subspace of M_{nn} and that dim $S_{nn} = [n(n + 1)]/2$.

27. Suppose that $v_1, v_2, \ldots, v_m$ are linearly independent vectors in a vector space V of dimension n and $m < n$. Show that $\{v_1, v_2, \ldots, v_m\}$ can be enlarged to a basis for V. That is, there exist vectors $v_{m+1}, v_{m+2}, \ldots, v_n$ such that $\{v_1, v_2, \ldots, v_n\}$ is a basis. [*Hint:* Look at the proof of Theorem 5.]

28. Let $\{v_1, v_2, \ldots, v_n\}$ be a basis for V. Let $u_1 = v_1$, $u_2 = v_1 + v_2$, $u_3 = v_1 + v_2 + v_3, \ldots, u_n = v_1 + v_2 + \cdots + v_n$. Show that $\{u_1, u_2, \ldots, u_n\}$ is also a basis for V.

29. Show that if $\{v_1, v_2, \ldots, v_n\}$ spans V, then dim $V \leq n$. [*Hint:* Use the result of Problem 4.5.49.]

30. Let H and K be subspaces of V such that $H \subseteq K$ and dim $H = $ dim $K < \infty$. Show that $H = K$.

31. Let H and K be subspaces of V and define $H + K = \{h + k: h \in H \text{ and } k \in K\}$.
   a.   Show that $H + K$ is a subspace of V.
   b.   If $H \cap K = \{0\}$, show that dim $(H + K) = $ dim $H + $ dim K.

*32. If H is a subspace of the finite dimensional vector space V, show that there exists a unique subspace K of V such that **(a)** $H \cap K = \{0\}$ and **(b)** $H + K = V$.

33. Show that two vectors $\mathbf{v}_1$ and $\mathbf{v}_2$ in $\mathbb{R}^2$ with endpoints at the origin are collinear if and only if dim span $\{\mathbf{v}_1, \mathbf{v}_2\} = 1$.

34. Show that three vectors $\mathbf{v}_1$, $\mathbf{v}_2$, and $\mathbf{v}_3$ in $\mathbb{R}^3$ with endpoints at the origin are coplanar if and only if dim span $\{\mathbf{v}_1, \mathbf{v}_2, \mathbf{v}_3\} \leq 2$.

35. Show that any n vectors which span an n-dimensional space V form a basis for V. [*Hint:* Show that if the n vectors are not linearly independent, then dim $V < n$.]

*36. Show that every subspace of a finite dimensional vector space has a basis.

37. Find two bases for $\mathbb{R}^4$ that contain $(1, 0, 1, 0)$ and $(0, 1, 0, 1)$ and have no other vectors in common.

38. For what values of the real number a do the vectors $(a, 1, 0)$, $(1, 0, a)$, and $(1 + a, 1, a)$ constitute a basis for $\mathbb{R}^3$?

4.7 THE RANK, NULLITY, ROW SPACE, AND COLUMN SPACE OF A MATRIX

In Section 4.5 we introduced the notion of linear independence. We showed that if A is an invertible $n \times n$ matrix, then the columns and rows of A form sets of linearly independent vectors. However, if A is not invertible (so that det $A = 0$), or if A is not a square matrix, then these results tell us nothing about the number of linearly independent rows or columns of A. In this section we fill in this gap. We also show how a basis for the span of a set of vectors can be obtained by row reduction.

Let A be an $m \times n$ matrix and let

$$\boxed{N_A = \{\mathbf{x} \in \mathbb{R}^n : A\mathbf{x} = \mathbf{0}\}} \tag{1}$$

Then, as we saw in Example 4.6.10 on page 249, N_A is a subspace of $\mathbb{R}^n$.

DEFINITION 1   **Kernel and Nullity of a Matrix** N_A is called the **kernel** of A and $\nu(A) = \dim N_A$ is called the **nullity** of A. If N_A contains only the zero vector, then $\nu(A) = 0$.

EXAMPLE 1   **The Kernel and Nullity of a 2 × 3 Matrix**   Let $A = \begin{pmatrix} 1 & 2 & -1 \\ 2 & -1 & 3 \end{pmatrix}$. Then, as we saw in Example 4.6.11 on page 249, N_A is spanned by $\begin{pmatrix} -1 \\ 1 \\ 1 \end{pmatrix}$ and $\nu(A) = 1$.   ■

EXAMPLE 2   **The Kernel and Nullity of a 3 × 3 Matrix**   Let $A = \begin{pmatrix} 2 & -1 & 3 \\ 4 & -2 & 6 \\ -6 & 3 & -9 \end{pmatrix}$. Then, by Example 4.6.12 on page 250, $\left\{ \begin{pmatrix} 1 \\ 2 \\ 0 \end{pmatrix}, \begin{pmatrix} 0 \\ 3 \\ 1 \end{pmatrix} \right\}$ is a basis for N_A and $\nu(A) = 2$.   ∎

THEOREM 1   Let A be an $n \times n$ matrix. Then A is invertible if and only if $\nu(A) = 0$.

Proof   By our Summing Up Theorem [Theorem 4.5.6, page 237, parts (i) and (ii)], A is invertible if and only if the homogeneous system $A\mathbf{x} = \mathbf{0}$ has only the trivial solution $\mathbf{x} = \mathbf{0}$. But, from equation (1), this means that A is invertible if and only if $N_A = \{\mathbf{0}\}$. Thus A is invertible if and only if $\nu(A) = \dim N_A = 0$.   ∎

DEFINITION 2   **Range of a Matrix**   Let A be an $m \times n$ matrix. Then the **range** of A, denoted by Range A, is given by

$$\text{Range } A = \{\mathbf{y} \in \mathbb{R}^m : A\mathbf{x} = \mathbf{y} \text{ for some } \mathbf{x} \in \mathbb{R}^n\} \tag{2}$$

THEOREM 2   Let A be an $m \times n$ matrix. Then Range A is a subspace of $\mathbb{R}^m$.

Proof   Suppose that $\mathbf{y}_1$ and $\mathbf{y}_2$ are in Range A. Then there are vectors $\mathbf{x}_1$ and $\mathbf{x}_2$ in $\mathbb{R}^n$ such that $\mathbf{y}_1 = A\mathbf{x}_1$ and $\mathbf{y}_2 = A\mathbf{x}_2$. Therefore

$$A(\alpha\mathbf{x}_1) = \alpha A\mathbf{x}_1 = \alpha\mathbf{y}_1 \quad \text{and} \quad A(\mathbf{x}_1 + \mathbf{x}_2) = A\mathbf{x}_1 + A\mathbf{x}_2 = \mathbf{y}_1 + \mathbf{y}_2$$

so $\alpha\mathbf{y}_1$ and $\mathbf{y}_1 + \mathbf{y}_2$ are in Range A. Thus, from Theorem 4.3.1, Range A is a subspace of $\mathbb{R}^m$.   ∎

DEFINITION 3   **Rank of a Matrix**   Let A be an $m \times n$ matrix. Then the **rank** of A, denoted by $\rho(A)$, is given by

$$\rho(A) = \dim \text{Range } A$$

We shall give two definitions and a theorem that make the calculation of rank relatively easy.

DEFINITION 4   **Row and Column Space of a Matrix** If A is an $m \times n$ matrix, let $\{\mathbf{r}_1, \mathbf{r}_2, \ldots, \mathbf{r}_m\}$ denote the rows of A and let $\{\mathbf{c}_1, \mathbf{c}_2, \ldots, \mathbf{c}_n\}$ denote the columns of A. Then we define

$$\boxed{R_A = \text{\bf row space } of\ A = \text{span }\{\mathbf{r}_1, \mathbf{r}_2, \ldots, \mathbf{r}_m\}} \tag{3}$$

and

$$\boxed{C_A = \text{\bf column space } of\ A = \text{span }\{\mathbf{c}_1, \mathbf{c}_2, \ldots, \mathbf{c}_n\}} \tag{4}$$

Note. R_A is a subspace of $\mathbb{R}^n$ and C_A is a subspace of $\mathbb{R}^m$.

We have introduced a lot of notation in just two pages. Let us stop for a moment to illustrate these ideas with an example.

EXAMPLE 3   **Finding N_A, $\nu(A)$, Range A, $\rho(A)$, R_A, and C_A for a 2 × 3 Matrix**

Let $A = \begin{pmatrix} 1 & 2 & -1 \\ 2 & -1 & 3 \end{pmatrix}$. A is a 2×3 matrix.

i. *The kernel of $A = N_A = \{\mathbf{x} \in \mathbb{R}^3\colon A\mathbf{x} = \mathbf{0}\}$*. As we saw in Example 1,

$$N_A = \text{span}\left\{\begin{pmatrix} -1 \\ 1 \\ 1 \end{pmatrix}\right\}.$$

ii. *The nullity of $A = \nu(A) = \dim N_A = 1$.*

iii. *The range of $A = $ Range $A = \{\mathbf{y} \in \mathbb{R}^2\colon A\mathbf{x} = \mathbf{y}$ for some $\mathbf{x} \in \mathbb{R}^3\}$*. Let $\mathbf{y} = \begin{pmatrix} y_1 \\ y_2 \end{pmatrix}$ be in $\mathbb{R}^2$. Then, if $\mathbf{y} \in $ Range A, there is an $\mathbf{x} \in \mathbb{R}^3$ such that $A\mathbf{x} = \mathbf{y}$.

Writing $\mathbf{x} = \begin{pmatrix} x_1 \\ x_2 \\ x_3 \end{pmatrix}$, we have

$$\begin{pmatrix} 1 & 2 & -1 \\ 2 & -1 & 3 \end{pmatrix}\begin{pmatrix} x_1 \\ x_2 \\ x_3 \end{pmatrix} = \begin{pmatrix} y_1 \\ y_2 \end{pmatrix}$$

or

$$\begin{aligned} x_1 + 2x_2 - x_3 &= y_1 \\ 2x_1 - x_2 + 3x_3 &= y_2. \end{aligned}$$

Row-reducing this system, we have

$$\begin{pmatrix} 1 & 2 & -1 & \big| & y_1 \\ 2 & -1 & 3 & \big| & y_2 \end{pmatrix} \xrightarrow{R_2 \to R_2 - 2R_1} \begin{pmatrix} 1 & 2 & -1 & \big| & y_1 \\ 0 & -5 & 5 & \big| & y_2 - 2y_1 \end{pmatrix}$$

$$\xrightarrow{R_2 \to -\frac{1}{5}R_2} \begin{pmatrix} 1 & 2 & -1 & \big| & y_1 \\ 0 & 1 & -1 & \big| & \dfrac{2y_1 - y_2}{5} \end{pmatrix} \xrightarrow{R_1 \to R_1 - 2R_2} \begin{pmatrix} 1 & 0 & 1 & \big| & \dfrac{y_1 + 2y_2}{5} \\ 0 & 1 & -1 & \big| & \dfrac{2y_1 - y_2}{5} \end{pmatrix}$$

Thus, if x_3 is chosen arbitrarily, we see that

$$x_1 = -x_3 + \frac{y_1 + 2y_2}{5} \quad \text{and} \quad x_2 = x_3 + \frac{2y_1 - y_2}{5}$$

That is, for every $\mathbf{y} = \begin{pmatrix} y_1 \\ y_2 \end{pmatrix} \in \mathbb{R}^2$, there are an infinite number of vectors $\mathbf{x} \in \mathbb{R}^3$ such that $A\mathbf{x} = \mathbf{y}$. Thus Range $A = \mathbb{R}^2$. Note, for example, that if $\mathbf{y} = \begin{pmatrix} 2 \\ -3 \end{pmatrix}$, then, choosing $x_3 = 0$ (the simplest choice), we have

$$x_1 = \frac{2 + 2(-3)}{5} = -\frac{4}{5} \qquad x_2 = \frac{2(2) - (-3)}{5} = \frac{7}{5}$$

and

$$A\mathbf{x} = \begin{pmatrix} 1 & 2 & -1 \\ 2 & -1 & 3 \end{pmatrix} \begin{pmatrix} -\frac{4}{5} \\ \frac{7}{5} \\ 0 \end{pmatrix} = \begin{pmatrix} \frac{10}{5} \\ -\frac{15}{5} \end{pmatrix} = \begin{pmatrix} 2 \\ -3 \end{pmatrix} = \mathbf{y}$$

iv. *The rank of* $A = \rho(A) = \dim$ Range $A = \dim \mathbb{R}^2 = 2$.

v. *The row space of* $A = R_A = \operatorname{span}\{(1, 2, -1), (2, -1, 3)\}$. Since these two vectors are linearly independent, we see that R_A is a two-dimensional subspace of $\mathbb{R}^3$. From Example 4.6.9 on page 249, we observe that R_A is a plane passing through the origin.

vi. *The column space of* $A =$

$$C_A = \operatorname{span}\left\{\begin{pmatrix} 1 \\ 2 \end{pmatrix}, \begin{pmatrix} 2 \\ -1 \end{pmatrix}, \begin{pmatrix} -1 \\ 3 \end{pmatrix}\right\} = \mathbb{R}^2$$

since $\begin{pmatrix} 1 \\ 2 \end{pmatrix}$ and $\begin{pmatrix} 2 \\ -1 \end{pmatrix}$, being linearly independent, constitute a basis for $\mathbb{R}^2$. ∎

In Example 3 we may observe that Range $A = C_A = \mathbb{R}^2$ and $\dim R_A = \dim C_A = \dim$ Range $A = \rho(A) = 2$. This is no coincidence.

THEOREM 3   If A is an $m \times n$ matrix, then:

   i. C_A = Range A

   ii. dim R_A = dim C_A = dim Range A = $\rho(A)$ ■

The proof of this theorem is not difficult, but it is quite long. We defer it to the end of the section.

EXAMPLE 4   **Finding Range A and $\rho(A)$ for a 3 × 3 Matrix**   Find a basis for Range A and

determine the rank of $A = \begin{pmatrix} 2 & -1 & 3 \\ 4 & -2 & 6 \\ -6 & 3 & -9 \end{pmatrix}$.

Solution   Since $\mathbf{r}_2 = 2\mathbf{r}_1$ and $\mathbf{r}_3 = -3\mathbf{r}_1$, we see that $\rho(A) = \dim R_A = 1$. Thus any column

in C_A is a basis for C_A = Range A. For example, $\begin{pmatrix} 2 \\ 4 \\ -6 \end{pmatrix}$ is a basis for Range A.   ■

The following theorem will simplify our computations.

THEOREM 4   If A is row (or column) equivalent to B, then $R_A = R_B$, $\rho(A) = \rho(B)$, and $\nu(A) = \nu(B)$.

Proof   Recall from Definition 1.9.3, page 73, that A is row equivalent to B if A can be "reduced" to B by elementary row operations. The definition for "column equivalent" is similar. Suppose that C is the matrix obtained by performing an elementary row operation on A. We first show that $R_A = R_C$. Since B is obtained by performing several elementary row operations on A, our first result, applied several times, will imply that $R_A = R_B$.

Case 1: Interchange two rows of A. Then $R_A = R_C$ because the rows of A and C are the same (just written in a different order).

Case 2: Multiply the ith row of A by $c \neq 0$. If the rows of A are $\{\mathbf{r}_1, \mathbf{r}_2, \ldots, \mathbf{r}_i, \ldots, \mathbf{r}_m\}$, then the rows of C are $\{\mathbf{r}_1, \mathbf{r}_2, \ldots, c\mathbf{r}_i, \ldots, \mathbf{r}_m\}$. Obviously, $c\mathbf{r}_i = c(\mathbf{r}_i)$ and $\mathbf{r}_i = (1/c)\mathbf{r}_i$. Thus each row of C is a multiple of one row of A and vice versa. This means that each row of C is in the span of the rows of A and vice versa. We have

$$R_A \subseteq R_C \quad \text{and} \quad R_C \subseteq R_A, \quad \text{so } R_C = R_A$$

Case 3: Multiply the ith row of A by $c \neq 0$ and add it to the jth row. Now the rows of C are $\{\mathbf{r}_1, \mathbf{r}_2, \ldots, \mathbf{r}_i, \ldots, c\mathbf{r}_i + \mathbf{r}_j, \ldots, \mathbf{r}_m\}$. Here

$$\mathbf{r}_j = \underbrace{(c\mathbf{r}_i + \mathbf{r}_j)}_{j\text{th row of } C} - \underset{\underset{i\text{th row of } C}{\uparrow}}{c\mathbf{r}_i}$$

so each row of A can be written as a linear combination of the rows of C and vice versa. Then, as before,

$$R_A \subseteq R_C \quad \text{and} \quad R_C \subseteq R_A, \quad \text{so } R_C = R_A$$

We have shown that $R_A = R_B$. Hence $\rho(R_A) = \rho(R_B)$. Finally, the set of solutions to $A\mathbf{x} = \mathbf{0}$ does not change under elementary row operations. Thus $N_A = N_B$, so $\nu(A) = \nu(B)$. ∎

EXAMPLE 5   **Finding $\rho(A)$ and R_A for a 3 × 3 matrix**   Determine the rank and row space of
$$A = \begin{pmatrix} 1 & -1 & 3 \\ 2 & 0 & 4 \\ -1 & -3 & 1 \end{pmatrix}.$$

Solution   We row-reduce to obtain a simpler matrix:

$$\begin{pmatrix} 1 & -1 & 3 \\ 2 & 0 & 4 \\ -1 & -3 & 1 \end{pmatrix} \xrightarrow[\substack{R_2 \to R_2 - 2R_1 \\ R_3 \to R_3 + R_1}]{} \begin{pmatrix} 1 & -1 & 3 \\ 0 & 2 & -2 \\ 0 & -4 & 4 \end{pmatrix}$$

$$\xrightarrow[R_2 \to \frac{1}{2}R_2]{} \begin{pmatrix} 1 & -1 & 3 \\ 0 & 1 & -1 \\ 0 & -4 & 4 \end{pmatrix} \xrightarrow[R_3 \to R_3 + 4R_2]{} \begin{pmatrix} 1 & -1 & 3 \\ 0 & 1 & -1 \\ 0 & 0 & 0 \end{pmatrix} = B$$

Since B has two independent rows, we have $\rho(B) = \rho(A) = 2$ and

$$R_A = \text{span} \{(1, -1, 3), (0, 1, -1)\}$$ ∎

Theorem 4 is useful when we want to find a basis for the span of a set of vectors.

EXAMPLE 6   **Finding a Basis for the Span of Four Vectors in $\mathbb{R}^3$**   Find a basis for the space spanned by

$$\mathbf{v}_1 = \begin{pmatrix} 1 \\ 2 \\ -3 \end{pmatrix}, \qquad \mathbf{v}_2 = \begin{pmatrix} -2 \\ 0 \\ 4 \end{pmatrix}, \qquad \mathbf{v}_3 = \begin{pmatrix} 0 \\ 4 \\ -2 \end{pmatrix}, \qquad \mathbf{v}_4 = \begin{pmatrix} -2 \\ -4 \\ 6 \end{pmatrix}$$

Solution   We write the vectors as rows of a matrix A and then reduce the matrix to row echelon form. The resulting matrix will have the same row space as A.

$$\begin{pmatrix} 1 & 2 & -3 \\ -2 & 0 & 4 \\ 0 & 4 & -2 \\ -2 & -4 & 6 \end{pmatrix} \xrightarrow[\substack{R_2 \to R_2 + 2R_1 \\ R_4 \to R_4 + 2R_1}]{} \begin{pmatrix} 1 & 2 & -3 \\ 0 & 4 & -2 \\ 0 & 4 & -2 \\ 0 & 0 & 0 \end{pmatrix}$$

$$\xrightarrow[R_3 \to R_3 - R_2]{} \begin{pmatrix} 1 & 2 & -3 \\ 0 & 4 & -2 \\ 0 & 0 & 0 \\ 0 & 0 & 0 \end{pmatrix} \xrightarrow[R_2 \to \frac{1}{4}R_2]{} \begin{pmatrix} 1 & 2 & -3 \\ 0 & 1 & -\frac{1}{2} \\ 0 & 0 & 0 \\ 0 & 0 & 0 \end{pmatrix}$$

Thus a basis for span $\{v_1, v_2, v_3, v_4\}$ is $\left\{ \begin{pmatrix} 1 \\ 2 \\ -3 \end{pmatrix}, \begin{pmatrix} 0 \\ 1 \\ -\frac{1}{2} \end{pmatrix} \right\}$. For example,

$$\begin{pmatrix} -2 \\ 0 \\ 4 \end{pmatrix} = -2 \begin{pmatrix} 1 \\ 2 \\ -3 \end{pmatrix} + 4 \begin{pmatrix} 0 \\ 1 \\ -\frac{1}{2} \end{pmatrix}.$$

■

The next theorem gives the relationship between rank and nullity.

THEOREM 5   Let A be an $m \times n$ matrix. Then

$$\boxed{\rho(A) + \nu(A) = n} \tag{5}$$

Proof   We assume that $k = \rho(A)$ and that the first k columns of A are linearly independent. Let c_i $(i > k)$ denote any other column of A. Since $c_1, c_2, \ldots, c_k$ form a basis for C_A, we have, for some scalars $a_1, a_2, \ldots, a_k$,

$$c_i = a_1 c_1 + a_2 c_2 + \cdots + a_k c_k \tag{6}$$

Thus, by adding $-a_1 c_1, -a_2 c_2, \ldots, -a_k c_k$ successively to the ith column of A, we obtain a new $m \times n$ matrix B with $\rho(B) = \rho(A)$ and $\nu(B) = \nu(A)$ with the ith column of $B = 0$. We do this to all other columns of A (except the first k) to obtain the matrix

$$D = \begin{pmatrix} a_{11} & a_{12} & \cdots & a_{1k} & 0 & 0 & \cdots & 0 \\ a_{21} & a_{22} & \cdots & a_{2k} & 0 & 0 & \cdots & 0 \\ \vdots & \vdots & & \vdots & \vdots & \vdots & & \vdots \\ a_{m1} & a_{m2} & \cdots & a_{mk} & 0 & 0 & \cdots & 0 \end{pmatrix} \tag{7}$$

where $\rho(D) = \rho(A)$ and $\nu(D) = \nu(A)$. By possibly rearranging the rows of D, we can assume that the first k rows of D are independent. Then we do the same thing to the rows (i.e., add multiples of the first k rows to the last $m - k$ rows) to obtain a new matrix:

$$F = \begin{pmatrix} a_{11} & a_{12} & \cdots & a_{1k} & 0 & \cdots & 0 \\ a_{21} & a_{22} & \cdots & a_{2k} & 0 & \cdots & 0 \\ \vdots & \vdots & & \vdots & \vdots & & \vdots \\ a_{k1} & a_{k2} & \cdots & a_{kk} & 0 & \cdots & 0 \\ 0 & 0 & \cdots & 0 & 0 & \cdots & 0 \\ \vdots & \vdots & & \vdots & \vdots & & \vdots \\ 0 & 0 & \cdots & 0 & 0 & \cdots & 0 \end{pmatrix}$$

where $\rho(F) = \rho(A)$ and $\nu(F) = \nu(A)$. It is now obvious that if $i > k$, then $F\mathbf{e}_i = \mathbf{0}$,† so $E_k = \{\mathbf{e}_{k+1}, \mathbf{e}_{k+2}, \ldots, \mathbf{e}_n\}$ is a linearly independent set of $n - k$ vectors in N_F. We now show that E_k spans N_F. Let the vector $\mathbf{x} \in N_F$ have the form

$$\mathbf{x} = \begin{pmatrix} x_1 \\ x_2 \\ \vdots \\ x_k \\ \vdots \\ x_n \end{pmatrix}$$

Then

$$\mathbf{0} = F\mathbf{x} = \begin{pmatrix} a_{11}x_1 + a_{12}x_2 + \cdots + a_{1k}x_k \\ a_{21}x_1 + a_{22}x_2 + \cdots + a_{2k}x_k \\ \vdots & \vdots & \vdots \\ a_{k1}x_1 + a_{k2}x_2 + \cdots + a_{kk}x_k \\ 0 \\ \vdots \\ 0 \end{pmatrix} = \begin{pmatrix} 0 \\ 0 \\ \vdots \\ 0 \end{pmatrix}$$

The determinant of the matrix of the $k \times k$ homogeneous system described above is nonzero, since the rows of this matrix are linearly independent. Thus the only solution to the system is $x_1 = x_2 = \cdots = x_k = 0$. Thus $\mathbf{x}$ has the form

$$(0, 0, \ldots, 0, x_{k+1}, x_{k+2}, \ldots, x_n) = x_{k+1}\mathbf{e}_{k+1} + x_{k+2}\mathbf{e}_{k+2} + \cdots + x_n\mathbf{e}_n$$

This means that E_k spans N_F so that $\nu(F) = n - k = n - \rho(F)$. This completes the proof. ∎

EXAMPLE 7 **Illustration that $\rho(A) + \nu(A) = n$** For $A = \begin{pmatrix} 1 & 2 & -1 \\ 2 & -1 & 3 \end{pmatrix}$ we calculated (in Examples 1 and 3) that $\rho(A) = 2$ and $\nu(A) = 1$; this illustrates that $\rho(A) + \nu(A) = n \ (=3)$. ∎

---

†Recall that $\mathbf{e}_i$ is the vector with a 1 in the ith position and a zero everywhere else.

EXAMPLE 8   **Illustration That $\rho(A) + \nu(A) = n$**   For $A = \begin{pmatrix} 1 & -1 & 3 \\ 2 & 0 & 4 \\ -1 & -3 & 1 \end{pmatrix}$, calculate $\nu(A)$.

Solution   In Example 5 we found that $\rho(A) = 2$. Thus $\nu(A) = 3 - 2 = 1$.   ∎

THEOREM 6   Let A be an $n \times n$ matrix. Then A is invertible if and only if $\rho(A) = n$.

Proof   By Theorem 1, A is invertible if and only if $\nu(A) = 0$. But, by Theorem 5, $\rho(A) = n - \nu(A)$. Thus A is invertible if and only if $\rho(A) = n - 0 = n$.   ∎

We next show how the notion of rank can be used to solve linear systems of equations. Again we consider the system of m equations in n unknowns

$$
\begin{aligned}
a_{11}x_1 + a_{12}x_2 + \cdots + a_{1n}x_n &= b_1 \\
a_{21}x_1 + a_{22}x_2 + \cdots + a_{2n}x_n &= b_2 \\
\vdots \qquad \vdots \qquad\qquad \vdots \qquad &\ \ \vdots \\
a_{m1}x_1 + a_{m2}x_2 + \cdots + a_{mn}x_n &= b_m
\end{aligned}
\tag{8}
$$

which we write as $A\mathbf{x} = \mathbf{b}$. We use the symbol $(A, \mathbf{b})$ to denote the $m \times (n + 1)$ augmented matrix obtained (as in Section 1.3) by adjoining the vector $\mathbf{b}$ to A.

THEOREM 7   The system $A\mathbf{x} = \mathbf{b}$ has at least one solution if and only if $\mathbf{b} \in C_A$. This will occur if and only if A and the augmented matrix $(A, \mathbf{b})$ have the same rank.

Proof   If $\mathbf{c}_1, \mathbf{c}_2, \ldots, \mathbf{c}_n$ are the columns of A, then we can write system (8) as

$$
x_1\mathbf{c}_1 + x_2\mathbf{c}_2 + \cdots + x_n\mathbf{c}_n = \mathbf{b}
\tag{9}
$$

System (9) will have a solution if and only if $\mathbf{b}$ can be written as a linear combination of the columns of A. That is, to have a solution we must have $\mathbf{b} \in C_A$. If $\mathbf{b} \in C_A$, then $(A, \mathbf{b})$ has the same number of linearly independent columns as A so that A and $(A, \mathbf{b})$ have the same rank. If $\mathbf{b} \notin C_A$, then $\rho(A, \mathbf{b}) = \rho(A) + 1$ and the system has no solutions. This completes the proof.   ∎

EXAMPLE 9   **Using Theorem 7 to Determine Whether a System Has Solutions**   Determine whether the system

$$
\begin{aligned}
2x_1 + 4x_2 + 6x_3 &= 18 \\
4x_1 + 5x_2 + 6x_3 &= 24 \\
2x_1 + 7x_2 + 12x_3 &= 40
\end{aligned}
$$

has solutions.

Solution   Let $A = \begin{pmatrix} 2 & 4 & 6 \\ 4 & 5 & 6 \\ 2 & 7 & 12 \end{pmatrix}$. Then we row-reduce to obtain, successively,

$$\xrightarrow{R_1 \to \frac{1}{2}R_1} \begin{pmatrix} 1 & 2 & 3 \\ 4 & 5 & 6 \\ 2 & 7 & 12 \end{pmatrix} \xrightarrow[R_3 \to R_3 - 2R_1]{R_2 \to R_2 - 4R_1} \begin{pmatrix} 1 & 2 & 3 \\ 0 & -3 & -6 \\ 0 & 3 & 6 \end{pmatrix}$$

$$\xrightarrow{R_2 \to -\frac{1}{3}R_2} \begin{pmatrix} 1 & 2 & 3 \\ 0 & 1 & 2 \\ 0 & 3 & 6 \end{pmatrix} \xrightarrow[R_3 \to R_3 - 3R_2]{R_1 \to R_1 - 2R_2} \begin{pmatrix} 1 & 0 & -1 \\ 0 & 1 & 2 \\ 0 & 0 & 0 \end{pmatrix}$$

Thus $\rho(A) = 2$. Similarly, we row-reduce $(A, \mathbf{b})$ to obtain

$$\begin{pmatrix} 2 & 4 & 6 & | & 18 \\ 4 & 5 & 6 & | & 24 \\ 2 & 7 & 12 & | & 40 \end{pmatrix} \xrightarrow{R_1 \to \frac{1}{2}R_1} \begin{pmatrix} 1 & 2 & 3 & | & 9 \\ 4 & 5 & 6 & | & 24 \\ 2 & 7 & 12 & | & 40 \end{pmatrix}$$

$$\xrightarrow[R_3 \to R_3 - 2R_1]{R_2 \to R_2 - 4R_1} \begin{pmatrix} 1 & 2 & 3 & | & 9 \\ 0 & -3 & -6 & | & -12 \\ 0 & 3 & 6 & | & 22 \end{pmatrix}$$

$$\xrightarrow{R_2 \to -\frac{1}{3}R_2} \begin{pmatrix} 1 & 2 & 3 & | & 9 \\ 0 & 1 & 2 & | & 4 \\ 0 & 3 & 6 & | & 22 \end{pmatrix} \xrightarrow[R_3 \to R_3 - 3R_2]{R_1 \to R_1 - 2R_2} \begin{pmatrix} 1 & 0 & -1 & | & 1 \\ 0 & 1 & 2 & | & 4 \\ 0 & 0 & 0 & | & 10 \end{pmatrix}$$

It is easy to see that the last three columns of the last matrix are linearly independent. Thus $\rho(A, \mathbf{b}) = 3$ and there are no solutions to the system. ∎

EXAMPLE 10   **Using Theorem 7 to Determine Whether a System Has Solutions**   Determine whether the system

$$x_1 - x_2 + 2x_3 = 4$$
$$2x_1 + x_2 - 3x_3 = -2$$
$$4x_1 - x_2 + x_3 = 6$$

has solutions.

Solution   Let $A = \begin{pmatrix} 1 & -1 & 2 \\ 2 & 1 & -3 \\ 4 & -1 & 1 \end{pmatrix}$. Then $\det A = 0$, so $\rho(A) < 3$. Since the first column is not a multiple of the second, we see that the first two columns are linearly independent; hence $\rho(A) = 2$. To compute $\rho(A, \mathbf{b})$, we row-reduce:

$$\begin{pmatrix} 1 & -1 & 2 & | & 4 \\ 2 & 1 & -3 & | & -2 \\ 4 & -1 & 1 & | & 6 \end{pmatrix} \xrightarrow[R_3 \to R_3 - 4R_1]{R_2 \to R_2 - 2R_1} \begin{pmatrix} 1 & -1 & 2 & | & 4 \\ 0 & 3 & -7 & | & -10 \\ 0 & 3 & -7 & | & -10 \end{pmatrix}$$

We see that $\rho(A, \mathbf{b}) = 2$ and there are an infinite number of solutions to the system. (If there were a unique solution, we would have det $A \neq 0$.) ∎

The results of this section allow us to improve on our Summing Up Theorem—last seen in Section 4.5 page 237.

THEOREM 8 **Summing Up Theorem—View 6** Let A be an $n \times n$ matrix. Then the following nine statements are equivalent: that is, each one implies the other eight (so if one is true, all are true.)

   i. A is invertible.

   ii. The only solution to the homogeneous system $A\mathbf{x} = \mathbf{0}$ is the trivial solution $(\mathbf{x} = \mathbf{0})$.

   iii. The system $A\mathbf{x} = \mathbf{b}$ has a unique solution for every n-vector $\mathbf{b}$.

   iv. A is row equivalent to the $n \times n$ identity matrix I_n.

   v. A can be written as the product of elementary matrices.

   vi. The rows (and columns) of A are linearly independent.

   vii. det $A \neq 0$.

   viii. $\nu(A) = 0$.

   ix. $\rho(A) = n$.

Moreover, if one of the above fails to hold, then for every vector $\mathbf{b} \in \mathbb{R}^n$, the system $A\mathbf{x} = \mathbf{b}$ has either no solution or an infinite number of solutions. It has an infinite number of solutions if and only if $\rho(A) = \rho((A, \mathbf{b}))$. ∎

We conclude this section with a proof of Theorem 3.

Proof of Theorem 3† We first show that $C_A = $ Range A. As before we let $\mathbf{e}_j$ denote the vector in $\mathbb{R}^n$ with a 1 in the jth position and zero everywhere else. We write A in the form

$$A = \begin{pmatrix} a_{11} & a_{12} & \cdots & a_{1j} & \cdots & a_{1n} \\ a_{21} & a_{22} & \cdots & a_{2j} & \cdots & a_{2n} \\ \vdots & \vdots & & \vdots & & \vdots \\ a_{m1} & a_{m2} & \cdots & a_{mj} & \cdots & a_{mn} \end{pmatrix}$$

Then $A\mathbf{e}_j$ is the jth column of A. Thus each column of A is in Range A so that

$$C_A \subseteq \text{Range } A \tag{10}$$

---

† If time permits.

Let $\{\mathbf{y}_1, \ldots, \mathbf{y}_k\}$ be a basis for Range A. Now let us look at one of the basis vectors, say $\mathbf{y}_i$. We have, by the definition of the range, $\mathbf{y}_i = A\mathbf{x}_i$ for some $\mathbf{x}_i \in \mathbb{R}^n$. But $\{\mathbf{e}_1, \mathbf{e}_2, \ldots, \mathbf{e}_n\}$ is a basis in $\mathbb{R}^n$, so there exist constants $c_1, c_2, \ldots, c_n$ such that

$$\mathbf{x}_i = c_1\mathbf{e}_1 + c_2\mathbf{e}_2 + \cdots + c_n\mathbf{e}_n \tag{11}$$

Then

$$\mathbf{y}_i = A\mathbf{x}_i = A(c_1\mathbf{e}_1 + c_2\mathbf{e}_2 + \cdots + c_n\mathbf{e}_n) = c_1A\mathbf{e}_1 + c_2A\mathbf{e}_2 + \cdots + c_nA\mathbf{e}_n \tag{12}$$

But $A\mathbf{e}_j$ is the jth column of A, so (12) shows that we can write $\mathbf{y}_i$ as a linear combination of the columns of A. Thus each basis vector in Range A is in the column space C_A of A so that

$$\text{Range } A \subseteq C_A \tag{13}$$

Combining (10) and (13), we see that Range $A = C_A$. To complete the proof, we must show that if R_A denotes the row space of A, then $\dim R_A = \dim C_A$. We denote the rows of A by $\mathbf{r}_1, \mathbf{r}_2, \ldots, \mathbf{r}_m$, and let $k = \dim R_A$. Let $S = \{\mathbf{s}_1, \mathbf{s}_2, \ldots, \mathbf{s}_k\}$ be a basis for R_A. Then every row vector of A can be written as a linear combination of the vectors in S, and we have, for some constants α_{ij},

$$
\begin{aligned}
\mathbf{r}_1 &= \alpha_{11}\mathbf{s}_1 + \alpha_{12}\mathbf{s}_2 + \cdots + \alpha_{1k}\mathbf{s}_k \\
\mathbf{r}_2 &= \alpha_{21}\mathbf{s}_1 + \alpha_{22}\mathbf{s}_2 + \cdots + \alpha_{2k}\mathbf{s}_k \\
&\ \ \vdots \qquad\quad \vdots \qquad\quad \vdots \qquad\qquad \vdots \\
\mathbf{r}_m &= \alpha_{m1}\mathbf{s}_1 + \alpha_{m2}\mathbf{s}_2 + \cdots + \alpha_{mk}\mathbf{s}_k
\end{aligned}
\tag{14}
$$

Now, the jth component of $\mathbf{r}_i$ is a_{ij}. Thus if we equate the jth components of both sides of (14), we obtain

$$
\begin{aligned}
a_{1j} &= \alpha_{11}s_{1j} + \alpha_{12}s_{2j} + \cdots + \alpha_{1k}s_{kj} \\
a_{2j} &= \alpha_{21}s_{1j} + \alpha_{22}s_{2j} + \cdots + \alpha_{2k}s_{kj} \\
&\ \ \vdots \qquad\quad \vdots \qquad\quad \vdots \qquad\qquad \vdots \\
a_{mj} &= \alpha_{m1}s_{1j} + \alpha_{m2}s_{2j} + \cdots + \alpha_{mk}s_{kj}
\end{aligned}
$$

or

$$
\begin{pmatrix} a_{1j} \\ a_{2j} \\ \vdots \\ a_{mj} \end{pmatrix} = s_{1j} \begin{pmatrix} \alpha_{11} \\ \alpha_{21} \\ \vdots \\ \alpha_{m1} \end{pmatrix} + s_{2j} \begin{pmatrix} \alpha_{12} \\ \alpha_{22} \\ \vdots \\ \alpha_{m2} \end{pmatrix} + \cdots + s_{kj} \begin{pmatrix} \alpha_{1k} \\ \alpha_{2k} \\ \vdots \\ \alpha_{mk} \end{pmatrix} \tag{15}
$$

Here $\mathbf{s}_i = (s_{i1}, s_{i2}, \ldots, s_{in})$. Let $\boldsymbol{\alpha}_i$ denote the vector $\begin{pmatrix} \alpha_{1i} \\ \alpha_{2i} \\ \vdots \\ \alpha_{mi} \end{pmatrix}$. Then since the

left-hand side of (15) is the jth column of A, we see that we can write every column of A as a linear combination of $\boldsymbol{\alpha}_1, \boldsymbol{\alpha}_2, \ldots, \boldsymbol{\alpha}_k$, which means that

$\alpha_1, \alpha_2, \ldots, \alpha_k$ span C_A and

$$\dim C_A \leq k = \dim R_A \tag{16}$$

But equation (16) holds for any matrix A. In particular, it holds for A^t. But $C_{A^t} = R_A$ and $R_{A^t} = C_A$. Thus, since, from (16), $\dim C_{A^t} \leq \dim R_{A^t}$, we have

$$\dim R_A \leq \dim C_A \tag{17}$$

Combining (16) and (17) completes the proof. ∎

PROBLEMS 4.7

In Problems 1–15 find the rank and nullity of the given matrix.

1. $\begin{pmatrix} 1 & 2 \\ 3 & 4 \end{pmatrix}$

2. $\begin{pmatrix} 1 & -1 & 2 \\ 3 & 1 & 0 \end{pmatrix}$

3. $\begin{pmatrix} -1 & 3 & 2 \\ 2 & -6 & -4 \end{pmatrix}$

4. $\begin{pmatrix} 1 & -1 & 2 \\ 3 & 1 & 4 \\ -1 & 0 & 4 \end{pmatrix}$

5. $\begin{pmatrix} 1 & -1 & 2 \\ 3 & 1 & 4 \\ 5 & -1 & 8 \end{pmatrix}$

6. $\begin{pmatrix} -1 & 2 & 1 \\ 2 & -4 & -2 \\ -3 & 6 & 3 \end{pmatrix}$

7. $\begin{pmatrix} 1 & -1 & 2 & 3 \\ 0 & 1 & 4 & 3 \\ 1 & 0 & 6 & 6 \end{pmatrix}$

8. $\begin{pmatrix} 1 & -1 & 2 & 3 \\ 0 & 1 & 4 & 3 \\ 1 & 0 & 6 & 5 \end{pmatrix}$

9. $\begin{pmatrix} 2 & 3 \\ -1 & 1 \\ 4 & 7 \end{pmatrix}$

10. $\begin{pmatrix} 1 & -1 & 2 & 3 \\ 0 & 1 & 0 & 1 \\ 1 & 0 & 1 & 0 \\ 0 & 0 & 0 & 1 \end{pmatrix}$

11. $\begin{pmatrix} 1 & -1 & 2 & 1 \\ -1 & 0 & 1 & 2 \\ 1 & -2 & 5 & 4 \\ 2 & -1 & 1 & -1 \end{pmatrix}$

12. $\begin{pmatrix} 1 & -1 & 2 & 3 \\ -2 & 2 & -4 & -6 \\ 2 & -2 & 4 & 6 \\ 3 & -3 & 6 & 9 \end{pmatrix}$

13. $\begin{pmatrix} -1 & -1 & 0 & 0 \\ 0 & 0 & 2 & 3 \\ 4 & 0 & -2 & 1 \\ 3 & -1 & 0 & 4 \end{pmatrix}$

14. $\begin{pmatrix} 3 & 0 & 0 \\ 0 & 0 & 0 \\ 0 & 0 & 6 \end{pmatrix}$

15. $\begin{pmatrix} 1 & 2 & 3 \\ 0 & 0 & 4 \\ 0 & 0 & 6 \end{pmatrix}$

In Problems 16–22 find a basis for the range and kernel of the given matrix.

16. The matrix of Problem 2

17. The matrix of Problem 5

18. The matrix of Problem 6

19. The matrix of Problem 8

20. The matrix of Problem 11

21. The matrix of Problem 12

22. The matrix of Problem 13

In Problems 23–26 find a basis for the span of the given set of vectors.

23. $\begin{pmatrix} 1 \\ 4 \\ -2 \end{pmatrix}, \begin{pmatrix} 2 \\ 1 \\ 2 \end{pmatrix}, \begin{pmatrix} -1 \\ 3 \\ -4 \end{pmatrix}$

24. $(1, -2, 3), (2, -1, 4), (3, -3, 3), (2, 1, 0)$

25. $(1, -1, 1, -1), (2, 0, 0, 1), (4, -2, 2, 1), (7, -3, 3, -1)$

26. $\begin{pmatrix} 1 \\ 0 \\ 0 \\ 1 \end{pmatrix}, \begin{pmatrix} 0 \\ 1 \\ 1 \\ 0 \end{pmatrix}, \begin{pmatrix} 1 \\ -2 \\ -2 \\ 1 \end{pmatrix}, \begin{pmatrix} 0 \\ 2 \\ 2 \\ 1 \end{pmatrix}$

In Problems 27–30 use Theorem 7 to determine whether the given system has any solutions.

27. $\begin{aligned} x_1 + x_2 - x_3 &= 7 \\ 4x_1 - x_2 + 5x_3 &= 4 \\ 6x_1 + x_2 + 3x_3 &= 20 \end{aligned}$

28. $\begin{aligned} x_1 + x_2 - x_3 &= 7 \\ 4x_1 - x_2 + 5x_3 &= 4 \\ 6x_1 + x_2 + 3x_3 &= 18 \end{aligned}$

29. $\begin{aligned} x_1 - 2x_2 + x_3 + x_4 &= 2 \\ 3x_1 \phantom{{}+ 2x_2} + 2x_3 - 2x_4 &= -8 \\ 4x_2 - x_3 - x_4 &= 1 \\ 5x_1 \phantom{{}+ 2x_2} + 3x_3 - x_4 &= -3 \end{aligned}$

30. $\begin{aligned} x_1 - 2x_2 + x_3 + x_4 &= 2 \\ 3x_1 \phantom{{}+ 2x_2} + 2x_3 - 2x_4 &= -8 \\ 4x_2 - x_3 - x_4 &= 1 \\ 5x_1 \phantom{{}+ 2x_2} + 3x_3 - x_4 &= 0 \end{aligned}$

31. Show that the rank of a diagonal matrix is equal to the number of nonzero components on the diagonal.

32. Let A be an upper triangular $n \times n$ matrix with zeros on the diagonal. Show that $\rho(A) < n$.

33. Show that for any matrix A, $\rho(A) = \rho(A^t)$.

34. Show that if A is an $m \times n$ matrix and $m < n$, then **(a)** $\rho(A) \leq m$ and **(b)** $\nu(A) \geq n - m$.

35. Let A be an $m \times n$ matrix and let B and C be invertible $m \times m$ and $n \times n$ matrices, respectively. Prove that $\rho(A) = \rho(BA) = \rho(AC)$. That is, multiplying a matrix by an invertible matrix does not change its rank.

36. Let A and B be $m \times n$ and $n \times p$ matrices, respectively. Show that $\rho(AB) \leq \min(\rho(A), \rho(B))$.

37. Let A be a 5×7 matrix with rank 5. Show that the linear system $A\mathbf{x} = \mathbf{b}$ has at least one solution for every 5-vector $\mathbf{b}$.

*38. Let A and B be $m \times n$ matrices. Show that if $\rho(A) = \rho(B)$, then there exist invertible matrices C and D such that $B = CAD$.

39. If $B = CAD$, where C and D are invertible, prove that $\rho(A) = \rho(B)$.

40. Suppose that any k rows of A are linearly independent while any $k + 1$ rows of A are linearly dependent. Show that $\rho(A) = k$.

41. If A is an $n \times n$ matrix, show that $\rho(A) < n$ if and only if there is a vector $\mathbf{x} \in \mathbb{R}^n$ such that $\mathbf{x} \neq \mathbf{0}$ and $A\mathbf{x} = \mathbf{0}$.

42. Let A be an $m \times n$ matrix. Suppose that for every $\mathbf{y} \in \mathbb{R}^m$ there is an $\mathbf{x} \in \mathbb{R}^n$ such that $A\mathbf{x} = \mathbf{y}$. Show that $\rho(A) = m$.

4.8   RANK AND DETERMINANTS OF SUBMATRICES (Optional)

In this section we show how the rank of a matrix can be calculated by computing certain determinants.

DEFINITION 1   ***k*-Square Submatrix** Let A be an $m \times n$ matrix. The matrix obtained as the intersection of k rows and k columns of A is called a ***k*-square submatrix** of A.

EXAMPLE 1   **Submatrices of a 3 × 4 Matrix**

$$A = \begin{pmatrix} 1 & 3 & -2 & 6 \\ 3 & 1 & 6 & 2 \\ 4 & .1 & 5 & 2 \end{pmatrix}$$

i. Each component of A is a 1-square submatrix of A.

ii. The intersection of rows 1 and 3 and columns 3 and 4, say, is a 2-square submatrix:

$$\begin{pmatrix} -2 & 6 \\ 5 & 2 \end{pmatrix}$$

There are 18 2-square submatrices. Two others are

$$\begin{pmatrix} 1 & 3 \\ 3 & 1 \end{pmatrix} \quad \text{and} \quad \begin{pmatrix} 3 & 6 \\ 1 & 2 \end{pmatrix}$$

rows 1 and 2   rows 1 and 3
columns 1 and 2   columns 2 and 4

iii. There are four 3-square submatrices. These are

$$\begin{pmatrix} 1 & 3 & -2 \\ 3 & 1 & 6 \\ 4 & 1 & 5 \end{pmatrix}, \quad \begin{pmatrix} 1 & 3 & 6 \\ 3 & 1 & 2 \\ 4 & 1 & 2 \end{pmatrix}, \quad \begin{pmatrix} 1 & -2 & 6 \\ 3 & 6 & 2 \\ 4 & 5 & 2 \end{pmatrix}, \quad \text{and} \quad \begin{pmatrix} 3 & -2 & 6 \\ 1 & 6 & 2 \\ 1 & 5 & 2 \end{pmatrix} \quad \blacksquare$$

DEFINITION 2   **Subdeterminant of Order *k*** Let A be an $m \times n$ matrix. A **subdeterminant of order *k*** of A is the determinant of a k-square submatrix of A.

EXAMPLE 2   **Subdeterminants of a 3 × 4 Matrix**   Let

$$A = \begin{pmatrix} 1 & 3 & -2 & 6 \\ 3 & 1 & 6 & 2 \\ 4 & 1 & 5 & 2 \end{pmatrix}$$

i. Each component of A is a subdeterminant of order 1.

ii. Three subdeterminants of order 2 are

$$\begin{pmatrix} -2 & 6 \\ 5 & 2 \end{pmatrix} = -34, \qquad \begin{pmatrix} 1 & 3 \\ 3 & 1 \end{pmatrix} = -8, \quad \text{and} \quad \begin{pmatrix} 3 & 6 \\ 1 & 2 \end{pmatrix} = 0$$

iii. Two subdeterminants of order 3 are

$$\begin{vmatrix} 1 & 3 & -2 \\ 3 & 1 & 6 \\ 4 & 1 & 5 \end{vmatrix} = 28 \quad \text{and} \quad \begin{vmatrix} 3 & -2 & 6 \\ 1 & 6 & 2 \\ 1 & 5 & 2 \end{vmatrix} = 0$$

$\blacksquare$

THEOREM 1   Let A be an $m \times n$ matrix. Then

$$\rho(A) \geq k$$

if and only if A has a nonzero subdeterminant of order k.

Proof   Suppose that $\rho(A) \geq k$. From Theorem 4.7.3, we know that

$$\dim R_A = \dim C_A = \dim \text{Range } A = \rho(A) \geq k$$

Thus A has at least k independent rows. Let B be a $k \times n$ matrix whose rows are k linearly independent rows of A. Then $k = \rho(B) = \dim C_B$; that is, B has k linearly independent columns. If E is the $k \times k$ matrix whose columns are k linearly independent columns of B, then E is a k-square submatrix of A. Since the columns of E are linearly independent, $\det E \neq 0$. We have shown that A has a nonzero subdeterminant of order k.

Now suppose that A has a nonzero subdeterminant of order k; that is, A has a k-square submatrix E with $\det E \neq 0$. Then the rows of E are linearly independent. Each of the k rows of E is contained in a row of A. Thus the corresponding k rows of A are linearly independent. Since A has at least k linearly independent rows, we conclude that $\rho(A) \geq k$. $\blacksquare$

THEOREM 2   Let A be an $m \times n$ matrix. Then

$\rho(A) = r > 0$ if and only if A has at least one nonzero subdeterminant of order r and every subdeterminant of order $r + 1$ is zero.

Proof   Suppose that $\rho(A) = r$. Then, by Theorem 1, A has a nonzero subdeterminant of order r. Let E be an $(r + 1)$-square submatrix. Then $\det E = 0$ because if $\det E \neq$

0, then $\rho(A) \geq k + 1$ by Theorem 1. Conversely, suppose that A has an r-square submatrix E with det $E \neq 0$ but that every subdeterminant of order $r + 1$ is zero. Thus, again by Theorem 1, $\rho(A) \geq r$. But $\rho(A) < r + 1$ because if $\rho(A) \geq r + 1$, then A has nonzero subdeterminant of order $r + 1$, in contradiction to the hypothesis. Thus $\rho(A) = k$ and the theorem is proved. ∎

EXAMPLE 3   **Using Subdeterminants to Compute the Rank of a Matrix**   Use Theorem 2 to find the rank of each matrix:

$$\textbf{(a) } A = \begin{pmatrix} 1 & 2 & 3 \\ 2 & 4 & 6 \\ -3 & -6 & -9 \end{pmatrix} \qquad \textbf{(b) } A = \begin{pmatrix} 1 & 2 & -1 & 4 \\ 3 & 1 & 5 & 2 \\ 1 & -3 & 7 & -6 \end{pmatrix}$$

Solution   (a) A has nine nonzero subdeterminants of order 1. The nine subdeterminants of order 2 are

$$\begin{vmatrix} 1 & 2 \\ 2 & 4 \end{vmatrix} = \begin{vmatrix} 1 & 3 \\ 2 & 6 \end{vmatrix} = \begin{vmatrix} 2 & 3 \\ 4 & 6 \end{vmatrix} = \begin{vmatrix} 1 & 2 \\ -3 & -6 \end{vmatrix} = \begin{vmatrix} 1 & 3 \\ -3 & -9 \end{vmatrix} = \begin{vmatrix} 2 & 3 \\ -6 & -9 \end{vmatrix}$$

$$= \begin{vmatrix} 2 & 4 \\ -3 & -6 \end{vmatrix} = \begin{vmatrix} 2 & 6 \\ -3 & -9 \end{vmatrix} = \begin{vmatrix} 4 & 6 \\ -6 & -9 \end{vmatrix} = 0$$

Thus $\rho(A) = 1$.

(b) $\begin{vmatrix} 1 & 2 \\ 3 & 1 \end{vmatrix} = -5 \neq 0$, so $\rho(A) \geq 2$. But

Check this

$$\begin{vmatrix} 1 & 2 & -1 \\ 3 & 1 & 5 \\ 1 & -3 & 7 \end{vmatrix} = \begin{vmatrix} 1 & 2 & 4 \\ 3 & 1 & 2 \\ 1 & -3 & -6 \end{vmatrix} = \begin{vmatrix} 2 & -1 & 4 \\ 1 & 5 & 2 \\ -3 & 7 & -6 \end{vmatrix} = 0$$

So $\rho(A) = 2$. ∎

PROBLEMS   4.8

In Problems 1–10 use Theorem 2 to determine the rank of the given matrix.

1. $\begin{pmatrix} 2 & -3 \\ -6 & 9 \end{pmatrix}$ **2.** $\begin{pmatrix} 4 & 1 \\ 1 & 4 \end{pmatrix}$ **3.** $\begin{pmatrix} 1 & 0 & 6 \\ 2 & -1 & 1 \end{pmatrix}$ **4.** $\begin{pmatrix} 1 & 0 & 2 \\ 3 & 0 & 6 \end{pmatrix}$

5. $\begin{pmatrix} 1 & 0 & 2 \\ 3 & 1 & 4 \\ 1 & 1 & 0 \end{pmatrix}$ **6.** $\begin{pmatrix} 1 & 0 & 2 \\ 3 & 1 & 4 \\ 1 & 1 & 1 \end{pmatrix}$ **7.** $\begin{pmatrix} 1 & 3 & -1 & 4 \\ 2 & 1 & 4 & 1 \\ 1 & 0 & 2 & 3 \end{pmatrix}$

8. $\begin{pmatrix} 1 & 3 & -1 & 4 \\ 2 & 1 & 4 & 1 \\ -1 & 2 & -5 & 3 \end{pmatrix}$ **9.** $\begin{pmatrix} 1 & 2 & 1 \\ 2 & -1 & 4 \\ 4 & 3 & 6 \\ 2 & 5 & 2 \end{pmatrix}$ **10.** $\begin{pmatrix} 1 & 2 & 1 \\ 2 & -1 & 4 \\ 4 & 3 & 6 \\ 7 & 4 & 4 \end{pmatrix}$

11. Let A be an $m \times n$ matrix and suppose that every subdeterminant of order k is zero. Prove that every subdeterminant of order $k + 1$ is zero.

4.9   CHANGE OF BASIS

In $\mathbb{R}^2$ we wrote vectors in terms of the standard basis $\mathbf{i} = \begin{pmatrix} 1 \\ 0 \end{pmatrix}$, $\mathbf{j} = \begin{pmatrix} 0 \\ 1 \end{pmatrix}$. In $\mathbb{R}^n$ we defined the standard basis $\{\mathbf{e}_1, \mathbf{e}_2, \dots, \mathbf{e}_n\}$. In P_n we defined the standard basis to be $\{1, x, x^2, \dots, x^n\}$. These bases are most commonly used because it is relatively easy to work with them. But it sometimes happens that some other basis is more convenient. There are infinitely many bases to choose from since in an n-dimensional vector space *any* n linearly independent vectors form a basis. In this section we shall see how to change from one basis to another by computing a certain matrix.

We start with a simple example. Let $\mathbf{u}_1 = \begin{pmatrix} 1 \\ 0 \end{pmatrix}$ and $\mathbf{u}_2 = \begin{pmatrix} 0 \\ 1 \end{pmatrix}$. Then $B_1 = \{\mathbf{u}_1, \mathbf{u}_2\}$ is the standard basis in $\mathbb{R}^2$. Let $\mathbf{v}_1 = \begin{pmatrix} 1 \\ 3 \end{pmatrix}$ and $\mathbf{v}_2 = \begin{pmatrix} -1 \\ 2 \end{pmatrix}$. Since $\mathbf{v}_1$ and $\mathbf{v}_2$ are linearly independent (because $\mathbf{v}_1$ is not a multiple of $\mathbf{v}_2$), $B_2 = \{\mathbf{v}_1, \mathbf{v}_2\}$ is a second basis in $\mathbb{R}^2$. Let $\mathbf{x} = \begin{pmatrix} x_1 \\ x_2 \end{pmatrix}$ be a vector in $\mathbb{R}^2$. This notation means that

$$\mathbf{x} = \begin{pmatrix} x_1 \\ x_2 \end{pmatrix} = x_1 \begin{pmatrix} 1 \\ 0 \end{pmatrix} + x_2 \begin{pmatrix} 0 \\ 1 \end{pmatrix} = x_1 \mathbf{u}_1 + x_2 \mathbf{u}_2$$

That is, $\mathbf{x}$ is written in terms of the vectors in the basis B_1. To emphasize this fact, we write

$$(\mathbf{x})_{B_1} = \begin{pmatrix} x_1 \\ x_2 \end{pmatrix}$$

Since B_2 is another basis in $\mathbb{R}^2$, there are scalars c_1 and c_2 such that

$$\mathbf{x} = c_1 \mathbf{v}_1 + c_2 \mathbf{v}_2 \tag{1}$$

Once these scalars are found, we write

$$(\mathbf{x})_{B_2} = \begin{pmatrix} c_1 \\ c_2 \end{pmatrix}$$

to indicate that $\mathbf{x}$ is now expressed in terms of the vectors in B_2. To find the numbers c_1 and c_2, we write the old basis vectors ($\mathbf{u}_1$ and $\mathbf{u}_2$) in terms of the new basis vectors ($\mathbf{v}_1$ and $\mathbf{v}_2$). It is easy to verify that

$$\mathbf{u}_1 = \begin{pmatrix} 1 \\ 0 \end{pmatrix} = \tfrac{2}{5}\begin{pmatrix} 1 \\ 3 \end{pmatrix} - \tfrac{3}{5}\begin{pmatrix} -1 \\ 2 \end{pmatrix} = \tfrac{2}{5}\mathbf{v}_1 - \tfrac{3}{5}\mathbf{v}_2 \tag{2}$$

and

$$\mathbf{u}_2 = \begin{pmatrix} 0 \\ 1 \end{pmatrix} = \tfrac{1}{5}\begin{pmatrix} 1 \\ 3 \end{pmatrix} + \tfrac{1}{5}\begin{pmatrix} -1 \\ 2 \end{pmatrix} = \tfrac{1}{5}\mathbf{v}_1 + \tfrac{1}{5}\mathbf{v}_2 \tag{3}$$

That is,

$$(\mathbf{u}_1)_{B_2} = \begin{pmatrix} \frac{2}{5} \\ -\frac{3}{5} \end{pmatrix} \quad \text{and} \quad (\mathbf{u}_2)_{B_2} = \begin{pmatrix} \frac{1}{5} \\ \frac{1}{5} \end{pmatrix}$$

Then

from (2) and (3)
↓

$$\mathbf{x} = x_1\mathbf{u}_1 + x_2\mathbf{u}_2 = x_1(\tfrac{2}{5}\mathbf{v}_1 - \tfrac{3}{5}\mathbf{v}_2) + x_2(\tfrac{1}{5}\mathbf{v}_1 + \tfrac{1}{5}\mathbf{v}_2)$$
$$= (\tfrac{2}{5}x_1 + \tfrac{1}{5}x_2)\mathbf{v}_1 + (-\tfrac{3}{5}x_1 + \tfrac{1}{5}x_2)\mathbf{v}_2$$

Thus, from (1),

$$c_1 = \tfrac{2}{5}x_1 + \tfrac{1}{5}x_2$$
$$c_2 = -\tfrac{3}{5}x_1 + \tfrac{1}{5}x_2$$

or

$$(\mathbf{x})_{B_2} = \begin{pmatrix} c_1 \\ c_2 \end{pmatrix} = \begin{pmatrix} \frac{2}{5}x_1 + \frac{1}{5}x_2 \\ -\frac{3}{5}x_1 + \frac{1}{5}x_2 \end{pmatrix} = \begin{pmatrix} \frac{2}{5} & \frac{1}{5} \\ -\frac{3}{5} & \frac{1}{5} \end{pmatrix}\begin{pmatrix} x_1 \\ x_2 \end{pmatrix}$$

For example, if $(\mathbf{x})_{B_1} = \begin{pmatrix} 3 \\ -4 \end{pmatrix}$, then

$$(\mathbf{x})_{B_2} = \begin{pmatrix} \frac{2}{5} & \frac{1}{5} \\ -\frac{3}{5} & \frac{1}{5} \end{pmatrix}\begin{pmatrix} 3 \\ -4 \end{pmatrix} = \begin{pmatrix} \frac{2}{5} \\ -\frac{13}{5} \end{pmatrix}$$

Check.

$$\tfrac{2}{5}\mathbf{v}_1 - \tfrac{13}{5}\mathbf{v}_2 = \tfrac{2}{5}\begin{pmatrix} 1 \\ 3 \end{pmatrix} - \tfrac{13}{5}\begin{pmatrix} -1 \\ 2 \end{pmatrix} = \begin{pmatrix} \frac{2}{5} + \frac{13}{5} \\ \frac{6}{5} - \frac{26}{5} \end{pmatrix} = \begin{pmatrix} 3 \\ -4 \end{pmatrix} = 3\begin{pmatrix} 1 \\ 0 \end{pmatrix} - 4\begin{pmatrix} 0 \\ 1 \end{pmatrix}$$
$$= 3\mathbf{u}_1 - 4\mathbf{u}_2$$

The matrix $A = \begin{pmatrix} \frac{2}{5} & \frac{1}{5} \\ -\frac{3}{5} & \frac{1}{5} \end{pmatrix}$ is called the **transition matrix** from B_1 to B_2, and we have shown that

$$(\mathbf{x})_{B_2} = A(\mathbf{x})_{B_1} \tag{4}$$

We illustrate the two bases $\left\{ \begin{pmatrix} 1 \\ 0 \end{pmatrix}, \begin{pmatrix} 0 \\ 1 \end{pmatrix} \right\}$ and $\left\{ \begin{pmatrix} 1 \\ 3 \end{pmatrix}, \begin{pmatrix} -1 \\ 2 \end{pmatrix} \right\}$ in Figure 4.4.

This example can be easily generalized, but first we need to extend our notation. Let $B_1 = \{\mathbf{u}_1, \mathbf{u}_2, \ldots, \mathbf{u}_n\}$ and $B_2 = \{\mathbf{v}_1, \mathbf{v}_2, \ldots, \mathbf{v}_n\}$ be two bases for an n-dimensional real vector space V. Let $\mathbf{x} \in V$. Then $\mathbf{x}$ can be written in terms of both bases:

$$\mathbf{x} = b_1\mathbf{u}_1 + b_2\mathbf{u}_2 + \cdots + b_n\mathbf{u}_n \tag{5}$$

and

$$\mathbf{x} = c_1\mathbf{v}_1 + c_2\mathbf{v}_2 + \cdots + c_n\mathbf{v}_n \tag{6}$$

where the b_i's and c_i's are real numbers. We then write $(\mathbf{x})_{B_1} = \begin{pmatrix} b_1 \\ b_2 \\ \vdots \\ b_n \end{pmatrix}$ to denote the

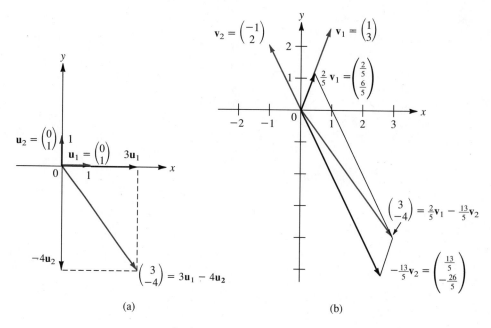

Figure 4.4   (a) Writing $\begin{pmatrix} 3 \\ -4 \end{pmatrix}$ in terms of standard basis $\left\{ \begin{pmatrix} 1 \\ 0 \end{pmatrix}, \begin{pmatrix} 0 \\ 1 \end{pmatrix} \right\}$.

(b) Writing $\begin{pmatrix} 3 \\ -4 \end{pmatrix}$ in terms of basis $\left\{ \begin{pmatrix} 1 \\ 3 \end{pmatrix}, \begin{pmatrix} -1 \\ 2 \end{pmatrix} \right\}$

representation of $\mathbf{x}$ in terms of the basis B_1. This is unambiguous because the coefficients b_i in (5) are unique by Theorem 4.6.1, page 245. Likewise $(\mathbf{x})_{B_2} = \begin{pmatrix} c_1 \\ c_2 \\ \vdots \\ c_n \end{pmatrix}$ has a similar meaning. Suppose that $\mathbf{w}_1 = a_1\mathbf{u}_1 + a_2\mathbf{u}_2 + \cdots + a_n\mathbf{u}_n$ and $\mathbf{w}_2 = b_1\mathbf{u}_1 + b_2\mathbf{u}_2 + \cdots + b_n\mathbf{u}_n$. Then $\mathbf{w}_1 + \mathbf{w}_2 = (a_1 + b_1)\mathbf{u}_1 + (a_2 + b_2)\mathbf{u}_2 + \cdots + (a_n + b_n)\mathbf{u}_n$, so that

$$(\mathbf{w}_1 + \mathbf{w}_2)_{B_1} = (\mathbf{w}_1)_{B_1} + (\mathbf{w}_2)_{B_1}$$

That is, in the new notation we can add vectors just as we add vectors in $\mathbb{R}^n$. Moreover, it is easy to show that

$$\alpha(\mathbf{w})_{B_1} = (\alpha\mathbf{w})_{B_1}$$

Now since B_2 is a basis, each $\mathbf{u}_j$ in B_1 can be written as a linear combination of the $\mathbf{v}_i$'s. Thus there exists a unique set of scalars $a_{1j}, a_{2j}, \ldots, a_{nj}$ such that for $j = 1, 2, \ldots, n$

$$\mathbf{u}_j = a_{1j}\mathbf{v}_1 + a_{2j}\mathbf{v}_2 + \cdots + a_{nj}\mathbf{v}_n \tag{7}$$

or

$$(\mathbf{u}_j)_{B_2} = \begin{pmatrix} a_{1j} \\ a_{2j} \\ \vdots \\ a_{nj} \end{pmatrix} \tag{8}$$

DEFINITION 1   **Transition Matrix**   The $n \times n$ matrix A whose columns are given by (8) is called the **transition matrix** from basis B_1 to basis B_2. That is,

$$A = \begin{pmatrix} a_{11} & a_{12} & a_{13} & \cdots & a_{1n} \\ a_{21} & a_{22} & a_{23} & \cdots & a_{2n} \\ \vdots & \vdots & \vdots & & \vdots \\ a_{n1} & a_{n2} & a_{n3} & \cdots & a_{nn} \end{pmatrix} \tag{9}$$

$$\uparrow \qquad \uparrow \qquad \uparrow \qquad\qquad \uparrow$$
$$(\mathbf{u}_1)_{B_2}(\mathbf{u}_2)_{B_2}(\mathbf{u}_3)_{B_2} \cdots (\mathbf{u}_n)_{B_2}$$

THEOREM 1   Let B_1 and B_2 be bases for a vector space V. Let A be the transition matrix from B_1 to B_2. Then, for every $\mathbf{x} \in V$,

$$\boxed{(\mathbf{x})_{B_2} = A(\mathbf{x})_{B_1}} \tag{10}$$

Proof   We use the representation of $\mathbf{x}$ given in (5) and (6):

from (5)
$\downarrow$

$$\mathbf{x} = b_1\mathbf{u}_1 + b_2\mathbf{u}_2 + \cdots + b_n\mathbf{u}_n$$

from (7)
$\downarrow$

$$= b_1(a_{11}\mathbf{v}_1 + a_{21}\mathbf{v}_2 + \cdots + a_{n1}\mathbf{v}_n) + b_2(a_{12}\mathbf{v}_1 + a_{22}\mathbf{v}_2 + \cdots + a_{n2}\mathbf{v}_n)$$
$$+ \cdots + b_n(a_{1n}\mathbf{v}_1 + a_{2n}\mathbf{v}_2 + \cdots + a_{nn}\mathbf{v}_n)$$
$$= (a_{11}b_1 + a_{12}b_2 + \cdots + a_{1n}b_n)\mathbf{v}_1 + (a_{21}b_1 + a_{22}b_2 + \cdots + a_{2n}b_n)\mathbf{v}_2 + \cdots$$
$$+ (a_{n1}b_1 + a_{n2}b_2 + \cdots + a_{nn}b_n)\mathbf{v}_n$$

from (6)
$\downarrow$

$$= c_1\mathbf{v}_1 + c_2\mathbf{v}_2 + \cdots + c_n\mathbf{v}_n \tag{11}$$

Thus

$$\text{from (11)}$$

$$(\mathbf{x})_{B_2} = \begin{pmatrix} c_1 \\ c_2 \\ \vdots \\ c_n \end{pmatrix} \stackrel{\downarrow}{=} \begin{pmatrix} a_{11}b_1 + a_{12}b_2 + \cdots + a_{1n}b_n \\ a_{21}b_1 + a_{22}b_2 + \cdots + a_{2n}b_n \\ \vdots \qquad \vdots \qquad \qquad \vdots \\ a_{n1}b_1 + a_{n2}b_2 + \cdots + a_{nn}b_n \end{pmatrix}$$

$$= \begin{pmatrix} a_{11} & a_{12} & \cdots & a_{1n} \\ a_{21} & a_{22} & \cdots & a_{2n} \\ \vdots & \vdots & & \vdots \\ a_{n1} & a_{n2} & \cdots & a_{nn} \end{pmatrix} \begin{pmatrix} b_1 \\ b_2 \\ \vdots \\ b_n \end{pmatrix} = A(\mathbf{x})_{B_1} \qquad (12)$$

■

Before doing any further examples, we prove a theorem that is very useful for computations.

THEOREM 2   If A is the transition matrix from B_1 to B_2, then A^{-1} is the transition matrix from B_2 to B_1.

Proof   Let C be the transition matrix from B_2 to B_1. Then, from (10), we have

$$(\mathbf{x})_{B_1} = C(\mathbf{x})_{B_2} \qquad (13)$$

But $(\mathbf{x})_{B_2} = A(\mathbf{x})_{B_1}$, and substituting this into (13) yields

$$(\mathbf{x})_{B_1} = CA(\mathbf{x})_{B_1} \qquad (14)$$

We leave it as an exercise (see Problem 39) to show that (14) can hold for every $\mathbf{x}$ in V only if $CA = I$. Thus, from Theorem 1.9.8 on page 79, $C = A^{-1}$, and the theorem is proven.   ■

Remark. This theorem makes it especially easy to find the transition matrix from the standard basis $B_1 = \{\mathbf{e}_1, \mathbf{e}_2, \ldots, \mathbf{e}_n\}$ in $\mathbb{R}^n$ to any other basis in $\mathbb{R}^n$. Let $B_2 = \{\mathbf{v}_1, \mathbf{v}_2, \ldots, \mathbf{v}_n\}$ be any other basis. Let C be the matrix whose columns are the vectors $\mathbf{v}_1, \mathbf{v}_2, \ldots, \mathbf{v}_n$. Then C is the transition matrix from B_2 to B_1 since each vector $\mathbf{v}_i$ is already written in terms of the standard basis. For example,

$$\begin{pmatrix} 1 \\ 3 \\ -2 \\ 4 \end{pmatrix}_{B_1} = \begin{pmatrix} 1 \\ 3 \\ -2 \\ 4 \end{pmatrix} = 1\begin{pmatrix} 1 \\ 0 \\ 0 \\ 0 \end{pmatrix} + 3\begin{pmatrix} 0 \\ 1 \\ 0 \\ 0 \end{pmatrix} - 2\begin{pmatrix} 0 \\ 0 \\ 1 \\ 0 \end{pmatrix} + 4\begin{pmatrix} 0 \\ 0 \\ 0 \\ 1 \end{pmatrix}$$

Thus the transition matrix from B_1 to B_2 is C^{-1}.

> **Procedure for Finding the Transition Matrix from the Standard Basis to the Basis $B_2 = \{v_1, v_2, \ldots, v_n\}$**
>
> **i.** Write the matrix C whose columns are $v_1, v_2, \ldots, v_n$.
>
> **ii.** Compute C^{-1}. This is the required transition matrix.

EXAMPLE 1   **Writing Vectors in $\mathbb{R}^3$ in Terms of a New Basis**   In $\mathbb{R}^3$ let $B_1 = \{i, j, k\}$ and let

$$B_2 = \left\{ \begin{pmatrix} 1 \\ 0 \\ 2 \end{pmatrix}, \begin{pmatrix} 3 \\ -1 \\ 0 \end{pmatrix}, \begin{pmatrix} 0 \\ 1 \\ -2 \end{pmatrix} \right\}. \text{ If } x = \begin{pmatrix} x \\ y \\ z \end{pmatrix} \in \mathbb{R}^3, \text{ write } x \text{ in terms of the vectors in } B_2.$$

Solution   We first verify that B_2 is a basis. This is evident since $\begin{vmatrix} 1 & 3 & 0 \\ 0 & -1 & 1 \\ 2 & 0 & -2 \end{vmatrix} = 8 \neq 0$.

Since $u_1 = \begin{pmatrix} 1 \\ 0 \\ 0 \end{pmatrix}$, $u_2 = \begin{pmatrix} 0 \\ 1 \\ 0 \end{pmatrix}$, and $u_3 = \begin{pmatrix} 0 \\ 0 \\ 1 \end{pmatrix}$, we immediately see that the transition matrix, C, from B_2 to B_1 is given by

$$C = \begin{pmatrix} 1 & 3 & 0 \\ 0 & -1 & 1 \\ 2 & 0 & -2 \end{pmatrix}$$

Thus, from Theorem 2, the transition matrix A from B_1 to B_2 is

$$A = C^{-1} = \tfrac{1}{8} \begin{pmatrix} 2 & 6 & 3 \\ 2 & -2 & -1 \\ 2 & 6 & -1 \end{pmatrix}$$

For example, if $(x)_{B_1} = \begin{pmatrix} 1 \\ -2 \\ 4 \end{pmatrix}$, then

$$(x)_{B_2} = \tfrac{1}{8} \begin{pmatrix} 2 & 6 & 3 \\ 2 & -2 & -1 \\ 2 & 6 & -1 \end{pmatrix} \begin{pmatrix} 1 \\ -2 \\ 4 \end{pmatrix} = \tfrac{1}{8} \begin{pmatrix} 2 \\ 2 \\ -14 \end{pmatrix} = \begin{pmatrix} \tfrac{1}{4} \\ \tfrac{1}{4} \\ -\tfrac{7}{4} \end{pmatrix}$$

As a check, note that

$$\tfrac{1}{4} \begin{pmatrix} 1 \\ 0 \\ 2 \end{pmatrix} + \tfrac{1}{4} \begin{pmatrix} 3 \\ -1 \\ 0 \end{pmatrix} - \tfrac{7}{4} \begin{pmatrix} 0 \\ 1 \\ -2 \end{pmatrix} = \begin{pmatrix} 1 \\ -2 \\ 4 \end{pmatrix} = 1 \begin{pmatrix} 1 \\ 0 \\ 0 \end{pmatrix} - 2 \begin{pmatrix} 0 \\ 1 \\ 0 \end{pmatrix} + 4 \begin{pmatrix} 0 \\ 0 \\ 1 \end{pmatrix} \qquad \blacksquare$$

EXAMPLE 2   **Writing Polynomials in P_2 in Terms of a New Basis**   In P_2 the standard basis is $B_1 = \{1, x, x^2\}$. Another basis is $B_2 = \{4x - 1, 2x^2 - x, 3x^2 + 3\}$. If $p = a_0 + a_1x + a_2x^2$, write p in terms of the polynomials in B_2.

Solution   We first verify that B_2 is a basis. If $c_1(4x - 1) + c_2(2x^2 - x) + c_3(3x^2 + 3) = 0$ for all x, then, rearranging terms, we obtain

$$(-c_1 + 3c_3)1 + (4c_1 - c_2)x + (2c_2 + 3c_3)x^2 = 0$$

But since $\{1, x, x^2\}$ is a linearly independent set, we must have

$$\begin{aligned} -c_1 \quad\quad\;\; + 3c_3 &= 0 \\ 4c_1 - \;\; c_2 \quad\quad\;\; &= 0 \\ 2c_2 + 3c_3 &= 0 \end{aligned}$$

The determinant of this homogeneous system is $\begin{vmatrix} -1 & 0 & 3 \\ 4 & -1 & 0 \\ 0 & 2 & 3 \end{vmatrix} = 27 \neq 0$, which

means that $c_1 = c_2 = c_3 = 0$ is the only solution. Now $(4x - 1)_{B_1} = \begin{pmatrix} -1 \\ 4 \\ 0 \end{pmatrix}$,

$(2x^2 - x)_{B_1} = \begin{pmatrix} 0 \\ -1 \\ 2 \end{pmatrix}$, and $(3 + 3x^2)_{B_1} = \begin{pmatrix} 3 \\ 0 \\ 3 \end{pmatrix}$. Hence

$$C = \begin{pmatrix} -1 & 0 & 3 \\ 4 & -1 & 0 \\ 0 & 2 & 3 \end{pmatrix}$$

is the transition matrix from B_2 to B_1 so that

$$A = C^{-1} = \tfrac{1}{27}\begin{pmatrix} -3 & 6 & 3 \\ -12 & -3 & 12 \\ 8 & 2 & 1 \end{pmatrix}$$

is the transition matrix from B_1 to B_2. Since $(a_0 + a_1x + a_2x^2)_{B_1} = \begin{pmatrix} a_0 \\ a_1 \\ a_2 \end{pmatrix}$, we have

$$\begin{aligned} (a_0 + a_1x + a_2x^2)_{B_2} &= \tfrac{1}{27}\begin{pmatrix} -3 & 6 & 3 \\ -12 & -3 & 12 \\ 8 & 2 & 1 \end{pmatrix}\begin{pmatrix} a_0 \\ a_1 \\ a_2 \end{pmatrix} \\[6pt] &= \begin{pmatrix} \tfrac{1}{27}[-3a_0 + 6a_1 + 3a_2] \\ \tfrac{1}{27}[-12a_0 - 3a_1 + 12a_2] \\ \tfrac{1}{27}[8a_0 + 2a_1 + a_2] \end{pmatrix} \end{aligned}$$

For example, if $p(x) = 5x^2 - 3x + 4$, then

$$(5x^2 - 3x + 4)_{B_2} = \tfrac{1}{27}\begin{pmatrix} -3 & 6 & 3 \\ -12 & -3 & 12 \\ 8 & 2 & 1 \end{pmatrix}\begin{pmatrix} 4 \\ -3 \\ 5 \end{pmatrix} = \begin{pmatrix} -\tfrac{15}{27} \\ \tfrac{21}{27} \\ \tfrac{31}{27} \end{pmatrix}$$

or

check this
↓

$$5x^2 - 3x + 4 = -\tfrac{15}{27}(4x - 1) + \tfrac{21}{27}(2x^2 - x) + \tfrac{31}{27}(3x^2 + 3)$$ ∎

EXAMPLE 3   **Converting from One Basis to Another in $\mathbb{R}^2$**   Let $B_1 = \left\{ \begin{pmatrix} 3 \\ 1 \end{pmatrix}, \begin{pmatrix} 2 \\ -1 \end{pmatrix} \right\}$ and $B_2 = \left\{ \begin{pmatrix} 2 \\ 4 \end{pmatrix}, \begin{pmatrix} -5 \\ 3 \end{pmatrix} \right\}$ be two bases in $\mathbb{R}^2$. If $(\mathbf{x})_{B_1} = \begin{pmatrix} b_1 \\ b_2 \end{pmatrix}$, write $\mathbf{x}$ in terms of the vectors in B_2.

Solution   This problem is a bit more difficult because neither basis is the standard basis. We must write the vectors in B_1 as linear combinations of the vectors in B_2. That is, we must find constants $a_{11}, a_{21}, a_{12}, a_{22}$ such that

$$\begin{pmatrix} 3 \\ 1 \end{pmatrix} = a_{11}\begin{pmatrix} 2 \\ 4 \end{pmatrix} + a_{21}\begin{pmatrix} -5 \\ 3 \end{pmatrix} \quad \text{and} \quad \begin{pmatrix} 2 \\ -1 \end{pmatrix} = a_{12}\begin{pmatrix} 2 \\ 4 \end{pmatrix} + a_{22}\begin{pmatrix} -5 \\ 3 \end{pmatrix}$$

This leads to the following systems:

$$\begin{array}{cc} 2a_{11} - 5a_{21} = 3 & 2a_{12} - 5a_{22} = 2 \\ & \text{and} \\ 4a_{11} + 3a_{21} = 1 & 4a_{12} + 3a_{22} = -1 \end{array}$$

The solutions are $a_{11} = \tfrac{7}{13}$, $a_{21} = -\tfrac{5}{13}$, $a_{12} = \tfrac{1}{26}$, and $a_{22} = -\tfrac{5}{13}$. Thus

$$A = \tfrac{1}{26}\begin{pmatrix} 14 & 1 \\ -10 & -10 \end{pmatrix}$$

and

$$(\mathbf{x})_{B_2} = \tfrac{1}{26}\begin{pmatrix} 14 & 1 \\ -10 & -10 \end{pmatrix}\begin{pmatrix} b_1 \\ b_2 \end{pmatrix} = \begin{pmatrix} \tfrac{1}{26}(14b_1 + b_2) \\ -\tfrac{10}{26}(b_1 + b_2) \end{pmatrix}$$

in standard basis
↓

For example, let $\mathbf{x} = \begin{pmatrix} 7 \\ 4 \end{pmatrix}$. Then

$$\begin{pmatrix} 7 \\ 4 \end{pmatrix}_{B_1} = b_1\begin{pmatrix} 3 \\ 1 \end{pmatrix} + b_2\begin{pmatrix} 2 \\ -1 \end{pmatrix} = 3\begin{pmatrix} 3 \\ 1 \end{pmatrix} - \begin{pmatrix} 2 \\ -1 \end{pmatrix}$$

so that

$$\begin{pmatrix} 7 \\ 4 \end{pmatrix}_{B_1} = \begin{pmatrix} 3 \\ -1 \end{pmatrix}$$

and

$$\begin{pmatrix} 7 \\ 4 \end{pmatrix}_{B_2} = \tfrac{1}{26}\begin{pmatrix} 14 & 1 \\ -10 & -10 \end{pmatrix}\begin{pmatrix} 3 \\ -1 \end{pmatrix} = \begin{pmatrix} \tfrac{41}{26} \\ -\tfrac{20}{26} \end{pmatrix}$$

That is,

check!
↓

$$\begin{pmatrix} 7 \\ 4 \end{pmatrix} = \tfrac{41}{26}\begin{pmatrix} 2 \\ 4 \end{pmatrix} - \tfrac{20}{26}\begin{pmatrix} -5 \\ 3 \end{pmatrix}$$ ∎

Using the notation of this section, we can derive a convenient way to determine whether a given set of vectors in any finite dimensional real vector space is linearly dependent or independent.

THEOREM 3   Let $B_1 = \{\mathbf{v}_1, \mathbf{v}_2, \ldots, \mathbf{v}_n\}$ be a basis for the n-dimensional vector space V. Suppose that

$$(\mathbf{x}_1)_{B_1} = \begin{pmatrix} a_{11} \\ a_{21} \\ \vdots \\ a_{n1} \end{pmatrix}, \; (\mathbf{x}_2)_{B_1} = \begin{pmatrix} a_{12} \\ a_{22} \\ \vdots \\ a_{n2} \end{pmatrix}, \; \ldots, \; (\mathbf{x}_n)_{B_1} = \begin{pmatrix} a_{1n} \\ a_{2n} \\ \vdots \\ a_{nn} \end{pmatrix}$$

Let

$$A = \begin{pmatrix} a_{11} & a_{12} & \cdots & a_{1n} \\ a_{21} & a_{22} & \cdots & a_{2n} \\ \vdots & \vdots & & \vdots \\ a_{n1} & a_{n2} & \cdots & a_{nn} \end{pmatrix}$$

Then $\mathbf{x}_1, \mathbf{x}_2, \ldots, \mathbf{x}_n$ are linearly independent if and only if $\det A \neq 0$.

Proof   Let $\mathbf{a}_1, \mathbf{a}_2, \ldots, \mathbf{a}_n$ denote the columns of A. Suppose that

$$c_1 \mathbf{x}_1 + c_2 \mathbf{x}_2 + \cdots + c_n \mathbf{x}_n = \mathbf{0} \tag{15}$$

Then, using the addition defined on page 272, we may write (15) as

$$(c_1 \mathbf{a}_1 + c_2 \mathbf{a}_2 + \cdots + c_n \mathbf{a}_n)_{B_1} = (\mathbf{0})_{B_1} \tag{16}$$

Equation (16) gives two representations of the zero vector in V in terms of the basis vectors in B_1. Since the representation of a vector in terms of basis vectors is unique (by Theorem 4.6.1, page 245) we conclude that

$$c_1 \mathbf{a}_1 + c_2 \mathbf{a}_2 + \cdots + c_n \mathbf{a}_n = \mathbf{0} \tag{17}$$

where the zero on the right-hand side is the zero vector in $\mathbb{R}^n$. But this proves the theorem, since equation (17) involves the columns of A, which are linearly independent if and only if $\det A \neq 0$.   ■

EXAMPLE 4   **Determining Whether Three Polynomials in P_2 Are Linearly Dependent or Independent**   In P_2 determine whether the polynomials $3 - x$, $2 + x^2$, and $4 + 5x - 2x^2$ are linearly dependent or independent.

Solution   Using the basis $B_1 = \{1, x, x^2\}$, we have $(3 - x)_{B_1} = \begin{pmatrix} 3 \\ -1 \\ 0 \end{pmatrix}$, $(2 + x^2)_{B_1} = \begin{pmatrix} 2 \\ 0 \\ 1 \end{pmatrix}$,

and $(4 + 5x - 2x^2)_{B_1} = \begin{pmatrix} 4 \\ 5 \\ -2 \end{pmatrix}$. Then $\det A = \begin{vmatrix} 3 & 2 & 4 \\ -1 & 0 & 5 \\ 0 & 1 & -2 \end{vmatrix} = -23 \neq 0$, so

the polynomials are independent.   ■

EXAMPLE 5 **Determining Whether Four 2 × 2 Matrices Are Linearly Dependent or Independent**   In M_{22} determine whether the matrices $\begin{pmatrix} 1 & 2 \\ 3 & 6 \end{pmatrix}$, $\begin{pmatrix} -1 & 3 \\ -1 & 1 \end{pmatrix}$, $\begin{pmatrix} 2 & -1 \\ 0 & 1 \end{pmatrix}$, and $\begin{pmatrix} 1 & 4 \\ 4 & 9 \end{pmatrix}$ are linearly dependent or independent.

Solution   Using the standard basis $B_1 = \left\{ \begin{pmatrix} 1 & 0 \\ 0 & 0 \end{pmatrix}, \begin{pmatrix} 0 & 1 \\ 0 & 0 \end{pmatrix}, \begin{pmatrix} 0 & 0 \\ 1 & 0 \end{pmatrix}, \begin{pmatrix} 0 & 0 \\ 0 & 1 \end{pmatrix} \right\}$, we obtain

$$\det A = \begin{vmatrix} 1 & -1 & 2 & 1 \\ 2 & 3 & -1 & 4 \\ 3 & -1 & 0 & 4 \\ 6 & 1 & 1 & 9 \end{vmatrix} = 0$$

so the matrices are dependent. Note that det $A = 0$ because the fourth row of A is the sum of the first three rows of A. Note also that

$$-29 \begin{pmatrix} 1 & 2 \\ 3 & 6 \end{pmatrix} - 7 \begin{pmatrix} -1 & 3 \\ -1 & 1 \end{pmatrix} + \begin{pmatrix} 2 & -1 \\ 0 & 1 \end{pmatrix} + 20 \begin{pmatrix} 1 & 4 \\ 4 & 9 \end{pmatrix} = \begin{pmatrix} 0 & 0 \\ 0 & 0 \end{pmatrix}$$

which illustrates that the four matrices are linearly dependent.   ■

PROBLEMS 4.9

In Problems 1–5 write $\begin{pmatrix} x \\ y \end{pmatrix} \in \mathbb{R}^2$ in terms of the given basis.

1. $\begin{pmatrix} 1 \\ 1 \end{pmatrix}, \begin{pmatrix} 1 \\ -1 \end{pmatrix}$ **2.** $\begin{pmatrix} 2 \\ -3 \end{pmatrix}, \begin{pmatrix} 3 \\ -2 \end{pmatrix}$ **3.** $\begin{pmatrix} 5 \\ 7 \end{pmatrix}, \begin{pmatrix} 3 \\ -4 \end{pmatrix}$ **4.** $\begin{pmatrix} -1 \\ -2 \end{pmatrix}, \begin{pmatrix} -1 \\ 2 \end{pmatrix}$

5. $\begin{pmatrix} a \\ c \end{pmatrix}, \begin{pmatrix} b \\ d \end{pmatrix}$, where $ad - bc \neq 0$

In Problems 6–10 write $\begin{pmatrix} x \\ y \\ z \end{pmatrix} \in \mathbb{R}^3$ in terms of the given basis.

6. $\begin{pmatrix} 1 \\ 0 \\ 0 \end{pmatrix}, \begin{pmatrix} 0 \\ 0 \\ 1 \end{pmatrix}, \begin{pmatrix} 1 \\ 1 \\ 1 \end{pmatrix}$ **7.** $\begin{pmatrix} 1 \\ 0 \\ 0 \end{pmatrix}, \begin{pmatrix} 1 \\ 1 \\ 0 \end{pmatrix}, \begin{pmatrix} 1 \\ 1 \\ 1 \end{pmatrix}$ **8.** $\begin{pmatrix} 1 \\ 0 \\ -1 \end{pmatrix}, \begin{pmatrix} -1 \\ 1 \\ 0 \end{pmatrix}, \begin{pmatrix} 0 \\ 1 \\ 1 \end{pmatrix}$

9. $\begin{pmatrix} 2 \\ 1 \\ 3 \end{pmatrix}, \begin{pmatrix} -1 \\ 4 \\ 5 \end{pmatrix}, \begin{pmatrix} 3 \\ -2 \\ -4 \end{pmatrix}$ **10.** $\begin{pmatrix} a \\ 0 \\ 0 \end{pmatrix}, \begin{pmatrix} b \\ d \\ 0 \end{pmatrix}, \begin{pmatrix} c \\ e \\ f \end{pmatrix}$, where $adf \neq 0$

In Problems 11–13 write the polynomial $a_0 + a_1 x + a_2 x^2$ in P_2 in terms of the given basis.

11. $1, x - 1, x^2 - 1$ **12.** $6, 2 + 3x, 3 + 4x + 5x^2$ **13.** $x + 1, x - 1, x^2 - 1$

14. In M_{22} write the matrix $\begin{pmatrix} 2 & -1 \\ 4 & 6 \end{pmatrix}$ in terms of the basis $\left\{ \begin{pmatrix} 1 & 1 \\ -1 & 0 \end{pmatrix}, \begin{pmatrix} 2 & 0 \\ 3 & 1 \end{pmatrix}, \begin{pmatrix} 0 & 1 \\ -1 & 0 \end{pmatrix}, \begin{pmatrix} 0 & -2 \\ 0 & 4 \end{pmatrix} \right\}$.

15. In P_3 write the polynomial $2x^3 - 3x^2 + 5x - 6$ in terms of the basis polynomials 1, $1 + x$, $x + x^2$, $x^2 + x^3$.

16. In P_3 write the polynomial $4x^2 - x + 5$ in terms of the basis polynomials 1, $1 - x$, $(1 - x)^2$, $(1 - x)^3$.

17. In $\mathbb{R}^2$ suppose that $(\mathbf{x})_{B_1} = \begin{pmatrix} 2 \\ -1 \end{pmatrix}$, where $B_1 = \left\{ \begin{pmatrix} 1 \\ 1 \end{pmatrix}, \begin{pmatrix} 2 \\ 3 \end{pmatrix} \right\}$. Write $\mathbf{x}$ in terms of the basis $B_2 = \left\{ \begin{pmatrix} 0 \\ 3 \end{pmatrix}, \begin{pmatrix} 5 \\ -1 \end{pmatrix} \right\}$.

18. In $\mathbb{R}^2$, $(\mathbf{x})_{B_1} = \begin{pmatrix} 4 \\ -1 \end{pmatrix}$, where $B_1 = \left\{ \begin{pmatrix} 2 \\ -5 \end{pmatrix}, \begin{pmatrix} 7 \\ 3 \end{pmatrix} \right\}$. Write $\mathbf{x}$ in terms of $B_2 = \left\{ \begin{pmatrix} -2 \\ 1 \end{pmatrix}, \begin{pmatrix} -3 \\ 2 \end{pmatrix} \right\}$.

19. In $\mathbb{R}^3$, $(\mathbf{x})_{B_1} = \begin{pmatrix} 2 \\ -1 \\ 4 \end{pmatrix}$, where $B_1 = \left\{ \begin{pmatrix} 1 \\ -1 \\ 0 \end{pmatrix}, \begin{pmatrix} 0 \\ 1 \\ -1 \end{pmatrix}, \begin{pmatrix} 1 \\ 0 \\ 1 \end{pmatrix} \right\}$. Write $\mathbf{x}$ in terms of $B_2 = \left\{ \begin{pmatrix} 3 \\ 0 \\ 0 \end{pmatrix}, \begin{pmatrix} 1 \\ 2 \\ -1 \end{pmatrix}, \begin{pmatrix} 0 \\ 1 \\ 5 \end{pmatrix} \right\}$.

20. In P_2, $(\mathbf{x})_{B_1} = \begin{pmatrix} 2 \\ 1 \\ 3 \end{pmatrix}$, where $B_1 = \{1 - x, 3x, x^2 - x - 1\}$. Write $\mathbf{x}$ in terms of $B_2 = \{3 - 2x, 1 + x, x + x^2\}$.

In Problems 21–28 use Theorem 2 to determine whether the given set of vectors is linearly dependent or independent.

21. In P_2: $2 + 3x + 5x^2$, $1 - 2x + x^2$, $-1 + 6x^2$

22. In P_2: $-3 + x^2$, $2 - x + 4x^2$, $4 + 2x$

23. In P_2: $x + 4x^2$, $-2 + 2x$, $2 + x + 12x^2$

24. In P_2: $-2 + 4x - 2x^2$, $3 + x$, $6 + 8x$

25. In P_3: $1 + x^2$, $-1 - 3x + 4x^2 + 5x^3$, $2 + 5x - 6x^3$, $4 + 6x + 3x^2 + 7x^3$

26. In M_{22}: $\begin{pmatrix} 2 & 0 \\ 3 & 4 \end{pmatrix}, \begin{pmatrix} -3 & -2 \\ 7 & 1 \end{pmatrix}, \begin{pmatrix} 1 & 0 \\ -1 & -3 \end{pmatrix}, \begin{pmatrix} 11 & 2 \\ -5 & -5 \end{pmatrix}$

27. In M_{22}: $\begin{pmatrix} 1 & -3 \\ 2 & 4 \end{pmatrix}, \begin{pmatrix} 1 & 4 \\ 5 & 0 \end{pmatrix}, \begin{pmatrix} -1 & 6 \\ -1 & 3 \end{pmatrix}, \begin{pmatrix} 0 & 0 \\ 3 & 0 \end{pmatrix}$

28. In M_{22}: $\begin{pmatrix} a & 0 \\ 0 & 0 \end{pmatrix}, \begin{pmatrix} b & c \\ 0 & 0 \end{pmatrix}, \begin{pmatrix} d & e \\ f & 0 \end{pmatrix}, \begin{pmatrix} g & h \\ j & k \end{pmatrix}$, where $acfk \neq 0$

29. In P_n, let $p_1, p_2, \ldots, p_{n+1}$ be $n + 1$ polynomials such that $p_i(0) = 0$ for $i = 1, 2, \ldots, n + 1$. Show that the polynomials are linearly dependent.

Calculus **30.** In Problem 29 suppose that $p_i^{(j)} = 0$ for $i = 1, 2, \ldots, n + 1$, and for some j with $1 \leq j \leq n$, and $p_i^{(j)}$ denotes the jth derivative of p_i. Show that the polynomials are linearly dependent in P_n.

31. In M_{mn} let $A_1, A_2, \ldots, A_{mn}$ be mn matrices each of whose components in the 1, 1 position is zero. Show that the matrices are linearly dependent.

*32. Suppose the x- and y-axes in the plane are rotated counterclockwise through an angle of θ (measure in degrees or radians). This gives us new axes which we denote (x', y'). What are the x- and y-coordinates of the now rotated basis vectors $\mathbf{i}$ and $\mathbf{j}$?

33. Show that the "change of coordinates" matrix in Problem 32 is given by
$$A^{-1} = \begin{pmatrix} \cos\theta & \sin\theta \\ -\sin\theta & \cos\theta \end{pmatrix}.$$

34. If, in Problems 32 and 33, $\theta = \pi/6 = 30°$, write the vector $\begin{pmatrix} -4 \\ 3 \end{pmatrix}$ in terms of the new coordinate axes x' and y'.

35. If $\theta = \pi/4 = 45°$, write $\begin{pmatrix} 2 \\ -7 \end{pmatrix}$ in terms of the new coordinate axes.

36. If $\theta = 2\pi/3 = 120°$, write $\begin{pmatrix} 4 \\ 5 \end{pmatrix}$ in terms of the new coordinate axes.

37. Let $C = (c_{ij})$ be an $n \times n$ invertible matrix and let $B_1 = \{\mathbf{v}_1, \mathbf{v}_2, \ldots, \mathbf{v}_n\}$ be a basis for a vector space V. Let

$$\mathbf{c}_1 = \begin{pmatrix} c_{11} \\ c_{21} \\ \vdots \\ c_{n1} \end{pmatrix}_{B_1}, \quad \mathbf{c}_2 = \begin{pmatrix} c_{12} \\ c_{22} \\ \vdots \\ c_{n2} \end{pmatrix}_{B_1}, \quad \ldots, \quad \mathbf{c}_n = \begin{pmatrix} c_{1n} \\ c_{2n} \\ \vdots \\ c_{nn} \end{pmatrix}_{B_1}$$

Show that $B_2 = \{\mathbf{c}_1, \mathbf{c}_2, \ldots, \mathbf{c}_n\}$ is a basis for V.

38. Let B_1 and B_2 be bases for the n-dimensional vector space V and let C be the transition matrix from B_1 to B_2. Show that C^{-1} is the transition matrix from B_2 to B_1.

39. Show that $(\mathbf{x})_{B_1} = CA(\mathbf{x})_{B_1}$ for every $\mathbf{x}$ in a vector space V if and only if $CA = I$ [*Hint:* Let $\mathbf{x}_i$ be the *ith* vector in B_1. Then $(\mathbf{x}_i)_{B_1}$ has a 1 in the *ith* position and a 0 everywhere else. What can you say about $CA(\mathbf{x}_i)_{B_1}$?]

4.10   ORTHONORMAL BASES AND PROJECTIONS IN $\mathbb{R}^n$

In $\mathbb{R}^n$ we saw that n linearly independent vectors constitute a basis. The most commonly used basis is the standard basis $E = \{\mathbf{e}_1, \mathbf{e}_2, \ldots, \mathbf{e}_n\}$. These vectors have two properties:

i. $\mathbf{e}_i \cdot \mathbf{e}_j = 0$ if $i \neq j$

ii. $\mathbf{e}_i \cdot \mathbf{e}_i = 1$

DEFINITION 1   **Orthonormal Set in $\mathbb{R}^n$**   The set of vectors $S = \{\mathbf{u}_1, \mathbf{u}_2, \ldots, \mathbf{u}_k\}$ in $\mathbb{R}^n$ is said to be an **orthonormal set** if

$$\mathbf{u}_i \cdot \mathbf{u}_j = 0 \qquad \text{if } i \neq j \tag{1}$$

$$\mathbf{u}_i \cdot \mathbf{u}_i = 1 \tag{2}$$

If only equation (1) is satisfied, the set is called **orthogonal.**

Since we shall be working with the scalar product extensively in this section, let us recall some basic facts (see Theorem 1.7.1, page 44). Without mentioning them again explicitly, we shall use these facts often in the rest of this section.

If **u**, **v**, and **w** are in $\mathbb{R}^n$ and α is a real number, then

$$\mathbf{u} \cdot \mathbf{v} = \mathbf{v} \cdot \mathbf{u} \tag{3}$$

$$(\mathbf{u} + \mathbf{v}) \cdot \mathbf{w} = \mathbf{u} \cdot \mathbf{w} + \mathbf{v} \cdot \mathbf{w} \tag{4}$$

$$\mathbf{u} \cdot (\mathbf{v} + \mathbf{w}) = \mathbf{u} \cdot \mathbf{v} + \mathbf{u} \cdot \mathbf{w} \tag{5}$$

$$(\alpha\mathbf{u}) \cdot \mathbf{v} = \alpha(\mathbf{u} \cdot \mathbf{v}) \tag{6}$$

$$\mathbf{u} \cdot (\alpha\mathbf{v}) = \alpha(\mathbf{u} \cdot \mathbf{v}) \tag{7}$$

DEFINITION 2   **Length or Norm of a Vector** We now give another useful definition. If $\mathbf{v} \in \mathbb{R}^n$, then the **length** or **norm** of **v**, written $|\mathbf{v}|$, is given by

$$|\mathbf{v}| = \sqrt{\mathbf{v} \cdot \mathbf{v}} \tag{8}$$

Note. If $\mathbf{v} = (x_1, x_2, \ldots, x_n)$, then $\mathbf{v} \cdot \mathbf{v} = x_1^2 + x_2^2 + \cdots + x_n^2$. This means that

$$\mathbf{v} \cdot \mathbf{v} \geq 0 \quad \text{and} \quad \mathbf{v} \cdot \mathbf{v} = 0 \qquad \text{if and only if } \mathbf{v} = \mathbf{0} \tag{9}$$

Thus we can take the square root in (8), and we have

$$|\mathbf{v}| = \sqrt{\mathbf{v} \cdot \mathbf{v}} \geq 0 \qquad \text{for every } \mathbf{v} \in \mathbb{R}^n \tag{10}$$

$$|\mathbf{v}| = 0 \qquad \text{if and only if } \mathbf{v} = \mathbf{0} \tag{11}$$

EXAMPLE 1   **The Norm of a Vector in $\mathbb{R}^2$**   Let $\mathbf{v} = (x, y) \in \mathbb{R}^2$. Then $|\mathbf{v}| = \sqrt{x^2 + y^2}$ conforms to our usual definition of length of a vector in the plane (see Equation 3.1.1, page 156). ∎

EXAMPLE 2   **The Norm of a Vector in $\mathbb{R}^3$**   If $\mathbf{v} = (x, y, z) \in \mathbb{R}^3$, then $|\mathbf{v}| = \sqrt{x^2 + y^2 + z^2}$ as in Section 3.3 ∎

EXAMPLE 3   **The Norm of a Vector in $\mathbb{R}^5$**   If $\mathbf{v} = (2, -1, 3, 4, -6) \in \mathbb{R}^5$, then $|\mathbf{v}| = \sqrt{4 + 1 + 9 + 16 + 36} = \sqrt{66}$. ∎

We can now restate Definition 1:

> A set of vectors is orthonormal if any pair of them is orthogonal and each has length 1.

Orthonormal sets of vectors are reasonably easy to work with. We shall see an example of this characteristic in Chapter 5. Now we prove that any finite orthogonal set of nonzero vectors is linearly independent.

THEOREM 1   If $S = \{\mathbf{v}_1, \mathbf{v}_2, \ldots, \mathbf{v}_k\}$ is an orthogonal set of nonzero vectors, then S is linearly independent.

Proof   Suppose that $c_1\mathbf{v}_1 + c_2\mathbf{v}_2 + \cdots + c_n\mathbf{v}_k = \mathbf{0}$. Then, for any $i = 1, 2, \ldots, k$,

$$0 = \mathbf{0} \cdot \mathbf{v}_i = (c_1\mathbf{v}_1 + c_2\mathbf{v}_2 + \cdots + c_i\mathbf{v}_i + \cdots + c_k\mathbf{v}_k) \cdot \mathbf{v}_i$$
$$= c_1(\mathbf{v}_1 \cdot \mathbf{v}_i) + c_2(\mathbf{v}_2 \cdot \mathbf{v}_i) + \cdots + c_i(\mathbf{v}_i \cdot \mathbf{v}_i) + \cdots + c_k(\mathbf{v}_k \cdot \mathbf{v}_i)$$
$$= c_1 0 + c_2 0 + \cdots + c_i |\mathbf{v}_i|^2 + \cdots + c_k 0 = c_i |\mathbf{v}_i|^2$$

Since $\mathbf{v}_i \neq 0$ by hypothesis, $|\mathbf{v}_i|^2 > 0$ and we have $c_i = 0$. This is true for $i = 1, 2, \ldots, k$ and the proof is complete.   ∎

We now see how *any* basis in $\mathbb{R}^n$ can be "turned into" an orthonormal basis. The method described below is called the **Gram-Schmidt orthonormalization process.**[†]

THEOREM 2   **Gram-Schmidt Orthonormalization Process**   Let H be an m-dimensional subspace of $\mathbb{R}^n$. Then H has an orthonormal basis.[‡]

Proof   Let $S = \{\mathbf{v}_1, \mathbf{v}_2, \ldots, \mathbf{v}_m\}$ be a basis for H. We shall prove the theorem by constructing an orthonormal basis from the vectors in S. Before giving the steps in this construction, we note the simple fact that a linearly independent set of vectors does *not* contain the zero vector (see Problem 21).

Step 1. Let

$$\mathbf{u}_1 = \frac{\mathbf{v}_1}{|\mathbf{v}_1|} \tag{12}$$

Then

$$\mathbf{u}_1 \cdot \mathbf{u}_1 = \left(\frac{\mathbf{v}_1}{|\mathbf{v}_1|}\right) \cdot \left(\frac{\mathbf{v}_1}{|\mathbf{v}_1|}\right) = \left(\frac{1}{|\mathbf{v}_1|^2}\right)(\mathbf{v}_1 \cdot \mathbf{v}_1) = 1$$

so that $|\mathbf{u}_1| = 1$.

[†] Jörgen Pederson Gram (1850–1916) was a Danish actuary who was very interested in the science of measurement. Erhardt Schmidt (1876–1959) was a German mathematician.

[‡] Note that H may be $\mathbb{R}^n$ in this theorem. That is, $\mathbb{R}^n$ itself has an orthonormal basis.

Step 2. Let

$$\mathbf{v}_2' = \mathbf{v}_2 - (\mathbf{v}_2 \cdot \mathbf{u}_1)\mathbf{u}_1 \tag{13}$$

Then

$$\mathbf{v}_2' \cdot \mathbf{u}_1 = \mathbf{v}_2 \cdot \mathbf{u}_1 - (\mathbf{v}_2 \cdot \mathbf{u}_1)(\mathbf{u}_1 \cdot \mathbf{u}_1) = \mathbf{v}_2 \cdot \mathbf{u}_1 - \mathbf{v}_2 \cdot \mathbf{u}_1 = 0,$$

so that $\mathbf{v}_2'$ is orthogonal to $\mathbf{u}_1$. Moreover, by Theorem 1, $\mathbf{u}_1$ and $\mathbf{v}_2'$ are linearly independent. $\mathbf{v}_2' \neq 0$ because otherwise $\mathbf{v}_2 = (\mathbf{v}_2 \cdot \mathbf{u}_1)\mathbf{u}_1 = \dfrac{(\mathbf{v}_2 \cdot \mathbf{u}_1)}{|\mathbf{v}_1|}\mathbf{v}_1$, contradicting the independence of $\mathbf{v}_1$ and $\mathbf{v}_2$.

Step 3. Let

$$\mathbf{u}_2 = \frac{\mathbf{v}_2'}{|\mathbf{v}_2'|} \tag{14}$$

Then clearly $\{\mathbf{u}_1, \mathbf{u}_2\}$ is an orthonormal set.

Suppose now that the vectors $\mathbf{u}_1, \mathbf{u}_2, \ldots, \mathbf{u}_k$ $(k < m)$ have been constructed and form an orthonormal set. We show how to construct $\mathbf{u}_{k+1}$.

Step 4. Let

$$\mathbf{v}_{k+1}' = \mathbf{v}_{k+1} - (\mathbf{v}_{k+1} \cdot \mathbf{u}_1)\mathbf{u}_1 - (\mathbf{v}_{k+1} \cdot \mathbf{u}_2)\mathbf{u}_2 - \cdots - (\mathbf{v}_{k+1} \cdot \mathbf{u}_k)\mathbf{u}_k \tag{15}$$

Then, for $i = 1, 2, \ldots, k$,

$$\mathbf{v}_{k+1}' \cdot \mathbf{u}_i = \mathbf{v}_{k+1} \cdot \mathbf{u}_i - (\mathbf{v}_{k+1} \cdot \mathbf{u}_1)(\mathbf{u}_1 \cdot \mathbf{u}_i) - (\mathbf{v}_{k+1} \cdot \mathbf{u}_2)(\mathbf{u}_2 \cdot \mathbf{u}_i)$$
$$- \cdots - (\mathbf{v}_{k+1} \cdot \mathbf{u}_i)(\mathbf{u}_i \cdot \mathbf{u}_i) - \cdots - (\mathbf{v}_{k+1} \cdot \mathbf{u}_k)(\mathbf{u}_k \cdot \mathbf{u}_i)$$

But $\mathbf{u}_j \cdot \mathbf{u}_i = 0$ if $j \neq i$ and $\mathbf{u}_i \cdot \mathbf{u}_i = 1$. Thus

$$\mathbf{v}_{k+1}' \cdot \mathbf{u}_i = \mathbf{v}_{k+1} \cdot \mathbf{u}_i - \mathbf{v}_{k+1} \cdot \mathbf{u}_i = 0$$

Hence $\{\mathbf{u}_1, \mathbf{u}_2, \ldots, \mathbf{u}_k, \mathbf{v}_{k+1}'\}$ is an orthogonal, linearly independent set and $\mathbf{v}_{k+1}' \neq \mathbf{0}$.

Step 5. Let $\mathbf{u}_{k+1} = \mathbf{v}_{k+1}'/|\mathbf{v}_{k+1}'|$. Then clearly $\{\mathbf{u}_1, \mathbf{u}_2, \ldots, \mathbf{u}_k, \mathbf{u}_{k+1}\}$ is an orthonormal set, and we continue in this manner until $k + 1 = m$ and the proof is complete. ∎

EXAMPLE 4   **Constructing an Orthonormal Basis in $\mathbb{R}^3$**   Construct an orthonormal basis in $\mathbb{R}^3$ starting with the basis $\{\mathbf{v}_1, \mathbf{v}_2, \mathbf{v}_3\} = \left\{ \begin{pmatrix} 1 \\ 1 \\ 0 \end{pmatrix}, \begin{pmatrix} 0 \\ 1 \\ 1 \end{pmatrix}, \begin{pmatrix} 1 \\ 0 \\ 1 \end{pmatrix} \right\}$.

Solution   We have $|\mathbf{v}_1| = \sqrt{2}$, so $\mathbf{u}_1 = \begin{pmatrix} 1/\sqrt{2} \\ 1/\sqrt{2} \\ 0 \end{pmatrix}$. Then

$$\mathbf{v}_2' = \mathbf{v}_2 - (\mathbf{v}_2 \cdot \mathbf{u}_1)\mathbf{u}_1 = \begin{pmatrix} 0 \\ 1 \\ 1 \end{pmatrix} - \frac{1}{\sqrt{2}}\begin{pmatrix} 1/\sqrt{2} \\ 1/\sqrt{2} \\ 0 \end{pmatrix} = \begin{pmatrix} 0 \\ 1 \\ 1 \end{pmatrix} - \begin{pmatrix} \frac{1}{2} \\ \frac{1}{2} \\ 0 \end{pmatrix} = \begin{pmatrix} -\frac{1}{2} \\ \frac{1}{2} \\ 1 \end{pmatrix}$$

Since $|\mathbf{v}_2'| = \sqrt{3/2}$, $\mathbf{u}_2 = \sqrt{2/3}\begin{pmatrix} -\frac{1}{2} \\ \frac{1}{2} \\ 1 \end{pmatrix} = \begin{pmatrix} -1/\sqrt{6} \\ 1/\sqrt{6} \\ 2/\sqrt{6} \end{pmatrix}$. Continuing, we have

$$\mathbf{v}_3' = \mathbf{v}_3 - (\mathbf{v}_3 \cdot \mathbf{u}_1)\mathbf{u}_1 - (\mathbf{v}_3 \cdot \mathbf{u}_2)\mathbf{u}_2$$

$$= \begin{pmatrix} 1 \\ 0 \\ 1 \end{pmatrix} - \frac{1}{\sqrt{2}}\begin{pmatrix} 1/\sqrt{2} \\ 1/\sqrt{2} \\ 0 \end{pmatrix} - \frac{1}{\sqrt{6}}\begin{pmatrix} -1/\sqrt{6} \\ 1/\sqrt{6} \\ 2/\sqrt{6} \end{pmatrix} = \begin{pmatrix} 1 \\ 0 \\ 1 \end{pmatrix} - \begin{pmatrix} \frac{1}{2} \\ \frac{1}{2} \\ 0 \end{pmatrix} - \begin{pmatrix} -\frac{1}{6} \\ \frac{1}{6} \\ \frac{2}{6} \end{pmatrix} = \begin{pmatrix} \frac{2}{3} \\ -\frac{2}{3} \\ \frac{2}{3} \end{pmatrix}$$

Finally, $|\mathbf{v}_3'| = \sqrt{12/9} = 2/\sqrt{3}$, so that $\mathbf{u}_3 = \dfrac{\sqrt{3}}{2}\begin{pmatrix} 2/3 \\ -2/3 \\ 2/3 \end{pmatrix} = \begin{pmatrix} 1/\sqrt{3} \\ -1/\sqrt{3} \\ 1/\sqrt{3} \end{pmatrix}$. Thus

the orthonormall basis is $\left\{ \begin{pmatrix} 1/\sqrt{2} \\ 1/\sqrt{2} \\ 0 \end{pmatrix}, \begin{pmatrix} -1/\sqrt{6} \\ 1/\sqrt{6} \\ 2/\sqrt{6} \end{pmatrix}, \begin{pmatrix} 1/\sqrt{3} \\ -1/\sqrt{3} \\ 1/\sqrt{3} \end{pmatrix} \right\}$. This result should be

checked.   ∎

EXAMPLE 5   **Finding an Orthonormal Basis for a Subspace of $\mathbb{R}^3$**   Find an orthonormal

basis for the set of vectors in $\mathbb{R}^3$ lying on the plane $\pi = \left\{ \begin{pmatrix} x \\ y \\ z \end{pmatrix} : 2x - y + 3z = 0 \right\}$.

Solution   As we saw in Example 4.6.3, page 244, a basis for this two-dimensional subspace is

$\mathbf{v}_1 = \begin{pmatrix} 1 \\ 2 \\ 0 \end{pmatrix}$ and $\mathbf{v}_2 = \begin{pmatrix} 0 \\ 3 \\ 1 \end{pmatrix}$. Then $|\mathbf{v}_1| = \sqrt{5}$ and $\mathbf{u}_1 = \mathbf{v}_1/|\mathbf{v}_1| = \begin{pmatrix} 1/\sqrt{5} \\ 2/\sqrt{5} \\ 0 \end{pmatrix}$. Continu-

ing, we define

$$\mathbf{v}_2' = \mathbf{v}_2 - (\mathbf{v}_2 \cdot \mathbf{u}_1)\mathbf{u}_1$$

$$= \begin{pmatrix} 0 \\ 3 \\ 1 \end{pmatrix} - (6/\sqrt{5})\begin{pmatrix} 1/\sqrt{5} \\ 2/\sqrt{5} \\ 0 \end{pmatrix} = \begin{pmatrix} 0 \\ 3 \\ 1 \end{pmatrix} - \begin{pmatrix} \frac{6}{5} \\ \frac{12}{5} \\ 0 \end{pmatrix} = \begin{pmatrix} -\frac{6}{5} \\ \frac{3}{5} \\ 1 \end{pmatrix}$$

Finally, $|\mathbf{v}_2'| = \sqrt{70/25} = \sqrt{70}/5$,   so that $\mathbf{u}_2 = \mathbf{v}_2'/|\mathbf{v}_2'| = (5/\sqrt{70})\begin{pmatrix} -\frac{6}{5} \\ \frac{3}{5} \\ 1 \end{pmatrix} =$

$\begin{pmatrix} -6/\sqrt{70} \\ 3/\sqrt{70} \\ 5/\sqrt{70} \end{pmatrix}$. Thus the orthonormal basis is $\left\{ \begin{pmatrix} 1/\sqrt{5} \\ 2/\sqrt{5} \\ 0 \end{pmatrix}, \begin{pmatrix} -6/\sqrt{70} \\ 3/\sqrt{70} \\ 5/\sqrt{70} \end{pmatrix} \right\}$. To check

this answer we note that (1) the vectors are orthogonal, (2) each has length 1, and
(3) each satisfies $2x - y + 3z = 0$.

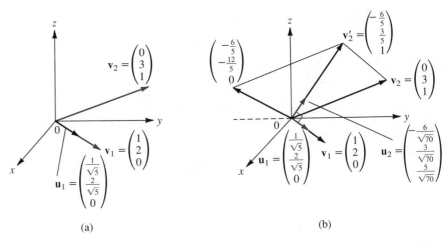

Figure 4.5 The vectors $\mathbf{u}_1$ and $\mathbf{u}_2$ form an orthonormal basis for the plane spanned by the vectors $\mathbf{v}_1$ and $\mathbf{v}_2$

In Figure 4.5a we draw the vectors $\mathbf{v}_1$, $\mathbf{v}_2$, and $\mathbf{u}_1$. In Figure 4.5b we draw the

vector $-\begin{pmatrix} \frac{6}{5} \\ \frac{12}{5} \\ 0 \end{pmatrix} = \begin{pmatrix} -\frac{6}{5} \\ -\frac{12}{5} \\ 0 \end{pmatrix}$ and add it to $\mathbf{v}_2$ using the parallelogram rule to obtain

$\mathbf{v}_2' = \begin{pmatrix} -\frac{6}{5} \\ \frac{3}{5} \\ 1 \end{pmatrix}$. Finally, $\mathbf{u}_2$ lies 1 unit along the vector $\mathbf{v}_2'$.

Remark. We can see why $\mathbf{u}_1$ and $\mathbf{u}_2$ must be orthogonal. We note that

$$\mathbf{v}_2' = \mathbf{v}_2 - (\mathbf{v}_2 \cdot \mathbf{u}_1)\mathbf{u}_1 = \mathbf{v}_2 - \left(\mathbf{v}_2 \cdot \frac{\mathbf{v}_1}{|\mathbf{v}_1|}\right)\left(\frac{\mathbf{v}_1}{|\mathbf{v}_1|}\right) = \mathbf{v}_2 - \frac{(\mathbf{v}_2 \cdot \mathbf{v}_1)}{|\mathbf{v}_1|^2}\mathbf{v}_1$$

But, from Definitions 3.2.4, page 169 (in $\mathbb{R}^2$) and 3.3.3, page 181 (in $\mathbb{R}^3$), $[(\mathbf{v}_2 \cdot \mathbf{v}_1)/|\mathbf{v}_1|^2]\mathbf{v}_1$ is the projection of $\mathbf{v}_2$ on $\mathbf{v}_1$. Moreover, from Figure 3.14, page 168, the vector $\mathbf{v}_2' = \mathbf{v}_2 - \text{proj}_{\mathbf{v}_1} \mathbf{v}_2$ is a vector orthogonal to $\mathbf{v}_1$.

Thus we see that in a certain sense the process we have described here is really a generalization of the notion of projection in $\mathbb{R}^2$ and $\mathbb{R}^3$. ∎

We now define a new kind of matrix that will be very useful in later chapters.

DEFINITION 3   **Orthogonal Matrix** The $n \times n$ matrix Q is called **orthogonal** if Q is invertible and

$$Q^{-1} = Q^t \tag{16}$$

Note that if $Q^{-1} = Q^t$, then $Q^t Q = I$.

Orthogonal matrices are not difficult to find, according to the next theorem.

THEOREM 3   The $n \times n$ matrix Q is orthogonal if and only if the columns of Q form an orthonormal basis for $\mathbb{R}^n$.

Proof   Let

$$Q = \begin{pmatrix} a_{11} & a_{12} & \cdots & a_{1n} \\ a_{21} & a_{22} & \cdots & a_{2n} \\ \vdots & \vdots & & \vdots \\ a_{n1} & a_{n2} & \cdots & a_{nn} \end{pmatrix}$$

Then

$$Q^t = \begin{pmatrix} a_{11} & a_{21} & \cdots & a_{n1} \\ a_{12} & a_{22} & \cdots & a_{n2} \\ \vdots & \vdots & & \vdots \\ a_{1n} & a_{2n} & \cdots & a_{nn} \end{pmatrix}$$

Let $B = (b_{ij}) = Q^t Q$. Then

$$b_{ij} = a_{1i}a_{1j} + a_{2i}a_{2j} + \cdots + a_{ni}a_{nj} = \mathbf{c}_i \cdot \mathbf{c}_j \tag{17}$$

where $\mathbf{c}_i$ denotes the ith column of Q. If the columns of Q are orthonormal, then

$$b_{ij} = \begin{cases} 0 & \text{if } i \neq j \\ 1 & \text{if } i = j \end{cases} \tag{18}$$

That is, $B = I$. Conversely, if $Q^t = Q^{-1}$, then $B = I$, so that (18) holds and (17) shows that the columns of Q are orthonormal. This completes the proof. ∎

EXAMPLE 6   **An Orthogonal Matrix**   From Example 4, the vectors $\begin{pmatrix} 1/\sqrt{2} \\ 1/\sqrt{2} \\ 0 \end{pmatrix}$, $\begin{pmatrix} -1/\sqrt{6} \\ 1/\sqrt{6} \\ 2/\sqrt{6} \end{pmatrix}$, $\begin{pmatrix} 1/\sqrt{3} \\ -1/\sqrt{3} \\ 1/\sqrt{3} \end{pmatrix}$ form an orthonormal basis in $\mathbb{R}^3$. Thus the matrix $Q = \begin{pmatrix} 1/\sqrt{2} & -1/\sqrt{6} & 1/\sqrt{3} \\ 1/\sqrt{2} & 1/\sqrt{6} & -1/\sqrt{3} \\ 0 & 2/\sqrt{6} & 1/\sqrt{3} \end{pmatrix}$ is an orthogonal matrix. To check this we note that

$$Q^t Q = \begin{pmatrix} 1/\sqrt{2} & 1/\sqrt{2} & 0 \\ -1/\sqrt{6} & 1/\sqrt{6} & 2/\sqrt{6} \\ 1/\sqrt{3} & -1/\sqrt{3} & 1/\sqrt{3} \end{pmatrix} \begin{pmatrix} 1/\sqrt{2} & -1/\sqrt{6} & 1/\sqrt{3} \\ 1/\sqrt{2} & 1/\sqrt{6} & -1/\sqrt{3} \\ 0 & 2/\sqrt{6} & 1/\sqrt{3} \end{pmatrix} = \begin{pmatrix} 1 & 0 & 0 \\ 0 & 1 & 0 \\ 0 & 0 & 1 \end{pmatrix}$$

∎

In the proof of Theorem 2 we defined $\mathbf{v}_2' = \mathbf{v}_2 - (\mathbf{v}_2 \cdot \mathbf{u}_1)\mathbf{u}_1$. But, as we have seen, $(\mathbf{v}_2 \cdot \mathbf{u}_1)\mathbf{u}_1 = \text{proj}_{\mathbf{u}_1} \mathbf{v}_2$ (since $|\mathbf{u}_1|^2 = 1$). We now extend this notion from projection onto a vector to projection onto a subspace.

DEFINITION 4   **Orthogonal Projection** Let H be a subspace of $\mathbb{R}^n$ with orthonormal basis $\{\mathbf{u}_1, \mathbf{u}_2, \ldots, \mathbf{u}_k\}$. If $\mathbf{v} \in \mathbb{R}^n$, then the **orthogonal projection** of $\mathbf{v}$ onto H, denoted by $\text{proj}_H \mathbf{v}$, is given by

$$\text{proj}_H \mathbf{v} = (\mathbf{v} \cdot \mathbf{u}_1)\mathbf{u}_1 + (\mathbf{v} \cdot \mathbf{u}_2)\mathbf{u}_2 + \cdots + (\mathbf{v} \cdot \mathbf{u}_k)\mathbf{u}_k \tag{19}$$

Note that $\text{proj}_H \mathbf{v} \in H$.

EXAMPLE 7   **The Orthogonal Projection of a Vector onto a Plane**   Find $\text{proj}_\pi \mathbf{v}$, where π is the plane $\left\{ \begin{pmatrix} x \\ y \\ z \end{pmatrix} : 2x - y + 3z = 0 \right\}$ and $\mathbf{v}$ is the vector $\begin{pmatrix} 3 \\ -2 \\ 4 \end{pmatrix}$.

Solution   From Example 5, an orthonormal basis for π is $\mathbf{u}_1 = \begin{pmatrix} 1/\sqrt{5} \\ 2/\sqrt{5} \\ 0 \end{pmatrix}$ and $\mathbf{u}_2 = \begin{pmatrix} -6/\sqrt{70} \\ 3/\sqrt{70} \\ 5/\sqrt{70} \end{pmatrix}$. Then

$$\text{proj}_\pi \mathbf{v} = \left[\begin{pmatrix} 3 \\ -2 \\ 4 \end{pmatrix} \cdot \begin{pmatrix} 1/\sqrt{5} \\ 2/\sqrt{5} \\ 0 \end{pmatrix} \right] \begin{pmatrix} 1/\sqrt{5} \\ 2/\sqrt{5} \\ 0 \end{pmatrix} + \left[\begin{pmatrix} 3 \\ -2 \\ 4 \end{pmatrix} \cdot \begin{pmatrix} -6/\sqrt{70} \\ 3/\sqrt{70} \\ 5/\sqrt{70} \end{pmatrix} \right] \begin{pmatrix} -6/\sqrt{70} \\ 3/\sqrt{70} \\ 5/\sqrt{70} \end{pmatrix}$$

$$= -\frac{1}{\sqrt{5}} \begin{pmatrix} 1/\sqrt{5} \\ 2/\sqrt{5} \\ 0 \end{pmatrix} - \frac{4}{\sqrt{70}} \begin{pmatrix} -6/\sqrt{70} \\ 3/\sqrt{70} \\ 5/\sqrt{70} \end{pmatrix} = \begin{pmatrix} -\frac{1}{5} \\ -\frac{2}{5} \\ 0 \end{pmatrix} + \begin{pmatrix} \frac{24}{70} \\ -\frac{12}{70} \\ -\frac{20}{70} \end{pmatrix} = \begin{pmatrix} \frac{1}{7} \\ -\frac{4}{7} \\ -\frac{2}{7} \end{pmatrix} \quad \blacksquare$$

The notion of projection gives us a convenient way to write a vector in $\mathbb{R}^n$ in terms of an orthonormal basis.

THEOREM 4   Let $B = \{\mathbf{u}_1, \mathbf{u}_2, \ldots, \mathbf{u}_n\}$ be an orthonormal basis for $\mathbb{R}^n$ and let $\mathbf{v} \in \mathbb{R}^n$. Then

$$\mathbf{v} = (\mathbf{v} \cdot \mathbf{u}_1)\mathbf{u}_1 + (\mathbf{v} \cdot \mathbf{u}_2)\mathbf{u}_2 + \cdots + (\mathbf{v} \cdot \mathbf{u}_n)\mathbf{u}_n \tag{20}$$

That is, $\mathbf{v} = \text{proj}_{\mathbb{R}^n} \mathbf{v}$.

Proof   Since B is a basis, we can write $\mathbf{v}$ in a unique way as $\mathbf{v} = c_1\mathbf{u}_1 + c_2\mathbf{u}_2 + \cdots + c_n\mathbf{u}_n$. Then

$$\mathbf{v} \cdot \mathbf{u}_i = c_1(\mathbf{u}_1 \cdot \mathbf{u}_i) + c_2(\mathbf{u}_2 \cdot \mathbf{u}_i) + \cdots + c_i(\mathbf{u}_i \cdot \mathbf{u}_i) + \cdots + c_n(\mathbf{u}_n \cdot \mathbf{u}_i) = c_i$$

since the $\mathbf{u}_i$'s are orthonormal. Since this is true for $i = 1, 2, \ldots, n$, the proof is complete.   ∎

EXAMPLE 8   **Writing a Vector in Terms of an Orthonormal Basis**   Write the vector $\begin{pmatrix} 2 \\ -1 \\ 3 \end{pmatrix}$ in

$\mathbb{R}^3$ in terms of the orthonormal basis $\left\{ \begin{pmatrix} 1/\sqrt{2} \\ 1/\sqrt{2} \\ 0 \end{pmatrix}, \begin{pmatrix} -1/\sqrt{6} \\ 1/\sqrt{6} \\ 2/\sqrt{6} \end{pmatrix}, \begin{pmatrix} 1/\sqrt{3} \\ -1/\sqrt{3} \\ 1/\sqrt{3} \end{pmatrix} \right\}.$

Solution $\begin{pmatrix} 2 \\ -1 \\ 3 \end{pmatrix} = \left[\begin{pmatrix} 2 \\ -1 \\ 3 \end{pmatrix} \cdot \begin{pmatrix} 1/\sqrt{2} \\ 1/\sqrt{2} \\ 0 \end{pmatrix} \right] \begin{pmatrix} 1/\sqrt{2} \\ 1/\sqrt{2} \\ 0 \end{pmatrix} + \left[\begin{pmatrix} 2 \\ -1 \\ 3 \end{pmatrix} \cdot \begin{pmatrix} -1/\sqrt{6} \\ 1/\sqrt{6} \\ 2/\sqrt{6} \end{pmatrix} \right] \begin{pmatrix} -1/\sqrt{6} \\ 1/\sqrt{6} \\ 2/\sqrt{6} \end{pmatrix}$

$+ \left[\begin{pmatrix} 2 \\ -1 \\ 3 \end{pmatrix} \cdot \begin{pmatrix} 1/\sqrt{3} \\ -1/\sqrt{3} \\ 1/\sqrt{3} \end{pmatrix} \right] \begin{pmatrix} 1/\sqrt{3} \\ -1/\sqrt{3} \\ 1/\sqrt{3} \end{pmatrix}$

$= \dfrac{1}{\sqrt{2}} \begin{pmatrix} 1/\sqrt{2} \\ 1/\sqrt{2} \\ 0 \end{pmatrix} + \dfrac{3}{\sqrt{6}} \begin{pmatrix} -1/\sqrt{6} \\ 1/\sqrt{6} \\ 2/\sqrt{6} \end{pmatrix} + \dfrac{6}{\sqrt{3}} \begin{pmatrix} 1/\sqrt{3} \\ -1/\sqrt{3} \\ 1/\sqrt{3} \end{pmatrix}$ ∎

Before continuing, we need to know that an orthogonal projection is well defined. By this we mean that the definition of $\text{proj}_H \mathbf{v}$ is independent of the orthonormal basis chosen in H. The following theorem takes care of this problem.

THEOREM 5   Let H be a subspace of $\mathbb{R}^n$. Suppose that H has two orthonormal bases, $\{\mathbf{u}_1, \mathbf{u}_2, \ldots, \mathbf{u}_k\}$ and $\{\mathbf{w}_1, \mathbf{w}_2, \ldots, \mathbf{w}_k\}$. Let $\mathbf{v}$ be a vector in $\mathbb{R}^n$. Then

$$(\mathbf{v} \cdot \mathbf{u}_1)\mathbf{u}_1 + (\mathbf{v} \cdot \mathbf{u}_2)\mathbf{u}_2 + \cdots + (\mathbf{v} \cdot \mathbf{u}_k)\mathbf{u}_k$$
$$= (\mathbf{v} \cdot \mathbf{w}_1)\mathbf{w}_1 + (\mathbf{v} \cdot \mathbf{w}_2)\mathbf{w}_2 + \cdots + (\mathbf{v} \cdot \mathbf{w}_k)\mathbf{w}_k \quad (21)$$

Proof   Choose vectors $\mathbf{u}_{k+1}, \mathbf{u}_{k+2}, \ldots, \mathbf{u}_n$ such that $B_1 = \{\mathbf{u}_1, \mathbf{u}_2, \ldots, \mathbf{u}_k, \mathbf{u}_{k+1}, \ldots, \mathbf{u}_n\}$ is an orthonormal basis for $\mathbb{R}^n$ (this can be done as in the proof of Theorem 2). Then $B_2 = \{\mathbf{w}_1, \mathbf{w}_2, \ldots, \mathbf{w}_k, \mathbf{u}_{k+1}, \mathbf{u}_{k+2}, \ldots, \mathbf{u}_n\}$ is also an orthonormal basis for $\mathbb{R}^n$. To see this, note first that none of the vectors $\mathbf{u}_{k+1}, \mathbf{u}_{k+2}, \ldots, \mathbf{u}_n$ can be written as a linear combination of $\mathbf{w}_1, \mathbf{w}_2, \ldots, \mathbf{w}_k$ because none of these vectors is in H and $\{\mathbf{w}_1, \mathbf{w}_2, \ldots, \mathbf{w}_k\}$ is a basis for H. Thus B_2 is a basis for $\mathbb{R}^n$ because it contains n linearly independent vectors. The orthonormality of the vectors in B_2 follows from the way these vectors were chosen ($\mathbf{u}_{k+j}$ is orthogonal to every vector in H for $j = 1, 2, \ldots, n - k$). Let $\mathbf{v}$ be a vector in $\mathbb{R}^n$.

Then, from Theorem 4 [equation (20)],

$$\mathbf{v} = (\mathbf{v} \cdot \mathbf{u}_1)\mathbf{u}_1 + (\mathbf{v} \cdot \mathbf{u}_2)\mathbf{u}_2 + \cdots + (\mathbf{v} \cdot \mathbf{u}_k)\mathbf{u}_k + (\mathbf{v} \cdot \mathbf{u}_{k+1})\mathbf{u}_{k+1} + \cdots + (\mathbf{v} \cdot \mathbf{u}_n)\mathbf{u}_n$$

$$= (\mathbf{v} \cdot \mathbf{w}_1)\mathbf{w}_1 + (\mathbf{v} \cdot \mathbf{w}_2)\mathbf{w}_2 + \cdots + (\mathbf{v} \cdot \mathbf{w}_k)\mathbf{w}_k + (\mathbf{v} \cdot \mathbf{u}_{k+1})\mathbf{u}_{k+1} + \cdots + (\mathbf{v} \cdot \mathbf{u}_n)\mathbf{u}_n \tag{22}$$

Equation (21) now follows from equation (22).   ∎

DEFINITION 5   **Orthogonal Complement** Let H be a subspace of $\mathbb{R}^n$. Then the **orthogonal complement** of H, denoted by $H^\perp$, is given by

$$H^\perp = \{\mathbf{x} \in \mathbb{R}^n \colon \mathbf{x} \cdot \mathbf{h} = 0 \quad \text{for every } \mathbf{h} \in H\}$$

THEOREM 6   If H is a subspace of $\mathbb{R}^n$, then:

  i. $H^\perp$ is a subspace of $\mathbb{R}^n$.

  ii. $H \cap H^\perp = \{\mathbf{0}\}$.

  iii. $\dim H^\perp = n - \dim H$.

Proof   **i.** If $\mathbf{x}$ and $\mathbf{y}$ are in $H^\perp$ and if $\mathbf{h} \in H$, then $(\mathbf{x} + \mathbf{y}) \cdot \mathbf{h} = \mathbf{x} \cdot \mathbf{h} + \mathbf{y} \cdot \mathbf{h} = 0 + 0 = 0$ and $(\alpha\mathbf{x} \cdot \mathbf{h}) = \alpha(\mathbf{x} \cdot \mathbf{h}) = 0$, so $H^\perp$ is a subspace.

  ii. If $\mathbf{x} \in H \cap H^\perp$, then $\mathbf{x} \cdot \mathbf{x} = 0$, so $\mathbf{x} = \mathbf{0}$, which shows that $H \cap H^\perp = \{\mathbf{0}\}$.

  iii. Let $\{\mathbf{u}_1, \mathbf{u}_2, \ldots, \mathbf{u}_k\}$ be an orthonormal basis for H. By the result of Problem 4.6.27, page 252, this can be expanded into a basis B for $\mathbb{R}^n$: $B = \{\mathbf{u}_1, \mathbf{u}_2, \ldots, \mathbf{u}_k, \mathbf{v}_{k+1}, \ldots, \mathbf{v}_n\}$. Using the Gram-Schmidt process, we can turn B into an orthonormal basis for $\mathbb{R}^n$. As in the proof of Theorem 2, the already orthonormal $\mathbf{u}_1, \mathbf{u}_2, \ldots, \mathbf{u}_k$ will remain unchanged in the process, and we obtain the orthonormal basis $B_1 = \{\mathbf{u}_1, \mathbf{u}_2, \ldots, \mathbf{u}_k, \mathbf{u}_{k+1}, \ldots, \mathbf{u}_n\}$. To complete the proof, we need only show that $\{\mathbf{u}_{k+1}, \ldots, \mathbf{u}_n\}$ is a basis for $H^\perp$. Since the $\mathbf{u}_i$'s are independent, we must show that they span $H^\perp$. Let $\mathbf{x} \in H^\perp$; then, by Theorem 4,

$$\mathbf{x} = (\mathbf{x} \cdot \mathbf{u}_1)\mathbf{u}_1 + (\mathbf{x} \cdot \mathbf{u}_2)\mathbf{u}_2 + \cdots + (\mathbf{x} \cdot \mathbf{u}_k)\mathbf{u}_k$$
$$+ (\mathbf{x} \cdot \mathbf{u}_{k+1})\mathbf{u}_{k+1} + \cdots + (\mathbf{x} \cdot \mathbf{u}_n)\mathbf{u}_n$$

But $(\mathbf{x} \cdot \mathbf{u}_i) = 0$ for $i = 1, 2, \ldots, k$, since $\mathbf{x} \in H^\perp$ and $\mathbf{u}_i \in H$. Thus $\mathbf{x} = (\mathbf{x} \cdot \mathbf{u}_{k+1})\mathbf{u}_{k+1} + \cdots + (\mathbf{x} \cdot \mathbf{u}_n)\mathbf{u}_n$. This shows that $\{\mathbf{u}_{k+1}, \ldots, \mathbf{u}_n\}$ is a basis for $H^\perp$ which means that $\dim H^\perp = n - k$.   ∎

The spaces H and $H^\perp$ allow us to "decompose" any vector in $\mathbb{R}^n$.

THEOREM 7   **Projection Theorem** Let H be a subspace of $\mathbb{R}^n$ and let $\mathbf{v} \in \mathbb{R}^n$. Then there exists a unique pair of vectors $\mathbf{h}$ and $\mathbf{p}$ such that $\mathbf{h} \in H$, $\mathbf{p} \in H^\perp$, and

$$\boxed{\mathbf{v} = \mathbf{h} + \mathbf{p} = \text{proj}_H\, \mathbf{v} + \text{proj}_{H^\perp}\, \mathbf{v}} \tag{23}$$

Proof   Let $\mathbf{h} = \text{proj}_H\, \mathbf{v}$ and let $\mathbf{p} = \mathbf{v} - \mathbf{h}$. By Definition 4 we have $\mathbf{h} \in H$. We now show that $\mathbf{p} \in H^\perp$. Let $\{\mathbf{u}_1, \mathbf{u}_2, \ldots, \mathbf{u}_k\}$ be a basis for H. Then

$$\mathbf{h} = (\mathbf{v} \cdot \mathbf{u}_1)\mathbf{u}_1 + (\mathbf{v} \cdot \mathbf{u}_2)\mathbf{u}_2 + \cdots + (\mathbf{v} \cdot \mathbf{u}_k)\mathbf{u}_k$$

Let $\mathbf{x}$ be a vector in H. There exist constants $\alpha_1, \alpha_2, \ldots, \alpha_k$ such that

$$\mathbf{x} = \alpha_1\mathbf{u}_1 + \alpha_2\mathbf{u}_2 + \cdots + \alpha_k\mathbf{u}_k$$

Then

$$\mathbf{p} \cdot \mathbf{x} = (\mathbf{v} - \mathbf{h}) \cdot \mathbf{x} = [\mathbf{v} - (\mathbf{v} \cdot \mathbf{u}_1)\mathbf{u}_1 - (\mathbf{v} \cdot \mathbf{u}_2)\mathbf{u}_2 - \cdots - (\mathbf{v} \cdot \mathbf{u}_k)\mathbf{u}_k]$$
$$\cdot [\alpha_1\mathbf{u}_1 + \alpha_2\mathbf{u}_2 + \cdots + \alpha_k\mathbf{u}_k] \tag{24}$$

Since $\mathbf{u}_i \cdot \mathbf{u}_j = \begin{cases} 0, & i \neq j \\ 1, & i = j \end{cases}$ it is easy to verify that the scalar product in (24) is given by

$$\mathbf{p} \cdot \mathbf{x} = \sum_{i=1}^{k} \alpha_i(\mathbf{v} \cdot \mathbf{u}_i) - \sum_{i=1}^{k} \alpha_i(\mathbf{v} \cdot \mathbf{u}_i) = 0$$

Thus $\mathbf{p} \cdot \mathbf{x} = 0$ for every $\mathbf{x} \in H$, which means that $\mathbf{p} \in H^\perp$. To show that $\mathbf{p} = \text{proj}_{H^\perp}\, \mathbf{v}$, we extend $\{\mathbf{u}_1, \mathbf{u}_2, \ldots, \mathbf{u}_k\}$ to an orthonormal basis for $\mathbb{R}^n$: $\{\mathbf{u}_1, \mathbf{u}_2, \ldots, \mathbf{u}_k, \mathbf{u}_{k+1}, \ldots, \mathbf{u}_n\}$. Then $\{\mathbf{u}_{k+1}, \ldots, \mathbf{u}_n\}$ is a basis for $H^\perp$ and, by Theorem 4,

$$\mathbf{v} = (\mathbf{v} \cdot \mathbf{u}_1)\mathbf{u}_1 + (\mathbf{v} \cdot \mathbf{u}_2)\mathbf{u}_2 + \cdots + (\mathbf{v} \cdot \mathbf{u}_k)\mathbf{u}_k + (\mathbf{v} \cdot \mathbf{u}_{k+1})\mathbf{u}_{k+1}$$
$$+ \cdots + (\mathbf{v} \cdot \mathbf{u}_n)\mathbf{u}_n$$
$$= \text{proj}_H\, \mathbf{v} + \text{proj}_{H^\perp}\, \mathbf{v} \qquad \text{(by Definition 4)}$$

This proves equation (23). To prove uniqueness, suppose that $\mathbf{v} = \mathbf{h}_1 - \mathbf{p}_1 = \mathbf{h}_2 - \mathbf{p}_2$, where $\mathbf{h}_1, \mathbf{h}_2 \in H$ and $\mathbf{p}_1, \mathbf{p}_2 \in H^\perp$. Then $\mathbf{h}_1 - \mathbf{h}_2 = \mathbf{p}_1 - \mathbf{p}_2$. But $\mathbf{h}_1 - \mathbf{h}_2 \in H$ and $\mathbf{p}_1 - \mathbf{p}_2 \in H^\perp$, so $\mathbf{h}_1 - \mathbf{h}_2 \in H \cap H^\perp = \{\mathbf{0}\}$. Thus $\mathbf{h}_1 - \mathbf{h}_2 = \mathbf{0}$ and $\mathbf{p}_1 - \mathbf{p}_2 = \mathbf{0}$, which completes the proof. ∎

EXAMPLE 9   **Decomposing a Vector in $\mathbb{R}^3$**   In $\mathbb{R}^3$ let $\pi = \left\{ \begin{pmatrix} x \\ y \\ z \end{pmatrix} : 2x - y + 3z = 0 \right\}$.

Write the vector $\begin{pmatrix} 3 \\ -2 \\ 4 \end{pmatrix}$ as $\mathbf{h} + \mathbf{p}$, where $\mathbf{h} \in \pi$ and $\mathbf{p} \in \pi^\perp$.

Solution   An orthonormal basis for π is $B_1 = \left\{ \begin{pmatrix} 1/\sqrt{5} \\ 2/\sqrt{5} \\ 0 \end{pmatrix}, \begin{pmatrix} -6/\sqrt{70} \\ 3/\sqrt{70} \\ 5/\sqrt{70} \end{pmatrix} \right\}$ and, from Example

7, $\mathbf{h} = \text{proj}_\pi \, \mathbf{v} = \begin{pmatrix} \frac{1}{7} \\ -\frac{4}{7} \\ -\frac{2}{7} \end{pmatrix} \in \pi$. Then

$$\mathbf{p} = \mathbf{v} - \mathbf{h} = \begin{pmatrix} 3 \\ -2 \\ 4 \end{pmatrix} - \begin{pmatrix} \frac{1}{7} \\ -\frac{4}{7} \\ -\frac{2}{7} \end{pmatrix} = \begin{pmatrix} \frac{20}{7} \\ -\frac{10}{7} \\ \frac{30}{7} \end{pmatrix} \in \pi^\perp$$

Note that $\mathbf{p} \cdot \mathbf{h} = 0$.   ∎

The following theorem is very useful in statistics and other applied areas. We shall provide one application of this theorem in the next section and apply an extended version of this result in Section 4.12.

THEOREM 8   **Norm Approximation Theorem** Let H be a subspace of $\mathbb{R}^n$ and let $\mathbf{v}$ be a vector in $\mathbb{R}^n$. Then in H, $\text{proj}_H \, \mathbf{v}$ is the best approximation to $\mathbf{v}$ in the following sense: If $\mathbf{h}$ is any other vector in H, then

$$\boxed{|\mathbf{v} - \text{proj}_H \, \mathbf{v}| < |\mathbf{v} - \mathbf{h}|} \tag{25}$$

Proof   From Theorem 7, $\mathbf{v} - \text{proj}_H \, \mathbf{v} \in H^\perp$. We write

$$\mathbf{v} - \mathbf{h} = (\mathbf{v} - \text{proj}_H \, \mathbf{v}) + (\text{proj}_H \, \mathbf{v} - \mathbf{h})$$

The first term on the right is in $H^\perp$, while the second is in H, so

$$(\mathbf{v} - \text{proj}_H \, \mathbf{v}) \cdot (\text{proj}_H \, \mathbf{v} - \mathbf{h}) = 0 \tag{26}$$

Now

$$
\begin{aligned}
|\mathbf{v} - \mathbf{h}|^2 &= (\mathbf{v} - \mathbf{h}) \cdot (\mathbf{v} - \mathbf{h}) \\
&= [(\mathbf{v} - \text{proj}_H \, \mathbf{v}) + (\text{proj}_H \, \mathbf{v} - \mathbf{h})] \cdot [(\mathbf{v} - \text{proj}_H \, \mathbf{v}) + (\text{proj}_H \, \mathbf{v} - \mathbf{h})] \\
&= |\mathbf{v} - \text{proj}_H \, \mathbf{v}|^2 + 2(\mathbf{v} - \text{proj}_H \, \mathbf{v}) \cdot (\text{proj}_H \, \mathbf{v} - \mathbf{h}) + |\text{proj}_H \, \mathbf{v} - \mathbf{h}|^2 \\
&= |\mathbf{v} - \text{proj}_H \, \mathbf{v}|^2 + |\text{proj}_H \, \mathbf{v} - \mathbf{h}|^2
\end{aligned}
$$

But $|\text{proj}_H \, \mathbf{v} - \mathbf{h}|^2 > 0$ because $\mathbf{h} \neq \text{proj}_H \, \mathbf{v}$. Hence

$$|\mathbf{v} - \mathbf{h}|^2 > |\mathbf{v} - \text{proj}_H \, \mathbf{v}|^2$$

or

$$|\mathbf{v} - \mathbf{h}| > |\mathbf{v} - \text{proj}_H \, \mathbf{v}|$$ ∎

Orthogonal Bases in $\mathbb{R}^3$ with Integer Coefficients and Integer Norms

It is sometimes useful to construct an orthogonal basis of vectors where the coordinates and norm of each vector is an integer. For example,

$$\left\{ \begin{pmatrix} 2 \\ 2 \\ -1 \end{pmatrix}, \quad \begin{pmatrix} 2 \\ -1 \\ 2 \end{pmatrix}, \quad \begin{pmatrix} -1 \\ 2 \\ 2 \end{pmatrix} \right\}$$

constitutes an orthogonal basis in $\mathbb{R}^3$ where each vector has norm 3. As another example,

$$\left\{ \begin{pmatrix} 12 \\ 4 \\ -3 \end{pmatrix}, \quad \begin{pmatrix} 0 \\ 3 \\ 4 \end{pmatrix}, \quad \begin{pmatrix} -25 \\ 48 \\ -36 \end{pmatrix} \right\}$$

is an orthogonal basis in $\mathbb{R}^3$ whose vectors have norms 13, 5, and 65, respectively. Finding bases like this in $\mathbb{R}^3$ turns out to be not so difficult as you might imagine. A discussion of this topic appears in the interesting paper "Orthogonal Bases of $\mathbb{R}^3$ with Integer Coordinates and Integer Lengths" by Anthony Osborne and Hans Liebeck in *The American Mathematical Monthly,* Volume 96, Number 1, January 1989, pp. 49–53.

We close this section with an important theorem.

THEOREM 9   **Cauchy-Schwarz Inequality in $\mathbb{R}^n$** Let $\mathbf{u}$ and $\mathbf{v}$ be vectors in $\mathbb{R}^n$. Then

  i. $|\mathbf{u} \cdot \mathbf{v}| \le |\mathbf{u}|\,|\mathbf{v}|$.   (27)

  ii. $|\mathbf{u} \cdot \mathbf{v}| = |\mathbf{u}|\,|\mathbf{v}|$ if and only if $\mathbf{v} = \lambda \mathbf{u}$ for some real number λ.

Proof   i. If $\mathbf{u} = \mathbf{0}$ or $\mathbf{v} = \mathbf{0}$ (or both), then (27) holds (both sides are equal to 0). We assume that $\mathbf{u} \ne \mathbf{0}$ and $\mathbf{v} \ne \mathbf{0}$. Then

$$0 \le \left| \frac{\mathbf{u}}{|\mathbf{u}|} - \frac{\mathbf{v}}{|\mathbf{v}|} \right|^2 = \left(\frac{\mathbf{u}}{|\mathbf{u}|} - \frac{\mathbf{v}}{|\mathbf{v}|} \right) \cdot \left(\frac{\mathbf{u}}{|\mathbf{u}|} - \frac{\mathbf{v}}{|\mathbf{v}|} \right) = \frac{\mathbf{u} \cdot \mathbf{u}}{|\mathbf{u}|^2} - \frac{2\mathbf{u} \cdot \mathbf{v}}{|\mathbf{u}|\,|\mathbf{v}|} + \frac{\mathbf{v} \cdot \mathbf{v}}{|\mathbf{v}|^2}$$

$$= \frac{|\mathbf{u}|^2}{|\mathbf{u}|^2} - \frac{2\mathbf{u} \cdot \mathbf{v}}{|\mathbf{u}|\,|\mathbf{v}|} + \frac{|\mathbf{v}|^2}{|\mathbf{v}|^2} = 2 - \frac{2\mathbf{u} \cdot \mathbf{v}}{|\mathbf{u}|\,|\mathbf{v}|}$$

which is ≥ 0. Thus $\dfrac{2\mathbf{u} \cdot \mathbf{v}}{|\mathbf{u}|\,|\mathbf{v}|} \le 2$, so $\dfrac{\mathbf{u} \cdot \mathbf{v}}{|\mathbf{u}|\,|\mathbf{v}|} \le 1$ and $\mathbf{u} \cdot \mathbf{v} \le |\mathbf{u}|\,|\mathbf{v}|$. Similarly, starting with $0 \le \left| \dfrac{\mathbf{u}}{|\mathbf{u}|} + \dfrac{\mathbf{v}}{|\mathbf{v}|} \right|$, we end up with $\dfrac{\mathbf{u} \cdot \mathbf{v}}{|\mathbf{u}|\,|\mathbf{v}|} \ge -1$ or $\mathbf{u} \cdot \mathbf{v} \ge -|\mathbf{u}|\,|\mathbf{v}|$. Putting these together, we obtain

$$-|\mathbf{u}|\,|\mathbf{v}| \le \mathbf{u} \cdot \mathbf{v} \le |\mathbf{u}|\,|\mathbf{v}| \quad \text{or} \quad |\mathbf{u} \cdot \mathbf{v}| \le |\mathbf{u}|\,|\mathbf{v}|$$

ii. If $\mathbf{u} = \lambda \mathbf{v}$, then $|\mathbf{u} \cdot \mathbf{v}| = |\lambda \mathbf{v} \cdot \mathbf{v}| = |\lambda| \, |\mathbf{v}|^2$ and $|\mathbf{u}| \, |\mathbf{v}| = |\lambda \mathbf{v}| \, |\mathbf{v}| = |\lambda| \, |\mathbf{v}| \, |\mathbf{v}| = |\lambda| \, |\mathbf{v}|^2 = |\mathbf{u} \cdot \mathbf{v}|$. Conversely, suppose that $|\mathbf{u} \cdot \mathbf{v}| = |\mathbf{u}| \, |\mathbf{v}|$ with $\mathbf{u} \neq 0$ and $\mathbf{v} \neq 0$. Then $\left| \dfrac{\mathbf{u} \cdot \mathbf{v}}{|\mathbf{u}| \, |\mathbf{v}|} \right| = 1$ so $\dfrac{\mathbf{u} \cdot \mathbf{v}}{|\mathbf{u}| \, |\mathbf{v}|} = \pm 1$

Case 1: $\dfrac{\mathbf{u} \cdot \mathbf{v}}{|\mathbf{u}| \, |\mathbf{v}|} = 1$. Then

$$\left| \frac{\mathbf{u}}{|\mathbf{u}|} - \frac{\mathbf{v}}{|\mathbf{v}|} \right|^2 = \left(\frac{\mathbf{u}}{|\mathbf{u}|} - \frac{\mathbf{v}}{|\mathbf{v}|} \right) \cdot \left(\frac{\mathbf{u}}{|\mathbf{u}|} - \frac{\mathbf{v}}{|\mathbf{v}|} \right) \overset{\text{as in (i)}}{=} 2 - \frac{2 \mathbf{u} \cdot \mathbf{v}}{|\mathbf{u}| \, |\mathbf{v}|} = 2 - 2 = 0.$$

Thus

$$\frac{\mathbf{u}}{|\mathbf{u}|} = \frac{\mathbf{v}}{|\mathbf{v}|} \quad \text{or} \quad \mathbf{u} = \frac{|\mathbf{u}|}{|\mathbf{v}|} \mathbf{v} = \lambda \mathbf{v}$$

Case 2: $\dfrac{\mathbf{u} \cdot \mathbf{v}}{|\mathbf{u}| \, |\mathbf{v}|} = -1$. Then

$$\left| \frac{\mathbf{u}}{|\mathbf{u}|} + \frac{\mathbf{v}}{|\mathbf{v}|} \right|^2 = 2 + \frac{2 \mathbf{u} \cdot \mathbf{v}}{|\mathbf{u}| \, |\mathbf{v}|} = 2 - 2 = 0$$

so

$$\frac{\mathbf{u}}{|\mathbf{u}|} = -\frac{\mathbf{v}}{|\mathbf{v}|} \quad \text{and} \quad \mathbf{u} = -\frac{|\mathbf{u}|}{|\mathbf{v}|} \mathbf{v} = \lambda \mathbf{v} \qquad \blacksquare$$

PROBLEMS 4.10

In Problems 1–13 construct an orthonormal basis for the given vector space or subspace.

1. In $\mathbb{R}^2$, starting with the basis vectors $\begin{pmatrix} 1 \\ 1 \end{pmatrix}$, $\begin{pmatrix} -1 \\ 1 \end{pmatrix}$

2. $H = \{(x, y) \in \mathbb{R}^2 : x + y = 0\}$ 3. $H = \{(x, y) \in \mathbb{R}^2 : ax + by = 0\}$

4. In $\mathbb{R}^2$, starting with $\begin{pmatrix} a \\ b \end{pmatrix}$, $\begin{pmatrix} c \\ d \end{pmatrix}$, where $ad - bc \neq 0$.

5. $\pi = \{(x, y, z) : 2x - y - z = 0\}$ 6. $\pi = \{(x, y, z) : 3x - 2y + 6z = 0\}$

7. $L = \{(x, y, z) : x/2 = y/3 = z/4\}$

8. $L = \{(x, y, z) : x = 3t, \ y = -2t, \ z = t; \ t \text{ real}\}$

9. $H = \{(x, y, z, w) \in \mathbb{R}^4 : 2x - y + 3z - w = 0\}$

10. $\pi = \{(x, y, z) : ax + by + cz = 0\}$, where $abc \neq 0$

11. $L = \{(x, y, z) : x/a = y/b = z/c\}$, where $abc \neq 0$.

12. $H = \{x_1, x_2, x_3, x_4, x_5\} \in \mathbb{R}^5 : 2x_1 - 3x_2 + x_3 + 4x_4 - x_5 = 0\}$

13. H is the solution space of

$$\begin{aligned} x - 3y + z &= 0 \\ -2x + 2y - 3z &= 0 \\ 4x - 8y + 5z &= 0 \end{aligned}$$

*14. Find an orthonormal basis in $\mathbb{R}^4$ that includes the vectors

$$\mathbf{u}_1 = \begin{pmatrix} 1/\sqrt{2} \\ 0 \\ 1/\sqrt{2} \\ 0 \end{pmatrix} \quad \text{and} \quad \mathbf{u}_2 = \begin{pmatrix} -\frac{1}{2} \\ \frac{1}{2} \\ \frac{1}{2} \\ -\frac{1}{2} \end{pmatrix}$$

[*Hint:* First find two vectors $\mathbf{v}_3$ and $\mathbf{v}_4$ to complete the basis.]

15. Show that $Q = \begin{pmatrix} \frac{2}{3} & \frac{1}{3} & \frac{2}{3} \\ \frac{1}{3} & \frac{2}{3} & -\frac{2}{3} \\ -\frac{2}{3} & \frac{2}{3} & \frac{1}{3} \end{pmatrix}$ is an orthogonal matrix.

16. Show that if P and Q are orthogonal $n \times n$ matrices, then PQ is orthogonal.

17. Verify the result of Problem 16 with

$$P = \begin{pmatrix} 1/\sqrt{2} & -1/\sqrt{2} \\ 1/\sqrt{2} & 1/\sqrt{2} \end{pmatrix} \quad \text{and} \quad Q = \begin{pmatrix} 1/3 & -\sqrt{8}/3 \\ \sqrt{8}/3 & 1/3 \end{pmatrix}$$

18. Show that if Q is a symmetric orthogonal matrix, then $Q^2 = I$.

19. Show that if Q is orthogonal, then $\det Q = \pm 1$.

20. Show that for any real number t, the matrix $A = \begin{pmatrix} \sin t & \cos t \\ \cos t & -\sin t \end{pmatrix}$ is orthogonal.

21. Let $\{\mathbf{v}_1, \mathbf{v}_2, \ldots, \mathbf{v}_k\}$ be a linearly independent set of vectors in $\mathbb{R}^n$. Prove that $\mathbf{v}_i \neq \mathbf{0}$ for $i = 1, 2, \ldots, k$. [*Hint:* If $\mathbf{v}_i = \mathbf{0}$, then it is easy to find constants $c_1, c_2, \ldots, c_k$ with $c_i \neq 0$ such that $c_1\mathbf{v}_1 + c_2\mathbf{v}_2 + \cdots + c_k\mathbf{v}_k = \mathbf{0}$.]

In Problems 22–28 a subspace H and a vector $\mathbf{v}$ are given. **(a)** Compute $\text{proj}_H \mathbf{v}$; **(b)** find an orthonormal basis for $H^\perp$; **(c)** write $\mathbf{v}$ as $\mathbf{h} + \mathbf{p}$, where $\mathbf{h} \in H$ and $\mathbf{p} \in H^\perp$.

22. $H = \left\{ \begin{pmatrix} x \\ y \end{pmatrix} \in \mathbb{R}^2 : x + y = 0 \right\}$; $\mathbf{v} = \begin{pmatrix} -1 \\ 2 \end{pmatrix}$

23. $H = \left\{ \begin{pmatrix} x \\ y \end{pmatrix} \in \mathbb{R}^2 : ax + by = 0 \right\}$; $\mathbf{v} = \begin{pmatrix} a \\ b \end{pmatrix}$

24. $H = \left\{ \begin{pmatrix} x \\ y \\ z \end{pmatrix} \in \mathbb{R}^3 : ax + by + cz = 0 \right\}$; $\mathbf{v} = \begin{pmatrix} a \\ b \\ c \end{pmatrix}$, $\mathbf{v} \neq \mathbf{0}$

25. $H = \left\{ \begin{pmatrix} x \\ y \\ z \end{pmatrix} \in \mathbb{R}^3 : 3x - 2y + 6z = 0 \right\}$; $\mathbf{v} = \begin{pmatrix} -3 \\ 1 \\ 4 \end{pmatrix}$

26. $H = \left\{ \begin{pmatrix} x \\ y \\ z \end{pmatrix} \in \mathbb{R}^3 : x/2 = y/3 = z/4 \right\}$; $\mathbf{v} = \begin{pmatrix} 1 \\ 1 \\ 1 \end{pmatrix}$

27. $H = \left\{ \begin{pmatrix} x \\ y \\ z \\ w \end{pmatrix} \in \mathbb{R}^4 : 2x - y + 3z - w = 0 \right\}$; $\mathbf{v} = \begin{pmatrix} 1 \\ -1 \\ 2 \\ 3 \end{pmatrix}$

28. $H = \left\{ \begin{pmatrix} x \\ y \\ z \\ w \end{pmatrix} \in \mathbb{R}^4 : x = y \text{ and } w = 3y \right\}; \; \mathbf{v} = \begin{pmatrix} -1 \\ 2 \\ 3 \\ 1 \end{pmatrix}$

29. Let $\mathbf{u}_1$ and $\mathbf{u}_2$ be two orthonormal vectors in $\mathbb{R}^n$. Show that $|\mathbf{u}_1 - \mathbf{u}_2| = \sqrt{2}$.

30. If $\mathbf{u}_1, \mathbf{u}_2, \ldots, \mathbf{u}_n$ are orthonormal, show that

$$|\mathbf{u}_1 + \mathbf{u}_2 + \cdots + \mathbf{u}_n|^2 = |\mathbf{u}_1|^2 + |\mathbf{u}_2|^2 + \cdots + |\mathbf{u}_n|^2 = n$$

31. Find a condition on the numbers a and b such that $\left\{ \begin{pmatrix} a \\ b \end{pmatrix}, \begin{pmatrix} b \\ -a \end{pmatrix} \right\}$ and $\left\{ \begin{pmatrix} a \\ b \end{pmatrix}, \begin{pmatrix} -b \\ a \end{pmatrix} \right\}$ form orthonormal bases in $\mathbb{R}^2$.

32. Show that *any* orthonormal basis in $\mathbb{R}^2$ has one of the forms of the bases in Problem 31.

33. Using the Cauchy-Schwarz inequality, prove that if $|\mathbf{u} + \mathbf{v}| = |\mathbf{u}| + |\mathbf{v}|$, then $\mathbf{u}$ and $\mathbf{v}$ are linearly dependent.

34. Using the Cauchy-Schwarz inequality, prove the **triangle inequality:**

$$|\mathbf{u} + \mathbf{v}| \leq |\mathbf{u}| + |\mathbf{v}|.$$

[*Hint:* Expand $|\mathbf{u} + \mathbf{v}|^2$.]

*35. Suppose that $\mathbf{x}_1, \mathbf{x}_2, \ldots, \mathbf{x}_k$ are vectors in $\mathbb{R}^n$ (not all zero) and

$$|\mathbf{x}_1 + \mathbf{x}_2 + \cdots + \mathbf{x}_k| = |\mathbf{x}_1| + |\mathbf{x}_2| + \cdots + |\mathbf{x}_k|$$

Show that dim span $\{\mathbf{x}_1, \mathbf{x}_2, \ldots, \mathbf{x}_k\} = 1$. [*Hint:* Use the results of Problem 33 and 34.]

36. Let $\{\mathbf{u}_1, \mathbf{u}_2, \ldots, \mathbf{u}_n\}$ be an orthonormal basis in $\mathbb{R}^n$ and let $\mathbf{v}$ be a vector in $\mathbb{R}^n$. Prove that $|\mathbf{v}|^2 = |\mathbf{v} \cdot \mathbf{u}_1|^2 + |\mathbf{v} \cdot \mathbf{u}_2|^2 + \cdots + |\mathbf{v} \cdot \mathbf{u}_n|^2$. This equality is called **Parseval's equality** in $\mathbb{R}^n$.

37. Show that for any subspace H of $\mathbb{R}^n$, $(H^\perp)^\perp = H$.

38. Let H_1 and H_2 be two subspaces of $\mathbb{R}^n$ and suppose that $H_1^\perp = H_2^\perp$. Show that $H_1 = H_2$.

39. If H_1 and H_2 are subspaces of $\mathbb{R}^n$, show that if $H_1 \subset H_2$, then $H_2^\perp \subset H_1^\perp$.

40. Prove the **generalized Pythagorean theorem:** Let $\mathbf{u}$ and $\mathbf{v}$ be vectors in $\mathbb{R}^n$ with $\mathbf{u} \perp \mathbf{v}$. Then

$$|\mathbf{u} + \mathbf{v}|^2 = |\mathbf{u}|^2 + |\mathbf{v}|^2$$

4.11 LEAST SQUARES APPROXIMATION

In many problems in the biological, physical, and social sciences it is useful to describe the relationship among the variables of the problem by means of a mathematical expression. Thus, for example, we may describe the relationship among cost, revenue, and profit by means of the simple formula

$$P = R - C$$

In a different vein, we may represent the relationship among the acceleration due to gravity, the time an object has been falling, and the height of the object by the physical law

$$s = s_0 - v_0 t - \tfrac{1}{2}g t^2$$

where s_0 is the initial height of the object and v_0 is its initial velocity.

Unfortunately, formulas like the ones above do not come easily. It is usually the task of the scientist or economist to sort through large amounts of data in order to find relationships among the variables in the problem. A common way to do this is to fit a curve among the various data points. This curve may be a straight line or a quadratic or a cubic, etc. The object is to find the curve of the given type that "best" fits the given data. In this section we show how to do this when there are two variables in the problem. In every case we assume that there are n data points $(x_1, y_1), (x_2, y_2), \ldots, (x_n, y_n)$.

In Figure 4.6 we can indicate three of the curves that can be used to fit data.

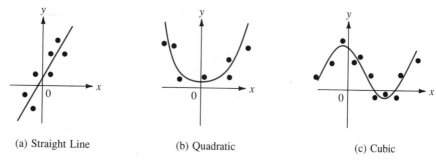

(a) Straight Line   (b) Quadratic   (c) Cubic

Figure 4.6   Three curves in the xy-plane

Straight-Line Approximation

Before continuing, we must be clear as to what we mean by the "best fit." Suppose we seek a straight line of the form $y = b + mx$ that best represents the n data points $(x_1, y_1), (x_2, y_2), \ldots, (x_n, y_n)$.

Figure 4.7 illustrates what is going on (using three data points). From the figure we see that if we assume that the x- and y-variables are related by the formula $y = b + mx$, then, for example, for $x = x_1$, the corresponding y-value is $b + mx_1$. This is different from the "true" y-value $y = y_1$.

In $\mathbb{R}^2$ the distance between the points (a_1, b_1) and (a_2, b_2) is given by $d = \sqrt{(a_1 - a_2)^2 + (b_1 - b_2)^2}$. Therefore, in determining how to choose the line $y = b + mx$ that best approximates the given data, it is reasonable to use the criterion of choosing the line that minimizes the sum of the squares of the distances between the points and the line. Note that since the distance between (x_1, y_1) and $(x_1, b + mx_1)$ is $y_1 - (b + mx_1)$, our problem (for n data points) can be stated as follows:

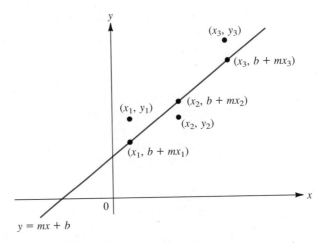

Figure 4.7   Points on the straight line have coordinates $(x, b + mx)$

The Least Squares Problem for a Line

Find numbers m and b such that the sum

$$[y_1 - (b + mx_1)]^2 + [y_2 - (b + mx_2)]^2 + \cdots + [y_n - (b + mx_n)]^2 \qquad (1)$$

is a minimum. For this choice of m and b, the line
$y = mx + b$ is called the **least squares straight-line
approximation to the data points** $(x_1, y_1), (x_2, y_2), \ldots, (x_n, y_n)$.

Having defined the problem, we now seek a method for finding the least squares approximation. This is most easily done by writing everything in matrix form. If the points $(x_1, y_1), (x_2, y_2), \ldots, (x_n, y_n)$ all lie on the line $y = b + mx$ (that is, if they are collinear), then we have

$$
\begin{aligned}
y_1 &= b & &+ mx_1 \\
y_2 &= b & &+ mx_2 \\
&\ \vdots & &\ \ \vdots \\
y_n &= b & &+ mx_n
\end{aligned}
$$

or

$$\mathbf{y} = A\mathbf{u} \qquad (2)$$

where

$$
\mathbf{y} = \begin{pmatrix} y_1 \\ y_2 \\ \vdots \\ y_n \end{pmatrix}, \quad
A = \begin{pmatrix} 1 & x_1 \\ 1 & x_2 \\ \vdots & \vdots \\ 1 & x_n \end{pmatrix}, \quad \text{and} \quad
\mathbf{u} = \begin{pmatrix} b \\ m \end{pmatrix} \qquad (3)
$$

If the points are not collinear, then $\mathbf{y} - A\mathbf{u} \neq \mathbf{0}$ and the problem becomes

> **Vector Form of the Least Squares Problem**
>
> find a vector $\mathbf{u}$ such that the Euclidean norm
>
> $$\left|\mathbf{y} - A\mathbf{u}\right|$$ (4)
>
> is a minimum.

Note that in $\mathbb{R}^2$, $|(x, y)| = \sqrt{x^2 + y^2}$; in $\mathbb{R}^3$, $|(x, y, z)| = \sqrt{x^2 + y^2 + z^2}$, etc. Thus minimizing (4) is equivalent to minimizing the sum of the squares in (1).

Finding the minimizing vector $\mathbf{u}$ is not so difficult as it seems. Since A is an $n \times 2$ matrix and $\mathbf{u}$ is a 2×1 matrix, the vector $A\mathbf{u}$ is a vector in $\mathbb{R}^n$ and belongs to the range of A. The range of A is a subspace of $\mathbb{R}^n$ of dimension at most two (since at most two of the columns of A are linearly independent). Thus, by the norm approximation theorem in $\mathbb{R}^n$ (Theorem 8 on page 292), (4) is a minimum when

$$A\mathbf{u} = \text{Proj}_H \, \mathbf{y}$$

where H is the range of A. We illustrate this graphically in the case $n = 3$.

In $\mathbb{R}^3$ the range of A will be a plane or a line passing through the origin (since these are the only subspaces of $\mathbb{R}^3$ of dimension one or two). Look at Figure 4.8.

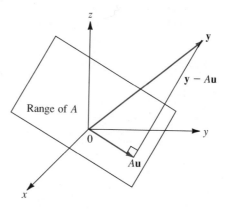

Figure 4.8 $\mathbf{y} - A\mathbf{u}$ is orthogonal to $A\mathbf{u}$

We denote the minimizing vector $\bar{\mathbf{u}}$. It follows from the figure (and the Pythagorean theorem) that $|\mathbf{y} - A\mathbf{u}|$ is minimized when $\mathbf{y} - A\mathbf{u}$ is orthogonal to the range of A. That is, if $\bar{\mathbf{u}}$ is the minimizing vector, then, for every vector $\mathbf{u} \in \mathbb{R}^2$,

$$A\mathbf{u} \perp (\mathbf{y} - A\bar{\mathbf{u}})$$ (5)

Using the definition of the scalar product in $\mathbb{R}^n$, we find that (5) becomes

$$A\mathbf{u} \cdot (\mathbf{y} - A\bar{\mathbf{u}}) = 0$$
$$(A\mathbf{u})^t(\mathbf{y} - A\bar{\mathbf{u}}) = 0 \qquad \text{formula (6) on page 85}$$
$$(\mathbf{u}^tA^t)(\mathbf{y} - A\bar{\mathbf{u}}) = 0 \qquad \text{Theorem 1 (ii) on page 84}$$

or

$$\mathbf{u}^t(A^t\mathbf{y} - A^tA\bar{\mathbf{u}}) = 0 \qquad\qquad\qquad (6)$$

Equation (6) can hold for every $\mathbf{u} \in \mathbb{R}^2$ only if

$$A^t\mathbf{y} - A^tA\bar{\mathbf{u}} = 0 \qquad\qquad\qquad (7)$$

Solving (7) for $\bar{\mathbf{u}}$, we obtain

**Solution to the Least Square Problem
for a Straight-Line Fit**

If A and $\mathbf{y}$ are as in (3), then the line $y = mx + b$
gives the best straight-line fit (in the least squares sense) to
the data points (x_1, y_1), (x_2, y_2), . . . , (x_n, y_n) when $\begin{pmatrix} b \\ m \end{pmatrix} = \bar{\mathbf{u}}$ and

$$\bar{\mathbf{u}} = (A^tA)^{-1}A^ty \qquad\qquad (8)$$

Here we have assumed that A^tA is invertible. This is always the case when the n data points are not collinear. The proof of this fact is left to the end of the section.

EXAMPLE 1   **Finding the Best Straight-Line Fit to Four Points**   Find the best straight-line fit to the data points $(1, 4)$, $(-2, 5)$, $(3, -1)$, and $(4, 1)$.

Solution   Here

$$A = \begin{pmatrix} 1 & 1 \\ 1 & -2 \\ 1 & 3 \\ 1 & 4 \end{pmatrix}, \qquad A^t = \begin{pmatrix} 1 & 1 & 1 & 1 \\ 1 & -2 & 3 & 4 \end{pmatrix} \quad \text{and} \quad \mathbf{y} = \begin{pmatrix} 4 \\ 5 \\ -1 \\ 1 \end{pmatrix}$$

Then

$$A^tA = \begin{pmatrix} 4 & 6 \\ 6 & 30 \end{pmatrix}, \qquad (A^tA)^{-1} = \tfrac{1}{84}\begin{pmatrix} 30 & -6 \\ -6 & 4 \end{pmatrix} \quad \text{and}$$

$$\bar{\mathbf{u}} = (A^tA)^{-1}A^ty = \tfrac{1}{84}\begin{pmatrix} 30 & -6 \\ -6 & 4 \end{pmatrix}\begin{pmatrix} 1 & 1 & 1 & 1 \\ 1 & -2 & 3 & 4 \end{pmatrix}\begin{pmatrix} 4 \\ 5 \\ -1 \\ 1 \end{pmatrix}$$

$$= \tfrac{1}{84}\begin{pmatrix} 30 & -6 \\ -6 & 4 \end{pmatrix}\begin{pmatrix} 9 \\ -5 \end{pmatrix} = \tfrac{1}{84}\begin{pmatrix} 300 \\ -74 \end{pmatrix} \approx \begin{pmatrix} 3.57 \\ -0.88 \end{pmatrix}$$

Therefore, the best straight-line fit is given by

$$y = 3.57 - 0.88x$$

This line and the four data points are sketched in Figure 4.9.

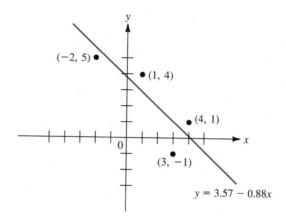

Figure 4.9   The best straight-line fit to the four points is $y = 3.57 - 0.88x$

Quadratic Approximation

Here we wish to fit a quadratic to our n data points. Recall that a quadratic in x is any expression of the form

$$y = a + bx + cx^2 \tag{9}$$

Equation (9) is the equation of a parabola in the plane. If the n data points were on the parabola, we would have

$$\begin{aligned} y_1 &= a + bx_1 + cx_1^2 \\ y_2 &= a + bx_2 + cx_2^2 \\ &\;\;\vdots \\ y_n &= a + bx_n + cx_n^2 \end{aligned} \tag{10}$$

For

$$\mathbf{y} = \begin{pmatrix} y_1 \\ y_2 \\ \vdots \\ y_n \end{pmatrix}, \qquad A = \begin{pmatrix} 1 & x_1 & x_1^2 \\ 1 & x_2 & x_2^2 \\ \vdots & \vdots & \vdots \\ 1 & x_n & x_n^2 \end{pmatrix} \quad \text{and} \quad \mathbf{u} = \begin{pmatrix} a \\ b \\ c \end{pmatrix} \tag{11}$$

(10) can be rewritten as

$$\mathbf{y} = A\mathbf{u}$$

as before. If the data points do not all lie on the same parabola, then $\mathbf{y} = A\mathbf{u} \neq \mathbf{0}$ for any vector $\mathbf{u}$, and our problem is, again,

---

Find a vector $\mathbf{u}$ in $\mathbb{R}^3$ such that $|\mathbf{y} - A\mathbf{u}|$ is a minimum.

Using reasoning similar to that used earlier, we can show that if the data points do not all lie on one parabola, then $A'A$ is invertible and the minimizing vector $\bar{\mathbf{u}}$ is given by

$$\bar{\mathbf{u}} = (A'A)^{-1}A'\mathbf{y} \qquad\qquad (12)$$

EXAMPLE 2   **Finding the Best Quadratic Fit to Four Points**   Find the best quadratic fit to the data points of Example 1.

Solution   Here

$$A = \begin{pmatrix} 1 & 1 & 1 \\ 1 & -2 & 4 \\ 1 & 3 & 9 \\ 1 & 4 & 16 \end{pmatrix}, \quad A' = \begin{pmatrix} 1 & 1 & 1 & 1 \\ 1 & -2 & 3 & 4 \\ 1 & 4 & 9 & 16 \end{pmatrix} \quad \text{and} \quad \mathbf{y} = \begin{pmatrix} 4 \\ 5 \\ -1 \\ 1 \end{pmatrix}$$

Then

$$A'A = \begin{pmatrix} 4 & 6 & 30 \\ 6 & 30 & 84 \\ 30 & 84 & 354 \end{pmatrix}, \quad (A'A)^{-1} = \tfrac{1}{4752}\begin{pmatrix} 3564 & 396 & -396 \\ 396 & 516 & -156 \\ -396 & -156 & 84 \end{pmatrix}$$

and

$$\bar{\mathbf{u}} = (A'A)^{-1}A'\mathbf{y} = \tfrac{1}{4752}\begin{pmatrix} 3565 & 396 & -396 \\ 396 & 516 & -156 \\ -396 & -156 & 84 \end{pmatrix}\begin{pmatrix} 1 & 1 & 1 & 1 \\ 1 & -2 & 3 & 4 \\ 1 & 4 & 9 & 16 \end{pmatrix}\begin{pmatrix} 4 \\ 5 \\ -1 \\ 1 \end{pmatrix}$$

$$= \tfrac{1}{4752}\begin{pmatrix} 3564 & 396 & -396 \\ 396 & 516 & -156 \\ -396 & -156 & 84 \end{pmatrix}\begin{pmatrix} 9 \\ -5 \\ 31 \end{pmatrix} = \tfrac{1}{4752}\begin{pmatrix} 17820 \\ -3852 \\ -180 \end{pmatrix} \approx \begin{pmatrix} 3.75 \\ -0.81 \\ -0.04 \end{pmatrix}$$

Thus the best quadratic fit to the data is given by the parabola

$$y = 3.75 - 0.81x - 0.04x^2$$

The parabola and data points are sketched in Figure 4.10.

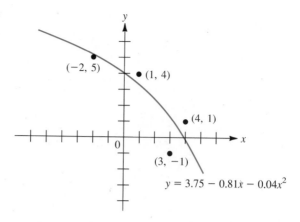

Figure 4.10 The quadratic $y = 3.75 - 0.81x - 0.04x^2$ is the best quadratic fit to the four points

■

EXAMPLE 3 **Finding the Best Quadratic Fit to Five Data Points Can Provide an Estimate for g**   The method of curve-fitting can be used to measure physical constants. Suppose, for example, that an object is dropped from a height of 200 meters. The following measurements are taken:

Elapsed Time	Height (in meters)
0	200
1	195
2	180
4	120
6	25

If an object at an initial height of 200 meters is dropped from rest, then its height after t seconds is given by

$$s = 200 - \tfrac{1}{2}gt^2.$$

To estimate g, we can fit a quadratic to the five data points given above. The coefficients of the t^2 term will, if our measurements are accurate, be a reasonable approximation to the number $-\tfrac{1}{2}g$. Using the earlier notation, we have

$$A = \begin{pmatrix} 1 & 0 & 0 \\ 1 & 1 & 1 \\ 1 & 2 & 4 \\ 1 & 4 & 16 \\ 1 & 6 & 36 \end{pmatrix}, \quad A^t = \begin{pmatrix} 1 & 1 & 1 & 1 & 1 \\ 0 & 1 & 2 & 4 & 6 \\ 0 & 1 & 4 & 16 & 36 \end{pmatrix} \quad \text{and} \quad \mathbf{y} = \begin{pmatrix} 200 \\ 195 \\ 180 \\ 120 \\ 25 \end{pmatrix}$$

Then

$$
A^t A = \begin{pmatrix} 5 & 13 & 57 \\ 13 & 57 & 289 \\ 57 & 289 & 1569 \end{pmatrix}, \qquad (A^t A)^{-1} = \tfrac{1}{7504}\begin{pmatrix} 5912 & -3924 & 508 \\ -3924 & 4596 & -704 \\ 508 & -704 & 116 \end{pmatrix}
$$

and

$$
\bar{u} = \tfrac{1}{7504}\begin{pmatrix} 5912 & -3924 & 508 \\ -3924 & 4596 & -704 \\ 508 & -704 & 116 \end{pmatrix}\begin{pmatrix} 1 & 1 & 1 & 1 & 1 \\ 0 & 1 & 2 & 4 & 6 \\ 0 & 1 & 4 & 16 & 36 \end{pmatrix}\begin{pmatrix} 200 \\ 195 \\ 180 \\ 120 \\ 25 \end{pmatrix}
$$

$$
= \tfrac{1}{7504}\begin{pmatrix} 5912 & -3924 & 508 \\ -3924 & 4596 & -704 \\ 508 & -704 & 116 \end{pmatrix}\begin{pmatrix} 720 \\ 1185 \\ 3735 \end{pmatrix} = \tfrac{1}{7504}\begin{pmatrix} 1504080 \\ -8460 \\ -35220 \end{pmatrix} \approx \begin{pmatrix} 200.44 \\ -1.13 \\ -4.69 \end{pmatrix}
$$

Thus the data points are fitted by the quadratic

$$
s(t) = 200.44 - 1.13t - 4.69t^2
$$

and we have $\tfrac{1}{2}g \approx 4.69$ or

$$
g \approx 2(4.69) = 9.38 \text{ m/sec}^2
$$

This is reasonably close to the correct value 9.81 m/sec^2. To obtain a more accurate approximation for g, we would need to obtain more accurate observations.   ■

We note here that higher-order polynomial approximations are carried out in a virtually identical manner. For details, see Problems 7 and 9.

We conclude this section by proving the result which guarantees that equation (8) will always be valid except when the data points lie on the same vertical line.

THEOREM 1   Let $(x_1, y_1), (x_2, y_2), \ldots, (x_n, y_n)$ be n points in $\mathbb{R}^2$, and suppose that not all the x_i's are equal. Then if A is given as in (3), the matrix $A^t A$ is an invertible 2×2 matrix.

Note. If $x_1 = x_2 = x_3 = \cdots = x_n$, then all the data points lie on the vertical line $x = x_1$, and the best linear approximation is, of course, this line.

Proof   We have

$$
A = \begin{pmatrix} 1 & x_1 \\ 1 & x_2 \\ \vdots & \vdots \\ 1 & x_n \end{pmatrix}
$$

Since not all the x_i's are equal, the columns of A are linearly independent. Now

$$A^t A = \begin{pmatrix} 1 & 1 & \cdots & 1 \\ x_1 & x_2 & \cdots & x_n \end{pmatrix} \begin{pmatrix} 1 & x_1 \\ 1 & x_2 \\ \vdots & \vdots \\ 1 & x_n \end{pmatrix} = \begin{pmatrix} n & \sum\limits_{i=1}^{n} x_i \\ \sum\limits_{i=1}^{n} x_i & \sum\limits_{i=1}^{n} x_i^2 \end{pmatrix}$$

If $A^t A$ is not invertible, then $\det A^t A = 0$. This means that

$$n \sum_{i=1}^{n} x_i^2 = \left(\sum_{i=1}^{n} x_i \right)^2 \tag{13}$$

Let $\mathbf{u} = \begin{pmatrix} 1 \\ 1 \\ \vdots \\ 1 \end{pmatrix}$ and $\mathbf{x} = \begin{pmatrix} x_1 \\ x_2 \\ \vdots \\ x_n \end{pmatrix}$. Then

$$|\mathbf{u}|^2 = \mathbf{u} \cdot \mathbf{u} = n, \qquad |\mathbf{x}|^2 = \sum_{i=1}^{n} x_i^2, \quad \text{and} \quad \mathbf{u} \cdot \mathbf{x} = \sum_{i=1}^{n} x_i$$

so that equation (13) can be restated as

$$|\mathbf{u}|^2 |\mathbf{x}|^2 = |\mathbf{u} \cdot \mathbf{x}|^2$$

or, taking square roots, we obtain

$$|\mathbf{u} \cdot \mathbf{x}| = |\mathbf{u}| \, |\mathbf{x}|$$

Now the Cauchy-Schwartz inequality (page 293) states that $|\mathbf{u} \cdot \mathbf{x}| \leq |\mathbf{u}| \, |\mathbf{x}|$ with equality if and only if $\mathbf{x}$ is a constant multiple of $\mathbf{u}$. But $\mathbf{u}$ and $\mathbf{x}$ are the columns of A which are linearly independent, by hypothesis. This contradiction proves the theorem. ∎

PROBLEMS   4.11

In Problems 1–3 find the best straight-line fit to the given data points.

1. $(1, 3), (-2, 4), (7, 0)$

2. $(-3, 7), (4, 9)$

3. $(1, -3), (4, 6), (-2, 5), (3, -1)$

In Problems 4–6 find the best quadratic fit to the given data points.

4. $(2, -5), (3, 0), (1, 1), (4, -2)$

5. $(-7, 3), (2, 8), (1, 5)$

6. $(1, -1)$, $(3, -6)$, $(5, 2)$, $(-3, 1)$, $(7, 4)$

7. The general cubic is given by

$$a + bx + cx^2 + dx^3$$

Show that the best cubic approximation to n data points is given by

$$\mathbf{u} = \begin{pmatrix} a \\ b \\ c \\ d \end{pmatrix} = (A^t A)^{-1} A^t \mathbf{y}$$

where $\mathbf{y}$ is as before, and

$$A = \begin{pmatrix} 1 & x_1 & x_1^2 & x_1^3 \\ 1 & x_2 & x_2^2 & x_2^3 \\ \vdots & \vdots & \vdots & \vdots \\ 1 & x_n & x_n^2 & x_n^3 \end{pmatrix}$$

8. Find the best cubic approximation to the data points $(3, -2)$, $(0, 3)$, $(-1, 4)$, $(2, -2)$, $(1, 2)$

9. The general kth-degree polynomial is given by

$$a_0 + a_1 x + a_2 x^2 + \cdots + a_k x^k$$

Show that the best kth-degree fit to n data points is given by

$$\bar{\mathbf{u}} = \begin{pmatrix} a_0 \\ a_1 \\ \vdots \\ a_k \end{pmatrix} = (A^t A)^{-1} A^t \mathbf{y}$$

where

$$A = \begin{pmatrix} 1 & x_1 & x_1^2 & \cdots & x_1^k \\ 1 & x_2 & x_2^2 & \cdots & x_2^k \\ \vdots & \vdots & \vdots & \cdots & \vdots \\ 1 & x_n & x_n^2 & \cdots & x_n^k \end{pmatrix}$$

10. The points $(1, 5.52)$, $(-1, 15.52)$, $(3, 11.28)$, and $(-2, 26.43)$ all lie on a parabola.
  a. Find the parabola.
  b. Show that $|\mathbf{y} - A\bar{\mathbf{u}}| = 0$.

11. A manufacturer buys large quantities of a certain machine replacement part. He finds that his cost depends on the number of cases bought at the same time and that the cost per unit decreases as the number of cases bought increases. He assumes that cost is a quadratic function of volume and, from past invoices, he obtains the following table:

Number of Cases Bought	Total Cost (dollars)
10	150
30	260
50	325
100	500
175	670

Find his total cost function.

12. A person throws a ball straight into the air. Its height is given by $s(t) = s_0 + v_0 t + \frac{1}{2}gt^2$. The following measurements are taken:

Elapsed Time (seconds)	Height (feet)
1	57
1.5	67
2.5	68
4	9.5

Using these data, estimate

a. The height at which the ball was released

b. Its initial velocity

c. g (in ft/sec^2)

4.12 INNER PRODUCT SPACES AND PROJECTIONS

This section makes use of a knowledge of elementary properties of complex numbers (summarized in Appendix 2) and some familiarity with material in the first year of calculus.

In Section 1.7 we saw how we could multiply two vectors in $\mathbb{R}^n$ to get a scalar. This scalar product is also called an *inner product*. Other vector spaces have inner products defined on them as well. Before giving a general definition, we note that in $\mathbb{R}^n$ the inner product of two vectors is a real scalar. In other spaces (see Example 2 below) the inner product gives us a complex scalar. To include all cases, therefore, we assume in the following definition that the inner product of two vectors is a complex number. Since every real number is a complex number, this definition includes the real case as well.

DEFINITION 1 **Inner Product Space** The complex vector space V is called an **inner product space** if for every pair of vectors $\mathbf{u}$ and $\mathbf{v}$ in V, there is a unique complex number $(\mathbf{u}, \mathbf{v})$, called the **inner product** of $\mathbf{u}$ and $\mathbf{v}$, such that if $\mathbf{u}$, $\mathbf{v}$, and $\mathbf{w}$ are in V and $\alpha \in \mathbb{C}$, then

i. $(\mathbf{v}, \mathbf{v}) \geq 0$

ii. $(\mathbf{v}, \mathbf{v}) = 0$ if and only if $\mathbf{v} = \mathbf{0}$

iii. $(\mathbf{u}, \mathbf{v} + \mathbf{w}) = (\mathbf{u}, \mathbf{v}) + (\mathbf{u}, \mathbf{w})$

iv. $(\mathbf{u} + \mathbf{v}, \mathbf{w}) = (\mathbf{u}, \mathbf{w}) + (\mathbf{v}, \mathbf{w})$

v. $(\mathbf{u}, \mathbf{v}) = \overline{(\mathbf{v}, \mathbf{u})}$

vi. $(\alpha\mathbf{u}, \mathbf{v}) = \alpha(\mathbf{u}, \mathbf{v})$

vii. $(\mathbf{u}, \alpha\mathbf{v}) = \overline{\alpha}(\mathbf{u}, \mathbf{v})$

The bar in conditions (*v*) and (*vii*) denotes the complex conjugate.

Note. If $(\mathbf{u}, \mathbf{v})$ is real, then $(\overline{\mathbf{u}, \mathbf{v}}) = (\mathbf{u}, \mathbf{v})$ and we can remove the bar in (*v*).

EXAMPLE 1   **An Inner Product in $\mathbb{R}^n$** $\mathbb{R}^n$ is an inner product space with $(\mathbf{u}, \mathbf{v}) = \mathbf{u} \cdot \mathbf{v}$. Conditions ($iii$–$vii$) are contained in Theorem 1.7.1 on page 44. Conditions (i) and (ii) are included in the result (4.10.9) on page 282.   ∎

EXAMPLE 2   **An Inner Product in $\mathbb{C}^n$**   We defined the space $\mathbb{C}^n$ in Example 4.2.13, page 216. Let $\mathbf{x} = (x_1, x_2, \ldots, x_n)$ and $\mathbf{y} = (y_1, y_2, \ldots, y_n)$ be in $\mathbb{C}^n$. (Remember— this means that the x_i's and y_i's are complex numbers.) Then we define

$$(\mathbf{x}, \mathbf{y}) = x_1 \overline{y}_1 + x_2 \overline{y}_2 + \cdots + x_n \overline{y}_n \tag{1}$$

To show that equation (1) defines an inner product, we need some facts about complex numbers. If these are unfamiliar, refer to Appendix 2. For (i),

$$(\mathbf{x}, \mathbf{x}) = x_1 \overline{x}_1 + x_2 \overline{x}_2 + \cdots + x_n \overline{x}_n = |x_1|^2 + |x_2|^2 + \cdots + |x_n|^2$$

Thus (i) and (ii) are satisfied since $|x_i|$ is a real number. Conditions (iii) and (iv) follow from the fact that $z_1(z_2 + z_3) = z_1 z_2 + z_1 z_3$ for any complex numbers z_1, z_2, and z_3. Condition (v) follows from the fact that $\overline{z_1 z_2} = \overline{z}_1 \, \overline{z}_2$ and $\overline{\overline{z}}_1 = z_1$ so that $\overline{x_1 \overline{y}_1} = \overline{x}_1 y_1$. Condition ($vi$) is obvious. For ($vii$), $(\mathbf{u}, \alpha\mathbf{v}) = \overline{(\alpha\mathbf{v}, \mathbf{u})} = \overline{(\alpha\mathbf{v}, \mathbf{u})} = \overline{\alpha}\overline{(\mathbf{v}, \mathbf{u})} = \overline{\alpha}(\mathbf{u}, \mathbf{v})$. Here we used ($vi$) and ($v$).   ∎

EXAMPLE 3   **The Inner Product of Two Vectors in $\mathbb{C}^3$**   In $\mathbb{C}^3$ let $\mathbf{x} = (1 + i, -3, 4 - 3i)$ and $\mathbf{y} = (2 - i, -i, 2 + i)$. Then

$$(\mathbf{x}, \mathbf{y}) = (1 + i)(\overline{2 - i}) + (-3)(\overline{-i}) + (4 - 3i)(\overline{2 + i})$$
$$= (1 + i)(2 + i) + (-3)(i) + (4 - 3i)(2 - i)$$
$$= (1 + 3i) - 3i + (5 - 10i) = 6 - 10i$$ ∎

EXAMPLE 4   **An Inner Product in $C[a, b]$**   Suppose that $a < b$; let $V = C[a, b]$, the space of

[Calculus]   functions which are continuous on the interval $[a, b]$, and define

$$(f, g) = \int_a^b f(t)g(t) \, dt \tag{2}$$

We shall see that this is also an inner product.†

(i) $(f, f) = \int_a^b f^2(t) \, dt \geq 0$. It is a basic theorem of calculus that if $f \in C[a, b]$, $f \geq 0$ on $[a, b]$, and $\int_a^b f(t) \, dt = 0$, then $f = 0$ on $[a, b]$. This proves (i) and (ii). (iii–vii) follow from basic facts about definite integrals.   ∎

Note. In $C[a, b]$ the scalars are assumed to be real numbers and the functions are real-valued so that we do not have to worry about complex conjugates. However if

---

†This is not the only way to define an inner product on $C[a, b]$, but it is the most common one.

the functions are complex-valued, then we can still define an inner product. See Problem 27 for details.

EXAMPLE 5   **The Inner Product of Two Functions in $C[0, 1]$**   Let $f(t) = t^2 \in C[0, 1]$ and
Calculus $g(t) = (4 - t) \in C[0, 1]$. Then

$$(f, g) = \int_0^1 t^2(4 - t) \, dt = \int_0^1 (4t^2 - t^3) \, dt = \left(\frac{4t^3}{3} - \frac{t^4}{4}\right)\Big|_0^1 = \frac{13}{12}$$ ∎

DEFINITION 2   Let V be an inner product space and suppose that $\mathbf{u}$ and $\mathbf{v}$ are in V. Then

   i. $\mathbf{u}$ and $\mathbf{v}$ are **orthogonal** if $(\mathbf{u}, \mathbf{v}) = 0$.

   ii. The **norm** of $\mathbf{u}$, denoted by $|\mathbf{u}|$, is given by

$$|\mathbf{u}| = \sqrt{(\mathbf{u}, \mathbf{u})} \tag{3}$$

Note. Equation (3) makes sense since $(\mathbf{u}, \mathbf{u}) \geq 0$.

EXAMPLE 6   **Two Orthogonal Vectors in $\mathbb{C}^2$**   In $\mathbb{C}^2$ the vectors $(3, -i)$ and $(2, 6i)$ are orthogonal because

$$((3, -i), (2, 6i)) = 3 \cdot \overline{2} + (-i)(\overline{6i}) = 6 + (-i)(-6i) = 6 - 6 = 0$$

Also $|(3, -i)| = \sqrt{3 \cdot 3 + (-i)(i)} = \sqrt{10}$. ∎

EXAMPLE 7   **Two Orthogonal Functions in $C[0, 2\pi]$**   In $C[0, 2\pi]$ the functions $\sin t$ and $\cos t$ are orthogonal since

$$(\sin t, \cos t) = \int_0^{2\pi} \sin t \cos t \, dt = \frac{1}{2}\int_0^{2\pi} \sin 2t \, dt = -\frac{\cos 2t}{4}\Big|_0^{2\pi} = 0$$

Also

$$|\sin t| = (\sin t, \sin t)^{1/2}$$
$$= \left[\int_0^{2\pi} \sin^2 t \, dt\right]^{1/2}$$
$$= \left[\frac{1}{2}\int_0^{2\pi} (1 - \cos 2t) \, dt\right]^{1/2}$$
$$= \left[\frac{1}{2}\left(t - \frac{\sin 2t}{2}\right)\Big|_0^{2\pi}\right]^{1/2}$$
$$= \sqrt{\pi}$$ ∎

If you look at the proofs of Theorems 4.10.1 and 4.10.2 on page 283, you will see that no use was made of the fact that $V = \mathbb{R}^n$. The same theorems are true in any inner product space V. We list them for convenience after giving a definition.

DEFINITION 3   **Orthonormal Set** The set of vectors $\{\mathbf{v}_1, \mathbf{v}_2, \ldots, \mathbf{v}_n\}$ is an **orthonormal set** in V if

$$(\mathbf{v}_i, \mathbf{v}_j) = 0 \qquad \text{for } i \neq j \tag{4}$$

and

$$|\mathbf{v}_i| = \sqrt{(\mathbf{v}_i, \mathbf{v}_i)} = 1 \tag{5}$$

If only (4) holds, the set is said to be **orthogonal.**

THEOREM 1   Any finite orthogonal set of nonzero vectors in an inner product space is linearly independent. ∎

THEOREM 2   Any finite, linearly independent set in an inner product space can be made into an orthonormal set by the Gram-Schmidt process. In particular, any finite dimensional inner product space has an orthonormal basis. ∎

EXAMPLE 8

Calculus

An Orthonormal Basis in $P_2[0, 1]$ Construct an orthonormal basis for $P_2[0, 1]$.

Solution   We start with the standard basis $\{1, x, x^2\}$. Since $P_2[0, 1]$ is a subspace of $C[0, 1]$, we may use the inner product of Example 4. Since $\int_0^1 1^2 \, dx = 1$, we let $\mathbf{u}_1 = 1$. Then $\mathbf{v}_1' = \mathbf{v}_2 - (\mathbf{v}_2, \mathbf{u}_1)\mathbf{u}_1$. Here $(\mathbf{v}_2, \mathbf{u}_1) = \int_0^1 (x \cdot 1) \, dx = \frac{1}{2}$. Thus $\mathbf{v}_2' = x - \frac{1}{2} \cdot 1 = x - \frac{1}{2}$. Next we compute

$$\left| x - \tfrac{1}{2} \right| = \left[\int_0^1 (x - \tfrac{1}{2})^2 \, dx \right]^{1/2} = \left[\int_0^1 (x^2 - x + \tfrac{1}{4}) \, dx \right]^{1/2} = \frac{1}{\sqrt{12}} = \frac{1}{2\sqrt{3}}$$

Hence $\mathbf{u}_2 = 2\sqrt{3}(x - \tfrac{1}{2}) = \sqrt{3}(2x - 1)$. Then

$$\mathbf{v}_3' = \mathbf{v}_3 - (\mathbf{v}_3, \mathbf{u}_1)\mathbf{u}_1 - (\mathbf{v}_3, \mathbf{u}_2)\mathbf{u}_2.$$

We have $(\mathbf{v}_3, \mathbf{u}_1) = \int_0^1 x^2 \, dx = \frac{1}{3}$ and

$$(\mathbf{v}_3, \mathbf{u}_2) = \sqrt{3} \int_0^1 x^2(2x - 1) \, dx = \sqrt{3} \int_0^1 (2x^3 - x^2) \, dx = \frac{\sqrt{3}}{6}$$

Thus

$$\mathbf{v}_3' = x^2 - \tfrac{1}{3} - \frac{\sqrt{3}}{6}[\sqrt{3}(2x - 1)] = x^2 - x + \tfrac{1}{6}$$

and

$$|\mathbf{v}_3'| = \left[\int_0^1 (x^2 - x + \tfrac{1}{6})^2 \, dx \right]^{1/2}$$

$$= \left[\int_0^1 \left(x^4 - 2x^3 + \tfrac{4}{3}x^2 - \frac{x}{3} + \tfrac{1}{36} \right) dx \right]^{1/2}$$

$$= \left[\left(\frac{x^5}{5} - \frac{x^4}{2} + \frac{4x^3}{9} - \frac{x^2}{6} + \frac{x}{36} \right) \Big|_0^1 \right]^{1/2}$$

$$= \frac{1}{\sqrt{180}} = \frac{1}{6\sqrt{5}}$$

Thus $\mathbf{u}_3 = 6\sqrt{5}(x^2 - x + \tfrac{1}{6}) = \sqrt{5}(6x^2 - 6x + 1)$. Finally, the orthonormal basis is $\{1, \sqrt{3}(2x - 1), \sqrt{5}(6x^2 - 6x + 1)\}$. ∎

EXAMPLE 9

Calculus

An Infinite Orthonormal Set in $C[0, 2\pi]$ In $C[0, 2\pi]$ the infinite set

$$S = \left\{ \frac{1}{\sqrt{2\pi}}, \frac{1}{\sqrt{\pi}} \sin x, \frac{1}{\sqrt{\pi}} \cos x, \frac{1}{\sqrt{\pi}} \sin 2x, \frac{1}{\sqrt{\pi}} \cos 2x, \dots, \right.$$

$$\left. \frac{1}{\sqrt{\pi}} \sin nx, \frac{1}{\sqrt{\pi}} \cos nx, \dots \right\}$$

is an orthonormal set. This follows since if $m \neq n$, then

$$\int_0^{2\pi} \sin mx \cos nx \, dx = \int_0^{2\pi} \sin mx \sin nx \, dx = \int_0^{2\pi} \cos mx \cos nx \, dx = 0$$

To prove one of these, we note that

$$\int_0^{2\pi} \sin mx \cos nx \, dx = \frac{1}{2} \int_0^{2\pi} [\sin (m + n)x + \sin (m - n)x] \, dx$$

$$= -\frac{1}{2} \left[\frac{\cos (m + n)x}{m + n} + \frac{\cos (m - n)x}{m - n} \right] \Big|_0^{2\pi}$$

$$= 0$$

since $\cos x$ is periodic of period 2π. We have seen that $|\sin x| = \sqrt{\pi}$. Thus $|(1/\sqrt{\pi}) \sin x| = 1$. The other facts follow in a similar fashion. This example provides a situation in which we have an *infinite* orthonormal set. In fact, although this is far beyond us in this elementary text, the functions in S are actually a basis in $C[0, 2\pi]$. Suppose $f \in C[0, 2\pi]$. Then if we write f as an infinite linear combination of the vectors in S, we obtain what is called the **Fourier series representation** of f. ∎

DEFINITION 4   **Orthogonal Projection** Let H be a subspace of an inner product space V with an orthonormal basis $\{\mathbf{u}_1, \mathbf{u}_2, \dots, \mathbf{u}_k\}$. If $\mathbf{v} \in V$, then the **orthogonal projection** of

v onto H, denoted by $\text{proj}_H \mathbf{v}$, is given by

$$\text{proj}_H \mathbf{v} = (\mathbf{v}, \mathbf{u}_1)\mathbf{u}_1 + (\mathbf{v}, \mathbf{u}_2)\mathbf{u}_2 + \cdots + (\mathbf{v}, \mathbf{u}_k)\mathbf{u}_k \tag{6}$$

The following theorems have proofs that are identical to their $\mathbb{R}^n$ counterparts proved in Section 4.10.

THEOREM 3   Let H be a subspace of the finite dimensional inner product space V. Suppose that H has two orthonormal bases $\{\mathbf{u}_1, \mathbf{u}_2, \ldots, \mathbf{u}_k\}$ and $\{\mathbf{w}_1, \mathbf{w}_2, \ldots, \mathbf{w}_k\}$. Let $\mathbf{v} \in V$. Then

$$(\mathbf{v}, \mathbf{u}_1)\mathbf{u}_1 + (\mathbf{v}, \mathbf{u}_2)\mathbf{u}_2 + \cdots + (\mathbf{v}, \mathbf{u}_k)\mathbf{u}_k$$
$$= (\mathbf{v}, \mathbf{w}_1)\mathbf{w}_1 + (\mathbf{v}, \mathbf{w}_2)\mathbf{w}_2 + \cdots + (\mathbf{v}, \mathbf{w}_k)\mathbf{w}_k \quad \blacksquare$$

DEFINITION 5   **Orthogonal Complement** Let H be a subspace of the inner product space V. Then the **orthogonal complement** of H, denoted by $H^\perp$, is given by

$$H^\perp = \{\mathbf{x} \in V \colon (\mathbf{x}, \mathbf{h}) = 0 \quad \text{for every } \mathbf{h} \in H\} \tag{7}$$

THEOREM 4   If H is a subspace of the inner product space V, then

i. $H^\perp$ is a subspace of V.

ii. $H \cap H^\perp = \{\mathbf{0}\}$.

iii. $\dim H^\perp = n - \dim H$ if $\dim V = n < \infty$. $\quad \blacksquare$

THEOREM 5   **Projection Theorem** Let H be a finite dimensional subspace of the inner product space V and suppose that $\mathbf{v} \in V$. Then there exists a unique pair of vectors $\mathbf{h}$ and $\mathbf{p}$ such that $\mathbf{h} \in H$, $\mathbf{p} \in H^\perp$, and

$$\mathbf{v} = \mathbf{h} + \mathbf{p} \tag{8}$$

where $\mathbf{h} = \text{proj}_H \mathbf{v}$.

If V is finite dimensional, then $\mathbf{p} = \text{proj}_{H^\perp} \mathbf{v}$. $\quad \blacksquare$

Remark. If you look at the proof of Theorem 4.10.7, you will notice that (8) holds even if V is infinite dimensional. The only difference is that if V is infinite dimensional, then $H^\perp$ is infinite dimensional (since H is finite dimensional), and so $\text{proj}_{H^\perp} \mathbf{v}$ is not defined.

THEOREM 6

Norm Approximation Theorem Let H be a finite dimensional subspace of the inner product space V and let $\mathbf{v}$ be a vector in V. Then, in H, $\text{proj}_H \mathbf{v}$ is the best approximation to $\mathbf{v}$ in the following sense: If $\mathbf{h}$ is any other vector in H, then

$$\boxed{|\mathbf{v} - \text{proj}_H \mathbf{v}| < |\mathbf{v} - \mathbf{h}|} \tag{9}$$

∎

EXAMPLE 10

Calculus

Computing a Projection onto $P_2[0, 1]$ Since $P_2[0, 1]$ is a finite dimensional subspace of $C[0, 1]$, we can talk about $\text{proj}_{P_2[0,1]} f$ if $f \in C[0, 1]$. If $f(x) = e^x$, for example, we compute $\text{proj}_{P_2[0,1]} e^x$. Since $\{\mathbf{u}_1, \mathbf{u}_2, \mathbf{u}_3\} = \{1, \sqrt{3}(2x - 1), \sqrt{5}(6x^2 - 6x + 1)\}$ is an orthonormal basis in $P_2[0, 1]$ by Example 8, we have

$$\text{proj}_{P_2[0,1]} e^x = (e^x, 1)1 + (e^x, \sqrt{3}(2x - 1))\sqrt{3}(2x - 1)$$
$$+ (e^x, \sqrt{5}(6x^2 - 6x + 1))\sqrt{5}(6x^2 - 6x + 1)$$

We shall spare you the computations. Using the fact that $\int_0^1 e^x \, dx = e - 1$, $\int_0^1 xe^x \, dx = 1$, and $\int_0^1 x^2 e^x \, dx = e - 2$, we obtain $(e^x, 1) = e - 1$, $(e^x, \sqrt{3}(2x - 1)) = \sqrt{3}(3 - e)$, and $(e^x, \sqrt{5}(6x^2 - 6x + 1)) = \sqrt{5}(7e - 19)$. Finally,

$$\text{proj}_{P_2[0,1]} e^x = (e - 1) + \sqrt{3}(3 - e)\sqrt{3}(2x - 1)$$
$$+ \sqrt{5}(7e - 19)(\sqrt{5})(6x^2 - 6x + 1)$$
$$= (e - 1) + (9 - 3e)(2x - 1)$$
$$+ 5(7e - 19)(6x^2 - 6x + 1)$$
$$\approx 1.01 + 0.85x + 0.84x^2$$

∎

We conclude this section with an application of the norm approximation theorem.

Mean Square Approximation of a Continuous Function
Let $f \in C[a, b]$. We wish to approximate f by an nth-degree polynomial. What is the polynomial that does this with the smallest error?

In order to answer this question, we must define what we mean by *error*. There are many different ways to define error. Three are given below:

$$\textbf{Maximum error} = \max |f(x) - g(x)| \text{ for } x \in [a, b] \tag{10}$$

$$\textbf{Area error} = \int_a^b |f(x) - g(x)| \, dx \tag{11}$$

$$\textbf{Mean square error} = \int_a^b |f(x) - g(x)|^2 \, dx \tag{12}$$

EXAMPLE 11

Calculus

Computing Errors   Let $f(x) = x^2$ and $g(x) = x^3$ on $[0, 1]$. On $[0, 1]$, $x^2 \geq x^3$ so $|x^2 - x^3| = x^2 - x^3$. Then

i. Maximum error $= \max (x^2 - x^3)$. To compute this we compute $d/dx \, (x^2 - x^3) = 2x - 3x^2 = x(2 - 3x) = 0$ when $x = 0$ and $x = 2/3$. The maximum error occurs when $x = 2/3$ and is given by $[(\frac{2}{3})^2 - (\frac{2}{3})^3] = \frac{4}{9} - \frac{8}{27} = \frac{4}{27} \approx 0.148$.

ii. Area error $= \int_0^1 (x^2 - x^3) \, dx = (x^3/3 - x^4/4)|_0^1 = \frac{1}{3} - \frac{1}{4} = \frac{1}{12} \approx 0.083$. This is sketched in Figure 4.11.

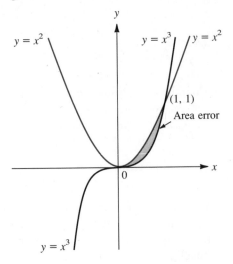

Figure 4.11   Illustration of area error

iii. Mean square error $= \int_0^1 (x^2 - x^3)^2 \, dx = \int_0^1 (x^4 - 2x^5 + x^6) \, dx = (x^5/5 - x^6/3 + x^7/7)|_0^1 = \frac{1}{5} - \frac{1}{3} + \frac{1}{7} = \frac{1}{105} \approx 0.00952$ ∎

Each of the three measurements of error is useful. The mean square error is used in statistics and other applications. We can use the norm approximation theorem to find the unique nth-degree polynomial that approximates a given continuous function with the smallest mean square error.

From Example 4, $C[a, b]$ is an inner product space with

$$(f, g) = \int_a^b f(t)g(t) \, dt \tag{13}$$

For every positive integer n, $P_n[a, b]$—the space of nth-degree polynomials defined on $[a, b]$—is a finite dimensional subspace of $C[a, b]$. We compute, for $f \in C[a, b]$ and $p_n \in P_n$,

$$|f - p_n|^2 = (f - p_n, f - p_n) = \int_a^b [(f(t) - p_n(t))(f(t) - p_n(t)] \, dt$$

$$= \int_a^b |f(t) - p_n(t)|^2 \, dt = \text{mean square error}$$

Thus, by Theorem 6,

> The nth-degree polynomial that approximates a continuous
> function with the smallest mean square error is given by
>
> $$p_n = \text{proj}_{P_n} f$$

(14)

EXAMPLE 12   **The Best Mean Square Quadratic Approximation to e^x**   From Example 10, the second-degree polynomial that best approximates e^x on $[0, 1]$ in the mean square sense is given by

$$p_2(x) \approx 1.01 + 0.85x + 0.84x^2$$ ∎

PROBLEMS 4.12

1. Let D_n denote the set of $n \times n$ diagonal matrices with real components under the usual matrix operations. If A and B are in D_n, define

$$(A, B) = a_{11}b_{11} + a_{22}b_{22} + \cdots + a_{nn}b_{nn}$$

Prove that D_n is an inner product space.

2. If $A \in D_n$, show that $|A| = 1$ if and only if $a_{11}^2 + a_{22}^2 + \cdots + a_{nn}^2 = 1$.

3. Find an orthonormal basis for D_n.

4. Find an orthonormal basis for D_2 starting with $A = \begin{pmatrix} 2 & 0 \\ 0 & 1 \end{pmatrix}$ and $B = \begin{pmatrix} -3 & 0 \\ 0 & 4 \end{pmatrix}$.

5. In $\mathbb{C}^2$ find an orthonormal basis starting with the basis $(1, i)$, $(2 - i, 3 + 2i)$.

Calculus 6. Find an orthonormal basis for $P_3[0, 1]$.

Calculus 7. Find an orthonormal basis for $P_2[-1, 1]$. The polynomials you obtain are called **normalized Legendre polynomials.**

*Calculus 8. Find an orthonormal basis for $P_2[a, b]$, $a < b$.

9. If $A = (a_{ij})$ is an $n \times n$ matrix, the **trace** of A, written tr A, is the sum of the diagonal components of A: tr $A = a_{11} + a_{22} + \cdots a_{nn}$. In M_{nn} define $(A, B) = $ tr (AB^t). Prove that with the preceding inner product, M_{nn} is an inner product space.

10. If $A \in M_{nn}$, show that $|A|^2$ is the sum of the squares of the elements of A. [*Note:* Here $|A| = (A, A)^{1/2}$, not det A.]

11. Find an orthonormal basis for M_{22}.

12. We can think of the complex plane as a vector space over the reals with basis vectors 1, i. If $z = a + ib$ and $w = c + id$, define $(z, w) = ac + bd$. Show that this is an inner product and that $|z|$ is the usual length of a complex number.

13. Let a, b, and c be three distinct real numbers. Let p and q be in P_2 and define $(p, q) = p(a)q(a) + p(b)q(b) + p(c)q(c)$.

   a. Prove that (p, q) is an inner product in P_2.
   b. Is $(p, q) = p(a)q(a) + p(b)q(b)$ an inner product?

14. In $\mathbb{R}^2$, if $\mathbf{x} = \begin{pmatrix} x_1 \\ x_2 \end{pmatrix}$ and $\mathbf{y} = \begin{pmatrix} y_1 \\ y_2 \end{pmatrix}$, let $(\mathbf{x}, \mathbf{y})_* = x_1 y_1 + 3 x_2 y_2$. Show that $(x, y)_*$ is an inner product on $\mathbb{R}^2$.

15. With the inner product of Problem 14, calculate $\left| \begin{pmatrix} 2 \\ -3 \end{pmatrix} \right|_*$.

16. In $\mathbb{R}^2$ let $(\mathbf{x}, \mathbf{y}) = x_1 y_1 - x_2 y_2$. Is this an inner product? If not, why not?

*17. Let V be an inner product space. Prove that $|(\mathbf{u}, \mathbf{v})| \le |\mathbf{u}| \, |\mathbf{v}|$. This is called the **Cauchy-Schwarz inequality**. [*Hint:* See Theorem 9 in Section 4.10.]

*18. Using the result of Problem 17, prove that $|\mathbf{u} + \mathbf{v}| \le |\mathbf{u}| + |\mathbf{v}|$. This is called the **triangle inequality**.

Calculus 19. In $P_3[0, 1]$ let H be the subspace spanned by $\{1, x^2\}$. Find $H^\perp$.

* Calculus 20. In $C[-1, 1]$ let H be the subspace of even functions. Show that $H^\perp$ consists of odd functions. [*Hint: f* is odd if $f(-x) = -f(x)$ and is even if $f(-x) = f(x)$.]

* Calculus 21. $H = P_2[0, 1]$ is a subspace of $P_3[0, 1]$. Write the polynomial $1 + 2x + 3x^2 - x^3$ as $h(x) + p(x)$, where $h(x) \in H$ and $p(x) \in H^\perp$.

*22. Find a second-degree polynomial that best approximates $\sin \dfrac{\pi}{2} x$ on the interval $[0, 1]$ in the mean square sense.

*23. Solve Problem 22 for the function $\cos \dfrac{\pi}{2} x$.

24. Let A be an $m \times n$ matrix with complex entries. Then the **conjugate transpose** of A, denoted by A^*, is defined by $(A^*)_{ij} = \overline{a_{ji}}$. Compute A^* if

$$A = \begin{pmatrix} 1 - 2i & 3 + 4i \\ 2i & -6 \end{pmatrix}$$

25. Let A be an invertible $n \times n$ matrix with complex entries. A is called **unitary** if $A^{-1} = A^*$. Show that the following matrix is unitary:

$$A = \begin{pmatrix} \dfrac{1}{\sqrt{2}} & -\dfrac{1}{2} + \dfrac{i}{2} \\ \dfrac{1}{\sqrt{2}} & \dfrac{1}{2} - \dfrac{i}{2} \end{pmatrix}$$

*26. Show that an $n \times n$ matrix with complex entries is unitary if and only if the columns of A constitute an orthonormal basis for $\mathbb{C}^n$.

* Calculus 27. A function f is said to be **complex-valued** on the (real) interval $[a, b]$ if $f(x)$ can be written

$$f(x) = f_1(x) + f_2(x)i, \qquad x \in [a, b]$$

where f_1 and f_2 are real-valued functions. The complex-valued function f is **continuous** if f_1 and f_2 are continuous. Let $CV[a, b]$ denote the set of complex-valued functions that are continuous on $[a, b]$. For f and g in $CV[a, b]$, define

$$(f, g) = \int_a^b f(x) \overline{g(x)} \, dx \qquad (15)$$

Show that (15) defines an inner product in $CV[a, b]$.

Calculus **28.** Show that $f(x) = \sin x + i \cos x$ and $g(x) = \sin x - i \cos x$ are orthogonal in $CV[0, \pi]$.

Calculus **29.** Compute $|\sin x + i \cos x|$ in $CV[0, \pi]$.

4.13   THE FOUNDATIONS OF VECTOR SPACE THEORY: THE EXISTENCE OF A BASIS (Optional)

In this section we prove one of the most important results in linear algebra: **Every vector space has a basis.** The proof is more difficult than any other proof in this book; it involves concepts that are part of the foundation of mathematics. It will take some hard work to go through the details of this proof. However, after you have done so, you should have a deeper appreciation of a fundamental mathematical idea.

We begin with some definitions.

DEFINITION 1   **Partial Ordering** Let S be a set. A **partial ordering** on S is a relation, denoted by $\leq$, which is defined for some of the ordered pairs of elements of S and satisfies three conditions:

  i. $x \leq x$ for all $x \in S$ **reflexive law**

  ii. If $x \leq y$ and $y \leq x$, then $x = y$ **antisymmetric law**

  iii. If $x \leq y$ and $y \leq z$, then $x \leq z$ **transitive law**

It may be the case that there are elements x and y in S such that neither $x \leq y$ nor $y \leq x$. However, if for every $x, y \in S$, either $x \leq y$ or $y \leq x$, then the ordering is said to be a **total ordering.** If $x \leq y$ or $y \leq x$, then x and y are said to be **comparable.**

Notation. $x < y$ means $x \leq y$ and $x \neq y$.

EXAMPLE 1   **A Partial Ordering on $\mathbb{R}$**   The real numbers are partially ordered by $\leq$ where $\leq$ stands for "less than or equal to." The ordering here is a total ordering.   ∎

EXAMPLE 2   **A Partial Order on a Set of Subsets**   Let S be a set and let $P(S)$, called the **power set** of S, denote the set of all subsets of S.

We say that $A \leq B$ if $A \subseteq B$. The inclusion relation is a partial ordering on $P(S)$. This is easy to prove. We have

  i. $A \subseteq A$ for every set A.

  ii. $A \subseteq B$ and $B \subseteq A$ if and only if $A = B$.

iii. Suppose $A \subseteq B$ and $B \subseteq C$. If $x \in A$, then $x \in B$, so $x \in C$. This means that $A \subseteq C$.

Except in unusual circumstances (for example, if S contains only one element), the ordering will not be a total ordering. This is illustrated in Figure 4.12.

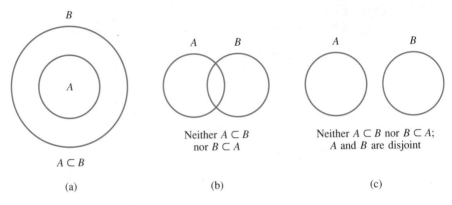

Figure 4.12   Three possibilities for set inclusion

■

DEFINITION 2   **Chain, Upper Bound, and Maximal Element**   Let S be a set partially ordered by $\leq$.

i. A subset T of S is called a **chain** if it is totally ordered; that is, if x and y are distinct elements of T, then $x \leq y$ or $y \leq x$.

ii. Let C be a subset of S. An element $u \in S$ is an **upper bound** for C if $c \leq u$ for every element $c \in C$.

iii. The element $m \in S$ is a **maximal element** for S if there is no $s \in S$ with $m < s$.

Remark 1. In (*ii*) the upper bound for C must be comparable with every element in C but it need not be in C (although it must be in S). For example, the number 1 is an upper bound for the set $(0, 1)$ but is not in $(0, 1)$. Any number greater than 1 is an upper bound. However, there is no number in $(0, 1)$ that is an upper bound for $(0, 1)$.

Remark 2. If m is a maximal element for S, it is not necessarily the case that $s \leq m$ for every $s \in S$. In fact, m may be comparable to very few elements of S. The only condition for maximality is that there is no element of S "larger than" m.

EXAMPLE 3   **A Chain of Subsets of $\mathbb{R}^2$**   Let $S = \mathbb{R}^2$. Then $P(S)$ consists of subsets of the xy-plane. Let $D_r = \{(x, y): x^2 + y^2 < r^2\}$; that is, D_r is an open disk of radius r—the interior of the circle of radius r centered at the origin. Let

$$T = \{D_r: \ r > 0\}$$

Clearly T is a chain. For if D_{r_1} and D_{r_2} are in T, then

$$D_{r_1} \subseteq D_{r_2} \text{ if } r_1 \le r_2 \text{ and } D_{r_2} \subseteq D_{r_1} \text{ if } r_2 \le r_1 \qquad \blacksquare$$

Before going further, we need some new notation. Let V be a vector space. We have seen that a linear combination of vectors in V is a finite sum $\sum_{i=1}^{n} \alpha_i v_i = \alpha_1 v_1 + \alpha_2 v_2 + \cdots + \alpha_n v_n$. If you have studied power series, you have seen infinite sums of the form $\sum_{n=0}^{\infty} a_n x^n$. For example,

$$e^x = \sum_{n=0}^{\infty} \frac{x^n}{n!} = 1 + x + \frac{x^2}{2!} + \frac{x^3}{3!} + \cdots$$

Here we need a different kind of sum. Let C be a set of vectors in V.† For each $\mathbf{v} \in C$, let $\alpha_{\mathbf{v}}$ denote a scalar (the set of scalars is given in the definition of V). Then when we write

$$\mathbf{x} = \sum_{\mathbf{v} \in C} \alpha_{\mathbf{v}} \, \mathbf{v} \qquad (1)$$

it will be understood that only a finite number of the scalars $\alpha_{\mathbf{v}}$ are nonzero and that all the terms with $\alpha_{\mathbf{v}} = 0$ are to be left out of the summation. We can describe the sum in (1) as follows:

For each $\mathbf{v} \in C$, assign a scalar $\alpha_{\mathbf{v}}$ and form the product $\alpha_{\mathbf{v}} \, \mathbf{v}$. Then $\mathbf{x}$ is the sum of the finite subset of the vectors $\alpha_{\mathbf{v}} \, \mathbf{v}$ for which $\alpha_{\mathbf{v}} \ne 0$.

DEFINITION 3   **Linear Combination, Spanning Set, Linear Independence, and Basis**

i. Let C be a subset of a vector space V. Then any vector that can be written in form (1) is called a **linear combination** of vectors in C. The set of linear combinations of vectors in C is denoted by L(C).

ii. The set C is said to **span** the vector space V if $V \subseteq \text{L}(C)$.

iii. A subset C of a vector space V is said to be **linearly independent** if

$$\sum_{\mathbf{v} \in C} \alpha_{\mathbf{v}} \, \mathbf{v} = \mathbf{0}$$

holds only when $\alpha_{\mathbf{v}} = 0$ for every $\mathbf{v} \in C$.

iv. The subset B of a vector space V is a **basis** for V if it spans V and is linearly independent.

Remark. If C contains only a finite number of vectors, then these definitions are precisely the ones we have seen earlier in this chapter.

---

†C is not necessarily a subspace of V.

THEOREM 1   Let B be a linearly independent subset of a vector space V. Then B is a basis if and only if it is maximal; that is, if $B \subsetneqq D$, then D is linearly dependent.

Proof   Suppose that B is a basis and that $B \subsetneqq D$. Choose $\mathbf{x}$ such that $\mathbf{x} \in D$ but $\mathbf{x} \notin B$. Since B is a basis, $\mathbf{x}$ can be written as a linear combination of vectors in B:

$$\mathbf{x} = \sum_{\mathbf{v} \in B} \alpha_{\mathbf{v}} \, \mathbf{v}$$

If $\alpha_{\mathbf{v}} = 0$ for every $\mathbf{v}$, then $\mathbf{x} = \mathbf{0}$ and D is dependent. Otherwise $\alpha_{\mathbf{v}} \neq 0$ for some $\mathbf{v}$ and so the sum

$$\mathbf{x} - \sum_{\mathbf{v} \in B} \alpha_{\mathbf{v}} \, \mathbf{v} = \mathbf{0}$$

shows that D is dependent. Thus B is maximal.

Conversely, suppose that B is maximal. Let $\mathbf{x}$ be a vector in V that is not in B. Let $D = B \cup \{\mathbf{x}\}$. Then D is dependent (since B is maximal) and there is an equation

$$\sum_{\mathbf{v} \in B} \alpha_{\mathbf{v}} \, \mathbf{v} + \beta \mathbf{x} = \mathbf{0}$$

in which not every coefficient is zero. But $\beta \neq 0$ since otherwise we would obtain a contradiction of the linear independence of B. Thus we can write

$$\mathbf{x} = -\beta^{-1} \sum_{\mathbf{v} \in B} \alpha_{\mathbf{v}} \, \mathbf{v} \dagger$$

Thus B is a spanning set and is therefore a basis for V.   ∎

Where is this all leading? Perhaps you can see the general direction. We have defined ordering on sets and maximal elements. We have shown that a linearly independent set is a basis if it is maximal. We now are lacking only a result that can help us prove the existence of a maximal element. That result is one of the basic assumptions of mathematics.

Many of you studied Euclidean geometry in high school. There, you had perhaps your first contact with mathematical proof. To prove things, Euclid made certain assumptions, which he called *axioms*. For example, he assumed that the shortest distance between two points is a straight line. Starting with these axioms, he, and students of geometry, were able to prove a number of theorems.

In all branches of mathematics it is necessary to have some axioms. If we

---

†If the scalars are real or complex numbers, then $\beta^{-1} = 1/\beta$.

assume nothing, we can prove nothing. To complete our proof we need the following axiom:

AXIOM   **Zorn's Lemma†** If S is a nonempty, partially ordered set such that every nonempty chain has an upper bound, then S has a maximal element.

Remark. The **axiom of choice** says, roughly, that given a number (finite or infinite) of nonempty sets, there is a function that chooses one element from each set. This axiom is equivalent to Zorn's lemma; that is, if you assume the axiom of choice, then you can prove Zorn's lemma and vice versa. For a proof of this equivalence and other interesting results, see the excellent book *Naive Set Theory* by Paul R. Halmos (New York: Van Nostrand, 1960), especially page 63.

We can now, finally, state and prove our main result.

THEOREM 2   Every vector space V has a basis.

Proof   We show that V has a maximal linearly independent subset. We do this in steps.

   i. Let S be the collection of all linearly independent subsets partially ordered by inclusion.

   ii. A chain in S is a subset T of S such that if A and B are in T, either $A \subseteq B$ or $B \subseteq A$.

   iii. Let T be a chain. Define

$$M(T) = \bigcup_{A \in T} A$$

   Clearly $M(T)$ is a subset of V and $A \subseteq M(T)$ for every $A \in T$. We want to show that $M(T)$ is an upper bound for T. Since $A \subseteq M(T)$ for every $A \in T$, we need only show that $M(T) \in S$; that is, we must show that $M(T)$ is linearly independent.

   iv. Suppose $\displaystyle\sum_{v \in M(T)} \alpha_v \, v = 0$ where only a finite number of the α_v's are nonzero. Denote these scalars by $\alpha_1, \alpha_2, \ldots, \alpha_n$ and the corresponding vectors by $v_1, v_2, \ldots, v_n$. For each i, $i = 1, 2, \ldots, n$, there is a set $A_i \in T$ such that $v_i \in A_i$ (because each v_i is in $M(T)$ and $M(T)$ is the union of sets in T). But T is totally ordered, so one of the A_i's contains all the others (see Problem 3); call

---

† Max A. Zorn (born 1906) spent a number of years at Indiana University where he is now a Professor Emeritus. He published his famous result in 1935 ("A Remark on Method in Transfinite Algebra," *Bulletin of the American Mathematical Society* 41 (1935): 667–670).

it A_k. (We can only draw this conclusion because $\{A_1, A_2, \ldots, A_n\}$ is finite.) Thus $A_i \subseteq A_k$ for $i = 1, 2, \ldots, n$ and $\mathbf{v}_1, \mathbf{v}_2, \ldots, \mathbf{v}_n \in A_k$. Since A_k is linearly independent and $\sum_{i=1}^{n} \alpha_i \mathbf{v}_i = \mathbf{0}$, it follows that $\alpha_1 = \alpha_2 = \cdots = \alpha_n = 0$.

Thus $M(T)$ is linearly independent.

v. S is nonempty because $\varnothing \in S$ ($\varnothing$ denotes the empty set). We have shown that every chain T in S has an upper bound $M(T)$ which is in S. By Zorn's lemma, S has a maximal element. But S consists of all linearly independent subsets of V. The maximal element $B \in S$ is therefore a maximal linearly independent subset of V. Thus, by Theorem 1, B is a basis for V. ∎

PROBLEMS   4.13

1. Show that every linearly independent set in a vector space V can be expanded to a basis.

2. Show that every spanning set in a vector space V has a subset that is a basis.

3. Let $A_1, A_2, \ldots, A_n$ be n sets in a chain T. Show that one of the sets contains all the others. [*Hint:* Because T is a chain, either $A_1 \subseteq A_2$ or $A_2 \subseteq A_1$. Thus the result is true if $n = 2$. Complete the proof by mathematical induction.]

SUMMARY

- A **real vector space** V is a set of objects, called **vectors,** together with two operations called **addition** (denoted by $\mathbf{x} + \mathbf{y}$) and **scalar multiplication** (denoted by $\alpha \mathbf{x}$) that satisfy the following ten axioms: (pp. 211, 212)

  i. If $\mathbf{x} \in V$ and $\mathbf{y} \in V$, then $\mathbf{x} + \mathbf{y} \in V$ (closure under addition).

  ii. For all $\mathbf{x}$, $\mathbf{y}$, and $\mathbf{z}$ in V, $(\mathbf{x} + \mathbf{y}) + \mathbf{z} = \mathbf{x} + (\mathbf{y} + \mathbf{z})$ (associative law of vector addition).

  iii. There is a vector $\mathbf{0} \in V$ such that for all $\mathbf{x} \in V$, $\mathbf{x} + \mathbf{0} = \mathbf{0} + \mathbf{x} = \mathbf{x}$ ($\mathbf{0}$ is called the additive identity).

  iv. If $\mathbf{x} \in V$, there is a vector $-\mathbf{x}$ in V such that $\mathbf{x} + (-\mathbf{x}) = \mathbf{0}$ ($-\mathbf{x}$ is called the additive inverse of $\mathbf{x}$).

  v. If $\mathbf{x}$ and $\mathbf{y}$ are in V, then $\mathbf{x} + \mathbf{y} = \mathbf{y} + \mathbf{x}$ (commutative law of vector addition).

  vi. If $\mathbf{x} \in V$ and α is a scalar, then $\alpha \mathbf{x} \in V$ (closure under scalar multiplication).

  vii. If $\mathbf{x}$ and $\mathbf{y}$ are in V and α is a scalar, then $\alpha(\mathbf{x} + \mathbf{y}) = \alpha \mathbf{x} + \alpha \mathbf{y}$ (first distributive law).

  viii. If $\mathbf{x} \in V$ and α and β are scalars, then $(\alpha + \beta)\mathbf{x} = \alpha \mathbf{x} + \beta \mathbf{x}$ (second distributive law).

  ix. If $\mathbf{x} \in V$ and α and β are scalars, then $\alpha(\beta \mathbf{x}) = \alpha \beta \mathbf{x}$ (associative law of scalar multiplication).

  x. For every vector $\mathbf{x} \in V$, $1\mathbf{x} = \mathbf{x}$ (the scalar 1 is called a multiplicative identity).

- **The space** $\mathbb{R}^n = \{(x_1, x_2, \ldots, x_n): x_i \in \mathbb{R} \text{ for } i = 1, 2, \ldots, n\}$.   (p. 212)
- **The space** $P_n = \{\text{polynomials of degree less than or equal to } n\}$.   (p. 215)
- **The space** $C[a, b] = \{\text{real-valued functions that are continuous on the interval } [a, b]\}$.   (p. 215)
- **The space** $M_{mn} = \{m \times n \text{ matrices with real coefficients}\}$.   (p. 215)
- **The space** $\mathbb{C}_n = \{(c_1, c_2, \ldots, c_n): c_i \in \mathbb{C} \text{ for } i = 1, 2, \ldots, n\}$. $\mathbb{C}$ denotes the set of complex numbers.   (p. 216)
- A **subspace** H of a vector space V is a subset of V that is itself a vector space.   (p. 219)
- A nonempty subset H of a vector space V is a subspace of V if the following two rules hold: (p. 219)

  i.   If $\mathbf{x} \in H$ and $\mathbf{y} \in H$, then $\mathbf{x} + \mathbf{y} \in H$.

  ii.   If $\mathbf{x} \in H$, then $\alpha\mathbf{x} \in H$ for every scalar α.

- A **proper subspace** of a vector space V is a subspace of V other than $\{\mathbf{0}\}$ or V.   (p. 220)
- A **linear combination** of the vectors $\mathbf{v}_1, \mathbf{v}_2, \ldots, \mathbf{v}_n$ in a vector space V is a sum of the form   (p. 224)

$$\alpha_1\mathbf{v}_1 + \alpha_2\mathbf{v}_2 + \cdots + \alpha_n\mathbf{v}_n$$

where $\alpha_1, \alpha_2, \ldots, \alpha_n$ are scalars.
- The vectors $\mathbf{v}_1, \mathbf{v}_2, \ldots, \mathbf{v}_n$ in a vector space V are said to **span** V if every vector in V can be written as a linear combination of $\mathbf{v}_1, \mathbf{v}_2, \ldots, \mathbf{v}_n$.   (p. 225)
- The **span of a set of vectors** $\mathbf{v}_1, \mathbf{v}_2, \ldots, \mathbf{v}_k$ in a vector space V is the set of linear combinations of $\mathbf{v}_1, \mathbf{v}_2, \ldots, \mathbf{v}_k$.   (p. 226)
- span $\{\mathbf{v}_1, \mathbf{v}_2, \ldots, \mathbf{v}_k\}$ is a subspace of V.   (p. 226)
- **Linear Dependence and Independence·**The vectors $\mathbf{v}_1, \mathbf{v}_2, \ldots, \mathbf{v}_n$ in a vector space V are said to be **linearly dependent** if there exist scalars $c_1, c_2, \ldots, c_n$ not all zero such that   (p. 230)

$$c_1\mathbf{v}_1 + c_2\mathbf{v}_2 + \cdots + c_n\mathbf{v}_n = \mathbf{0}$$

If the vectors are not linearly dependent, they are said to be **linearly independent**.
- Two vectors in a vector space V are linearly dependent if and only if one is a scalar multiple of the other.   (p. 231)
- Any set of n linearly independent vectors in $\mathbb{R}^n$ spans $\mathbb{R}^n$.   (p. 237)
- A set of n vectors in $\mathbb{R}^m$ is linearly dependent if $n > m$.   (p. 234)
- **Basis·**A set of vectors $\mathbf{v}_1, \mathbf{v}_2, \ldots, \mathbf{v}_n$ is a **basis** for a vector space V if   (p. 244)

  i. $\{\mathbf{v}_1, \mathbf{v}_2, \ldots, \mathbf{v}_n\}$ is linearly independent

  ii. $\{\mathbf{v}_1, \mathbf{v}_2, \ldots, \mathbf{v}_n\}$ spans V

- Every set of n linearly independent vectors in $\mathbb{R}^n$ is a basis in $\mathbb{R}^n$ (p. 244)
- The **standard basis** in $\mathbb{R}^n$ consists of the n vectors   (p. 244)

$$\mathbf{e}_1 = \begin{pmatrix} 1 \\ 0 \\ 0 \\ \vdots \\ 0 \end{pmatrix}, \mathbf{e}_2 = \begin{pmatrix} 0 \\ 1 \\ 0 \\ \vdots \\ 0 \end{pmatrix}, \mathbf{e}_3 = \begin{pmatrix} 0 \\ 0 \\ 1 \\ \vdots \\ 0 \end{pmatrix}, \ldots, \mathbf{e}_n = \begin{pmatrix} 0 \\ 0 \\ 0 \\ \vdots \\ 1 \end{pmatrix}$$

- **Dimension·**If the vector space V has a finite basis, then the **dimension** of V is the number of vectors in every basis and V is called a **finite dimensional vector space**. Otherwise V is called an **infinite dimensional vector space**. If $V = \{\mathbf{0}\}$, then V is said to be **zero dimensional**.   (p. 247)

We write the dimension of V as dim V.

- If H is a subspace of the finite dimensional space V, then dim $H \le$ dim V (p. 248)
- The only proper subspaces of $\mathbb{R}^3$ are sets of vectors lying on a single line or a single plane passing through the origin.   (p. 249)
- The **kernel** of an $n \times n$ matrix A is the subspace of $\mathbb{R}^n$ given by   (p. 253)

$$N_A = \{\mathbf{x} \in \mathbb{R}^n: A\mathbf{x} = \mathbf{0}\}$$

- The **nullity** of an $n \times n$ matrix A is the dimension of N_A and is denoted by $\nu(A)$.   (p. 253)
- Let A be an $m \times n$ matrix. The **range of A,** denoted by Range A, is the subspace of $\mathbb{R}^m$ given by   (p. 254)

$$\text{Range } A = \{\mathbf{y} \in \mathbb{R}^m: A\mathbf{x} = \mathbf{y} \text{ for some } \mathbf{x} \in \mathbb{R}^n\}$$

- The **rank of A,** denoted by $\rho(A)$, is the dimension of Range A.   (p. 254)
- The **row space of A,** denoted by R_A, is the span of the rows of A and is a subspace of $\mathbb{R}^n$.   (p. 255)
- The **column space of A,** denoted by C_A, is the span of the columns of A and is a subspace of $\mathbb{R}^m$ (p. 255)
- If A is an $m \times n$ matrix, then

$$C_A = \text{Range } A \quad \text{and} \quad \dim R_A = \dim C_A = \dim \text{Range } A = \rho(A)$$ (pp. 257, 259)

Moreover,

$$\rho(A) + \nu(A) = n$$

- The system $A\mathbf{x} = \mathbf{b}$ has at least one solution if and only if $\rho(A) = \rho(A, \mathbf{b})$ where $(A, \mathbf{b})$ is the augmented matrix obtained by adjoining the column vector $\mathbf{b}$ to A.   (p. 261)
- **Summing Up Theorem·**Let A be an $n \times n$ matrix. Then the following are equivalent: (p. 263)

   i. A is invertible.

   ii. The only solution to the homogeneous system $A\mathbf{x} = \mathbf{0}$ is the trivial solution ($\mathbf{x} = \mathbf{0}$).

   iii. The system $A\mathbf{x} = \mathbf{b}$ has a unique solution for every n-vector $\mathbf{b}$.

   iv. A is row equivalent to the $n \times n$ identity matrix I_n.

   v. A can be written as the product of elementary matrices.

   vi. det $A \ne 0$.

   vii. The columns (and rows) of A are linearly independent.

   viii. $\nu(A) = 0$

   ix. $\rho(A) = n$.

- Let A be an $m \times n$ matrix.   (p. 267)
   A **k-square submatrix** of A is the $k \times k$ matrix obtained as the intersection of k rows and k columns of A.
   A **subdeterminant of order k** is the determinant of a k-square submatrix of A.
- Let A be an $m \times n$ matrix. Then $\rho(A) = r > 0$ if and only if A has at least one nonzero subdeterminant of order r and every subdeterminant of order $r + 1$ is zero.   (p. 268)
- Let $B_1 = \{\mathbf{u}_1, \mathbf{u}_2, \dots, \mathbf{u}_n\}$ and $B_2 = \{\mathbf{v}_1, \mathbf{v}_2, \dots, \mathbf{u}_n\}$ be two bases for the vector space V. If $\mathbf{x} \in V$ and   (pp. 271–273)

$$\mathbf{x} = b_1\mathbf{u}_1 + b_2\mathbf{u}_2 + \cdots + b_n\mathbf{u}_n = c_1\mathbf{v}_1 + c_2\mathbf{v}_2 + c_n\mathbf{v}_n$$

then we write $(\mathbf{x})_{B_1} = \begin{pmatrix} b_1 \\ b_2 \\ \vdots \\ b_n \end{pmatrix}$ and $(\mathbf{x})_{B_2} = \begin{pmatrix} c_1 \\ c_2 \\ \vdots \\ c_n \end{pmatrix}$.

Suppose that $(\mathbf{u}_j)_{B_2} = \begin{pmatrix} a_{1j} \\ a_{2j} \\ \vdots \\ a_{nj} \end{pmatrix}$. Then the **transition matrix** from B_1 to B_2 is the $n \times n$ matrix   (p. 273)

$$A = \begin{pmatrix} a_{11} & a_{12} & \cdots & a_{1n} \\ a_{21} & a_{22} & \cdots & a_{2n} \\ \vdots & \vdots & & \vdots \\ a_{n1} & a_{n2} & \cdots & a_{nn} \end{pmatrix}$$

Moreover, $(\mathbf{x})_{B_2} = A(\mathbf{x})_{B_1}$.

- If A is the transition matrix from B_1 to B_2, then A^{-1} is the transition matrix from B_2 to B_1.   (p. 274)

- If $(\mathbf{x}_j)_{B_1} = \begin{pmatrix} a_{1j} \\ a_{2j} \\ \vdots \\ a_{nj} \end{pmatrix}$ for $j = 1, 2, \ldots, n$, then $\mathbf{x}_1, \mathbf{x}_2, \ldots, \mathbf{x}_n$ are linearly independent if and

only if det $A \neq 0$, where   (p. 278)

$$A = \begin{pmatrix} a_{11} & a_{12} & \cdots & a_{1n} \\ a_{21} & a_{22} & \cdots & a_{2n} \\ \vdots & \vdots & & \vdots \\ a_{n1} & a_{n2} & \cdots & a_{nn} \end{pmatrix}$$

- The vectors $\mathbf{u}_1, \mathbf{u}_2, \ldots, \mathbf{u}_k$ in $\mathbb{R}^n$ form an **orthogonal set** if $\mathbf{u}_i \cdot \mathbf{u}_j = 0$ for $i \neq j$. If, in addition, $\mathbf{u}_i \cdot \mathbf{u}_i = 1$ for $i = 1, 2, \ldots, k$, the set is said to be **orthonormal**.   (p. 281)
- $|\mathbf{v}| = |\mathbf{v} \cdot \mathbf{v}|^{1/2}$ is called the **length** or **norm** of $\mathbf{v}$.   (p. 282)
- Every subspace of $\mathbb{R}^n$ has an orthonormal basis. **The Gram-Schmidt orthonormalization process** can be used to construct such a basis.   (p. 283)
- An **orthogonal matrix** is an invertible $n \times n$ matrix Q such that $Q^{-1} = Q^t$.   (p. 286)
- An $n \times n$ matrix is orthogonal if and only if its columns form an orthonormal basis for $\mathbb{R}^n$.   (p. 287)
- Let H be a subspace of $\mathbb{R}^n$ with orthonormal basis $\{\mathbf{u}_1, \mathbf{u}_2, \ldots, \mathbf{u}_k\}$. If $\mathbf{v} \in \mathbb{R}^n$, then the **orthogonal projection** of $\mathbf{v}$ onto H, denoted by $\text{proj}_H \mathbf{v}$, is given by   (p. 288)

$$\text{proj}_H \mathbf{v} = (\mathbf{v} \cdot \mathbf{u}_1)\mathbf{u}_1 + (\mathbf{v} \cdot \mathbf{u}_2)\mathbf{u}_2 + \cdots + (\mathbf{v} \cdot \mathbf{u}_k)\mathbf{u}_k$$

- Let H be a subspace of $\mathbb{R}^n$. Then the **orthogonal complement** of H, denoted by $H^\perp$, is given by   (p. 290)

$$H^\perp = \{\mathbf{x} \in \mathbb{R}^n: \mathbf{x} \cdot \mathbf{h} = 0 \quad \text{for every } \mathbf{h} \in H\}$$

- **Projection Theorem** · Let H be a subspace of $\mathbb{R}^n$ and let $\mathbf{v} \in \mathbb{R}^n$. Then there exists a unique pair of vectors $\mathbf{h}$ and $\mathbf{p}$ such that $\mathbf{h} \in H$, $\mathbf{p} \in H^\perp$, and   (p. 291)

$$\mathbf{v} = \mathbf{h} + \mathbf{p} = \text{proj}_H \mathbf{v} + \text{proj}_{H^\perp} \mathbf{v}$$

- **Norm Approximation Theorem** · Let H be a subspace of $\mathbb{R}^n$ and let $\mathbf{v}$ be a vector in $\mathbb{R}^n$. Then, in H, $\text{proj}_H \mathbf{v}$ is the best approximation to $\mathbf{v}$ in the following sense: If $\mathbf{h}$ is any other vector in H, then   (p. 292)

$$|\mathbf{v} - \text{proj}_H \mathbf{v}| < |\mathbf{v} - \mathbf{h}|$$

- Let $(x_1, y_1), (x_2, y_2), \ldots, (x_n, y_n)$ be a set of data points. If we wish to represent these data by the straight line $y = mx + b$, then the **least squares problem** is to find the m and b that minimizes the sum of squares   (p. 298)

$$[y_1 - (b + mx_1)]^2 + [y_2 - (b + mx_2)]^2 + \cdots + [y_n - (b + mx_n)]^2$$

The solution to this problem is to set   (p. 300)

$$\begin{pmatrix} b \\ m \end{pmatrix} = \mathbf{u} = (A^t A)^{-1} A^t \mathbf{y}$$

$$\text{where } \mathbf{y} = \begin{pmatrix} y_1 \\ y_2 \\ \vdots \\ y_n \end{pmatrix} \text{ and } A = \begin{pmatrix} 1 & x_1 \\ 1 & x_2 \\ \vdots & \vdots \\ 1 & x_n \end{pmatrix}.$$

Similar results apply when we attempt to represent the data using a polynomial of degree > 1.

- **Inner Product Space·**The complex vector space V is called an **inner product space** if for every pair of vectors $\mathbf{u}$ and $\mathbf{v}$ in V, there is a unique complex number $(\mathbf{u}, \mathbf{v})$, called the **inner product** of $\mathbf{u}$ and $\mathbf{v}$, such that if $\mathbf{u}$, $\mathbf{v}$, and $\mathbf{w}$ are in V and $\alpha \in \mathbb{C}$, then   (p. 307)

   i. $(\mathbf{v}, \mathbf{v}) \geq 0$

  ii. $(\mathbf{v}, \mathbf{v}) = 0$ if and only if $\mathbf{v} = 0$

 iii. $(\mathbf{u}, \mathbf{v} + \mathbf{w}) = (\mathbf{u}, \mathbf{v}) + (\mathbf{u}, \mathbf{w})$

 iv. $(\mathbf{u} + \mathbf{v}, \mathbf{w}) = (\mathbf{u}, \mathbf{w}) + (\mathbf{v}, \mathbf{w})$

  v. $(\mathbf{u}, \mathbf{v}) = \overline{(\mathbf{v}, \mathbf{u})}$

 vi. $(\alpha\mathbf{u}, \mathbf{v}) = \alpha(\mathbf{u}, \mathbf{v})$

vii. $(\mathbf{u}, \alpha\mathbf{v}) = \overline{\alpha}(\mathbf{u}, \mathbf{v})$

- **Inner Product in $\mathbb{C}^n$**   (p. 308)

$$(\mathbf{x}, \mathbf{y}) = x_1\overline{y}_1 + x_2\overline{y}_2 + \cdots + x_n\overline{y}_n$$

- Let V be an inner product space and suppose that $\mathbf{u}$ and $\mathbf{v}$ are in $\mathbf{V}$. Then   (p. 309)

$$\mathbf{u} \text{ and } \mathbf{v} \text{ are } \textbf{orthogonal} \text{ if } (\mathbf{u}, \mathbf{v}) = 0$$

The **norm** of $\mathbf{u}$, denoted by $|\mathbf{u}|$, is given by

$$|\mathbf{u}| = \sqrt{(u, u)}$$

- **Orthonormal Set·**The set of vectors $\{\mathbf{v}_1, \mathbf{v}_2, \ldots, \mathbf{v}_n\}$ is an **orthonormal set** in V if   (p. 310)

$$(\mathbf{v}_i, \mathbf{v}_j) = 0 \qquad \text{for } i \neq j$$

and

$$|\mathbf{v}_i| = \sqrt{(\mathbf{v}_i, \mathbf{v}_i)} = 1$$

If only the first condition holds, then the set is said to be **orthogonal**.

- **Orthogonal Projection·**Let H be a subspace of an inner product space V with an orthonormal basis $\{\mathbf{u}_1, \mathbf{u}_2, \ldots, \mathbf{u}_k\}$. If $\mathbf{v} \in V$, then the **orthogonal projection** of $\mathbf{v}$ onto H, denoted by $\text{proj}_H \mathbf{v}$, is given by   (pp. 312, 313)

$$\text{proj}_H \mathbf{v} = (\mathbf{v}, \mathbf{u}_1)\mathbf{u}_1 + (\mathbf{v}, \mathbf{u}_2)\mathbf{u}_2 + \cdots + (\mathbf{v}, \mathbf{u}_k)\mathbf{u}_k$$

- **Orthogonal Complement**·Let H be a subspace of the inner product space V. Then the **orthogonal complement** of H, denoted by $H^\perp$, is given by

$$H^\perp = \{x \in V: (x, h) = 0 \quad \text{for every } h \in H\}$$

(p. 312)

- If H is a subspace of the inner product space V, then

  i. $H^\perp$ is a subspace of **v**.

  ii. $H \cap H^\perp = \{0\}$.

  iii. $\dim H^\perp = n - \dim H$ if $\dim V = n < \infty$.

(p. 312)

- **Projection Theorem**·Let H be a finite dimensional subspace of the inner product space V and suppose that $v \in V$. Then there exists a unique pair of vectors h and p such that $h \in H, p \in H^\perp$, and

$$v = h + p$$

(p. 312)

where $h = \text{proj}_H v$.

If V is finite dimensional, then $p = \text{proj}_{H^\perp} v$.

- **Norm Approximation Theorem**·Let H be a finite dimensional subspace of the inner product space V and let v be a vector in V. Then, in H, $\text{proj}_H v$ is the best approximation to v in the following sense: If h is any other vector in H, then

$$|v - \text{proj}_H v| < |v - h|$$

(p. 313)

REVIEW EXERCISES

In Exercises 1–10 determine whether the given set is a vector space. If so, determine its dimension. If it is finite dimensional, find a basis for it.

1. The vectors (x, y, z) in $\mathbb{R}^3$ satisfying $x + 2y - z = 0$

2. The vectors (x, y, z) in $\mathbb{R}^3$ satisfying $x + 2y - z \le 0$

3. The vectors (x, y, z, w) in $\mathbb{R}^4$ satisfying $x + y + z + w = 0$

4. The vectors in $\mathbb{R}^3$ satisfying $x - 2 = y + 3 = z - 4$

5. The set of upper triangular $n \times n$ matrices under the operations of matrix addition and scalar multiplication

6. Th set of polynomials of degree ≤ 5

7. The set of polynomials of degree 5

8. The set of 3×2 matrices $A = (a_{ij})$, with $a_{12} = 0$, under the operations of matrix addition and scalar multiplication

9. The set in Exercise 8 except that $a_{12} = 1$

10. The set $S = \{f \in C[0, 2]: f(2) = 0\}$

In Exercises 11–19 determine whether the given set of vectors is linearly dependent or independent.

11. $\begin{pmatrix} 2 \\ 3 \end{pmatrix}; \begin{pmatrix} 4 \\ -6 \end{pmatrix}$

12. $\begin{pmatrix} 2 \\ 3 \end{pmatrix}; \begin{pmatrix} 4 \\ 6 \end{pmatrix}$

13. $\begin{pmatrix} 1 \\ -1 \\ 2 \end{pmatrix}; \begin{pmatrix} 3 \\ 0 \\ 1 \end{pmatrix}; \begin{pmatrix} 0 \\ 0 \\ 0 \end{pmatrix}$

14. $\begin{pmatrix} 1 \\ -4 \\ 2 \end{pmatrix}; \begin{pmatrix} 0 \\ 2 \\ -1 \end{pmatrix}; \begin{pmatrix} 2 \\ -10 \\ 5 \end{pmatrix}$

15. $\begin{pmatrix} 1 \\ 0 \\ 0 \\ 0 \end{pmatrix}; \begin{pmatrix} 0 \\ 1 \\ 0 \\ 0 \end{pmatrix}; \begin{pmatrix} 0 \\ 0 \\ 1 \\ 0 \end{pmatrix}; \begin{pmatrix} 0 \\ 0 \\ 0 \\ 1 \end{pmatrix}$

16. In P_3: $1, 2 - x^2, 3 - x, 7x^2 - 8x$

17. In P_3: $1, 2 + x^3, 3 - x, 7x^2 - 8x$

18. In M_{22}: $\begin{pmatrix} 1 & -1 \\ 0 & 0 \end{pmatrix}, \begin{pmatrix} 1 & 1 \\ 0 & 0 \end{pmatrix}, \begin{pmatrix} 0 & 0 \\ 1 & 1 \end{pmatrix}, \begin{pmatrix} 0 & 0 \\ 1 & -1 \end{pmatrix}$

19. In M_{22}: $\begin{pmatrix} 1 & 1 \\ 0 & 0 \end{pmatrix}, \begin{pmatrix} 1 & -1 \\ 0 & 0 \end{pmatrix}, \begin{pmatrix} 0 & 0 \\ 1 & 1 \end{pmatrix}, \begin{pmatrix} 0 & 0 \\ 1 & -1 \end{pmatrix}$

20. Using determinants, determine whether each set of vectors is linearly dependent or independent.

a. $\begin{pmatrix} 1 \\ 5 \\ 2 \end{pmatrix}; \begin{pmatrix} 3 \\ 0 \\ 4 \end{pmatrix}; \begin{pmatrix} -5 \\ 5 \\ 6 \end{pmatrix}$ **b.** $(2, 1, 4); (3, -2, 6); (-1, -4, -2)$

In Exercises 21–26 find a basis for the given vector space and determine its dimension.

21. The vectors in $\mathbb{R}^3$ lying on the plane $2x + 3y - 4z = 0$

22. $H = \{(x, y): 2x - 3y = 0\}$

23. $\{v \in \mathbb{R}^4: 3x - y - z + w = 0\}$

24. $\{p \in P_3: p(0) = 0\}$

25. The set of diagonal 4×4 matrices

26. M_{32}

In Exercises 27–32 find the kernel, range, nullity, and rank of the given matrix.

27. $A = \begin{pmatrix} 1 & -2 \\ -2 & 4 \end{pmatrix}$

28. $A = \begin{pmatrix} 1 & -1 & 3 \\ 2 & 0 & 4 \\ 0 & -2 & 2 \end{pmatrix}$

29. $A = \begin{pmatrix} 1 & -1 & 2 \\ 0 & 1 & 4 \\ 1 & -1 & 0 \end{pmatrix}$

30. $A = \begin{pmatrix} 2 & 4 & -2 \\ -1 & -2 & 1 \end{pmatrix}$

31. $A = \begin{pmatrix} 2 & 3 \\ -1 & 2 \\ 4 & 6 \end{pmatrix}$

32. $A = \begin{pmatrix} 1 & -1 & 2 & 3 \\ 0 & 1 & -1 & 0 \\ 1 & -2 & 3 & 3 \\ 2 & -3 & 5 & 6 \end{pmatrix}$

In Exercises 33–36 write the given vector in terms of the new given basis vectors.

33. In $\mathbb{R}^2$: $\mathbf{x} = \begin{pmatrix} 2 \\ -1 \end{pmatrix}; \begin{pmatrix} 1 \\ 2 \end{pmatrix}, \begin{pmatrix} -1 \\ 2 \end{pmatrix}$

34. In $\mathbb{R}^3$: $\mathbf{x} = \begin{pmatrix} -3 \\ 4 \\ 2 \end{pmatrix}; \begin{pmatrix} 1 \\ 0 \\ 1 \end{pmatrix}, \begin{pmatrix} 1 \\ 1 \\ 0 \end{pmatrix}, \begin{pmatrix} 0 \\ 2 \\ 3 \end{pmatrix}$

35. In P_2: $\mathbf{x} = 4 + x^2$; $1 + x^2, 1 + x, 1$

36. In M_{22}: $\mathbf{x} = \begin{pmatrix} 3 & 1 \\ 0 & 1 \end{pmatrix}; \begin{pmatrix} 1 & 1 \\ 0 & 0 \end{pmatrix}, \begin{pmatrix} 1 & -1 \\ 0 & 0 \end{pmatrix}, \begin{pmatrix} 0 & 0 \\ 1 & 1 \end{pmatrix}, \begin{pmatrix} 0 & 0 \\ 1 & -1 \end{pmatrix}$

In Exercises 37 and 38 determine the rank of the given matrix by considering subdeterminants.

37. $\begin{pmatrix} 1 & 3 & -1 & 4 \\ 2 & 1 & 6 & 2 \\ 1 & -2 & 7 & -2 \end{pmatrix}$

38. $\begin{pmatrix} 2 & 3 & -1 \\ 4 & 1 & 6 \\ 0 & -5 & 8 \\ -2 & -3 & 3 \end{pmatrix}$

In Exercises 39–42 find an orthonormal basis for the given vector space.

39. $\mathbb{R}^2$ starting with the basis $\begin{pmatrix} 2 \\ 3 \end{pmatrix}, \begin{pmatrix} -1 \\ 4 \end{pmatrix}$

40. $\{(x, y, z) \in \mathbb{R}^3: x - y - z = 0\}$ **41.** $\{(x, y, z) \in \mathbb{R}^3: x = y = z\}$

42. $\{(x, y, z, w) \in \mathbb{R}^4: x = z \text{ and } y = w\}$

In Exercises 43–45: **(a)** compute $\text{proj}_H \mathbf{v}$; **(b)** find an orthonormal basis for $H^{\perp}$; **(c)** write $\mathbf{v}$ as $\mathbf{h} + \mathbf{p}$, where $\mathbf{h} \in H$ and $\mathbf{p} \in H^{\perp}$.

43. H is the subspace of Problem 40; $\mathbf{v} = \begin{pmatrix} -1 \\ 2 \\ 4 \end{pmatrix}$.

44. H is the subspace of Problem 41; $\mathbf{v} = \begin{pmatrix} 1 \\ 0 \\ -1 \end{pmatrix}$.

45. H is the subspace of Problem 42; $\mathbf{v} = \begin{pmatrix} 1 \\ 0 \\ 0 \\ 1 \end{pmatrix}$.

[Calculus] **46.** Find an orthonormal basis for $P_2[0, 2]$.

[Calculus] **47.** Use the result of Exercise 46 to find the polynomial that gives the best mean square approximation to e^x on the interval $[0, 2]$.

48. Find the best straight-line fit to the points $(2, 5)$, $(-1, -3)$, $(1, 0)$

49. Find the best quadratic fit to the points in Exercise 48.

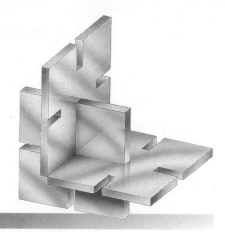

5

Linear Transformations

5.1 DEFINITION AND EXAMPLES

In this chapter we discuss a special class of functions, called *linear transformations,* which occur with great frequency in linear algebra and other branches of mathematics. They are also important in a wide variety of applications. Before defining a linear transformation, let us study two simple examples to see what can happen.

EXAMPLE 1   **Reflection About the *x*-Axis**   In $\mathbb{R}^2$ define a function T by the formula $T\begin{pmatrix} x \\ y \end{pmatrix} = \begin{pmatrix} x \\ -y \end{pmatrix}$. Geometrically, T takes a vector in $\mathbb{R}^2$ and reflects it about the *x*-axis. This is illustrated in Figure 5.1. Once we have given our basic definition, we shall see that T is a linear transformation from $\mathbb{R}^2$ into $\mathbb{R}^2$.

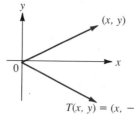

Figure 5.1   The vector $(x, -y)$ is the reflection about the *x*-axis of the vector (x, y)

EXAMPLE 2   **Transforming a Production Vector into a Raw Material Vector**   A manufacturer makes four different products, each of which requires three raw materials. We denote the four products by P_1, P_2, P_3, and P_4 and denote the raw materials by R_1, R_2, and R_3. The accompanying table gives the number of units of each raw material required to manufacture 1 unit of each product.

		Needed to Produce 1 Unit of			
		P_1	P_2	P_3	P_4
Number of	R_1	2	1	3	4
Units of					
Raw	R_2	4	2	2	1
Material					
	R_3	3	3	1	2

A natural question arises: If certain numbers of the four products are produced, how many units of each raw material are needed? We let p_1, p_2, p_3, and p_4 denote the number of items of the four products manufactured and let r_1, r_2, and r_3 denote the number of units of the three raw materials needed. Then we define

$$\mathbf{p} = \begin{pmatrix} p_1 \\ p_2 \\ p_3 \\ p_4 \end{pmatrix} \qquad \mathbf{r} = \begin{pmatrix} r_1 \\ r_2 \\ r_3 \end{pmatrix} \qquad A = \begin{pmatrix} 2 & 1 & 3 & 4 \\ 4 & 2 & 2 & 1 \\ 3 & 3 & 1 & 2 \end{pmatrix}$$

For example, suppose that $\mathbf{p} = \begin{pmatrix} 10 \\ 30 \\ 20 \\ 50 \end{pmatrix}$. How many units of R_1 are needed to produce these numbers of units of the four products? From the table, we find that

$$r_1 = p_1 \cdot 2 + p_2 \cdot 1 + p_3 \cdot 3 + p_4 \cdot 4$$
$$= 10 \cdot 2 + 30 \cdot 1 + 20 \cdot 3 + 50 \cdot 4 = 310 \text{ units}$$

Similarly,

$$r_2 = 10 \cdot 4 + 30 \cdot 2 + 20 \cdot 2 + 50 \cdot 1 = 190 \text{ units}$$

and

$$r_3 = 10 \cdot 3 + 30 \cdot 3 + 20 \cdot 1 + 50 \cdot 2 = 240 \text{ units}$$

In general, we see that

$$\begin{pmatrix} 2 & 1 & 3 & 4 \\ 4 & 2 & 2 & 1 \\ 3 & 3 & 1 & 2 \end{pmatrix} \begin{pmatrix} p_1 \\ p_2 \\ p_3 \\ p_4 \end{pmatrix} = \begin{pmatrix} r_1 \\ r_2 \\ r_3 \end{pmatrix}$$

or

$$\mathbf{r} = A\mathbf{p}$$

We can look at this in another way. If $\mathbf{p}$ is called the **production vector** and $\mathbf{r}$ the **raw material vector,** we define the function T by $\mathbf{r} = T\mathbf{p} = A\mathbf{p}$. That is, T is the function that "transforms" the production vector into the raw material vector. It is defined by ordinary matrix multiplication. As we shall see, this function is also a linear transformation.   ■

Before defining a linear transformation, let us say a bit about functions. In Section 1.8 we wrote a system of equations as

$$A\mathbf{x} = \mathbf{b}$$

where A is an $m \times n$ matrix, $\mathbf{x} \in \mathbb{R}^n$ and $\mathbf{b} \in \mathbb{R}^m$. We were asked to find $\mathbf{x}$ when A and $\mathbf{b}$ were known. However, we can look at this equation in another way: Suppose A is given. Then the equation $A\mathbf{x} = \mathbf{b}$ "says": Give me an $\mathbf{x}$ in $\mathbb{R}^n$ and I'll give you a $\mathbf{b}$ in $\mathbb{R}^m$; that is, A represents a *function* with domain $\mathbb{R}^n$ and range in $\mathbb{R}^m$.

The function defined above has the property that $A(\alpha\mathbf{x}) = \alpha A\mathbf{x}$ if α is a scalar and $A(\mathbf{x} + \mathbf{y}) = A\mathbf{x} + A\mathbf{y}$. This property characterizes linear transformations.

DEFINITION 1   **Linear Transformation** Let V and W be vector spaces. A **linear transformation** T from V into W is a function that assigns to each vector $\mathbf{v} \in V$ a unique vector $T\mathbf{v} \in W$ and that satisfies, for each $\mathbf{u}$ and $\mathbf{v}$ in V and each scalar α,

and

$$T(\mathbf{u} + \mathbf{v}) = T\mathbf{u} + T\mathbf{v} \tag{1}$$

$$T(\alpha\mathbf{v}) = \alpha T\mathbf{v} \tag{2}$$

Notation. We write $T: V \to W$ to indicate that T takes the vector space V into the vector space W.

Terminology. Linear transformations are often called **linear operators.**

EXAMPLE 3   **A Linear Transformation from $\mathbb{R}^2$ to $\mathbb{R}^3$**   Let $T: \mathbb{R}^2 \to \mathbb{R}^3$ be defined by

$$T\begin{pmatrix} x \\ y \end{pmatrix} = \begin{pmatrix} x + y \\ x - y \\ 3y \end{pmatrix}. \text{ For example, } T\begin{pmatrix} 2 \\ -3 \end{pmatrix} = \begin{pmatrix} -1 \\ 5 \\ -9 \end{pmatrix}. \text{ Then}$$

$$T\left[\begin{pmatrix} x_1 \\ y_1 \end{pmatrix} + \begin{pmatrix} x_2 \\ y_2 \end{pmatrix}\right] = T\begin{pmatrix} x_1 + x_2 \\ y_1 + y_2 \end{pmatrix} = \begin{pmatrix} x_1 + x_2 + y_1 + y_2 \\ x_1 + x_2 - y_1 - y_2 \\ 3y_1 + 3y_2 \end{pmatrix}$$

$$= \begin{pmatrix} x_1 + y_1 \\ x_1 - y_1 \\ 3y_1 \end{pmatrix} + \begin{pmatrix} x_2 + y_2 \\ x_2 - y_2 \\ 3y_2 \end{pmatrix}$$

But

$$\begin{pmatrix} x_1 + y_1 \\ x_1 - y_1 \\ 3y_1 \end{pmatrix} = T\begin{pmatrix} x_1 \\ y_1 \end{pmatrix} \quad \text{and} \quad \begin{pmatrix} x_2 + y_2 \\ x_2 - y_2 \\ 3y_2 \end{pmatrix} = T\begin{pmatrix} x_2 \\ y_2 \end{pmatrix}$$

Thus

$$T\left[\begin{pmatrix} x_1 \\ y_1 \end{pmatrix} + \begin{pmatrix} x_2 \\ y_2 \end{pmatrix}\right] = T\begin{pmatrix} x_1 \\ y_1 \end{pmatrix} + T\begin{pmatrix} x_2 \\ y_2 \end{pmatrix}$$

Similarly,

$$T\left[\alpha\begin{pmatrix} x \\ y \end{pmatrix}\right] = T\begin{pmatrix} \alpha x \\ \alpha y \end{pmatrix} = \begin{pmatrix} \alpha x + \alpha y \\ \alpha x - \alpha y \\ 3\alpha y \end{pmatrix} = \alpha\begin{pmatrix} x + y \\ x - y \\ 3y \end{pmatrix} = \alpha T\begin{pmatrix} x \\ y \end{pmatrix}$$

Thus T is a linear transformation. ■

EXAMPLE 4   **The Zero Transformation**   Let V and W be vector spaces and define T: $V \to W$ by $T\mathbf{v} = \mathbf{0}$ for every $\mathbf{v}$ in V. Then $T(\mathbf{v}_1 + \mathbf{v}_2) = \mathbf{0} = \mathbf{0} + \mathbf{0} = T\mathbf{v}_1 + T\mathbf{v}_2$ and $T(\alpha\mathbf{v}) = \mathbf{0} = \alpha\mathbf{0} = \alpha T\mathbf{v}$. Here T is called the **zero transformation.** ■

EXAMPLE 5   **The Identity Transformation**   Let V be a vector space and define I: $V \to V$ by $I\mathbf{v} = \mathbf{v}$ for every $\mathbf{v}$ in V. Here I is obviously a linear transformation. It is called the **identity transformation** or **identity operator.** ■

EXAMPLE 6   **A Reflection Transformation**   Let T: $\mathbb{R}^2 \to \mathbb{R}^2$ be defined by $T\begin{pmatrix} x \\ y \end{pmatrix} = \begin{pmatrix} -x \\ y \end{pmatrix}$.
It is easy to verify that T is linear. Geometrically, T takes a vector in $\mathbb{R}^2$ and reflects it about the y-axis (see Figure 5.2).

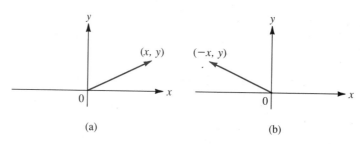

(a)   (b)

Figure 5.2   The vector $(-x, y)$ is the reflection about the y-axis of the vector (x, y)

■

EXAMPLE 7   **A Transformation from $\mathbb{R}^n \to \mathbb{R}^m$ Given by Multiplication by an $m \times n$ Matrix**
Let A be an $m \times n$ matrix and define T: $\mathbb{R}^n \to \mathbb{R}^m$ by $T\mathbf{x} = A\mathbf{x}$. Since $A(\mathbf{x} + \mathbf{y}) =$

$A\mathbf{x} + A\mathbf{y}$ and $A(\alpha\mathbf{x}) = \alpha A\mathbf{x}$ if $\mathbf{x}$ and $\mathbf{y}$ are in $\mathbb{R}^n$, we see that T is a linear transformation. Thus: *Every $m \times n$ matrix A gives rise to a linear transformation from $\mathbb{R}^n$ into $\mathbb{R}^m$*. In Section 5.3 we shall see that a certain converse is true: *Every linear transformation between finite dimensional vector spaces can be represented by a matrix.* ∎

EXAMPLE 8   **A Rotation Transformation**   Suppose the vector $\mathbf{v} = \begin{pmatrix} x \\ y \end{pmatrix}$ in the xy-plane is rotated through an angle of θ (measured in degrees or radians) in the counterclockwise direction. Call the new rotated vector $\mathbf{v}' = \begin{pmatrix} x' \\ y' \end{pmatrix}$. Then, as in Figure 5.3, if r

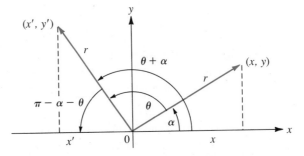

Figure 5.3 (x', y') is obtained by rotating (x, y) through the angle θ

denotes the length of $\mathbf{v}$ (which is unchanged by rotation),

$$x = r \cos \alpha \qquad\qquad y = r \sin \alpha$$
$$x' = r \cos (\theta + \alpha) \qquad y' = r \sin (\theta + \alpha)\dagger$$

But $r \cos (\theta + \alpha) = r \cos \theta \cos \alpha - r \sin \theta \sin \alpha$, so that

$$\boxed{x' = x \cos \theta - y \sin \theta} \tag{3}$$

Similarly, $r \sin (\theta + \alpha) = r \sin \theta \cos \alpha + r \cos \theta \sin \alpha$ or

$$\boxed{y' = x \sin \theta + y \cos \theta} \tag{4}$$

Let

$$A_\theta = \begin{pmatrix} \cos \theta & -\sin \theta \\ \sin \theta & \cos \theta \end{pmatrix} \tag{5}$$

---

† These follow from the standard definitions of $\cos \theta$ and $\sin \theta$ as the x- and y-coordinates of a point on the unit circle. If (x, y) is a point on the circle centered at the origin of radius r, then $x = r \cos \varphi$ and $y = r \sin \varphi$, where φ is the angle the vector (x, y) makes with the positive x-axis.

Then, from (3) and (4), we see that $A_\theta \begin{pmatrix} x \\ y \end{pmatrix} = \begin{pmatrix} x' \\ y' \end{pmatrix}$. The linear transformation $T: \mathbb{R}^2 \rightarrow \mathbb{R}^2$ defined by $T\mathbf{v} = A_\theta \mathbf{v}$, where A_θ is given by (5), is called a **rotation transformation.** ∎

EXAMPLE 9 **An Orthogonal Projection Transformation**   Let H be a subspace of $\mathbb{R}^n$. We define the **orthogonal projection transformation** $P: V \rightarrow H$ by

$$P\mathbf{v} = \text{proj}_H \mathbf{v} \tag{6}$$

Let $\{\mathbf{u}_1, \mathbf{u}_2, \ldots, \mathbf{u}_k\}$ be an orthonormal basis for H. Then from Definition 4.10.4 on page 288 we have

$$P\mathbf{v} = (\mathbf{v} \cdot \mathbf{u}_1)\mathbf{u}_1 + (\mathbf{v} \cdot \mathbf{u}_2)\mathbf{u}_2 + \cdots + (\mathbf{v} \cdot \mathbf{u}_k)\mathbf{u}_k \tag{7}$$

Since $(\mathbf{v}_1 + \mathbf{v}_2) \cdot \mathbf{u} = \mathbf{v}_1 \cdot \mathbf{u} + \mathbf{v}_2 \cdot \mathbf{u}$ and $(\alpha\mathbf{v}) \cdot \mathbf{u} = \alpha(\mathbf{v} \cdot \mathbf{u})$, we see that P is a linear transformation. ∎

EXAMPLE 10 **Two Projection Operators**   Let $T: \mathbb{R}^3 \rightarrow \mathbb{R}^3$ be defined by $T\begin{pmatrix} x \\ y \\ z \end{pmatrix} = \begin{pmatrix} x \\ y \\ 0 \end{pmatrix}$.

Then T is the projection operator taking a vector in space and projecting it into the xy-plane. Similarly, $T\begin{pmatrix} x \\ y \\ z \end{pmatrix} = \begin{pmatrix} x \\ 0 \\ z \end{pmatrix}$ projects a vector in space into the xz-plane. These two transformations are depicted in Figure 5.4.

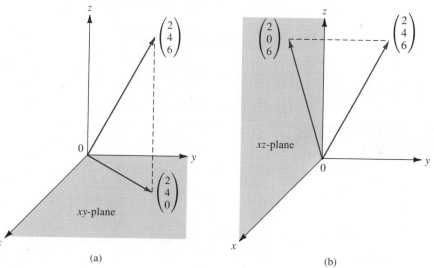

(a)   (b)

Figure 5.4   (a) Projection onto xy-plane: $T\begin{pmatrix} x \\ y \\ z \end{pmatrix} = \begin{pmatrix} x \\ y \\ 0 \end{pmatrix}$. (b) Projection onto xz-plane: $T\begin{pmatrix} x \\ y \\ z \end{pmatrix} = \begin{pmatrix} x \\ 0 \\ z \end{pmatrix}$

∎

EXAMPLE 11   **A Transpose Operator**   Define $T: M_{mn} \to M_{nm}$ by $T(A) = A^t$. Since $(A + B)^t = A^t + B^t$ and $(\alpha A)^t = \alpha A^t$, we see that T, called the **transpose operator,** is a linear transformation. ∎

EXAMPLE 12   **An Integral Operator**   Let $J: C[0, 1] \to \mathbb{R}$ be defined by $Jf = \int_0^1 f(x)\, dx$.
Calculus   Since $\int_0^1 [f(x) + g(x)]\, dx = \int_0^1 f(x)\, dx + \int_0^1 g(x)\, dx$ and $\int_0^1 \alpha f(x)\, dx = \alpha \int_0^1 f(x)\, dx$ if f and g are continuous, we see that J is linear. For example, $J(x^3) = \frac{1}{4}$. J is called an **integral operator.** ∎

EXAMPLE 13   **A Differential Operator**   Let $D: C^1[0, 1] \to C[0, 1]$ be defined by $Df = f'$.
Calculus   Since $(f + g)' = f' + g'$ and $(\alpha f)' = \alpha f'$ if f and g are differentiable, we see that D is linear. D is called a **differential operator.** ∎

WARNING   Not every transformation that looks linear actually is linear. For example, define $T: \mathbb{R} \to \mathbb{R}$ by $Tx = 2x + 3$. Then $\{(x, Tx): x \in \mathbb{R}\}$ is a straight line in the xy-plane. But T is not linear since $T(x + y) = 2(x + y) + 3 = 2x + 2y + 3$ and $Tx + Ty = (2x + 3) + (2y + 3) = 2x + 2y + 6$. The only linear transformations from $\mathbb{R}$ to $\mathbb{R}$ are functions of the form $f(x) = mx$ for some real number m. Thus among all straight-line functions, the only ones that are linear are those that pass through the origin. In algebra and calculus, a **linear function** with domain $\mathbb{R}$ is defined as a function having the form $f(x) = mx + b$. Thus we can say that a linear function is a linear transformation from $\mathbb{R}$ to $\mathbb{R}$ if and only if b (the y-intercept) is zero. □

EXAMPLE 14   **A Transformation That Is Not Linear**   Let $T: C[0, 1] \to \mathbb{R}$ be defined by $Tf = f(0) + 1$. Then T is not linear. To see this, we compute

$$T[f + g] = (f + g)(0) + 1 = f(0) + g(0) + 1$$
$$Tf + Tg = [f(0) + 1] + [g(0) + 1] = f(0) + g(0) + 2$$

This provides another example of a transformation that might look linear but in fact is not. ∎

PROBLEMS 5.1

In Problems 1–29 determine whether the given transformation from V to W is linear.

1. $T: \mathbb{R}^2 \to \mathbb{R}^2$; $T\begin{pmatrix} x \\ y \end{pmatrix} = \begin{pmatrix} x \\ 0 \end{pmatrix}$

2. $T: \mathbb{R}^2 \to \mathbb{R}^2$; $T\begin{pmatrix} x \\ y \end{pmatrix} = \begin{pmatrix} 1 \\ y \end{pmatrix}$

3. $T: \mathbb{R}^3 \to \mathbb{R}^2$; $T\begin{pmatrix} x \\ y \\ z \end{pmatrix} = \begin{pmatrix} x \\ y \end{pmatrix}$

4. $T: \mathbb{R}^3 \to \mathbb{R}^2$; $T\begin{pmatrix} x \\ y \\ z \end{pmatrix} = \begin{pmatrix} 0 \\ y \end{pmatrix}$

5. $T: \mathbb{R}^3 \to \mathbb{R}^2$; $T\begin{pmatrix} x \\ y \\ z \end{pmatrix} = \begin{pmatrix} 1 \\ z \end{pmatrix}$

6. $T: \mathbb{R}^2 \to \mathbb{R}^2$; $T\begin{pmatrix} x \\ y \end{pmatrix} = \begin{pmatrix} x^2 \\ y^2 \end{pmatrix}$

7. $T: \mathbb{R}^2 \to \mathbb{R}^2$; $T\begin{pmatrix} x \\ y \end{pmatrix} = \begin{pmatrix} y \\ x \end{pmatrix}$

8. $T: \mathbb{R}^2 \to \mathbb{R}^2$; $T\begin{pmatrix} x \\ y \end{pmatrix} = \begin{pmatrix} x + y \\ x - y \end{pmatrix}$

9. $T: \mathbb{R}^2 \to \mathbb{R}$; $T\begin{pmatrix} x \\ y \end{pmatrix} = xy$

10. $T: \mathbb{R}^n \to \mathbb{R}$; $T\begin{pmatrix} x_1 \\ x_2 \\ \vdots \\ x_n \end{pmatrix} = x_1 + x_2 + \cdots + x_n$

11. $T: \mathbb{R} \to \mathbb{R}^n$; $T(x) = \begin{pmatrix} x \\ x \\ \vdots \\ x \end{pmatrix}$

12. $T: \mathbb{R}^4 \to \mathbb{R}^2$; $T\begin{pmatrix} x \\ y \\ z \\ w \end{pmatrix} = \begin{pmatrix} x + z \\ y + w \end{pmatrix}$

13. $T: \mathbb{R}^4 \to \mathbb{R}^2$; $T = \begin{pmatrix} x \\ y \\ z \\ w \end{pmatrix} = \begin{pmatrix} xz \\ yw \end{pmatrix}$

14. $T: M_{nn} \to M_{nn}$; $T(A) = AB$, where B is a fixed $n \times n$ matrix

15. $T: M_{nn} \to M_{nn}$; $T(A) = A^t A$

16. $T: M_{mn} \to M_{mp}$; $T(A) = AB$, where B is a fixed $n \times p$ matrix

17. $T: D_n \to D_n$; $T(D) = D^2$ (D_n is the set of $n \times n$ diagonal matrices)

18. $T: D_n \to D_n$; $T(D) = I + D$

19. $T: P_2 \to P_1$; $T(a_0 + a_1 x + a_2 x^2) = a_0 + a_1 x$

20. $T: P_2 \to P_1$; $T(a_0 + a_1 x + a_2 x^2) = a_1 + a_2 x$

21. $T: \mathbb{R} \to P_n$; $T(a) = a + ax + ax^2 + \cdots + ax^n$

22. $T: P_2 \to P_4$; $T(p(x)) = [p(x)]^2$

23. $T: C[0, 1] \to C[0, 1]$; $Tf(x) = f^2(x)$

24. $T: C[0, 1] \to C[0, 1]$; $Tf(x) = f(x) + 1$

Calculus **25.** $T: C[0, 1] \to \mathbb{R}$; $Tf = \int_0^1 f(x)g(x)\, dx$, where g is a fixed function in $C[0, 1]$

Calculus **26.** $T: C^1[0, 1] \to C[0, 1]$; $Tf = (fg)'$, where g is a fixed function in $C^1[0, 1]$

27. $T: C[0, 1] \to C[1, 2]$; $Tf(x) = f(x - 1)$

28. $T: C[0, 1] \to \mathbb{R}$; $Tf = f(\tfrac{1}{2})$

29. $T: M_{nn} \to \mathbb{R}$; $T(A) = \det A$

30. Let $T: \mathbb{R}^2 \to \mathbb{R}^2$ be given by $T(x, y) = (-x, -y)$. Describe T geometrically.

31. Let T be a linear transformation from $\mathbb{R}^2 \to \mathbb{R}^3$ such that $T\begin{pmatrix} 1 \\ 0 \end{pmatrix} = \begin{pmatrix} 1 \\ 2 \\ 3 \end{pmatrix}$ and $T\begin{pmatrix} 0 \\ 1 \end{pmatrix} = \begin{pmatrix} -4 \\ 0 \\ 5 \end{pmatrix}$. Find: **(a)** $T\begin{pmatrix} 2 \\ 4 \end{pmatrix}$ and **(b)** $T\begin{pmatrix} -3 \\ 7 \end{pmatrix}$.

32. In Example 8:
   a. Find the rotation matrix A_θ when $\theta = \pi/6$.
   b. What happens to the vector $\begin{pmatrix} -3 \\ 4 \end{pmatrix}$ if it is rotated through an angle of $\pi/6$ in the counterclockwise direction?

33. Let $A_\theta = \begin{pmatrix} \cos\theta & -\sin\theta & 0 \\ \sin\theta & \cos\theta & 0 \\ 0 & 0 & 1 \end{pmatrix}$. Describe geometrically the linear transformation $T: \mathbb{R}^3 \to \mathbb{R}^3$ given by $Tx = A_\theta x$.

34. Answer the questions in Problem 33 for $A_\theta = \begin{pmatrix} \cos\theta & 0 & -\sin\theta \\ 0 & 1 & 0 \\ \sin\theta & 0 & \cos\theta \end{pmatrix}$.

35. Suppose that in a real vector space V, T satisfies $T(\mathbf{x} + \mathbf{y}) = T\mathbf{x} + T\mathbf{y}$ and $T(\alpha\mathbf{x}) = \alpha T\mathbf{x}$ for $\alpha \geq 0$. Show that T is linear.

36. Find a linear transformation $T: M_{33} \to M_{22}$.

37. If T is a linear transformation from V to W, show that $T(\mathbf{x} - \mathbf{y}) = T\mathbf{x} - T\mathbf{y}$.

38. If T is a linear transformation from V to W, show that $T\mathbf{0} = \mathbf{0}$. Are the two zero vectors here the same?

39. Let V be an inner product space and let $\mathbf{u}_0 \in V$ be fixed. Let $T: V \to \mathbb{R}$ (or $\mathbb{C}$) be defined by $T\mathbf{v} = (\mathbf{v}, \mathbf{u}_0)$. Show that T is linear.

*40. Show that if V is a complex inner product space and $T: V \to \mathbb{C}$ is defined by $T\mathbf{v} = (\mathbf{u}_0, \mathbf{v})$ for a fixed vector $\mathbf{u}_0 \in V$, then T is not linear.

41. Let V be an inner product space with the finite dimensional subspace H. Let $\{\mathbf{u}_1, \mathbf{u}_2, \ldots, \mathbf{u}_k\}$ be a basis for H. Show that $T: V \to H$ defined by $T\mathbf{v} = (\mathbf{v}, \mathbf{u}_1)\mathbf{u}_1 + (\mathbf{v}, \mathbf{u}_2)\mathbf{u}_2 + \cdots + (\mathbf{v}, \mathbf{u}_k)\mathbf{u}_k$ is a linear transformation.

42. Let V and W be vector spaces. Let $L(V, W)$ denote the set of linear transformations from V to W. If T_1 and T_2 are in $L(V, W)$, define αT_1 and $T_1 + T_2$ by $(\alpha T_1)\mathbf{v} = \alpha(T_1\mathbf{v})$ and $(T_1 + T_2)\mathbf{v} = T_1\mathbf{v} + T_2\mathbf{v}$. Prove that $L(V, W)$ is a vector space.

5.2 PROPERTIES OF LINEAR TRANSFORMATIONS: RANGE AND KERNEL

In this section we develop some of the basic properties of linear transformations.

THEOREM 1   Let $T: V \to W$ be a linear transformation. Then, for all vectors $\mathbf{u}$, $\mathbf{v}$, $\mathbf{v}_1$, $\mathbf{v}_2$, $\ldots$, $\mathbf{v}_n$ in V and all scalars $\alpha_1, \alpha_2, \ldots, \alpha_n$:

i. $T(\mathbf{0}) = \mathbf{0}$

ii. $T(\mathbf{u} - \mathbf{v}) = T\mathbf{u} - T\mathbf{v}$

iii. $T(\alpha_1\mathbf{v}_1 + \alpha_2\mathbf{v}_2 + \cdots + \alpha_n\mathbf{v}_n) = \alpha_1 T\mathbf{v}_1 + \alpha_2 T\mathbf{v}_2 + \cdots + \alpha_n T\mathbf{v}_n$

Note. In part (i) the $\mathbf{0}$ on the left is the zero vector in V, whereas the $\mathbf{0}$ on the right is the zero vector in W.

Proof   i. $T(\mathbf{0}) = T(\mathbf{0} + \mathbf{0}) = T(\mathbf{0}) + T(\mathbf{0})$. Thus

$$\mathbf{0} = T(\mathbf{0}) - T(\mathbf{0}) = T(\mathbf{0}) + T(\mathbf{0}) - T(\mathbf{0}) = T(\mathbf{0}).$$

ii. $T(\mathbf{u} - \mathbf{v}) = T[\mathbf{u} + (-1)\mathbf{v}] = T\mathbf{u} + T[(-1)\mathbf{v}] = T\mathbf{u} + (-1)T\mathbf{v} = T\mathbf{u} - T\mathbf{v}$.

iii. We prove this part by induction (see Appendix 1). For $n = 2$, we get $T(\alpha_1\mathbf{v}_1 + \alpha_2\mathbf{v}_2) = T(\alpha_1\mathbf{v}_1) + T(\alpha_2\mathbf{v}_2) = \alpha_1 T\mathbf{v}_1 + \alpha_2 T\mathbf{v}_2$. Thus the equation holds for $n = 2$. We assume that it holds for $n = k$ and prove it for $n = k + 1$: $T(\alpha_1\mathbf{v}_1 + \alpha_2\mathbf{v}_2 + \cdots + \alpha_k\mathbf{v}_k + \alpha_{k+1}\mathbf{v}_{k+1}) = T(\alpha_1\mathbf{v}_1 + \alpha_2\mathbf{v}_2 + \cdots + \alpha_k\mathbf{v}_k) + T(\alpha_{k+1}\mathbf{v}_{k+1})$, and using the equation in part (iii) for $n = k$, this is equal to $(\alpha_1 T\mathbf{v}_1 + \alpha_2 T\mathbf{v}_2 + \cdots + \alpha_k T\mathbf{v}_k) + \alpha_{k+1} T\mathbf{v}_{k+1}$, which is what we wanted to show. This completes the proof. ∎

Remark. Note that parts (i) and (ii) of Theorem 1 are special cases of part (iii).

An important fact about linear transformations is that they are completely determined by what they do to basis vectors.

THEOREM 2   Let V be a finite dimensional vector space with basis $B = \{\mathbf{v}_1, \mathbf{v}_2, \ldots, \mathbf{v}_n\}$. Let $\mathbf{w}_1, \mathbf{w}_2, \ldots, \mathbf{w}_n$ be n vectors in W. Suppose that T_1 and T_2 are two linear transformations from V to W such that $T_1\mathbf{v}_i = T_2\mathbf{v}_i = \mathbf{w}_i$ for $i = 1, 2, \ldots, n$. Then for any vector $\mathbf{v} \in V$, $T_1\mathbf{v} = T_2\mathbf{v}$. That is, $T_1 = T_2$.

Proof   Since B is a basis for V, there exists a unique set of scalars $\alpha_1, \alpha_2, \ldots, \alpha_n$ such that $\mathbf{v} = \alpha_1\mathbf{v}_1 + \alpha_2\mathbf{v}_2 + \cdots + \alpha_n\mathbf{v}_n$. Then, from part (iii) of Theorem 1,

$$T_1\mathbf{v} = T_1(\alpha_1\mathbf{v}_1 + \alpha_2\mathbf{v}_2 + \cdots + \alpha_n\mathbf{v}_n) = \alpha_1 T_1\mathbf{v}_1 + \alpha_2 T_1\mathbf{v}_2 + \cdots + \alpha_n T_1\mathbf{v}_n$$
$$= \alpha_1\mathbf{w}_1 + \alpha_2\mathbf{w}_2 + \cdots + \alpha_n\mathbf{w}_n$$

Similarly,

$$T_2\mathbf{v} = T_2(\alpha_1\mathbf{v}_1 + \alpha_2\mathbf{v}_2 + \cdots + \alpha_n\mathbf{v}_n) = \alpha_1 T_2\mathbf{v}_1 + \alpha_2 T_2\mathbf{v}_2 + \cdots + \alpha_n T_2\mathbf{v}_n$$
$$= \alpha_1\mathbf{w}_1 + \alpha_2\mathbf{w}_2 + \cdots + \alpha_n\mathbf{w}_n$$

Thus $T_1\mathbf{v} = T_2\mathbf{v}$. ∎

Theorem 2 tells us that if $T: V \to W$ and V is finite dimensional, then we need to know only what T does to basis vectors in V. This determines T completely. To see this, let $\mathbf{v}_1, \mathbf{v}_2, \ldots, \mathbf{v}_n$ be a basis in V and let $\mathbf{v}$ be another vector in V. Then, as in the proof of Theorem 2,

$$T\mathbf{v} = \alpha_1 T\mathbf{v}_1 + \alpha_2 T\mathbf{v}_2 + \cdots + \alpha_n T\mathbf{v}_n$$

Thus we can compute $T\mathbf{v}$ for any vector $\mathbf{v} \in V$ if we know $T\mathbf{v}_1, T\mathbf{v}_2, \ldots, T\mathbf{v}_n$.

EXAMPLE 1   **If You Know What a Linear Transformation Does to Basis Vectors, Then You Know What It Does to Any Other Vector**   Let T be a linear transformation from $\mathbb{R}^3$ into $\mathbb{R}^2$ and suppose that $T\begin{pmatrix} 1 \\ 0 \\ 0 \end{pmatrix} = \begin{pmatrix} 2 \\ 3 \end{pmatrix}$, $T\begin{pmatrix} 0 \\ 1 \\ 0 \end{pmatrix} = \begin{pmatrix} -1 \\ 4 \end{pmatrix}$, and $T\begin{pmatrix} 0 \\ 0 \\ 1 \end{pmatrix} =$

$\begin{pmatrix} 5 \\ -3 \end{pmatrix}$. Compute $T\begin{pmatrix} 3 \\ -4 \\ 5 \end{pmatrix}$.

Solution   We have $\begin{pmatrix} 3 \\ -4 \\ 5 \end{pmatrix} = 3\begin{pmatrix} 1 \\ 0 \\ 0 \end{pmatrix} - 4\begin{pmatrix} 0 \\ 1 \\ 0 \end{pmatrix} + 5\begin{pmatrix} 0 \\ 0 \\ 1 \end{pmatrix}$.

Thus

$$T\begin{pmatrix} 3 \\ -4 \\ 5 \end{pmatrix} = 3T\begin{pmatrix} 1 \\ 0 \\ 0 \end{pmatrix} - 4T\begin{pmatrix} 0 \\ 1 \\ 0 \end{pmatrix} + 5T\begin{pmatrix} 0 \\ 0 \\ 1 \end{pmatrix}$$

$$= 3\begin{pmatrix} 2 \\ 3 \end{pmatrix} - 4\begin{pmatrix} -1 \\ 4 \end{pmatrix} + 5\begin{pmatrix} 5 \\ -3 \end{pmatrix} = \begin{pmatrix} 6 \\ 9 \end{pmatrix} + \begin{pmatrix} 4 \\ -16 \end{pmatrix} + \begin{pmatrix} 25 \\ -15 \end{pmatrix} = \begin{pmatrix} 35 \\ -22 \end{pmatrix}$$ ■

Another question arises: If $\mathbf{w}_1, \mathbf{w}_2, \ldots, \mathbf{w}_n$ are n vectors in W, does there exist a linear transformation T such that $T\mathbf{v}_i = \mathbf{w}_i$ for $i = 1, 2, \ldots, n$? The answer is yes, as the next theorem shows.

THEOREM 3   Let V be a finite dimensional vector space with basis $B = \{\mathbf{v}_1, \mathbf{v}_2, \ldots, \mathbf{v}_n\}$. Let W be a vector space containing the n vectors $\mathbf{w}_1, \mathbf{w}_2, \ldots, \mathbf{w}_n$. Then there exists a unique linear transformation $T: V \to W$ such that $T\mathbf{v}_i = \mathbf{w}_i$ for $i = 1, 2, \ldots, n$.

Proof   Define a function T as follows:

i. $T\mathbf{v}_i = \mathbf{w}_i$

ii. If $\mathbf{v} = \alpha_1\mathbf{v}_1 + \alpha_2\mathbf{v}_2 + \cdots + \alpha_n\mathbf{v}_n$, then

$$T\mathbf{v} = \alpha_1\mathbf{w}_1 + \alpha_2\mathbf{w}_2 + \cdots + \alpha_n\mathbf{w}_n \qquad (1)$$

Because B is a basis for V, T is defined for every $\mathbf{v} \in V$; and since W is a vector space, $T\mathbf{v} \in W$. Thus it only remains to show that T is linear. But this follows directly from equation (1). For if $\mathbf{u} = \alpha_1\mathbf{v}_1 + \alpha_2\mathbf{v}_2 + \cdots + \alpha_n\mathbf{v}_n$ and $\mathbf{v} = \beta_1\mathbf{v}_1 + \beta_2\mathbf{v}_2 + \cdots + \beta_n\mathbf{v}_n$, then

$$T(\mathbf{u} + \mathbf{v}) = T[(\alpha_1 + \beta_1)\mathbf{v}_1 + (\alpha_2 + \beta_2)\mathbf{v}_2 + \cdots + (\alpha_n + \beta_n)\mathbf{v}_n]$$
$$= (\alpha_1 + \beta_1)\mathbf{w}_1 + (\alpha_2 + \beta_2)\mathbf{w}_2 + \cdots + (\alpha_n + \beta_n)\mathbf{w}_n$$
$$= (\alpha_1\mathbf{w}_1 + \alpha_2\mathbf{w}_2 + \cdots + \alpha_n\mathbf{w}_n) + (\beta_1\mathbf{w}_1 + \beta_2\mathbf{w}_2 + \cdots + \beta_n\mathbf{w}_n)$$
$$= T\mathbf{u} + T\mathbf{v}$$

Similarly, $T(\alpha\mathbf{v}) = \alpha T\mathbf{v}$, so T is linear. The uniqueness of T follows from Theorem 2 and the theorem is proved. ■

Remark. In Theorems 2 and 3 the vectors $\mathbf{w}_1, \mathbf{w}_2, \ldots, \mathbf{w}_n$ need not be independent and, in fact, need not even be distinct. Moreover, we emphasize that the theorems are true if V is any finite dimensional vector space, not just $\mathbb{R}^n$. Note also that W does not have to be finite dimensional.

EXAMPLE 2 **Finding a Linear Transformation from $\mathbb{R}^2$ into a Subspace of $\mathbb{R}^3$** Find a linear transformation from $\mathbb{R}^2$ into the plane

$$W = \left\{ \begin{pmatrix} x \\ y \\ z \end{pmatrix} : 2x - y + 3z = 0 \right\}$$

Solution From Example 4.6.3, page 244, we know that W is a two-dimensional subspace of $\mathbb{R}^3$ with basis vectors $\mathbf{w}_1 = \begin{pmatrix} 1 \\ 2 \\ 0 \end{pmatrix}$ and $\mathbf{w}_2 = \begin{pmatrix} 0 \\ 3 \\ 1 \end{pmatrix}$. Using the standard basis in $\mathbb{R}^2$, $\mathbf{v}_1 = \begin{pmatrix} 1 \\ 0 \end{pmatrix}$ and $\mathbf{v}_2 = \begin{pmatrix} 0 \\ 1 \end{pmatrix}$, we define the linear transformation T by $T\begin{pmatrix} 1 \\ 0 \end{pmatrix} = \begin{pmatrix} 1 \\ 2 \\ 0 \end{pmatrix}$ and $T\begin{pmatrix} 0 \\ 1 \end{pmatrix} = \begin{pmatrix} 0 \\ 3 \\ 1 \end{pmatrix}$. Then, as the discussion following Theorem 2 shows, T is completely determined. For example,

$$T\begin{pmatrix} 5 \\ -7 \end{pmatrix} = T\left[5\begin{pmatrix} 1 \\ 0 \end{pmatrix} - 7\begin{pmatrix} 0 \\ 1 \end{pmatrix} \right] = 5T\begin{pmatrix} 1 \\ 0 \end{pmatrix} - 7T\begin{pmatrix} 0 \\ 1 \end{pmatrix} = 5\begin{pmatrix} 1 \\ 2 \\ 0 \end{pmatrix} - 7\begin{pmatrix} 0 \\ 3 \\ 1 \end{pmatrix} = \begin{pmatrix} 5 \\ -11 \\ -7 \end{pmatrix}. \blacksquare$$

We now turn to two important definitions in the theory of linear transformations.

DEFINITION 1 **Kernel and Range of a Linear Transformation** Let V and W be vector spaces and let $T: V \rightarrow W$ be a linear transformation. Then

i. The **kernel** of T, denoted by ker T, is given by

$$\boxed{\ker T = \{\mathbf{v} \in V : T\mathbf{v} = \mathbf{0}\}} \tag{2}$$

ii. The **range** of T, denoted by range T, is given by

$$\boxed{\text{Range } T = \{\mathbf{w} \in W : \mathbf{w} = T\mathbf{v} \text{ for some } \mathbf{v} \in V\}} \tag{3}$$

Remark 1. Note that ker T is nonempty because, by Theorem 1, $T(\mathbf{0}) = \mathbf{0}$ so that $\mathbf{0} \in$ ker T for any linear transformation T. We shall be interested in finding other vectors in V that get "mapped to zero." Again note that when we write $T(\mathbf{0}) = \mathbf{0}$, the $\mathbf{0}$ on the left is in V and the $\mathbf{0}$ on the right is in W.

Remark 2. Range T is simply the set of "images" of vectors in V under the transformation T. In fact, if $\mathbf{w} = T\mathbf{v}$, we say that $\mathbf{w}$ is the **image** of $\mathbf{v}$ under T.

Image

Before giving examples of kernels and ranges, we prove a theorem that will be very useful.

THEOREM 4   If $T\colon V \to W$ is a linear transformation, then

   i. ker T is a subspace of V.

   ii. range T is a subspace of W.

Proof   **i.** Let $\mathbf{u}$ and $\mathbf{v}$ be in ker T; then $T(\mathbf{u} + \mathbf{v}) = T\mathbf{u} + T\mathbf{v} = \mathbf{0} + \mathbf{0} = \mathbf{0}$ and $T(\alpha\mathbf{u}) = \alpha T\mathbf{u} = \alpha\mathbf{0} = \mathbf{0}$ so that $\mathbf{u} + \mathbf{v}$ and $\alpha\mathbf{u}$ are in ker T.

   ii. Let $\mathbf{w}$ and $\mathbf{x}$ be in range T. Then $\mathbf{w} = T\mathbf{u}$ and $\mathbf{x} = T\mathbf{v}$ for two vectors $\mathbf{u}$ and $\mathbf{v}$ in V. This means that $T(\mathbf{u} + \mathbf{v}) = T\mathbf{u} + T\mathbf{v} = \mathbf{w} + \mathbf{x}$ and $T(\alpha\mathbf{u}) = \alpha T\mathbf{u} = \alpha\mathbf{w}$. Thus $\mathbf{w} + \mathbf{x}$ and $\alpha\mathbf{w}$ are in range T.   ∎

EXAMPLE 3   **Kernel and Range of the Zero Transformation**   Let $T\mathbf{v} = \mathbf{0}$ for every $\mathbf{v} \in V$. (T is the zero transformation.) Then ker $T = V$ and range $T = \{\mathbf{0}\}$.   ∎

EXAMPLE 4   **Kernel and Range of the Identity Transformation**   Let $T\mathbf{v} = \mathbf{v}$ for every $\mathbf{v} \in V$. (T is the identity transformation.) Then ker $T = \{\mathbf{0}\}$ and range $T = V$.   ∎

The zero and identity transformations provide two extremes. In the first, everything is in the kernel. In the second, only the zero vector is in the kernel. The cases in between are more interesting.

EXAMPLE 5   **Kernel and Range of a Projection Operator**   Let $T\colon \mathbb{R}^3 \to \mathbb{R}^3$ be defined by $T\begin{pmatrix} x \\ y \\ z \end{pmatrix} = \begin{pmatrix} x \\ y \\ 0 \end{pmatrix}$. That is (see Example 5.1.10, page 335), T is the projection operator from $\mathbb{R}^3$ into the xy-plane. If $T\begin{pmatrix} x \\ y \\ z \end{pmatrix} = \begin{pmatrix} x \\ y \\ 0 \end{pmatrix} = \mathbf{0} = \begin{pmatrix} 0 \\ 0 \\ 0 \end{pmatrix}$, then $x = y = 0$. Thus ker $T = \{(x, y, z)\colon x = y = 0\} =$ the z-axis, and range $T = \{(x, y, z)\colon z = 0\} =$ the xy-plane. Note that dim ker $T = 1$ and dim range $T = 2$.   ∎

DEFINITION 2   **Nullity and Rank of a Linear Transformation**   If T is a linear transformation from V to W, then we define

> **Nullity** of $T = \nu(T) = \dim \ker T$ (4)
>
> **Rank** of $T = \rho(T) = \dim \text{range } T$ (5)

Remark. In Section 4.7 we defined the rank and nullity of a matrix. According to Example 5.1.7, every $m \times n$ matrix A gives rise to a linear transformation $T: \mathbb{R}^n \to \mathbb{R}^m$ defined by $T\mathbf{x} = A\mathbf{x}$. Evidently, $\ker T = N_A$, range $T = $ range $A = C_A$, $\nu(T) = \nu(A)$ and $\rho(T) = \rho(A)$. Thus we see that the definitions of kernel, range, nullity, and rank of a linear transformation are generalizations of the same concepts applied to matrices.

EXAMPLE 6   **Kernel and Nullity of a Projection Operator**   Let H be a subspace of $\mathbb{R}^n$ and define $T\mathbf{v} = \text{proj}_H \mathbf{v}$. Clearly range $T = H$. From Theorem 4.10.7, page 291, we can write any $\mathbf{v} \in V$ as $\mathbf{v} = \mathbf{h} + \mathbf{p} = \text{proj}_H \mathbf{v} + \text{proj}_{H^\perp} \mathbf{v}$. If $T\mathbf{v} = \mathbf{0}$, then $\mathbf{h} = \mathbf{0}$, which means that $\mathbf{v} = \mathbf{p} \in H^\perp$. Thus $\ker T = H^\perp$, $\rho(T) = \dim H$, and $\nu(T) = \dim H^\perp = n - \rho(T)$.   ∎

EXAMPLE 7   **The Kernel and Range of a Transpose Operator**   Let $V = M_{mn}$ and define $T: M_{mn} \to M_{mn}$ by $T(A) = A^t$ (see Example 5.1.11, page 336). If $TA = A^t = 0$, then A^t is the $n \times m$ zero matrix so that A is the $m \times n$ zero matrix. Thus $\ker T = \{\mathbf{0}\}$, and clearly range $T = M_{nm}$. This means that $\nu(T) = 0$ and $\rho(T) = nm$.   ∎

EXAMPLE 8   **The Kernel and Range of a Transformation from P_3 to P_2**   Let $T: P_3 \to P_2$ be defined by $T(p) = T(a_0 + a_1 x + a_2 x^2 + a_3 x^3) = a_0 + a_1 x + a_2 x^2$. Then if $T(p) = 0$, $a_0 + a_1 x + a_2 x^2 = 0$ for every x, which implies that $a_0 = a_1 = a_2 = 0$. Thus $\ker T = \{p \in P_3: p(x) = a_3 x^3\}$ and range $T = P_2$, $\nu(T) = 1$, and $\rho(T) = 3$.   ∎

EXAMPLE 9   [Calculus]   **The Kernel and Range of an Integral Operator**   Let $V = C[0, 1]$ and define $J: C[0, 1] \to \mathbb{R}$ by $Jf = \int_0^1 f(x)\, dx$ (see Example 5.1.12, page 336). Then $\ker J = \{f \in C[0, 1]: \int_0^1 f(x)\, dx = 0\}$. Let α be a real number. Then the constant function $f(x) = \alpha$ for $x \in [0, 1]$ is in $C[0, 1]$ and $\int_0^1 \alpha\, dx = \alpha$. Since this is true for every real number α, we have range $J = \mathbb{R}$.   ∎

In the next section we shall see how every linear transformation from one finite dimensional vector space to another can be represented by a matrix. This will enable us to compute the kernel and range of any linear transformation between finite dimensional vector spaces by finding the kernel and range of a corresponding matrix.

PROBLEMS   5.2

In Problems 1–10 find the kernel, range, rank, and nullity of the given linear transformation.

1. $T\colon \mathbb{R}^2 \to \mathbb{R}^2$; $T\begin{pmatrix} x \\ y \end{pmatrix} = \begin{pmatrix} x \\ 0 \end{pmatrix}$

2. $T\colon \mathbb{R}^3 \to \mathbb{R}^2$; $T\begin{pmatrix} x \\ y \\ z \end{pmatrix} = \begin{pmatrix} z \\ y \end{pmatrix}$

3. $T\colon \mathbb{R}^2 \to \mathbb{R}$; $T\begin{pmatrix} x \\ y \end{pmatrix} = x + y$

4. $T\colon \mathbb{R}^4 \to \mathbb{R}^2$; $T\begin{pmatrix} x \\ y \\ z \\ w \end{pmatrix} = \begin{pmatrix} x + z \\ y + w \end{pmatrix}$

5. $T\colon M_{22} \to M_{22}$: $T(A) = AB$, where $B = \begin{pmatrix} 1 & 2 \\ 0 & 1 \end{pmatrix}$

6. $T\colon \mathbb{R} \to P_3$: $T(a) = a + ax + ax^2 + ax^3$

* 7. $T\colon M_{nn} \to M_{nn}$: $T(A) = A^t + A$ $\boxed{\text{Calculus}}$ **8.** $T\colon C^1[0, 1] \to C[0, 1]$; $Tf = f'$

9. $T\colon C[0, 1] \to \mathbb{R}$; $Tf = f(\tfrac{1}{2})$

10. $T\colon \mathbb{R}^2 \to \mathbb{R}^2$: T is a rotation through an angle of $\pi/3$

11. Let $T\colon V \to W$ be a linear transformation, let $\{\mathbf{v}_1, \mathbf{v}_2, \ldots, \mathbf{v}_n\}$ be a basis for V, and suppose that $T\mathbf{v}_i = \mathbf{0}$ for $i = 1, 2, \ldots, n$. Show that T is the zero transformation.

12. In Problem 11 suppose that $W = V$ and $T\mathbf{v}_i = \mathbf{v}_i$ for $i = 1, 2, \ldots, n$. Show that T is the identity operator.

13. Let $T\colon V \to \mathbb{R}^3$. Prove that range T is either **(a)** $\{\mathbf{0}\}$, **(b)** a line through the origin, **(c)** a plane through the origin, or **(d)** $\mathbb{R}^3$.

14. Let $T\colon \mathbb{R}^3 \to V$. Show that ker T is one of four spaces listed in Problem 13.

15. Find all linear transformations from $\mathbb{R}^2$ into $\mathbb{R}^2$ such that the line $y = 0$ is carried into the line $x = 0$.

16. Find all linear transformations from $\mathbb{R}^2$ into $\mathbb{R}^2$ that carry the line $y = ax$ into the line $y = bx$.

17. Find a linear transformation T from $\mathbb{R}^3 \to \mathbb{R}^3$ such that

$$\ker T = \{(x, y, z)\colon 2x - y + z = 0\}.$$

18. Find a linear transformation T from $\mathbb{R}^3 \to \mathbb{R}^3$ such that

$$\text{range } T = \{(x, y, z)\colon 2x - y + z = 0\}.$$

19. Let $T\colon M_{nn} \to M_{nn}$ be defined by $TA = A - A^t$. Show that ker $T = \{$symmetric $n \times n$ matrices$\}$ and range of $T = \{$skew-symmetric $n \times n$ matrices$\}$.

* $\boxed{\text{Calculus}}$ **20.** Let $T\colon C^1[0, 1] \to C[0, 1]$ be defined by $Tf(x) = xf'(x)$. Find the kernel and range of T.

*21. In Problem 5.1.42 you were asked to show that the set of linear transformations from a vector space V to a vector space W, denoted by $L(V, W)$, is a vector space. Suppose that dim $V = n < \infty$ and dim $W = m < \infty$. Find dim $L(V, W)$.

22. Let H be a subspace of V where dim $H = k$ and dim $V = n$. Let U be the subset of $L(V, V)$ having the property that if $T \in L(V, V)$, then $T\mathbf{h} = \mathbf{0}$ for every $\mathbf{h} \in H$.
   a. Prove that U is a subspace of $L(V, V)$.
   b. Find dim U.

*23. Let S and T be in $L(V, V)$ such that ST is the zero transformation. Prove or disprove: TS is the zero transformation.

5.3 THE MATRIX REPRESENTATION OF A LINEAR TRANSFORMATION

If A is an $m \times n$ matrix and $T: \mathbb{R}^n \to \mathbb{R}^m$ is defined by $T\mathbf{x} = A\mathbf{x}$, then, as we saw in Example 5.1.7 on page 333, T is a linear transformation. We shall now see that for *every* linear transformation from $\mathbb{R}^n$ into $\mathbb{R}^m$, there exists an $m \times n$ matrix A such that $T\mathbf{x} = A\mathbf{x}$ for every $\mathbf{x} \in \mathbb{R}^n$. This fact is extremely useful. As we saw in the remark on page 343, if $T\mathbf{x} = A\mathbf{x}$, then ker $T = N_A$ and range $T = R_A$. Moreover, $\nu(T) = \dim \ker T = \nu(A)$ and $\rho(T) = \dim \text{range } T = \rho(A)$. Thus we can determine the kernel, range, nullity, and rank of a linear transformation from $\mathbb{R}^n \to \mathbb{R}^m$ by determining the kernel and range space of a corresponding matrix. Moreover, once we know that $T\mathbf{x} = A\mathbf{x}$, we can evaluate $T\mathbf{x}$ for any $\mathbf{x}$ in $\mathbb{R}^n$ by simple matrix multiplication.

But this is not all. As we shall see, any linear transformation between finite dimensional vector spaces can be represented by a matrix.

THEOREM 1   Let $T: \mathbb{R}^n \to \mathbb{R}^m$ be a linear transformation. Then there exists a unique $m \times n$ matrix A_T such that

$$T\mathbf{x} = A_T\mathbf{x} \qquad \text{for every } \mathbf{x} \in \mathbb{R}^n \qquad (1)$$

Proof   Let $\mathbf{w}_1 = T\mathbf{e}_1, \mathbf{w}_2 = T\mathbf{e}_2, \ldots, \mathbf{w}_n = T\mathbf{e}_n$. Let A_T be the matrix whose columns are $\mathbf{w}_1, \mathbf{w}_2, \ldots, \mathbf{w}_n$ and let A_T also denote the transformation from $\mathbb{R}^n \to \mathbb{R}^m$ which multiplies a vector in $\mathbb{R}^n$ on the left by A_T. If

$$\mathbf{w}_i = \begin{pmatrix} a_{1i} \\ a_{2i} \\ \vdots \\ a_{mi} \end{pmatrix} \qquad \text{for } i = 1, 2, \ldots, n$$

then

$$A_T e_i = \begin{pmatrix} a_{11} & a_{12} & \cdots & a_{1i} & \cdots & a_{1n} \\ a_{21} & a_{22} & \cdots & a_{2i} & \cdots & a_{2n} \\ \vdots & \vdots & & \vdots & & \vdots \\ a_{m1} & a_{m2} & \cdots & a_{mi} & \cdots & a_{mn} \end{pmatrix} \begin{pmatrix} 0 \\ 0 \\ \vdots \\ 1 \\ 0 \\ \vdots \\ 0 \end{pmatrix} \begin{matrix} \\ \\ \\ \leftarrow \\ \\ \\ i\text{th} \end{matrix} = \begin{pmatrix} a_{1i} \\ a_{2i} \\ \vdots \\ a_{mi} \end{pmatrix} = \mathbf{w}_i.$$

position

Thus $A_T e_i = \mathbf{w}_i$ for $i = 1, 2, \ldots, n$. By Theorem 5.2.2 on page 339, T and the transformation A_T are the same because they agree on basis vectors.

We can now show that A_T is unique. Suppose that $Tx = A_T x$ and $Tx = B_T x$ for every $\mathbf{x} \in \mathbb{R}^n$. Then $A_T x = B_T x$ or, setting $C_T = A_T - B_T$, we have $C_T x = \mathbf{0}$ for every $\mathbf{x} \in \mathbb{R}^n$. In particular, $C_T e_i = \mathbf{0}$ for $i = 1, 2, \ldots, n$. But, as we see from the proof of the first part of the theorem, $C_T e_i$ is the ith column of C_T. Thus each of the n columns of C_T is the m-zero vector and $C_T = 0$, the $m \times n$ zero matrix. This shows that $A_T = B_T$ and the theorem is proved. ∎

Remark 1. In this theorem we assumed that every vector in $\mathbb{R}^n$ and $\mathbb{R}^m$ is written in terms of the standard basis vectors in those spaces. If we choose other bases for $\mathbb{R}^n$ and $\mathbb{R}^m$, we shall, of course, get a different matrix A_T. See, for instance, Example 4.9.1 on page 275 or Example 8 below.

Remark 2. The proof of the theorem shows us that A_T is easily obtained as the matrix whose columns are the vectors Te_i.

DEFINITION 1   **Transformation Matrix**   The matrix A_T in Theorem 1 is called the **transformation matrix** corresponding to T.

In Section 5.2 we defined the range, rank, kernel, and nullity of a linear transformation. In Section 4.7 we defined the range, rank, kernel, and nullity of a matrix. The proof of the following theorem follows from Theorem 1 and is left as an exercise (see Problem 36).

THEOREM 2   Let A_T be the transformation matrix corresponding to the linear transformation T. Then:

  i. range $T = R_{A_T} = C_{A_T}$

ii. $\rho(T) = \rho(A_T)$

iii. $\ker T = N_{A_T}$

iv. $\nu(T) = \nu(A_T)$ ∎

EXAMPLE 1 **The Matrix Representation of a Projection Transformation** Find the transformation matrix A_T corresponding to the projection of a vector in $\mathbb{R}^3$ onto the xy-plane.

Solution Here $T\begin{pmatrix} x \\ y \\ z \end{pmatrix} = \begin{pmatrix} x \\ y \\ 0 \end{pmatrix}$. In particular, $T\begin{pmatrix} 1 \\ 0 \\ 0 \end{pmatrix} = \begin{pmatrix} 1 \\ 0 \\ 0 \end{pmatrix}$, $T\begin{pmatrix} 0 \\ 1 \\ 0 \end{pmatrix} = \begin{pmatrix} 0 \\ 1 \\ 0 \end{pmatrix}$, and $T\begin{pmatrix} 0 \\ 0 \\ 1 \end{pmatrix} = \begin{pmatrix} 0 \\ 0 \\ 0 \end{pmatrix}$.

Thus $A_T = \begin{pmatrix} 1 & 0 & 0 \\ 0 & 1 & 0 \\ 0 & 0 & 0 \end{pmatrix}$. Note that $A_T\begin{pmatrix} x \\ y \\ z \end{pmatrix} = \begin{pmatrix} 1 & 0 & 0 \\ 0 & 1 & 0 \\ 0 & 0 & 0 \end{pmatrix}\begin{pmatrix} x \\ y \\ z \end{pmatrix} = \begin{pmatrix} x \\ y \\ 0 \end{pmatrix}$. ∎

EXAMPLE 2 **The Matrix Representation of a Transformation from $\mathbb{R}^3$ to $\mathbb{R}^4$**
Let $T: \mathbb{R}^3 \to \mathbb{R}^4$ be defined by

$$T\begin{pmatrix} x \\ y \\ z \end{pmatrix} = \begin{pmatrix} x - y \\ y + z \\ 2x - y - z \\ -x + y + 2z \end{pmatrix}$$

Find A_T, $\ker T$, range T, $\nu(T)$, and $\rho(T)$.

Solution $T\begin{pmatrix} 1 \\ 0 \\ 0 \end{pmatrix} = \begin{pmatrix} 1 \\ 0 \\ 2 \\ -1 \end{pmatrix}$, $T\begin{pmatrix} 0 \\ 1 \\ 0 \end{pmatrix} = \begin{pmatrix} -1 \\ 1 \\ -1 \\ 1 \end{pmatrix}$, and $T\begin{pmatrix} 0 \\ 0 \\ 1 \end{pmatrix} = \begin{pmatrix} 0 \\ 1 \\ -1 \\ 2 \end{pmatrix}$. Thus

$$A_T = \begin{pmatrix} 1 & -1 & 0 \\ 0 & 1 & 1 \\ 2 & -1 & -1 \\ -1 & 1 & 2 \end{pmatrix}$$

Note (as a check) that

$$\begin{pmatrix} 1 & -1 & 0 \\ 0 & 1 & 1 \\ 2 & -1 & -1 \\ -1 & 1 & 2 \end{pmatrix}\begin{pmatrix} x \\ y \\ z \end{pmatrix} = \begin{pmatrix} x - y \\ x + z \\ 2x - y - z \\ -x + y + 2z \end{pmatrix}$$

Next we compute the kernel and range of A. Row-reducing, we obtain

$$\begin{pmatrix} 1 & -1 & 0 \\ 0 & 1 & 1 \\ 2 & -1 & -1 \\ -1 & 1 & 2 \end{pmatrix} \xrightarrow[\substack{R_3 \to R_3 - 2R_1 \\ R_4 \to R_4 + R_1}]{} \begin{pmatrix} 1 & -1 & 0 \\ 0 & 1 & 1 \\ 0 & 1 & -1 \\ 0 & 0 & 2 \end{pmatrix}$$

$$\xrightarrow[\substack{R_1 \to R_1 + R_2 \\ R_3 \to R_3 - R_2}]{} \begin{pmatrix} 1 & 0 & 1 \\ 0 & 1 & 1 \\ 0 & 0 & -2 \\ 0 & 0 & 2 \end{pmatrix} \xrightarrow[R_4 \to R_4 + R_3]{} \begin{pmatrix} 1 & 0 & 1 \\ 0 & 1 & 1 \\ 0 & 0 & -2 \\ 0 & 0 & 0 \end{pmatrix}$$

The first three rows of the last matrix are linearly independent (because their determinant is -2) so

since $\rho(A) + \nu(A) = 3$
$$\downarrow$$
$$\rho(A) = 3 \text{ and } \nu(A) = 3 - 3 = 0.$$

This means that $\ker T = \{\mathbf{0}\}$, range $T = \text{span}\left\{ \begin{pmatrix} 1 \\ 0 \\ 2 \\ -1 \end{pmatrix}, \begin{pmatrix} -1 \\ 1 \\ -1 \\ 1 \end{pmatrix}, \begin{pmatrix} 0 \\ 1 \\ -1 \\ 2 \end{pmatrix} \right\}$, $\nu(T) = 0$,

and $\rho(T) = 3$.   ∎

EXAMPLE 3   **The Matrix Representation of a Transformation from $\mathbb{R}^3$ to $\mathbb{R}^3$**

Let $T\colon \mathbb{R}^3 \to \mathbb{R}^3$ be defined by $T\begin{pmatrix} x \\ y \\ z \end{pmatrix} = \begin{pmatrix} 2x - y + 3z \\ 4x - 2y + 6z \\ -6x + 3y - 9z \end{pmatrix}$. Find A_T, $\ker T$,

range T, $\nu(T)$, and $\rho(T)$.

Solution   Since $T\begin{pmatrix} 1 \\ 0 \\ 0 \end{pmatrix} = \begin{pmatrix} 2 \\ 4 \\ -6 \end{pmatrix}$, $T\begin{pmatrix} 0 \\ 1 \\ 0 \end{pmatrix} = \begin{pmatrix} -1 \\ -2 \\ 3 \end{pmatrix}$, and $T\begin{pmatrix} 0 \\ 0 \\ 1 \end{pmatrix} = \begin{pmatrix} 3 \\ 6 \\ -9 \end{pmatrix}$, we have

$$A_T = \begin{pmatrix} 2 & -1 & 3 \\ 4 & -2 & 6 \\ -6 & 3 & -9 \end{pmatrix}$$

Theorem 2(*ii*)
$$\downarrow$$

From Example 4.7.4 on page 257 we see that $\rho(A) = \rho(T) = 1$ and

range $T = \text{span}\left\{ \begin{pmatrix} 2 \\ 4 \\ -6 \end{pmatrix} \right\}$. Then $\nu(T) = 2$.

Theorem 2(*iii*)
↓

To find $N_A = \ker T$, we row-reduce to solve the system $Ax = 0$:

$$\begin{pmatrix} 2 & -1 & 3 & | & 0 \\ 4 & -2 & 6 & | & 0 \\ -6 & 3 & -9 & | & 0 \end{pmatrix} \xrightarrow[R_3 \to R_3 + 3R_1]{R_2 \to R_2 - 2R_1} \begin{pmatrix} 2 & -1 & 3 & | & 0 \\ 0 & 0 & 0 & | & 0 \\ 0 & 0 & 0 & | & 0 \end{pmatrix}.$$

This means that $\begin{pmatrix} x \\ y \\ z \end{pmatrix} \in N_A$ if $2x - y + 3z = 0$ or $y = 2x + 3z$. First setting $x = 1$, $z = 0$ and then $x = 0$, $z = 1$, we obtain a basis for N_A:

$$\ker T = N_A = \operatorname{span}\left\{ \begin{pmatrix} 1 \\ 2 \\ 0 \end{pmatrix}, \begin{pmatrix} 0 \\ 3 \\ 1 \end{pmatrix} \right\}.$$

∎

EXAMPLE 4   **The Matrix Representation of a Zero Transformation**   It is easy to verify that if T is the zero transformation from $\mathbb{R}^n \to \mathbb{R}^m$, then A_T is the $m \times n$ zero matrix. Similarly, if T is the identity transformation from $\mathbb{R}^n \to \mathbb{R}^n$, then $A_T = I_n$.   ∎

EXAMPLE 5   **The Matrix Representation of a Rotation Transformation**   We saw in Example 5.1.8 on page 334 that if T is the function that rotates every vector in $\mathbb{R}^2$ through an angle of θ, then $A_T = \begin{pmatrix} \cos\theta & -\sin\theta \\ \sin\theta & \cos\theta \end{pmatrix}$.   ∎

We now generalize the notion of matrix representation to arbitrary finite dimensional vector spaces.

THEOREM 3   Let V be an n-dimensional vector space, W be an m-dimensional vector space, and $T: V \to W$ be a linear transformation. Let $B_1 = \{v_1, v_2, \ldots, v_n\}$ be a basis for V and let $B_2 = \{w_1, w_2, \ldots, w_m\}$ be a basis for W. Then there is a unique $m \times n$ matrix A_T such that

$$\boxed{(Tx)_{B_2} = A_T(x)_{B_1}} \tag{2}$$

Remark 1. The notation in (2) is the notation of Section 4.9 (see page 270). If $x \in V = c_1v_1 + c_2v_2 + \cdots + c_nv_n$, then $(x)_{B_1} = \begin{pmatrix} c_1 \\ c_2 \\ \vdots \\ c_n \end{pmatrix}$. Let $c = \begin{pmatrix} c_1 \\ c_2 \\ \vdots \\ c_n \end{pmatrix}$. Then A_Tc is an m-vector that we denote by $\mathbf{d} = \begin{pmatrix} d_1 \\ d_2 \\ \vdots \\ d_m \end{pmatrix}$. Equation (2) says that $(T\mathbf{x})_{B_2} = \begin{pmatrix} d_1 \\ d_2 \\ \vdots \\ d_m \end{pmatrix}$.

That is,

$$T\mathbf{x} = d_1\mathbf{w}_1 + d_2\mathbf{w}_2 + \cdots + d_m\mathbf{w}_m$$

Remark 2. As in Theorem 1, the uniqueness of A_T is relative to the bases B_1 and B_2. If we change the bases, we change A_T (see Examples 8 and 9, and Theorem 5).

Proof   Let $T\mathbf{v}_1 = \mathbf{y}_1$, $T\mathbf{v}_2 = \mathbf{y}_2$, . . . , $T\mathbf{v}_n = \mathbf{y}_n$. Since $\mathbf{y}_i \in W$, we have, for $i = 1, 2, \ldots, n$,

$$\mathbf{y}_i = a_{1i}\mathbf{w}_1 + a_{2i}\mathbf{w}_2 + \cdots + a_{mi}\mathbf{w}_m$$

for some (unique) set of scalars $a_{1i}, a_{2i}, \ldots, a_{mi}$, and we write

$$(\mathbf{y}_1)_{B_2} = \begin{pmatrix} a_{11} \\ a_{21} \\ \vdots \\ a_{m1} \end{pmatrix}, \ (\mathbf{y}_2)_{B_2} = \begin{pmatrix} a_{12} \\ a_{22} \\ \vdots \\ a_{m2} \end{pmatrix}, \ \ldots, \ (\mathbf{y}_n)_{B_2} = \begin{pmatrix} a_{1n} \\ a_{2n} \\ \vdots \\ a_{mn} \end{pmatrix}$$

This means, for example, that $\mathbf{y}_1 = a_{11}\mathbf{w}_1 + a_{21}\mathbf{w}_2 + \cdots + a_{m1}\mathbf{w}_m$. We now define

$$A_T = \begin{pmatrix} a_{11} & a_{12} & \cdots & a_{1n} \\ a_{21} & a_{22} & \cdots & a_{2n} \\ \vdots & \vdots & & \vdots \\ a_{m1} & a_{m2} & \cdots & a_{mn} \end{pmatrix}$$

Since

$$(\mathbf{v}_1)_{B_1} = \begin{pmatrix} 1 \\ 0 \\ \vdots \\ 0 \end{pmatrix}, \quad (\mathbf{v}_2)_{B_1} = \begin{pmatrix} 0 \\ 1 \\ 0 \\ \vdots \\ 0 \end{pmatrix}, \ \ldots, \ (\mathbf{v}_n)_{B_1} = \begin{pmatrix} 0 \\ 0 \\ \vdots \\ 1 \end{pmatrix}$$

we have, as in the proof of Theorem 1,

$$A_T(\mathbf{v}_i)_{B_1} = \begin{pmatrix} a_{11} & a_{12} & \cdots & a_{1n} \\ a_{21} & a_{22} & \cdots & a_{2n} \\ \vdots & \vdots & & \vdots \\ a_{i1} & a_{i2} & \cdots & a_{in} \\ \vdots & \vdots & & \vdots \\ a_{m1} & a_{m2} & \cdots & a_{mn} \end{pmatrix} \overset{i\text{th position}}{\begin{pmatrix} 0 \\ 0 \\ \vdots \\ 1 \\ 0 \\ \vdots \\ 0 \end{pmatrix}} = \begin{pmatrix} a_{1i} \\ a_{2i} \\ \vdots \\ a_{mi} \end{pmatrix} = (\mathbf{y}_i)_{B_2}$$

If $\mathbf{x}$ is in V, then

$$\mathbf{x} = c_1\mathbf{v}_1 + c_2\mathbf{v}_2 + \cdots + c_n\mathbf{v}_n$$

$$(\mathbf{x})_{B_1} = \begin{pmatrix} c_1 \\ c_2 \\ \vdots \\ c_n \end{pmatrix}$$

and

$$(A_T(\mathbf{x})_{B_1})_{B_2} = \begin{pmatrix} a_{11} & a_{12} & \cdots & a_{1n} \\ a_{21} & a_{22} & \cdots & a_{2n} \\ \vdots & \vdots & & \vdots \\ a_{m1} & a_{m2} & \cdots & a_{mn} \end{pmatrix} \begin{pmatrix} c_1 \\ c_2 \\ \vdots \\ c_n \end{pmatrix}$$

$$= \begin{pmatrix} a_{11}c_1 + a_{12}c_2 + \cdots + a_{1n}c_n \\ a_{21}c_1 + a_{22}c_2 + \cdots + a_{2n}c_n \\ \vdots & \vdots & \vdots \\ a_{m1}c_1 + a_{m2}c_2 + \cdots + a_{mn}c_n \end{pmatrix}$$

$$= c_1 \begin{pmatrix} a_{11} \\ a_{21} \\ \vdots \\ a_{m1} \end{pmatrix} + c_2 \begin{pmatrix} a_{12} \\ a_{22} \\ \vdots \\ a_{m2} \end{pmatrix} + \cdots + c_n \begin{pmatrix} a_{1n} \\ a_{2n} \\ \vdots \\ a_{mn} \end{pmatrix}$$

$$= c_1(\mathbf{y}_1)_{B_2} + c_2(\mathbf{y}_2)_{B_2} + \cdots + c_n(\mathbf{y}_n)_{B_2}$$

Similarly, $T\mathbf{x} = T(c_1\mathbf{v}_1 + c_2\mathbf{v}_2 + \cdots + c_n\mathbf{v}_n) = c_1T\mathbf{v}_1 + c_2T\mathbf{v}_2 + \cdots + c_nT\mathbf{v}_n = c_1\mathbf{y}_1 + c_2\mathbf{y}_2 + \cdots + c_n\mathbf{y}_n$. Thus $(T\mathbf{x})_{B_2} = A_T(\mathbf{x})_{B_1}$. The proof of uniqueness is exactly as in the proof of uniqueness in Theorem 1. ∎

The following useful result follows immediately from Theorem 4.7.5 on page 259 and generalizes Theorem 2. Its proof is left as an exercise (see Problem 37).

THEOREM 4   Let V and W be finite dimensional vector spaces with dim $V = n$. Let $T: V \rightarrow W$ be a linear transformation and let A_T be a matrix representation of T. Then:

  i. $\rho(T) = \rho(A_T)$

  ii. $\nu(T) = \nu(A_T)$

  iii. $\nu(T) + \rho(T) = n$

  ∎

EXAMPLE 6   **The Matrix Representation of a Transformation from P_2 to P_3**   Let $T: P_2 \rightarrow P_3$ be defined by $(Tp)(x) = xp(x)$. Find A_T and use it to determine the kernel and range of T.

Solution   Using the standard basis $B_1 = \{1, x, x^2\}$ in P_2 and $B_2 = \{1, x, x^2, x^3\}$ in P_3, we

have $(T(1))_{B_2} = (x)_{B_2} = \begin{pmatrix} 0 \\ 1 \\ 0 \\ 0 \end{pmatrix}$, $(T(x))_{B_2} = (x^2)_{B_2} = \begin{pmatrix} 0 \\ 0 \\ 1 \\ 0 \end{pmatrix}$, and $(T(x^2))_{B_2} = (x^3)_{B_2} =$

$\begin{pmatrix} 0 \\ 0 \\ 0 \\ 1 \end{pmatrix}$. Thus $A_T = \begin{pmatrix} 0 & 0 & 0 \\ 1 & 0 & 0 \\ 0 & 1 & 0 \\ 0 & 0 & 1 \end{pmatrix}$. Clearly $\rho(A) = 3$ and a basis for R_A is

$\left\{ \begin{pmatrix} 0 \\ 1 \\ 0 \\ 0 \end{pmatrix}, \begin{pmatrix} 0 \\ 0 \\ 1 \\ 0 \end{pmatrix}, \begin{pmatrix} 0 \\ 0 \\ 0 \\ 1 \end{pmatrix} \right\}$. Therefore range $T = \text{span } \{x, x^2, x^3\}$. Since

$$\nu(A) = 3 - \rho(A) = 0, \text{ we see that ker } T = \{0\}. \qquad \blacksquare$$

EXAMPLE 7   **The Matrix Representation of a Transformation from P_3 to P_2**
Let $T: P_3 \rightarrow P_2$ be defined by $T(a_0 + a_1x + a_2x^2 + a_3x^3) = a_1 + a_2x^2$. Compute A_T and use it to find the kernel and range of T.

Solution   Using the standard bases $B_1 = (1, x, x^2, x^3\}$ in P_3 and $B_2 = \{1, x, x^2\}$ in P_2, we

immediately see that $(T(1))_{B_2} = \begin{pmatrix} 0 \\ 0 \\ 0 \end{pmatrix}$, $(T(x))_{B_2} = \begin{pmatrix} 1 \\ 0 \\ 0 \end{pmatrix}$, $(T(x^2))_{B_2} = \begin{pmatrix} 0 \\ 0 \\ 1 \end{pmatrix}$, and

$(T(x^3))_{B_2} = \begin{pmatrix} 0 \\ 0 \\ 0 \end{pmatrix}$. Thus $A_T = \begin{pmatrix} 0 & 1 & 0 & 0 \\ 0 & 0 & 0 & 0 \\ 0 & 0 & 1 & 0 \end{pmatrix}$. Clearly $\rho(A) = 2$ and a basis for R_A

is $\left\{ \begin{pmatrix} 1 \\ 0 \\ 0 \end{pmatrix}, \begin{pmatrix} 0 \\ 0 \\ 1 \end{pmatrix} \right\}$ so that range $T = \text{span } \{1, x^2\}$. Then $\nu(A) = 4 - 2 = 2$; and if

$A_T \begin{pmatrix} a_0 \\ a_1 \\ a_2 \\ a_3 \end{pmatrix} = \begin{pmatrix} 0 \\ 0 \\ 0 \end{pmatrix}$, then $a_1 = 0$ and $a_2 = 0$. Hence a_0 and a_3 are arbitrary and a basis

for N_A is $\left\{ \begin{pmatrix} 1 \\ 0 \\ 0 \\ 0 \end{pmatrix}, \begin{pmatrix} 0 \\ 0 \\ 0 \\ 1 \end{pmatrix} \right\}$, so that a basis for ker T is $\{1, x^3\}$.

$\blacksquare$

In all the examples of this section we have obtained the matrix A_T by using the standard basis in each vector space. However, Theorem 3 holds for any bases in V and W. The next example illustrates this.

EXAMPLE 8   **Finding a Matrix Representation Relative to Two Nonstandard Bases in $\mathbb{R}^2$**

Let $T: \mathbb{R}^2 \to \mathbb{R}^2$ be defined by $T\begin{pmatrix} x \\ y \end{pmatrix} = \begin{pmatrix} x + y \\ x - y \end{pmatrix}$. Using the bases $B_1 = B_2 = \left\{ \begin{pmatrix} 1 \\ -1 \end{pmatrix}, \begin{pmatrix} -3 \\ 2 \end{pmatrix} \right\}$, compute A_T.

Solution   We have $T\begin{pmatrix} 1 \\ -1 \end{pmatrix} = \begin{pmatrix} 0 \\ 2 \end{pmatrix}$ and $T\begin{pmatrix} -3 \\ 2 \end{pmatrix} = \begin{pmatrix} -1 \\ -5 \end{pmatrix}$. Since $\begin{pmatrix} 0 \\ 2 \end{pmatrix} = -6\begin{pmatrix} 1 \\ -1 \end{pmatrix} - 2\begin{pmatrix} -3 \\ 2 \end{pmatrix}$, we find that $\begin{pmatrix} 0 \\ 2 \end{pmatrix}_{B_2} = \begin{pmatrix} -6 \\ -2 \end{pmatrix}$. Similarly, $\begin{pmatrix} -1 \\ -5 \end{pmatrix} = 17\begin{pmatrix} 1 \\ -1 \end{pmatrix} + 6\begin{pmatrix} -3 \\ 2 \end{pmatrix}$ so that $\begin{pmatrix} -1 \\ -5 \end{pmatrix}_{B_2} = \begin{pmatrix} 17 \\ 6 \end{pmatrix}$. Thus $A_T = \begin{pmatrix} -6 & 17 \\ -2 & 6 \end{pmatrix}$. To compute $T\begin{pmatrix} -4 \\ 7 \end{pmatrix}$, for example, we first write $\begin{pmatrix} -4 \\ 7 \end{pmatrix} = -13\begin{pmatrix} 1 \\ -1 \end{pmatrix} - 3\begin{pmatrix} -3 \\ 2 \end{pmatrix}$ so $\begin{pmatrix} -4 \\ 7 \end{pmatrix}_{B_1} = \begin{pmatrix} -13 \\ -3 \end{pmatrix}$. Then $\left(T\begin{pmatrix} -4 \\ 7 \end{pmatrix} \right)_{B_2} = A_T\begin{pmatrix} -4 \\ 7 \end{pmatrix}_{B_1} = A_T\begin{pmatrix} -13 \\ -3 \end{pmatrix} = \begin{pmatrix} -6 & 17 \\ -2 & 6 \end{pmatrix}\begin{pmatrix} -13 \\ -3 \end{pmatrix} = \begin{pmatrix} 27 \\ 8 \end{pmatrix}$. Hence $T\begin{pmatrix} -4 \\ 7 \end{pmatrix} = 27\begin{pmatrix} 1 \\ -1 \end{pmatrix} + 8\begin{pmatrix} -3 \\ 2 \end{pmatrix} = \begin{pmatrix} 3 \\ -11 \end{pmatrix}$. Note that $T\begin{pmatrix} -4 \\ 7 \end{pmatrix} = \begin{pmatrix} -4 + 7 \\ -4 - 7 \end{pmatrix} = \begin{pmatrix} 3 \\ -11 \end{pmatrix}$, which verifies our calculations.   ∎

To avoid confusion, we shall, unless explicitly stated otherwise, always compute the matrix A_T with respect to the standard basis.† If $T: V \to V$ is a linear transformation and some other basis B is used, then we refer to A_T as *the transformation matrix of T with respect to the basis B.* Thus, in the last example, $A_T = \begin{pmatrix} -6 & 17 \\ -2 & 6 \end{pmatrix}$ is the transformation matrix of T with respect to the basis $\left\{ \begin{pmatrix} 1 \\ -1 \end{pmatrix}, \begin{pmatrix} -3 \\ 2 \end{pmatrix} \right\}$.

Before leaving this section, we must answer an obvious question. Why bother to use a basis other than the standard basis since the computations are, as in Example 8, a good deal more complicated? The answer is that it is often possible to find a basis B^* in $\mathbb{R}^n$ so that the transformation matrix with respect to B^* is a diagonal matrix. Diagonal matrices are very easy to work with, and, as we shall see in Chapter 6, there are numerous advantages to writing a matrix in a diagonal form.

EXAMPLE 9   **The Matrix Representation of a Linear Transformation Relative to Two Nonstandard Bases in $\mathbb{R}^2$ May Be Diagonal**   Let $T: \mathbb{R}^2 \to \mathbb{R}^2$ be defined by $T\begin{pmatrix} x \\ y \end{pmatrix} = \begin{pmatrix} 12x + 10y \\ -15x - 13y \end{pmatrix}$. Find A_T with respect to the basis $B_1 = B_2 = \left\{ \begin{pmatrix} 1 \\ -1 \end{pmatrix}, \begin{pmatrix} 2 \\ -3 \end{pmatrix} \right\}$.

---

† That is, in any space where we have defined a standard basis.

Solution $T\begin{pmatrix} 1 \\ -1 \end{pmatrix} = \begin{pmatrix} 2 \\ -2 \end{pmatrix}$ and $T\begin{pmatrix} 2 \\ -3 \end{pmatrix} = \begin{pmatrix} -6 \\ 9 \end{pmatrix}$. Then $\begin{pmatrix} 2 \\ -2 \end{pmatrix} = 2\begin{pmatrix} 1 \\ -1 \end{pmatrix} + 0\begin{pmatrix} 2 \\ -3 \end{pmatrix}$ so

$\begin{pmatrix} 2 \\ -2 \end{pmatrix}_{B_2} = \begin{pmatrix} 2 \\ 0 \end{pmatrix}$. Similarly, $\begin{pmatrix} -6 \\ 9 \end{pmatrix} = 0\begin{pmatrix} 1 \\ -1 \end{pmatrix} - 3\begin{pmatrix} 2 \\ -3 \end{pmatrix}$ so $\begin{pmatrix} -6 \\ 9 \end{pmatrix}_{B_2} = \begin{pmatrix} 0 \\ -3 \end{pmatrix}$.

Thus $A_T = \begin{pmatrix} 2 & 0 \\ 0 & -3 \end{pmatrix}$.

There is another way to solve this problem. The vectors $\begin{pmatrix} 1 \\ -1 \end{pmatrix}$ and $\begin{pmatrix} 2 \\ -3 \end{pmatrix}$ are

written in terms of the standard basis $S = \left\{ \begin{pmatrix} 1 \\ 0 \end{pmatrix}, \begin{pmatrix} 0 \\ 1 \end{pmatrix} \right\}$. That is, $\begin{pmatrix} 1 \\ -1 \end{pmatrix} = 1\begin{pmatrix} 1 \\ 0 \end{pmatrix} +$

$(-1)\begin{pmatrix} 0 \\ 1 \end{pmatrix}$ and $\begin{pmatrix} 2 \\ -3 \end{pmatrix} = 2\begin{pmatrix} 1 \\ 0 \end{pmatrix} + (-3)\begin{pmatrix} 0 \\ 1 \end{pmatrix}$. Thus the matrix $A = \begin{pmatrix} 1 & 2 \\ -1 & -3 \end{pmatrix}$ is the

matrix whose first and second columns represent the expansions of the vectors in B_1 in terms of the standard basis. Then, from the procedure outlined on page 275, the

matrix $A^{-1} = \begin{pmatrix} 3 & 2 \\ -1 & -1 \end{pmatrix}$ is the transition matrix from S to B_1. Similarly, the

matrix A is the transition matrix from B_1 to S (see Problem 4.9.38, page 281). Now suppose that $\mathbf{x}$ is written in terms of B_1. Then $A\mathbf{x}$ is the same vector now written in

terms of S. Let $C = \begin{pmatrix} 12 & 10 \\ -15 & -13 \end{pmatrix}$. Then $CA\mathbf{x} = T(A\mathbf{x})$ is the image of $A\mathbf{x}$ written

in terms of S. Finally, since we want $T(A\mathbf{x})$ in terms of B_1 (that was the problem), we multiply on the left by the transition matrix A^{-1} to obtain $(T\mathbf{x})_{B_1} = (A^{-1}CA)(\mathbf{x})_{B_1}$. That is,

$$A_T = A^{-1}CA = \begin{pmatrix} 3 & 2 \\ -1 & -1 \end{pmatrix}\begin{pmatrix} 12 & 10 \\ -15 & -13 \end{pmatrix}\begin{pmatrix} 1 & 2 \\ -1 & -3 \end{pmatrix} = \begin{pmatrix} 3 & 2 \\ -1 & -1 \end{pmatrix}\begin{pmatrix} 2 & -6 \\ -2 & 9 \end{pmatrix}$$

$$= \begin{pmatrix} 2 & 0 \\ 0 & -3 \end{pmatrix}$$

as before. We summarize this result below. ∎

THEOREM 5   Let $T: \mathbb{R}^n \to \mathbb{R}^m$ be a linear transformation. Suppose that C is the transformation matrix of T with respect to the standard bases S_n and S_m in $\mathbb{R}^n$ and $\mathbb{R}^m$, respectively. Let A_1 be the transition matrix from S_n to the basis B_1 in $\mathbb{R}^n$ and let A_2 be the transition matrix from S_m to the basis B_2 in $\mathbb{R}^m$. If A_T denotes the transformation matrix of T with respect to the bases B_1 and B_2, then

$$\boxed{A_T = A_2^{-1}CA_1} \tag{3}$$
∎

In Example 9 we saw that by looking at the linear transformation T with respect to a new basis, the transformation matrix A_T turned out to be a diagonal matrix. We shall return to this "diagonalizing" procedure in Section 6.3. Given a linear transformation from $\mathbb{R}^n$ to $\mathbb{R}^n$, we shall see that it is often possible to find a basis B such that the transformation matrix of T with respect to B will be diagonal.

The Geometry of Linear Transformations from $\mathbb{R}^2$ to $\mathbb{R}^2$

Let $T: \mathbb{R}^2 \to \mathbb{R}^2$ be a linear transformation with matrix representation A_T. We now show that if A_T is invertible, then T can be written as a succession of one or more of four special transformations, called **expansions, compressions, reflections,** and **shears.**

Expansions Along the x- or y-Axis. An **expansion along the x-axis** is a linear transformation that multiplies the x-coordinate of a vector in $\mathbb{R}^2$ by a constant $c > 1$. That is,

$$T\binom{x}{y} = \binom{cx}{y}$$

Then $T\binom{1}{0} = \binom{c}{0}$ and $T\binom{0}{1} = \binom{0}{1}$, so if $A_T = \begin{pmatrix} c & 0 \\ 0 & 1 \end{pmatrix}$, we have

$$T\binom{x}{y} = A\binom{x}{y} = \begin{pmatrix} c & 0 \\ 0 & 1 \end{pmatrix}\binom{x}{y} = \binom{cx}{y}$$

Similarly, **an expansion along the y-axis** is a linear transformation that multiplies the y-coordinate of every vector in $\mathbb{R}^2$ by a constant $c > 1$. As above,

If $T\binom{x}{y} = \binom{x}{cy}$, then the matrix representation of T is $A_T = \begin{pmatrix} 1 & 0 \\ 0 & c \end{pmatrix}$ so that
$\begin{pmatrix} 1 & 0 \\ 0 & c \end{pmatrix}\binom{x}{y} = \binom{x}{cy}$.

In Figure 5.5 we depict an expansion along each axis.

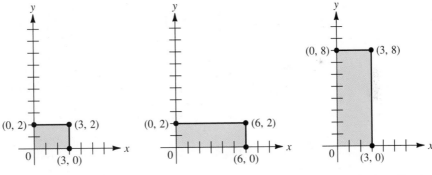

(a) We start with   (b) Expansion in the x-direction (c) Expansion in the y-direction
this rectangle   with $c = 2$ with $c = 4$

Figure 5.5   Two expansions

Compression Along the x-Axis or y-Axis. A **compression** along the x- or y-axis is a linear transformation that multiplies the x- or y-coordinate of a vector in $\mathbb{R}^2$ by a positive constant $c < 1$. The matrix representations of a compression are the same as for an expansion except that for a compression $0 < c < 1$, whereas for an expansion $c > 1$. Two compressions are illustrated in Figure 5.6.

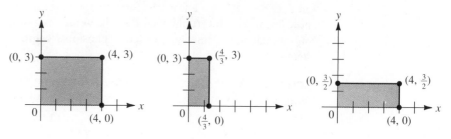

(a) We start with this rectangle   (b) Compression along the x-axis with $c = \frac{1}{3}$ (c) Compression along the y-axis with $c = \frac{1}{2}$

Figure 5.6   Two compressions

Reflections. There are three kinds of reflections that will be of interest to us. In Example 5.1.1 on page 330 we saw that the transformation

$$T\begin{pmatrix} x \\ y \end{pmatrix} = \begin{pmatrix} x \\ -y \end{pmatrix}$$

reflects a vector in $\mathbb{R}^2$ about the x-axis (see Figure 5.1). In Example 5.1.6 on page 333 we saw that the transformation

$$T\begin{pmatrix} x \\ y \end{pmatrix} = \begin{pmatrix} -x \\ y \end{pmatrix}$$

reflects a vector in $\mathbb{R}^2$ about the y-axis (see Figure 5.2). Now

$$\begin{pmatrix} 1 & 0 \\ 0 & -1 \end{pmatrix}\begin{pmatrix} x \\ y \end{pmatrix} = \begin{pmatrix} x \\ -y \end{pmatrix} \quad \text{and} \quad \begin{pmatrix} -1 & 0 \\ 0 & 1 \end{pmatrix}\begin{pmatrix} x \\ y \end{pmatrix} = \begin{pmatrix} -x \\ y \end{pmatrix}$$

so

$\begin{pmatrix} 1 & 0 \\ 0 & -1 \end{pmatrix}$ is the matrix representation of the reflection about the x-axis

and

$\begin{pmatrix} -1 & 0 \\ 0 & 1 \end{pmatrix}$ is the matrix representation of the reflection about the y-axis.

Finally, the mapping $T\begin{pmatrix} x \\ y \end{pmatrix} = \begin{pmatrix} y \\ x \end{pmatrix}$, which interchanges x and y, has the effect of reflecting a vector in $\mathbb{R}^2$ **about the line** $y = x$ (see Figure 5.7).

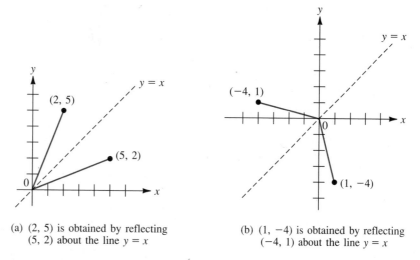

(a) (2, 5) is obtained by reflecting
    (5, 2) about the line $y = x$

(b) (1, −4) is obtained by reflecting
    (−4, 1) about the line $y = x$

Figure 5.7   Reflecting a vector in $\mathbb{R}^2$ about the line $y = x$

If $T\begin{pmatrix} x \\ y \end{pmatrix} = \begin{pmatrix} y \\ x \end{pmatrix}$, then $T\begin{pmatrix} 1 \\ 0 \end{pmatrix} = \begin{pmatrix} 0 \\ 1 \end{pmatrix}$ and $T\begin{pmatrix} 0 \\ 1 \end{pmatrix} = \begin{pmatrix} 1 \\ 0 \end{pmatrix}$, so the matrix representation of the linear transformation that reflects a vector in $\mathbb{R}^2$ about the line $y = x$ is $A = \begin{pmatrix} 0 & 1 \\ 1 & 0 \end{pmatrix}$.

Shears. A **shear along the x-axis** is a transformation that takes a vector $\begin{pmatrix} x \\ y \end{pmatrix}$ into a new vector $\begin{pmatrix} x + cy \\ y \end{pmatrix}$, where c is a constant that may be positive or negative. Two shears along the x-axis are illustrated in Figure 5.8.

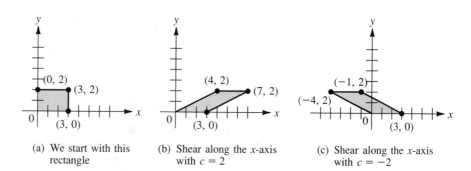

(a) We start with this
    rectangle

(b) Shear along the x-axis
    with $c = 2$

(c) Shear along the x-axis
    with $c = -2$

Figure 5.8   Two shears along the x-axis

Let T be a shear along the x-axis. Then

$$T\begin{pmatrix} 1 \\ 0 \end{pmatrix} = \begin{pmatrix} 1 \\ 0 \end{pmatrix} \quad \text{and} \quad T\begin{pmatrix} 0 \\ 1 \end{pmatrix} = \begin{pmatrix} 0 + c \cdot 1 \\ 1 \end{pmatrix} = \begin{pmatrix} c \\ 1 \end{pmatrix}$$

so the matrix representation of T is $\begin{pmatrix} 1 & c \\ 0 & 1 \end{pmatrix}$. For example, in Figure 5.8b, $c = 2$, so

$$A_T = \begin{pmatrix} 1 & 2 \\ 0 & 1 \end{pmatrix} \quad \text{and} \quad A_T\begin{pmatrix} 3 \\ 2 \end{pmatrix} = \begin{pmatrix} 1 & 2 \\ 0 & 1 \end{pmatrix}\begin{pmatrix} 3 \\ 2 \end{pmatrix} = \begin{pmatrix} 7 \\ 2 \end{pmatrix}$$

In Figure 5.8c, $c = -2$, so $A_T = \begin{pmatrix} 1 & -2 \\ 0 & 1 \end{pmatrix}$,

$$A_T\begin{pmatrix} 3 \\ 2 \end{pmatrix} = \begin{pmatrix} 1 & -2 \\ 0 & 1 \end{pmatrix}\begin{pmatrix} 3 \\ 2 \end{pmatrix} = \begin{pmatrix} -1 \\ 2 \end{pmatrix} \quad \text{and}$$

$$A_T\begin{pmatrix} 0 \\ 2 \end{pmatrix} = \begin{pmatrix} 1 & -2 \\ 0 & 1 \end{pmatrix}\begin{pmatrix} 0 \\ 2 \end{pmatrix} = \begin{pmatrix} -4 \\ 2 \end{pmatrix}$$

Note that $A_T\begin{pmatrix} 3 \\ 0 \end{pmatrix} = \begin{pmatrix} 1 & -2 \\ 0 & 1 \end{pmatrix}\begin{pmatrix} 3 \\ 0 \end{pmatrix} = \begin{pmatrix} 3 \\ 0 \end{pmatrix}$. That is, vectors with y-coordinate zero are left unchanged by shears along the x-axis.

A **shear along the y-axis** is a transformation that takes a vector $\begin{pmatrix} x \\ y \end{pmatrix}$ into a new vector $\begin{pmatrix} x \\ y + cx \end{pmatrix}$, where c is a constant that may be positive or negative. Two shears along the y-axis are illustrated in Figure 5.9.

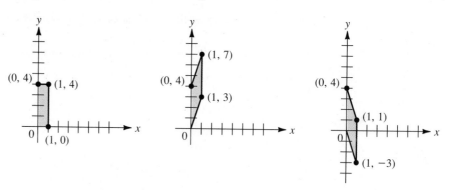

(a) We start with this rectangle   (b) Shear along the y-axis with $c = 3$ (c) Shear along the y-axis with $c = -3$

Figure 5.9   Two shears along the y-axis

If T is a shear along the y-axis, then

$$T\begin{pmatrix} 1 \\ 0 \end{pmatrix} = \begin{pmatrix} 1 \\ c \end{pmatrix} \quad \text{and} \quad T\begin{pmatrix} 0 \\ 1 \end{pmatrix} = \begin{pmatrix} 0 \\ 1 \end{pmatrix}, \text{ so } A_T = \begin{pmatrix} 1 & 0 \\ c & 1 \end{pmatrix}.$$

For example, in Figure 5.9b $c = 3$, so

$$A_T = \begin{pmatrix} 1 & 0 \\ 3 & 1 \end{pmatrix} \quad \text{and} \quad A_T\begin{pmatrix} 1 \\ 4 \end{pmatrix} = \begin{pmatrix} 1 & 0 \\ 3 & 1 \end{pmatrix}\begin{pmatrix} 1 \\ 4 \end{pmatrix} = \begin{pmatrix} 1 \\ 7 \end{pmatrix}$$

In Figure 5.9c $c = -3$, so

$$A_T\begin{pmatrix} 1 \\ 4 \end{pmatrix} = \begin{pmatrix} 1 & 0 \\ -3 & 1 \end{pmatrix}\begin{pmatrix} 1 \\ 4 \end{pmatrix} = \begin{pmatrix} 1 \\ 1 \end{pmatrix} \quad \text{and} \quad A_T\begin{pmatrix} 1 \\ 0 \end{pmatrix} = \begin{pmatrix} 1 & 0 \\ -3 & 1 \end{pmatrix}\begin{pmatrix} 1 \\ 0 \end{pmatrix} = \begin{pmatrix} 1 \\ -3 \end{pmatrix}$$

Note that $A_T\begin{pmatrix} 0 \\ 4 \end{pmatrix} = \begin{pmatrix} 1 & 0 \\ -3 & 1 \end{pmatrix}\begin{pmatrix} 0 \\ 4 \end{pmatrix} = \begin{pmatrix} 0 \\ 4 \end{pmatrix}$. That is, vectors with x-coordinate zero are left unchanged by shears along the y-axis.

In Table 5.1 we summarize these types of linear transformations.

Table 5.1   Special linear transformations from $\mathbb{R}^2$ to $\mathbb{R}^2$

Transformation	Matrix Representation of Transformation; A_T
Expansion along the x-axis	$\begin{pmatrix} c & 0 \\ 0 & 1 \end{pmatrix}, c > 1$
Expansion along the y-axis	$\begin{pmatrix} 1 & 0 \\ 0 & c \end{pmatrix}, c > 1$
Compression along the x-axis	$\begin{pmatrix} c & 0 \\ 0 & 1 \end{pmatrix}, 0 < c < 1$
Compression along the y-axis	$\begin{pmatrix} 1 & 0 \\ 0 & c \end{pmatrix}, 0 < c < 1$
Reflection about the line $y = x$	$\begin{pmatrix} 0 & 1 \\ 1 & 0 \end{pmatrix}$
Reflection about the x-axis	$\begin{pmatrix} 1 & 0 \\ 0 & -1 \end{pmatrix}$
Reflection about the y-axis	$\begin{pmatrix} -1 & 0 \\ 0 & 1 \end{pmatrix}$
Shear along the x-axis	$\begin{pmatrix} 1 & c \\ 0 & 1 \end{pmatrix}$
Shear along the y-axis	$\begin{pmatrix} 1 & 0 \\ c & 1 \end{pmatrix}$

In Section 1.11 we discussed elementary matrices. Multiplication of a matrix by an elementary matrix has the effect of performing an elementary row operation on that matrix. Table 5.2 lists the elementary matrices in $\mathbb{R}^2$.

Table 5.2   The Elementary Matrices in $\mathbb{R}^2$

Elementary Row Operation	Elementary Matrix	Illustration
$R_1 \rightarrow cR_1$	$\begin{pmatrix} c & 0 \\ 0 & 1 \end{pmatrix}$	$\begin{pmatrix} c & 0 \\ 0 & 1 \end{pmatrix}\begin{pmatrix} x & y \\ z & w \end{pmatrix} = \begin{pmatrix} cx & cy \\ z & w \end{pmatrix}$
$R_2 \rightarrow cR_2$	$\begin{pmatrix} 1 & 0 \\ 0 & c \end{pmatrix}$	$\begin{pmatrix} 1 & 0 \\ 0 & c \end{pmatrix}\begin{pmatrix} x & y \\ z & w \end{pmatrix} = \begin{pmatrix} x & y \\ cz & cw \end{pmatrix}$
$R_1 \rightarrow R_1 + cR_2$	$\begin{pmatrix} 1 & c \\ 0 & 1 \end{pmatrix}$	$\begin{pmatrix} 1 & c \\ 0 & 1 \end{pmatrix}\begin{pmatrix} x & y \\ z & w \end{pmatrix} = \begin{pmatrix} x + cz & y + cw \\ z & w \end{pmatrix}$
$R_2 \rightarrow R_2 + cR_1$	$\begin{pmatrix} 1 & 0 \\ c & 1 \end{pmatrix}$	$\begin{pmatrix} 1 & 0 \\ c & 1 \end{pmatrix}\begin{pmatrix} x & y \\ z & w \end{pmatrix} = \begin{pmatrix} x & y \\ z + cx & w + cy \end{pmatrix}$
$R_1 \rightleftarrows R_2$	$\begin{pmatrix} 0 & 1 \\ 1 & 0 \end{pmatrix}$	$\begin{pmatrix} 0 & 1 \\ 1 & 0 \end{pmatrix}\begin{pmatrix} x & y \\ z & w \end{pmatrix} = \begin{pmatrix} z & w \\ x & y \end{pmatrix}$

THEOREM 6   Every 2×2 elementary matrix E is one of the following:

  i.   The matrix representation of an expansion along the x- or y-axis
  ii. The matrix representation of a compression along the x- or y-axis
  iii. The matrix representation of a reflection about the line $y = x$
  iv. The matrix representation of a shear along the x- or y-axis
  v.   The matrix representation of a reflection about the x- or y-axis
  vi. The product of the matrix representation of a reflection about the x- or y-axis and the matrix representation of an expansion or compression

Proof   We refer to Tables 5.1 and 5.2.

Case 1: $E = \begin{pmatrix} c & 0 \\ 0 & 1 \end{pmatrix}$, $c > 0$ This is the matrix representation of an expansion along the x-axis if $c > 1$ or a compression along the x-axis if $0 < c < 1$.

Case 2: $E = \begin{pmatrix} c & 0 \\ 0 & 1 \end{pmatrix}$, $c < 0$

  Case 2a: $c = -1$ Then $E = \begin{pmatrix} -1 & 0 \\ 0 & 1 \end{pmatrix}$, which is the matrix representation of a reflection about the y-axis.

  Case 2b: $c < 0$, $c \neq -1$.   Then $-c > 0$ and

$$E = \begin{pmatrix} c & 0 \\ 0 & 1 \end{pmatrix} = \begin{pmatrix} -1 & 0 \\ 0 & 1 \end{pmatrix}\begin{pmatrix} -c & 0 \\ 0 & 1 \end{pmatrix}$$

which is the product of the matrix representation of a reflection about the y-axis and the matrix representation of an expansion (if $-c > 1$) or compression (if $0 < -c < 1$) along the x-axis.

Case 3: $E = \begin{pmatrix} 1 & 0 \\ 0 & c \end{pmatrix}$, $c > 0$ Same as case 1 with the y-axis replacing the x-axis.

Case 4: $E = \begin{pmatrix} 1 & 0 \\ 0 & c \end{pmatrix}$, $c < 0$ Same as case 2 with the axes interchanged.

Case 5: $E = \begin{pmatrix} 1 & c \\ 0 & 1 \end{pmatrix}$ This is the matrix representation of a shear along the x-axis.

Case 6: $E = \begin{pmatrix} 1 & 0 \\ c & 1 \end{pmatrix}$ This is the matrix representation of a shear along the y-axis.

Case 7: $E = \begin{pmatrix} 0 & 1 \\ 1 & 0 \end{pmatrix}$ This is the matrix representation of a reflection about the line $y = x$. ∎

In Theorem 1.11.3 on page 89 we showed that every invertible matrix can be written as the product of elementary matrices. In Theorem 6 we showed that every elementary matrix in $\mathbb{R}^2$ can be written as a product of matrix representations of expansions, compressions, shears, and reflections. Thus we have the following result.

THEOREM 7   Let $T: \mathbb{R}^2 \to \mathbb{R}^2$ be a linear transformation such that its matrix representation is invertible. Then T can be obtained as a succession of expansions, compressions, shears, and reflections. ∎

EXAMPLE 10   **Decomposing a Linear Transformation in $\mathbb{R}^2$ into a Succession of Expansions, Compressions, Shears, and Reflections**   Consider the transformation $T: \mathbb{R}^2 \to \mathbb{R}^2$ with matrix representation $A_T = \begin{pmatrix} 1 & 2 \\ 3 & 4 \end{pmatrix}$. Using the technique of Section 1.11 (look at Example 3 on page 90), we can write A_T as a product of three elementary matrices:

$$\begin{pmatrix} 1 & 2 \\ 3 & 4 \end{pmatrix} = \begin{pmatrix} 1 & 0 \\ 3 & 1 \end{pmatrix}\begin{pmatrix} 1 & 0 \\ 0 & -2 \end{pmatrix}\begin{pmatrix} 1 & 2 \\ 0 & 1 \end{pmatrix} \tag{4}$$

Now

$$\begin{pmatrix} 1 & 0 \\ 3 & 1 \end{pmatrix}$$ represents a shear along the y-axis (with $c = 3$)

$$\begin{pmatrix} 1 & 2 \\ 0 & 1 \end{pmatrix}$$ represents a shear along the x-axis (with $c = 2$)

$$\begin{pmatrix} 1 & 0 \\ 0 & -2 \end{pmatrix} = \begin{pmatrix} 1 & 0 \\ 0 & -1 \end{pmatrix}\begin{pmatrix} 1 & 0 \\ 0 & 2 \end{pmatrix}$$ represents an expansion along the y-axis (with $c = 2$) followed by a reflection about the x-axis

Thus to apply T to a vector in $\mathbb{R}^2$, we:

i. Shear along the x-axis with $c = 2$.

ii. Expand along the y-axis with $c = 2$.

iii. Reflect about the x-axis.

iv. Shear along the y-axis with $c = 3$.

Note that we do these operations in the reverse order in which we write the matrices in (4).

To illustrate this, suppose that $\mathbf{v} = \begin{pmatrix} 3 \\ -2 \end{pmatrix}$.

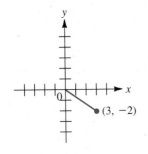

Then

$$T\mathbf{v} = A_T\mathbf{v} = \begin{pmatrix} 1 & 2 \\ 3 & 4 \end{pmatrix}\begin{pmatrix} 3 \\ -2 \end{pmatrix} = \begin{pmatrix} -1 \\ 1 \end{pmatrix}$$

(a) We start with this vector

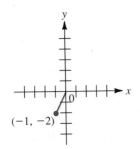

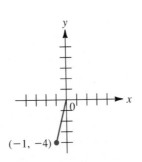

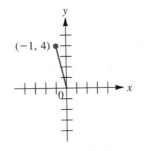

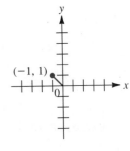

(b) Vector obtained by shearing along the x-axis with $c = 2$

(c) Vector obtained by expanding along the y-axis with $c = 2$

(d) Vector obtained by reflecting about the x-axis

(e) Vector obtained by shearing along the y-axis with $c = 3$

Figure 5.10   Decomposing the linear transformation $T\begin{pmatrix} 3 \\ -2 \end{pmatrix} = \begin{pmatrix} 1 & 2 \\ 3 & 4 \end{pmatrix}\begin{pmatrix} 3 \\ -2 \end{pmatrix}$ into a succession of shears, expansions, and reflections

Using the operations (i)–(iv), we have

$$\begin{pmatrix} 3 \\ -2 \end{pmatrix} \xrightarrow{\text{Shear}} \begin{pmatrix} 1 & 2 \\ 0 & 1 \end{pmatrix}\begin{pmatrix} 3 \\ -2 \end{pmatrix} = \begin{pmatrix} -1 \\ -2 \end{pmatrix} \xrightarrow{\text{Expansion}} \begin{pmatrix} 1 & 0 \\ 0 & 2 \end{pmatrix}\begin{pmatrix} -1 \\ -2 \end{pmatrix} = \begin{pmatrix} -1 \\ -4 \end{pmatrix}$$

$$\xrightarrow{\text{Reflection}} \begin{pmatrix} 1 & 0 \\ 0 & -1 \end{pmatrix}\begin{pmatrix} -1 \\ -4 \end{pmatrix} = \begin{pmatrix} -1 \\ 4 \end{pmatrix} \xrightarrow{\text{Shear}} \begin{pmatrix} 1 & 0 \\ 3 & 1 \end{pmatrix}\begin{pmatrix} -1 \\ 4 \end{pmatrix} = \begin{pmatrix} -1 \\ 1 \end{pmatrix}$$

We sketch these steps in Figure 5.10.

■

PROBLEMS   5.3

In Problems 1–30 find the matrix representation A_T of the linear transformation T, ker T, range T, $\nu(T)$, and $\rho(T)$. Unless otherwise stated, assume that B_1 and B_2 are standard bases.

1. $T: \mathbb{R}^2 \to \mathbb{R}^2$; $T\begin{pmatrix} x \\ y \end{pmatrix} = \begin{pmatrix} x - 2y \\ -x + y \end{pmatrix}$

2. $T: \mathbb{R}^2 \to \mathbb{R}^3$; $T\begin{pmatrix} x \\ y \end{pmatrix} = \begin{pmatrix} x + y \\ x - y \\ 2x + 3y \end{pmatrix}$

3. $T: \mathbb{R}^3 \to \mathbb{R}^2$; $T\begin{pmatrix} x \\ y \\ z \end{pmatrix} = \begin{pmatrix} x - y + z \\ -2x + 2y - 2z \end{pmatrix}$

4. $T: \mathbb{R}^2 \to \mathbb{R}^2$; $T\begin{pmatrix} x \\ y \end{pmatrix} = \begin{pmatrix} ax + by \\ cx + dy \end{pmatrix}$

5. $T: \mathbb{R}^3 \to \mathbb{R}^3$; $T\begin{pmatrix} x \\ y \\ z \end{pmatrix} = \begin{pmatrix} x - y + 2z \\ 3x + y + 4z \\ 5x - y + 8z \end{pmatrix}$

6. $T: \mathbb{R}^3 \to \mathbb{R}^3$; $T\begin{pmatrix} x \\ y \\ z \end{pmatrix} = \begin{pmatrix} -x + 2y + z \\ 2x - 4y - 2z \\ -3x + 6y + 3z \end{pmatrix}$

7. $T: \mathbb{R}^4 \to \mathbb{R}^3$; $T\begin{pmatrix} x \\ y \\ z \\ w \end{pmatrix} = \begin{pmatrix} x - y + 2z + 3w \\ y + 4z + 3w \\ x + 6z + 6w \end{pmatrix}$

8. $T: \mathbb{R}^4 \to \mathbb{R}^4$; $T\begin{pmatrix} x \\ y \\ z \\ w \end{pmatrix} = \begin{pmatrix} x - y + 2z + w \\ -x + z + 2w \\ x - 2y + 5z + 4w \\ 2x - y + z - w \end{pmatrix}$

9. $T: \mathbb{R}^2 \to \mathbb{R}^2$; $T\begin{pmatrix} x \\ y \end{pmatrix} = \begin{pmatrix} x - 2y \\ 2x + y \end{pmatrix}$; $B_1 = B_2 = \left\{ \begin{pmatrix} 1 \\ -2 \end{pmatrix}, \begin{pmatrix} 3 \\ 2 \end{pmatrix} \right\}$

10. $T: \mathbb{R}^2 \to \mathbb{R}^2$; $T\begin{pmatrix} x \\ y \end{pmatrix} = \begin{pmatrix} 4x - y \\ 3x + 2y \end{pmatrix}$; $B_1 = B_2 = \left\{ \begin{pmatrix} -1 \\ 1 \end{pmatrix}, \begin{pmatrix} 4 \\ 3 \end{pmatrix} \right\}$

11. $T: \mathbb{R}^3 \to \mathbb{R}^2; T\begin{pmatrix} x \\ y \\ z \end{pmatrix} = \begin{pmatrix} 2x + y + z \\ y - 3z \end{pmatrix};$

$$B_1 = \left\{ \begin{pmatrix} 1 \\ 0 \\ 1 \end{pmatrix}, \begin{pmatrix} 1 \\ 1 \\ 0 \end{pmatrix}, \begin{pmatrix} 1 \\ 1 \\ 1 \end{pmatrix} \right\}; B_2 = \left\{ \begin{pmatrix} 1 \\ -1 \end{pmatrix}, \begin{pmatrix} 2 \\ 3 \end{pmatrix} \right\}$$

12. $T: \mathbb{R}^2 \to \mathbb{R}^3; T\begin{pmatrix} x \\ y \end{pmatrix} = \begin{pmatrix} x - y \\ 2x + y \\ y \end{pmatrix}; B_1 = \left\{ \begin{pmatrix} 2 \\ 1 \end{pmatrix}, \begin{pmatrix} 1 \\ 2 \end{pmatrix} \right\}; B_2 = \left\{ \begin{pmatrix} 1 \\ -1 \\ 0 \end{pmatrix}, \begin{pmatrix} 0 \\ 2 \\ 0 \end{pmatrix}, \begin{pmatrix} 0 \\ 2 \\ 5 \end{pmatrix} \right\}$

13. $T: P_2 \to P_3; T(a_0 + a_1 x + a_2 x^2) = a_1 - a_1 x + a_0 x^3$

14. $T: \mathbb{R} \to P_3; T(a) = a + ax + ax^2 + ax^3$

15. $T: P_3 \to \mathbb{R}; T(a_0 + a_1 x + a_2 x^2 + a_3 x^3) = a_2$

16. $T: P_3 \to P_1; T(a_0 + a_1 x + a_2 x^2 + a_3 x^3) = (a_1 + a_3)x - a_2$

17. $T: P_3 \to P_2; T(a_0 + a_1 x + a_2 x^2 + a_3 x^3) = (a_0 - a_1 + 2a_2 + 3a_3)$
$+ (a_1 + 4a_2 + 3a_3)x + (a_0 + 6a_2 + 5a_3)x^2$

18. $T: M_{22} \to M_{22}; T\begin{pmatrix} a & b \\ c & d \end{pmatrix} = \begin{pmatrix} a - b + 2c + d & -a + 2c + 2d \\ a - 2b + 5c + 4d & 2a - b + c - d \end{pmatrix}$

19. $T: M_{22} \to M_{22}; T\begin{pmatrix} a & b \\ c & d \end{pmatrix} = \begin{pmatrix} a + b + c + d & a + b + c \\ a + b & a \end{pmatrix}$

20. $T: P_2 \to P_3; T[p(x)] = xp(x); B_1 = \{1, x, x^2\}; B_2 = \{1, (1 + x), (1 + x)^2, (1 + x)^3\}$

Calculus **21.** $D: P_4 \to P_3; Dp(x) = p'(x)$ Calculus **22.** $T: P_4 \to P_4; Tp(x) = xp'(x) - p(x)$

*Calculus **23.** $D: P_n \to P_{n-1}; Dp(x) = p'(x)$ Calculus **24.** $D: P_4 \to P_2; Dp(x) = p''(x)$

*Calculus **25.** $T: P_4 \to P_4; Tp(x) = p''(x) + xp'(x) + 2p(x)$

*Calculus **26.** $D: P_n \to P_{n-k}; Dp(x) = p^{(k)}(x)$

*Calculus **27.** $T: P_n \to P_n; Tp(x) = x^n p^{(n)}(x) + x^{n-1}p^{(n-1)}(x) + \cdots + xp'(x) + p(x)$

Calculus **28.** $J: P_n \to \mathbb{R}; Jp = \int_0^1 p(x)\, dx$ **29.** $T: \mathbb{R}^3 \to P_2; T\begin{pmatrix} a \\ b \\ c \end{pmatrix} = a + bx + cx^2$

30. $T: P_3 \to \mathbb{R}^3; T(a_0 + a_1 x + a_2 x^2 + a_3 x^3) = \begin{pmatrix} a_3 - a_2 \\ a_1 + a_3 \\ a_2 - a_1 \end{pmatrix}$

31. Let $T: M_{mn} \to M_{nm}$ be given by $TA = A^t$. Find A_T with respect to the standard bases in M_{mn} and M_{nm}.

*32.** Let $T: \mathbb{C}^2 \to \mathbb{C}^2$ be given by $T\begin{pmatrix} x \\ y \end{pmatrix} = \begin{pmatrix} x + iy \\ (1 + i)y - x \end{pmatrix}$. Find A_T.

Calculus **33.** Let $V = \text{span}\{1, \sin x, \cos x\}$. Find A_D, where $D: V \to V$ is defined by $Df(x) = f'(x)$. Find range D and ker D.

Calculus **34.** Answer the questions of Problems 33 given $V = \text{span}\{e^x, xe^x, x^2 e^x\}$.

35. Let $T: \mathbb{C}^2 \to \mathbb{C}^2$ be given by $T\mathbf{x} = \text{proj}_H \mathbf{x}$, where $H = \text{span}\{(1/\sqrt{2})(1, i)\}$. Find A_T.

36. Prove Theorem 2.

37. Prove Theorem 4.

In Problems 38–45 describe in words the linear transformation $T: \mathbb{R}^2 \to \mathbb{R}^2$ with the given matrix representation A_T.

38. $A_T = \begin{pmatrix} 4 & 0 \\ 0 & 1 \end{pmatrix}$

39. $A_T = \begin{pmatrix} 1 & 0 \\ 0 & \frac{1}{4} \end{pmatrix}$

40. $A_T = \begin{pmatrix} 1 & 0 \\ 0 & -1 \end{pmatrix}$

41. $A_T = \begin{pmatrix} 1 & 2 \\ 0 & 1 \end{pmatrix}$

42. $A_T = \begin{pmatrix} 1 & -3 \\ 0 & 1 \end{pmatrix}$

43. $A_T = \begin{pmatrix} 1 & 0 \\ \frac{1}{2} & 1 \end{pmatrix}$

44. $A_T = \begin{pmatrix} 1 & 0 \\ -5 & 1 \end{pmatrix}$

45. $A_T = \begin{pmatrix} 0 & 1 \\ 1 & 0 \end{pmatrix}$

In Problems 46–55 write the 2×2 matrix representation of the given linear transformation and draw a sketch of the region obtained when the transformation is applied to the given rectangle.

46. Expansion along the y-axis with $c = 2$

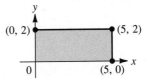

47. Compression along the x-axis with $c = \frac{1}{4}$

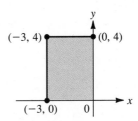

48. Shear along the x-axis with $c = -2$

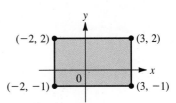

49. Shear along the y-axis with $c = 3$

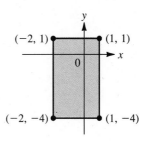

50. Shear along the y-axis with $c = -\frac{1}{2}$

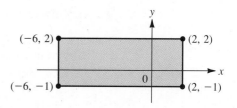

51. Shear along the x-axis with $c = \frac{1}{5}$

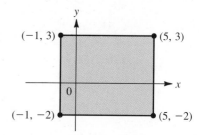

52. Reflection about the x-axis

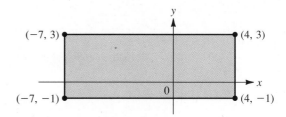

53. Reflection about the y-axis

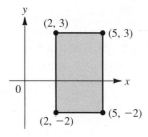

54. Reflection about the line $y = x$

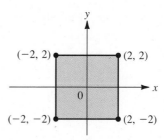

55. Reflection about the line $y = x$

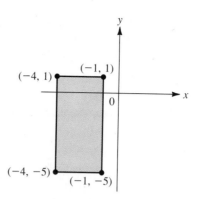

In Problems 56–63 write each linear transformation with given transformation matrix A_T as a succession of expansions, compressions, reflections, and shears.

56. $A_T = \begin{pmatrix} 2 & -1 \\ 5 & 0 \end{pmatrix}$ **57.** $A_T = \begin{pmatrix} 3 & 2 \\ -1 & 4 \end{pmatrix}$ **58.** $A_T = \begin{pmatrix} 0 & -2 \\ 3 & -5 \end{pmatrix}$

59. $A_T = \begin{pmatrix} 3 & 6 \\ 4 & 2 \end{pmatrix}$ **60.** $A_T = \begin{pmatrix} 0 & 3 \\ 1 & -2 \end{pmatrix}$ **61.** $A_T = \begin{pmatrix} 0 & -2 \\ 5 & 7 \end{pmatrix}$

62. $A_T = \begin{pmatrix} 3 & 7 \\ -4 & -8 \end{pmatrix}$ **63.** $A_T = \begin{pmatrix} -1 & 10 \\ 6 & 2 \end{pmatrix}$

5.4   ISOMORPHISMS

In this section we introduce some important terminology and then prove a theorem which says that all n-dimensional vector spaces are "essentially" the same.

DEFINITION 1	**One-to-One Transformation** Let $T: V \to W$ be a linear transformation. Then T is **one-to-one,** written 1–1, if

$$\boxed{T\mathbf{v}_1 = T\mathbf{v}_2 \quad \text{implies that} \quad \mathbf{v}_1 = \mathbf{v}_2} \tag{1}$$

That is, T is 1–1 if every vector $\mathbf{w}$ in the range of T is the image of exactly one vector in V.

THEOREM 1

Proof

Let $T: V \to W$ be a linear transformation. Then T is 1–1 if and only if ker $T = \{\mathbf{0}\}$.

Suppose ker $T = \{\mathbf{0}\}$ and $T\mathbf{v}_1 = T\mathbf{v}_2$. Then $T\mathbf{v}_1 - T\mathbf{v}_2 = T(\mathbf{v}_1 - \mathbf{v}_2) = \mathbf{0}$, which means that $(\mathbf{v}_1 - \mathbf{v}_2) \in$ ker $T = \{\mathbf{0}\}$. Thus $\mathbf{v}_1 - \mathbf{v}_2 = \mathbf{0}$, so $\mathbf{v}_1 = \mathbf{v}_2$, which shows that T is 1–1. Now suppose that T is 1–1 and $\mathbf{v} \in$ ker T. Then $T\mathbf{v} = \mathbf{0}$. But $T\mathbf{0} = \mathbf{0}$ also. Thus, since T is 1–1, $\mathbf{v} = \mathbf{0}$. This completes the proof.   ∎

EXAMPLE 1 **A 1–1 Transformation from $\mathbb{R}^2$ to $\mathbb{R}^2$** Let $T\colon \mathbb{R}^2 \to \mathbb{R}^2$ be defined by $T\begin{pmatrix} x \\ y \end{pmatrix} = \begin{pmatrix} x - y \\ 2x + y \end{pmatrix}$. We easily find $A_T = \begin{pmatrix} 1 & -1 \\ 2 & 1 \end{pmatrix}$ and $\rho(A_T) = 2$; hence $\nu(A_T) = 0$ and $N_{A_T} = \ker T = \{\mathbf{0}\}$. Thus T is 1–1. ∎

EXAMPLE 2 **A Transformation from $\mathbb{R}^2$ to $\mathbb{R}^2$ That Is Not 1–1** Let $T\colon \mathbb{R}^2 \to \mathbb{R}^2$ be defined by $T\begin{pmatrix} x \\ y \end{pmatrix} = \begin{pmatrix} x - y \\ 2x - 2y \end{pmatrix}$. Then $A_T = \begin{pmatrix} 1 & -1 \\ 2 & -2 \end{pmatrix}$, $\rho(A_T) = 1$, and $\nu(A_T) = 1$; hence $\nu(T) = 1$ and T is not 1–1. Note, for example, that $T\begin{pmatrix} 1 \\ 1 \end{pmatrix} = \mathbf{0} = T\begin{pmatrix} 0 \\ 0 \end{pmatrix}$. ∎

DEFINITION 2 **Onto Transformation** Let $T\colon V \to W$ be a linear transformation. Then T is said to be **onto** W or simply **onto,** if for every $\mathbf{w} \in W$ there is at least one $\mathbf{v} \in V$ such that $T\mathbf{v} = \mathbf{w}$. That is: *T is onto W if and only if range $T = W$.*

EXAMPLE 3 **Determining Whether a Transformation Is Onto** In Example 1, $\rho(A_T) = 2$; hence range $T = \mathbb{R}^2$ and T is onto. In Example 2, $\rho(A_T) = 1$ and range $T = \text{span} \left\{ \begin{pmatrix} 1 \\ 2 \end{pmatrix} \right\} \neq \mathbb{R}^2$; hence T is not onto. ∎

THEOREM 2 Let $T\colon V \to W$ be a linear transformation and suppose that $\dim V = \dim W = n$.

> **i.** If T is 1–1, then T is onto.
>
> **ii.** If T is onto, then T is 1–1.

Proof Let A_T be the matrix representation of T. Then if T is 1–1, $\ker T = \{\mathbf{0}\}$ and $\nu(A_T) = 0$, which means that $\rho(T) = \rho(A_T) = n - 0 = n$ so that range $T = W$. If T is onto, then $\rho(A_T) = n$ so that $\nu(T) = \nu(A_T) = 0$ and T is 1–1. ∎

THEOREM 3 Let $T\colon V \to W$ be a linear transformation. Suppose that $\dim V = n$ and $\dim W = m$. Then:

i. If $n > m$, T is not 1–1.

ii. If $m > n$, T is not onto.

Proof **i.** Let $\{\mathbf{v}_1, \mathbf{v}_2, \ldots, \mathbf{v}_n\}$ be a basis for V. Let $\mathbf{w}_i = T\mathbf{v}_i$ for $i = 1, 2, \ldots, n$ and look at the set $S = \{\mathbf{w}_1, \mathbf{w}_2, \ldots, \mathbf{w}_n\}$. Since $m = \dim W < n$, the set S is linearly dependent. Thus there exist scalars not all zero such that $c_1\mathbf{w}_1 + c_2\mathbf{w}_2 + \cdots + c_n\mathbf{w}_n = \mathbf{0}$. Let $\mathbf{v} = c_1\mathbf{v}_1 + c_2\mathbf{v}_2 + \cdots + c_n\mathbf{v}_n$. Since the $\mathbf{v}_i$'s are linearly independent and since not all the c_i's are zero, we see that $\mathbf{v} \neq \mathbf{0}$. But $T\mathbf{v} = T(c_1\mathbf{v}_1 + c_2\mathbf{v}_2 + \cdots + c_n\mathbf{v}_n) = c_1 T\mathbf{v}_1 + c_2 T\mathbf{v}_2 + \cdots + c_n T\mathbf{v}_n = c_1\mathbf{w}_1 + c_2\mathbf{w}_2 + \cdots + c_n\mathbf{w}_n = \mathbf{0}$. Thus $\mathbf{v} \in \ker T$ and $\ker T \neq \{\mathbf{0}\}$.

ii. If $\mathbf{v} \in V$, then $\mathbf{v} = a_1\mathbf{v}_1 + a_2\mathbf{v}_2 + \cdots + a_n\mathbf{v}_n$ for some scalars $a_1, a_2, \ldots, a_n$ and $T\mathbf{v} = a_1 T\mathbf{v}_1 + a_2 T\mathbf{v}_2 + \cdots + a_n T\mathbf{v}_n = a_1\mathbf{w}_1 + a_2\mathbf{w}_2 + \cdots + a_n\mathbf{w}_n$. Thus $\{\mathbf{w}_1, \mathbf{w}_2, \ldots, \mathbf{w}_n\} = \{T\mathbf{v}_1, T\mathbf{v}_2, \ldots, T\mathbf{v}_n\}$ spans range T. Then, from Problem 4.6.29 on page 252, $\rho(T) = \dim$ range $T \leq n$. Since $m > n$, this shows that range $T \neq W$. Thus T is not onto. ∎

EXAMPLE 4   **A Transformation from $\mathbb{R}^3 \to \mathbb{R}^2$ is not 1–1**   Let $T \colon \mathbb{R}^3 \to \mathbb{R}^2$ be given by

$$T\begin{pmatrix} x \\ y \\ z \end{pmatrix} = \begin{pmatrix} 1 & 2 & 3 \\ 4 & 5 & 6 \end{pmatrix}\begin{pmatrix} x \\ y \\ z \end{pmatrix}.$$ Here $n = 3$ and $m = 2$, so T is not 1–1. To see this,

observe that

$$T\begin{pmatrix} -1 \\ 2 \\ 0 \end{pmatrix} = \begin{pmatrix} 1 & 2 & 3 \\ 4 & 5 & 6 \end{pmatrix}\begin{pmatrix} -1 \\ 2 \\ 0 \end{pmatrix} = \begin{pmatrix} 3 \\ 6 \end{pmatrix} \quad \text{and} \quad T\begin{pmatrix} 2 \\ -4 \\ 3 \end{pmatrix} = \begin{pmatrix} 1 & 2 & 3 \\ 4 & 5 & 6 \end{pmatrix}\begin{pmatrix} 2 \\ -4 \\ 3 \end{pmatrix} = \begin{pmatrix} 3 \\ 6 \end{pmatrix}$$

That is, two different vectors in $\mathbb{R}^3$ have the same image in $\mathbb{R}^2$. ∎

EXAMPLE 5   **A Transformation from $\mathbb{R}^2$ to $\mathbb{R}^3$ Is Not Onto**   Let $T \colon \mathbb{R}^2 \to \mathbb{R}^3$ be given by

$$T\begin{pmatrix} x \\ y \end{pmatrix} = \begin{pmatrix} 1 & 2 \\ 3 & 4 \\ 5 & 6 \end{pmatrix}\begin{pmatrix} x \\ y \end{pmatrix}.$$ Here $n = 2$ and $m = 3$, so T is not onto. To show this, we

must find a vector in $\mathbb{R}^3$ which is not in the range of T. One such vector is $\begin{pmatrix} 0 \\ 0 \\ 1 \end{pmatrix}$. That

is, there is no vector $\mathbf{x} = \begin{pmatrix} x \\ y \end{pmatrix}$ in $\mathbb{R}^2$ such that $T\mathbf{x} = \begin{pmatrix} 0 \\ 0 \\ 1 \end{pmatrix}$. We prove this by assuming

that $T\begin{pmatrix} x \\ y \end{pmatrix} = \begin{pmatrix} 0 \\ 0 \\ 1 \end{pmatrix}$. That is,

$$\begin{pmatrix} 1 & 2 \\ 3 & 4 \\ 5 & 6 \end{pmatrix}\begin{pmatrix} x \\ y \end{pmatrix} = \begin{pmatrix} 0 \\ 0 \\ 1 \end{pmatrix} \quad \text{or} \quad \begin{pmatrix} x + 2y \\ 3x + 4y \\ 5x + 6y \end{pmatrix} = \begin{pmatrix} 0 \\ 0 \\ 1 \end{pmatrix}$$

Row-reducing, we have

$$\begin{pmatrix} 1 & 2 & | & 0 \\ 3 & 4 & | & 0 \\ 5 & 6 & | & 1 \end{pmatrix} \xrightarrow[\substack{R_2 \to R_2 - 3R_1 \\ R_3 \to R_3 - 5R_1}]{} \begin{pmatrix} 1 & 2 & | & 0 \\ 0 & -2 & | & 0 \\ 0 & -4 & | & 1 \end{pmatrix} \xrightarrow[\substack{R_3 \to R_3 - 2R_2}]{} \begin{pmatrix} 1 & 2 & | & 0 \\ 0 & -2 & | & 0 \\ 0 & 0 & | & 1 \end{pmatrix}$$

The last line reads $0 \cdot x + 0 \cdot y = 1$, so the system is inconsistent and $\begin{pmatrix} 0 \\ 0 \\ 1 \end{pmatrix}$ is not in

the range of T.   ∎

DEFINITION 3   **Isomorphism** Let $T: V \to W$ be a linear transformation. Then T is an **isomorphism** if T is 1–1 and onto.

DEFINITION 4   **Isomorphic Vector Spaces** The vector spaces V and W are said to be **isomorphic** if there exists an isomorphism T from V onto W. In this case we write $V \cong W$.

Remark. The word "isomorphism" comes from the Greek *isomorphos* meaning "of equal form" (*iso* = equal; *morphos* = form). After a few examples we shall see how closely related are the "forms" of isomorphic vector spaces.

Let $T: \mathbb{R}^n \to \mathbb{R}^n$ and let A_T be the matrix representation of T. Now T is 1–1 if and only if ker $T = \{0\}$, which is true if and only if $\nu(A_T) = 0$ if and only if det $A_T \neq 0$. Thus we can extend our Summing Up Theorem (last seen on page 263) in another direction.

THEOREM 4   **Summing Up Theorem—View 7** Let A be an $n \times n$ matrix. Then the following ten statements are equivalent; that is, each one implies the other nine (so that if one is true, all are true):

    i. A is invertible.

    ii. The only solution to the homogeneous system $A\mathbf{x} = \mathbf{0}$ is the trivial solution ($\mathbf{x} = \mathbf{0}$).

    iii. The system $A\mathbf{x} = \mathbf{b}$ has a unique solution for every n-vector $\mathbf{b}$.

    iv. A is row equivalent to the $n \times n$ identity matrix I_n.

    v. A can be written as the product of elementary matrices.

    vi. The rows (and columns) of A are linearly independent.

    vii. det $A \neq 0$.

    viii. $\nu(A) = 0$.

ix. $\rho(A) = n$.

x. The linear transformation T from $\mathbb{R}^n$ to $\mathbb{R}^n$ defined by $T\mathbf{x} = A\mathbf{x}$ is an isomorphism. ∎

We now look at some examples of isomorphisms between other pairs of vector spaces.

EXAMPLE 6   **An Isomorphism Between $\mathbb{R}^3$ and P_2**   Let $T: \mathbb{R}^3 \to P_2$ be defined by $T\begin{pmatrix} a \\ b \\ c \end{pmatrix} = a + bx + cx^2$. It is easy to verify that T is linear. Suppose that $T\begin{pmatrix} a \\ b \\ c \end{pmatrix} = \mathbf{0} = 0 + 0x + 0x^2$. Then $a = b = c = 0$. That is, ker $T = \{\mathbf{0}\}$ and T is 1–1. If $p(x) = a_0 + a_1x + a_2x^2$, then $p(x) = T\begin{pmatrix} a_0 \\ a_1 \\ a_2 \end{pmatrix}$. This means that range $T = P_2$ and T is onto. Thus $\mathbb{R}^3 \cong P_2$. ∎

Note. dim $\mathbb{R}^3$ = dim P_2 = 3. Thus, by Theorem 2, once we know that T is 1–1, we also know that it is onto. We verified that it was onto; but it was unnecessary to do so.

EXAMPLE 7   **An Isomorphism Between Two Infinite Dimensional Vector Spaces**

Calculus

Let $V = \{ f \in C^1[0, 1]: f(0) = 0 \}$ and $W = C[0, 1]$. Let $D: V \to W$ be given by $Df = f'$. Suppose that $Df = Dg$. Then $f' = g'$ or $(f - g)' = 0$ and $f(x) - g(x) = c$, a constant. But $f(0) = g(0) = 0$, so $c = 0$ and $f = g$. Thus D is 1–1. Let $g \in C[0, 1]$ and let $f(x) = \int_0^x g(t)\ dt$. Then, from the fundamental theorem of calculus, $f \in C^1[0, 1]$ and $f'(x) = g(x)$ for every $x \in [0, 1]$. Moreover, since $\int_0^0 g(t)\ dt = 0$, we have $f(0) = 0$. Thus, for every g in W, there is an $f \in V$ such that $Df = g$. Hence D is onto and we have shown that $V \cong W$. ∎

The following theorem illustrates the similarity of two isomorphic vector spaces.

THEOREM 5   Let $T: V \to W$ be an isomorphism:

i. If $\mathbf{v}_1, \mathbf{v}_2, \ldots, \mathbf{v}_n$ span V, then $T\mathbf{v}_1, T\mathbf{v}_2, \ldots, T\mathbf{v}_n$ span W.

ii. If $\mathbf{v}_1, \mathbf{v}_2, \ldots, \mathbf{v}_n$ are linearly independent in V, then $T\mathbf{v}_1, T\mathbf{v}_2, \ldots, T\mathbf{v}_n$ are linearly independent in W.

iii. If $\{\mathbf{v}_1, \mathbf{v}_2, \ldots, \mathbf{v}_n\}$ is a basis in V, then $\{T\mathbf{v}_1, T\mathbf{v}_2, \ldots, T\mathbf{v}_n\}$ is a basis in W.

iv. If V is finite dimensional, then W is finite dimensional and dim V = dim W.

Proof   i. Let $\mathbf{w} \in W$. Then, since T is onto, there is a $\mathbf{v} \in V$ such that $T\mathbf{v} = \mathbf{w}$. Since the $\mathbf{v}_i$'s span V, we can write $\mathbf{v} = a_1\mathbf{v}_1 + a_2\mathbf{v}_2 + \cdots + a_n\mathbf{v}_n$ so that $\mathbf{w} = T\mathbf{v} = a_1 T\mathbf{v}_1 + a_2 T\mathbf{v}_2 + \cdots + a_n T\mathbf{v}_n$ and this shows that $\{T\mathbf{v}_1, T\mathbf{v}_2, \ldots, T\mathbf{v}_n\}$ spans W.

ii. Suppose $c_1 T\mathbf{v}_1 + c_2 T\mathbf{v}_2 + \cdots + c_n T\mathbf{v}_n = \mathbf{0}$. Then $T(c_1\mathbf{v}_1 + c_2\mathbf{v}_2 + \cdots + c_n\mathbf{v}_n) = \mathbf{0}$. Thus, since T is 1–1, $c_1\mathbf{v}_1 + c_2\mathbf{v}_2 + \cdots + c_n\mathbf{v}_n = \mathbf{0}$, which implies that $c_1 = c_2 = \cdots = c_n = 0$ since the $\mathbf{v}_i$'s are independent.

iii. This follows from parts (i) and (ii).

iv. This follows from part (iii).   ∎

In general, it is difficult to show that two infinite dimensional vector spaces are isomorphic. For finite dimensional spaces, however, it is remarkably easy. Theorem 3 shows that if dim $V \neq$ dim W, then V and W are not isomorphic. The next theorem shows that if dim $V =$ dim W, and if V and W are real vector spaces, then V and W are isomorphic. That is,

---

Two real finite dimensional spaces of the same dimension are isomorphic.

THEOREM 6   Let V and W be two real† finite dimensional vector spaces with dim $V =$ dim W. Then $V \cong W$.

Proof   Let $\{\mathbf{v}_1, \mathbf{v}_2, \ldots, \mathbf{v}_n\}$ be a basis for V and let $\{\mathbf{w}_1, \mathbf{w}_2, \ldots, \mathbf{w}_n\}$ be a basis for W. Define the linear transformation T by

$$T\mathbf{v}_i = \mathbf{w}_i \qquad \text{for } i = 1, 2, \ldots, n \tag{2}$$

By Theorem 5.2.2 on page 339 there is exactly one linear transformation that satisfies equation (2). Suppose $\mathbf{v} \in V$ and $T\mathbf{v} = \mathbf{0}$. Then if $\mathbf{v} = c_1\mathbf{v}_1 + c_2\mathbf{v}_2 + \cdots + c_n\mathbf{v}_n$, we have $T\mathbf{v} = c_1 T\mathbf{v}_1 + \cdots + c_n T\mathbf{v}_n = c_1\mathbf{w}_1 + c_2\mathbf{w}_2 + \cdots + c_n\mathbf{w}_n = \mathbf{0}$. But, since $\mathbf{w}_1, \mathbf{w}_2, \ldots, \mathbf{w}_n$ are linearly independent, $c_1 = c_2 = \cdots = c_n = 0$. Thus $\mathbf{v} = \mathbf{0}$ and T is 1–1. Since V and W are finite dimensional and dim $V =$ dim W, T is onto by Theorem 2 and the proof is complete.   ∎

---

† We need the word "real" here because it is important that the sets of scalars in V and W be the same. Otherwise, the condition $T(\alpha\mathbf{v}) = \alpha T\mathbf{v}$ might not hold because $\mathbf{v} \in V$, $T\mathbf{v} \in W$, and either $\alpha\mathbf{v}$ or $\alpha T\mathbf{v}$ might not be defined. Theorem 6 is true if the word "real" is omitted and, instead, we impose the conditions that V and W be defined with the same set of scalars (like $\mathbb{C}$, for example).

This last result is one of the central results of linear algebra. Loosely speaking, it says that if you know one real n-dimensional vector space, you know all real vector spaces of dimension n. That is, loosely speaking, $\mathbb{R}^n$ is the only n-dimensional vector space over the reals.

PROBLEMS 5.4

  1. Show that $T: M_{mn} \to M_{nm}$ defined by $TA = A^t$ is an isomorphism.

  2. Show that $T: \mathbb{R}^n \to \mathbb{R}^n$ is an isomorphism if and only if A_T is invertible.

  * **3.** Let V and W be n-dimensional real vector spaces and let B_1 and B_2 be bases for V and W, respectively. Let A_T be the transformation matrix relative to the bases B_1 and B_2. Show that $T: V \to W$ is an isomorphism if and only if $\det A_T \neq 0$.

  4. Find an isomorphism between D_n, the $n \times n$ diagonal matrices with real entries, and $\mathbb{R}^n$. [*Hint:* Look first at the case $n = 2$.]

  5. For what value of m is the set of $n \times n$ symmetric matrices isomorphic to $\mathbb{R}^m$?

  6. Show that the set of $n \times n$ symmetric matrices is isomorphic to the set of $n \times n$ upper triangular matrices.

  7. Let $V = P_4$ and $W = \{p \in P_5: p(0) = 0\}$. Show that $V \cong W$.

  Calculus **8.** Define $T: P_n \to P_n$ by $Tp = p + p'$. Show that T is an isomorphism.

  9. Find a condition on the numbers m, n, p, q such that $M_{mn} \cong M_{pq}$.

  10. Show that $D_n \cong P_{n-1}$.

  11. Prove that any two finite dimensional complex vector spaces V and W with $\dim V = \dim W$ are isomorphic.

  12. Define $T: C[0, 1] \to C[3, 4]$ by $Tf(x) = f(x - 3)$. Show that T is an isomorphism.

  13. Let B be an invertible $n \times n$ matrix. Show that $T: M_{nm} \to M_{nm}$ defined by $TA = AB$ is an isomorphism.

  Calculus **14.** Show that the transformation $Tp(x) = xp'(x)$ is not an isomorphism from P_n into P_n.

  15. Let H be a subspace of the finite dimensional inner product space V. Show that $T: V \to H$ defined by $Tv = \text{proj}_H v$ is onto. Under what circumstances will it be 1–1?

  16. Show that if $T: V \to W$ is an isomorphism, then there exists an isomorphism $S: W \to V$ such that $S(Tv) = v$. Here S is called the **inverse transformation** of T and is denoted by T^{-1}.

  17. Show that if $T: \mathbb{R}^n \to \mathbb{R}^n$ is defined by $Tx = Ax$ and if T is an isomorphism, then A is invertible and the inverse transformation T^{-1} is given by $T^{-1}x = A^{-1}x$.

  18. Find T^{-1} for the isomorphism of Problem 7.

  ***19.** Consider the space $C = \{z = a + ib$, where a and b are real numbers and $i^2 = -1\}$. Show that if the scalars are taken to be the reals, then $C \cong \mathbb{R}^2$.

  ***20.** Consider the space $\mathbb{C}_\mathbb{R}^n = \{(c_1, c_2, \ldots, c_n): c_i \in C$ and the scalars are the reals$\}$. Show that $\mathbb{C}_\mathbb{R}^n \cong \mathbb{R}^{2n}$. [*Hint:* See Problem 19.]

5.5   ISOMETRIES

In this section we describe a special kind of linear transformation between vector spaces. We begin with a very useful result.

THEOREM 1 Let A be an $m \times n$ matrix with real entries.† Then for any vectors $\mathbf{x} \in \mathbb{R}^n$ and $\mathbf{y} \in \mathbb{R}^m$:

$$(A\mathbf{x}) \cdot \mathbf{y} = \mathbf{x} \cdot (A^t\mathbf{y}) \tag{1}$$

Proof

$$\begin{array}{ccc} \text{equation (6)} & \text{Theorem 1(ii)} & \text{Associative law} \\ \text{on p. 85} & \text{on p. 84} & \text{for matrix multiplication} \\ \downarrow & \downarrow & \downarrow \end{array}$$

$$A\mathbf{x} \cdot \mathbf{y} = (A\mathbf{x})^t\mathbf{y} = (\mathbf{x}^tA^t)\mathbf{y} = \mathbf{x}^t(A^t\mathbf{y})$$

$$\begin{array}{cc} \text{equation (6)} & \text{Theorem 1(i)} \\ \text{on p. 85} & \text{on p. 84} \\ \downarrow & \downarrow \end{array}$$

$$= (\mathbf{x}^t)^t \cdot (A^t\mathbf{y}) = \mathbf{x} \cdot (A^t\mathbf{y}) \qquad \blacksquare$$

Recall from Section 4.10, page 286, that a matrix Q with real entries is **orthogonal** if Q is invertible and $Q^{-1} = Q^t$. In Theorem 4.10.3 on page 287 we proved that Q is orthogonal if and only if the columns of Q form an orthonormal basis for $\mathbb{R}^n$. Now let Q be an $n \times n$ orthogonal matrix and let $T: \mathbb{R}^n \to \mathbb{R}^n$ be the linear transformation defined by $T\mathbf{x} = Q\mathbf{x}$. Then, using equation (1), we compute

$$(T\mathbf{x} \cdot T\mathbf{y}) = Q\mathbf{x} \cdot Q\mathbf{y} = \mathbf{x} \cdot (Q^tQ\mathbf{y}) = \mathbf{x} \cdot (I\mathbf{y}) = \mathbf{x} \cdot \mathbf{y}.$$

In particular, if $\mathbf{x} = \mathbf{y}$, we see that $T\mathbf{x} \cdot T\mathbf{x} = \mathbf{x} \cdot \mathbf{x}$ or

$$|T\mathbf{x}| = |\mathbf{x}|$$

for every $\mathbf{x}$ in $\mathbb{R}^n$.

DEFINITION 1 **Isometry** A linear transformation $T: \mathbb{R}^n \to \mathbb{R}^n$ is called an **isometry** if, for every $\mathbf{x}$ in $\mathbb{R}^n$,

$$|T\mathbf{x}| = |\mathbf{x}| \tag{2}$$

---

†This result can easily be extended to matrices with complex components. See Problem 21.

Because of equation (2) we can say: An isometry in $\mathbb{R}^n$ is a linear transformation that preserves length in $\mathbb{R}^n$. Note that (2) implies that

$$\boxed{|T\mathbf{x} - T\mathbf{y}| = |\mathbf{x} - \mathbf{y}|} \tag{3}$$

[since $T\mathbf{x} - T\mathbf{y} = T(\mathbf{x} - \mathbf{y})$].

THEOREM 2 Let T be an isometry from $\mathbb{R}^n \to \mathbb{R}^n$ and suppose that $\mathbf{x}$ and $\mathbf{y}$ are in $\mathbb{R}^n$. Then

$$\boxed{T\mathbf{x} \cdot T\mathbf{y} = \mathbf{x} \cdot \mathbf{y}} \tag{4}$$

That is, an isometry in $\mathbb{R}^n$ preserves the scalar product.

Proof
$$|T\mathbf{x} - T\mathbf{y}|^2 = (T\mathbf{x} - T\mathbf{y}) \cdot (T\mathbf{x} - T\mathbf{y}) = |T\mathbf{x}|^2 - 2T\mathbf{x} \cdot T\mathbf{y} + |T\mathbf{y}|^2 \tag{5}$$
$$|\mathbf{x} - \mathbf{y}|^2 = (\mathbf{x} - \mathbf{y}) \cdot (\mathbf{x} - \mathbf{y}) = |\mathbf{x}|^2 - 2\mathbf{x} \cdot \mathbf{y} + |\mathbf{y}|^2 \tag{6}$$

Since $|T\mathbf{x} - T\mathbf{y}|^2 = |\mathbf{x} - \mathbf{y}|^2$, $|T\mathbf{x}|^2 = |\mathbf{x}|^2$ and $|T\mathbf{y}|^2 = |\mathbf{y}|^2$, equations (5) and (6) show that

$$-2T\mathbf{x} \cdot T\mathbf{y} = -2\mathbf{x} \cdot \mathbf{y} \quad \text{or} \quad T\mathbf{x} \cdot T\mathbf{y} = \mathbf{x} \cdot \mathbf{y} \qquad \blacksquare$$

In the derivation of equation (2) we showed that if the matrix representation of T is an orthogonal matrix, then T is an isometry. Conversely, suppose that T is an isometry. If A is the matrix representation of T, then for any $\mathbf{x}$ and $\mathbf{y}$ in $\mathbb{R}^n$

$$\overset{\text{from (4)}}{\downarrow} \qquad \overset{\text{from (1)}}{\downarrow}$$
$$\mathbf{x} \cdot \mathbf{y} = T\mathbf{x} \cdot T\mathbf{y} = A\mathbf{x} \cdot A\mathbf{y} = \mathbf{x} \cdot A^t A\mathbf{y}$$
$$\mathbf{x} \cdot \mathbf{y} - \mathbf{x} \cdot A^t A\mathbf{y} = 0 \quad \text{or} \quad \mathbf{x} \cdot (\mathbf{y} - A^t A\mathbf{y}) = 0$$

Thus (see page 290)

$$\mathbf{y} - A^t A\mathbf{y} \in (\mathbb{R}^n)^\perp = \{\mathbf{0}\}$$

We see that for every $\mathbf{y} \in \mathbb{R}^n$

$$\mathbf{y} = A^t A\mathbf{y} \tag{7}$$

This implies that $A^t A = I$, so A is orthogonal.

We have proved the following theorem.

THEOREM 3 A linear transformation $T: \mathbb{R}^n \to \mathbb{R}^n$ is an isometry if and only if the matrix representation of T is orthogonal. $\blacksquare$

Isometries of $\mathbb{R}^2$

Let T be an isometry from $\mathbb{R}^2$ to $\mathbb{R}^2$. Let

$$\mathbf{u}_1 = T\begin{pmatrix} 1 \\ 0 \end{pmatrix} \quad \text{and} \quad \mathbf{u}_2 = T\begin{pmatrix} 0 \\ 1 \end{pmatrix}$$

Then $\mathbf{u}_1$ and $\mathbf{u}_2$ are unit vectors [because of (2)] and

$$\overset{\text{from (4)}}{\downarrow}$$

$$\mathbf{u}_1 \cdot \mathbf{u}_2 = \begin{pmatrix} 1 \\ 0 \end{pmatrix} \cdot \begin{pmatrix} 0 \\ 1 \end{pmatrix} = 0$$

Thus $\mathbf{u}_1$ and $\mathbf{u}_2$ are orthogonal. From equation (3.1.7) on page 162, there exists a number θ, with $0 \le \theta < 2\pi$ such that

$$\mathbf{u}_1 = \begin{pmatrix} \cos \theta \\ \sin \theta \end{pmatrix}$$

Since $\mathbf{u}_1$ and $\mathbf{u}_2$ are orthogonal,

$$\text{Direction of } \mathbf{u}_2 = \text{direction of } \mathbf{u}_1 \pm \frac{\pi}{2}$$

In the first case,

$$\mathbf{u}_2 = \begin{pmatrix} \cos\left(\theta + \dfrac{\pi}{2}\right) \\ \sin\left(\theta + \dfrac{\pi}{2}\right) \end{pmatrix} = \begin{pmatrix} -\sin\theta \\ \cos\theta \end{pmatrix}$$

In the second case,

$$\mathbf{u}_2 = \begin{pmatrix} \cos\left(\theta - \dfrac{\pi}{2}\right) \\ \sin\left(\theta - \dfrac{\pi}{2}\right) \end{pmatrix} = \begin{pmatrix} \sin\theta \\ -\cos\theta \end{pmatrix}$$

So the matrix representation of T is either

$$Q_1 = \begin{pmatrix} \cos\theta & -\sin\theta \\ \sin\theta & \cos\theta \end{pmatrix} \quad \text{or} \quad Q_2 = \begin{pmatrix} \cos\theta & \sin\theta \\ \sin\theta & -\cos\theta \end{pmatrix}$$

From Example 5.1.8 on page 334, we see that Q_1 is the matrix representation of a rotation transformation (counterclockwise through an angle of θ). Now, as is easily verified,

$$\begin{pmatrix} \cos\theta & \sin\theta \\ \sin\theta & -\cos\theta \end{pmatrix} = \begin{pmatrix} \cos\theta & -\sin\theta \\ \sin\theta & \cos\theta \end{pmatrix}\begin{pmatrix} 1 & 0 \\ 0 & -1 \end{pmatrix}$$

But the transformation $T\colon \mathbb{R}^2 \to \mathbb{R}^2$ given by

$$T\begin{pmatrix} x \\ y \end{pmatrix} = \begin{pmatrix} 1 & 0 \\ 0 & -1 \end{pmatrix}\begin{pmatrix} x \\ y \end{pmatrix} = \begin{pmatrix} x \\ -y \end{pmatrix}$$

is a reflection of $\begin{pmatrix} x \\ y \end{pmatrix}$ about the x-axis (see Example 5.1.1. on page 330). Thus we have the following.

THEOREM 4   Let $T\colon \mathbb{R}^2 \to \mathbb{R}^2$ be an isometry. Then T is either:

  i.   a rotation transformation

  or

  ii.   a reflection about the x-axis followed by a rotation transformation.   ∎

Isometries have some interesting properties.

THEOREM 5   Let $T\colon \mathbb{R}^n \to \mathbb{R}^n$ be an isometry. Then:

  i.   If $\mathbf{u}_1, \mathbf{u}_2, \ldots, \mathbf{u}_n$ is an orthogonal set, then $T\mathbf{u}_1, T\mathbf{u}_2, \ldots, T\mathbf{u}_n$ is an orthogonal set.

  ii. T is an isomorphism.

Proof   **i.**   If $i \neq j$ and $\mathbf{u}_i \cdot \mathbf{u}_j = 0$, then $(T\mathbf{u}_i) \cdot (T\mathbf{u}_j) = \mathbf{u}_i \cdot \mathbf{u}_j = 0$, which proves (i).

  ii.   Let $\mathbf{u}_1, \mathbf{u}_2, \ldots, \mathbf{u}_n$ be an orthonormal basis for $\mathbb{R}^n$. Then, by part (i) and the fact that $|T\mathbf{u}_i| = |\mathbf{u}_i| = 1$, we find that $T\mathbf{u}_1, T\mathbf{u}_2, \ldots, T\mathbf{u}_n$ is an orthonormal set in $\mathbb{R}^n$. By Theorem 4.10.1 on page 283 these vectors are linearly independent and hence form a basis for $\mathbb{R}^n$. Thus range $T = \mathbb{R}^n$, which proves that ker $T = \{\mathbf{0}\}$ (since $\nu(T) + \rho(T) = n$).   ∎

We conclude this section by outlining how we can extend the notion of isometry to an arbitrary inner product space. Recall from page 309 that in an inner product space V,

$$|\mathbf{v}| = (\mathbf{v}, \mathbf{v})^{1/2}$$

DEFINITION 2   **Isometry**   Let V and W be real (or complex) inner product spaces and let $T\colon V \to W$ be a linear transformation. Then T is an **isometry** if, for every $\mathbf{v} \in V$,

$$\boxed{|\mathbf{v}_1|_V = |T\mathbf{v}_1|_W}$$ (7)

Two facts follow immediately: First, since $T(\mathbf{v}_1 - \mathbf{v}_2) = T\mathbf{v}_1 - T\mathbf{v}_2$, we have for every $\mathbf{v}_1$ and $\mathbf{v}_2$ in V,

$$\boxed{|T\mathbf{v}_1 - T\mathbf{v}_2|_W = |\mathbf{v}_1 - \mathbf{v}_2|_V}$$

THEOREM 6   Let $T: V \to W$ be an isometry. Then, for every $\mathbf{v}_1$ and $\mathbf{v}_2$ in V,

$$\boxed{(T\mathbf{v}_1, T\mathbf{v}_2) = (\mathbf{v}_1, \mathbf{v}_2)} \qquad \text{(8)}$$

That is, an isometry preserves inner products.

The proof of Theorem 6 is identical to the proof of Theorem 2 with inner products in V and W replacing the scalar product in $\mathbb{R}^n$. ∎

DEFINITION 3   **Isometrically Isomorphic Vector Spaces** Two vector spaces V and W are said to be **isometrically isomorphic** if there exists a linear transformation $T: V \to W$ that is both an isometry and an isomorphism.

THEOREM 7   Any two n-dimensional real inner product spaces are isometrically isomorphic.

Proof   Let $\{\mathbf{u}_1, \mathbf{u}_2, \ldots, \mathbf{u}_n\}$ and $\{\mathbf{w}_1, \mathbf{w}_2, \ldots, \mathbf{w}_n\}$ be orthonormal bases for V and W, respectively. Let $T: V \to W$ be the linear transformation defined by $T\mathbf{u}_i = \mathbf{w}_i$, $i = 1, 2, \ldots, n$. If we can show that T is an isometry, then we shall be done since reasoning as in the proof of Theorem 5 shows us that T is also an isomorphism. Let $\mathbf{x}$ and $\mathbf{y}$ be in V. Then there exist sets of real numbers $c_1, c_2, \ldots, c_n$ and $d_1, d_2, \ldots, d_n$ such that $\mathbf{x} = c_1\mathbf{u}_1 + c_2\mathbf{u}_2 + \cdots + c_n\mathbf{u}_n$ and $\mathbf{y} = d_1\mathbf{u}_1 + d_2\mathbf{u}_2 + \cdots + d_n\mathbf{u}_n$. Since the $\mathbf{u}_i$'s are orthonormal, $(\mathbf{x}, \mathbf{y}) = ((c_1\mathbf{u}_1 + c_2\mathbf{u}_2 + \cdots + c_n\mathbf{u}_n), (d_1\mathbf{u}_1 + d_2\mathbf{u}_2 + \cdots + d_n\mathbf{u}_n)) = c_1d_1 + c_2d_2 + \cdots + c_nd_n$. Similarly, since $T\mathbf{x} = c_1T\mathbf{u}_1 + c_2T\mathbf{u}_2 + \cdots + c_nT\mathbf{u}_n = c_1\mathbf{w}_1 + c_2\mathbf{w}_2 + \cdots + c_n\mathbf{w}_n$, we obtain $(T\mathbf{x}, T\mathbf{y}) = ((c_1\mathbf{w}_1 + c_2\mathbf{w}_2 + \cdots + c_n\mathbf{w}_n), (d_1\mathbf{w}_1 + d_2\mathbf{w}_2 + \cdots + d_n\mathbf{w}_n)) = c_1d_1 + c_2d_2 + \cdots + c_nd_n$ because the $\mathbf{w}_i$'s are orthonormal. This completes the proof. ∎

EXAMPLE 1   **An Isometry Between $\mathbb{R}^3$ and $P_2[0, 1]$**   We illustrate this theorem by showing that $\mathbb{R}^3$ and $P_2[0, 1]$ are isometrically isomorphic. In $\mathbb{R}^3$ we use the standard basis $\left\{ \begin{pmatrix} 1 \\ 0 \\ 0 \end{pmatrix}, \begin{pmatrix} 0 \\ 1 \\ 0 \end{pmatrix}, \begin{pmatrix} 0 \\ 0 \\ 1 \end{pmatrix} \right\}$. In P_2 we use the orthonormal basis $\{1, \sqrt{3}(2x - 1),$

$\sqrt{5}(6x^2 - 6x + 1)\}$. (See Example 4.12.8 on page 310.) Let $\mathbf{x} = \begin{pmatrix} a_1 \\ b_1 \\ c_1 \end{pmatrix}$ and $\mathbf{y} = \begin{pmatrix} a_2 \\ b_2 \\ c_2 \end{pmatrix}$ be in $\mathbb{R}^3$. Then $(\mathbf{x}, \mathbf{y}) = \mathbf{x} \cdot \mathbf{y} = a_1a_2 + b_1b_2 + c_1c_2$. Recall that in $P_2[0, 1]$

we defined $(p, q) = \int_0^1 p(x)q(x) \, dx$. We define $T\begin{pmatrix} 1 \\ 0 \\ 0 \end{pmatrix} = 1$, $T\begin{pmatrix} 0 \\ 1 \\ 0 \end{pmatrix} = \sqrt{3}(2x - 1)$,

and $T\begin{pmatrix} 0 \\ 0 \\ 1 \end{pmatrix} = \sqrt{5}(6x^2 - 6x + 1)$; hence

$$T\begin{pmatrix} a \\ b \\ c \end{pmatrix} = a + b\sqrt{3}(2x - 1) + c\sqrt{5}(6x^2 - 6x + 1)$$

and

$$(T\mathbf{x}, T\mathbf{y}) = \int_0^1 [a_1 + b_1\sqrt{3}(2x - 1) + c_1\sqrt{5}(6x^2 - 6x + 1)]$$
$$\times [a_2 + b_2\sqrt{3}(2x - 1) + c_2\sqrt{5}(6x^2 - 6x + 1)] \, dx$$
$$= a_1a_2 \int_0^1 dx + \int_0^1 b_1b_23(2x - 1)^2 \, dx + \int_0^1 c_1c_2[5(6x^2 - 6x + 1)^2] \, dx$$
$$+ (a_1b_2 + a_2b_1) \int_0^1 \sqrt{3}(2x - 1) \, dx$$
$$+ (a_1c_2 + a_2c_1) \int_0^1 \sqrt{5}(6x^2 - 6x + 1) \, dx$$
$$+ (b_1c_2 + b_2c_1) \int_0^1 [\sqrt{3}(2x - 1)][\sqrt{5}(6x^2 - 6x + 1)] \, dx$$
$$= a_1a_2 + b_1b_2 + c_1c_2$$

Here we saved time by using the fact that $\{1, \sqrt{3}(2x - 1), \sqrt{5}(6x^2 - 6x + 1)\}$ is an orthonormal set. Thus $T: \mathbb{R}^3 \to P_2[0, 1]$ is an isometry. ∎

PROBLEMS   5.5

1. Show that for any real number θ, the transformation $T: \mathbb{R}^3 \to \mathbb{R}^3$ defined by $T\mathbf{x} = A\mathbf{x}$, where

$$A = \begin{pmatrix} \sin\theta & \cos\theta & 0 \\ \cos\theta & -\sin\theta & 0 \\ 0 & 0 & 1 \end{pmatrix}$$

is an isometry.

2. Do the same for the transformation T, where

$$A = \begin{pmatrix} \cos\theta & 0 & -\sin\theta \\ 0 & 1 & 0 \\ \sin\theta & 0 & \cos\theta \end{pmatrix}$$

3. Let A and B be orthogonal $n \times n$ matrices. Show that $T: \mathbb{R}^n \to \mathbb{R}^n$ defined by $T\mathbf{x} = AB\mathbf{x}$ is an isometry.

4. Find A_T if T is the transformation from $\mathbb{R}^3 \to \mathbb{R}^3$ defined by

$$T\begin{pmatrix} 2/3 \\ 1/3 \\ -2/3 \end{pmatrix} = \begin{pmatrix} 1/\sqrt{2} \\ 1/\sqrt{2} \\ 0 \end{pmatrix} \quad T\begin{pmatrix} 1/3 \\ 2/3 \\ 2/3 \end{pmatrix} = \begin{pmatrix} -1/\sqrt{6} \\ 1/\sqrt{6} \\ 2/\sqrt{6} \end{pmatrix} \quad T\begin{pmatrix} 2/3 \\ -2/3 \\ 1/3 \end{pmatrix} = \begin{pmatrix} 1/\sqrt{3} \\ -1/\sqrt{3} \\ 1/\sqrt{3} \end{pmatrix}$$

Show that A_T is orthogonal.

5. Prove Theorem 6.

6. Let $T: \mathbb{R}^2 \to \mathbb{R}^2$ be an isometry. Show that T preserves angles. That is: (angle between $\mathbf{x}$ and $\mathbf{y}$) = (angle between $T\mathbf{x}$ and $T\mathbf{y}$).

7. Give an example of a linear transformation from $\mathbb{R}^2$ onto $\mathbb{R}^2$ that preserves angles and is *not* an isometry.

8. For $\mathbf{x}, \mathbf{y} \in \mathbb{R}^n$ and $\mathbf{x}$ and $\mathbf{y} \neq \mathbf{0}$, define: (angle between $\mathbf{x}$ and $\mathbf{y}$) = $\angle(\mathbf{x}, \mathbf{y}) = \cos^{-1}[(\mathbf{x} \cdot \mathbf{y})/|\mathbf{x}|\,|\mathbf{y}|]$. Show that if $T: \mathbb{R}^n \to \mathbb{R}^n$ is an isometry, then T preserves angles.

9. Let $T: \mathbb{R}^n \to \mathbb{R}^n$ be an isometry and let $T\mathbf{x} = A\mathbf{x}$. Show that $S\mathbf{x} = A^{-1}\mathbf{x}$ is an isometry.

In Problems 10–14 find an isometry between the given pair of spaces.

Calculus **10.** $P_1[-1, 1]$, $\mathbb{R}^2$ *Calculus **11.** $P_3[-1, 1]$, $\mathbb{R}^4$

*12. M_{22}, $\mathbb{R}^4$ *Calculus **13.** M_{22}, $P_3[-1, 1]$

14. D_n and $\mathbb{R}^n$ (D_n = set of diagonal $n \times n$ matrices)

15. Let A be an $n \times n$ matrix with complex components. Then the **conjugate transpose** of A, denoted by A^*, is defined by $(A^*)_{ij} = \overline{a}_{ji}$. Compute A^* if $A = \begin{pmatrix} 1 + i & -4 + 2i \\ 3 & 6 - 3i \end{pmatrix}$.

16. The $n \times n$ complex matrix A is called **hermitian** if $A^* = A$. Show that the matrix $A = \begin{pmatrix} 4 & 3 - 2i \\ 3 + 2i & 6 \end{pmatrix}$ is hermitian.

17. Show that if A is hermitian, then the diagonal components of A are real.

18. The $n \times n$ complex matrix A is called **unitary** if $A^* = A^{-1}$. Show that the matrix

$$A = \begin{pmatrix} \dfrac{1 + i}{2} & \dfrac{3 - 2i}{\sqrt{26}} \\ \dfrac{1 + i}{2} & \dfrac{-3 + 2i}{\sqrt{26}} \end{pmatrix}$$

is unitary.

19. Show that A is unitary if and only if the columns of A form an orthonormal basis in $\mathbb{C}^n$.

20. Show that if A is unitary, then $|\det A| = 1$.

21. Let A be an $n \times n$ matrix with complex components. In $\mathbb{C}^n$, if $\mathbf{x} = (c_1, c_2, \ldots, c_n)$ and $y = (d_1, d_2, \ldots, d_n)$, define the inner product $(\mathbf{x}, \mathbf{y}) = c_1\overline{d_1} + c_2\overline{d_2} + \cdots + c_n\overline{d_n}$. (See Example 4.12.2.) Prove that $(A\mathbf{x}, \mathbf{y}) = (\mathbf{x}, A^*\mathbf{y})$.

*22. Show that any two complex inner product spaces of the same (finite) dimension are isometrically isomorphic.

SUMMARY

- **Linear Transformation**·Let V and W be vector spaces. A **linear transformation** T from V into W is a function that assigns to each vector $\mathbf{v} \in V$ a unique vector $T\mathbf{v} \in W$ and that satisfies, for each $\mathbf{u}$ and $\mathbf{v}$ in V and each scalar α,   (p. 332)

$$T(\mathbf{u} + \mathbf{v}) = T\mathbf{u} + T\mathbf{v}$$

and

$$T(\alpha\mathbf{v}) = \alpha T\mathbf{v}$$

- **Basic Properties of Linear Transformations**·Let $T: V \to W$ be a linear transformation. Then for all vectors $\mathbf{u}, \mathbf{v}, \mathbf{v}_1, \mathbf{v}_2, \ldots, \mathbf{v}_n$ in V and all scalars $\alpha_1, \alpha_2, \ldots, \alpha_n$:   (p. 338)

  i. $T(\mathbf{0}) = \mathbf{0}$

  ii. $T(\mathbf{u} - \mathbf{v}) = T\mathbf{u} - T\mathbf{v}$

  iii. $T(\alpha_1\mathbf{v}_1 + \alpha_2\mathbf{v}_2 + \cdots + \alpha_n\mathbf{v}_n) = \alpha_1 T\mathbf{v}_1 + \alpha_2 T\mathbf{v}_2 + \cdots + \alpha_n T\mathbf{v}_n$

- **Kernel and Range of a Linear Transformation**·Let V and W be vector spaces and let $T: V \to W$ be a linear transformation. Then the **kernel** of T, denoted by ker T, is given by

$$\ker T = \{\mathbf{v} \in V: T\mathbf{v} = \mathbf{0}\}$$

  (pp. 341, 342)

  The **range** of T, denoted by range T, is given by

$$\text{Range } T = \{\mathbf{w} \in W: \mathbf{w} = T\mathbf{v} \text{ for some } \mathbf{v} \in V\}$$

  Ker T is a subspace of V and range T is a subspace of W.

- **Nullity and Rank of a Linear Transformation**·If T is a linear transformation from V to W, then   (p. 343)

$$\textbf{Nullity of } T = \nu(T) = \dim \ker T$$

$$\textbf{Rank of } T = \rho(T) = \dim \text{range } T$$

- **Transformation Matrix**·Let $T: \mathbb{R}^n \to \mathbb{R}^m$ be a linear transformation. Then there is a unique $m \times n$ matrix A_T such that   (pp. 345, 346)

$$T\mathbf{x} = A_T\mathbf{x} \quad \text{for every } \mathbf{x} \in \mathbb{R}^n$$

  The matrix A_T is called the **transformation matrix** of T.

- Let A_T be the transformation matrix corresponding to the linear transformation T. Then: (pp. 346, 347)

  i. range $T = R_{A_T} = C_{A_T}$

  ii. $\rho(T) = \rho(A_T)$

  iii. ker $T = N_{A_T}$

  iv. $\nu(T) = \nu(A_T)$

- **Matrix Representation of a Linear Transformation** · Let V be a real n-dimensional vector space, W be a real m-dimensional vector space, and $T: V \rightarrow W$ be a linear transformation. Let $B_1 = \{\mathbf{v}_1, \mathbf{v}_2, \ldots, \mathbf{v}_n\}$ be a basis for V and let $B_2 = \{\mathbf{w}_1, \mathbf{w}_2, \ldots, \mathbf{w}_m\}$ be a basis for W. Then there is a unique $m \times n$ matrix A_T such that   (p. 349)

$$(T\mathbf{x})_{B_2} = A_T(\mathbf{x})_{B_1}$$

A_T is called the **matrix representation** of T

- Let V and W be finite dimensional vector spaces with dim $V = n$. Let $T: V \rightarrow W$ be a linear transformation and let A_T be a matrix representation of T. Then: (p. 351)

    i. $\rho(T) = \rho(A_T)$

    ii. $\nu(T) = \nu(A_T)$

    iii. $\nu(T) + \rho(T) = n$

- **One-to-One Transformation** · Let $T: V \rightarrow W$ be a linear transformation. Then T is **one-to-one,**   (p. 367) written 1–1, if $T\mathbf{v}_1 = T\mathbf{v}_2$ implies that $\mathbf{v}_1 = \mathbf{v}_2$. That is, T is 1–1 if every vector $\mathbf{w}$ in the range of T is the image of exactly one vector in V.
- Let $T: V \rightarrow W$ be a linear transformation. Then T is 1–1 if and only if ker $T = \{\mathbf{0}\}$.   (p. 367)
- **Onto Transformation** · Let $T: V \rightarrow W$ be a linear transformation. Then T is said to be **onto** W or   (p. 368) simply **onto,** if for every $\mathbf{w} \in W$ there is at least one $\mathbf{v} \in V$ such that $T\mathbf{v} = \mathbf{w}$. That is: *T is onto W if and only if range T = W.*
- Let $T: V \rightarrow W$ be a linear transformation and suppose that dim $V =$ dim $W = n$:   (p. 368)

    i. If T is 1–1, then T is onto.

    ii. If T is onto, then T is 1–1.

- Let $T: V \rightarrow W$ be a linear transformation. Suppose that dim $V = n$ and dim $W = m$. Then: (p. 368)

    i. If $n > m$, T is not 1–1.

    ii. If $m > n$, T is not onto.

- **Isomorphism** · Let $T: V \rightarrow W$ be a linear transformation. Then T is an **isomorphism** if T is 1–1 and onto.   (p. 370)
- **Isomorphic Vector Spaces** · The vector spaces V and W are said to be **isomorphic** if there exists an isomorphism T from V onto W. In this case we write $V \cong W$.   (p. 370)
- Any two real finite dimensional vector spaces of the same dimension are isomorphic.   (p. 372)
- **Summing Up Theorem** · Let A be an $n \times n$ matrix. Then the following ten statements are equivalent: (p. 370)

    i. A is invertible.

    ii. The only solution to the homogeneous system $A\mathbf{x} = \mathbf{0}$ is the trivial solution ($\mathbf{x} = \mathbf{0}$).

    iii. The system $A\mathbf{x} = \mathbf{b}$ has a unique solution for every n-vector $\mathbf{b}$.

    iv. A is row equivalent to the $n \times n$ identity matrix I_n.

    v. A can be written as the product of elementary matrices.

    vi. The rows (and columns) of A are linearly independent.

    vii. det $A \neq 0$.

    viii. $\nu(A) = 0$.

    ix. $\rho(A) = n$.

    x. The linear transformation T from $\mathbb{R}^n$ to $\mathbb{R}^n$ defined by $T\mathbf{x} = A\mathbf{x}$ is an isomorphism.

- Let $T: V \to W$ be an isomorphism: (p. 371)

  i. If $\mathbf{v}_1, \mathbf{v}_2, \ldots, \mathbf{v}_n$ span V, then $T\mathbf{v}_1, T\mathbf{v}_2, \ldots, T\mathbf{v}_n$ span W.

  ii. If $\mathbf{v}_1, \mathbf{v}_2, \ldots, \mathbf{v}_n$ are linearly independent in V, then $T\mathbf{v}_1, T\mathbf{v}_2, \ldots, T\mathbf{v}_n$ are linearly independent in W.

  iii. If $\{\mathbf{v}_1, \mathbf{v}_2, \ldots, \mathbf{v}_n\}$ is a basis in V, then $\{T\mathbf{v}_1, T\mathbf{v}_2, \ldots, T\mathbf{v}_n\}$ is a basis in W.

  iv. If V is finite dimensional, then W is finite dimensional and $\dim V = \dim W$.

- **Isometry**·A linear transformation $T: \mathbb{R}^n \to \mathbb{R}^n$ is called an **isometry** if, for every $\mathbf{x}$ in $\mathbb{R}^n$,   (p. 374)

$$|T\mathbf{x}| = |\mathbf{x}|$$

- If T is an isometry from $\mathbb{R}^n \to \mathbb{R}^n$, then, for every $\mathbf{x}$ and $\mathbf{y}$ in $\mathbb{R}^n$,   (p. 375)

$$|T\mathbf{x} - T\mathbf{y}| = |\mathbf{x} - \mathbf{y}| \quad \text{and} \quad T\mathbf{x} \cdot T\mathbf{y} = \mathbf{x} \cdot \mathbf{y}$$

- Let $T: \mathbb{R}^n \to \mathbb{R}^n$ be an isometry. Then: (p. 377)

  i. If $\mathbf{u}_1, \mathbf{u}_2, \ldots, \mathbf{u}_n$ is an orthogonal set, then $T\mathbf{u}_1, T\mathbf{u}_2, \ldots, T\mathbf{u}_n$ is an orthogonal set.

  ii. T is an isomorphism.

- A linear transformation $T: \mathbb{R}^n \to \mathbb{R}^n$ is an isometry if and only if the matrix representation of T is orthogonal.   (p. 375)
- **Isometry**·Let V and W be real (or complex) inner product spaces and let $T: V \to W$ be a linear transformation. Then T is an **isometry** if, for every $\mathbf{v}_1, \mathbf{v}_2 \in V$,   (p. 377)

$$|\mathbf{v}_1|_V = |T\mathbf{v}_1|_W$$

- **Isometrically Isomorphic Vector Spaces**·Two vector spaces V and W are said to be **isometrically isomorphic** if there exists a linear transformation $T: V \to W$ that is both an isometry and an isomorphism.   (p. 378)
- Any two n-dimensional real inner product spaces are isometrically isomorphic.   (p. 378)

REVIEW EXERCISES

In Exercises 1–6 determine whether the given transformation from V to W is linear.

1. $T: \mathbb{R}^2 \to \mathbb{R}^2$; $T(x, y) = (0, -y)$ **2.** $T: \mathbb{R}^3 \to \mathbb{R}^3$; $T(x, y, z) = (1, y, z)$

3. $T: \mathbb{R}^2 \to \mathbb{R}$; $T(x, y) = x/y$ **4.** $T: P_1 \to P_2$; $(Tp)(x) = xp(x)$

5. $T: P_2 \to P_2$; $(Tp)(x) = 1 + p(x)$ **6.** $T: C[0, 1] \to C[0, 1]$; $Tf(x) = f(1)$

In Exercises 7–12 find the kernel, range, rank, and nullity of the given linear transformation.

7. $T: \mathbb{R}^2 \to \mathbb{R}^2$; $T\begin{pmatrix} x \\ y \end{pmatrix} = \begin{pmatrix} 2 & -1 \\ 4 & 7 \end{pmatrix}\begin{pmatrix} x \\ y \end{pmatrix}$

8. $T: \mathbb{R}^3 \to \mathbb{R}^3$; $T\begin{pmatrix} x \\ y \\ z \end{pmatrix} = \begin{pmatrix} 1 & 2 & -1 \\ 2 & 4 & 3 \\ 1 & 2 & -6 \end{pmatrix}\begin{pmatrix} x \\ y \\ z \end{pmatrix}$

9. $T: \mathbb{R}^3 \to \mathbb{R}^2$; $T\begin{pmatrix} x \\ y \\ z \end{pmatrix} = \begin{pmatrix} y \\ -x \end{pmatrix}$

10. $T: P_2 \to P_4$; $Tp(x) = x^2 p(x)$

11. $T: M_{22} \to M_{22}$; $T(A) = AB$, where $B = \begin{pmatrix} 1 & 1 \\ -1 & 1 \end{pmatrix}$

12. $T: C[0, 1] \to \mathbb{R}$; $Tf = f(1)$

In Exercises 13–18 find the matrix representation of the given linear transformation and find the kernel, range, nullity, and rank of the transformation.

13. $T: \mathbb{R}^2 \to \mathbb{R}^2$; $T(x, y) = (0, -y)$ **14.** $T: \mathbb{R}^3 \to \mathbb{R}^2$; $T(x, y, z) = (y, z)$

15. $T: \mathbb{R}^4 \to \mathbb{R}^2$; $T(x, y, z, w) = (x - 2z, 2y + 3w)$

16. $T: P_3 \to P_4$; $(Tp)(x) = xp(x)$

17. $T: M_{22} \to M_{22}$; $TA = AB$, where $B = \begin{pmatrix} -1 & 0 \\ 1 & 2 \end{pmatrix}$

18. $T: \mathbb{R}^2 \to \mathbb{R}^2$; $T(x, y) = (x - y, 2x + 3y)$; $B_1 = \left\{ \begin{pmatrix} 1 \\ 1 \end{pmatrix}, \begin{pmatrix} 1 \\ 2 \end{pmatrix} \right\}$; $B_2 = \left\{ \begin{pmatrix} -1 \\ 3 \end{pmatrix}, \begin{pmatrix} 4 \\ 1 \end{pmatrix} \right\}$

In Exercises 19–22 describe in words the linear transformation $T: \mathbb{R}^2 \to \mathbb{R}^2$ with the given matrix representation A_T.

19. $A_T = \begin{pmatrix} 3 & 0 \\ 0 & 1 \end{pmatrix}$ **20.** $A_T = \begin{pmatrix} 1 & 0 \\ 0 & \frac{1}{3} \end{pmatrix}$ **21.** $A_T = \begin{pmatrix} 1 & 0 \\ -2 & 1 \end{pmatrix}$ **22.** $A_T = \begin{pmatrix} 1 & -5 \\ 0 & 1 \end{pmatrix}$

In Exercises 23–26 write the 2×2 matrix representation of the given linear transformation and draw a sketch of the region obtained when the transformation is applied to the given rectangle.

23. Expansion along the x-axis with $c = 3$

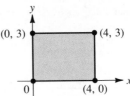

24. Compression along the y-axis with $c = \frac{1}{3}$

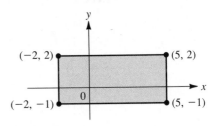

25. Reflection about the line $y = x$

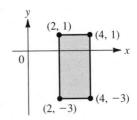

26. Shear along the x-axis with $c = -3$

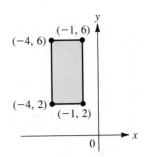

In Exercises 27–30 write each linear transformation matrix A_T as a succession of expansions, compressions, reflections, and shears.

27. $A_T = \begin{pmatrix} 1 & 3 \\ -2 & 2 \end{pmatrix}$ **28.** $A_T = \begin{pmatrix} 0 & 5 \\ -3 & 2 \end{pmatrix}$ **29.** $A_T = \begin{pmatrix} -6 & 4 \\ 1 & 3 \end{pmatrix}$ **30.** $A_T = \begin{pmatrix} 2 & 1 \\ 1 & 5 \end{pmatrix}$

31. Find an isomorphism $T: P_2 \to \mathbb{R}^3$.

Calculus **32.** Find an isometry $T: \mathbb{R}^2 \to P_1[-1, 1]$.

6

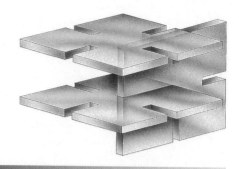

Eigenvalues, Eigenvectors, and Canonical Forms

6.1 EIGENVALUES AND EIGENVECTORS

Let $T: V \to V$ be a linear transformation. In a great variety of applications (one of which is given in the next section) it is useful to find a vector $\mathbf{v}$ in V such that $T\mathbf{v}$ and $\mathbf{v}$ are parallel. That is, we seek a vector $\mathbf{v}$ and a scalar λ such that

$$T\mathbf{v} = \lambda\mathbf{v} \tag{1}$$

If $\mathbf{v} \neq \mathbf{0}$ and λ satisfy (1), then λ is called an *eigenvalue* of T and $\mathbf{v}$ is called an *eigenvector* of T corresponding to the eigenvalue λ. The purpose of this chapter is to investigate properties of eigenvalues and eigenvectors. If V is finite dimensional, then T can be represented by a matrix A_T. For that reason we shall discuss eigenvalues and eigenvectors of $n \times n$ matrices.

DEFINITION 1   **Eigenvalue and Eigenvector** Let A be an $n \times n$ matrix with real† components. The number λ (real or complex) is called an **eigenvalue** of A if there is a *nonzero* vector $\mathbf{v}$ in $\mathbb{C}^n$ such that

$$\boxed{A\mathbf{v} = \lambda\mathbf{v}} \tag{2}$$

The vector $\mathbf{v} \neq \mathbf{0}$ is called an **eigenvector** *of A corresponding to the eigenvalue λ.*

---

† This definition is also valid if A has complex components; but as the matrices we shall be dealing with will, for the most part, have real components, the definition is sufficient for our purposes.

Note. The word "eigen" is the German word for "own" or "proper." Eigenvalues are also called **proper values** or **characteristic values** and eigenvectors are called **proper vectors** or **characteristic vectors.**

Remark. As we shall see (for instance, in Example 6) a matrix with real components can have complex eigenvalues and eigenvectors. That is why, in the definition, we have asserted that $\mathbf{v} \in \mathbb{C}^n$. We shall not be using many facts about complex numbers in this book. For a discussion of those few facts we do need, see Appendix 2.

EXAMPLE 1   **Eigenvalues and Eigenvectors of a 2 × 2 Matrix**   Let $A = \begin{pmatrix} 10 & -18 \\ 6 & -11 \end{pmatrix}$. Then $A\begin{pmatrix} 2 \\ 1 \end{pmatrix} = \begin{pmatrix} 10 & -18 \\ 6 & -11 \end{pmatrix}\begin{pmatrix} 2 \\ 1 \end{pmatrix} = \begin{pmatrix} 2 \\ 1 \end{pmatrix}$. Thus $\lambda_1 = 1$ is an eigenvalue of A with corresponding eigenvector $\mathbf{v}_1 = \begin{pmatrix} 2 \\ 1 \end{pmatrix}$. Similarly, $A\begin{pmatrix} 3 \\ 2 \end{pmatrix} = \begin{pmatrix} 10 & -18 \\ 6 & -11 \end{pmatrix}\begin{pmatrix} 3 \\ 2 \end{pmatrix} = \begin{pmatrix} -6 \\ -4 \end{pmatrix} = -2\begin{pmatrix} 3 \\ 2 \end{pmatrix}$ so that $\lambda_2 = -2$ is an eigenvalue of A with corresponding eigenvector $\mathbf{v}_2 = \begin{pmatrix} 3 \\ 2 \end{pmatrix}$. As we soon shall see, these are the only eigenvalues of A.   ■

EXAMPLE 2   **Eigenvalues and Eigenvectors of the Identity Matrix**   Let $A = I$. Then for any $\mathbf{v} \in \mathbb{C}^n$, $A\mathbf{v} = I\mathbf{v} = \mathbf{v}$. Thus 1 is the only eigenvalue of A and every $\mathbf{v} \neq \mathbf{0} \in \mathbb{C}^n$ is an eigenvector of I.   ■

We shall compute the eigenvalues and eigenvectors of many matrices in this section. But first we need to prove some facts that will simplify our computations.

Suppose that λ is an eigenvalue of A. Then there exists a nonzero vector $\mathbf{v} = \begin{pmatrix} x_1 \\ x_2 \\ \vdots \\ x_n \end{pmatrix} \neq \mathbf{0}$ such that $A\mathbf{v} = \lambda\mathbf{v} = \lambda I\mathbf{v}$. Rewriting this, we have

$$(A - \lambda I)\mathbf{v} = \mathbf{0} \tag{3}$$

If A is an $n \times n$ matrix, equation (3) is a homogeneous system of n equations in the unknowns $x_1, x_2, \ldots, x_n$. Since, by assumption, the system has nontrivial solutions, we conclude that $\det (A - \lambda I) = 0$. Conversely, if $\det (A - \lambda I) = 0$, then equation (3) has nontrivial solutions and λ is an eigenvalue of A. On the other hand, if $\det (A - \lambda I) \neq 0$, then (3) has only the solution $\mathbf{v} = \mathbf{0}$ so that λ is *not* an eigenvalue of A. Summing up these facts, we have the following theorem.

THEOREM 1   Let A be an $n \times n$ matrix. Then λ is an eigenvalue of A if and only if

$$p(\lambda) = \det (A - \lambda I) = 0 \qquad \qquad (4)$$

∎

DEFINITION 2   **Characteristic Equation and Polynomial** Equation (4) is called the **characteristic equation** of A; $p(\lambda)$ is called the **characteristic polynomial** of A.

As will become apparent in the examples, $p(\lambda)$ is a polynomial of degree n in λ. For example, if $A = \begin{pmatrix} a & b \\ c & d \end{pmatrix}$, then $A - \lambda I = \begin{pmatrix} a & b \\ c & d \end{pmatrix} - \begin{pmatrix} \lambda & 0 \\ 0 & \lambda \end{pmatrix} = \begin{pmatrix} a - \lambda & b \\ c & d - \lambda \end{pmatrix}$ and $p(\lambda) = \det (A - \lambda I) = (a - \lambda)(d - \lambda) - bc = \lambda^2 - (a + d)\lambda + (ad - bc)$.

According to the **fundamental theorem of algebra,** any polynomial of degree n with real or complex coefficients has exactly n roots (counting multiplicities). By this we mean, for example, that the polynomial $(\lambda - 1)^5$ has five roots, all equal to the number 1. Since any eigenvalue of A is a root of the characteristic equation of A, we conclude that

> *Counting multiplicities, every $n \times n$ matrix has exactly n eigenvalues.*

THEOREM 2   Let λ be an eigenvalue of the $n \times n$ matrix A and let $E_\lambda = \{ \mathbf{v} : A\mathbf{v} = \lambda \mathbf{v} \}$. Then E_λ is a subspace of $\mathbb{C}^n$.

Proof   If $A\mathbf{v} = \lambda \mathbf{v}$, then $(A - \lambda I)\mathbf{v} = \mathbf{0}$. Thus E_λ is the kernel of the matrix $A - \lambda I$, which, by Example 4.6.10 on page 249, is a subspace† of $\mathbb{C}^n$.   ∎

DEFINITION 3   **Eigenspace** Let λ be an eigenvalue of A. The subspace E_λ is called the **eigenspace‡** of A corresponding to the eigenvalue λ.

We now prove another useful result.

---

† In Example 4.6.10 on page 249 we saw that ker A is a subspace of $\mathbb{R}^n$ if A is a real matrix. The extension of this result to $\mathbb{C}^n$ presents no difficulties.

‡ Note that $\mathbf{0} \in E_\lambda$ since E_λ is a subspace. However, $\mathbf{0}$ is *not* an eigenvector.

THEOREM 3 Let A be an $n \times n$ matrix and let $\lambda_1, \lambda_2, \ldots, \lambda_m$ be distinct eigenvalues of A with corresponding eigenvectors $\mathbf{v}_1, \mathbf{v}_2, \ldots, \mathbf{v}_m$. Then $\mathbf{v}_1, \mathbf{v}_2, \ldots, \mathbf{v}_m$ are linearly independent. That is: *Eigenvectors corresponding to distinct eigenvalues are linearly independent.*

Proof We prove this by mathematical induction. We start with $m = 2$. Suppose that

$$c_1\mathbf{v}_1 + c_2\mathbf{v}_2 = \mathbf{0} \tag{5}$$

Then, multiplying both sides of (5) by A, we have

$$\mathbf{0} = A(c_1\mathbf{v}_1 + c_2\mathbf{v}_2) = c_1A\mathbf{v}_1 + c_2A\mathbf{v}_2$$

or (since $A\mathbf{v}_i = \lambda_i\mathbf{v}_i$ for $i = 1, 2$)

$$c_1\lambda_1\mathbf{v}_1 + c_2\lambda_2\mathbf{v}_2 = \mathbf{0} \tag{6}$$

We then multiply (5) by λ_1 and subtract it from (6) to obtain

$$(c_1\lambda_1\mathbf{v}_1 + c_2\lambda_2\mathbf{v}_2) - (c_1\lambda_1\mathbf{v}_1 + c_2\lambda_1\mathbf{v}_2) = \mathbf{0}$$

or

$$c_2(\lambda_2 - \lambda_1)\mathbf{v}_2 = \mathbf{0}$$

Since $\mathbf{v}_2 \neq \mathbf{0}$ (by the definition of an eigenvector) and since $\lambda_2 \neq \lambda_1$, we conclude that $c_2 = 0$. Then inserting $c_2 = 0$ in (5), we see that $c_1 = 0$, which proves the theorem in the case $m = 2$. Now suppose that the theorem is true for $m = k$. That is, we assume that any k eigenvectors corresponding to distinct eigenvalues are linearly independent. We prove the theorem for $m = k + 1$. So we assume that

$$c_1\mathbf{v}_1 + c_2\mathbf{v}_2 + \cdots + c_k\mathbf{v}_k + c_{k+1}\mathbf{v}_{k+1} = \mathbf{0} \tag{7}$$

Then, multiplying both sides of (7) by A and using the fact that $A\mathbf{v}_i = \lambda_i\mathbf{v}_i$, we obtain

$$c_1\lambda_1\mathbf{v}_1 + c_2\lambda_2\mathbf{v}_2 + \cdots + c_k\lambda_k\mathbf{v}_k + c_{k+1}\lambda_{k+1}\mathbf{v}_{k+1} = \mathbf{0} \tag{8}$$

We multiply both sides of (7) by λ_{k+1} and subtract it from (8):

$$c_1(\lambda_1 - \lambda_{k+1})\mathbf{v}_1 + c_2(\lambda_2 - \lambda_{k+1})\mathbf{v}_2 + \cdots + c_k(\lambda_k - \lambda_{k+1})\mathbf{v}_k = \mathbf{0}$$

But, by the induction assumption, $\mathbf{v}_1, \mathbf{v}_2, \ldots, \mathbf{v}_k$ are linearly independent. Thus $c_1(\lambda_1 - \lambda_{k+1}) = c_2(\lambda_2 - \lambda_{k+1}) = \cdots = c_k(\lambda_k - \lambda_{k+1}) = 0$; and, since $\lambda_i \neq \lambda_{k+1}$ for $i = 1, 2, \ldots, k$, we conclude that $c_1 = c_2 = \cdots = c_k = 0$. But, from (7), this means that $c_{k+1} = 0$. Thus the theorem is true for $m = k + 1$ and the proof is complete. ∎

If

$$A = \begin{pmatrix} a_{11} & a_{12} & \cdots & a_{1n} \\ a_{21} & a_{22} & \cdots & a_{2n} \\ \vdots & \vdots & & \vdots \\ a_{n1} & a_{n2} & \cdots & a_{nn} \end{pmatrix}$$

then

$$p(\lambda) = \det (A - \lambda I) = \begin{vmatrix} a_{11} - \lambda & a_{12} & \cdots & a_{1n} \\ a_{21} & a_{22} - \lambda & \cdots & a_{2n} \\ \vdots & \vdots & & \vdots \\ a_{n1} & a_{n2} & \cdots & a_{nn} - \lambda \end{vmatrix}$$

and $p(\lambda) = 0$ can be written in the form

$$p(\lambda) = \lambda^n + b_{n-1}\lambda^{n-1} + \cdots + b_1\lambda + b_0 = 0 \tag{9}$$

Equation (9) has n roots, some of which may be repeated. If $\lambda_1, \lambda_2, \ldots, \lambda_m$ are the distinct roots of (9) with multiplicities $r_1, r_2, \ldots, r_m$, respectively, then (9) may be factored to obtain

$$p(\lambda) = (\lambda - \lambda_1)^{r_1}(\lambda - \lambda_2)^{r_2} \cdots (\lambda - \lambda_m)^{r_m} = 0 \tag{10}$$

Algebraic multiplicity

The numbers $r_1, r_2, \ldots, r_m$ are called the **algebraic multiplicities** of the eigenvalues $\lambda_1, \lambda_2, \ldots, \lambda_m$, respectively.

We now calculate eigenvalues and corresponding eigenspaces. We do this in a three-step procedure:

Procedure for Computing Eigenvalues and Eigenvectors

i. Find $p(\lambda) = \det (A - \lambda I)$.

ii. Find the roots $\lambda_1, \lambda_2, \ldots, \lambda_m$ of $p(\lambda) = 0$.

iii. Corresponding to each eigenvalue λ_i, solve the homogeneous system $(A - \lambda_i I)\mathbf{v} = \mathbf{0}$.

Remark 1. Step (ii) is usually the most difficult one.

Remark 2. A relatively easy way to find eigenvalues and eigenvectors of 2×2 matrices is suggested in Problems 35 and 36.

EXAMPLE 3   **Computing Eigenvalues and Eigenvectors**   Let $A = \begin{pmatrix} 4 & 2 \\ 3 & 3 \end{pmatrix}$. Then

$\det (A - \lambda I) = \begin{vmatrix} 4 - \lambda & 2 \\ 3 & 3 - \lambda \end{vmatrix} = (4 - \lambda)(3 - \lambda) - 6 = \lambda^2 - 7\lambda + 6 = (\lambda - 1)(\lambda - 6) = 0$. Thus the eigenvalues of A are $\lambda_1 = 1$ and $\lambda_2 = 6$. For $\lambda_1 = 1$, we solve $(A - I)\mathbf{v} = \mathbf{0}$ or $\begin{pmatrix} 3 & 2 \\ 3 & 2 \end{pmatrix}\begin{pmatrix} x_1 \\ x_2 \end{pmatrix} = \begin{pmatrix} 0 \\ 0 \end{pmatrix}$. Clearly any eigenvector corresponding to $\lambda_1 = 1$ satisfies $3x_1 + 2x_2 = 0$. One such eigenvector is $\mathbf{v}_1 = \begin{pmatrix} 2 \\ -3 \end{pmatrix}$. Thus $E_1 = \text{span} \left\{ \begin{pmatrix} 2 \\ -3 \end{pmatrix} \right\}$. Similarly, the equation $(A - 6I)\mathbf{v} = \mathbf{0}$ means that $\begin{pmatrix} -2 & 2 \\ 3 & -3 \end{pmatrix} \begin{pmatrix} x_1 \\ x_2 \end{pmatrix} = \begin{pmatrix} 0 \\ 0 \end{pmatrix}$ or $x_1 = x_2$. Thus $\mathbf{v}_2 = \begin{pmatrix} 1 \\ 1 \end{pmatrix}$ is an eigenvector corresponding to $\lambda_2 = 6$ and $E_6 = \text{span} \left\{ \begin{pmatrix} 1 \\ 1 \end{pmatrix} \right\}$. Note that $\mathbf{v}_1$ and $\mathbf{v}_2$ are linearly independent since one is not a multiple of the other. ■

EXAMPLE 4   **A 3 × 3 Matrix with Distinct Eigenvalues**   Let $A = \begin{pmatrix} 1 & -1 & 4 \\ 3 & 2 & -1 \\ 2 & 1 & -1 \end{pmatrix}$. Then

$$\det(A - \lambda I) = \begin{vmatrix} 1-\lambda & -1 & 4 \\ 3 & 2-\lambda & -1 \\ 2 & 1 & -1-\lambda \end{vmatrix}$$

$$= -(\lambda^3 - 2\lambda^2 - 5\lambda + 6) = -(\lambda - 1)(\lambda + 2)(\lambda - 3) = 0$$

Thus the eigenvalues of A are $\lambda_1 = 1$, $\lambda_2 = -2$, and $\lambda_3 = 3$. Corresponding to $\lambda_1 = 1$ we have

$$(A - I)\mathbf{v} = \begin{pmatrix} 0 & -1 & 4 \\ 3 & 1 & -1 \\ 2 & 1 & -2 \end{pmatrix} \begin{pmatrix} x_1 \\ x_2 \\ x_3 \end{pmatrix} = \begin{pmatrix} 0 \\ 0 \\ 0 \end{pmatrix}$$

Solving by row reduction, we obtain, successively,

$$\begin{pmatrix} 0 & -1 & 4 & | & 0 \\ 3 & 1 & -1 & | & 0 \\ 2 & 1 & -2 & | & 0 \end{pmatrix} \xrightarrow[\substack{R_2 \to R_2 + R_1 \\ R_3 \to R_3 + R_1}]{} \begin{pmatrix} 0 & -1 & 4 & | & 0 \\ 3 & 0 & 3 & | & 0 \\ 2 & 0 & 2 & | & 0 \end{pmatrix}$$

$$\xrightarrow[R_2 \to \frac{1}{3}R_2]{} \begin{pmatrix} 0 & -1 & 4 & | & 0 \\ 1 & 0 & 1 & | & 0 \\ 2 & 0 & 2 & | & 0 \end{pmatrix} \xrightarrow[R_3 \to R_3 - 2R_2]{} \begin{pmatrix} 0 & -1 & 4 & | & 0 \\ 1 & 0 & 1 & | & 0 \\ 0 & 0 & 0 & | & 0 \end{pmatrix}$$

Thus $x_1 = -x_3$, $x_2 = 4x_3$, an eigenvector is $\mathbf{v}_1 = \begin{pmatrix} -1 \\ 4 \\ 1 \end{pmatrix}$, and $E_1 = \text{span} \left\{ \begin{pmatrix} -1 \\ 4 \\ 1 \end{pmatrix} \right\}$.

For $\lambda_2 = -2$, we have $[A - (-2I)]\mathbf{v} = (A + 2I)\mathbf{v} = \mathbf{0}$ or

$$\begin{pmatrix} 3 & -1 & 4 \\ 3 & 4 & -1 \\ 2 & 1 & 1 \end{pmatrix} \begin{pmatrix} x_1 \\ x_2 \\ x_3 \end{pmatrix} = \begin{pmatrix} 0 \\ 0 \\ 0 \end{pmatrix}.$$

This leads to

$$\begin{pmatrix} 3 & -1 & 4 & | & 0 \\ 3 & 4 & -1 & | & 0 \\ 2 & 1 & 1 & | & 0 \end{pmatrix} \xrightarrow[\begin{subarray}{l} R_2 \to R_2 + 4R_1 \\ R_3 \to R_3 + R_1 \end{subarray}]{} \begin{pmatrix} 3 & -1 & 4 & | & 0 \\ 15 & 0 & 15 & | & 0 \\ 5 & 0 & 5 & | & 0 \end{pmatrix}$$

$$\xrightarrow[R_2 \to \frac{1}{15}R_2]{} \begin{pmatrix} 3 & -1 & 4 & | & 0 \\ 1 & 0 & 1 & | & 0 \\ 5 & 0 & 5 & | & 0 \end{pmatrix} \xrightarrow[\begin{subarray}{l} R_1 \to R_1 - 4R_2 \\ R_3 \to R_3 - 5R_2 \end{subarray}]{} \begin{pmatrix} -1 & -1 & 0 & | & 0 \\ 1 & 0 & 1 & | & 0 \\ 0 & 0 & 0 & | & 0 \end{pmatrix}$$

Thus $x_2 = -x_1$, $x_3 = -x_1$, and an eigenvector is $\mathbf{v}_2 = \begin{pmatrix} 1 \\ -1 \\ -1 \end{pmatrix}$. Thus

$E_{-2} = \text{span} \left\{ \begin{pmatrix} 1 \\ -1 \\ -1 \end{pmatrix} \right\}$. Finally, for $\lambda_3 = 3$, we have

$$(A - 3I)\mathbf{v} = \begin{pmatrix} -2 & -1 & 4 \\ 3 & -1 & -1 \\ 2 & 1 & -4 \end{pmatrix} \begin{pmatrix} x_1 \\ x_2 \\ x_3 \end{pmatrix} = \begin{pmatrix} 0 \\ 0 \\ 0 \end{pmatrix}$$

and

$$\begin{pmatrix} -2 & -1 & 4 & | & 0 \\ 3 & -1 & -1 & | & 0 \\ 2 & 1 & -4 & | & 0 \end{pmatrix} \xrightarrow[\begin{subarray}{l} R_2 \to R_2 - R_1 \\ R_3 \to R_3 + R_1 \end{subarray}]{} \begin{pmatrix} -2 & -1 & 4 & | & 0 \\ 5 & 0 & -5 & | & 0 \\ 0 & 0 & 0 & | & 0 \end{pmatrix}$$

$$\xrightarrow[R_2 \to \frac{1}{5}R_2]{} \begin{pmatrix} -2 & -1 & 4 & | & 0 \\ 1 & 0 & -1 & | & 0 \\ 0 & 0 & 0 & | & 0 \end{pmatrix} \xrightarrow[R_1 \to R_1 + 4R_2]{} \begin{pmatrix} 2 & -1 & 0 & | & 0 \\ 1 & 0 & -1 & | & 0 \\ 0 & 0 & 0 & | & 0 \end{pmatrix}$$

Hence $x_3 = x_1$, $x_2 = 2x_1$, and $\mathbf{v}_3 = \begin{pmatrix} 1 \\ 2 \\ 1 \end{pmatrix}$ so that $E_3 = \text{span} \left\{ \begin{pmatrix} 1 \\ 2 \\ 1 \end{pmatrix} \right\}$.   ∎

Remark. In this and every other example there is always an infinite number of choices for each eigenvector. We arbitrarily choose a simple one by setting one or more of the x_i's equal to a convenient number. Here we have set one of the x_i's equal to 1.

EXAMPLE 5   **A 2 × 2 Matrix with One of Its Eigenvalues Equal to Zero**   Let $A = \begin{pmatrix} 2 & -1 \\ -4 & 2 \end{pmatrix}$. Then det $(A - \lambda I) = \begin{vmatrix} 2 - \lambda & -1 \\ -4 & 2 - \lambda \end{vmatrix} = \lambda^2 - 4\lambda = \lambda(\lambda - 4)$. Thus the eigenvalues are $\lambda_1 = 0$ and $\lambda_2 = 4$. The eigenspace corresponding to zero is simply the kernel of A. We calculate $\begin{pmatrix} 2 & -1 \\ -4 & 2 \end{pmatrix}\begin{pmatrix} x_1 \\ x_2 \end{pmatrix} = \begin{pmatrix} 0 \\ 0 \end{pmatrix}$ or $2x_1 = x_2$ and an

eigenvector is $v_1 = \begin{pmatrix} 1 \\ 2 \end{pmatrix}$. Thus $\ker A = E_0 = \text{span}\left\{ \begin{pmatrix} 1 \\ 2 \end{pmatrix} \right\}$. Corresponding to $\lambda_2 = $

4 we have $\begin{pmatrix} -2 & -1 \\ -4 & -2 \end{pmatrix}\begin{pmatrix} x_1 \\ x_2 \end{pmatrix} = \begin{pmatrix} 0 \\ 0 \end{pmatrix}$, so $E_4 = \text{span}\left\{ \begin{pmatrix} 1 \\ -2 \end{pmatrix} \right\}$. ∎

EXAMPLE 6 **A 2×2 Matrix with Complex Conjugate Eigenvalues** Let $A = \begin{pmatrix} 3 & -5 \\ 1 & -1 \end{pmatrix}$.

Then $\det (A - \lambda I) = \begin{vmatrix} 3 - \lambda & -5 \\ 1 & -1 - \lambda \end{vmatrix} = \lambda^2 - 2\lambda + 2 = 0$ and

$$\lambda = \frac{-(-2) \pm \sqrt{4 - 4(1)(2)}}{2} = \frac{2 \pm \sqrt{-4}}{2} = \frac{2 \pm 2i}{2} = 1 \pm i$$

Thus $\lambda_1 = 1 + i$ and $\lambda_2 = 1 - i$. We compute

$$[A - (1 + i)I]v = \begin{pmatrix} 2 - i & -5 \\ 1 & -2 - i \end{pmatrix}^\dagger \begin{pmatrix} x_1 \\ x_2 \end{pmatrix} = \begin{pmatrix} 0 \\ 0 \end{pmatrix}$$

and we obtain $(2 - i)x_1 - 5x_2 = 0$ and $x_1 + (-2 - i)x_2 = 0$. Thus $x_1 = (2 + i)x_2$, which yields the eigenvector $v_1 = \begin{pmatrix} 2 + i \\ 1 \end{pmatrix}$ and $E_{1+i} = $ span $\left\{ \begin{pmatrix} 2 + i \\ 1 \end{pmatrix} \right\}$. Similarly, $[A - (1 - i)I]v = \begin{pmatrix} 2 + i & -5 \\ 1 & -2 + i \end{pmatrix}\begin{pmatrix} x_1 \\ x_2 \end{pmatrix} = \begin{pmatrix} 0 \\ 0 \end{pmatrix}$ or $x_1 + (-2 + i)x_2 = 0$, which yields $x_1 = (2 - i)x_2$, $v_2 = \begin{pmatrix} 2 - i \\ 1 \end{pmatrix}$, and $E_{1-i} = $ span $\left\{ \begin{pmatrix} 2 - i \\ 1 \end{pmatrix} \right\}$. ∎

Remark 1. This example illustrates that a real matrix may have complex eigenvalues and eigenvectors. Some texts define eigenvalues of real matrices to be *real* roots of the characteristic equation. With this definition the matrix of the last example has *no* eigenvalues. This might make the computations simpler, but it also greatly reduces the usefulness of the theory of eigenvalues and eigenvectors. We shall see a significant illustration of the use of complex eigenvalues in Section 6.7.

Remark 2. Note that $\lambda_2 = 1 - i$ is the complex conjugate of $\lambda_1 = 1 + i$. Also, the components of v_2 are complex conjugates of the components of v_1. This is no coincidence. In Problem 33 you are asked to prove that

---

† Note that the columns of this matrix are linearly dependent because $\begin{pmatrix} -5 \\ -2 - i \end{pmatrix} = (-2 - i)\begin{pmatrix} 2 - i \\ 1 \end{pmatrix}$.

> The eigenvalues of a real matrix occur in complex conjugate pairs
> and
> corresponding eigenvectors are complex conjugates of one another.

EXAMPLE 7   **A 2 × 2 Matrix with One Eigenvalue and Two Linearly Independent Eigenvectors**   Let $A = \begin{pmatrix} 4 & 0 \\ 0 & 4 \end{pmatrix}$. Then $\det (A - \lambda I) = \begin{vmatrix} 4 - \lambda & 0 \\ 0 & 4 - \lambda \end{vmatrix} = (\lambda - 4)^2 = 0$; hence $\lambda = 4$ is an eigenvalue of algebraic multiplicity 2. It is obvious that $A\mathbf{v} = 4\mathbf{v}$ for every vector $\mathbf{v} \in \mathbb{R}^2$ so that $E_4 = \mathbb{R}^2 = \text{span} \left\{ \begin{pmatrix} 1 \\ 0 \end{pmatrix}, \begin{pmatrix} 0 \\ 1 \end{pmatrix} \right\}$.   ∎

EXAMPLE 8   **A 2 × 2 Matrix with One Eigenvalue and Only One Linearly Independent Eigenvector**   Let $A = \begin{pmatrix} 4 & 1 \\ 0 & 4 \end{pmatrix}$. Then $\det (A - \lambda I) = \begin{vmatrix} 4 - \lambda & 1 \\ 0 & 4 - \lambda \end{vmatrix} = (\lambda - 4)^2 = 0$; thus $\lambda = 4$ is again an eigenvalue of algebraic multiplicity 2. But this time we have $(A - 4I)\mathbf{v} = \begin{pmatrix} 0 & 1 \\ 0 & 0 \end{pmatrix} \begin{pmatrix} x_1 \\ x_2 \end{pmatrix} = \begin{pmatrix} x_2 \\ 0 \end{pmatrix}$. Thus $x_2 = 0$, $\mathbf{v}_1 = \begin{pmatrix} 1 \\ 0 \end{pmatrix}$ is an eigenvector, and $E_4 = \text{span} \left\{ \begin{pmatrix} 1 \\ 0 \end{pmatrix} \right\}$.   ∎

EXAMPLE 9   **A 3 × 3 Matrix with Two Distinct Eigenvalues and Three Linearly Independent Eigenvectors**   Let $A = \begin{pmatrix} 3 & 2 & 4 \\ 2 & 0 & 2 \\ 4 & 2 & 3 \end{pmatrix}$. Then $\det (A - \lambda I) =$

$$\begin{vmatrix} 3 - \lambda & 2 & 4 \\ 2 & -\lambda & 2 \\ 4 & 2 & 3 - \lambda \end{vmatrix} = -\lambda^3 + 6\lambda^2 + 15\lambda + 8 = -(\lambda + 1)^2(\lambda - 8) = 0 \text{ so that}$$

the eigenvalues are $\lambda_1 = 8$ and $\lambda_2 = -1$ (with algebraic multiplicity 2). For $\lambda_1 = 8$, we obtain

$$(A - 8I)\mathbf{v} = \begin{pmatrix} -5 & 2 & 4 \\ 2 & -8 & 2 \\ 4 & 2 & -5 \end{pmatrix} \begin{pmatrix} x_1 \\ x_2 \\ x_3 \end{pmatrix} = \begin{pmatrix} 0 \\ 0 \\ 0 \end{pmatrix}$$

or, row reducing,

$$\begin{pmatrix} -5 & 2 & 4 & | & 0 \\ 2 & -8 & 2 & | & 0 \\ 4 & 2 & -5 & | & 0 \end{pmatrix} \xrightarrow[R_3 \to R_3 - R_1]{R_2 \to R_2 + 4R_1} \begin{pmatrix} -5 & 2 & 4 & | & 0 \\ -18 & 0 & 18 & | & 0 \\ 9 & 0 & -9 & | & 0 \end{pmatrix}$$

$$\xrightarrow{R_2 \to \frac{1}{18}R_2} \begin{pmatrix} -5 & 2 & 4 & | & 0 \\ -1 & 0 & 1 & | & 0 \\ 9 & 0 & -9 & | & 0 \end{pmatrix} \xrightarrow[R_3 \to R_3 + 9R_2]{R_1 \to R_1 - 5R_2} \begin{pmatrix} 0 & 2 & -1 & | & 0 \\ -1 & 0 & 1 & | & 0 \\ 0 & 0 & 0 & | & 0 \end{pmatrix}$$

Hence $x_3 = 2x_2$ and $x_1 = x_3$, we obtain the eigenvector $\mathbf{v}_1 = \begin{pmatrix} 2 \\ 1 \\ 2 \end{pmatrix}$, and $E_8 =$

span $\left\{ \begin{pmatrix} 2 \\ 1 \\ 2 \end{pmatrix} \right\}$. For $\lambda_2 = -1$, we have $(A + I)\mathbf{v} = \begin{pmatrix} 4 & 2 & 4 \\ 2 & 1 & 2 \\ 4 & 2 & 4 \end{pmatrix} \begin{pmatrix} x_1 \\ x_2 \\ x_3 \end{pmatrix} = \begin{pmatrix} 0 \\ 0 \\ 0 \end{pmatrix}$, which

gives us the single equation $2x_1 + x_2 + 2x_3 = 0$ or $x_2 = -2x_1 - 2x_3$. If $x_1 = 1$

and $x_3 = 0$, we obtain $\mathbf{v}_2 = \begin{pmatrix} 1 \\ -2 \\ 0 \end{pmatrix}$. If $x_1 = 0$ and $x_3 = 1$, we obtain $\mathbf{v}_3 = \begin{pmatrix} 0 \\ -2 \\ 1 \end{pmatrix}$.

Thus $E_{-1} = $ span $\left\{ \begin{pmatrix} 1 \\ -2 \\ 0 \end{pmatrix}, \begin{pmatrix} 0 \\ -2 \\ 1 \end{pmatrix} \right\}$. There are other convenient choices for eigen-

vectors. For example, $\mathbf{v} = \begin{pmatrix} 1 \\ 0 \\ -1 \end{pmatrix}$ is in E_{-1} since $\mathbf{v} = \mathbf{v}_2 - \mathbf{v}_3$. ∎

EXAMPLE 10 **A 3 × 3 Matrix with One Eigenvalue and Only One Linearly Independent Eigenvector** Let $A = \begin{pmatrix} -5 & -5 & -9 \\ 8 & 9 & 18 \\ -2 & -3 & -7 \end{pmatrix}$. Then $\det (A - \lambda I) =$

$$\begin{vmatrix} -5 - \lambda & -5 & -9 \\ 8 & 9 - \lambda & 18 \\ -2 & -3 & -7 - \lambda \end{vmatrix} = -\lambda^3 - 3\lambda^2 - 3\lambda - 1 = -(\lambda + 1)^3 = 0.$$ Thus

$\lambda = -1$ is an eigenvalue of algebraic multiplicity 3. To compute E_{-1}, we set

$(A + I)\mathbf{v} = \begin{pmatrix} -4 & -5 & -9 \\ 8 & 10 & 18 \\ -2 & -3 & -6 \end{pmatrix} \begin{pmatrix} x_1 \\ x_2 \\ x_3 \end{pmatrix} = \begin{pmatrix} 0 \\ 0 \\ 0 \end{pmatrix}$ and row-reduce to obtain, successively,

$$\begin{pmatrix} -4 & -5 & -9 & | & 0 \\ 8 & 10 & 18 & | & 0 \\ -2 & -3 & -6 & | & 0 \end{pmatrix} \xrightarrow[R_2 \to R_2 + 4R_3]{R_1 \to R_1 - 2R_3} \begin{pmatrix} 0 & 1 & 3 & | & 0 \\ 0 & -2 & -6 & | & 0 \\ -2 & -3 & -6 & | & 0 \end{pmatrix}$$

$$\xrightarrow[R_3 \to R_3 + 3R_1]{R_2 \to R_2 + 2R_1} \begin{pmatrix} 0 & 1 & 3 & | & 0 \\ 0 & 0 & 0 & | & 0 \\ -2 & 0 & 3 & | & 0 \end{pmatrix}$$

This yields $x_2 = -3x_3$ and $2x_1 = 3x_3$. Setting $x_3 = 2$, we obtain only one linearly independent eigenvector: $\mathbf{v}_1 = \begin{pmatrix} 3 \\ -6 \\ 2 \end{pmatrix}$. Thus $E_{-1} = \text{span} \left\{ \begin{pmatrix} 3 \\ -6 \\ 2 \end{pmatrix} \right\}$.

■

EXAMPLE 11   **A 3 × 3 Matrix with One Eigenvalue and Two Linearly Independent Eigenvectors**   Let $A = \begin{pmatrix} -1 & -3 & -9 \\ 0 & 5 & 18 \\ 0 & -2 & -7 \end{pmatrix}$.   Then $\det (A - \lambda I) =$

$$\begin{vmatrix} -1 - \lambda & -3 & -9 \\ 0 & 5 - \lambda & 18 \\ 0 & -2 & -7 - \lambda \end{vmatrix} = -(\lambda + 1)^3 = 0.$$ Thus, as in Example 10, $\lambda = -1$

is an eigenvalue of algebraic multiplicity 3. To find E_{-1}, we compute $(A + I)\mathbf{v} =$
$\begin{pmatrix} 0 & -3 & -9 \\ 0 & 6 & 18 \\ 0 & -2 & -6 \end{pmatrix} \begin{pmatrix} x_1 \\ x_2 \\ x_3 \end{pmatrix} = \begin{pmatrix} 0 \\ 0 \\ 0 \end{pmatrix}$. Thus $-2x_2 - 6x_3 = 0$ or $x_2 = -3x_3$, and x_1 is arbi-

trary. Setting $x_1 = 0$, $x_3 = 1$, we obtain $\mathbf{v}_1 = \begin{pmatrix} 0 \\ -3 \\ 1 \end{pmatrix}$. Setting $x_1 = 1$, $x_3 = 1$ yields

$\mathbf{v}_2 = \begin{pmatrix} 1 \\ -3 \\ 1 \end{pmatrix}$. Thus $E_{-1} = \text{span} \left\{ \begin{pmatrix} 0 \\ -3 \\ 1 \end{pmatrix}, \begin{pmatrix} 1 \\ -3 \\ 1 \end{pmatrix} \right\}$.

■

In each of the last five examples we found an eigenvalue with an algebraic multiplicity of 2 or more. But, as we saw in Examples 8, 10, and 11, the number of linearly independent eigenvectors is not necessarily equal to the algebraic multiplicity of the eigenvalue (as was the case in Examples 7 and 9). This observation leads to the following definition.

DEFINITION 4   **Geometric Multiplicity**   Let λ be an eigenvalue of the matrix A. Then the **geometric multiplicity** of λ is the dimension of the eigenspace corresponding to λ (which is the nullity of the matrix $A - \lambda I$). That is,

$$\text{Geometric multiplicity of } \lambda = \dim E_\lambda = \nu(A - \lambda I)$$

In Examples 7 and 9 we saw that for the eigenvalues of algebraic multiplicity 2, the geometric multiplicities were also 2. In Example 8 the geometric multiplicity of $\lambda = 4$ was 1 while the algebraic multiplicity was 2. In Example 10 the algebraic multiplicity was 3 and the geometric multiplicity was 1. In Example 11 the algebraic multiplicity was 3 and the geometric multiplicity was 2. These examples illustrate the fact that if the algebraic multiplicity of λ is greater than 1, then we cannot predict the geometric multiplicity of λ without additional information.

If A is a 2×2 matrix and λ is an eigenvalue with algebraic multiplicity 2, then the geometric multiplicity of λ is ≤ 2 since there can be at most two linearly independent vectors in a two-dimensional space. Let A be a 3×3 matrix having two eigenvalues λ_1 and λ_2 with algebraic multiplicities 1 and 2, respectively. Then the geometric multiplicity of λ_2 is ≤ 2 because otherwise we would have at least four linearly independent vectors in a three-dimensional space. Intuitively, it seems that the geometric multiplicity of an eigenvalue is always less than or equal to its algebraic multiplicity. The proof of the following theorem is not difficult if additional facts about determinants are proved. Since this would take us too far afield, we omit the proof.†

THEOREM 4   Let λ be an eigenvalue of A. Then

$$\boxed{\text{Geometric multiplicity of } \lambda \leq \text{algebraic multiplicity of } \lambda}$$

Note. The geometric multiplicity of an eigenvalue is never zero. This follows from Definition 1, which states that if λ is an eigenvalue, then there exists a *nonzero* eigenvector corresponding to λ.

In the rest of this chapter an important problem for us will be to determine whether a given $n \times n$ matrix does or does not have n linearly independent eigenvectors. From what we have already discussed in this section, the following theorem is apparent.

THEOREM 5   Let A be an $n \times n$ matrix. Then A has n linearly independent eigenvectors if and only if the geometric multiplicity of every eigenvalue is equal to its algebraic multiplicity. In particular, A has n linearly independent eigenvectors if all the eigenvalues are distinct (since then the algebraic multiplicity of every eigenvalue is 1).   ■

In Example 5 we saw a matrix for which zero was an eigenvalue. In fact, from Theorem 1, it is evident that zero is an eigenvalue of A if and only if det $A =$ det $(A - 0I) = 0$. This enables us to extend, for the last time, our Summing Up Theorem (see Theorem 5.4.4, page 370).

---

†For a proof see Theorem 11.2.6 in C. R. Wylie's book *Advanced Engineering Mathematics* (New York: McGraw-Hill, 1975).

THEOREM 6   **Summing Up Theorem—View 8** Let A be an $n \times n$ matrix. Then the following eleven statements are equivalent; that is, each one implies the other ten (so that if one is true, all are true):

    i. A is invertible.

    ii. The only solution to the homogeneous system $A\mathbf{x} = \mathbf{0}$ is the trivial solution ($\mathbf{x} = \mathbf{0}$).

    iii. The system $A\mathbf{x} = \mathbf{b}$ has a unique solution for every n-vector $\mathbf{b}$.

    iv. A is row equivalent to the $n \times n$ identity matrix I_n.

    v. A can be written as the product of elementary matrices.

    vi. The rows (and columns) of A are linearly independent.

    vii. $\det A \neq 0$.

    viii. $\nu(A) = 0$.

    ix. $\rho(A) = n$.

    x. The linear transformation T from $\mathbb{R}^n$ to $\mathbb{R}^n$ defined by $T\mathbf{x} = A\mathbf{x}$ is an isomorphism.

    xi. Zero is *not* an eigenvalue of A.   ■

PROBLEMS 6.1

In Problems 1–20 calculate the eigenvalues and eigenspaces of the given matrix. If the algebraic multiplicity of an eigenvalue is greater than 1, calculate its geometric multiplicity.

1. $\begin{pmatrix} -2 & -2 \\ -5 & 1 \end{pmatrix}$ **2.** $\begin{pmatrix} -12 & 7 \\ -7 & 2 \end{pmatrix}$ **3.** $\begin{pmatrix} 2 & -1 \\ 5 & -2 \end{pmatrix}$

4. $\begin{pmatrix} -3 & 0 \\ 0 & -3 \end{pmatrix}$ **5.** $\begin{pmatrix} -3 & 2 \\ 0 & -3 \end{pmatrix}$ **6.** $\begin{pmatrix} 3 & 2 \\ -5 & 1 \end{pmatrix}$

7. $\begin{pmatrix} 1 & -1 & 0 \\ -1 & 2 & -1 \\ 0 & -1 & 1 \end{pmatrix}$ **8.** $\begin{pmatrix} 1 & 1 & -2 \\ -1 & 2 & 1 \\ 0 & 1 & -1 \end{pmatrix}$ **9.** $\begin{pmatrix} 5 & 4 & 2 \\ 4 & 5 & 2 \\ 2 & 2 & 2 \end{pmatrix}$

10. $\begin{pmatrix} 1 & 2 & 2 \\ 0 & 2 & 1 \\ -1 & 2 & 2 \end{pmatrix}$ **11.** $\begin{pmatrix} 0 & 1 & 0 \\ 0 & 0 & 1 \\ 1 & -3 & 3 \end{pmatrix}$ **12.** $\begin{pmatrix} -3 & -7 & -5 \\ 2 & 4 & 3 \\ 1 & 2 & 2 \end{pmatrix}$

13. $\begin{pmatrix} 1 & -1 & -1 \\ 1 & -1 & 0 \\ 1 & 0 & -1 \end{pmatrix}$ **14.** $\begin{pmatrix} 7 & -2 & -4 \\ 3 & 0 & -2 \\ 6 & -2 & -3 \end{pmatrix}$ **15.** $\begin{pmatrix} 4 & 6 & 6 \\ 1 & 3 & 2 \\ -1 & -5 & -2 \end{pmatrix}$

16. $\begin{pmatrix} 4 & 1 & 0 & 1 \\ 2 & 3 & 0 & 1 \\ -2 & 1 & 2 & -3 \\ 2 & -1 & 0 & 5 \end{pmatrix}$ **17.** $\begin{pmatrix} a & 0 & 0 & 0 \\ 0 & a & 0 & 0 \\ 0 & 0 & a & 0 \\ 0 & 0 & 0 & a \end{pmatrix}$

18. $\begin{pmatrix} a & b & 0 & 0 \\ 0 & a & 0 & 0 \\ 0 & 0 & a & 0 \\ 0 & 0 & 0 & a \end{pmatrix}; b \neq 0$

19. $\begin{pmatrix} a & b & 0 & 0 \\ 0 & a & c & 0 \\ 0 & 0 & a & 0 \\ 0 & 0 & 0 & a \end{pmatrix}; bc \neq 0$

20. $\begin{pmatrix} a & b & 0 & 0 \\ 0 & a & c & 0 \\ 0 & 0 & a & d \\ 0 & 0 & 0 & a \end{pmatrix}; bcd \neq 0$

21. Show that for any real numbers a and b, the matrix $A = \begin{pmatrix} a & b \\ -b & a \end{pmatrix}$ has the eigenvectors $\begin{pmatrix} 1 \\ i \end{pmatrix}$ and $\begin{pmatrix} 1 \\ -i \end{pmatrix}$.

In Problems 22–28 assume that the matrix A has the eigenvalues $\lambda_1, \lambda_2, \ldots, \lambda_k$.

22. Show that the eigenvalues of A^t are $\lambda_1, \lambda_2, \ldots, \lambda_k$.

23. Show that the eigenvalues of αA are $\alpha\lambda_1, \alpha\lambda_2, \ldots, \alpha\lambda_k$.

24. Show that A^{-1} exists if and only if $\lambda_1\lambda_2 \cdots \lambda_k \neq 0$.

*25. If A^{-1} exists, show that the eigenvalues of A^{-1} are $1/\lambda_1, 1/\lambda_2, \ldots, 1/\lambda_k$.

26. Show that the matrix $A - \alpha I$ has the eigenvalues $\lambda_1 - \alpha, \lambda_2 - \alpha, \ldots, \lambda_k - \alpha$.

*27. Show that the eigenvalues of A^2 are $\lambda_1^2, \lambda_2^2, \ldots, \lambda_k^2$.

*28. Show that the eigenvalues of A^m are $\lambda_1^m, \lambda_2^m, \ldots, \lambda_k^m$ for $m = 1, 2, 3, \ldots$.

29. Let λ be an eigenvalue of A with corresponding eigenvector $\mathbf{v}$. Let $p(\lambda) = a_0 + a_1\lambda + a_2\lambda^2 + \cdots + a_n\lambda^n$. Define the matrix $p(A)$ by $p(A) = a_0 I + a_1 A + a_2 A^2 + \cdots + a_n A^n$. Show that $p(A)\mathbf{v} = p(\lambda)\mathbf{v}$.

30. Using the result of Problem 29, show that if $\lambda_1, \lambda_2, \ldots, \lambda_k$ are eigenvalues of A, then $p(\lambda_1), p(\lambda_2), \ldots, p(\lambda_k)$ are eigenvalues of $p(A)$.

31. Show that if A is an upper triangular matrix, then the eigenvalues of A are the diagonal components of A.

32. Let $A_1 = \begin{pmatrix} 2 & 0 & 0 & 0 \\ 0 & 2 & 0 & 0 \\ 0 & 0 & 2 & 0 \\ 0 & 0 & 0 & 2 \end{pmatrix}$, $A_2 = \begin{pmatrix} 2 & 1 & 0 & 0 \\ 0 & 2 & 0 & 0 \\ 0 & 0 & 2 & 0 \\ 0 & 0 & 0 & 2 \end{pmatrix}$, $A_3 = \begin{pmatrix} 2 & 1 & 0 & 0 \\ 0 & 2 & 1 & 0 \\ 0 & 0 & 2 & 0 \\ 0 & 0 & 0 & 2 \end{pmatrix}$, and

$A_4 = \begin{pmatrix} 2 & 1 & 0 & 0 \\ 0 & 2 & 1 & 0 \\ 0 & 0 & 2 & 1 \\ 0 & 0 & 0 & 2 \end{pmatrix}$. Show that, for each matrix, $\lambda = 2$ is an eigenvalue of algebraic multiplicity 4. In each case compute the geometric multiplicity of $\lambda = 2$.

*33. Let A be a real $n \times n$ matrix. Show that if λ_1 is a complex eigenvalue of A with eigenvector $\mathbf{v}_1$, then $\bar{\lambda}_1$ is an eigenvalue of A with eigenvector $\bar{\mathbf{v}}_1$.

*34. A **probability matrix** is an $n \times n$ matrix having two properties:

   i. $a_{ij} \geq 0$ for every i and j.
   ii. The sum of the components in every column is 1.

Prove that 1 is an eigenvalue of every probability matrix.

*35. Let $A = \begin{pmatrix} a & b \\ c & d \end{pmatrix}$ be a 2×2 matrix. Suppose that $b \neq 0$. Let m be a root (real or complex) of the equation

$$bm^2 + (a - d)m - c = 0$$

show that $a + bm$ is an eigenvalue of A with corresponding eigenvector $\mathbf{v} = \begin{pmatrix} 1 \\ m \end{pmatrix}$. This gives us an easy way to compute eigenvalues and eigenvectors of 2×2 matrices. [This procedure appeared in the paper "A Simple Algorithm for Finding Eigenvalues and Eigenvectors for 2×2 Matrices" by Tyre A. Newton in The *American Mathematical Monthly*, 97(1), January, 1990, 57–60.]

36. Let $A = \begin{pmatrix} a & 0 \\ c & d \end{pmatrix}$ be a 2×2 matrix. Show that d is an eigenvalue of A with corresponding eigenvector $\begin{pmatrix} 0 \\ 1 \end{pmatrix}$.

6.2   A MODEL OF POPULATION GROWTH (Optional)

In this section we show how the theory of eigenvalues and eigenvectors can be used to analyze a model of the growth of a bird population.† We begin by discussing a simple model of population growth. We assume that a certain species grows at a constant rate; that is, the population of the species after one time period (which could be an hour, a week, a month, a year, etc.) is a constant multiple of the population in the previous time period. One way this could happen, for example, is that each generation is distinct and each organism produces r offspring and then dies. If p_n denotes the population after the nth time period, we would have

$$p_n = rp_{n-1}$$

For example, this model might describe a bacteria population, where, at a given time, an organism splits into two separate organisms. Then $r = 2$. Let p_0 denote the initial population. Then $p_1 = rp_0$, $p_2 = rp_1 = r(rp_0) = r^2p_0$, $p_3 = rp_2 = r(r^2p_0) = r^3p_0$, and so on, so that

$$p_n = r^np_0 \tag{1}$$

From this model we see that the population increases without bound if $r > 1$ and decreases to zero if $r < 1$. If $r = 1$ the population remains at the constant value p_0.

This model is, evidently, very simplistic. One obvious objection is that the number of offspring produced depends, in many cases, on the ages of the adults. For example, in a human population the average female adult over 50 would certainly produce fewer children than the average 21-year-old female. To deal with this difficulty, we introduce a model which allows for age groupings with different fertility rates.

---

† The material in this section is based on a paper by D. Cooke: "A 2×2 Matrix Model of Population Growth," *Mathematical Gazette* 61(416): 120–123.

We look at a model of population growth for a species of birds. In this bird population we assume that the number of female birds equals the number of males. Let $p_{j,n-1}$ denote the population of juvenile (immature) females in the $(n-1)$st year and let $p_{a,n-1}$ denote the number of adult females in the $(n-1)$st year. Some of the juvenile birds will die during the year. We assume that a certain proportion α of the juvenile birds survive to become adults in the spring of the nth year. Each surviving female bird produces eggs later in the spring, which hatch to produce, on the average, k juvenile female birds in the following spring. Adults also die, and the proportion of adults that survive from one spring to the next is β.

This constant survival rate of birds is not just a simplistic assumption. It appears to be the case with most of the natural bird populations that have been studied. This means that the adult survival rate of many bird species is independent of age. Perhaps few birds in the wild survive long enough to exhibit the effects of old age. Moreover, in many species the number of offspring seems to be uninfluenced by the age of the mother.

In the notation introduced above $p_{j,n}$ and $p_{a,n}$ represent, respectively, the populations of juvenile and adult females in the nth year. Putting together all the information given, we arrive at the following 2×2 system:

$$
\begin{aligned}
p_{j,n} &= kp_{a,n-1} \\
p_{a,n} &= \alpha p_{j,n-1} + \beta p_{a,n-1}
\end{aligned}
\tag{2}
$$

or

$$
\mathbf{p}_n = A\mathbf{p}_{n-1}
\tag{3}
$$

where $\mathbf{p}_n = \begin{pmatrix} p_{j,n} \\ p_{a,n} \end{pmatrix}$ and $A = \begin{pmatrix} 0 & k \\ \alpha & \beta \end{pmatrix}$. It is clear, from (3), that $\mathbf{p}_1 = A\mathbf{p}_0$, $\mathbf{p}_2 = A\mathbf{p}_1 = A(A\mathbf{p}_0) = A^2\mathbf{p}_0, \ldots$, and so on. Hence

$$
\boxed{\mathbf{p}_n = A^n\mathbf{p}_0}
\tag{4}
$$

where $\mathbf{p}_0$ is the vector of initial populations of juvenile and adult females.

Equation (4) is like equation (1), but now we are able to distinguish between the survival rates of juvenile and adult birds.

EXAMPLE 1   **An Illustration of the Model Carried Through 20 Generations**   Let $A = \begin{pmatrix} 0 & 2 \\ 0.3 & 0.5 \end{pmatrix}$. This means that each adult female produces two female offspring and, since the number of males is assumed equal to the number of females, at least four eggs—and probably many more, since losses among fledglings are likely to be high. From the model, it is apparent that α and β lie in the interval $[0, 1]$. Since juvenile birds are not so likely as adults to survive, we must have $\alpha < \beta$.

In Table 6.1 we assume that, initially, there are 10 female (and 10 male) adults and no juveniles. The computations were done on a computer, but the work would not be too onerous if done on a hand calculator. For example, $\mathbf{p}_1 = \begin{pmatrix} 0 & 2 \\ 0.3 & 0.5 \end{pmatrix}\begin{pmatrix} 0 \\ 10 \end{pmatrix} = \begin{pmatrix} 20 \\ 5 \end{pmatrix}$ so that $p_{j,1} = 20$, $p_{a,1} = 5$, the total female population after 1 year is 25, and the ratio of juvenile to adult females is 4 to 1. In the second year $\mathbf{p}_2 = \begin{pmatrix} 0 & 2 \\ 0.3 & 0.5 \end{pmatrix}\begin{pmatrix} 20 \\ 5 \end{pmatrix} = \begin{pmatrix} 10 \\ 8.5 \end{pmatrix}$, which we round down to $\begin{pmatrix} 10 \\ 8 \end{pmatrix}$ since we cannot have $8\frac{1}{2}$ adult birds. Table 6.1 tabulates the ratios $p_{j,n}/p_{a,n}$ and the ratios T_n/T_{n-1} of the total number of females in successive years.

Table 6.1

Year n	No. of juveniles $p_{j,n}$	No. of adults $p_{a,n}$	Total female population T_n in nth year	$p_{j,n}/p_{a,n}$†	T_n/T_{n-1}†
0	0	10	10	0	—
1	20	5	25	4.00	2.50
2	10	8	18	1.18	0.74
3	17	7	24	2.34	1.31
4	14	8	22	1.66	0.96
5	17	8	25	2.00	1.13
10	22	12	34	1.87	1.06
11	24	12	36	1.88	1.07
12	25	13	38	1.88	1.06
20	42	22	64	1.88	1.06

†The figures in these columns were obtained before the numbers in the previous columns were rounded. Thus, for example, in year 2, $p_{j,2}/p_{a,2} = 10/8.5 \approx 1.176470588 \approx 1.18$. ∎

In Table 6.1 it seems as if the ratio $p_{j,n}/p_{a,n}$ is approaching the constant 1.88 while the total population seems to be increasing at a constant rate of 6 percent a year. Let us see if we can determine why this is the case.

First, we return to the general case (equation (4)). Suppose that A has the real distinct eigenvalues λ_1 and λ_2 with corresponding eigenvectors $\mathbf{v}_1$ and $\mathbf{v}_2$. Since $\mathbf{v}_1$ and $\mathbf{v}_2$ are linearly independent, they form a basis for $\mathbb{R}^2$ and we can write

$$\mathbf{p}_0 = a_1\mathbf{v}_1 + a_2\mathbf{v}_2 \tag{5}$$

for some real numbers a_1 and a_2. Then (4) becomes

$$\mathbf{p}_n = A^n(a_1\mathbf{v}_1 + a_2\mathbf{v}_2) \tag{6}$$

But $A\mathbf{v}_1 = \lambda_1\mathbf{v}_1$ and $A^2\mathbf{v}_1 = A(A\mathbf{v}_1) = A(\lambda_1\mathbf{v}_1) = \lambda_1 A\mathbf{v}_1 = \lambda_1(\lambda_1\mathbf{v}_1) = \lambda_1^2\mathbf{v}_1$. Thus we can see that $A^n\mathbf{v}_1 = \lambda_1^n\mathbf{v}_1$, $A^n\mathbf{v}_2 = \lambda_2^n\mathbf{v}_2$, and, from (6),

$$\mathbf{p}_n = a_1\lambda_1^n\mathbf{v}_1 + a_2\lambda_2^n\mathbf{v}_2 \tag{7}$$

The characteristic equation of A is $\begin{vmatrix} -\lambda & k \\ \alpha & \beta - \lambda \end{vmatrix} = \lambda^2 - \beta\lambda - k\alpha = 0$ or

$\lambda = (\beta \pm \sqrt{\beta^2 + 4\alpha k})/2$. By assumption, $k > 0$, $0 < \alpha < 1$, and $0 < \beta < 1$. Hence $4\alpha k > 0$ and $\beta^2 + 4\alpha k > 0$, which means that the eigenvalues are, indeed, real and distinct and that one eigenvalue λ_1 is positive, one λ_2 is negative, and $|\lambda_1| > |\lambda_2|$. We can write (7) as

$$\mathbf{p}_n = \lambda_1^n \left[a_1 \mathbf{v}_1 + \left(\frac{\lambda_2}{\lambda_1} \right)^n a_2 \mathbf{v}_2 \right] \tag{8}$$

Since $|\lambda_2/\lambda_1| < 1$, it is apparent that $(\lambda_2/\lambda_1)^n$ gets very small as n gets large. Thus, for n large,

$$\mathbf{p}_n \approx a_1 \lambda_1^n \mathbf{v}_1 \tag{9}$$

This means that, in the long run, the age distribution stabilizes and is proportional to $\mathbf{v}_1$. Each age group will change by a factor of λ_1 each year. Thus—in the long run—equation (4) acts just like equation (1). In the short term—that is, before "stability" is reached—the numbers oscillate. The magnitude of this oscillation depends on the magnitude of λ_2/λ_1 (which is negative, thus explaining the oscillation).

EXAMPLE 1
(continued)

The Eigenvalues and Eigenvectors of A Determine the Behavior in Future Generations   For $A = \begin{pmatrix} 0 & 2 \\ 0.3 & 0.5 \end{pmatrix}$, we have $\lambda^2 - 0.5\lambda - 0.6 = 0$ or $\lambda = (0.5 \pm \sqrt{0.25 + 2.4})/2 = (0.5 \pm \sqrt{2.65})/2$ so that $\lambda_1 \approx 1.06$ and $\lambda_2 \approx -0.56$. This explains the 6 percent increase in population noted in the last column of Table 6.1. Corresponding to the eigenvalue $\lambda_1 = 1.06$, we compute $(A - 1.06I)\mathbf{v}_1 = \begin{pmatrix} -1.06 & 2 \\ 0.3 & -0.56 \end{pmatrix}\begin{pmatrix} x_1 \\ x_2 \end{pmatrix} = \begin{pmatrix} 0 \\ 0 \end{pmatrix}$ or $1.06x_1 = 2x_2$ so that $\mathbf{v}_1 = \begin{pmatrix} 1 \\ 0.53 \end{pmatrix}$ is an eigenvector. Similarly, $(A + 0.56)\mathbf{v}_2 = \begin{pmatrix} 0.56 & 2 \\ 0.3 & 1.06 \end{pmatrix}\begin{pmatrix} x_1 \\ x_2 \end{pmatrix} = \begin{pmatrix} 0 \\ 0 \end{pmatrix}$ so that $0.56x_1 + 2x_2 = 0$ and $\mathbf{v}_2 = \begin{pmatrix} 1 \\ -0.28 \end{pmatrix}$ is a second eigenvector. Note that in $\mathbf{v}_1$ we have $1/0.53 \approx 1.88$. This explains the ratio $p_{j,n}/p_{a,n}$ in the fifth column of the table. ∎

Remark. In the preceding computations precision was lost because we rounded to only two decimal places of accuracy. Much greater accuracy is obtained by using a hand calculator or computer. For example, using a hand calculator, we easily calculate $\lambda_1 = 1.06394103$, $\lambda_2 = -0.5639410298$, $\mathbf{v}_1 = \begin{pmatrix} 1 \\ 0.531970515 \end{pmatrix}$, $\mathbf{v}_2 = \begin{pmatrix} 1 \\ -0.2819705149 \end{pmatrix}$, and the ratio of $p_{j,n}$ to $p_{a,n}$ is seen to be $1/0.5319710515 \approx 1.879801537$.

It is remarkable just how much information is available from a simple computation of eigenvalues. It is of great interest to know whether a population will ulti-mately increase or decrease. It will increase if $\lambda_1 > 1$, and the condition for that is $(\beta + \sqrt{\beta^2 + 4\alpha k})/2 > 1$ or $\sqrt{\beta^2 + 4\alpha k} > 2 - \beta$ or $\beta^2 + 4\alpha k > (2 - \beta)^2 = 4 - 4\beta + \beta^2$. This leads to $4\alpha k > 4 - 4\beta$ or

$$k > \frac{1 - \beta}{\alpha} \tag{10}$$

In Example 1 we had $\beta = 0.5$, $\alpha = 0.3$; thus (10) is satisfied if $k > 0.5/0.3 \approx 1.67$.

Before we close this section we indicate two limitations of this model:

i.   Birth and death rates often change from year to year and are particularly depen-dent on the weather. This model assumes a constant environment.

ii. Ecologists have found that for many species birth and death rates vary with the size of the population. In particular, a population cannot grow when it reaches a certain size due to the effects of limited food resources and overcrowding. It is obvious that a population cannot grow indefinitely at a constant rate. Otherwise that population would overrun the earth.

PROBLEMS   6.2

In Problems 1–3 find the numbers of juvenile and adult female birds after 1, 2, 5, 10, 19, and 20 years. Then find the long-term ratios of $p_{j,n}$ to $p_{a,n}$ and T_n to T_{n-1}. [*Hint:* Use equations (7) and (9) and a calculator and round to three decimals.]

1. $\mathbf{p}_0 = \begin{pmatrix} 0 \\ 12 \end{pmatrix}$; $k = 3$, $\alpha = 0.4$, $\beta = 0.6$

2. $\mathbf{p}_0 = \begin{pmatrix} 0 \\ 15 \end{pmatrix}$; $k = 1$, $\alpha = 0.3$, $\beta = 0.4$

3. $\mathbf{p}_0 = \begin{pmatrix} 0 \\ 20 \end{pmatrix}$; $k = 4$, $\alpha = 0.7$, $\beta = 0.8$

4.   Show that if $\alpha = \beta$ and $\alpha > \frac{1}{2}$, then the bird population will always increase in the long run if at least one female offspring on the average is produced by each female adult.

5.   Show that, in the long run, the ratio $p_{j,n}/p_{a,n}$ approaches the limiting value k/λ_1.

6.   Suppose we divide the adult birds into two age groups: those 1–5 years old and those more than 5 years old. Assume that the survival rate for birds in the first group is β, whereas in the second group it is γ (and $\beta > \gamma$). Assume that the birds in the first group are equally divided as to age. (That is, if there are 100 birds in the group, then 20 are 1 year old, 20 are 2 years old, and so on.) Formulate a 3 × 3 matrix model for this situation.

6.3   SIMILAR MATRICES AND DIAGONALIZATION

In this section we describe an interesting and useful relationship that can hold between two matrices.

DEFINITION 1   **Similar Matrices**   Two $n \times n$ matrices A and B are said to be **similar** if there exists an invertible $n \times n$ matrix C such that

$$\boxed{B = C^{-1}AC} \qquad (1)$$

Similarity transformation

The function defined by (1) which takes the matrix A into the matrix B is called a **similarity transformation.** We can write this linear transformation as

$$T(A) = C^{-1}AC$$

Note. $C^{-1}(A_1 + A_2)C = C^{-1}A_1C + C^{-1}A_2C$ and $C^{-1}(\alpha A)C = \alpha C^{-1}AC$, so that the function defined by (1) is, in fact, a linear transformation. This explains the use of the word "transformation" in Definition 1.

The purpose of this section is to show that (1) similar matrices have several important properties in common and (2) most matrices are similar to diagonal matrices. (See the remark on page 409.)

EXAMPLE 1   **Two   Similar   Matrices**   Let $A = \begin{pmatrix} 2 & 1 \\ 0 & -1 \end{pmatrix}$, $B = \begin{pmatrix} 4 & -2 \\ 5 & -3 \end{pmatrix}$,   and

$C = \begin{pmatrix} 2 & -1 \\ -1 & 1 \end{pmatrix}$.   Then $CB = \begin{pmatrix} 2 & -1 \\ -1 & 1 \end{pmatrix}\begin{pmatrix} 4 & -2 \\ 5 & -3 \end{pmatrix} = \begin{pmatrix} 3 & -1 \\ 1 & -1 \end{pmatrix}$ and

$AC = \begin{pmatrix} 2 & 1 \\ 0 & -1 \end{pmatrix}\begin{pmatrix} 2 & -1 \\ -1 & 1 \end{pmatrix} = \begin{pmatrix} 3 & -1 \\ 1 & -1 \end{pmatrix}$. Thus $CB = AC$. Since $\det C = 1 \neq 0$, C is invertible; and since $CB = AC$, we have $C^{-1}CB = C^{-1}AC$ or $B = C^{-1}AC$. This shows that A and B are similar.   ∎

EXAMPLE 2   **A Matrix That Is Similar to a Diagonal Matrix**   Let $D = \begin{pmatrix} 1 & 0 & 0 \\ 0 & -1 & 0 \\ 0 & 0 & 2 \end{pmatrix}$,

$A = \begin{pmatrix} -6 & -3 & -25 \\ 2 & 1 & 8 \\ 2 & 2 & 7 \end{pmatrix}$,   and $C = \begin{pmatrix} 2 & 4 & 3 \\ 0 & 1 & -1 \\ 3 & 5 & 7 \end{pmatrix}$. C is   invertible   because $\det C = 3 \neq 0$. We then compute:

$$CA = \begin{pmatrix} 2 & 4 & 3 \\ 0 & 1 & -1 \\ 3 & 5 & 7 \end{pmatrix} \begin{pmatrix} -6 & -3 & -25 \\ 2 & 1 & 8 \\ 2 & 2 & 7 \end{pmatrix} = \begin{pmatrix} 2 & 4 & 3 \\ 0 & -1 & 1 \\ 6 & 10 & 14 \end{pmatrix}$$

$$DC = \begin{pmatrix} 1 & 0 & 0 \\ 0 & -1 & 0 \\ 0 & 0 & 2 \end{pmatrix} \begin{pmatrix} 2 & 4 & 3 \\ 0 & 1 & -1 \\ 3 & 5 & 7 \end{pmatrix} = \begin{pmatrix} 2 & 4 & 3 \\ 0 & -1 & 1 \\ 6 & 10 & 14 \end{pmatrix}$$

Thus $CA = DC$ and $A = C^{-1}DC$, so A and D are similar.   ∎

Note. In Examples 1 and 2 it was not necessary to compute C^{-1}. It was only necessary to know that C was nonsingular.

THEOREM 1   If A and B are similar $n \times n$ matrices, then A and B have the same characteristic equation, and therefore have the same eigenvalues.

Proof   Since A and B are similar, $B = C^{-1}AC$ and

$$\det (B - \lambda I) = \det (C^{-1}AC - \lambda I) = \det [C^{-1}AC - C^{-1}(\lambda I)C]$$
$$= \det [C^{-1}(A - \lambda I)C] = \det (C^{-1}) \det (A - \lambda I) \det (C)$$
$$= \det (C^{-1}) \det (C) \det (A - \lambda I) = \det (C^{-1}C) \det (A - \lambda I)$$
$$= \det I \det (A - \lambda I) = \det (A - \lambda I)$$

This means that A and B have the same characteristic equation, and, since eigenvalues are roots of the characteristic equation, they have the same eigenvalues.   ∎

EXAMPLE 3   **Eigenvalues of Similar Matrices Are the Same**   In Example 2 it is obvious that the eigenvalues of $D = \begin{pmatrix} 1 & 0 & 0 \\ 0 & -1 & 0 \\ 0 & 0 & 2 \end{pmatrix}$ are 1, -1, and 2. Thus these are the eigenvalues of $A = \begin{pmatrix} -6 & -3 & -25 \\ 2 & 1 & 8 \\ 2 & 2 & 7 \end{pmatrix}$. Check this by verifying that $\det (A - I) = \det (A + I) = \det (A - 2I) = 0$.   ∎

In a variety of applications it is quite useful to "diagonalize" a matrix A—that is, to find a diagonal matrix similar to A.

DEFINITION 2   **Diagonalizable Matrix**   An $n \times n$ matrix A is **diagonalizable** if there is a diagonal matrix D such that A is similar to D.

Remark. If D is a diagonal matrix, then its eigenvalues are its diagonal components. If A is similar to D, then A and D have the same eigenvalues (by Theorem 1). Putting these two facts together, we observe that if A is diagonalizable, then A is similar to a diagonal matrix whose diagonal components are the eigenvalues of A.

The next theorem tells us when a matrix is diagonalizable.

THEOREM 2   An $n \times n$ matrix A is diagonalizable if and only if it has n linearly independent eigenvectors. In that case the diagonal matrix D similar to A is given by

$$D = \begin{pmatrix} \lambda_1 & 0 & 0 & \cdots & 0 \\ 0 & \lambda_2 & 0 & \cdots & 0 \\ 0 & 0 & \lambda_3 & \cdots & 0 \\ \vdots & \vdots & \vdots & & \vdots \\ 0 & 0 & 0 & \cdots & \lambda_n \end{pmatrix} \tag{2}$$

where $\lambda_1, \lambda_2, \ldots, \lambda_n$ are the eigenvalues of A. If C is a matrix whose columns are linearly independent eigenvectors of A, then

$$\boxed{D = C^{-1}AC} \tag{3}$$

Proof   We first assume that A has n linearly independent eigenvectors $\mathbf{v}_1, \mathbf{v}_2, \ldots, \mathbf{v}_n$ corresponding to the (not necessarily distinct) eigenvalues $\lambda_1, \lambda_2, \ldots, \lambda_n$.

Let

$$\mathbf{v}_1 = \begin{pmatrix} c_{11} \\ c_{21} \\ \vdots \\ c_{n1} \end{pmatrix}, \; \mathbf{v}_2 = \begin{pmatrix} c_{12} \\ c_{22} \\ \vdots \\ c_{n2} \end{pmatrix}, \; \ldots, \; \mathbf{v}_n = \begin{pmatrix} c_{1n} \\ c_{2n} \\ \vdots \\ c_{nn} \end{pmatrix}$$

and let

$$C = \begin{pmatrix} c_{11} & c_{12} & \cdots & c_{1n} \\ c_{21} & c_{22} & \cdots & c_{2n} \\ \vdots & \vdots & & \vdots \\ c_{n1} & c_{n2} & \cdots & c_{nn} \end{pmatrix}$$

Then C is invertible since its columns are linearly independent. Now

$$AC = \begin{pmatrix} a_{11} & a_{12} & \cdots & a_{1n} \\ a_{21} & a_{22} & \cdots & a_{2n} \\ \vdots & \vdots & & \vdots \\ a_{n1} & a_{n2} & \cdots & a_{nn} \end{pmatrix} \begin{pmatrix} c_{11} & c_{12} & \cdots & c_{1n} \\ c_{21} & c_{22} & \cdots & c_{2n} \\ \vdots & \vdots & & \vdots \\ c_{n1} & c_{n2} & \cdots & c_{nn} \end{pmatrix}$$

and we see that the ith column of AC is $A\begin{pmatrix} c_{1i} \\ c_{2i} \\ \vdots \\ c_{ni} \end{pmatrix} = A\mathbf{v}_i = \lambda_i \mathbf{v}_i$. Thus AC is the matrix

whose ith column is $\lambda_i \mathbf{v}_i$ and

$$AC = \begin{pmatrix} \lambda_1 c_{11} & \lambda_2 c_{12} & \cdots & \lambda_n c_{1n} \\ \lambda_1 c_{21} & \lambda_2 c_{22} & \cdots & \lambda_n c_{2n} \\ \vdots & \vdots & & \vdots \\ \lambda_1 c_{n1} & \lambda_2 c_{n2} & \cdots & \lambda_n c_{nn} \end{pmatrix}$$

But

$$CD = \begin{pmatrix} c_{11} & c_{12} & \cdots & c_{1n} \\ c_{21} & c_{22} & \cdots & c_{2n} \\ \vdots & \vdots & & \vdots \\ c_{n1} & c_{n2} & \cdots & c_{nn} \end{pmatrix}\begin{pmatrix} \lambda_1 & 0 & \cdots & 0 \\ 0 & \lambda_2 & \cdots & 0 \\ \vdots & \vdots & & \vdots \\ 0 & 0 & \cdots & \lambda_n \end{pmatrix}$$

$$= \begin{pmatrix} \lambda_1 c_{11} & \lambda_2 c_{12} & \cdots & \lambda_n c_{1n} \\ \lambda_1 c_{21} & \lambda_2 c_{22} & \cdots & \lambda_n c_{2n} \\ \vdots & \vdots & & \vdots \\ \lambda_1 c_{n1} & \lambda_2 c_{n2} & \cdots & \lambda_n c_{nn} \end{pmatrix}$$

Thus

$$AC = CD \tag{4}$$

and, since C is invertible, we can multiply both sides of (4) on the left by C^{-1} to obtain

$$D = C^{-1}AC \tag{5}$$

This proves that if A has n linearly independent eigenvectors, then A is diagonalizable. Conversely, suppose that A is diagonalizable. That is, suppose that (5) holds for some invertible matrix C. Let $\mathbf{v}_1, \mathbf{v}_2 \ldots, \mathbf{v}_n$ be the columns of C. Then $AC = CD$, and, reversing the arguments above, we immediately see that $A\mathbf{v}_i = \lambda_i \mathbf{v}_i$ for $i = 1, 2, \ldots, n$. Thus $\mathbf{v}_1, \mathbf{v}_2, \ldots, \mathbf{v}_n$ are eigenvectors of A and are linearly independent because C is invertible. ∎

Notation. To indicate that D is a diagonal matrix with diagonal components $\lambda_1, \lambda_2, \ldots, \lambda_n$, we write $D = \text{diag}(\lambda_1, \lambda_2, \ldots, \lambda_n)$.

Theorem 2 has a useful corollary that follows immediately from Theorem 6.1.3 on page 389.

COROLLARY   If the $n \times n$ matrix A has n distinct eigenvalues, then A is diagonalizable. ∎

Remark. If the real coefficients of a polynomial of degree n are picked at random, then, with probability 1, the polynomial will have n distinct roots. It is not difficult to see, intuitively, why this is so. If $n = 2$, for example, then the equation $\lambda^2 + a\lambda + b = 0$ has equal roots if and only if $a^2 = 4b$—a highly unlikely event if a and b are chosen at random. We can, of course, write down polynomials having roots of algebraic multiplicity greater than 1, but these polynomials are exceptional. Thus, without attempting to be mathematically precise, it is fair to say that *most* polynomials have distinct roots. Hence *most* matrices have distinct eigenvalues, and, as we stated at the beginning of the section, *most* matrices are diagonalizable.

EXAMPLE 4   **Diagonalizing a 2×2 Matrix**   Let $A = \begin{pmatrix} 4 & 2 \\ 3 & 3 \end{pmatrix}$. In Example 6.1.3 on page 390 we found the two linearly independent eigenvectors $\mathbf{v}_1 = \begin{pmatrix} 2 \\ -3 \end{pmatrix}$ and $\mathbf{v}_2 = \begin{pmatrix} 1 \\ 1 \end{pmatrix}$. Then, setting $C = \begin{pmatrix} 2 & 1 \\ -3 & 1 \end{pmatrix}$, we find that

$$C^{-1}AC = \frac{1}{5}\begin{pmatrix} 1 & -1 \\ 3 & 2 \end{pmatrix}\begin{pmatrix} 4 & 2 \\ 3 & 3 \end{pmatrix}\begin{pmatrix} 2 & 1 \\ -3 & 1 \end{pmatrix}$$

$$= \frac{1}{5}\begin{pmatrix} 1 & -1 \\ 3 & 2 \end{pmatrix}\begin{pmatrix} 2 & 6 \\ -3 & 6 \end{pmatrix} = \frac{1}{5}\begin{pmatrix} 5 & 0 \\ 0 & 30 \end{pmatrix} = \begin{pmatrix} 1 & 0 \\ 0 & 6 \end{pmatrix}$$

which is the matrix whose diagonal components are the eigenvalues of A. ∎

EXAMPLE 5   **Diagonalizing a 3×3 Matrix with Three Distinct Eigenvalues**   Let $A = \begin{pmatrix} 1 & -1 & 4 \\ 3 & 2 & -1 \\ 2 & 1 & -1 \end{pmatrix}$. In Example 6.1.4 on page 391 we computed the three linearly independent eigenvectors $\mathbf{v}_1 = \begin{pmatrix} -1 \\ 4 \\ 1 \end{pmatrix}$, $\mathbf{v}_2 = \begin{pmatrix} 1 \\ -1 \\ -1 \end{pmatrix}$, and $\mathbf{v}_3 = \begin{pmatrix} 1 \\ 2 \\ 1 \end{pmatrix}$. Then

$$C = \begin{pmatrix} -1 & 1 & 1 \\ 4 & -1 & 2 \\ 1 & -1 & 1 \end{pmatrix} \text{ and }$$

$$C^{-1}AC = -\frac{1}{6}\begin{pmatrix} 1 & -2 & 3 \\ -2 & -2 & 6 \\ -3 & 0 & -3 \end{pmatrix}\begin{pmatrix} 1 & -1 & 4 \\ 3 & 2 & -1 \\ 2 & 1 & -1 \end{pmatrix}\begin{pmatrix} -1 & 1 & 1 \\ 4 & -1 & 2 \\ 1 & -1 & 1 \end{pmatrix}$$

$$= -\frac{1}{6}\begin{pmatrix} 1 & -2 & 3 \\ -2 & -2 & 6 \\ -3 & 0 & -3 \end{pmatrix}\begin{pmatrix} -1 & -2 & 3 \\ 4 & 2 & 6 \\ 1 & 2 & 3 \end{pmatrix}$$

$$= -\frac{1}{6} \begin{pmatrix} -6 & 0 & 0 \\ 0 & 12 & 0 \\ 0 & 0 & -18 \end{pmatrix} = \begin{pmatrix} 1 & 0 & 0 \\ 0 & -2 & 0 \\ 0 & 0 & 3 \end{pmatrix}$$

with eigenvalues 1, −2, and 3.   ∎

Remark. Since there are an infinite number of ways to choose an eigenvector, there are an infinite number of ways to choose the diagonalizing matrix C. The only advice is to choose the eigenvectors and matrix C that are, arithmetically, the easiest to work with. This usually means that you should insert as many 0's and 1's as possible.

EXAMPLE 6   **Diagonalizing a 3 × 3 Matrix with Two Distinct Eigenvalues and Three Linearly Independent Eigenvectors**   Let $A = \begin{pmatrix} 3 & 2 & 4 \\ 2 & 0 & 2 \\ 4 & 2 & 3 \end{pmatrix}$. Then, from Example

6.1.9 on page 394, we have the three linearly independent eigenvectors $\mathbf{v}_1 = \begin{pmatrix} 2 \\ 1 \\ 2 \end{pmatrix}$,

$\mathbf{v}_2 = \begin{pmatrix} 1 \\ -2 \\ 0 \end{pmatrix}$, and $\mathbf{v}_3 = \begin{pmatrix} 0 \\ -2 \\ 1 \end{pmatrix}$. Setting $C = \begin{pmatrix} 2 & 1 & 0 \\ 1 & -2 & -2 \\ 2 & 0 & 1 \end{pmatrix}$, we obtain

$$C^{-1}AC = -\frac{1}{9} \begin{pmatrix} -2 & -1 & -2 \\ -5 & 2 & 4 \\ 4 & 2 & -5 \end{pmatrix} \begin{pmatrix} 3 & 2 & 4 \\ 2 & 0 & 2 \\ 4 & 2 & 3 \end{pmatrix} \begin{pmatrix} 2 & 1 & 0 \\ 1 & -2 & -2 \\ 2 & 0 & 1 \end{pmatrix}$$

$$= -\frac{1}{9} \begin{pmatrix} -2 & -1 & -2 \\ -5 & 2 & 4 \\ 4 & 2 & -5 \end{pmatrix} \begin{pmatrix} 16 & -1 & 0 \\ 8 & 2 & 2 \\ 16 & 0 & -1 \end{pmatrix}$$

$$= -\frac{1}{9} \begin{pmatrix} -72 & 0 & 0 \\ 0 & 9 & 0 \\ 0 & 0 & 9 \end{pmatrix} = \begin{pmatrix} 8 & 0 & 0 \\ 0 & -1 & 0 \\ 0 & 0 & -1 \end{pmatrix}$$

This example illustrates that A is diagonalizable even though its eigenvalues are not distinct.   ∎

EXAMPLE 7   **A 2 × 2 Matrix with Only One Linearly Independent Eigenvector Cannot Be Diagonalized**   Let $A = \begin{pmatrix} 4 & 1 \\ 0 & 4 \end{pmatrix}$. In Example 6.1.8 on page 394 we saw that A did *not* have two linearly independent eigenvectors. Suppose that A were diagonalizable (in contradiction to Theorem 2). Then $D = \begin{pmatrix} 4 & 0 \\ 0 & 4 \end{pmatrix}$ and there would be an invertible matrix C such that $C^{-1}AC = D$. Multiplying this equation on the left by C and on the right by C^{-1}, we find that $A = CDC^{-1} = C\begin{pmatrix} 4 & 0 \\ 0 & 4 \end{pmatrix}C^{-1} = C(4I)C^{-1} = 4CIC^{-1} = 4CC^{-1} = 4I = \begin{pmatrix} 4 & 0 \\ 0 & 4 \end{pmatrix} = D$. But $A \neq D$, so no such C exists.   ∎

We have seen that many matrices are similar to diagonal matrices. However, two questions remain:

i. Is it possible to determine whether a given matrix is diagonalizable without computing eigenvalues and eigenvectors?

ii. What do we do if A is not diagonalizable?

We shall find a partial answer to the first question in the next section and a complete answer to the second in Section 6.6. In Section 6.7 we shall see an important application of the diagonalizing procedure.

At the beginning of this chapter we defined eigenvectors and eigenvalues for a linear transformation $T: V \to V$, where dim $V = n$. We stated then that as T can be represented by an $n \times n$ matrix, we would limit our discussion to eigenvalues and eigenvectors of $n \times n$ matrices.

However, the linear transformation can be represented by many different $n \times n$ matrices—one for each basis chosen. Do these matrices have the same eigenvalues and eigenvectors? The answer is yes, as proved in the next theorem.

THEOREM 3 Let V be a finite dimensional vector space with bases $B_1 = \{v_1, v_2, \ldots, v_n\}$ and $B_2 = \{w_1, w_2, \ldots, w_n\}$. Let $T: V \to V$ be a linear transformation. If A_T is the matrix representation of T with respect to the basis B_1 and if C_T is the matrix representation of T with respect to the basis B_2, then A_T and C_T are similar.

Proof T is a linear transformation from V to itself. From Theorem 5.3.3 on page 349, we have

$$(Tx)_{B_1} = A_T(x)_{B_1} \tag{6}$$

and

$$(Tx)_{B_2} = C_T(x)_{B_2} \tag{7}$$

Let M denote the transition matrix from B_1 to B_2. Then, by Theorem 4.9.1 on page 273,

$$(x)_{B_2} = M(x)_{B_1} \tag{8}$$

for every x in V. Also,

$$(Tx)_{B_2} = M(Tx)_{B_1} \tag{9}$$

Inserting (8) and (9) in (7) yields

$$M(Tx)_{B_1} = C_T M(x)_{B_1} \tag{10}$$

The matrix M is invertible by the result of Theorem 4.9.2 on page 274. If we multiply both sides of (10) by M^{-1} (which is the transition matrix from B_2 to B_1), we obtain

$$(Tx)_{B_1} = M^{-1} C_T M(x)_{B_1} \tag{11}$$

Comparing (6) and (11), we have

$$A_T(x)_{B_1} = M^{-1} C_T M(x)_{B_1} \tag{12}$$

Since (12) holds for every $x \in V$, we conclude that

$$A_T = M^{-1} C_T M$$

That is, A_T and C_T are similar.   ∎

PROBLEMS   6.3

In Problems 1–15 determine whether the given matrix A is diagonalizable. If it is, find a matrix C such that $C^{-1}AC = D$.

1. $\begin{pmatrix} -2 & -2 \\ -5 & 1 \end{pmatrix}$
   2. $\begin{pmatrix} 3 & -1 \\ -2 & 4 \end{pmatrix}$
   3. $\begin{pmatrix} 2 & -1 \\ 5 & -2 \end{pmatrix}$

4. $\begin{pmatrix} 3 & -5 \\ 1 & -1 \end{pmatrix}$
   5. $\begin{pmatrix} 3 & 2 \\ -5 & 1 \end{pmatrix}$
   6. $\begin{pmatrix} 1 & -1 & 0 \\ -1 & 2 & -1 \\ 0 & -1 & 1 \end{pmatrix}$

7. $\begin{pmatrix} 1 & 1 & -2 \\ -1 & 2 & 1 \\ 0 & 1 & -1 \end{pmatrix}$
   8. $\begin{pmatrix} 2 & 1 & 0 \\ 0 & 0 & 1 \\ 0 & 0 & 0 \end{pmatrix}$
   9. $\begin{pmatrix} 3 & 0 & 0 \\ 0 & 0 & 1 \\ 0 & 0 & 2 \end{pmatrix}$

10. $\begin{pmatrix} 3 & -1 & -1 \\ 1 & 1 & -1 \\ 1 & -1 & 1 \end{pmatrix}$
   11. $\begin{pmatrix} 7 & -2 & -4 \\ 3 & 0 & -2 \\ 6 & -2 & -3 \end{pmatrix}$
   12. $\begin{pmatrix} 4 & 6 & 6 \\ 1 & 3 & 2 \\ -1 & -5 & -2 \end{pmatrix}$

13. $\begin{pmatrix} -3 & -7 & -5 \\ 2 & 4 & 3 \\ 1 & 2 & 2 \end{pmatrix}$
   14. $\begin{pmatrix} -2 & -2 & 0 & 0 \\ -5 & 1 & 0 & 0 \\ 0 & 0 & 2 & -1 \\ 0 & 0 & 5 & -2 \end{pmatrix}$
   15. $\begin{pmatrix} 4 & 1 & 0 & 1 \\ 2 & 3 & 0 & 1 \\ -2 & 1 & 2 & -3 \\ 2 & -1 & 0 & 5 \end{pmatrix}$

16. Show that if A is similar to B and B is similar to C, then A is similar to C.

17. If A is similar to B, show that A^n is similar to B^n for any positive integer n.

*18. If A is similar to B, show that $\rho(A) = \rho(B)$ and $\nu(A) = \nu(B)$. [*Hint:* First prove that if C is invertible, then $\nu(CA) = \nu(A)$ by showing that $x \in N_A$ if and only if $x \in N_{CA}$. Next prove that $\rho(AC) = \rho(A)$ by showing that $R_A = R_{AC}$. Conclude that $\rho(AC) = \rho(CA) = \rho(A)$. Finally, use the fact that C^{-1} is invertible to show that $\rho(C^{-1}AC) = \rho(A)$.]

19. Let $D = \begin{pmatrix} 1 & 0 \\ 0 & -1 \end{pmatrix}$. Compute D^{20}.

20. If A is similar to B, show that $\det A = \det B$.

21. Suppose that $C^{-1}AC = D$. Show that for any integer n, $A^n = CD^nC^{-1}$. This gives an easy way to compute powers of a diagonalizable matrix.

22. Let $A = \begin{pmatrix} 3 & -4 \\ 2 & -3 \end{pmatrix}$. Compute A^{20}. [*Hint:* Find a C such that $A = CDC^{-1}$]

*23. Let A be an $n \times n$ matrix whose characteristic equation is $(\lambda - c)^n = 0$. Show that A is diagonalizable if and only if $A = cI$.

24. Use the result of Problem 21 and Example 6 to compute A^{10}, where $A = \begin{pmatrix} 3 & 2 & 4 \\ 2 & 0 & 2 \\ 4 & 2 & 3 \end{pmatrix}$.

*25. Let A and B be real $n \times n$ matrices with distinct eigenvalues. Prove that $AB = BA$ if and only if A and B have the same eigenvectors.

26. If A is diagonalizable, show that $\det A = \lambda_1\lambda_2\cdots\lambda_n$, where $\lambda_1, \lambda_2, \ldots, \lambda_n$ are the eigenvalues of A.

6.4 SYMMETRIC MATRICES AND ORTHOGONAL DIAGONALIZATION

In this section we shall see that symmetric matrices† have a number of important properties. In particular, we show that any symmetric matrix has n linearly independent real eigenvectors and therefore, by Theorem 6.3.2, is diagonalizable. We begin by proving that the eigenvalues of a real symmetric matrix are real.

THEOREM 1 Let A be a real $n \times n$ symmetric matrix. Then the eigenvalues of A are real.

Proof‡ Let λ be an eigenvalue of A with eigenvector $\mathbf{v}$; that is, $A\mathbf{v} = \lambda\mathbf{v}$. Now $\mathbf{v}$ is a vector in $\mathbb{C}^n$, and an inner product in $\mathbb{C}^n$ (see Definition 4.12.1, page 413, and Example 4.12.2) satisfies

$$(\alpha\mathbf{x}, \mathbf{y}) = \alpha(\mathbf{x}, \mathbf{y}) \quad \text{and} \quad (\mathbf{x}, \alpha\mathbf{y}) = \bar{\alpha}(\mathbf{x}, \mathbf{y}) \tag{1}$$

Then

$$(A\mathbf{v}, \mathbf{v}) = (\lambda\mathbf{v}, \mathbf{v}) = \lambda(\mathbf{v}, \mathbf{v}) \tag{2}$$

Moreover, by Theorem 5.5.1 on page 374 and the fact that $A^t = A$,

$$(A\mathbf{v}, \mathbf{v}) = (\mathbf{v}, A^t\mathbf{v}) = (\mathbf{v}, A\mathbf{v}) = (\mathbf{v}, \lambda\mathbf{v}) = \bar{\lambda}(\mathbf{v}, \mathbf{v}) \tag{3}$$

---

†Recall that A is symmetric if and only if $A^t = A$.

‡This proof uses material in Sections 4.12 and 5.5 and should be omitted if those sections were not covered.

Thus, equating (2) and (3), we have

$$\lambda(\mathbf{v}, \mathbf{v}) = \bar{\lambda}(\mathbf{v}, \mathbf{v}) \tag{4}$$

But $(\mathbf{v}, \mathbf{v}) = |\mathbf{v}|^2 \neq 0$, since $\mathbf{v}$ is an eigenvector. Thus we can divide both sides of (4) by $(\mathbf{v}, \mathbf{v})$ to obtain

$$\lambda = \bar{\lambda} \tag{5}$$

If $\lambda = a + ib$, then $\bar{\lambda} = a - ib$ and, from (5), we have

$$a + ib = a - ib \tag{6}$$

which can hold only if $b = 0$. This shows that $\lambda = a$; hence λ is real and the proof is complete. ∎

We saw in Theorem 6.1.3 on page 389 that eigenvectors corresponding to distinct eigenvalues are linearly independent. For symmetric matrices the result is stronger: *Eigenvectors of a symmetric matrix corresponding to distinct eigenvalues are orthogonal.*

THEOREM 2   Let A be a real symmetric $n \times n$ matrix. If λ_1 and λ_2 are distinct eigenvalues with corresponding real eigenvectors $\mathbf{v}_1$ and $\mathbf{v}_2$, then $\mathbf{v}_1$ and $\mathbf{v}_2$ are orthogonal.

Proof   We compute

$$A\mathbf{v}_1 \cdot \mathbf{v}_2 = \lambda_1 \mathbf{v}_1 \cdot \mathbf{v}_2 = \lambda_1(\mathbf{v}_1 \cdot \mathbf{v}_2) \tag{7}$$

and

$$A\mathbf{v}_1 \cdot \mathbf{v}_2 = \mathbf{v}_1 \cdot A^t\mathbf{v}_2 = \mathbf{v}_1 \cdot A\mathbf{v}_2 = \mathbf{v}_1 \cdot (\lambda_2\mathbf{v}_2) = \lambda_2(\mathbf{v}_1 \cdot \mathbf{v}_2) \tag{8}$$

Combining (7) and (8), we have $\lambda_1(\mathbf{v}_1 \cdot \mathbf{v}_2) = \lambda_2(\mathbf{v}_1 \cdot \mathbf{v}_2)$ and since $\lambda_1 \neq \lambda_2$, we conclude that $\mathbf{v}_1 \cdot \mathbf{v}_2 = 0$. This is what we wanted to show. ∎

We now state the main result of this section. Its proof, which is difficult (and optional), is given at the end of this section.

THEOREM 3   Let A be a real symmetric $n \times n$ matrix. Then A has n real orthonormal eigenvectors. ∎

Remark. It follows from this theorem that the geometric multiplicity of each eigenvalue of A is equal to its algebraic multiplicity.

Theorem 3 tells us that if A is symmetric, then $\mathbb{R}^n$ has a basis $B = \{\mathbf{u}_1, \mathbf{u}_2, \ldots, \mathbf{u}_n\}$ consisting of orthonormal eigenvectors of A. Let Q be the matrix whose columns are $\mathbf{u}_1, \mathbf{u}_2, \ldots, \mathbf{u}_n$. Then, by Theorem 4.10.3 on page 287, Q is an orthogonal matrix. This leads to the following definition.

DEFINITION 1 **Orthogonally Diagonalizable Matrix** An $n \times n$ matrix A is said to be **orthogonally diagonalizable** if there exists an orthogonal matrix Q such that

$$Q^t A Q = D \qquad (9)$$

where $D = \text{diag}\,(\lambda_1, \lambda_2, \ldots, \lambda_n)$ and $\lambda_1, \lambda_2, \ldots, \lambda_n$ are the eigenvalues of A.

Note. Remember that Q is orthogonal if $Q^t = Q^{-1}$; hence (9) could be written as $Q^{-1}AQ = D$.

THEOREM 4 Let A be a real $n \times n$ matrix. Then A is orthogonally diagonalizable if and only if A is symmetric.

Proof Let A be symmetric. Then, by Theorems 2 and 3, A is orthogonally diagonalizable with Q the matrix whose columns are the orthonormal eigenvectors given in Theorem 3. Conversely, suppose that A is orthogonally diagonalizable. Then there exists an orthogonal matrix Q such that $Q^t A Q = D$. Multiplying this equation on the left by Q and on the right by Q^t and using the fact that $Q^t Q = Q Q^t = I$, we obtain

$$A = Q D Q^t \qquad (10)$$

Then $A^t = (QDQ^t)^t = (Q^t)^t D^t Q^t = QDQ^t = A$. Thus A is symmetric and the theorem is proved. In the last series of equations we used the facts that $(AB)^t = B^t A^t$ [part (*ii*) of Theorem 1.10.1, page 84], $(A^t)^t = A$ [part (*i*) of Theorem 1.10.1], and $D^t = D$ for any diagonal matrix D. ∎

Before giving examples, we provide the following three-step procedure for finding the orthogonal matrix Q that diagonalizes the symmetric matrix A.

Procedure for Finding a Diagonalizing Matrix Q

  i. Find a basis for each eigenspace of A.

  ii. Find an orthonormal basis for each eigenspace of A by using the Gram-Schmidt process.

  iii. Write Q as the matrix whose columns are the orthonormal eigenvectors obtained in step (*ii*).

EXAMPLE 1   **Diagonalizing a 2 × 2 Symmetric Matrix Using an Orthogonal Matrix**   Let

$A = \begin{pmatrix} 1 & -2 \\ -2 & 3 \end{pmatrix}$. Then the characteristic equation of A is $\det(A - \lambda I) =$

$\begin{vmatrix} 1 - \lambda & -2 \\ -2 & 3 - \lambda \end{vmatrix} = \lambda^2 - 4\lambda - 1 = 0$, which has the roots $\lambda = (4 \pm \sqrt{20})/2 =$

$(4 \pm 2\sqrt{5})/2 = 2 \pm \sqrt{5}$.   For $\lambda_1 = 2 - \sqrt{5}$,   we   obtain $(A - \lambda I)\mathbf{v} =$

$\begin{pmatrix} -1 + \sqrt{5} & -2 \\ -2 & 1 + \sqrt{5} \end{pmatrix}\begin{pmatrix} x_1 \\ x_2 \end{pmatrix} = \begin{pmatrix} 0 \\ 0 \end{pmatrix}$. An eigenvector is $\mathbf{v}_1 = \begin{pmatrix} 2 \\ -1 + \sqrt{5} \end{pmatrix}$ and

$|\mathbf{v}_1| = \sqrt{2^2 + (-1 + \sqrt{5})^2} = \sqrt{10 - 2\sqrt{5}}$. Thus

$$\mathbf{u}_1 = \frac{1}{\sqrt{10 - 2\sqrt{5}}}\begin{pmatrix} 2 \\ -1 + \sqrt{5} \end{pmatrix}.$$

Next, for $\lambda_2 = 2 + \sqrt{5}$, we compute $(A - \lambda I)\mathbf{v} = \begin{pmatrix} -1 - \sqrt{5} & -2 \\ -2 & 1 - \sqrt{5} \end{pmatrix}\begin{pmatrix} x_1 \\ x_2 \end{pmatrix} =$

$\begin{pmatrix} 0 \\ 0 \end{pmatrix}$ and $\mathbf{v}_2 = \begin{pmatrix} 1 - \sqrt{5} \\ 2 \end{pmatrix}$. Note that $\mathbf{v}_1 \cdot \mathbf{v}_2 = 0$ (which must be true according to

Theorem 2). Then $|\mathbf{v}_2| = \sqrt{10 - 2\sqrt{5}}$ so that $\mathbf{u}_2 = \frac{1}{\sqrt{10 - 2\sqrt{5}}}\begin{pmatrix} 1 - \sqrt{5} \\ 2 \end{pmatrix}$.

Finally,

$$Q = \frac{1}{\sqrt{10 - 2\sqrt{5}}}\begin{pmatrix} 2 & 1 - \sqrt{5} \\ -1 + \sqrt{5} & 2 \end{pmatrix}$$

$$Q^t = \frac{1}{\sqrt{10 - 2\sqrt{5}}}\begin{pmatrix} 2 & -1 + \sqrt{5} \\ 1 - \sqrt{5} & 2 \end{pmatrix}$$

and

$$Q^t A Q = \frac{1}{10 - 2\sqrt{5}}\begin{pmatrix} 2 & -1 + \sqrt{5} \\ 1 - \sqrt{5} & 2 \end{pmatrix}\begin{pmatrix} 1 & -2 \\ -2 & 3 \end{pmatrix}\begin{pmatrix} 2 & 1 - \sqrt{5} \\ -1 + \sqrt{5} & 2 \end{pmatrix}$$

$$= \frac{1}{10 - 2\sqrt{5}}\begin{pmatrix} 2 & -1 + \sqrt{5} \\ 1 - \sqrt{5} & 2 \end{pmatrix}\begin{pmatrix} 4 - 2\sqrt{5} & -3 - \sqrt{5} \\ -7 + 3\sqrt{5} & 4 + 2\sqrt{5} \end{pmatrix}$$

$$= \frac{1}{10 - 2\sqrt{5}}\begin{pmatrix} 30 - 14\sqrt{5} & 0 \\ 0 & 10 + 6\sqrt{5} \end{pmatrix} = \begin{pmatrix} 2 - \sqrt{5} & 0 \\ 0 & 2 + \sqrt{5} \end{pmatrix} \quad ■$$

EXAMPLE 2   **Diagonalizing a 3 × 3 Symmetric Matrix Using an Orthogonal Matrix**   Let

$A = \begin{pmatrix} 5 & 4 & 2 \\ 4 & 5 & 2 \\ 2 & 2 & 2 \end{pmatrix}$.   Then A is   symmetric   and $\det(A - \lambda I) =$

$\begin{pmatrix} 5 - \lambda & 4 & 2 \\ 4 & 5 - \lambda & 2 \\ 2 & 2 & 2 - \lambda \end{pmatrix} = -(\lambda - 1)^2(\lambda - 10)$. Corresponding to $\lambda = 1$ we compute the linearly independent eigenvectors $\mathbf{v}_1 = \begin{pmatrix} -1 \\ 1 \\ 0 \end{pmatrix}$ and $\mathbf{v}_2 = \begin{pmatrix} -1 \\ 0 \\ 2 \end{pmatrix}$. Corre-

sponding to $\lambda = 10$ we find that $\mathbf{v}_3 = \begin{pmatrix} 2 \\ 2 \\ 1 \end{pmatrix}$. To find Q, we apply the Gram-Schmidt

process to $\{\mathbf{v}_1, \mathbf{v}_2\}$, a basis for E_1. Since $|\mathbf{v}_1| = \sqrt{2}$, we set $\mathbf{u}_1 = \begin{pmatrix} -1/\sqrt{2} \\ 1/\sqrt{2} \\ 0 \end{pmatrix}$. Next

$$\mathbf{v}_2' = \mathbf{v}_2 - (\mathbf{v}_2 \cdot \mathbf{u}_1)\mathbf{u}_1 = \begin{pmatrix} -1 \\ 0 \\ 2 \end{pmatrix} - \frac{1}{\sqrt{2}}\begin{pmatrix} -1/\sqrt{2} \\ 1/\sqrt{2} \\ 0 \end{pmatrix}$$

$$= \begin{pmatrix} -1 \\ 0 \\ 2 \end{pmatrix} - \begin{pmatrix} -1/2 \\ 1/2 \\ 0 \end{pmatrix} = \begin{pmatrix} -1/2 \\ -1/2 \\ 2 \end{pmatrix}$$

Then $|\mathbf{v}_2| = \sqrt{18/4} = 3\sqrt{2}/2$ and $\mathbf{u}_2 = \dfrac{2}{3\sqrt{2}}\begin{pmatrix} -1/2 \\ -1/2 \\ 2 \end{pmatrix} = \begin{pmatrix} -1/3\sqrt{2} \\ -1/3\sqrt{2} \\ 4/3\sqrt{2} \end{pmatrix}$. We check

this by noting that $\mathbf{u}_1 \cdot \mathbf{u}_2 = 0$. Finally, we have $\mathbf{u}_3 = \mathbf{v}_3/|\mathbf{v}_3| = \frac{1}{3}\mathbf{v}_3 = \begin{pmatrix} 2/3 \\ 2/3 \\ 1/3 \end{pmatrix}$. We

can check this too by noting that $\mathbf{u}_1 \cdot \mathbf{u}_3 = 0$ and $\mathbf{u}_2 \cdot \mathbf{u}_3 = 0$. Thus

$$Q = \begin{pmatrix} -1/\sqrt{2} & -1/3\sqrt{2} & 2/3 \\ 1/\sqrt{2} & -1/3\sqrt{2} & 2/3 \\ 0 & 4/3\sqrt{2} & 1/3 \end{pmatrix}$$

and

$$Q^t A Q = \begin{pmatrix} -1/\sqrt{2} & 1/\sqrt{2} & 0 \\ -1/3\sqrt{2} & -1/3\sqrt{2} & 4/3\sqrt{2} \\ 2/3 & 2/3 & 1/3 \end{pmatrix}\begin{pmatrix} 5 & 4 & 2 \\ 4 & 5 & 2 \\ 2 & 2 & 2 \end{pmatrix}\begin{pmatrix} -1/\sqrt{2} & -1/3\sqrt{2} & 2/3 \\ 1/\sqrt{2} & -1/3\sqrt{2} & 2/3 \\ 0 & 4/3\sqrt{2} & 1/3 \end{pmatrix}$$

$$= \begin{pmatrix} -1/\sqrt{2} & 1/\sqrt{2} & 0 \\ -1/3\sqrt{2} & -1/3\sqrt{2} & 4/3\sqrt{2} \\ 2/3 & 2/3 & 1/3 \end{pmatrix}\begin{pmatrix} -1/\sqrt{2} & -1/3\sqrt{2} & 20/3 \\ 1/\sqrt{2} & -1/3\sqrt{2} & 20/3 \\ 0 & 4/3\sqrt{2} & 10/3 \end{pmatrix}$$

$$= \begin{pmatrix} 1 & 0 & 0 \\ 0 & 1 & 0 \\ 0 & 0 & 10 \end{pmatrix} \qquad \blacksquare$$

Conjugate   In this section we have proved results for real symmetric matrices. If $A = (a_{ij})$
transpose   is a complex matrix, then the **conjugate transpose** of A, denoted by A^*, is defined

Hermitian matrix

Unitary matrix

by: the ijth element of $A^* = \overline{a_{ji}}$. The matrix A is called **hermitian** if $A^* = A$. It turns out that Theorems 1, 2, and 3 are also true for hermitian matrices. Moreover, if we define a **unitary** matrix to be a complex matrix U with $U^* = U^{-1}$, then, using the proof of Theorem 4, we can show that a hermitian matrix is unitarily diagonalizable. We leave all these facts as exercises (see Problems 15–17).

We conclude this section with a proof of Theorem 3.

Proof of Theorem 3†

We prove that to every eigenvalue λ of algebraic multiplicity k, there correspond k orthonormal eigenvectors. This step, combined with Theorem 2, will be sufficient. Let $\mathbf{u}_1$ be an eigenvector of A corresponding to λ_1. We can assume that $|\mathbf{u}_1| = 1$. We can also assume that $\mathbf{u}_1$ is real because λ_1 is real and $\mathbf{u}_1 \in N_{A-\lambda_1 I}$, the kernel of the real matrix $A - \lambda_1 I$. This kernel is a subspace of $\mathbb{R}^n$ by Example 4.6.10 on page 249. Next we note that $\{\mathbf{u}_1\}$ can be expanded into a basis $\{\mathbf{u}_1, \mathbf{v}_2, \mathbf{v}_3, \ldots, \mathbf{v}_n\}$ for $\mathbb{R}^n$ and, by the Gram-Schmidt process, we can turn this basis into the orthonormal basis $\{\mathbf{u}_1, \mathbf{u}_2, \ldots, \mathbf{u}_n\}$. Let Q be the orthogonal matrix whose columns are $\mathbf{u}_1, \mathbf{u}_2, \ldots, \mathbf{u}_n$. For convenience of notation we write $Q = (\mathbf{u}_1, \mathbf{u}_2, \ldots, \mathbf{u}_n)$. Now Q is invertible and $Q^t = Q^{-1}$, so A is similar to $Q^t A Q$ and, by Theorem 6.3.1 on page 406, $Q^t A Q$ and A have the same characteristic polynomial: $|Q^t A Q - \lambda I| = |A - \lambda I|$. Then

$$Q^t = \begin{pmatrix} \mathbf{u}_1^t \\ \mathbf{u}_2^t \\ \vdots \\ \mathbf{u}_n^t \end{pmatrix}$$

so that

$$Q^t A Q = \begin{pmatrix} \mathbf{u}_1^t \\ \mathbf{u}_2^t \\ \vdots \\ \mathbf{u}_n^t \end{pmatrix} A(\mathbf{u}_1 \mathbf{u}_2 \cdots \mathbf{u}_n) = \begin{pmatrix} \mathbf{u}_1^t \\ \mathbf{u}_2^t \\ \vdots \\ \mathbf{u}_n^t \end{pmatrix} (A\mathbf{u}_1 A\mathbf{u}_2 \cdots A\mathbf{u}_n)$$

$$= \begin{pmatrix} \mathbf{u}_1^t \\ \mathbf{u}_2^t \\ \vdots \\ \mathbf{u}_n^t \end{pmatrix} (\lambda_1 \mathbf{u}_1, A\mathbf{u}_2 \cdots A\mathbf{u}_n) = \begin{pmatrix} \lambda_1 & \mathbf{u}_1^t A\mathbf{u}_2 & \cdots & \mathbf{u}_1^t A\mathbf{u}_n \\ 0 & \mathbf{u}_2^t A\mathbf{u}_2 & \cdots & \mathbf{u}_2^t A\mathbf{u}_n \\ \vdots & \vdots & & \vdots \\ 0 & \mathbf{u}_n^t A\mathbf{u}_2 & \cdots & \mathbf{u}_n^t A\mathbf{u}_n \end{pmatrix}$$

The zeros appear because $\mathbf{u}_1^t \mathbf{u}_j = \mathbf{u}_1 \cdot \mathbf{u}_j = 0$ if $j \neq 1$. Now $[Q^t A Q]^t = Q^t A^t (Q^t)^t = Q^t A Q$. Thus $Q^t A Q$ is symmetric, which means that there must be zeros in the first row of $Q^t A Q$ to match the zeros in the first column. Thus

$$Q^t A Q = \begin{pmatrix} \lambda_1 & 0 & 0 & \cdots & 0 \\ 0 & q_{22} & q_{23} & \cdots & q_{2n} \\ 0 & q_{32} & q_{33} & \cdots & q_{3n} \\ \vdots & \vdots & \vdots & & \vdots \\ 0 & q_{n2} & q_{n3} & \cdots & q_{nn} \end{pmatrix}$$

---

† If time permits.

and

$$|Q^tAQ - \lambda I| = \begin{vmatrix} \lambda_1 - \lambda & 0 & 0 & \cdots & 0 \\ 0 & q_{22} - \lambda & q_{23} & \cdots & q_{2n} \\ 0 & q_{32} & q_{33} - \lambda & \cdots & q_{3n} \\ \vdots & \vdots & \vdots & & \vdots \\ 0 & q_{n2} & q_{n3} & \cdots & q_{nn} - \lambda \end{vmatrix}$$

$$= (\lambda_1 - \lambda) \begin{vmatrix} q_{22} - \lambda & q_{23} & \cdots & q_{2n} \\ q_{32} & q_{33} - \lambda & \cdots & q_{3n} \\ \vdots & \vdots & & \vdots \\ q_{n2} & q_{n3} & \cdots & q_{nn} - \lambda \end{vmatrix} = (\lambda - \lambda_1)|M_{11}(\lambda)|$$

where $M_{11}(\lambda)$ is the 1, 1 minor of $Q^tAQ - \lambda I$. If $k = 1$, there is nothing to prove. If $k > 1$, then $|A - \lambda I|$ contains the factor $(\lambda - \lambda_1)^2$, and therefore $|Q^tAQ - \lambda I|$ also contains the factor $(\lambda - \lambda_1)^2$. Thus $|M_{11}(\lambda)|$ contains the factor $\lambda - \lambda_1$, which means that $|M_{11}(\lambda_1)| = 0$. This means that the last $n - 1$ columns of $Q^tAQ - \lambda_1 I$ are linearly dependent. Since the first column of $Q^tAQ - \lambda_1 I$ is the zero vector, this means that $Q^tAQ - \lambda_1 I$ contains at most $n - 2$ linearly independent columns. In other words, $\rho(Q^tAQ - \lambda_1 I) \le n - 2$. But $Q^tAQ - \lambda_1 I$ and $A - \lambda_1 I$ are similar; hence, by Problem 6.3.18, $\rho(A - \lambda_1 I) \le n - 2$. Therefore $\nu(A - \lambda_1 I) \ge 2$, which means that $E_\lambda = $ kernel of $(A - \lambda_1 I)$ contains at least two linearly independent eigenvectors. If $k = 2$, we are done. If $k > 2$, then we take two ortho-normal vectors $\mathbf{u}_1, \mathbf{u}_2$ in E_λ and expand them into a new orthonormal basis $\{\mathbf{u}_1, \mathbf{u}_2, \ldots, \mathbf{u}_n\}$ for $\mathbb{R}^n$ and define $P = \{\mathbf{u}_1, \mathbf{u}_2, \ldots, \mathbf{u}_n\}$. Then, exactly as before, we show that

$$P^tAP - \lambda I = \begin{pmatrix} \lambda_1 - \lambda & 0 & 0 & 0 & \cdots & 0 \\ 0 & \lambda_1 - \lambda & 0 & 0 & \cdots & 0 \\ 0 & 0 & \beta_{33} - \lambda & \beta_{34} & \cdots & \beta_{3n} \\ 0 & 0 & \beta_{43} & \beta_{44} - \lambda & \cdots & \beta_{4n} \\ \vdots & \vdots & \vdots & \vdots & & \vdots \\ 0 & 0 & \beta_{n3} & \beta_{n4} & \cdots & \beta_{nn} - \lambda \end{pmatrix}$$

Since $k > 2$, we show, as before, that the determinant of the matrix in brackets is zero when $\lambda = \lambda_1$—which shows that $\rho(P^tAP - \lambda_1 I) \le n - 3$ so that $\nu(P^tAP - \lambda_1 I) = \nu(A - \lambda_1 I) \ge 3$. Then dim $E_{\lambda_1} \ge 3$, and so on. We can clearly continue this process to show that dim $E_{\lambda_1} = k$. Finally, in each E_{λ_1} we can find an orthonor-mal basis. This completes the proof. ∎

PROBLEMS 6.4

In Problems 1–8 find an orthogonal matrix that diagonalizes the given symmetric matrix.

1. $\begin{pmatrix} 3 & 4 \\ 4 & -3 \end{pmatrix}$
2. $\begin{pmatrix} 2 & 1 \\ 1 & 2 \end{pmatrix}$
3. $\begin{pmatrix} 1 & -1 \\ -1 & 1 \end{pmatrix}$

4. $\begin{pmatrix} 1 & -1 & -1 \\ -1 & 1 & -1 \\ -1 & -1 & 1 \end{pmatrix}$ **5.** $\begin{pmatrix} -1 & 2 & 2 \\ 2 & -1 & 2 \\ 2 & 2 & 1 \end{pmatrix}$ **6.** $\begin{pmatrix} 1 & -1 & 0 \\ -1 & 2 & -1 \\ 0 & -1 & 1 \end{pmatrix}$

7. $\begin{pmatrix} 3 & 2 & 2 \\ 2 & 2 & 0 \\ 2 & 0 & 4 \end{pmatrix}$ **8.** $\begin{pmatrix} 1 & -1 & 0 & 0 \\ -1 & 0 & 0 & 0 \\ 0 & 0 & 0 & 0 \\ 0 & 0 & 0 & 2 \end{pmatrix}$

9. Let Q be a symmetric orthogonal matrix. Show that if λ is an eigenvalue of Q, then $\lambda = \pm 1$.

10. A is **orthogonally similar** to B if there exists an orthogonal matrix Q such that $B = Q^t A Q$. Suppose that A is orthogonally similar to B and that B is orthogonally similar to C. Show that A is orthogonally similar to C.

11. Show that if $Q = \begin{pmatrix} a & b \\ c & d \end{pmatrix}$ is orthogonal, then $b = \pm c$. [*Hint:* Write out the equations that result from the equation $Q^t Q = I$.]

12. Suppose that A is a real symmetric matrix every one of whose eigenvalues is zero. Show that A is the zero matrix.

13. Show that if a real 2×2 matrix A has eigenvectors that are orthogonal, then A is symmetric.

14. Let A be a real skew-symmetric matrix ($A^t = -A$). Prove that every eigenvalue of A is of the form $i\alpha$, where α is a real number. That is, prove that every eigenvalue of A is a **pure imaginary** number.

*15. Show that the eigenvalues of a complex $n \times n$ hermitian matrix are real. [*Hint:* Use the fact that in $\mathbb{C}^n$, $(A\mathbf{x}, \mathbf{y}) = (\mathbf{x}, A^*\mathbf{y})$.]

*16. If A is an $n \times n$ hermitian matrix, show that eigenvectors corresponding to different eigenvalues are orthogonal.

**17. By repeating the proof of Theorem 3, except that $\bar{\mathbf{v}}_i^t$ replaces $\mathbf{v}_i^t$ where appropriate, show that any $n \times n$ hermitian matrix has n orthonormal eigenvectors.

18. Find a unitary matrix U such that U^*AU is diagonal, where $A = \begin{pmatrix} 1 & 1-i \\ 1+i & 0 \end{pmatrix}$.

19. Do the same for $A = \begin{pmatrix} 2 & 3-3i \\ 3+3i & 5 \end{pmatrix}$.

20. Prove that the determinant of a hermitian matrix is real.

6.5 QUADRATIC FORMS AND CONIC SECTIONS

In this section we use the material of Section 6.4 to discover information about the graphs of quadratic equations. Quadratic equations and quadratic forms, which are defined below, arise in a variety of ways. For example, we can use quadratic forms to obtain information about the conic sections in $\mathbb{R}^2$ (circles, parabolas, ellipses, hyperbolas) and extend this theory to describe certain surfaces, called *quadric surfaces*, in $\mathbb{R}^3$. These topics are discussed later in the section. Although we shall not discuss it in this text, quadratic forms arise in a number of applications ranging from a description of cost functions in economics to an analysis of the control of a rocket traveling in space.

DEFINITION 1   **Quadratic Equation and Quadratic Form**

   i. A **quadratic equation in two variables with no linear terms** is an equation of the form

$$ax^2 + bxy + cy^2 = d \qquad\qquad (1)$$

where $|a| + |b| + |c| \neq 0$. That is, at least one of the numbers a, b, and c is nonzero.

   ii. A **quadratic form in two variables** is an expression of the form

$$F(x, y) = ax^2 + bxy + cy^2 \qquad\qquad (2)$$

where $|a| + |b| + |c| \neq 0$.

Obviously, quadratic equations and quadratic forms are closely related. We begin our analysis of quadratic forms with a simple example.

Consider the quadratic form $F(x, y) = x^2 - 4xy + 3y^2$. Let $\mathbf{v} = \begin{pmatrix} x \\ y \end{pmatrix}$ and $A = \begin{pmatrix} 1 & -2 \\ -2 & 3 \end{pmatrix}$. Then

$$
\begin{aligned}
A\mathbf{v} \cdot \mathbf{v} &= \begin{pmatrix} 1 & -2 \\ -2 & 3 \end{pmatrix}\begin{pmatrix} x \\ y \end{pmatrix} \cdot \begin{pmatrix} x \\ y \end{pmatrix} = \begin{pmatrix} x - 2y \\ -2x + 3y \end{pmatrix} \cdot \begin{pmatrix} x \\ y \end{pmatrix} \\
&= (x^2 - 2xy) + (-2xy + 3y^2) = x^2 - 4xy + 3y^2 = F(x, y)
\end{aligned}
$$

Thus we have "represented" the quadratic form $F(x, y)$ by the symmetric matrix A in the sense that

$$F(x, y) = A\mathbf{v} \cdot \mathbf{v} \qquad\qquad (3)$$

Conversely, if A is a symmetric matrix, then equation (3) defines a quadratic form $F(x, y) = A\mathbf{v} \cdot \mathbf{v}$.

We can represent $F(x, y)$ by many matrices but only one symmetric matrix. To see this, let $A = \begin{pmatrix} 1 & a \\ b & 3 \end{pmatrix}$, where $a + b = -4$. Then $A\mathbf{v} \cdot \mathbf{v} = F(x, y)$. If $A =$

$\begin{pmatrix} 1 & 3 \\ -7 & 3 \end{pmatrix}$, for example, then $A\mathbf{v} = \begin{pmatrix} x + 3y \\ -7x + 3y \end{pmatrix}$ and $A\mathbf{v} \cdot \mathbf{v} = x^2 - 4xy + 3y^2$. If, however, we insist that A be symmetric, then we must have $a + b = -4$ and $a = b$. This pair of equations has the unique solution $a = b = -2$.

If $F(x, y) = ax^2 + bxy + cy^2$ is a quadratic form, let

$$A = \begin{pmatrix} a & b/2 \\ b/2 & c \end{pmatrix} \tag{4}$$

Then

$$A\mathbf{v} \cdot \mathbf{v} = \left[\begin{pmatrix} a & b/2 \\ b/2 & c \end{pmatrix} \begin{pmatrix} x \\ y \end{pmatrix} \right] \cdot \begin{pmatrix} x \\ y \end{pmatrix} = \begin{pmatrix} ax + (b/2)y \\ (b/2)x + cy \end{pmatrix} \cdot \begin{pmatrix} x \\ y \end{pmatrix}$$
$$= ax^2 + bxy + cy^2 = F(x, y)$$

Now let us return to the quadratic equation (1). Using (3), we can write (1) as

$$A\mathbf{v} \cdot \mathbf{v} = d \tag{5}$$

where A is symmetric. By Theorem 6.4.4 on page 415, there is an orthogonal matrix Q such that $Q^t A Q = D$, where $D = \text{diag}(\lambda_1, \lambda_2)$ and λ_1 and λ_2 are the eigenvalues of A. Then $A = QDQ^t$ (remember that $Q^t = Q^{-1}$) and (5) can be written

$$(QDQ^t\mathbf{v}) \cdot \mathbf{v} = d \tag{6}$$

But, from Theorem 5.5.1 on page 374, $A\mathbf{v} \cdot \mathbf{y} = \mathbf{v} \cdot A^t\mathbf{y}$. Thus

$$Q(DQ^t\mathbf{v}) \cdot \mathbf{v} = DQ^t\mathbf{v} \cdot Q^t\mathbf{v} \tag{7}$$

so that (6) reads

$$[DQ^t\mathbf{v}] \cdot Q^t\mathbf{v} = d \tag{8}$$

Let $\mathbf{v}' = Q^t\mathbf{v}$. Then $\mathbf{v}'$ is a 2-vector and (8) becomes

$$D\mathbf{v}' \cdot \mathbf{v}' = d \tag{9}$$

Let us look at (9) more closely. We can write $\mathbf{v}' = \begin{pmatrix} x' \\ y' \end{pmatrix}$. Since a diagonal matrix is symmetric, (9) defines a quadratic form $\overline{F}(x', y')$ in the variables x' and y'. If $D = \begin{pmatrix} a' & 0 \\ 0 & c' \end{pmatrix}$, then $D\mathbf{v}' = \begin{pmatrix} a' & 0 \\ 0 & c' \end{pmatrix} \begin{pmatrix} x' \\ y' \end{pmatrix} = \begin{pmatrix} a'x' \\ c'y' \end{pmatrix}$ and

$$\overline{F}(x', y') = D\mathbf{v}' \cdot \mathbf{v}' = \begin{pmatrix} a'x' \\ c'y' \end{pmatrix} \cdot \begin{pmatrix} x' \\ y' \end{pmatrix} = a'x'^2 + c'y'^2$$

That is: $\overline{F}(x', y')$ *is a quadratic form with the* $x'y'$ *term missing.* Hence equation (9) is a quadratic equation in the new variables x', y' with the $x'y'$ term missing.

EXAMPLE 1   **Writing a Quadratic Form in New Variables** x' **and** y' **with the** $x'y'$ **Term Missing**   Consider the quadratic equation $x^2 - 4xy + 3y^2 = 6$. Then, as we have seen, the equation can be written in the form $A\mathbf{x} \cdot \mathbf{x} = 6$, where $A = \begin{pmatrix} 1 & -2 \\ -2 & 3 \end{pmatrix}$. In Example 6.4.1 on page 416 we saw that A can be diagonalized to $D = \begin{pmatrix} 2 - \sqrt{5} & 0 \\ 0 & 2 + \sqrt{5} \end{pmatrix}$ by using the orthogonal matrix

$$Q = \frac{1}{\sqrt{10 - 2\sqrt{5}}} \begin{pmatrix} 2 & 1 - \sqrt{5} \\ -1 + \sqrt{5} & 2 \end{pmatrix}$$

Then

$$\mathbf{x}' = \begin{pmatrix} x' \\ y' \end{pmatrix} = Q^t\mathbf{x} = \frac{1}{\sqrt{10 - 2\sqrt{5}}} \begin{pmatrix} 2 & -1 + \sqrt{5} \\ 1 - \sqrt{5} & 2 \end{pmatrix} \begin{pmatrix} x \\ y \end{pmatrix}$$

$$= \frac{1}{\sqrt{10 - 2\sqrt{5}}} \begin{pmatrix} 2x + (-1 + \sqrt{5})y \\ (1 - \sqrt{5})x + 2y \end{pmatrix}$$

and in the new variables the equation can be written as

$$(2 - \sqrt{5})x'^2 + (2 + \sqrt{5})y'^2 = 6 \qquad\blacksquare$$

Let us take another look at the matrix Q. Since Q is real and orthogonal, $1 = \det QQ^{-1} = \det QQ^t = \det Q \det Q^t = \det Q \det Q = (\det Q)^2$. Thus $\det Q = \pm 1$. If $\det Q = -1$, we can interchange the rows of Q to make the determinant of this new Q equal to 1. Then it can be shown (see Problem 36) that $Q = \begin{pmatrix} \cos\theta & -\sin\theta \\ \sin\theta & \cos\theta \end{pmatrix}$ for some number θ with $0 \le \theta < 2\pi$. But, from Example 5.1.8 on page 334, this means that Q is a rotation matrix. We have therefore proved the following theorem.

THEOREM 1   **Principal Axes Theorem in** $\mathbb{R}^2$ Let $ax^2 + bxy + cy^2 = d$ (10)
be a quadratic equation in the variables x and y. Then there exists a unique number θ in $[0, 2\pi)$ such that equation (10) can be written in the form

$$a'x'^2 + c'y'^2 = d \qquad (11)$$

where x', y' are the axes obtained by rotating the x- and y-axes through an angle of

θ in the counterclockwise direction. Moreover, the numbers a' and c' are the eigenvalues of the matrix $A = \begin{pmatrix} a & b/2 \\ b/2 & c \end{pmatrix}$. The x'- and y'-axes are called the **principal axes** of the graph of the quadratic equation (10). ■

We can use Theorem 1 to identify three important conic sections. Recall that the **standard equations** of a circle, ellipse, and hyperbola are

Circle: $\qquad\qquad x^2 + y^2 = r^2 \qquad\qquad$ **(12)**

Ellipse: $\qquad\qquad \dfrac{x^2}{a^2} + \dfrac{y^2}{b^2} = 1 \qquad\qquad$ **(13)**

Hyperbola: $\quad$ or $\begin{cases} \dfrac{x^2}{a^2} - \dfrac{y^2}{b^2} = 1 & \textbf{(14)} \\[2mm] \dfrac{y^2}{a^2} - \dfrac{x^2}{b^2} = 1 & \textbf{(15)} \end{cases}$

EXAMPLE 2   **A Hyperbola**   Identify the conic section whose equation is

$$x^2 - 4xy + 3y^2 = 6 \qquad\qquad \textbf{(16)}$$

Solution   In Example 1 we found that this can be written as $(2 - \sqrt{5})x'^2 + (2 + \sqrt{5})y'^2 = 6$ or

$$\frac{y'^2}{6/(2 + \sqrt{5})} - \frac{x'^2}{6/(\sqrt{5} - 2)} = 1$$

This is equation (15) with $a = \sqrt{6/(2 + \sqrt{5})} \approx 1.19$ and $b = \sqrt{6/(\sqrt{5} - 2)} \approx 5.04$. Since

$$Q = \frac{1}{\sqrt{10 - 2\sqrt{5}}} \begin{pmatrix} 2 & 1 - \sqrt{5} \\ -1 + \sqrt{5} & 2 \end{pmatrix}$$

and det $Q = 1$, we have, using Problem 36 and the fact that 2 and $-1 + \sqrt{5}$ are positive,

$$\cos \theta = \frac{2}{\sqrt{10 - 2\sqrt{5}}} \approx 0.85065$$

Thus θ is in the first quadrant and, using a table (or a calculator), we find that $\theta \approx 0.5536$ rad $\approx 31.7°$. Thus (16) is the equation of a standard hyperbola rotated through an angle of $31.7°$ (see Figure 6.1).

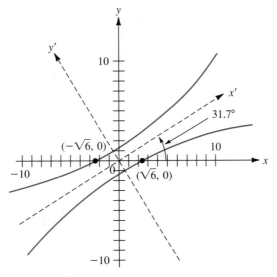

Figure 6.1   The hyperbola $x^2 - 4xy + 3y^2 = 6$

■

EXAMPLE 3   **An Ellipse**   Identify the conic section whose equation is

$$5x^2 - 2xy + 5y^2 = 4 \tag{17}$$

Solution   Here $A = \begin{pmatrix} 5 & -1 \\ -1 & 5 \end{pmatrix}$, the eigenvalues of A are $\lambda_1 = 4$ and $\lambda_2 = 6$, and two or-

thonormal eigenvectors are $\mathbf{v}_1 = \begin{pmatrix} 1/\sqrt{2} \\ 1/\sqrt{2} \end{pmatrix}$ and $\mathbf{v}_2 = \begin{pmatrix} 1/\sqrt{2} \\ -1/\sqrt{2} \end{pmatrix}$. Then $Q = $

$\begin{pmatrix} 1/\sqrt{2} & 1/\sqrt{2} \\ 1/\sqrt{2} & -1/\sqrt{2} \end{pmatrix}$. Before continuing, we note that det $Q = -1$. For Q to be a

rotation matrix, we need det $Q = 1$. This is easily accomplished by reversing the

eigenvectors. Thus we set $\lambda_1 = 6$, $\lambda_2 = 4$, $\mathbf{v}_1 = \begin{pmatrix} 1/\sqrt{2} \\ -1/\sqrt{2} \end{pmatrix}$, $\mathbf{v}_2 = \begin{pmatrix} 1/\sqrt{2} \\ 1/\sqrt{2} \end{pmatrix}$, and

$Q = \begin{pmatrix} 1/\sqrt{2} & 1/\sqrt{2} \\ -1/\sqrt{2} & 1/\sqrt{2} \end{pmatrix}$; now det $Q = 1$. Then $D = \begin{pmatrix} 6 & 0 \\ 0 & 4 \end{pmatrix}$ and (17) can be writ-

ten as $D\mathbf{v} \cdot \mathbf{v} = 4$ or

$$6x'^2 + 4y'^2 = 4 \tag{18}$$

where

$$\begin{pmatrix} x' \\ y' \end{pmatrix} = Q^t \begin{pmatrix} x \\ y \end{pmatrix} = \begin{pmatrix} 1/\sqrt{2} & -1/\sqrt{2} \\ 1/\sqrt{2} & 1/\sqrt{2} \end{pmatrix} \begin{pmatrix} x \\ y \end{pmatrix} = \begin{pmatrix} 1/\sqrt{2}\,x - 1/\sqrt{2}\,y \\ 1/\sqrt{2}\,x + 1/\sqrt{2}\,y \end{pmatrix}$$

Rewriting (18), we obtain $x'^2/(\frac{4}{6}) + y'^2/1 = 1$, which is equation (13) with $a = \sqrt{\frac{2}{3}}$
and $b = 1$. Moreover, since $1/\sqrt{2} > 0$ and $-1/\sqrt{2} < 0$, we have, from

Problem 36, $\theta = 2\pi - \cos^{-1}(1/\sqrt{2}) = 2\pi - \pi/4 = 7\pi/4 = 315°$. Thus (17) is the equation of a standard ellipse rotated through an angle of 315° (or 45° in the clockwise direction). (See Figure 6.2.)

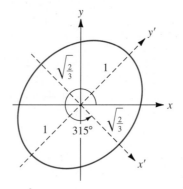

Figure 6.2   The ellipse $5x^2 - 2xy + 5y^2 = 4$

EXAMPLE 4   **A Degenerate Conic Section**   Identify the conic section whose equation is

$$-5x^2 + 2xy - 5y^2 = 4 \tag{19}$$

Solution   Referring to Example 3, equation (19) can be rewritten as

$$-6x'^2 - 4y'^2 = 4 \tag{20}$$

Since, for any real numbers x' and y', $-6x'^2 - 4y'^2 \leq 0$, we see that there are no real numbers x and y which satisfy (19). The conic section defined by (19) is called a **degenerate conic section.**   ∎

There is an easy way to identify the conic section defined by

$$ax^2 + bxy + cy^2 = d \tag{21}$$

If $A = \begin{pmatrix} a & b/2 \\ b/2 & c \end{pmatrix}$, then the characteristic equation of A is

$$\lambda^2 - (a + c)\lambda + (ac - b^2/4) = 0 = (\lambda - \lambda_1)(\lambda - \lambda_2)$$

This means that $\lambda_1\lambda_2 = ac - b^2/4$. But equation (21) can, as we have seen, be rewritten as

$$\lambda_1 x'^2 + \lambda_2 y'^2 = d \tag{22}$$

If λ_1 and λ_2 have the same sign, then (21) defines an ellipse (or a circle) or a degenerate conic as in Examples 3 and 4. If λ_1 and λ_2 have opposite signs, then (21) is the equation of a hyperbola (as in Example 2). We can therefore prove the following.

THEOREM 2   If $A = \begin{pmatrix} a & b/2 \\ b/2 & c \end{pmatrix}$, then the quadratic equation (21) with $d \neq 0$ is the equation of:

    i. A hyperbola if $\det A < 0$.

    ii. An ellipse, circle, or degenerate conic section if $\det A > 0$.

    iii. A pair of straight lines or a degenerate conic section if $\det A = 0$.

    iv. If $d = 0$, then (21) is the equation of two straight lines if $\det A \neq 0$ and the equation of a single line if $\det A = 0$.

Proof   We have already shown why (*i*) and (*ii*) are true. To prove part (*iii*), suppose that $\det A = 0$. Then, by our Summing Up Theorem (Theorem 6.1.6), $\lambda = 0$ is an eigenvalue of A and equation (22) reads $\lambda_1 x'^2 = d$ or $\lambda_2 y'^2 = d$. If $\lambda_1 x'^2 = d$ and $d/\lambda_1 > 0$, then $x_1' = \pm\sqrt{d/\lambda_1}$ is the equation of two straight lines in the xy-plane. If $d/\lambda_1 < 0$, then we have $x'^2 < 0$ (which is impossible) and we obtain a degenerate conic. The same facts hold if $\lambda_2 y'^2 = d$. Part (*iv*) is left as an exercise (see Problem 37). ∎

Note. In Example 2 we had $\det A = ac - b^2/4 = -1$. In Examples 3 and 4 we had $\det A = 24$.

The methods described above can be used to analyze quadratic equations in more than two variables. We give one example below.

EXAMPLE 5   **An Ellipsoid**   Consider the quadratic equation

$$5x^2 + 8xy + 5y^2 + 4xz + 4yz + 2z^2 = 100 \tag{23}$$

If $A = \begin{pmatrix} 5 & 4 & 2 \\ 4 & 5 & 2 \\ 2 & 2 & 2 \end{pmatrix}$ and $\mathbf{v} = \begin{pmatrix} x \\ y \\ z \end{pmatrix}$, then (23) can be written in the form

$$A\mathbf{v} \cdot \mathbf{v} = 100 \tag{24}$$

From Example 6.4.2 on page 416, $Q^t A Q = D = \begin{pmatrix} 1 & 0 & 0 \\ 0 & 1 & 0 \\ 0 & 0 & 10 \end{pmatrix}$, where $Q = \begin{pmatrix} -1/\sqrt{2} & -1/3\sqrt{2} & 2/3 \\ 1/\sqrt{2} & -1/3\sqrt{2} & 2/3 \\ 0 & 4/3\sqrt{2} & 1/3 \end{pmatrix}$.

Let

$$\mathbf{v}' = \begin{pmatrix} x' \\ y' \\ z' \end{pmatrix} = Q^t \mathbf{v} = \begin{pmatrix} -1/\sqrt{2} & 1/\sqrt{2} & 0 \\ -1/3\sqrt{2} & -1/3\sqrt{2} & 4/3\sqrt{2} \\ 2/3 & 2/3 & 1/3 \end{pmatrix} \begin{pmatrix} x \\ y \\ z \end{pmatrix}$$

$$= \begin{pmatrix} (-1/\sqrt{2})x + (1/\sqrt{2})y \\ -(1/3\sqrt{2})x - (1/3\sqrt{2})y + (4/3\sqrt{2})z \\ (2/3)x + (2/3)y + (1/3)z \end{pmatrix}$$

Then, as before, $A = QDQ^t$ and $A\mathbf{v} \cdot \mathbf{v} = QDQ^t\mathbf{v} \cdot \mathbf{v} = DQ^t\mathbf{v} \cdot Q^t\mathbf{v} = D\mathbf{v}' \cdot \mathbf{v}'$. Thus (24) can be written in the new variables x', y', z' as $D\mathbf{v}' \cdot \mathbf{v}' = 100$ or

$$x'^2 + y'^2 + 10z'^2 = 100 \tag{25}$$

In $\mathbb{R}^3$ the surface defined by (25) is called an **ellipsoid** (see Figure 6.3).

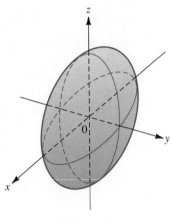

$$x'^2 + y'^2 + 10z'^2 = 100$$

Figure 6.3   The ellipsoid $5x^2 + 8xy + 5y^2 + 4xz + 4yz + 2z^2 = 100$, which can be written in new variables as $x'^2 + y'^2 + 10z'^2 = 100$

■

There is a great variety of three-dimensional surfaces of the form $A\mathbf{v} \cdot \mathbf{v} = d$, where $\mathbf{v} \in \mathbb{R}^2$. Such surfaces are called **quadric surfaces.**

We close this section by noting that quadratic forms can be defined in any number of variables.

DEFINITION 2   **Quadratic Form** Let $\mathbf{v} = \begin{pmatrix} x_1 \\ x_2 \\ \vdots \\ x_n \end{pmatrix}$ and let A be a symmetric $n \times n$ matrix. Then a

quadratic form in $x_1, x_2, \ldots, x_n$ is an expression of the form

$$F(x_1, x_2, \ldots, x_n) = A\mathbf{v} \cdot \mathbf{v} \tag{26}$$

EXAMPLE 6   **A Quadratic Form in Four Variables**

$$A = \begin{pmatrix} 2 & 1 & 2 & -2 \\ 1 & -4 & 6 & 5 \\ 2 & 6 & 7 & -1 \\ -2 & 5 & -1 & 3 \end{pmatrix}$$

Then

$$A\mathbf{v} \cdot \mathbf{v} = \left[\begin{pmatrix} 2 & 1 & 2 & -2 \\ 1 & -4 & 6 & 5 \\ 2 & 6 & 7 & -1 \\ -2 & 5 & -1 & 3 \end{pmatrix} \begin{pmatrix} x_1 \\ x_2 \\ x_3 \\ x_4 \end{pmatrix} \right] \cdot \begin{pmatrix} x_1 \\ x_2 \\ x_3 \\ x_4 \end{pmatrix}$$

$$= \begin{pmatrix} 2x_1 + x_2 + 2x_3 - 2x_4 \\ x_1 - 4x_2 + 6x_3 + 5x_4 \\ 2x_1 + 6x_2 + 7x_3 - x_4 \\ -2x_1 + 5x_2 - x_3 + 3x_4 \end{pmatrix} \cdot \begin{pmatrix} x_1 \\ x_2 \\ x_3 \\ x_4 \end{pmatrix}$$

$$= 2x_1^2 + 2x_1x_2 - 4x_2^2 + 4x_1x_3 + 12x_2x_3$$
$$+ 7x_3^2 - 4x_1x_4 + 10x_2x_4 - 2x_3x_4 + 3x_4^2$$

(after simplification)   ∎

EXAMPLE 7   **Finding a Symmetric Matrix That Corresponds to a Quadratic Form in Four Variables**   Find the symmetric matrix A corresponding to the quadratic form

$$5x_1^2 - 3x_1x_2 + 4x_2^2 + 8x_1x_3 - 9x_2x_3 + 2x_3^2 - x_1x_4 + 7x_2x_4 + 6x_3x_4 + 9x_4^2$$

Solution   If $A = (a_{ij})$, then, by looking at the earlier examples in this section, we see that a_{ii} is the coefficient of the x_i^2 term and $a_{ij} + a_{ji}$ is the coefficient of the x_ix_j term. Since A is symmetric, $a_{ij} = a_{ji}$; hence $a_{ij} = a_{ji} = \frac{1}{2} \cdot$ (coefficient of x_ix_j term). Putting this all together, we obtain

$$A = \begin{pmatrix} 5 & -\frac{3}{2} & 4 & -\frac{1}{2} \\ -\frac{3}{2} & 4 & -\frac{9}{2} & \frac{7}{2} \\ 4 & -\frac{9}{2} & 2 & 3 \\ -\frac{1}{2} & \frac{7}{2} & 3 & 9 \end{pmatrix}$$

∎

PROBLEMS   **6.5**

In Problems 1–13 write the quadratic equation in the form $A\mathbf{v} \cdot \mathbf{v} = d$ (where A is a symmetric matrix) and eliminate the xy-term by rotating the axes through an angle of θ. Write the equation in terms of the new variables and identify the conic section obtained.

1. $3x^2 - 2xy - 5 = 0$ **2.** $4x^2 + 4xy + y^2 = 9$ **3.** $4x^2 + 4xy - y^2 = 9$

4. $xy = 1$ **5.** $xy = a; a > 0$ **6.** $4x^2 + 2xy + 3y^2 + 2 = 0$

7. $xy = a;\ a < 0$ **8.** $x^2 + 4xy + 4y^2 - 6 = 0$ **9.** $-x^2 + 2xy - y^2 = 0$

10. $2x^2 + xy + y^2 = 4$ **11.** $3x^2 - 6xy + 5y^2 = 36$ **12.** $x^2 - 3xy + 4y^2 = 1$

13. $6x^2 + 5xy - 6y^2 + 7 = 0$

14. What are the possible forms of the graph of $ax^2 + bxy + cy^2 = 0$?

In Problems 15–18 write the quadratic form in new variables x', y', and z' so that no cross-product terms (xy, xz, yz) are present.

15. $x^2 - 2xy + y^2 - 2xz - 2yz + z^2$ **16.** $-x^2 + 4xy - y^2 + 4xz + 4yz + z^2$

17. $3x^2 + 4xy + 2y^2 + 4xz + 4z^2$ **18.** $x^2 - 2xy + 2y^2 - 2yz + z^2$

In Problems 19–21 find a symmetric matrix A such that the quadratic form can be written in the form $A\mathbf{x} \cdot \mathbf{x}$.

19. $x_1^2 + 2x_1x_2 + x_2^2 + 4x_1x_3 + 6x_2x_3 + 3x_3^2 + 7x_1x_4 - 2x_2x_4 + x_4^2$

20. $x_1^2 - x_2^2 + x_1x_3 - x_2x_4 + x_4^2$

21. $3x_1^2 - 7x_1x_2 - 2x_2^2 + x_1x_3 - x_2x_3 + 3x_3^2 - 2x_1x_4 + x_2x_4 - 4x_3x_4 - 6x_4^2 + 3x_1x_5$
$\quad - 5x_3x_5 + x_4x_5 - x_5^2$

22. Suppose that for some nonzero value of d, the graph of $ax^2 + bxy + cy^2 = d$ is a hyperbola. Show that the graph is a hyperbola for any other nonzero value of d.

23. Show that if $a \neq c$, the xy-term in quadratic equation (1) will be eliminated by rotation through an angle θ if θ is given by $\cot 2\theta = (a - c)/b$.

24. Show that if $a = c$ in Problem 23, then the xy-term will be eliminated by a rotation through an angle of either $\pi/4$ or $-\pi/4$.

*25. Suppose that a rotation converts $ax^2 + bxy + cy^2$ into $a'(x')^2 + b'(x'y') + c'(y')^2$. Show that:
  a. $a + c = a' + c'$ **b.** $b^2 - 4ac = b'^2 - 4a'c'$

*26. A quadratic form $F(\mathbf{x}) = F(x_1, x_2, \ldots, x_n)$ is said to be **positive definite** if $F(\mathbf{x}) \geq 0$ for every $\mathbf{x} \in \mathbb{R}^n$ and $F(\mathbf{x}) = 0$ if and only if $\mathbf{x} = \mathbf{0}$. Show that F is positive definite if and only if the symmetric matrix A associated with F has positive eigenvalues.

27. A quadratic form $F(\mathbf{x})$ is said to be **positive semidefinite** if $F(\mathbf{x}) \geq 0$ for every $\mathbf{x} \in \mathbb{R}^n$. Show that F is positive semidefinite if and only if the eigenvalues of the symmetric matrix associated with F are all nonnegative.

The definitions of **negative definite** and **negative semidefinite** are the definitions in Problems 26 and 27 with ≤ 0 replacing ≥ 0. A quadratic form is **indefinite** if it is none of the above. In Problems 28–35 determine whether the given quadratic form is positive definite, positive semidefinite, negative definite, negative semidefinite, or indefinite.

28. $3x^2 + 2y^2$ **29.** $-3x^2 - 3y^2$ **30.** $3x^2 - 2y^2$ **31.** $x^2 + 2xy + 2y^2$

32. $x^2 - 2xy + 2y^2$ **33.** $x^2 - 4xy + 3y^2$ **34.** $-x^2 + 4xy - 3y^2$ **35.** $-x^2 + 2xy - 2y^2$

*36. Let $Q = \begin{pmatrix} a & b \\ c & d \end{pmatrix}$ be a real orthogonal matrix with $\det Q = 1$. Define the number

$\theta \in [0, 2\pi)$:
  a. If $a \geq 0$ and $c > 0$, then $\theta = \cos^{-1} a$ $(0 < \theta \leq \pi/2)$.
  b. If $a \geq 0$ and $c < 0$, then $\theta = 2\pi - \cos^{-1} a$ $(3\pi/2 \leq \theta < 2\pi)$.
  c. If $a \leq 0$ and $c > 0$, then $\theta = \cos^{-1} a$ $(\pi/2 \leq \theta < \pi)$.
  d. If $a \leq 0$ and $c < 0$, then $\theta = 2\pi - \cos^{-1} a$ $(\pi < \theta \leq 3\pi/2)$.

e. If $a = 1$ and $c = 0$, then $\theta = 0$.

f. If $a = -1$ and $c = 0$, then $\theta = \pi$.

(Here $\cos^{-1} x \in [0, \pi]$ for $x \in [-1, 1]$.) With θ chosen as above, show that

$$Q = \begin{pmatrix} \cos\theta & -\sin\theta \\ \sin\theta & \cos\theta \end{pmatrix}.$$

37. Prove, using formula (22), that equation (21) is the equation of two straight lines in the xy-plane when $d = 0$ and $\det A \neq 0$. If $\det A = d = 0$, show that equation (21) is the equation of a single line.

38. Let A be the symmetric matrix representation of quadratic equation (1) with $d \neq 0$. Let λ_1 and λ_2 be the eigenvalues of A. Show that (1) is the equation of **(a)** a hyperbola if $\lambda_1\lambda_2 < 0$ and **(b)** a circle, ellipse, or degenerate conic section if $\lambda_1\lambda_2 > 0$.

6.6 JORDAN CANONICAL FORM

As we have seen, $n \times n$ matrices with n linearly independent eigenvectors can be brought into an especially nice form by a similarity transformation. Fortunately, as "most" polynomials have distinct roots, "most" matrices will have distinct eigenvalues. As we shall see in Section 6.7, however, matrices that are not diagonalizable (that is, that do not have n linearly independent eigenvectors) do arise in applications. In this case it is still possible to show that the matrix is similar to another, simpler matrix, but the new matrix is not diagonal and the transforming matrix C is harder to obtain.

To discuss this case fully, we define the matrix N_k to be the $k \times k$ matrix

$$N_k = \begin{pmatrix} 0 & 1 & 0 & \cdots & 0 \\ 0 & 0 & 1 & \cdots & 0 \\ \vdots & \vdots & \vdots & & \vdots \\ 0 & 0 & 0 & \cdots & 1 \\ 0 & 0 & 0 & \cdots & 0 \end{pmatrix} \tag{1}$$

Note that N_k is the matrix with 1's above the main diagonal and 0's everywhere else. We next define a $k \times k$ **Jordan**† **block matrix** $B(\lambda)$ by

Jordan block matrix

$$B(\lambda) = \lambda I + N_k = \begin{pmatrix} \lambda & 1 & 0 & \cdots & 0 & 0 \\ 0 & \lambda & 1 & \cdots & 0 & 0 \\ \vdots & \vdots & \vdots & & \vdots & \vdots \\ 0 & 0 & & \cdots & \lambda & 1 \\ 0 & 0 & & \cdots & 0 & \lambda \end{pmatrix} \tag{2}$$

---

†Named for the French mathematician Camille Jordan (1838–1922). The results in this section first appeared in Jordan's brilliant *Traité des substitutions et des équations algebriques* (Treatise on substitutions and algebraic equations), which was published in 1870.

That is, $B(\lambda)$ is the $k \times k$ matrix with the fixed number λ on the diagonal, 1's above the diagonal, and 0's everywhere else.

Note. We can (and often will) have a 1×1 Jordan block matrix. Such a matrix takes the form $B(\lambda) = (\lambda)$.

Jordan matrix   Finally, a **Jordan matrix** J has the form

$$
J = \begin{pmatrix}
B_1(\lambda_1) & 0 & \cdots & 0 \\
0 & B_2(\lambda_2) & \cdots & 0 \\
\vdots & \vdots & & \vdots \\
0 & 0 & \cdots & B_r(\lambda_r)
\end{pmatrix}
$$

where each $B_j(\lambda_j)$ is a Jordan block matrix. Thus: *a Jordan matrix is a matrix with Jordan block matrices down the diagonal and zeros everywhere else.*

EXAMPLE 1   **Three Jordan Matrices**   The following are examples of Jordan matrices. The Jordan blocks are outlined by the dotted lines:

i.
$$
\begin{pmatrix}
2 & 1 & 0 \\
0 & 2 & 0 \\
0 & 0 & 4
\end{pmatrix}
$$

ii.
$$
\begin{pmatrix}
-3 & 0 & 0 & 0 & 0 \\
0 & -3 & 1 & 0 & 0 \\
0 & 0 & -3 & 1 & 0 \\
0 & 0 & 0 & -3 & 0 \\
0 & 0 & 0 & 0 & 7
\end{pmatrix}
$$

iii.
$$
\begin{pmatrix}
4 & 1 & 0 & 0 & 0 & 0 & 0 \\
0 & 4 & 0 & 0 & 0 & 0 & 0 \\
0 & 0 & 3 & 1 & 0 & 0 & 0 \\
0 & 0 & 0 & 3 & 1 & 0 & 0 \\
0 & 0 & 0 & 0 & 3 & 0 & 0 \\
0 & 0 & 0 & 0 & 0 & 5 & 1 \\
0 & 0 & 0 & 0 & 0 & 0 & 5
\end{pmatrix}
$$

EXAMPLE 2   **The 2 × 2 Jordan Matrices**   The only 2×2 Jordan matrices are $\begin{pmatrix} \lambda_1 & 0 \\ 0 & \lambda_2 \end{pmatrix}$ and $\begin{pmatrix} \lambda & 1 \\ 0 & \lambda \end{pmatrix}$. In the first matrix the numbers λ_1 and λ_2 could be equal.

EXAMPLE 3   **The 3 × 3 Jordan Matrices**   The only 3×3 Jordan matrices are

$$
\begin{pmatrix} \lambda_1 & 0 & 0 \\ 0 & \lambda_2 & 0 \\ 0 & 0 & \lambda_3 \end{pmatrix} \quad
\begin{pmatrix} \lambda_1 & 0 & 0 \\ 0 & \lambda_2 & 1 \\ 0 & 0 & \lambda_2 \end{pmatrix} \quad
\begin{pmatrix} \lambda_1 & 1 & 0 \\ 0 & \lambda_1 & 0 \\ 0 & 0 & \lambda_2 \end{pmatrix} \quad
\begin{pmatrix} \lambda_1 & 1 & 0 \\ 0 & \lambda_1 & 1 \\ 0 & 0 & \lambda_1 \end{pmatrix}
$$

where λ_1, λ_2, and λ_3 are not necessarily distinct.   ∎

The following result is one of the most important theorems in matrix theory. Although its proof is beyond the scope of this book,[†] we shall prove this theorem in the 2×2 case (see Theorem 3) and suggest a proof for the 3×3 case in Problem 19.

THEOREM 1   Let A be an $n \times n$ matrix. Then there exists an invertible $n \times n$ matrix C such that

$$
\boxed{C^{-1}AC = J} \tag{3}
$$

where J is a Jordan matrix whose diagonal elements are the eigenvalues of A. Moreover, J is unique except for the order in which the Jordan blocks appear.   ∎

Remark. By the last sentence of the theorem we mean, for example, that if A is similar to

$$
J_1 = \begin{pmatrix}
2 & 1 & 0 & 0 & 0 & 0 \\
0 & 2 & 0 & 0 & 0 & 0 \\
0 & 0 & 3 & 1 & 0 & 0 \\
0 & 0 & 0 & 3 & 1 & 0 \\
0 & 0 & 0 & 0 & 3 & 0 \\
0 & 0 & 0 & 0 & 0 & 4
\end{pmatrix},
$$

then A is also similar to

$$
J_2 = \begin{pmatrix}
3 & 1 & 0 & 0 & 0 & 0 \\
0 & 3 & 1 & 0 & 0 & 0 \\
0 & 0 & 3 & 0 & 0 & 0 \\
0 & 0 & 0 & 4 & 0 & 0 \\
0 & 0 & 0 & 0 & 2 & 1 \\
0 & 0 & 0 & 0 & 0 & 2
\end{pmatrix}
\quad \text{and} \quad
J_3 = \begin{pmatrix}
4 & 0 & 0 & 0 & 0 & 0 \\
0 & 2 & 1 & 0 & 0 & 0 \\
0 & 0 & 2 & 0 & 0 & 0 \\
0 & 0 & 0 & 3 & 1 & 0 \\
0 & 0 & 0 & 0 & 3 & 1 \\
0 & 0 & 0 & 0 & 0 & 3
\end{pmatrix}
$$

[†] For a proof see G. Birkhoff and S. MacLane, *A Survey of Modern Algebra*, (New York: Macmillan, 1953), p. 334.

and three other Jordan matrices. That is, the actual Jordan blocks remain the same but we can change the order in which they are written.

DEFINITION 1   **Jordan Canonical Form**   The matrix J is called the **Jordan canonical form** of A.

Remark. If A is diagonalizable, then $J = D = \operatorname{diag}(\lambda_1, \lambda_2, \ldots, \lambda_n)$, where λ_1, $\lambda_2, \ldots, \lambda_n$ are the (not necessarily distinct) eigenvalues of A. Each diagonal component is a 1×1 Jordan block matrix.

We shall now see how to compute the Jordan canonical form of any 2×2 matrix. If A has two linearly independent eigenvectors, we already know what to do. Therefore the only case of interest occurs when A has a single eigenvalue λ of algebraic multiplicity 2 and geometric multiplicity 1. That is, we assume that, corresponding to λ, A has the single independent eigenvector $\mathbf{v}_1$. That is: *Any vector that is not a multiple of $\mathbf{v}_1$ is not an eigenvector.*

THEOREM 2   Let the 2×2 matrix A have an eigenvalue λ of algebraic multiplicity 2 and geometric multiplicity 1. Let $\mathbf{v}_1$ be an eigenvector corresponding to λ. Then there exists a vector $\mathbf{v}_2$ that satisfies the equation

$$(A - \lambda I)\mathbf{v}_2 = \mathbf{v}_1 \qquad (4)$$

Proof   Let $\mathbf{x} \in \mathbb{C}^2$ be a fixed vector that is *not* a multiple of $\mathbf{v}_1$ so that $\mathbf{x}$ is not an eigenvector of A. We first show that

$$\mathbf{w} = (A - \lambda I)\mathbf{x} \qquad (5)$$

is an eigenvector of A. That is, we shall show that $\mathbf{w} = c\mathbf{v}_1$ for some constant c. Since $\mathbf{w} \in \mathbb{C}^2$ and $\mathbf{v}_1$ and $\mathbf{x}$ are linearly independent, there exist constants c_1 and c_2 such that

$$\mathbf{w} = c_1\mathbf{v}_1 + c_2\mathbf{x} \qquad (6)$$

To show that $\mathbf{w}$ is an eigenvector of A, we must show that $c_2 = 0$. From (5) and (6), we find that

$$(A - \lambda I)\mathbf{x} = c_1\mathbf{v}_1 + c_2\mathbf{x} \qquad (7)$$

Let $B = A - (\lambda + c_2)I$. Then, from (7),

$$B\mathbf{x} = [A - (\lambda + c_2)I]\mathbf{x} = c_1\mathbf{v}_1 \qquad (8)$$

If we assume that $c_2 \neq 0$, then $\lambda + c_2 \neq \lambda$ and $\lambda + c_2$ is not an eigenvalue of A (since λ is the only eigenvalue of A). Thus $\det B = \det[A - (\lambda + c_2)I] \neq 0$, which means that B is invertible. Hence (8) can be written as

$$\mathbf{x} = B^{-1}c_1\mathbf{v}_1 = c_1B^{-1}\mathbf{v}_1 \qquad (9)$$

Then, multiplying both sides of (9) by λ, we have

$$\lambda\mathbf{x} = \lambda c_1B^{-1}\mathbf{v}_1 = c_1B^{-1}\lambda\mathbf{v}_1 = c_1B^{-1}A\mathbf{v}_1 \qquad (10)$$

But $B = A - (\lambda + c_2)I$, so

$$A = B + (\lambda + c_2)I \qquad (11)$$

Inserting (11) into (10), we have

$$\begin{aligned}\lambda\mathbf{x} &= c_1B^{-1}[B + (\lambda + c_2)I]\mathbf{v}_1 \\ &= c_1[I + (\lambda + c_2)B^{-1}]\mathbf{v}_1 \\ &= c_1\mathbf{v}_1 + (\lambda + c_2)c_1B^{-1}\mathbf{v}_1 \end{aligned} \qquad (12)$$

But, again using (8), $c_1B^{-1}\mathbf{v}_1 = \mathbf{x}$ so that (12) becomes

$$\lambda\mathbf{x} = c_1\mathbf{v}_1 + (\lambda + c_2)\mathbf{x} = c_1\mathbf{v}_1 + c_2\mathbf{x} + \lambda\mathbf{x}$$

or

$$\mathbf{0} = c_1\mathbf{v}_1 + c_2\mathbf{x} \qquad (13)$$

But $\mathbf{v}_1$ and $\mathbf{x}$ are linearly independent, so $c_1 = c_2 = 0$. This contradicts the assumption that $c_2 \neq 0$. Thus $c_2 = 0$ and, by (6), $\mathbf{w}$ is a multiple of $\mathbf{v}_1$ so that $\mathbf{w} = c_1\mathbf{v}_1$ is an eigenvector of A. Moreover, $\mathbf{w} \neq \mathbf{0}$ since if $\mathbf{w} = \mathbf{0}$, then (5) tells us that $\mathbf{x}$ is an eigenvector of A. Therefore $c_1 \neq 0$. Let

$$\mathbf{v}_2 = \frac{1}{c_1}\mathbf{x} \qquad (14)$$

Then $(A - \lambda I)\mathbf{v}_2 = (1/c_1)(A - \lambda I)\mathbf{x} = (1/c_1)\mathbf{w} = \mathbf{v}_1$. This proves the theorem. ∎

DEFINITION 2   **Generalized Eigenvector**   Let A be a 2×2 matrix with the single eigenvalue λ having geometric multiplicity 1. Let $\mathbf{v}_1$ be an eigenvector of A. Then the vector $\mathbf{v}_2$ defined by $(A - \lambda I)\mathbf{v}_2 = \mathbf{v}_1$ is called a **generalized eigenvector** of A corresponding to the eigenvalue λ.

EXAMPLE 4   **Finding a Generalized Eigenvector**   Let $A = \begin{pmatrix} 3 & -2 \\ 8 & -5 \end{pmatrix}$. The characteristic equation of A is $\lambda^2 + 2\lambda + 1 = (\lambda + 1)^2 = 0$, so $\lambda = -1$ is an eigenvalue of algebraic multiplicity 2. Then

$$(A - \lambda I)\mathbf{v} = (A + I)\mathbf{v} = \begin{pmatrix} 4 & -2 \\ 8 & -4 \end{pmatrix}\begin{pmatrix} x_1 \\ x_2 \end{pmatrix} = \begin{pmatrix} 0 \\ 0 \end{pmatrix}$$

This yields the eigenvector $\mathbf{v}_1 = \begin{pmatrix} 1 \\ 2 \end{pmatrix}$. There is no other linearly independent eigenvector. To find a generalized eigenvector $\mathbf{v}_2$, we compute $(A + I)\mathbf{v}_2 = \mathbf{v}_1$ or $\begin{pmatrix} 4 & -2 \\ 8 & -4 \end{pmatrix}\begin{pmatrix} x_1 \\ x_2 \end{pmatrix} = \begin{pmatrix} 1 \\ 2 \end{pmatrix}$, which yields the system

$$4x_1 - 2x_2 = 1$$
$$8x_1 - 4x_2 = 2$$

The second equation is double the first, so x_2 can be chosen arbitrarily and $x_1 = (1 + 2x_2)/4$. Therefore a possible choice for $\mathbf{v}_2$ is $\mathbf{v}_2 = \begin{pmatrix} \frac{1}{4} \\ 0 \end{pmatrix}$.   ∎

The reason for finding generalized eigenvectors is given in the following theorem.

THEOREM 3   Let A, λ, $\mathbf{v}_1$, and $\mathbf{v}_2$ be as in Theorem 2 and let C be the matrix whose columns are $\mathbf{v}_1$ and $\mathbf{v}_2$. Then $C^{-1}AC = J$, where $J = \begin{pmatrix} \lambda & 1 \\ 0 & \lambda \end{pmatrix}$ is the Jordan canonical form of A.

Proof   Since $\mathbf{v}_1$ and $\mathbf{v}_2$ are linearly independent, we see that C is invertible. Next note that $AC = A(\mathbf{v}_1, \mathbf{v}_2) = (A\mathbf{v}_1, A\mathbf{v}_2) = (\lambda\mathbf{v}_1, A\mathbf{v}_2)$. But, from equation (4), $A\mathbf{v}_2 = \mathbf{v}_1 + \lambda\mathbf{v}_2$ so that $AC = (\lambda\mathbf{v}_1, \mathbf{v}_1 + \lambda\mathbf{v}_2)$. But $CJ = (\mathbf{v}_1, \mathbf{v}_2)\begin{pmatrix} \lambda & 1 \\ 0 & \lambda \end{pmatrix} = (\lambda\mathbf{v}_1, \mathbf{v}_1 + \lambda\mathbf{v}_2)$. Thus $AC = CJ$, which means that $C^{-1}AC = J$ and the theorem is proved.   ∎

EXAMPLE 5   **Finding the Jordan Canonical Form of a 2 × 2 Matrix**   In Example 4 $\mathbf{v}_1 = \begin{pmatrix} 1 \\ 2 \end{pmatrix}$ and $\mathbf{v}_2 = \begin{pmatrix} \frac{1}{4} \\ 0 \end{pmatrix}$. Then $C = \begin{pmatrix} 1 & \frac{1}{4} \\ 2 & 0 \end{pmatrix}$, $C^{-1} = -2\begin{pmatrix} 0 & -\frac{1}{4} \\ -2 & 1 \end{pmatrix} = \begin{pmatrix} 0 & \frac{1}{2} \\ 4 & -2 \end{pmatrix}$, and

$$C^{-1}AC = \begin{pmatrix} 0 & \frac{1}{2} \\ 4 & -2 \end{pmatrix}\begin{pmatrix} 3 & -2 \\ 8 & -5 \end{pmatrix}\begin{pmatrix} 1 & \frac{1}{4} \\ 2 & 0 \end{pmatrix}$$

$$= \begin{pmatrix} 0 & \frac{1}{2} \\ 4 & -2 \end{pmatrix}\begin{pmatrix} -1 & \frac{3}{4} \\ -2 & 2 \end{pmatrix} = \begin{pmatrix} -1 & 1 \\ 0 & -1 \end{pmatrix} = J$$ ∎

The method described above can be generalized to obtain the Jordan canonical form of every matrix. We shall not do this, but one generalization is suggested in Problem 19. Although we shall not prove this fact, it is always possible to determine the number of 1's above the diagonal in the Jordan canonical form of an $n \times n$ matrix A. Let λ_i be an eigenvalue of A with algebraic multiplicity r_i and geometric multiplicity s_i. If $\lambda_1, \lambda_2, \ldots, \lambda_k$ are the eigenvalues of A, then

The number of 1's above the diagonal of the Jordan canonical form of A

$$= (r_1 - s_1) + (r_2 - s_2) + \cdots + (r_k - s_k) \tag{15}$$

$$= \sum_{i=1}^{k} r_i - \sum_{i=1}^{k} s_i = n - \sum_{i=1}^{k} s_i$$

If we know the characteristic equation of a matrix A, then we can determine the possible Jordan canonical forms of A.

EXAMPLE 6 **Determining the Possible Jordan Canonical Forms of a 4 × 4 Matrix with Given Characteristic Equation** If the characteristic equation of A is $(\lambda - 2)^3(\lambda + 3)$, then the possible Jordan canonical forms for A are

$$J = \begin{pmatrix} 2 & 0 & 0 & 0 \\ 0 & 2 & 0 & 0 \\ 0 & 0 & 2 & 0 \\ 0 & 0 & 0 & -3 \end{pmatrix}, \quad \begin{pmatrix} 2 & 1 & 0 & 0 \\ 0 & 2 & 0 & 0 \\ 0 & 0 & 2 & 0 \\ 0 & 0 & 0 & -3 \end{pmatrix}, \quad \begin{pmatrix} 2 & 1 & 0 & 0 \\ 0 & 2 & 1 & 0 \\ 0 & 0 & 2 & 0 \\ 0 & 0 & 0 & -3 \end{pmatrix}$$

or any matrix obtained by rearranging the Jordan blocks in J. The first matrix corresponds to a geometric multiplicity of 3 (for $\lambda = 2$); the second corresponds to a geometric multiplicity of 2; and the third corresponds to a geometric multiplicity of 1. ∎

PROBLEMS 6.6

In Problems 1–14 determine whether the given matrix is a Jordan matrix.

1. $\begin{pmatrix} 1 & 1 \\ 0 & -6 \end{pmatrix}$ **2.** $\begin{pmatrix} 1 & 0 \\ 0 & 0 \end{pmatrix}$ **3.** $\begin{pmatrix} 1 & 2 \\ 0 & 1 \end{pmatrix}$ **4.** $\begin{pmatrix} 1 & 0 & 0 \\ 0 & 3 & 1 \\ 0 & 0 & 3 \end{pmatrix}$

5. $\begin{pmatrix} 3 & 1 & 0 \\ 0 & 3 & 1 \\ 0 & 0 & 3 \end{pmatrix}$ **6.** $\begin{pmatrix} 3 & 1 & 0 \\ 0 & 3 & 1 \\ 0 & 0 & 2 \end{pmatrix}$ **7.** $\begin{pmatrix} 1 & 0 & 0 \\ 0 & 3 & 1 \\ 0 & 0 & 4 \end{pmatrix}$ **8.** $\begin{pmatrix} 1 & 1 & 0 \\ 0 & 3 & 1 \\ 0 & 0 & 3 \end{pmatrix}$

9. $\begin{pmatrix} 1 & 1 & 0 \\ 0 & 1 & 1 \\ 0 & 0 & 1 \end{pmatrix}$ **10.** $\begin{pmatrix} 1 & 0 & 0 & 0 & 0 \\ 0 & 2 & 1 & 0 & 0 \\ 0 & 0 & 2 & 1 & 0 \\ 0 & 0 & 0 & 2 & 0 \\ 0 & 0 & 0 & 0 & 2 \end{pmatrix}$ **11.** $\begin{pmatrix} 1 & 0 & 0 & 0 & 0 \\ 0 & 1 & 2 & 0 & 0 \\ 0 & 0 & 1 & 2 & 0 \\ 0 & 0 & 0 & 1 & 0 \\ 0 & 0 & 0 & 0 & 1 \end{pmatrix}$

12. $\begin{pmatrix} 2 & 0 & 0 & 0 & 0 \\ 0 & 3 & 1 & 0 & 0 \\ 0 & 0 & 3 & 0 & 0 \\ 0 & 0 & 0 & 5 & 1 \\ 0 & 0 & 0 & 0 & 5 \end{pmatrix}$ **13.** $\begin{pmatrix} a & 0 & 0 & 0 & 0 \\ 0 & b & 0 & 0 & 0 \\ 0 & 0 & c & 0 & 0 \\ 0 & 0 & 0 & d & 0 \\ 0 & 0 & 0 & 0 & e \end{pmatrix}$ **14.** $\begin{pmatrix} a & 1 & 0 & 0 & 0 \\ 0 & a & 0 & 0 & 0 \\ 0 & 0 & c & 1 & 0 \\ 0 & 0 & 0 & c & 1 \\ 0 & 0 & 0 & 0 & c \end{pmatrix}$

In Problems 15–18 find an invertible matrix C that transforms the 2×2 matrix to its Jordan canonical form.

15. $\begin{pmatrix} 6 & 1 \\ 0 & 6 \end{pmatrix}$ **16.** $\begin{pmatrix} -12 & 7 \\ -7 & 2 \end{pmatrix}$ **17.** $\begin{pmatrix} -10 & -7 \\ 7 & 4 \end{pmatrix}$ **18.** $\begin{pmatrix} 4 & -1 \\ 1 & 2 \end{pmatrix}$

*19. Let A be a 3×3 matrix. Assume that λ is an eigenvalue of A with algebraic multiplicity 3 and geometric multiplicity 1 and let v_1 be the corresponding eigenvector.
   a. Show that there is a solution, v_2, to the system $(A - \lambda I)v_2 = v_1$ such that v_1 and v_2 are linearly independent.
   b. With v_2 defined by part (a), show that there is a solution, v_3, to the system $(A - \lambda I)v_3 = v_2$ such that v_1, v_2, and v_3 are linearly independent.
   c. Show that if C is a matrix whose columns are v_1, v_2, and v_3, then

$$C^{-1}AC = \begin{pmatrix} \lambda & 1 & 0 \\ 0 & \lambda & 1 \\ 0 & 0 & \lambda \end{pmatrix}.$$

20. Apply the procedure described in Problem 19 to reduce the matrix $A = \begin{pmatrix} -2 & 1 & 0 \\ -2 & 1 & -1 \\ -1 & 1 & -2 \end{pmatrix}$
by a similarity transformation to its Jordan canonical form.

21. Do the same for $A = \begin{pmatrix} -1 & -2 & -1 \\ -1 & -1 & -1 \\ 2 & 3 & 2 \end{pmatrix}$.

22. Do the same for $A = \begin{pmatrix} -1 & -18 & -7 \\ 1 & -13 & -4 \\ -1 & 25 & 8 \end{pmatrix}$.

23. An $n \times n$ matrix A is **nilpotent** if there is an integer k such that $A^k = 0$. If k is the smallest such integer, then k is called the **index of nilpotency** of A. Prove that if k is the index of nilpotency of A and if $m \geq k$, then $A^m = 0$.

*24. Let N_k be the matrix defined by equation (1). Prove that N_k is nilpotent with index of nilpotency k.

25. Write down all possible 4×4 Jordan matrices.

In Problems 26–31 the characteristic polynomial of a matrix A is given. Write the possible Jordan canonical forms for A.

26. $(\lambda + 1)^2(\lambda - 2)^2$ **27.** $(\lambda - 3)^3(\lambda + 4)$ **28.** $(\lambda - 3)^4$

29. $(\lambda - 4)^3(\lambda + 3)^2$ **30.** $(\lambda - 6)(\lambda + 7)^4$ **31.** $(\lambda + 7)^5$

32. Using the Jordan canonical form, show that for any $n \times n$ matrix A, $\det A = \lambda_1\lambda_2\cdots\lambda_n$, where $\lambda_1, \lambda_2, \ldots, \lambda_n$ are the eigenvalues of A.

6.7 AN IMPORTANT APPLICATION: MATRIX DIFFERENTIAL EQUATIONS

Let $x = f(t)$ represent some physical quantity such as the volume of a substance, the population of a certain species, the mass of a decaying radioactive substance, or the number of dollars invested in bonds. Then the rate of growth of $f(t)$ is given by its derivative $f'(t) = dx/dt$. If $f(t)$ is growing at a constant rate, then $dx/dt = k$ and $x = kt + C$; that is, $x = f(t)$ is a straight-line function.

It is often more interesting and more appropriate to consider the **relative rate of growth** defined by

$$\text{Relative rate of growth} = \frac{\text{actual size of growth}}{\text{size of } f(t)} = \frac{f'(t)}{f(t)} = \frac{x'(t)}{x(t)} \qquad (1)$$

If the relative rate of growth is constant, then we have

$$\frac{x'(t)}{x(t)} = a \qquad (2)$$

or

$$x'(t) = ax(t) \qquad (3)$$

Differential equation

Equation (3) is called a **differential equation** because it is an equation involving a derivative. It is not difficult to prove that the only solutions to (3) are of the form

$$x(t) = ce^{at} \qquad (4)$$

Initial value

where c is an arbitrary constant. If, however, $x(t)$ represents some physical quantity, then it is the usual practice to specify an **initial value** $x_0 = x(0)$ of the quantity. Then, substituting $t = 0$ in (4), we have $x_0 = x(0) = ce^{a \cdot 0} = c$ or

$$x(t) = x_0 e^{at} \qquad (5)$$

The function $x(t)$ given by (5) is the unique solution to (3) satisfying the initial condition $x(0) = x_0$.

Equation (3) arises in a number of interesting applications. Some of these are undoubtedly given in your calculus text—in the chapter introducing the exponential function. In this section we consider a generalization of equation (3).

In the model discussed above we seek one unknown function. It often occurs that there are several functions linked by several differential equations. Examples are given later in the section. Consider the following system of n differential equations in n unknown functions:

$$
\begin{aligned}
x_1'(t) &= a_{11}x_1(t) + a_{12}x_2(t) + \cdots + a_{1n}x_n(t) \\
x_2'(t) &= a_{21}x_1(t) + a_{22}x_2(t) + \cdots + a_{2n}x_n(t) \\
&\vdots \qquad\quad \vdots \qquad\quad \vdots \qquad\qquad \vdots \\
x_n'(t) &= a_{n1}x_1(t) + a_{n2}x_2(t) + \cdots + a_{nn}x_n(t)
\end{aligned}
\qquad (6)
$$

where the a_{ij}'s are real numbers. System (6) is called an $n \times n$ **first-order system of linear differential equations.** The term "first order" means that only first derivatives occur in the system.

Now, let

$$\mathbf{x}(t) = \begin{pmatrix} x_1(t) \\ x_2(t) \\ \vdots \\ x_n(t) \end{pmatrix}$$

Vector function   Here $\mathbf{x}(t)$ is called a **vector function.** We define

$$\mathbf{x}'(t) = \begin{pmatrix} x_1'(t) \\ x_2'(t) \\ \vdots \\ x_n'(t) \end{pmatrix}$$

Then, if we define the $n \times n$ matrix

$$A = \begin{pmatrix} a_{11} & a_{12} & \cdots & a_{1n} \\ a_{21} & a_{22} & \cdots & a_{2n} \\ \vdots & \vdots & & \vdots \\ a_{n1} & a_{n2} & \cdots & a_{nn} \end{pmatrix}$$

system (6) can be written as

$$\boxed{\mathbf{x}'(t) = A\mathbf{x}(t)} \tag{7}$$

Note that equation (7) is almost identical to equation (3). The only difference is that now we have a vector function and a matrix whereas before we had a "scalar" function and a number (1×1 matrix).

To solve equation (7), we might guess that a solution would have the form e^{At}. But what does e^{At} mean? We shall answer that question in a moment. First, let us recall the series expansion of the function e^t:

$$e^t = 1 + t + \frac{t^2}{2!} + \frac{t^3}{3!} + \frac{t^4}{4!} + \cdots \tag{8}$$

This series converges for every real number t. Then, for any real number a,

$$e^{at} = 1 + at + \frac{(at)^2}{2!} + \frac{(at)^3}{3!} + \frac{(at)^4}{4!} + \cdots \tag{9}$$

DEFINITION 1   **The Matrix e^A** Let A be an $n \times n$ matrix with real (or complex) entries. Then e^A is an $n \times n$ matrix defined by

$$e^A = I + A + \frac{A^2}{2!} + \frac{A^3}{3!} + \frac{A^4}{4!} + \cdots = \sum_{k=0}^{\infty} \frac{A^k}{k!} \qquad (10)$$

Norm of a matrix

Remark. It is not difficult to prove that the series of matrices in equation (10) converges for every matrix A, but to do so would take us too far afield. We can, however, give an indication of why it is so. We first define $|A|_i$ to be the sum of the absolute values of the components in the ith row of A. We then define the **norm**† of A, denoted by $|A|$, by

$$|A| = \max_{1 \le i \le n} |A|_i \qquad (11)$$

It can be shown that

$$|AB| \le |A|\, |B| \qquad (12)$$

and

$$|A + B| \le |A| + |B| \qquad (13)$$

Then, using (12) and (13) in (10), we obtain

$$|e^A| \le 1 + |A| + \frac{|A|^2}{2!} + \frac{|A|^3}{3!} + \frac{|A|^4}{4!} + \cdots = e^{|A|}$$

Since $|A|$ is a real number, $e^{|A|}$ is finite. This shows that the series in (10) converges for any matrix A.

We shall now see the usefulness of the series in equation (10).

THEOREM 1   For any constant vector $\mathbf{c}$, $\mathbf{x}(t) = e^{At}\mathbf{c}$ is a solution of (7). Moreover, the solution of (7) given by $\mathbf{x}(t) = e^{At}\mathbf{x}_0$ satisfies $\mathbf{x}(0) = \mathbf{x}_0$.

Proof   We compute, using (10):

$$\mathbf{x}(t) = e^{At}\mathbf{c} = \left[I + At + A^2\frac{t^2}{2!} + A^3\frac{t^3}{3!} + \cdots \right]\mathbf{c} \qquad (14)$$

But since A is a constant matrix, we have

$$\frac{d}{dt} A^k \frac{t^k}{k!} = \frac{d}{dt} \frac{t^k}{k!} A^k = \frac{kt^{k-1}}{k!} A^k$$

$$= \frac{A^k t^{k-1}}{(k-1)!} = A\left[A^{k-1} \frac{t^{k-1}}{(k-1)!} \right] \qquad (15)$$

---

†This is called the **max-row sum norm** of A.

Then, combining (14) and (15), we obtain (since $\mathbf{c}$ is a constant vector)

$$\mathbf{x}'(t) = \frac{d}{dt}\, e^{At}\mathbf{c} = A\left[I + At + A^2\frac{t^2}{2!} + A^3\frac{t^3}{3!} + \cdots\right]\mathbf{c} = Ae^{At}\mathbf{c} = A\mathbf{x}(t)$$

Finally, since $e^{A\cdot 0} = e^0 = I$, we have

$$\mathbf{x}(0) = e^{A\cdot 0}\mathbf{x}_0 = I\mathbf{x}_0 = \mathbf{x}_0.$$ ∎

DEFINITION 2   **Principal Matrix Solution** The matrix e^{At} is called the **principal matrix solution** of the system $\mathbf{x}' = A\mathbf{x}$.

A major (and obvious) problem remains: How do we compute e^{At} in a practical way? We begin with two examples.

EXAMPLE 1   **Computing e^{At} When A Is a Diagonal Matrix**   Let $A = \begin{pmatrix} 1 & 0 & 0 \\ 0 & 2 & 0 \\ 0 & 0 & 3 \end{pmatrix}$. Then

$$A^2 = \begin{pmatrix} 1 & 0 & 0 \\ 0 & 2^2 & 0 \\ 0 & 0 & 3^2 \end{pmatrix},\ A^3 = \begin{pmatrix} 1 & 0 & 0 \\ 0 & 2^3 & 0 \\ 0 & 0 & 3^3 \end{pmatrix},\ \ldots,\ A^m = \begin{pmatrix} 1 & 0 & 0 \\ 0 & 2^m & 0 \\ 0 & 0 & 3^m \end{pmatrix}$$

and

$$e^{At} = I + At + \frac{A^2 t^2}{2!} + \frac{A^3 t^3}{3!} + \cdots = \begin{pmatrix} 1 & 0 & 0 \\ 0 & 1 & 0 \\ 0 & 0 & 1 \end{pmatrix} + \begin{pmatrix} t & 0 & 0 \\ 0 & 2t & 0 \\ 0 & 0 & 3t \end{pmatrix}$$

$$+ \begin{pmatrix} \dfrac{t^2}{2!} & 0 & 0 \\ 0 & \dfrac{2^2 t^2}{2!} & 0 \\ 0 & 0 & \dfrac{3^2 t^2}{2!} \end{pmatrix} + \begin{pmatrix} \dfrac{t^3}{3!} & 0 & 0 \\ 0 & \dfrac{2^3 t^3}{3!} & 0 \\ 0 & 0 & \dfrac{3^3 t^3}{3!} \end{pmatrix} + \cdots$$

$$= \begin{pmatrix} 1 + t + \dfrac{t^2}{2!} + \dfrac{t^3}{3!} + \cdots & 0 & 0 \\[2ex] 0 & 1 + (2t) + \dfrac{(2t)^2}{2!} + \dfrac{(2t)^3}{3!} + \cdots & 0 \\[2ex] 0 & 0 & 1 + (3t) + \dfrac{(3t)^2}{2!} + \dfrac{(3t)^3}{3!} + \cdots \end{pmatrix}$$

$$= \begin{pmatrix} e^t & 0 & 0 \\ 0 & e^{2t} & 0 \\ 0 & 0 & e^{3t} \end{pmatrix}$$ ∎

EXAMPLE 2   **Computing e^{At} When A is a 2 × 2 Matrix That Is Not Diagonalizable**   Let $A = \begin{pmatrix} a & 1 \\ 0 & a \end{pmatrix}$. Then, as is easily verified,

$$A^2 = \begin{pmatrix} a^2 & 2a \\ 0 & a^2 \end{pmatrix}, \; A^3 = \begin{pmatrix} a^3 & 3a^2 \\ 0 & a^3 \end{pmatrix}, \; \ldots, \; A^m = \begin{pmatrix} a^m & ma^{m-1} \\ 0 & a^m \end{pmatrix}, \ldots$$

so that

$$e^{At} = \begin{pmatrix} \displaystyle\sum_{m=0}^{\infty} \frac{(at)^m}{m!} & \displaystyle\sum_{m=1}^{\infty} \frac{ma^{m-1}t^m}{m!} \\ 0 & \displaystyle\sum_{m=0}^{\infty} \frac{(at)^m}{m!} \end{pmatrix}$$

Now

$$\sum_{m=1}^{\infty} \frac{ma^{m-1}t^m}{m!} = \sum_{m=1}^{\infty} \frac{a^{m-1}t^m}{(m-1)!} = t + at^2 + \frac{a^2t^3}{2!} + \frac{a^3t^4}{3!} + \cdots$$

$$= t\left(1 + at + \frac{a^2t^2}{2!} + \frac{a^3t^3}{3!} + \cdots\right) = te^{at}$$

Thus

$$e^{At} = \begin{pmatrix} e^{at} & te^{at} \\ 0 & e^{at} \end{pmatrix}$$ ∎

As Example 1 illustrates, it is easy to calculate e^{At} if A is a diagonal matrix. Example 1 shows that if $D = \text{diag}\,(\lambda_1, \lambda_2, \ldots, \lambda_n)$, then

$$e^{Dt} = \text{diag}\,(e^{\lambda_1 t}, e^{\lambda_2 t}, \ldots, e^{\lambda_n t}).$$

In Example 2 we calculated e^{At} for a matrix A in Jordan canonical form. It turns out that this is really all we need to be able to do, as the next theorem suggests.

THEOREM 2   Let J be the Jordan canonical form of a matrix A and let $J = C^{-1}AC$. Then $A = CJC^{-1}$ and

$$\boxed{e^{At} = Ce^{Jt}C^{-1}}$$ (16)

Proof   We first note that

$$A^n = (CJC^{-1})^n = \overbrace{(CJC^{-1})(CJC^{-1})\cdots(CJC^{-1})}^{n \text{ times}}$$
$$= CJ(C^{-1}C)J(C^{-1}C)J(C^{-1}C)\cdots(C^{-1}C)JC^{-1}$$
$$= CJ^nC^{-1}$$

It then follows that

$$(At)^n = C(Jt)^nC^{-1}$$ (17)

Thus

$$e^{At} = I + (At) + \frac{(At)^2}{2!} + \cdots = CIC^{-1} + C(Jt)C^{-1} + C\frac{(Jt)^2}{2!}C^{-1} + \cdots$$

$$= C\left[I + (Jt) + \frac{(Jt)^2}{2!} + \cdots\right]C^{-1} = Ce^{Jt}C^{-1} \qquad \blacksquare$$

Theorem 2 tells us that to calculate e^{At} we really need only to calculate e^{Jt}. When J is a diagonal (as is most often the case), then we know how to calculate e^{Jt}. If A is a 2×2 matrix that is not diagonalizable, then $J = \begin{pmatrix} \lambda & 1 \\ 0 & \lambda \end{pmatrix}$ and $e^{Jt} = \begin{pmatrix} e^{\lambda t} & te^{\lambda t} \\ 0 & e^{\lambda t} \end{pmatrix}$ as we calculated in Example 2. In fact, it is not difficult to calculate e^{Jt} where J is any Jordan matrix. It is first necessary to compute e^{Bt} for a Jordan block matrix B. A method for doing this is given in Problems 20–22.

We now apply our computations to a simple biological model of population growth. Suppose that in an ecosystem there are two interacting species S_1 and S_2. We denote the populations of the species at time t by $x_1(t)$ and $x_2(t)$. One system governing the relative growth of the two species is

$$\begin{aligned} x_1'(t) &= ax_1(t) + bx_2(t) \\ x_2'(t) &= cx_1(t) + dx_2(t) \end{aligned} \qquad (18)$$

We can interpret the constants a, b, c, and d as follows: If the species are competing, then it is reasonable to have $b < 0$ and $c < 0$. This is true because increases in the population of one species will slow the growth of the other. A second model is a *predator-prey* relationship. If S_1 is the prey and S_2 is the predator (S_2 eats S_1), then it is reasonable to have $b < 0$ and $c > 0$ since an increase in the predator species will cause a decrease in the prey species, while an increase in the prey species will cause an increase in the predator species (since it will have more food). Finally, in a *symbiotic* relationship (each species lives off the other), we would likely have $b > 0$ and $c > 0$. Of course, the constants a, b, c, and d depend on a wide variety of factors including available food, time of year, climate, limits due to overcrowding, other competing species, and so on. We shall analyze four different models by using the material in this section. We assume that t is measured in years.

EXAMPLE 3   **A Competitive Model**   Consider the system

$$\begin{aligned} x_1'(t) &= 3x_1(t) - x_2(t) \\ x_2'(t) &= -2x_1(t) + 2x_2(t) \end{aligned}$$

Here an increase in the population of one species causes a decline in the growth rate of another. Suppose that the initial populations are $x_1(0) = 90$ and $x_2(0) = 150$. Find the populations of both species for $t > 0$.

Solution We have $A = \begin{pmatrix} 3 & -1 \\ -2 & 2 \end{pmatrix}$. The eigenvalues of A are $\lambda_1 = 1$ and $\lambda_2 = 4$ with corresponding eigenvectors $\mathbf{v}_1 = \begin{pmatrix} 1 \\ 2 \end{pmatrix}$ and $\mathbf{v}_2 = \begin{pmatrix} 1 \\ -1 \end{pmatrix}$. Then

$$C = \begin{pmatrix} 1 & 1 \\ 2 & -1 \end{pmatrix} \quad C^{-1} = -\frac{1}{3}\begin{pmatrix} -1 & -1 \\ -2 & 1 \end{pmatrix} \quad J = D = \begin{pmatrix} 1 & 0 \\ 0 & 4 \end{pmatrix} \quad e^{Jt} = \begin{pmatrix} e^t & 0 \\ 0 & e^{4t} \end{pmatrix}$$

$$e^{At} = Ce^{Jt}C^{-1} = -\frac{1}{3}\begin{pmatrix} 1 & 1 \\ 2 & -1 \end{pmatrix}\begin{pmatrix} e^t & 0 \\ 0 & e^{4t} \end{pmatrix}\begin{pmatrix} -1 & -1 \\ -2 & 1 \end{pmatrix}$$

$$= -\frac{1}{3}\begin{pmatrix} 1 & 1 \\ 2 & -1 \end{pmatrix}\begin{pmatrix} -e^t & -e^t \\ -2e^{4t} & e^{4t} \end{pmatrix}$$

$$= -\frac{1}{3}\begin{pmatrix} -e^t - 2e^{4t} & -e^t + e^{4t} \\ -2e^t + 2e^{4t} & -2e^t - e^{4t} \end{pmatrix}$$

Finally, the solution to the system is given by

$$\mathbf{x}(t) = \begin{pmatrix} x_1(t) \\ x_2(t) \end{pmatrix} = e^{At}\mathbf{x}_0 = -\frac{1}{3}\begin{pmatrix} -e^t - 2e^{4t} & -e^t + e^{4t} \\ -2e^t + 2e^{4t} & -2e^t - e^{4t} \end{pmatrix}\begin{pmatrix} 90 \\ 150 \end{pmatrix}$$

$$= -\frac{1}{3}\begin{pmatrix} -240e^t - 30e^{4t} \\ -480e^t + 30e^{4t} \end{pmatrix} = \begin{pmatrix} 80e^t + 10e^{4t} \\ 160e^t - 10e^{4t} \end{pmatrix}$$

For example, after 6 months ($t = \frac{1}{2}$ year), $x_1(t) = 80e^{1/2} + 10e^2 \approx 206$ individuals, whereas $x_2(t) = 160e^{1/2} - 10e^2 \approx 190$ individuals. More significantly, $160e^t - 10e^{4t} = 0$ when $16e^t = e^{4t}$ or $16 = e^{3t}$ or $3t = \ln 16$ and $t = (\ln 16)/3 \approx 2.77/3 \approx 0.92$ years ≈ 11 months. Thus the second species will be eliminated after only 11 months even though it started with a larger population. In Problems 10 and 11 you are asked to show that neither population will be eliminated if $x_2(0) = 2x_1(0)$ and that the first population will be eliminated if $x_2(0) > 2x_1(0)$. Thus, as was well known to Darwin, survival in this very simple model depends on the relative sizes of the competing species when competition begins. ∎

EXAMPLE 4 **A Predator-Prey Model** We consider the following system in which species 1 is the prey and species 2 is the predator:

$$x_1'(t) = 2x_1(t) - x_2(t)$$
$$x_2'(t) = x_1(t) + 4x_2(t)$$

Find the populations of the two species for $t > 0$ if the initial populations are $x_1(0) = 500$ and $x_2(0) = 100$.

Solution Here $A = \begin{pmatrix} 2 & -1 \\ 1 & 4 \end{pmatrix}$ and the only eigenvalue is $\lambda = 3$ with the single eigenvector

$\begin{pmatrix} 1 \\ -1 \end{pmatrix}$. One solution to the equation $(A - 3I)\mathbf{v}_2 = \mathbf{v}_1$ (see Theorem 6.6.2 on page 434) is $\mathbf{v}_2 = \begin{pmatrix} 1 \\ -2 \end{pmatrix}$. Then

$$C = \begin{pmatrix} 1 & 1 \\ -1 & -2 \end{pmatrix} \qquad C^{-1} = \begin{pmatrix} 2 & 1 \\ -1 & -1 \end{pmatrix} \qquad J = \begin{pmatrix} 3 & 1 \\ 0 & 3 \end{pmatrix}$$

$$e^{Jt} = \begin{pmatrix} e^{3t} & te^{3t} \\ 0 & e^{3t} \end{pmatrix} = e^{3t}\begin{pmatrix} 1 & t \\ 0 & 1 \end{pmatrix} \qquad \text{(from Example 2)}$$

and

$$e^{At} = Ce^{Jt}C^{-1} = e^{3t}\begin{pmatrix} 1 & 1 \\ -1 & -2 \end{pmatrix}\begin{pmatrix} 1 & t \\ 0 & 1 \end{pmatrix}\begin{pmatrix} 2 & 1 \\ -1 & -1 \end{pmatrix}$$

$$= e^{3t}\begin{pmatrix} 1 & 1 \\ -1 & -2 \end{pmatrix}\begin{pmatrix} 2-t & 1-t \\ -1 & -1 \end{pmatrix}$$

$$= e^{3t}\begin{pmatrix} 1-t & -t \\ t & 1+t \end{pmatrix}$$

Thus the solution to the system is

$$\mathbf{x}(t) = \begin{pmatrix} x_1(t) \\ x_2(t) \end{pmatrix} = e^{At}\mathbf{x}_0 = e^{3t}\begin{pmatrix} 1-t & -t \\ t & 1+t \end{pmatrix}\begin{pmatrix} 500 \\ 100 \end{pmatrix} = e^{3t}\begin{pmatrix} 500 - 600t \\ 100 + 600t \end{pmatrix}$$

It is apparent that the prey species will be eliminated after $\frac{5}{6}$ year $= 10$ months—even though it started with a population five times as great as the predator species. In fact, it is easy to show (see Problem 12) that no matter how great the initial advantage of the prey species, the prey species will be eliminated in less than 1 year.   ∎

EXAMPLE 5 **Another Predator-Prey Model**   Consider the predator-prey model governed by the system

$$x_1'(t) = x_1(t) + x_2(t)$$
$$x_2'(t) = -x_1(t) + x_2(t)$$

If the initial populations are $x_1(0) = x_2(0) = 1000$, determining the populations of the two species for $t > 0$.

Solution   Here $A = \begin{pmatrix} 1 & 1 \\ -1 & 1 \end{pmatrix}$ with characteristic equation $\lambda^2 - 2\lambda + 2 = 0$, complex roots $\lambda_1 = 1 + i$ and $\lambda_2 = 1 - i$, and eigenvectors $\mathbf{v}_1 = \begin{pmatrix} 1 \\ i \end{pmatrix}$ and $\mathbf{v}_2 = \begin{pmatrix} 1 \\ -i \end{pmatrix}$.† Then

---

† Note that $\lambda_2 = \overline{\lambda_1}$ and $\mathbf{v}_2 = \overline{\mathbf{v}_1}$. This should be no surprise, because, according to the result of Problem 6.1.33 on page 399, eigenvalues of real matrices occur in complex conjugate pairs and their corresponding eigenvectors are complex conjugates.

$$C = \begin{pmatrix} 1 & 1 \\ i & -i \end{pmatrix}, \quad C^{-1} = -\frac{1}{2i}\begin{pmatrix} -i & -1 \\ -i & 1 \end{pmatrix} = \frac{1}{2}\begin{pmatrix} 1 & -i \\ 1 & i \end{pmatrix},$$

$$J = D = \begin{pmatrix} 1+i & 0 \\ 0 & 1-i \end{pmatrix}$$

and

$$e^{Jt} = \begin{pmatrix} e^{(1+i)t} & 0 \\ 0 & e^{(1-i)t} \end{pmatrix}$$

Now, by Euler's formula (see Appendix 2), $e^{it} = \cos t + i \sin t$. Thus

$$e^{(1+i)t} = e^t e^{it} = e^t(\cos t + i \sin t).$$

Similarly,

$$e^{(1-i)t} = e^t e^{-it} = e^t(\cos t - i \sin t).$$

Thus

$$e^{Jt} = e^t\begin{pmatrix} \cos t + i \sin t & 0 \\ 0 & \cos t - i \sin t \end{pmatrix}$$

and

$$e^{At} = Ce^{Jt}C^{-1} = \frac{e^t}{2}\begin{pmatrix} 1 & 1 \\ i & -i \end{pmatrix}\begin{pmatrix} \cos t + i \sin t & 0 \\ 0 & \cos t - i \sin t \end{pmatrix}\begin{pmatrix} 1 & -i \\ 1 & i \end{pmatrix}$$

$$= \frac{e^t}{2}\begin{pmatrix} 1 & 1 \\ i & -i \end{pmatrix}\begin{pmatrix} \cos t + i \sin t & -i \cos t + \sin t \\ \cos t - i \sin t & i \cos t + \sin t \end{pmatrix}$$

$$= \frac{e^t}{2}\begin{pmatrix} 2\cos t & 2\sin t \\ -2\sin t & 2\cos t \end{pmatrix} = e^t\begin{pmatrix} \cos t & \sin t \\ -\sin t & \cos t \end{pmatrix}$$

Finally,

$$\mathbf{x}(t) = e^{At}\mathbf{x}(0) = e^t\begin{pmatrix} \cos t & \sin t \\ -\sin t & \cos t \end{pmatrix}\begin{pmatrix} 1000 \\ 1000 \end{pmatrix} = \begin{pmatrix} 1000e^t(\cos t + \sin t) \\ 1000e^t(\cos t - \sin t) \end{pmatrix}$$

The prey species is eliminated when $1000e^t(\cos t - \sin t) = 0$ or when $\sin t = \cos t$. The first positive solution of this last equation is $t = \pi/4 \approx 0.7854$ year $\approx$ 9.4 months.   ∎

EXAMPLE 6   **A Model of Species Cooperation (Symbiosis)**   Consider the symbiotic model governed by the system

$$x_1'(t) = -\tfrac{1}{2}x_1(t) + x_2(t)$$
$$x_2'(t) = \tfrac{1}{4}x_1(t) - \tfrac{1}{2}x_2(t)$$

Note that in this model the population of each species increases proportionally to the population of the other and decreases proportionally to its own population. Suppose that $x_1(0) = 200$ and $x_2(0) = 500$. Determine the population of each species for $t > 0$.

Solution   Here $A = \begin{pmatrix} -\frac{1}{2} & 1 \\ \frac{1}{4} & -\frac{1}{2} \end{pmatrix}$ with eigenvalues $\lambda_1 = 0$ and $\lambda_2 = -1$ and corresponding

eigenvectors $\mathbf{v}_1 = \begin{pmatrix} 2 \\ 1 \end{pmatrix}$ and $\mathbf{v}_2 = \begin{pmatrix} 2 \\ -1 \end{pmatrix}$. Then

$$C = \begin{pmatrix} 2 & 2 \\ 1 & -1 \end{pmatrix}, \qquad C^{-1} = -\frac{1}{4}\begin{pmatrix} -1 & -2 \\ -1 & 2 \end{pmatrix}, \qquad J = D = \begin{pmatrix} 0 & 0 \\ 0 & -1 \end{pmatrix}$$

and

$$e^{Jt} = \begin{pmatrix} e^{0t} & 0 \\ 0 & e^{-t} \end{pmatrix} = \begin{pmatrix} 1 & 0 \\ 0 & e^{-t} \end{pmatrix}.$$

Thus

$$e^{At} = -\frac{1}{4}\begin{pmatrix} 2 & 2 \\ 1 & -1 \end{pmatrix}\begin{pmatrix} 1 & 0 \\ 0 & e^{-t} \end{pmatrix}\begin{pmatrix} -1 & -2 \\ -1 & 2 \end{pmatrix}$$

$$= -\frac{1}{4}\begin{pmatrix} 2 & 2 \\ 1 & -1 \end{pmatrix}\begin{pmatrix} -1 & -2 \\ -e^{-t} & 2e^{-t} \end{pmatrix}$$

$$= -\frac{1}{4}\begin{pmatrix} -2 - 2e^{-t} & -4 + 4e^{-t} \\ -1 + e^{-t} & -2 - 2e^{-t} \end{pmatrix}$$

and

$$\mathbf{x}(t) = e^{At}\mathbf{x}(0) = -\frac{1}{4}\begin{pmatrix} -2 - 2e^{-t} & -4 + 4e^{-t} \\ -1 + e^{-t} & -2 - 2e^{-t} \end{pmatrix}\begin{pmatrix} 200 \\ 500 \end{pmatrix}$$

$$= -\frac{1}{4}\begin{pmatrix} -2400 + 1600e^{-t} \\ -1200 - 800e^{-t} \end{pmatrix}$$

$$= \begin{pmatrix} 600 - 400e^{-t} \\ 300 + 200e^{-t} \end{pmatrix}$$

Note that $e^{-t} \to 0$ as $t \to \infty$. This means that as time goes on, the two cooperating species approach the **equilibrium** populations 600 and 300, respectively. Neither population is eliminated. ∎

PROBLEMS 6.7

In Problems 1–9 find the principal matrix solution e^{At} of the system $\mathbf{x}'(t) = A\mathbf{x}(t)$.

1. $A = \begin{pmatrix} -2 & -2 \\ -5 & 1 \end{pmatrix}$ **2.** $A = \begin{pmatrix} 3 & -1 \\ -2 & 4 \end{pmatrix}$ **3.** $A = \begin{pmatrix} 2 & -1 \\ 5 & -2 \end{pmatrix}$

4. $A = \begin{pmatrix} 3 & -5 \\ 1 & -1 \end{pmatrix}$ **5.** $A = \begin{pmatrix} -10 & -7 \\ 7 & 4 \end{pmatrix}$ **6.** $A = \begin{pmatrix} -2 & 1 \\ 5 & 2 \end{pmatrix}$

7. $A = \begin{pmatrix} -12 & 7 \\ -7 & 2 \end{pmatrix}$ **8.** $A = \begin{pmatrix} 1 & 1 & -2 \\ -1 & 2 & 1 \\ 0 & 1 & -1 \end{pmatrix}$ **9.** $A = \begin{pmatrix} 4 & 6 & 6 \\ 1 & 3 & 2 \\ -1 & -5 & -2 \end{pmatrix}$

10. In Example 3 show that if the initial vector $\mathbf{x}(0) = \begin{pmatrix} a \\ 2a \end{pmatrix}$, where a is a constant, then both populations grow at a rate proportional to e^t.

11. In Example 3 show that if $x_2(0) > 2x_1(0)$, then the first population will be eliminated.

12. In Example 4 show that the first population will become extinct in α years, where $\alpha = x_1(0)/[x_1(0) + x_2(0)]$.

*13. In a water desalinization plant there are two tanks of water. Suppose that tank 1 contains 1000 liters of brine in which 1000 kg of salt is dissolved and tank 2 contains 1000 liters of pure water. Suppose that water flows into tank 1 at the rate of 20 liters per minute and the mixture flows from tank 1 into tank 2 at a rate of 30 liters per minute. From tank 2, 10 liters is pumped back to tank 1 (establishing *feedback*) while 20 liters is flushed away. Find the amount of salt in both tanks at all times t. [*Hint:* Write the information as a 2×2 system and let $x_1(t)$ and $x_2(t)$ denote the amount of salt in each tank.]

14. A community of n individuals is exposed to an infectious disease.† At any given time t, the community is divided into three groups: group 1 with population $x_1(t)$ is the suscepti- ble group; group 2 with a population of $x_2(t)$ is the group of infected individuals in circulation; and group 3, population $x_3(t)$, consists of those who are isolated, dead, or immune. It is reasonable to assume that initially $x_2(t)$ and $x_3(t)$ will be small compared to $x_1(t)$. Let α and β be positive constants denoting the rates at which susceptibles become infected and infected individuals join group 3, respectively. Then a reasonable model for the spread of the disease is given by the system

$$
\begin{aligned}
x_1'(t) &= -\alpha x_1(0)x_2 \\
x_2'(t) &= \alpha x_1(0)x_2 - \beta x_2 \\
x_3'(t) &= \beta x_2
\end{aligned}
$$

   a. Write this system in the form $\mathbf{x}' = A\mathbf{x}$ and find the solution in terms of $x_1(0)$, $x_2(0)$, and $x_3(0)$. Note that $x_1(0) + x_2(0) + x_3(0) = n$.
   b. Show that if $\alpha x(0) < \beta$, then the disease will not produce an epidemic.
   c. What will happen if $\alpha x(0) > \beta$?

15. Consider the **second-order differential equation** $x''(t) + ax'(t) + bx(t) = 0$.
   a. Letting $x_1(t) = x(t)$ and $x_2(t) = x'(t)$, write the preceding equation as a first-order system in the form of equation (7), where A is a 2×2 matrix.
   b. Show that the characteristic equation of A is $\lambda^2 + a\lambda + b = 0$.

In Problems 16–19 use the result of Problem 15 to solve the given equation.

16. $x'' + 5x' + 6x = 0$; $x(0) = 1$, $x'(0) = 0$

17. $x'' + 6x' + 9x = 0$; $x(0) = 1$, $x'(0) = 2$

18. $x'' + 4x = 0$; $x(0) = 0$, $x'(0) = 1$

19. $x'' - 3x' - 10x = 0$; $x(0) = 3$, $x'(0) = 2$

20. Let $N_3 = \begin{pmatrix} 0 & 1 & 0 \\ 0 & 0 & 1 \\ 0 & 0 & 0 \end{pmatrix}$. Show that $N_3^3 = 0$, the zero matrix.

---

† For a discussion of this model, see N. Bailey, "The Total Size of a General Stochastic Epidemic," *Biometrika* 40 (1953): 177–185.

21. Show that $e^{N_3 t} = \begin{pmatrix} 1 & t & t^2/2 \\ 0 & 1 & t \\ 0 & 0 & 1 \end{pmatrix}$. [*Hint:* Write down the series for $e^{N_3 t}$ and use the result of Problem 20.]

22. Let $J = \begin{pmatrix} \lambda & 1 & 0 \\ 0 & \lambda & 1 \\ 0 & 0 & \lambda \end{pmatrix}$. Show that $e^{Jt} = e^{\lambda t} \begin{pmatrix} 1 & t & t^2/2 \\ 0 & 1 & t \\ 0 & 0 & 1 \end{pmatrix}$. [*Hint:* $Jt = \lambda It + N_3 t$. Use the fact that $e^{A+B} = e^A e^B$ if $AB = BA$.]

23. Using the result of Problem 22, compute e^{At}, where $A = \begin{pmatrix} -2 & 1 & 0 \\ -2 & 1 & -1 \\ -1 & 1 & -2 \end{pmatrix}$. [*Hint:* See Problem 6.6.20 on page 438.]

24. Compute e^{At}, where $A = \begin{pmatrix} -1 & -18 & -7 \\ 1 & -13 & -4 \\ -1 & 25 & 8 \end{pmatrix}$.

25. Compute e^{Jt}, where $J = \begin{pmatrix} \lambda & 1 & 0 & 0 \\ 0 & \lambda & 1 & 0 \\ 0 & 0 & \lambda & 1 \\ 0 & 0 & 0 & \lambda \end{pmatrix}$.

26. Compute e^{At}, where $A = \begin{pmatrix} 2 & 1 & 0 & 0 \\ 0 & 2 & 0 & 0 \\ 0 & 0 & 3 & 1 \\ 0 & 0 & 0 & 3 \end{pmatrix}$.

27. Compute e^{At}, where $A = \begin{pmatrix} -4 & 1 & 0 & 0 \\ 0 & -4 & 1 & 0 \\ 0 & 0 & -4 & 0 \\ 0 & 0 & 0 & 3 \end{pmatrix}$.

6.8   A DIFFERENT PERSPECTIVE: THE THEOREMS OF CAYLEY-HAMILTON AND GERSHGORIN

There are many interesting results concerning the eigenvalues of a matrix. In this section we discuss two of the more useful ones. The first says that any matrix satisfies its own characteristic equation. The second shows how to locate, crudely, the eigenvalues of any matrix with practically no computation.

Let $p(x) = x^n + a_{n-1}x^{n-1} + \cdots + a_1 x + a_0$ be a polynomial and let A be an $n \times n$ matrix. Then powers of A are defined and we define

$$p(A) = A^n + a_{n-1}A^{n-1} + \cdots + a_1 A + a_0 I \tag{1}$$

EXAMPLE 1   **Evaluating $P(A)$**   Let $A = \begin{pmatrix} -1 & 4 \\ 3 & 7 \end{pmatrix}$ and $p(x) = x^2 - 5x + 3$. Then

$$P(A) = A^2 - 5A + 3I = \begin{pmatrix} 13 & 24 \\ 18 & 61 \end{pmatrix} + \begin{pmatrix} 5 & -20 \\ -15 & -35 \end{pmatrix} + \begin{pmatrix} 3 & 0 \\ 0 & 3 \end{pmatrix} = \begin{pmatrix} 21 & 4 \\ 3 & 29 \end{pmatrix}. \blacksquare$$

Expression (1) is a polynomial with scalar coefficients defined for a matrix variable. We can also define a polynomial with *square matrix* coefficients by

$$Q(\lambda) = B_0 + B_1\lambda + B_2\lambda^2 + \cdots + B_n\lambda^n \tag{2}$$

If A is a matrix of the same size as B, then we define

$$Q(A) = B_0 + B_1A + B_2A^2 + \cdots + B_nA^n \tag{3}$$

We must be careful in (3) since matrices do not commute under multiplication.

THEOREM 1   If $P(\lambda)$ and $Q(\lambda)$ are polynomials in the scalar variable λ with square matrix coefficients and if $P(\lambda) = Q(\lambda)(A - \lambda I)$, then $P(A) = 0$.

Proof   If $Q(\lambda)$ is given by equation (2), then

$$P(\lambda) = (B_0 + B_1\lambda + B_2\lambda^2 + \cdots + B_n\lambda^n)(A - \lambda I)$$
$$= B_0A + B_1A\lambda + B_2A\lambda^2 + \cdots + B_nA\lambda^n$$
$$- B_0\lambda - B_1\lambda^2 - B_2\lambda^3 - \cdots - B_n\lambda^{n+1} \tag{4}$$

Then, substituting A for λ in (4), we obtain

$$P(A) = B_0A + B_1A^2 + B_2A^3 + \cdots + B_nA^{n+1}$$
$$- B_0A - B_1A^2 - B_2A^3 - \cdots - B_nA^{n+1} = 0 \qquad \blacksquare$$

Note. We cannot prove this theorem by substituting $\lambda = A$ to obtain $P(A) = Q(A)(A - A) = 0$. This is because it is possible to find polynomials $P(\lambda)$ and $Q(\lambda)$ with matrix coefficients such that $F(\lambda) = P(\lambda)Q(\lambda)$ but $F(A) \neq P(A)Q(A)$. (See Problem 17.)

We can now state the first main theorem.

THEOREM 2   **The Cayley-Hamilton Theorem**[†]   Every square matrix satisfies its own characteristic equation. That is, if $p(\lambda) = 0$ is the characteristic equation of A, then $p(A) = 0$.

[†] Named after Sir William Rowan Hamilton and Arthur Cayley (1821–1895) (see page 33 and 53). Cayley published the first discussion of this famous theorem in 1858. Independently, Hamilton discovered the result in his work on quaternions.

Proof   We have

$$p(\lambda) = \det(A - \lambda I) = \begin{vmatrix} a_{11} - \lambda & a_{12} & \cdots & a_{1n} \\ a_{21} & a_{22} - \lambda & \cdots & a_{2n} \\ \vdots & \vdots & & \vdots \\ a_{n1} & a_{n2} & \cdots & a_{nn} - \lambda \end{vmatrix}$$

Clearly any cofactor of $(A - \lambda I)$ is a polynomial in λ. Thus the adjoint of $A - \lambda I$ (see Definition 2.4.1, page 142) is an $n \times n$ matrix each of whose components is a polynomial in λ. That is,

$$\text{adj}(A - \lambda I) = \begin{pmatrix} p_{11}(\lambda) & p_{12}(\lambda) & \cdots & p_{1n}(\lambda) \\ p_{21}(\lambda) & p_{22}(\lambda) & \cdots & p_{2n}(\lambda) \\ \vdots & \vdots & & \vdots \\ p_{n1}(\lambda) & p_{22}(\lambda) & \cdots & p_{nn}(\lambda) \end{pmatrix}$$

This means that we can think of $\text{adj}(A - \lambda I)$ as a polynomial, $Q(\lambda)$, in λ with $n \times n$ matrix coefficients. To see this, look at the following:

$$\begin{pmatrix} -\lambda^2 - 2\lambda + 1 & 2\lambda^2 - 7\lambda - 4 \\ 4\lambda^2 + 5\lambda - 2 & -3\lambda^2 - \lambda + 3 \end{pmatrix} = \begin{pmatrix} -1 & 2 \\ 4 & -3 \end{pmatrix}\lambda^2 + \begin{pmatrix} -2 & -7 \\ 5 & -1 \end{pmatrix}\lambda + \begin{pmatrix} 1 & -4 \\ -2 & 3 \end{pmatrix}$$

Now, from Theorem 2.4.2 on page 143,

$$\det(A - \lambda I)I = [\text{adj}(A - \lambda I)][A - \lambda I] = Q(\lambda)(A - \lambda I) \qquad \textbf{(5)}$$

But $\det(A - \lambda I)I = p(\lambda)I$. If

$$p(\lambda) = \lambda^n + a_{n-1}\lambda^{n-1} + \cdots + a_1\lambda + a_0$$

then we define

$$P(\lambda) = p(\lambda)I = \lambda^n I + a_{n-1}\lambda^{n-1}I + \cdots + a_1\lambda I + a_0 I$$

Thus, from (5), we have $P(\lambda) = Q(\lambda)(A - \lambda I)$. Finally, from Theorem 1, $P(A) = 0$. This completes the proof. ∎

EXAMPLE 2   **Illustration of the Cayley-Hamilton Theorem**   Let $A = \begin{pmatrix} 1 & -1 & 4 \\ 3 & 2 & -1 \\ 2 & 1 & -1 \end{pmatrix}$. In Example 6.1.4 on page 391 we computed the characteristic equation $\lambda^3 - 2\lambda^2 - 5\lambda + 6 = 0$. Now we compute

$$A^2 = \begin{pmatrix} 6 & 1 & 1 \\ 7 & 0 & 11 \\ 3 & -1 & 8 \end{pmatrix}, \; A^3 = \begin{pmatrix} 11 & -3 & 22 \\ 29 & 4 & 17 \\ 16 & 3 & 5 \end{pmatrix}$$

and

$$A^3 - 2A^2 - 5A + 6I = \begin{pmatrix} 11 & -3 & 22 \\ 29 & 4 & 17 \\ 16 & 3 & 5 \end{pmatrix} + \begin{pmatrix} -12 & -2 & -2 \\ -14 & 0 & -22 \\ -6 & 2 & -16 \end{pmatrix}$$

$$+ \begin{pmatrix} -5 & 5 & -20 \\ -15 & -10 & 5 \\ -10 & -5 & 5 \end{pmatrix} + \begin{pmatrix} 6 & 0 & 0 \\ 0 & 6 & 0 \\ 0 & 0 & 6 \end{pmatrix}$$

$$= \begin{pmatrix} 0 & 0 & 0 \\ 0 & 0 & 0 \\ 0 & 0 & 0 \end{pmatrix} \qquad \blacksquare$$

In some situations the Cayley-Hamilton theorem is useful in calculating the inverse of a matrix. If A^{-1} exists and $p(A) = 0$, then $A^{-1}p(A) = 0$. To illustrate, if $p(\lambda) = \lambda^n + a_{n-1}\lambda^{n-1} + \cdots + a_1\lambda + a_0$, then

$$p(A) = A^n + a_{n-1}A^{n-1} + \cdots + a_1A + a_0I = 0$$

and

$$A^{-1}p(A) = A^{n-1} + a_{n-1}A^{n-2} + \cdots + a_2A + a_1I + a_0A^{-1} = 0$$

Thus

$$A^{-1} = \frac{1}{a_0}(-A^{n-1} - a_{n-1}A^{n-2} - \cdots - a_2A - a_1I) \qquad \text{(6)}$$

Note that $a_0 \neq 0$ because $a_0 = \det A$ (why?) and we assumed that A was invertible.

EXAMPLE 3   **Using the Cayley-Hamilton Theorem to Compute A^{-1}**

Let $A = \begin{pmatrix} 1 & -1 & 4 \\ 3 & 2 & -1 \\ 2 & 1 & -1 \end{pmatrix}$. Then $p(\lambda) = \lambda^3 - 2\lambda^2 - 5\lambda + 6$. Here $n = 3$, $a_2 = -2$, $a_1 = -5$, $a_0 = 6$, and

$$A^{-1} = \frac{1}{6}(-A^2 + 2A + 5I)$$

$$= \frac{1}{6}\left[\begin{pmatrix} -6 & -1 & -1 \\ -7 & 0 & -11 \\ -3 & 1 & -8 \end{pmatrix} + \begin{pmatrix} 2 & -2 & 8 \\ 6 & 4 & -2 \\ 4 & 2 & -2 \end{pmatrix} + \begin{pmatrix} 5 & 0 & 0 \\ 0 & 5 & 0 \\ 0 & 0 & 5 \end{pmatrix} \right]$$

$$= \frac{1}{6}\begin{pmatrix} 1 & -3 & 7 \\ -1 & 9 & -13 \\ 1 & 3 & -5 \end{pmatrix}$$

Note that we computed A^{-1} with a single division and with only one calculation of a determinant (in order to find $p(\lambda) = \det (A - \lambda I)$). This method is sometimes very efficient on a computer.   ■

Gershgorin's Circle Theorem

We now turn to the second important result of this section. Let A be an $n \times n$ matrix. We write, as usual,

$$A = \begin{pmatrix} a_{11} & a_{12} & \cdots & a_{1n} \\ a_{21} & a_{22} & \cdots & a_{2n} \\ \vdots & \vdots & & \vdots \\ a_{n1} & a_{n2} & \cdots & a_{nn} \end{pmatrix}$$

Define the number

$$r_1 = |a_{12}| + |a_{13}| + \cdots + |a_{1n}| = \sum_{j=2}^{n} |a_{1j}| \tag{7}$$

Similarly, define

$$r_i = |a_{i1}| + |a_{i2}| + \cdots + |a_{i,i-1}| + |a_{i,i+1}| + \cdots + |a_{i,n}|$$

$$= \sum_{\substack{j=1 \\ j \neq i}}^{n} |a_{ij}| \tag{8}$$

That is, r_i is the sum of the absolute values of the numbers on the ith row of A that are not on the main diagonal of A. Let

$$D_i = \{z \in \mathbb{C}: |z - a_{ii}| \leq r_i\} \tag{9}$$

Here D_i is a disk in the complex plane centered at a_{ii} with radius r_i (see Figure 6.4).

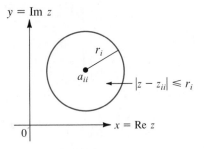

Figure 6.4   A circle of radius r_i centered at a_{ii}

The disk D_i consists of all points in the complex plane on and inside the circle $C_i = \{z \in \mathbb{C} : |z - a_{ii}| = r_i\}$. The circles C_i, $i = 1, 2, \ldots , n$, are called **Gershgorin circles.**

THEOREM 3

Gershgorin's Circle Theorem† Let A be an $n \times n$ matrix and let D_i be defined by equation (9). Then each eigenvalue of A is contained in at least one of the D_i's. That is, if the eigenvalues of A are $\lambda_1, \lambda_2, \ldots , \lambda_k$, then

$$\{\lambda_1, \lambda_2, \ldots , \lambda_k\} \subset \bigcup_{i=1}^{n} D_i \tag{10}$$

Proof

Let λ be an eigenvalue of A with eigenvector $\mathbf{v} = \begin{pmatrix} x_1 \\ x_2 \\ \vdots \\ x_n \end{pmatrix}$. Let $m = \max \{|x_1|, |x_2|,$

$\ldots , |x_n|\}$. Then $(1/m)\mathbf{v} = \begin{pmatrix} y_1 \\ y_2 \\ \vdots \\ y_n \end{pmatrix}$ is an eigenvector of A corresponding to λ and

$\max \{|y_1|, |y_2|, \ldots , |y_n|\} = 1$. Let y_i be the component of $\mathbf{y}$ with $|y_i| = 1$. Now $A\mathbf{y} = \lambda \mathbf{y}$. The ith component of the n-vector $A\mathbf{y}$ is $a_{i1}y_1 + a_{i2}y_2 + \cdots + a_{in}y_n$. The ith component of $\lambda \mathbf{y}$ is λy_i. Thus

$$a_{i1}y_1 + a_{i2}y_2 + \cdots + a_{in}y_n = \lambda y_i,$$

which we write as

$$\sum_{j=1}^{n} a_{ij}y_j = \lambda y_i \tag{11}$$

By subtracting $a_{ii}y_i$ from both sides, equation (11) can be rewritten as

$$\sum_{\substack{j=1 \\ j \neq i}}^{n} a_{ij}y_j = \lambda y_i - a_{ii}y_i = (\lambda - a_{ii})y_i \tag{12}$$

Next, taking the absolute value of both sides of (12) and using the triangle inequality ($|a + b| \leq |a| + |b|$), we obtain

$$|(a_{ii} - \lambda)y_i| = \left| -\sum_{\substack{j=1 \\ j \neq i}}^{n} a_{ij}y_j \right| \leq \sum_{\substack{j=1 \\ j \neq i}}^{n} |a_{ij}| \, |y_j| \tag{13}$$

---

†The Russian mathematician S. Gershgorin published this result in 1931.

We divide both sides of (13) by $|y_i|$ (which is equal to 1) to obtain

$$|a_{ii} - \lambda| \le \sum_{\substack{j=1 \\ j \ne i}}^{n} |a_{ij}| \frac{|y_j|}{|y_i|} \le \sum_{\substack{j=1 \\ j \ne 1}}^{n} |a_{ij}| = r_i \tag{14}$$

The last step followed the fact that $|y_j| \le |y_i|$ (by the way we chose y_i). But this proves the theorem since (14) shows that $\lambda \in D_i$. ∎

EXAMPLE 4   **Using Gershgorin's Theorem**   Let $A = \begin{pmatrix} 1 & -1 & 4 \\ 3 & 2 & -1 \\ 2 & 1 & -1 \end{pmatrix}$. Then $a_{11} = 1$, $a_{22} = 2$, $a_{33} = -1$, $r_1 = |-1| + |4| = 5$, $r_2 = |3| + |-1| = 4$, and $r_3 = |2| + |1| = 3$. Thus the eigenvalues of A lie within the boundaries of the three circles drawn in Figure 6.5. We can verify this since we know by Example 6.1.4 on page 391 that the eigenvalues of A are 1, -2, and 3, which lie within the three circles. Note that the Gershgorin circles can intersect one another.

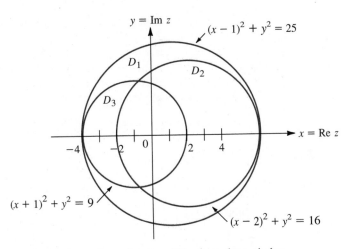

Figure 6.5   All the eigenvalues of A lie within these three circles

EXAMPLE 5   **Using Gershgorin's Theorem**   Find bounds on the eigenvalues of the matrix

$$A = \begin{pmatrix} 3 & 0 & -1 & -\frac{1}{4} & \frac{1}{4} \\ 0 & 5 & \frac{1}{2} & 0 & 1 \\ -\frac{1}{4} & 0 & 6 & \frac{1}{4} & \frac{1}{2} \\ 0 & -1 & \frac{1}{2} & -3 & \frac{1}{4} \\ \frac{1}{6} & -\frac{1}{6} & \frac{1}{3} & \frac{1}{3} & 4 \end{pmatrix}$$

Solution Here $a_{11} = 3$, $a_{22} = 5$, $a_{33} = 6$, $a_{44} = -3$, $a_{55} = 4$, $r_1 = \frac{3}{2}$, $r_2 = \frac{3}{2}$, $r_3 = 1$, $r_4 = \frac{7}{4}$, and $r_5 = 1$. The Gershgorin circles are drawn in Figure 6.6. It is clear from Theorem 3 and Figure 6.6 that if λ is an eigenvalue of A, then $|\lambda| \leq 7$ and Re $\lambda \geq -\frac{19}{4}$. Note the power of the Gershgorin theorem to find the approximate location of eigenvalues after doing very little work.

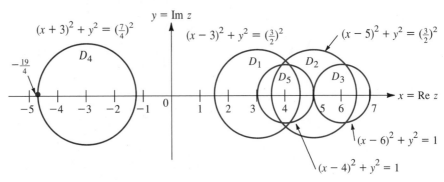

Figure 6.6 All the eigenvalues of A lie within these five circles

PROBLEMS 6.8

In Problems 1–9: **(a)** Find the characteristic equation $p(\lambda) = 0$ of the given matrix; **(b)** verify that $p(A) = 0$; **(c)** use part (*b*) to compute A^{-1}.

1. $\begin{pmatrix} -2 & -2 \\ -5 & 1 \end{pmatrix}$
2. $\begin{pmatrix} 2 & -1 \\ 5 & -2 \end{pmatrix}$
3. $\begin{pmatrix} 1 & -1 & 0 \\ -1 & 2 & -1 \\ 0 & -1 & 1 \end{pmatrix}$

4. $\begin{pmatrix} 1 & 2 & 2 \\ 0 & 2 & 1 \\ -1 & 2 & 2 \end{pmatrix}$
5. $\begin{pmatrix} 0 & 1 & 0 \\ 0 & 0 & 1 \\ 1 & -3 & 3 \end{pmatrix}$
6. $\begin{pmatrix} -3 & -7 & -5 \\ 2 & 4 & 3 \\ 1 & 2 & 2 \end{pmatrix}$

7. $\begin{pmatrix} 2 & -1 & 3 \\ 4 & 1 & 6 \\ 1 & 5 & 3 \end{pmatrix}$
8. $\begin{pmatrix} 1 & 0 & 1 & 0 \\ 2 & -1 & 0 & 2 \\ -1 & 0 & 0 & 1 \\ 4 & 1 & -1 & 0 \end{pmatrix}$
9. $\begin{pmatrix} a & b & 0 & 0 \\ 0 & a & c & 0 \\ 0 & 0 & a & d \\ 0 & 0 & 0 & a \end{pmatrix}$; $bcd \neq 0$

In Problems 10–14 draw the Gershgorin circles for the given matrix A and find a bound for $|\lambda|$ if λ is an eigenvalue of A.

10. $\begin{pmatrix} 2 & 1 & 0 \\ \frac{1}{2} & 5 & \frac{1}{2} \\ 1 & 0 & 6 \end{pmatrix}$
11. $\begin{pmatrix} 3 & -\frac{1}{2} & -\frac{1}{3} & 0 \\ 0 & 6 & 1 & 0 \\ \frac{1}{3} & -\frac{1}{3} & 5 & \frac{1}{3} \\ -\frac{1}{2} & \frac{1}{4} & -\frac{1}{4} & 4 \end{pmatrix}$
12. $\begin{pmatrix} 1 & 3 & -1 & 4 \\ 2 & 5 & 0 & -7 \\ 3 & -1 & 6 & 1 \\ 0 & 2 & 3 & 4 \end{pmatrix}$

13.
$$\begin{pmatrix} -7 & \frac{1}{5} & -\frac{1}{5} & \frac{2}{5} \\ -\frac{1}{10} & -10 & \frac{1}{10} & \frac{3}{10} \\ -\frac{1}{4} & \frac{1}{4} & 5 & \frac{1}{4} \\ 0 & -1 & 0 & 4 \end{pmatrix}$$

14.
$$\begin{pmatrix} 3 & 0 & -\frac{1}{3} & \frac{1}{3} & 0 & \frac{1}{3} \\ \frac{1}{2} & 5 & -\frac{1}{2} & 0 & 1 & 0 \\ \frac{1}{10} & -\frac{1}{5} & 4 & \frac{3}{5} & -\frac{1}{5} & \frac{1}{10} \\ -1 & 0 & 0 & -3 & 0 & 0 \\ \frac{1}{2} & 0 & -\frac{1}{2} & 0 & 2 & \frac{1}{2} \\ -\frac{1}{4} & \frac{1}{4} & \frac{1}{4} & 0 & -\frac{1}{4} & 0 \end{pmatrix}$$

15. Let $A = \begin{pmatrix} 2 & \frac{1}{2} & -\frac{1}{3} & \frac{1}{4} \\ \frac{1}{2} & 3 & \frac{1}{2} & 1 \\ -\frac{1}{3} & \frac{1}{2} & 5 & 2 \\ \frac{1}{4} & 1 & 2 & 4 \end{pmatrix}$. Prove that the eigenvalues of A are positive real numbers.

16. Let $A = \begin{pmatrix} -4 & 1 & 1 & 1 \\ 1 & -6 & 2 & 1 \\ 1 & 2 & -5 & 1 \\ 1 & 1 & 1 & -4 \end{pmatrix}$. Prove that the eigenvalues of A are negative and real.

17. Let $P(\lambda) = B_0 + B_1\lambda$ and $Q(\lambda) = C_0 + C_1\lambda$, where B_0, B_1, C_0, and C_1 are $n \times n$ matrices.

    a. Compute $F(\lambda) = P(\lambda)Q(\lambda)$.

    b. Let A be an $n \times n$ matrix. Show that $F(A) = P(A)Q(A)$ if and only if A commutes with both C_0 and C_1.

18. Let the $n \times n$ matrix A have the eigenvalues $\lambda_1, \lambda_2, \ldots, \lambda_n$ and let $r(A) = \max\limits_{1 \le i \le n} |\lambda_i|$. If $|A|$ is the max-row sum norm defined in Section 6.7, show that $r(A) \le |A|$.

19. The $n \times n$ matrix A is said to be **strictly diagonally dominant** if $|a_{ii}| > r_i$ for $i = 1, 2, \ldots, n$, where r_i is defined by equation (8). Show that if A is a strictly diagonally dominant matrix, then $\det A \neq 0$.

SUMMARY

- **Eigenvalue and Eigenvector**·Let A be an $n \times n$ matrix with real components. The number λ (real or complex) is called an **eigenvalue** of A if there is a *nonzero* vector $\mathbf{v}$ in $\mathbb{C}^n$ such that   (p. 386)

$$A\mathbf{v} = \lambda\mathbf{v}$$

The vector $\mathbf{v} \neq 0$ is called an **eigenvector** of A corresponding to the eigenvalue λ.
- Let A be an $n \times n$ matrix. Then λ is an eigenvalue of A if and only if   (p. 388)

$$p(\lambda) = \det(A - \lambda I) = 0$$

The equation $p(\lambda) = 0$ is called the **characteristic equation** of A; $p(\lambda)$ is called the **characteristic polynomial** of A.
- Counting multiplicities, every $n \times n$ matrix has exactly n eigenvalues.   (p. 388)
- Eigenvectors corresponding to different eigenvalues are linearly independent.   (p. 389)
- **Algebraic Multiplicity**·If $p(\lambda) = (\lambda - \lambda_1)^{r_1}(\lambda - \lambda_2)^{r_2} \cdots (\lambda - \lambda_m)^{r_m}$, then r_i is the **algebraic multiplicity** of λ_i.   (p. 390)
- The eigenvalues of a real matrix occur in complex conjugate pairs.   (p. 394)

- **Eigenspace**·If λ is an eigenvalue of the $n \times n$ matrix A, then $E_\lambda = \{v: Av = \lambda v\}$ is a subspace of $\mathbb{R}^n$ called the **eigenspace** of A corresponding to λ. It is denoted by E_λ. (p. 388)
- **Geometric Multiplicity**·The **geometric multiplicity** of an eigenvalue λ of the matrix A is equal to dim $E_\lambda = \nu(A - \lambda I)$. (p. 396)
- For any eigenvalue λ, geometric multiplicity $\leq$ algebraic multiplicity. (p. 397)
- Let A be an $n \times n$ matrix. Then A has n linearly independent eigenvectors if and only if the geometric multiplicity of every eigenvalue is equal to its algebraic multiplicity. In particular, A has n linearly independent eigenvectors if all the eigenvalues are distinct (since then the algebraic multiplicity of every eigenvalue is 1). (p. 397)
- **Summing Up Theorem**·Let A be an $n \times n$ matrix. Then the following eleven statements are equivalent; that is each one implies the other ten (so that if one is true, all are true): (p. 398)

  i. A is invertible.

  ii. The only solution to the homogeneous system $Ax = 0$ is the trivial solution ($x = 0$).

  iii. The system $Ax = b$ has a unique solution for every n-vector b.

  iv. A is row equivalent to the $n \times n$ identity matrix I_n.

  v. A can be written as the product of elementary matrices.

  vi. The rows (and columns) of A are linearly independent.

  vii. det $A \neq 0$.

  viii. $\nu(A) = 0$.

  ix. $\rho(A) = n$.

  x. The linear transformation T from $\mathbb{R}^n$ to $\mathbb{R}^n$ defined by $Tx = Ax$ is an isomorphism.

  xi. Zero is *not* an eigenvalue of A.

- **Similar Matrices**·Two $n \times n$ matrices A and B are said to be **similar** if there exists an invertible $n \times n$ matrix C such that (p. 405)
$$B = C^{-1}AC$$

  The function defined above which takes the matrix A into the matrix B is called a **similarity transformation.**
- Similar matrices have the same eigenvalues. (p. 406)
- **Diagonalizable Matrix**·An $n \times n$ matrix A is **diagonalizable** if there is a diagonal matrix D such that A is similar to D. (p. 406)
- An $n \times n$ matrix A is diagonalizable if and only if it has n linearly independent eigenvectors. In that case the diagonal matrix D similar to A is given by (p. 407)

$$D = \begin{pmatrix} \lambda_1 & 0 & 0 & \cdots & 0 \\ 0 & \lambda_2 & 0 & \cdots & 0 \\ 0 & 0 & \lambda_3 & \cdots & 0 \\ \vdots & \vdots & \vdots & & \vdots \\ 0 & 0 & 0 & \cdots & \lambda_n \end{pmatrix}$$

  where $\lambda_1, \lambda_2, \ldots, \lambda_n$ are the eigenvalues of A. If C is a matrix whose columns are linearly independent eigenvectors of A, then
$$D = C^{-1}AC$$

- If the $n \times n$ matrix A has n distinct eigenvalues, then A is diagonalizable. (p. 408)

- The eigenvalues of a real symmetric matrix are real.   (p. 413)
- Eigenvectors of a real symmetric matrix corresponding to distinct eigenvalues are orthogonal.   (p. 414)
- A real symmetric $n \times n$ matrix has n real orthonormal eigenvectors.   (p. 414)
- **Orthogonally Diagonalizable Matrix·**An $n \times n$ matrix A is said to be **orthogonally diagonalizable** if there exists an orthogonal matrix Q such that   (p. 415)

$$Q^t A Q = D$$

  where $D = \text{diag}(\lambda_1, \lambda_2, \ldots, \lambda_n)$ and $\lambda_1, \lambda_2, \ldots, \lambda_n$ are the eigenvalues of A.
- **Procedure for finding an orthogonal diagonalizing matrix Q for a real symmetric matrix A:**   (p. 415)

  i.   Find a basis for each eigenspace of A.

  ii.   Find an orthonormal basis for each eigenspace of A by using the Gram-Schmidt process.

  iii.   Write Q as the matrix whose columns are the orthonormal eigenvectors obtained in step (*ii*).

- The **conjugate transpose** of an $m \times n$ matrix $A = (a_{ij})$, denoted by A^*, is the $n \times m$ matrix whose ijth component is $\bar{a}_{ji}$.   (p. 417)
- A complex $n \times n$ matrix A is **hermitian** if $A^* = A$.   (p. 418)
- A complex $n \times n$ matrix U is **unitary** if $U^* = U^{-1}$.   (p. 418)
- **Quadratic Equation and Quadratic Form·**A **quadratic equation in two variables with no linear term** is an equation of the form   (p. 421)

$$ax^2 + bxy + cy^2 = d$$

  where $|a| + |b| + |c| \neq 0$ and a, b, c are real numbers.

  A **quadratic form in two variables** is an expression of the form

$$F(x, y) = ax^2 + bxy + cy^2 = d$$

  where $|a| + |b| + |c| \neq 0$ and a, b, c are real numbers.
- A quadratic form can be written as   (p. 422)

$$F(x, y) = A\mathbf{v} \cdot \mathbf{v}$$

  where $A = \begin{pmatrix} a & b/2 \\ b/2 & c \end{pmatrix}$ is a symmetric matrix.
- If the eigenvalues of A are a' and c', then the quadratic form can be written as   (p. 423)

$$F(x', y') = a'x'^2 + c'y'^2$$

  where $\begin{pmatrix} x' \\ y' \end{pmatrix} = Q^t \begin{pmatrix} x \\ y \end{pmatrix}$ and Q is the orthogonal matrix that diagonalizes A.
- **Principal Axes Theorem in $\mathbb{R}^2$·**Let

$$ax^2 + bxy + cy^2 = d \qquad\qquad (*) \qquad \text{(pp. 423, 424)}$$

be a quadratic equation in the variables x and y. Then there exists a unique number θ in $[0, 2\pi)$ such that equation ($*$) can be written in the form

$$a'x'^2 + c'y'^2 = d$$

where x', y' are the axes obtained by rotating the x- and y-axes through an angle of θ in the counterclockwise direction. Moreover, the numbers a' and c' are the eigenvalues of the matrix $A = \begin{pmatrix} a & b/2 \\ b/2 & c \end{pmatrix}$. The x'- and y'-axes are called the **principal axes** of the graph of the quadratic equation.

- If $A = \begin{pmatrix} a & b/2 \\ b/2 & c \end{pmatrix}$, then the quadratic equation ($*$) with $d \neq 0$ is the equation of: (p. 427)

  i.   A hyperbola if $\det A < 0$.

  ii. An ellipse, circle, or degenerate conic section if $\det A > 0$.

  iii. A pair of straight lines or a degenerate conic section if $\det A = 0$.

  iv. If $d = 0$, then ($*$) is the equation of two straight lines if $\det A \neq 0$ and the equation of a single line if $\det A = 0$.

- **Quadratic Form in $\mathbb{R}^n$·** Let $\mathbf{v} = \begin{pmatrix} x_1 \\ x_2 \\ \vdots \\ x_n \end{pmatrix}$ and let A be a symmetric $n \times n$ matrix. Then a **quadratic**

  form in $x_1, x_2, \ldots, x_n$ is an expression of the form   (p. 428)

  $$F(x_1, x_2, \ldots, x_n) = A\mathbf{v} \cdot \mathbf{v}$$

- The matrix N_k is the $k \times k$ matrix   (p. 431)

  $$N_k = \begin{pmatrix} 0 & 1 & 0 & \cdots & 0 \\ 0 & 0 & 1 & \cdots & 0 \\ \vdots & \vdots & \vdots & & \vdots \\ 0 & 0 & 0 & \cdots & 1 \\ 0 & 0 & 0 & \cdots & 0 \end{pmatrix}$$

- The $k \times k$ **Jordan Block** matrix $B(\lambda)$ is given by   (p. 431)

  $$B(\lambda) = \lambda I + N_k = \begin{pmatrix} \lambda & 1 & 0 & \cdots & 0 & 0 \\ 0 & \lambda & 1 & \cdots & 0 & 0 \\ \vdots & \vdots & \vdots & & \vdots & \vdots \\ 0 & 0 & & \cdots & \lambda & 1 \\ 0 & 0 & & \cdots & 0 & \lambda \end{pmatrix}$$

- A **Jordan matrix** J has the form   (p. 432)

  $$J = \begin{pmatrix} B_1(\lambda_1) & 0 & \cdots & 0 \\ 0 & B_2(\lambda_2) & \cdots & 0 \\ \vdots & \vdots & & \vdots \\ 0 & 0 & \cdots & B_r(\lambda_r) \end{pmatrix}$$

  where each $B_j(\lambda_j)$ is a Jordan block matrix.

- **Jordan Canonical Form·** Let A be an $n \times n$ matrix. Then there exists an invertible $n \times n$ matrix C such that   (pp. 433, 434)

  $$C^{-1}AC = J$$

  where J is a Jordan matrix whose diagonal elements are the eigenvalues of A. Moreover, J is unique except for the order in which the Jordan blocks appear.

  The matrix J is called the **Jordan canonical form** of A.

- Suppose A is a 2×2 matrix with one eigenvalue λ of geometric multiplicity 1. Then the Jordan canonical form of A is   (pp. 435, 436)

$$J = \begin{pmatrix} \lambda & 1 \\ 0 & \lambda \end{pmatrix}$$

The matrix C has columns $\mathbf{v}_1$ and $\mathbf{v}_2$, where $\mathbf{v}_1$ is an eigenvector and $\mathbf{v}_2$ is a **generalized eigenvector** of A. That is, $\mathbf{v}_2$ satisfies

$$(A - \lambda I)\mathbf{v}_2 = \mathbf{v}_1$$

- Let A be an $n \times n$ matrix. Then e^A is defined by   (pp. 440, 441)

$$e^A = I + A + \frac{A^2}{2!} + \frac{A^3}{3!} + \cdots = \sum_{k=0}^{\infty} \frac{A^k}{k!}$$

- The **principal matrix solution** to the vector differential equation $\mathbf{x}'(t) = A\mathbf{x}(t)$ is e^{At}.   (p. 442)
- The unique solution to the differential equation $\mathbf{x}'(t) = A\mathbf{x}(t)$ that satisfies $\mathbf{x}(0) = \mathbf{x}_0$
  is $\mathbf{x}(t) = e^{At}\mathbf{x}_0$.   (p. 441)
- If J is the Jordan canonical form of the matrix A and if $J = C^{-1}AC$, then   (p. 443)

$$e^{At} = Ce^{Jt}C^{-1}$$

- **The Cayley-Hamilton Theorem**·Every square matrix satisfies its own characteristic equation.   (p. 451)
  That is, if $p(\lambda) = 0$ is the characteristic equation of A, then $p(A) = 0$.
- **Gershgorin Circles**·Let

$$A = \begin{pmatrix} a_{11} & a_{12} & \cdots & a_{1n} \\ a_{21} & a_{22} & \cdots & a_{2n} \\ \vdots & \vdots & & \vdots \\ a_{n1} & a_{n2} & \cdots & a_{nn} \end{pmatrix}$$

and define the numbers   (p. 454)

$$r_1 = |a_{12}| + |a_{13}| + \cdots + |a_{1n}| = \sum_{j=2}^{n} |a_{1j}|$$

$$r_i = |a_{i1}| + |a_{i2}| + \cdots + |a_{i,i-1}| + |a_{i,i+1}| + \cdots + |a_{i,n}|$$

$$= \sum_{\substack{j=1 \\ j \neq i}}^{n} |a_{ij}|$$

The **Gershgorin circles** are the circles that bound the disks

$$D_i = \{z \in \mathbb{C} : |z - a_{ii}| \le r_i\} \qquad (**)$$

- **Gershgorin Circle Theorem**·Let A be an $n \times n$ matrix and let D_i be defined by equation $(**)$.
  Then each eigenvalue of A is contained in at least one of the D_i's. That is, if the eigenvalues of
  A are $\lambda_1, \lambda_2, \ldots, \lambda_k$, then   (p. 455)

$$\{\lambda_1, \lambda_2, \ldots, \lambda_k\} \subset \bigcup_{i=1}^{n} D_i$$

REVIEW EXERCISES

In Exercises 1–6 calculate the eigenvalues and eigenspaces of the given matrix.

1. $\begin{pmatrix} -8 & 12 \\ -6 & 10 \end{pmatrix}$
   2. $\begin{pmatrix} 2 & 5 \\ 0 & 2 \end{pmatrix}$
   3. $\begin{pmatrix} 1 & 0 & 0 \\ 3 & 7 & 0 \\ -2 & 4 & -5 \end{pmatrix}$

4. $\begin{pmatrix} 1 & -1 & 0 \\ 1 & 2 & 1 \\ -2 & 1 & -1 \end{pmatrix}$
   5. $\begin{pmatrix} 5 & -2 & 0 & 0 \\ 4 & -1 & 0 & 0 \\ 0 & 0 & 3 & -1 \\ 0 & 0 & 2 & 3 \end{pmatrix}$
   6. $\begin{pmatrix} -2 & 1 & 0 \\ 0 & -2 & 1 \\ 0 & 0 & -2 \end{pmatrix}$

In Exercises 7–15 determine whether the given matrix A is diagonalizable. If it is, find a matrix C such that $C^{-1}AC = D$. If A is symmetric, find an orthogonal matrix Q such that $Q^t A Q = D$.

7. $\begin{pmatrix} -18 & -15 \\ 20 & 17 \end{pmatrix}$
   8. $\begin{pmatrix} \frac{17}{2} & \frac{9}{2} \\ -15 & -8 \end{pmatrix}$
   9. $\begin{pmatrix} 1 & 1 & 1 \\ -1 & -1 & 0 \\ -1 & 0 & -1 \end{pmatrix}$

10. $\begin{pmatrix} 4 & 2 & 0 \\ 2 & 4 & 0 \\ 0 & 0 & -3 \end{pmatrix}$
   11. $\begin{pmatrix} -3 & 2 & 1 \\ -7 & 4 & 2 \\ -5 & 3 & 2 \end{pmatrix}$
   12. $\begin{pmatrix} 8 & 0 & 12 \\ 0 & -2 & 0 \\ 12 & 0 & -2 \end{pmatrix}$

13. $\begin{pmatrix} 2 & 2 & 0 \\ 2 & 2 & 0 \\ 0 & 0 & -3 \end{pmatrix}$
   14. $\begin{pmatrix} 4 & 2 & -2 & 2 \\ 1 & 3 & 1 & -1 \\ 0 & 0 & 2 & 0 \\ 1 & 1 & -3 & 5 \end{pmatrix}$
   15. $\begin{pmatrix} 3 & 4 & -4 & 0 \\ 0 & -1 & 0 & 0 \\ 0 & 0 & -1 & 0 \\ 0 & -4 & 4 & 3 \end{pmatrix}$

In Exercises 16–20 identify the conic section and write it in new variables with the xy term absent.

16. $xy = -4$
   17. $4x^2 + 2xy + 2y^2 = 8$
   18. $4x^2 - 3xy + y^2 = 1$

19. $3y^2 - 2xy - 5 = 0$
   20. $x^2 - 4xy + 4y^2 + 1 = 0$

21. Write the quadratic form $2x^2 + 4xy + 2y^2 - 3z^2$ in new variables x', y', and z' so that no cross-product terms are present.

In Exercises 22–24 find a matrix C such that $C^{-1}AC = J$, the Jordan canonical form of the matrix.

22. $\begin{pmatrix} -9 & 4 \\ -25 & 11 \end{pmatrix}$
   23. $\begin{pmatrix} -4 & 4 \\ -1 & 0 \end{pmatrix}$
   24. $\begin{pmatrix} 0 & -18 & -7 \\ 1 & -12 & -4 \\ -1 & 25 & 9 \end{pmatrix}$

In Exercises 25–27 compute e^{At}.

25. $A = \begin{pmatrix} -3 & 4 \\ -2 & 3 \end{pmatrix}$
   26. $A = \begin{pmatrix} -4 & 4 \\ -1 & 0 \end{pmatrix}$
   27. $A = \begin{pmatrix} -3 & -4 \\ 2 & 1 \end{pmatrix}$

28. Using the Cayley-Hamilton theorem, compute the inverse of

$$A = \begin{pmatrix} 2 & 3 & 1 \\ -1 & 1 & 0 \\ -2 & -1 & 4 \end{pmatrix}.$$

29. Use the Gershgorin circle theorem to find a bound on the eigenvalues of

$$A = \begin{pmatrix} 3 & \frac{1}{2} & -\frac{1}{2} & 0 \\ 0 & 4 & \frac{1}{3} & -\frac{1}{3} \\ 1 & 0 & 2 & -1 \\ \frac{1}{2} & -\frac{1}{2} & 1 & -3 \end{pmatrix}$$

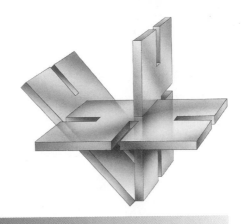

7

Numerical Methods

7.1 THE ERROR IN NUMERICAL COMPUTATIONS AND COMPUTATIONAL COMPLEXITY

In every chapter of this book we have performed numerical computations. We have, among other things, solved linear equations, multiplied and inverted matrices, found bases, and computed eigenvalues and eigenvectors. With few exceptions, the examples involved 2×2 and 3×3 matrices—not because most applications have only two or three variables but because the computations would have been too tedious otherwise.

With the recent and widespread use of calculators and computers, the situation has been altered. The remarkable strides made in the last few years in the theory of numerical methods for solving certain computational problems have made it possible to perform, quickly and accurately, the calculations mentioned in the first paragraph on high-order matrices.

The use of the computer presents new difficulties, however. Computers do not store numbers such as $\frac{2}{3}$, $7\frac{3}{7}$, $\sqrt{2}$, and π. Rather, every computer uses what is called *floating-point arithmetic*. In this system every number is represented in the form

$$x = \pm 0.d_1 d_2 \cdots d_k \times 10^n \qquad (1)$$

where $d_1, d_2, \ldots, d_k$ are single-digit positive integers and n is an integer. Any number written in this form is called a *floating-point number*. In equation (1) the number $\pm 0.d_1 d_2 \cdots d_k$ is called the *mantissa* and the number n is called the *exponent*. The number k is called the *number of significant digits* in the expression.

Different computers have different capabilities in the range of numbers expressible in the form of equation (1). Digits are usually represented in binary rather than decimal form. One popular computer, for example, carries 28 binary digits. Since $2^{28} = 268{,}435{,}456$, we can use the 28 binary digits to represent any eight-digit number. Hence $k = 8$.

EXAMPLE 1   **The Floating-Point Form of Four Numbers**   The following numbers are expressed in floating-point form:

  i. $\frac{1}{4} = 0.25$
  ii. $2378 = 0.2378 \times 10^4$
  iii. $-0.000816 = -0.816 \times 10^{-3}$
  iv. $83.27 = 0.8327 \times 10^2$ ■

If the number of significant digits were unlimited, then we would have no problem. Almost every time numbers are introduced into the computer, however, errors begin to accumulate. This can happen in one of two ways:

  i. **Truncation.** All significant digits after k digits are simply "cut off." For example, if truncation is used, $\frac{2}{3} = 0.666666\ldots$ is stored (with $k = 8$) as $\frac{2}{3} = 0.66666666 \times 10^0$.

  ii. **Rounding.** If $d_{k+1} \geq 5$, then 1 is added to d_k and the resulting number is truncated. Otherwise the number is simply truncated. For example, with rounding (and $k = 8$), $\frac{2}{3}$ is stored as $\frac{2}{3} = 0.66666667 \times 10^0$.

EXAMPLE 2   **Illustration of Truncation and Rounding**   We can illustrate how some numbers are stored with truncation and rounding by using eight significant digits:

Number	Truncated Number	Rounded Number
$\frac{8}{3}$	0.26666666×10^1	0.26666667×10^1
π	0.31415926×10^1	0.31415927×10^1
$-\frac{1}{57}$	$-0.17543859 \times 10^{-1}$	$-0.17543860 \times 10^{-1}$

■

Individual round-off or truncation errors do not seem very significant. When thousands of computational steps are involved, however, the *accumulated* round-off error can be devastating. Thus, in discussing any numerical scheme, it is necessary to know not only whether you will get the right answer, theoretically, but also how badly the round-off errors will accumulate. To keep track of things, we define two types of error. If x is the actual value of a number and if x^* is the number that appears in the computer, then the **absolute error** ϵ_a is defined by

Absolute error

$$\epsilon_a = |x^* - x| \tag{2}$$

Relative error   More interesting in most situations is the **relative error** ϵ_r, defined by

$$\epsilon_r = \left| \frac{x^* - x}{x} \right|$$

(3)

EXAMPLE 3   **Illustration of Relative Error**   Let $x = 2$ and $x^* = 2.1$. Then $\epsilon_a = 0.1$ and $\epsilon_r = 0.1/2 = 0.05$. If $x_1 = 2000$ and $x_1^* = 2000.1$, then again $\epsilon_a = 0.1$. But now $\epsilon_r = 0.1/2000 = 0.00005$. Most people would agree that the 0.1 error in the first case is more significant than the 0.1 error in the second.   ■

Much of numerical analysis is concerned with questions of **convergence** and **stability.** If x is the answer to a problem and our computational method gives us approximating values x_n, then the method converges if, theoretically, x_n approaches x as n gets large. If, moreover, it can be shown that the round-off errors will not accumulate in such a way as to make the answer very inaccurate, then the method is **stable.**

It is easy to give an example of a procedure in which round-off error can be quite large. Suppose we wish to compute $y = 1/(x - 0.66666665)$. For $x = \frac{2}{3}$, if the computer truncates, then $x = 0.66666666$ and $y = 1/0.00000001 = 10^8 = 10 \times 10^7$. If the computer rounds, then $x = 0.66666667$ and $y = 1/0.00000002 = 5 \times 10^7$. The difference here is enormous. The correct answer is $1/\left(\frac{2}{3} - \frac{66666665}{100,000,000} \right) = 1/\left(\frac{200,000,000}{300,000,000} - \frac{199,999,995}{300,000,000} \right) = 1/\frac{5}{300,000,000} = \frac{300,000,000}{5} = 60,000,000 = 6 \times 10^7$.

Computational Complexity

In solving problems on a computer, two questions naturally arise:

How accurate are my answers?

How much time will it take?

(After all, you pay by the hour on a computer.)

We discussed the first question in the first part of this section. To answer the second one, we must estimate the number of steps required to carry out a certain computation. The **computational complexity** of a problem is a measure of the number of arithmetic operations needed to solve the problem and the time needed to carry out all the needed operations.

There are two basic arithmetic operations carried out on a computer:

Operation	Average Time in microseconds†
Addition or subtraction	$\frac{1}{2}$ microsecond
Multiplication or division	2 microseconds

†1 microsecond = 1 millionth of a second = 10^{-6} second.

Thus in order to estimate the time needed to solve a problem on a computer, it is first necessary to count the number of additions, subtractions, multiplications, and divisions involved in solving the problem.

Counting the number of operations needed to solve a problem is often very difficult. We illustrate how it can be done in the case of Gauss-Jordan elimination. To simplify matters, we treat addition and subtraction as the same operation and multiplication and division as the same (although, in fact, each division takes three times as long as a multiplication; the average time for both is 2 microseconds).

EXAMPLE 4   **Counting Additions and Multiplications in Gauss-Jordan Elimination**   Let A be an invertible $n \times n$ matrix. Estimate the number of additions and multiplications needed to solve the system $A\mathbf{x} = \mathbf{b}$ by Gauss-Jordan elimination.

Solution   We begin, as in Section 1.3, by writing the system in the augmented-matrix form

$$\begin{pmatrix} a_{11} & a_{12} & \cdots & a_{1n} & \bigm| & b_1 \\ a_{21} & a_{22} & \cdots & a_{2n} & \bigm| & b_2 \\ \vdots & \vdots & & \vdots & \bigm| & \vdots \\ a_{n1} & a_{n2} & \cdots & a_{nn} & \bigm| & b_n \end{pmatrix}$$

Since A is invertible by assumption, its reduced row echelon form is the $n \times n$ identity matrix. We assume that in the row reduction no rows are permuted (interchanged), since such an interchange does not involve any additions or multiplications. Moreover, keeping track of row numbers is a bookkeeping task that requires considerably less time than an addition.

To keep track of which numbers are being computed during a given step, we write the augmented matrix with C's and L's. A C denotes a number just computed. An L denotes a number to be left alone.

Step 1. Divide each number in the first row by a_{11} $\left(\text{multiply by } \dfrac{1}{a_{11}}\right)$ to obtain

$n + 1$ columns   **Total for Step 1**

$$\begin{pmatrix} 1 & C & C & \cdots & C & C & \bigm| & C \\ L & L & L & \cdots & L & L & \bigm| & L \\ \vdots & \vdots & \vdots & & \vdots & \vdots & \bigm| & \vdots \\ L & L & L & \cdots & L & L & \bigm| & L \end{pmatrix}$$

n multiplications

$\left(\dfrac{a_{11}}{a_{11}} = 1 \text{ requires no computation, the}\right.$

1 is simply inserted in the 1,1 position.)

no additions

Step 2. Multiply row 1 by $-a_{i1}$ and add it to the ith row for $i = 2, 3, \ldots, n$:

$$\begin{pmatrix} 1 & L & L & \cdots & L & L & \bigm| & L \\ 0 & C & C & \cdots & C & C & \bigm| & C \\ 0 & C & C & \cdots & C & C & \bigm| & C \\ \vdots & \vdots & \vdots & & \vdots & \vdots & \bigm| & \vdots \\ 0 & C & C & \cdots & C & C & \bigm| & C \end{pmatrix}$$

Let us count the operations.

To obtain the new second row:

The 0 in the 2,1 position requires no work. We know that the number in the 2,1 position will be 0, so we simply place it there. There are $(n + 1) - 1 = n$ numbers in the second row that must be changed. For example, if we denote the new a_{22} by a'_{22}, then

$$a'_{22} = a_{22} - a_{21}a_{12}$$

This requires 1 multiplication and 1 addition. Since there are n numbers to be changed in the second row, there are n multiplications and n additions needed in the second row. The same is true in each of the $n - 1$ rows 2 through n. Thus

Total for Step 2

$(n - 1)n$ multiplications

$(n - 1)n$ additions

Notation From now on a'_{ij} will now denote the latest entry in the ith row and jth column.

Step 3. Divide everything in the second row by a'_{22}:

$$\begin{pmatrix} 1 & L & L & \cdots & L & L & | & L \\ 0 & 1 & C & \cdots & C & C & | & C \\ 0 & L & L & \cdots & L & L & | & L \\ \vdots & \vdots & \vdots & & \vdots & \vdots & | & \vdots \\ 0 & L & L & \cdots & L & L & | & L \end{pmatrix}$$

Total for Step 3

$n - 1$ multiplications. (As before, the 1 in the 2,2 position is simply placed there.)

no additions

Step 4. Multiply row 2 by $-a'_{i2}$ and add it to the ith row, for $i = 1, 3, 4, \ldots, n$:

$$\begin{pmatrix} 1 & 0 & C & \cdots & C & C & | & C \\ 0 & 1 & L & \cdots & L & L & | & L \\ 0 & 0 & C & \cdots & C & C & | & C \\ \vdots & \vdots & \vdots & & \vdots & \vdots & | & \vdots \\ 0 & C & C & \cdots & C & C & | & C \end{pmatrix}$$

Total for Step 4

$(n - 1)(n - 1)$ multiplications
$(n - 1)(n - 1)$ additions

As in Step 2, each change requires 1 multiplication and 1 addition. But now the first two components in each row require no computation; that is, $(n + 1) - 2 = n - 1$ numbers in each row are computed. As before, computations are done in $n - 1$ rows. This explains the numbers above.

You should now see the pattern. In step 5 we shall have $n - 2$ multiplications (divide each component in the third row, besides the first three, by a'_{33}). In step 6 there will be $n - 2$ multiplications and $n - 2$ additions needed in each of $n - 1$ rows for a total of $(n - 1)(n - 2)$ multiplications and $(n - 1)(n - 2)$ additions. We continue in this manner until there are four steps to go. Here is how the augmented matrix will appear:

$$\begin{pmatrix}
1 & 0 & 0 & \cdots & a'_{1,n-1} & a'_{1n} & \bigg| & b'_1 \\
0 & 1 & 0 & \cdots & a'_{2,n-1} & a'_{2n} & \bigg| & b'_2 \\
0 & 0 & 1 & \cdots & a'_{3,n-1} & a'_{3n} & \bigg| & b'_3 \\
\vdots & \vdots & \vdots & & \vdots & \vdots & & \vdots \\
0 & 0 & 0 & \cdots & a'_{n-1,n-1} & a'_{n-1,n} & \bigg| & b'_{n-1} \\
0 & 0 & 0 & \cdots & a'_{n,n-1} & a'_{nn} & \bigg| & b'_n
\end{pmatrix}$$

Last Step Minus 3. Divide the $(n-1)$st row by $a'_{n-1,n-1}$:

$$\begin{pmatrix}
1 & 0 & 0 & \cdots & L & L & \bigg| & L \\
0 & 1 & 0 & \cdots & L & L & \bigg| & L \\
0 & 0 & 1 & \cdots & L & L & \bigg| & L \\
\vdots & \vdots & \vdots & & \vdots & \vdots & & \vdots \\
0 & 0 & 0 & \cdots & 1 & C & \bigg| & C \\
0 & 0 & 0 & \cdots & L & L & \bigg| & L
\end{pmatrix}$$ 2 multiplications
no additions

Last Step Minus 2. Multiply the $(n-1)$st row by $-a'_{i,n-1}$ and add it to the ith row, for $i = 1, 2, \ldots, n-2, n$

$$\begin{pmatrix}
1 & 0 & 0 & \cdots & 0 & C & \bigg| & C \\
0 & 1 & 0 & \cdots & 0 & C & \bigg| & C \\
0 & 0 & 1 & \cdots & 0 & C & \bigg| & C \\
\vdots & \vdots & \vdots & & \vdots & \vdots & & \vdots \\
0 & 0 & 0 & \cdots & 1 & L & \bigg| & L \\
0 & 0 & 0 & \cdots & 0 & C & \bigg| & C
\end{pmatrix}$$ $2(n-1)$ multiplications
$2(n-1)$ additions

Last Step Minus 1. Divide the nth row by a'_{nn}:

$$\begin{pmatrix}
1 & 0 & 0 & \cdots & 0 & L & \bigg| & L \\
0 & 1 & 0 & \cdots & 0 & L & \bigg| & L \\
0 & 0 & 1 & \cdots & 0 & L & \bigg| & L \\
\vdots & \vdots & \vdots & & \vdots & \vdots & & \vdots \\
0 & 0 & 0 & \cdots & 1 & L & \bigg| & L \\
0 & 0 & 0 & \cdots & 0 & 1 & \bigg| & C
\end{pmatrix}$$ 1 multiplication
no additions

Last Step. Multiply the nth row by $-a'_{in}$ and add it to the ith row, for $i = 1, 2, \ldots, n-1$:

$$\begin{pmatrix}
1 & 0 & 0 & \cdots & 0 & 0 & \bigg| & C \\
0 & 1 & 0 & \cdots & 0 & 0 & \bigg| & C \\
0 & 0 & 1 & \cdots & 0 & 0 & \bigg| & C \\
\vdots & \vdots & \vdots & & \vdots & \vdots & & \vdots \\
0 & 0 & 0 & \cdots & 1 & 0 & \bigg| & C \\
0 & 0 & 0 & \cdots & 0 & 1 & \bigg| & L
\end{pmatrix}$$ $1(n-1)$ multiplications
$1(n-1)$ additions

We now find the totals:

For the odd-numbered steps there are

$$n + (n - 1) + (n - 2) + \cdots + 3 + 2 + 1 \text{ multiplications}$$

and

$$\text{no additions}$$

For the even-numbered steps there are

$$(n - 1)[n + (n - 1) + (n - 2) + \cdots + 3 + 2 + 1] \text{ multiplications}$$

and

$$(n - 1)[n + (n - 1) + (n - 2) + \cdots + 3 + 2 + 1] \text{ additions}$$

In Example 1 in Appendix 1 (page 504) we prove that

$$1 + 2 + 3 + \cdots + n = \frac{n(n + 1)}{2} \tag{4}$$

Thus the total number of multiplications is

From odd-numbered steps   From even-numbered steps

$$\frac{n(n + 1)}{2} \quad + \quad (n - 1)\left[\frac{n(n + 1)}{2}\right]$$

$$= \left[\frac{n(n + 1)}{2}\right][1 + (n - 1)] = n^2\left(\frac{n + 1}{2}\right) = \frac{n^3}{2} + \frac{n^2}{2}$$

and the total number of additions is $(n - 1)\left[\dfrac{n(n + 1)}{2}\right] = \dfrac{n^3 - n}{2} = \dfrac{n^3}{2} - \dfrac{n}{2}$ ∎

A Modification of Gauss-Jordan Elimination

There is a more efficient way to row-reduce A to the identity matrix: First reduce A to its row echelon form to obtain the matrix

$$\left(\begin{array}{cccccc|c}
1 & a'_{12} & a'_{13} & \cdots & a'_{1,n-1} & a'_{1n} & b'_1 \\
0 & 1 & a'_{23} & \cdots & a'_{2,n-1} & a'_{2n} & b'_2 \\
\vdots & \vdots & \vdots & & \vdots & \vdots & \vdots \\
0 & 0 & 0 & \cdots & 1 & a'_{n-1,n} & b'_{n-1} \\
0 & 0 & 0 & \cdots & 0 & 1 & b'_n
\end{array}\right)$$

The next step is to make zero all entries in the nth column above the 1 in the n,n position. This results in

$$\left(\begin{array}{cccccc|c}
1 & L & L & \cdots & L & 0 & C \\
0 & 1 & L & \cdots & L & 0 & C \\
\vdots & \vdots & \vdots & & \vdots & \vdots & \vdots \\
0 & 0 & 0 & \cdots & 1 & 0 & C \\
0 & 0 & 0 & \cdots & 0 & 1 & L
\end{array}\right)$$

Finally, working from right to left, make all the remaining entries above the diagonal equal to 0. In Problem 22 you are asked to show that with this modification, the number of multiplications is $\frac{1}{3}n^3 + n^2 - \frac{1}{3}n$ and the number of additions is $\frac{1}{3}n^3 + \frac{1}{2}n^2 - \frac{5}{6}n$.

For n large,

$$\frac{n^3}{2} + \frac{n^2}{2} \approx \frac{n^3}{2}$$

For example, when $n = 10,000$

$$\frac{n^3}{2} + \frac{n^2}{2} = 500,050,000,000 = 5.0005 \times 10^{11}$$

and

$$\frac{n^3}{2} = 500,000,000,000 = 5 \times 10^{11}$$

Similarly, for n large,

$$\frac{1}{3}n^3 + n^2 - \frac{1}{3}n \approx \frac{n^3}{3}$$

Since $\dfrac{n^3}{3}$ is less than $\dfrac{n^3}{2}$, we see that the modification described above is more efficient when n is large. (In fact, it is better when $n \geq 3$.)

In Table 7.1 we give the number of additions and multiplications required for several of the processes discussed in Chapters 1 and 2.

Table 7.1   Number of Arithmetic Operations for an Invertible $n \times n$ Matrix A

Technique	Number of Multiplications	Approximate Number of Multiplications for n large	Number of Additions	Approximate Number of Additions for n large
1. Solve $A\mathbf{x} = \mathbf{b}$ by Gauss-Jordan elimination	$\dfrac{n^3}{2} + \dfrac{n^2}{2}$	$\dfrac{n^3}{2}$	$\dfrac{n^3}{2} - \dfrac{n}{2}$	$\dfrac{n^3}{2}$
2. Solve $A\mathbf{x} = \mathbf{b}$ by modified Gauss-Jordan elimination	$\dfrac{n^3}{3} + n^2 - \dfrac{n}{3}$	$\dfrac{n^3}{3}$	$\dfrac{n^3}{3} + \dfrac{n^2}{2} - \dfrac{5n}{6}$	$\dfrac{n^3}{3}$
3. Solve $A\mathbf{x} = \mathbf{b}$ by Gaussian elimination with back substitution	$\dfrac{n^3}{3} + n^2 - \dfrac{n}{3}$	$\dfrac{n^3}{3}$	$\dfrac{n^3}{3} + \dfrac{n^2}{2} - \dfrac{5n}{6}$	$\dfrac{n^3}{3}$
4. Find A^{-1} by Gauss-Jordan elimination	n^3	n^3	$n^3 - 2n^2 + n$	n^3
5. Compute $\det A$ by reducing A to a triangular matrix and multiplying the diagonal components	$\dfrac{n^3}{3} + \dfrac{2n}{3} - 1$	$\dfrac{n^3}{3}$	$\dfrac{n^3}{3} - \dfrac{n^2}{2} + \dfrac{n}{6}$	$\dfrac{n^3}{3}$

You are asked to derive these formulas in Problems 22–25.

In this chapter we examine several schemes for solving systems of equations and computing eigenvalues and eigenvectors. Wherever possible, we also discuss convergence and stability. This is but a very superficial view of numerical methods, however; entire books and courses are devoted to the subject. For a more exhaustive view, you are encouraged to consult the following references:

Numerical Linear Algebra References

1. Blum, E. K. *Numerical Analysis and Computation: Theory and Practice.* Reading, Mass.: Addison-Wesley, 1972.

2. Burden, R. L., J. D. Faires, and A. C. Reynolds. *Numerical Analysis.* Boston: Prindle, Weber and Schmidt, 1978.

3. Conte, S. D. *Elementary Numerical Analysis,* 2nd Ed. New York: McGraw-Hill, 1972.

4. Faddeev, D. K., and V. N. Faddeeva. *Computational Methods of Linear Algebra.* San Francisco: Freeman, 1963.

5. Fox, L. *An Introduction to Numerical Linear Algebra.* New York: Oxford University Press, 1965.

PROBLEMS 7.1

In Problems 1–13 convert the number to a floating-point number with eight decimal places of accuracy. Either truncate (T) or round off (R) as indicated.

1. $\frac{1}{3}$ (T) 2. $\frac{7}{8}$ 3. -0.000035 4. $\frac{7}{9}$ (R)

5. $\frac{7}{9}$ (T) 6. $\frac{33}{7}$ (T) 7. $\frac{85}{11}$ (R) 8. $-18\frac{5}{6}$ (T)

9. $-18\frac{5}{6}$ (R) 10. 237,059,628 (T) 11. 237,059,628 (R)

12. -23.7×10^{15} 13. 8374.2×10^{-24}

In Problems 14–21 the number x and an approximation x^* are given. Find the absolute and relative errors ϵ_a and ϵ_r.

14. $x = 5; x^* = 0.49 \times 10^1$ 15. $x = 500; x^* = 0.4999 \times 10^3$

16. $x = 3720; x^* = 0.3704 \times 10^4$ 17. $x = \frac{1}{8}; x^* = 0.12 \times 10^0$

18. $x = \frac{1}{800}; x^* = 0.12 \times 10^{-2}$ 19. $x = -5\frac{5}{6}; x^* = -0.583 \times 10^1$

20. $x = 0.70465; x^* = 0.70466 \times 10^0$

21. $x = 70465; x^* = 0.70466 \times 10^5$

22. Derive the formulas in row 2 of Table 7.1. [*Hint:* You will need the following formula which is proved in Example 2 in Appendix 1:

$$1^2 + 2^2 + 3^2 + \cdots + n^2 = \frac{n(n + 1)(2n + 1)}{6}]$$

23. Derive the formulas in row 3 of Table 7.1.

24. Derive the formulas in row 4 of Table 7.1.

*25. Derive the formulas in row 5 of Table 7.1.

26. How many seconds would it take, on average, to solve $A\mathbf{x} = \mathbf{b}$ on a computer using Gauss-Jordan elimination if A is a 20×20 matrix?

27. Answer Problem 26 if the modification described in the text is used.

28. How many seconds would it take, on average, to invert a 50×50 matrix? a 200×200 matrix? a $10,000 \times 10,000$ matrix?

29. Derive a formula for the number of multiplications and additions required to compute the product AB where A is an $m \times n$ and B is an $n \times q$ matrix.

7.2 SOLVING LINEAR SYSTEMS I: GAUSSIAN ELIMINATION WITH PIVOTING

It is not difficult to program a computer to solve a system of linear equations by the Gaussian or Gauss-Jordan elimination method used throughout this text. There is, however, a variation of the method that was designed to reduce the accumulated round-off error in solving an $n \times n$ system of equations. Before describing this method, recall the definition of the row echelon form of a matrix (see Definition 1.3.2, page 13).

DEFINITION 1   **Row Echelon Form** Let A be an $m \times n$ matrix. Then A is in **row echelon form** if

i. All rows consisting entirely of zeros appear at the bottom of the matrix.

ii. The first (starting from the left) number in any row not consisting entirely of zeros is a 1.

iii. If two successive rows do not consist entirely of zeros, then the first 1 in the lower row occurs farther to the right than the first 1 in the higher row.

Note. If A is an $n \times n$ upper triangular matrix with 1's on the main diagonal, then it is easy to verify that A is in row echelon form.

From Chapter 1 it is apparent that any matrix can be reduced to row echelon form by Gaussian elimination. There is a computational problem with this method, however. If we divide by a small number that has been rounded, the result could contain a significant round-off error. For example, $1/0.00074 \approx 1351$ while $1/0.0007 \approx 1429$. To avoid this problem, we use a method called **Gaussian elimination with partial pivoting.** The idea is always to divide by the largest component in a column, thereby avoiding, so far as is possible, the type of error illustrated above. We describe the method with a simple example.

EXAMPLE 1   **Solving a System by Gaussian Elimination with Partial Pivoting**   Solve the following system by Gaussian elimination with partial pivoting:

$$\begin{aligned} x_1 - x_2 + x_3 &= 1 \\ -3x_1 + 2x_2 - 3x_3 &= -6 \\ 2x_1 - 5x_2 + 4x_3 &= 5 \end{aligned}$$

Solution

Pivot

Step 1. Write the system in augmented-matrix form. From the first column with nonzero components (called the **pivot column**), select the component with the *largest absolute value*. This component is called the **pivot**:

$$\text{pivot} \rightarrow \left(\begin{array}{ccc|c} 1 & -1 & 1 & 1 \\ \boxed{-3} & 2 & -3 & -6 \\ 2 & -5 & 4 & 5 \end{array}\right)$$

Step 2. Rearrange the rows to move the pivot to the top:

$$\left(\begin{array}{ccc|c} \boxed{-3} & 2 & -3 & -6 \\ 1 & -1 & 1 & 1 \\ 2 & -5 & 4 & 5 \end{array}\right) \quad \begin{array}{l}\text{(first and second rows} \\ \text{were interchanged)}\end{array}$$

Step 3. Divide the first row by the pivot:

$$\left(\begin{array}{ccc|c} 1 & -\frac{2}{3} & 1 & 2 \\ 1 & -1 & 1 & 1 \\ 2 & -5 & 4 & 5 \end{array}\right) \quad \begin{array}{l}\text{(first row divided} \\ \text{by} -3)\end{array}$$

Step 4. Add multiples of the first row to the other rows to make all the other components in the pivot column equal to zero:

$$\left(\begin{array}{ccc|c} 1 & -\frac{2}{3} & 1 & 2 \\ 0 & -\frac{1}{3} & 0 & -1 \\ 0 & -\frac{11}{3} & 2 & 1 \end{array}\right) \quad \begin{array}{l}\text{(first row multiplied by} -1 \\ \text{and} -2 \text{ and added to the second} \\ \text{and third rows)}\end{array}$$

Step 5. Cover the first row and perform steps 1–4 on the resulting *submatrix:*

$$\left(\begin{array}{ccc|c} 1 & -\frac{2}{3} & 1 & 2 \\ 0 & -\frac{1}{3} & 0 & -1 \\ 0 & \boxed{-\frac{11}{3}} & 2 & 1 \end{array}\right)$$

new pivot

$$\left(\begin{array}{ccc|c} 1 & -\frac{2}{3} & 1 & 2 \\ 0 & \boxed{-\frac{11}{3}} & 2 & 1 \\ 0 & -\frac{1}{3} & 0 & -1 \end{array}\right) \quad \begin{array}{l}\text{(first and second rows} \\ \text{of submatrix} \\ \text{interchanged)}\end{array}$$

$$\left(\begin{array}{ccc|c} 1 & -\frac{2}{3} & 1 & 2 \\ 0 & 1 & -\frac{6}{11} & -\frac{3}{11} \\ 0 & -\frac{1}{3} & 0 & -1 \end{array}\right) \quad \begin{array}{l}\text{(new first row divided} \\ \text{by pivot)}\end{array}$$

$$\left(\begin{array}{ccc|c} 1 & -\frac{2}{3} & 1 & 2 \\ 0 & 1 & -\frac{6}{11} & -\frac{3}{11} \\ 0 & 0 & -\frac{2}{11} & -\frac{12}{11} \end{array}\right) \quad \begin{array}{l}\text{(new first row multiplied by} \\ \frac{1}{3} \text{ and added to new} \\ \text{second row)}\end{array}$$

Step 6. Continue in this manner until the matrix is in row echelon form:

$$\begin{pmatrix} 1 & -\frac{2}{3} & 1 & \bigg| & 2 \\ 0 & 1 & -\frac{6}{11} & \bigg| & -\frac{3}{11} \\ 0 & 0 & \boxed{-\frac{2}{11}} & \bigg| & -\frac{12}{11} \end{pmatrix}$$

new pivot ⟶

$$\begin{pmatrix} 1 & -\frac{2}{3} & 1 & \bigg| & 2 \\ 0 & 1 & -\frac{6}{11} & \bigg| & -\frac{3}{11} \\ 0 & 0 & 1 & \bigg| & 6 \end{pmatrix} \quad \begin{array}{l}\text{(divided new first} \\ \text{row by pivot)}\end{array}$$

Step 7. Use **back substitution** to find the solution (if any) to the system. Evidently, we have $x_3 = 6$. Then $x_2 - \frac{6}{11}x_3 = -\frac{3}{11}$ or

$$x_2 = -\frac{3}{11} + \frac{6}{11}x_3 = -\frac{3}{11} + \frac{6}{11}(6) = 3$$

Finally, $x_1 - \frac{2}{3}x_2 + x_3 = 2$ or

$$x_1 = 2 + \frac{2}{3}x_2 - x_3 = 2 + \frac{2}{3}(3) - 6 = -2$$

The unique solution is given by the vector $(-2, 3, 6)$.   ■

Remark. **Complete pivoting** involves finding the component in A with largest absolute value, not just the component in the first nonzero column. The problem with this method is that it usually involves the relabeling of variables when the columns are interchanged to bring the pivot to the first column. In most problems complete pivoting is not much more accurate than partial pivoting—at least not enough to justify the extra time involved. For this reason the partial pivoting method described above is more popular.

We now examine the method applied to a computationally more difficult system. Calculations were done on a hand calculator and were rounded to six significant digits.

EXAMPLE 2   **Solving a System by Gaussian Elimination with Partial Pivoting**   Solve the system

$$\begin{aligned} 2x_1 - 3.5x_2 + \quad x_3 &= 22.35 \\ -5x_1 + \quad 3x_2 + 3.3x_3 &= -9.08 \\ 12x_1 + 7.8x_2 + 4.6x_3 &= 21.38 \end{aligned}$$

Solution   Using the steps outlined above, we obtain, successively,

$$\begin{pmatrix} 2 & -3.5 & 1 & \bigg| & 22.35 \\ -5 & 3 & 3.3 & \bigg| & -9.08 \\ \boxed{12} & 7.8 & 4.6 & \bigg| & 21.38 \end{pmatrix} \xrightarrow{R_1 \rightleftarrows R_3} \begin{pmatrix} \boxed{12} & 7.8 & 4.6 & \bigg| & 21.38 \\ -5 & 3 & 3.3 & \bigg| & -9.08 \\ 2 & -3.5 & 1 & \bigg| & 22.35 \end{pmatrix}$$

pivot

$$\xrightarrow{R_1 \to \frac{1}{12}R_1} \begin{pmatrix} 1 & 0.65 & 0.383333 & 1.78167 \\ -5 & 3 & 3.3 & -9.08 \\ 2 & -3.5 & 1 & 22.35 \end{pmatrix}$$

$$\xrightarrow[R_3 \to R_3 - 2R_1]{R_2 \to R_2 + 5R_1} \begin{pmatrix} 1 & 0.65 & 0.383333 & 1.78167 \\ 0 & \boxed{6.25} & 5.21667 & -0.17165 \\ 0 & -4.8 & 0.233334 & 18.7867 \end{pmatrix}$$

new pivot

$$\xrightarrow{R_2 \to \frac{1}{6.25}R_2} \begin{pmatrix} 1 & 0.65 & 0.383333 & 1.78167 \\ 0 & 1 & 0.834667 & -0.027464 \\ 0 & -4.8 & 0.233334 & 18.7867 \end{pmatrix}$$

$$\xrightarrow{R_3 \to R_3 + 4.8R_2} \begin{pmatrix} 1 & 0.65 & 0.383333 & 1.78167 \\ 0 & 1 & 0.834667 & -0.027464 \\ 0 & 0 & \boxed{4.23974} & 18.6549 \end{pmatrix}$$

new pivot

$$\xrightarrow{R_3 \to \frac{1}{4.23974}R_3} \begin{pmatrix} 1 & 0.65 & 0.383333 & 1.78167 \\ 0 & 1 & 0.834667 & -0.027464 \\ 0 & 0 & 1 & 4.40001 \end{pmatrix}$$

The matrix is now in row echelon form. Using back substitution, we obtain

$$x_3 \approx 4.40001$$
$$x_2 \approx -0.027464 - 0.834667x_3 = -0.027464 - (0.834667)(4.40001)$$
$$= -3.70001$$
$$x_1 \approx 1.78167 - (0.65)(x_2) - (0.383333)x_3 = 1.78167 - (0.65)(-3.70001)$$
$$- (0.383333)(4.40001) = 2.50001$$

The correct solution is $x_1 = 2.5$, $x_2 = -3.7$, and $x_3 = 4.4$. Our answers are very accurate indeed. ∎

Remark. Example 2 illustrates the fact that it is tedious and inefficient to use this method without a calculator—especially if several significant digits of accuracy are required.

The next example shows how pivoting can significantly improve the answers. Here we round to only three significant digits, thereby introducing greater round-off errors.

EXAMPLE 3   Partial Pivoting Can Make a Significant Difference   Consider the system

$$0.0002x_1 - 0.00031x_2 + 0.0017x_3 = 0.00609$$
$$5x_1 \qquad - 7x_2 \qquad + 6x_3 = 7$$
$$8x_1 \qquad + 6x_2 \qquad + 3x_3 = 2$$

The exact solution is $x_1 = -2$, $x_2 = 1$, $x_3 = 4$. Let us first solve the system by Gaussian elimination without pivoting, rounding to three significant figures.

$$
\begin{pmatrix}
0.0002 & -0.00031 & 0.0017 & \bigm| & 0.00609 \\
5 & -7 & 6 & \bigm| & 7 \\
8 & 6 & 3 & \bigm| & 2
\end{pmatrix}
\xrightarrow{R_1 \to \frac{1}{0.0002}R_1}
\begin{pmatrix}
1 & -1.55 & 8.5 & \bigm| & 30.5 \\
5 & -7 & 6 & \bigm| & 7 \\
8 & 6 & 3 & \bigm| & 2
\end{pmatrix}
$$

$$
\xrightarrow[R_3 \to R_3 - 8R_1]{R_2 \to R_2 - 5R_1}
\begin{pmatrix}
1 & -1.55 & 8.5 & \bigm| & 30.5 \\
0 & 0.75 & -36.5 & \bigm| & -146 \\
0 & 18.4 & -65 & \bigm| & -242
\end{pmatrix}
\xrightarrow{R_2 \to \frac{1}{0.75}R_2}
\begin{pmatrix}
1 & -1.55 & 8.5 & \bigm| & 30.5 \\
0 & 1 & -48.7 & \bigm| & -195 \\
0 & 18.4 & -65 & \bigm| & -242
\end{pmatrix}
$$

$$
\xrightarrow{R_3 \to R_3 - 18.4R_1}
\begin{pmatrix}
1 & -1.55 & 8.5 & \bigm| & 30.5 \\
0 & 1 & -48.7 & \bigm| & -195 \\
0 & 0 & 831 & \bigm| & 3350
\end{pmatrix}
\xrightarrow{R_3 \to \frac{1}{831}R_3}
\begin{pmatrix}
1 & -1.55 & 8.5 & \bigm| & 30.5 \\
0 & 1 & -48.7 & \bigm| & -195 \\
0 & 0 & 1 & \bigm| & 4.03
\end{pmatrix}
$$

This yields

$$
x_3 \approx 4.03
$$
$$
x_2 \approx -195 + (48.7)(4.03) = 1.26
$$
$$
x_1 \approx 30.5 + (1.55)(1.26) - 8.5(4.03) = -1.8
$$

Here the errors are significant. The relative errors, given as percentages, are

$$
x_1: \; \epsilon_r = \left| \frac{-0.2}{2} \right| = 10\%
$$

$$
x_2: \; \epsilon_r = \left| \frac{0.26}{1} \right| = 26\%
$$

$$
x_3: \; \epsilon_r = \left| \frac{0.03}{4} \right| = 0.75\%
$$

Let us now repeat the procedure *with* pivoting. We obtain (with the pivots circled)

$$
\begin{pmatrix}
0.0002 & -0.00031 & 0.0017 & \bigm| & 0.00609 \\
5 & -7 & 6 & \bigm| & 7 \\
\circledR{8} & 6 & 3 & \bigm| & 2
\end{pmatrix}
$$

$$
\xrightarrow{R_1 \rightleftarrows R_3}
\begin{pmatrix}
\circledR{8} & 6 & 3 & \bigm| & 2 \\
5 & -7 & 6 & \bigm| & 7 \\
0.0002 & -0.00031 & 0.0017 & \bigm| & 0.00609
\end{pmatrix}
$$

$$
\xrightarrow{R_1 \to \frac{1}{8}R_1}
\begin{pmatrix}
1 & 0.75 & 0.375 & \bigm| & 0.25 \\
5 & -7 & 6 & \bigm| & 7 \\
0.0002 & -0.00031 & 0.0017 & \bigm| & 0.00609
\end{pmatrix}
$$

$$
\xrightarrow[R_3 \to R_3 - 0.0002R_1]{R_2 \to R_2 - 5R_1}
\begin{pmatrix}
1 & 0.75 & 0.375 & \bigm| & 0.25 \\
0 & \circledR{-10.8} & 4.13 & \bigm| & 5.75 \\
0 & -0.00046 & 0.00163 & \bigm| & 0.00604
\end{pmatrix}
$$

$$\xrightarrow{R_2 \to -\frac{1}{10.8}R_2} \begin{pmatrix} 1 & 0.75 & 0.375 & 0.25 \\ 0 & 1 & -0.382 & -0.532 \\ 0 & -0.00046 & 0.00163 & 0.00604 \end{pmatrix}$$

$$\xrightarrow{R_3 \to R_3 + 0.00046R_2} \begin{pmatrix} 1 & 0.75 & 0.375 & 0.25 \\ 0 & 1 & -0.382 & -0.532 \\ 0 & 0 & \boxed{0.00145} & 0.0058 \end{pmatrix}$$

$$\xrightarrow{R_3 \to \frac{1}{0.00145}R_3} \begin{pmatrix} 1 & 0.75 & 0.375 & 0.25 \\ 0 & 1 & -0.382 & -0.532 \\ 0 & 0 & 1 & 4.00 \end{pmatrix}$$

Hence

$$x_3 = 4.00$$
$$x_2 = -0.532 + (0.382)(4.00) = 0.996$$
$$x_1 = 0.25 - 0.75(0.996) - (0.375)(4.00) = -2.00$$

Thus, with pivoting and three-significant-digit rounding, x_1 and x_3 are obtained exactly and x_2 is obtained with the relative error of $0.004/1 = 0.4$ percent. ∎

Before leaving this section, we note that there are some matrices for which a small round-off error can lead to a large error in the final result. Such matrices are called **ill-conditioned.**

EXAMPLE 4   **An Ill-Conditioned System**   Consider the system

$$\begin{aligned} x_1 + \quad x_2 &= 1 \\ x_1 + 1.005x_2 &= 0 \end{aligned}$$

The solution is easily seen to be $x_1 = 201, x_2 = -200$. If, with or without pivoting, the coefficients are rounded to three significant digits, we obtain the system

$$\begin{aligned} x_1 + \quad x_2 &= 1 \\ x_1 + 1.01x_2 &= 0 \end{aligned}$$

with the solution $x_1 = 101, x_2 = -100$. Note that by rounding we introduced a relative error of $0.005/1.005 \approx 0.5$ percent in one coefficient but this induced an error of about 50 percent in the final answer! ∎

There are techniques for recognizing and dealing with ill-conditioned matrices. Some of these are discussed in the references listed in the first section.

PROBLEMS 7.2

In Problems 1–4 solve the given system by Gaussian elimination with partial pivoting. Use a hand calculator and round to six significant digits at every step.

1.
$$2x_1 - x_2 + x_3 = 0.3$$
$$-4x_1 + 3x_2 - 2x_3 = -1.4$$
$$3x_1 - 8x_2 + 3x_3 = 0.1$$

2.
$$4.7x_1 + 1.81x_2 + 2.6x_3 = -5.047$$
$$-3.4x_1 - 0.25x_2 + 1.1x_3 = 11.495$$
$$12.3x_1 + 0.06x_2 + 0.77x_3 = 7.9684$$

3.
$$-7.4x_1 + 3.61x_2 + 8.04x_3 = 25.1499$$
$$12.16x_1 - 2.7x_2 - 0.891x_3 = 3.2157$$
$$-4.12x_1 + 6.63x_2 - 4.38x_3 = -36.1383$$

4.
$$4.1x_1 - 0.7x_2 + 8.3x_3 + 3.9x_4 = -4.22$$
$$2.6x_1 + 8.1x_2 + 0.64x_3 - 0.8x_4 = 37.452$$
$$-5.3x_1 - 0.2x_2 + 7.4x_3 - 0.55x_4 = -25.73$$
$$0.8x_1 - 1.3x_2 + 3.6x_3 + 1.6x_4 = -7.7$$

In Problems 5 and 6 solve the system by Gaussian elimination with and without pivoting by rounding to three significant figures. Then solve the system exactly and compute the relative errors of all six computed values.

5.
$$0.1x_1 + 0.05x_2 + 0.2x_3 = 1.3$$
$$12x_1 + 25x_2 - 3x_3 = 10$$
$$-7x_1 + 8x_2 + 15x_3 = 2$$

6.
$$0.02x_1 + 0.03x_2 - 0.04x_3 = -0.04$$
$$16x_1 + 2x_2 + 4x_3 = 0$$
$$50x_1 + 10x_2 + 8x_3 = 6$$

7. Show that the system

$$x_1 + x_2 = 50$$
$$x_1 + 1.026x_2 = 20$$

is ill-conditioned if rounding is done to three significant figures. What is the approximate relative error in each answer induced by rounding?

8. Do the same for the system

$$-0.0001x_1 + x_2 = 2$$
$$-x_1 + x_2 = 3$$

7.3   SOLVING LINEAR SYSTEMS II: ITERATIVE METHODS

In the last section we developed a method for solving linear systems *directly*. That is, we carried out a fixed number of steps that led to a single answer. In numerical analysis this procedure is the exception rather than the rule. A much more common procedure is called *iteration*. With iteration, the idea is to come up with a *sequence* of approximations to the answer. If things work well, this sequence will converge to the correct answer in the sense that each term or *iterate* in the sequence is a better approximation to the answer than the ones that precede it.

To give you a taste of what iteration is all about, consider the following *algorithm* for computing $\sqrt{2}$.†

$$x_{n+1} = \frac{1}{2}\left(x_n + \frac{2}{x_n}\right) \tag{1}$$

This means that we start with a value x_0 and use equation (1) to compute x_1; we then use (1) to compute x_2; and so on. In Table 7.2, computations were carried out on a hand calculator with 10 significant digits of precision. It is apparent that the method converges very quickly to the correct answer.

Table 7.2

n	x_n	$2/x_n$	$x_n + (2/x_n)$	$x_{n+1} = \frac{1}{2}[x_n + (2/x_n)]$
0	1.0	2.0	3.0	1.5
1	1.5	1.333333333	2.833333333	1.416666667
2	1.416666667	1.411764706	2.828431373	1.414215686
3	1.414215686	1.414211438	2.828427125	1.414213562
4	1.414213562	1.414213562	2.828427125	1.414213562

There are two commonly used iterative techniques for solving a system of equations $A\mathbf{x} = \mathbf{b}$: the **Jacobi†† method** and the **Gauss-Seidel‡ method.** These methods are used under certain special circumstances. If A is ill-conditioned, for example, then, as we have seen, certain direct techniques fail. If the matrix A has a large number of zeros (A is then called a *sparse matrix*), iterative techniques will often provide better results with less work. The two methods do not always converge, however. After describing them, we shall examine some conditions under which the methods always converge. In the following discussion we assume that $\det A \neq 0$ so that the system has a unique solution.

Jacobi Iteration

Let us illustrate the method by solving a particular system. First we note that since $\det A \neq 0$, A has no zero columns. Thus, by possibly rearranging the rows of A, we can get a new coefficient matrix A' with nonzero diagonal components (see Problem 14). Hence we assume that for the $n \times n$ matrix $A = (a_{ij})$, $a_{ii} \neq 0$ for $i = 1, 2, \ldots, n$.

---

† The method used here is a special case of **Newton's method**—It was discovered in the seventeenth century by Sir Isaac Newton. Actually, the special case given by (1) was known to the ancient Babylonians. For more details, see *A History of Mathematics* by Carl Boyer, Wiley, 1968, page 31.

†† See the biographical sketch on page 490.

‡ We encountered the great Carl Friedrich Gauss in Chapter 1. P. L. V. Seidel (1821–1896) was a German mathematician.

EXAMPLE 1   **Solving a System by Jacobi Iteration**   Solve the system

$$4.4x_1 - 2.3x_2 + 0.7x_3 = -7.43$$
$$0.8x_1 + 2.5x_2 + 1.1x_3 = 12.17 \tag{2}$$
$$-1.6x_1 + 0.4x_2 - 5.2x_3 = 26.12$$

Solution   The following computations are carried out to five significant figures.

Step 1. Rewrite system (2) so that, in the ith equation, x_i is written in terms of the other variables:

$$x_1 = -\frac{7.43}{4.4} + \frac{2.3}{4.4}x_2 - \frac{0.7}{4.4}x_3 = -1.6886 + 0.52273x_2 - 0.15909x_3$$

$$x_2 = \frac{12.17}{2.5} - \frac{0.8}{2.5}x_1 - \frac{1.1}{2.5}x_3 = 4.868 - 0.32x_1 - 0.44x_3 \tag{3}$$

$$x_3 = -\frac{26.12}{5.2} - \frac{1.6}{5.2}x_1 + \frac{0.4}{5.2}x_2 = -5.0231 - 0.30769x_1 + 0.076923x_2$$

Step 2. Arbitrarily choose an initial approximation to the solution: $x_1^{(0)}, x_2^{(0)}, x_3^{(0)}$. If no other information is available, choose $x_1^{(0)} = x_2^{(0)} = x_3^{(0)} = 0$.

Step 3. Substitute these initial values into the right-hand side of (3) to obtain the new approximation $x_1^{(1)}, x_2^{(1)}, x_3^{(1)}$:

$$x_1^{(1)} = -1.6886 + 0 - 0 = -1.6886$$
$$x_2^{(1)} = 4.868 - 0 - 0 = 4.868$$
$$x_3^{(1)} = -5.0231 - 0 + 0 = -5.0231$$

Step 4. Use the values computed in Step 3 to compute $x_1^{(2)}, x_2^{(2)}, x_3^{(2)}$ and continue in this fashion to generate the sequences $\{x_1^{(n)}\}, \{x_2^{(n)}\}, \{x_3^{(n)}\}$:

$$x_1^{(2)} = -1.6886 + 0.52273x_2^{(1)} - 0.15909x_3^{(1)}$$
$$= -1.6886 + 0.52273(4.868) - 0.15909(-5.0231) = 1.6552$$
$$x_2^{(2)} = 4.868 - 0.32x_1^{(1)} - 0.44x_3^{(1)}$$
$$= 4.868 - 0.32(-1.6886) - 0.44(-5.0231) = 7.6185$$
$$x_3^{(2)} = -5.0231 - 0.30769x_1^{(1)} + 0.076923x_2^{(1)}$$
$$= -5.0231 - 0.30769(-1.6886) + 0.076923(4.868) = -4.1291$$

Continuing in this fashion, we obtain Table 7.3 (rounded to five significant figures).

Table 7.3

Iterate	$x_1^{(n)}$	$x_2^{(n)}$	$x_3^{(n)}$
0	0	0	0
1	−1.6886	4.868	−5.0231
2	1.6552	7.6185	−4.1291
3	2.9507	6.1551	−4.9464
4	2.3158	6.1002	−5.4575
5	2.3684	6.5282	−5.2664
6	2.5617	6.4273	−5.2497
7	2.5063	6.3581	−5.3169
8	2.4808	6.4054	−5.3052
9	2.5037	6.4084	−5.2937
10	2.5034	6.3960	−5.3005
11	2.4980	6.3991	−5.3014
12	2.4998	6.4013	−5.2995
13	2.5006	6.3998	−5.2999
14	2.4999	6.3998	−5.3002
15	2.5000	6.4001	−5.3000

It appears (as we could have seen as early as iterate 8) that the sequences are converging to the values $x_1 = 2.5$, $x_2 = 6.4$, $x_3 = -5.3$. This can be verified by direct substitution. Note that, at least for this problem, the Jacobi iterates converge, but they converge rather slowly. The Gauss-Seidel method can improve the speed of convergence. ∎

Gauss-Seidel Iteration

If you look closely at the steps in Jacobi iteration, you will note some inefficiency in step 3. In computing $x_2^{(n)}$, we have already computed a new value for $x_1^{(n)}$ but have instead used the old value $x_1^{(n-1)}$. Since the iterates are converging, it makes sense to use the latest available information.

EXAMPLE 2   **Solving a System by Gauss-Seidel Iteration**   Solve the system of Example 1 by using the Gauss-Seidel method.

Solution   We start, as before, with $x_1^{(0)} = x_2^{(0)} = x_3^{(0)} = 0$. Then, as before,

$$x_1^{(1)} = -1.6886 + 0 + 0 = -1.6886$$

But the next step is different. Using this new approximation to x_1, we obtain

$$x_2^{(1)} = 4.868 - 0.32x_1^{(1)} - 0.44x_3^{(0)} = 4.868 - 0.32(-1.6886) = 5.4084$$

Now we have new approximations for both x_1 and x_2. Using these values in system (3), we have

$$x_3^{(1)} = -5.0231 - 0.30769x_1^{(1)} + 0.076923x_2^{(1)}$$
$$= -5.0231 - 0.30769(-1.6886) + 0.076923(5.4084) = -4.0875$$

Continuing in this way (always using the latest approximations), we obtain Table 7.4.

Table 7.4

Iterate	$x_1^{(n)}$	$x_2^{(n)}$	$x_3^{(n)}$
0	0	0	0
1	−1.6886	5.4084	−4.0875
2	1.7888	6.0941	−5.1047
3	2.3091	6.3752	−5.2432
4	2.4780	6.3820	−5.2946
5	2.4898	6.4009	−5.2968
6	2.5000	6.3986	−5.3001
7	2.4993	6.4003	−5.2998
8	2.5002	6.3998	−5.3001
9	2.5000	6.4000	−5.3000
10	2.5000	6.4000	−5.3000

Again we conclude that $x_1 = 2.5$, $x_2 = 6.4$, $x_3 = -5.3$. Note that the Gauss-Seidel iterations converge more rapidly than the Jacobi iterations. ∎

WARNING

It is usually (but not always) true that the Gauss-Seidel method is more efficient than the Jacobi method. In Example 7 we encounter a system for which the Jacobi iterates converge (slowly) but the Gauss-Seidel iterates diverge. The converse is also possible (see Problem 13). □

Convergence. As mentioned above, these two methods do not always give a converging sequence of iterates. We cite below several conditions which ensure that the iterates converge. The proofs are beyond the scope of this text; for a good discussion of this problem, consult the book by Fox cited earlier.

DEFINITION 1   **Strictly Diagonally Dominant Matrix** Let

$$A = \begin{pmatrix} a_{11} & a_{12} & \cdots & a_{1n} \\ a_{21} & a_{22} & \cdots & a_{2n} \\ \vdots & \vdots & & \vdots \\ a_{n1} & a_{n2} & \cdots & a_{nn} \end{pmatrix}$$

Then A is **strictly diagonally dominant** if, in every row, the absolute value of the diagonal component is greater than the sum of the absolute values of the off-diagonal components. That is,

$$|a_{ii}| > |a_{i1}| + |a_{i2}| + \cdots + |a_{i,i-1}| + |a_{i,i+1}| + \cdots + |a_{in}| = \sum_{\substack{j=1 \\ j \neq i}}^{n} |a_{ij}| \qquad (4)$$

for $i = 1, 2, \ldots, n$

EXAMPLE 3   **A Strictly Diagonally Dominant System**   The matrix
$$A = \begin{pmatrix} 4.4 & -2.3 & 0.7 \\ 0.8 & 2.5 & 1.1 \\ -1.6 & 0.4 & -5.2 \end{pmatrix} \text{ is strictly diagonally dominant because}$$

$$|4.4| > |-2.3| + |0.7| = 3$$
$$|2.5| > |0.8| + |1.1| = 1.9$$

and

$$|-5.2| > |-1.6| + |0.4| = 2 \qquad \blacksquare$$

EXAMPLE 4   **Interchanging Rows May Make a Matrix Strictly Diagonally Dominant**
Consider the system

$$\begin{aligned} x_1 + 3x_2 - x_3 &= 6 \\ 4x_1 - x_2 + x_3 &= 5 \\ x_1 + x_2 - 7x_3 &= -9 \end{aligned} \qquad (5)$$

The matrix $A = \begin{pmatrix} 1 & 3 & -1 \\ 4 & -1 & 1 \\ 1 & 1 & -7 \end{pmatrix}$ is *not* strictly diagonally dominant. However, if we interchange the first two equations in system (5) (which, of course, does not change the solutions), then the matrix of the rearranged system is $\begin{pmatrix} 4 & -1 & 1 \\ 1 & 3 & -1 \\ 1 & 1 & -7 \end{pmatrix}$, which *is* strictly diagonally dominant. $\qquad \blacksquare$

The importance of systems with strictly diagonally dominant coefficient matrices is given in the following theorem. Its proof is suggested in Problem 19.

THEOREM 1   If A is strictly diagonally dominant, then both the Jacobi and the Gauss-Seidel iterations converge to the unique solution to $A\mathbf{x} = \mathbf{b}$ for any vector $\mathbf{b}$.   ∎

Note. Since the matrix of Examples 1 and 2 is strictly diagonally dominant, we know before doing any computations that both sequences of iterates will converge.

Remark. As we shall see in Example 6, there are matrices that are *not* strictly diagonally dominant but for which both sequences of iterates will converge.

Let A be an $n \times n$ matrix. Let L denote the $n \times n$ matrix consisting of the components of A below the main diagonal and zero everywhere else; D is the $n \times n$ matrix with the same diagonal components as A and zero everywhere else; U is the matrix consisting of the components of A above the main diagonal and zeros everywhere else. Here L, D, and U are called the **lower triangular,** the **diagonal,** and the **upper triangular** parts of A, respectively. We have, clearly,

$$A = L + D + U \tag{6}$$

EXAMPLE 5   **Writing A as $L + D + U$**   Let $A = \begin{pmatrix} 1 & 2 & 3 \\ 4 & 5 & 6 \\ 7 & 8 & 9 \end{pmatrix}$. Then $L = \begin{pmatrix} 0 & 0 & 0 \\ 4 & 0 & 0 \\ 7 & 8 & 0 \end{pmatrix}$,

$D = \begin{pmatrix} 1 & 0 & 0 \\ 0 & 5 & 0 \\ 0 & 0 & 9 \end{pmatrix}$, and $U = \begin{pmatrix} 0 & 2 & 3 \\ 0 & 0 & 6 \\ 0 & 0 & 0 \end{pmatrix}$.   ∎

THEOREM 2   Let $r(A)$ denote the absolute value of the eigenvalue of A with largest absolute value. Then, referring to the system $A\mathbf{x} = \mathbf{b}$ with $\det A \neq 0$:

i. The Jacobi iterates will converge if and only if

$$\boxed{r[D^{-1}(L + U)] < 1} \tag{7}$$

ii. The Gauss-Seidel iterates will converge if and only if

$$\boxed{r[(D + L)^{-1}U] < 1} \tag{8}$$

EXAMPLE 6   **A Matrix for Which Both Methods Converge**   Let $A = \begin{pmatrix} 2 & 3 \\ 1 & 4 \end{pmatrix}$; then $L =$

$\begin{pmatrix} 0 & 0 \\ 1 & 0 \end{pmatrix}, D = \begin{pmatrix} 2 & 0 \\ 0 & 4 \end{pmatrix}, U = \begin{pmatrix} 0 & 3 \\ 0 & 0 \end{pmatrix}, D^{-1} = \begin{pmatrix} \frac{1}{2} & 0 \\ 0 & \frac{1}{4} \end{pmatrix}, D^{-1}(L + U) = \begin{pmatrix} 0 & \frac{3}{2} \\ \frac{1}{4} & 0 \end{pmatrix},$

and the eigenvalues of $D^{-1}(L + U)$ are $\pm\sqrt{\frac{3}{8}}$. Thus $r[D^{-1}(L + U)] = \sqrt{\frac{3}{8}}$. Similarly, we find that $(D + L)^{-1}U = \begin{pmatrix} 0 & \frac{3}{2} \\ 0 & -\frac{3}{8} \end{pmatrix}$ with eigenvalues 0 and $-\frac{3}{8}$. Thus $r[(D + L)^{-1}U] = \frac{3}{8}$. This provides an example of a matrix that is not strictly diagonally dominant but for which both the Jacobi and the Gauss-Seidel iterates will converge. ∎

EXAMPLE 7   **A Matrix for Which the Jacobi Iterates Converge and the Gauss-Seidel Iterates Diverge**   Let $A = \begin{pmatrix} 1 & 0 & 1 \\ -1 & 1 & 0 \\ 1 & 2 & -3 \end{pmatrix}$.   Then $L = \begin{pmatrix} 0 & 0 & 0 \\ -1 & 0 & 0 \\ 1 & 2 & 0 \end{pmatrix}$,

$D = \begin{pmatrix} 1 & 0 & 0 \\ 0 & 1 & 0 \\ 0 & 0 & -3 \end{pmatrix}$, and $U = \begin{pmatrix} 0 & 0 & 1 \\ 0 & 0 & 0 \\ 0 & 0 & 0 \end{pmatrix}$. We find that $D^{-1} = \begin{pmatrix} 1 & 0 & 0 \\ 0 & 1 & 0 \\ 0 & 0 & -\frac{1}{3} \end{pmatrix}$ and

$D^{-1}(L + U) = \begin{pmatrix} 1 & 0 & 0 \\ 0 & 1 & 0 \\ 0 & 0 & -\frac{1}{3} \end{pmatrix}\begin{pmatrix} 0 & 0 & 1 \\ -1 & 0 & 0 \\ 1 & 2 & 0 \end{pmatrix} = \begin{pmatrix} 0 & 0 & 1 \\ -1 & 0 & 0 \\ -\frac{1}{3} & -\frac{2}{3} & 0 \end{pmatrix}$.

The characteristic equation of $D^{-1}(L + U)$ is $\lambda^3 + \lambda/3 - \frac{2}{3} = 0$ with approximate roots $\lambda_1 \approx 0.747$, $\lambda_2 \approx -0.374 + 0.867i$, and $\lambda_3 \approx -0.374 - 0.867i$. We have $|\lambda_1| = 0.747$, $|\lambda_2| = |\lambda_3| = \sqrt{0.374^2 + 0.867^2} \approx 0.944$. Thus $r[D^{-1}(L + U)] \approx 0.944$ and *the Jacobi iterates will converge*. On the other hand,

we find that $D + L = \begin{pmatrix} 1 & 0 & 0 \\ -1 & 1 & 0 \\ 1 & 2 & -3 \end{pmatrix}$ and

$$(D + L)^{-1}U = \begin{pmatrix} 1 & 0 & 0 \\ 1 & 1 & 0 \\ 1 & \frac{2}{3} & -\frac{1}{3} \end{pmatrix}\begin{pmatrix} 0 & 0 & 1 \\ 0 & 0 & 0 \\ 0 & 0 & 0 \end{pmatrix} = \begin{pmatrix} 0 & 0 & 1 \\ 0 & 0 & 1 \\ 0 & 0 & 1 \end{pmatrix}$$

The characteristic equation of $(D + L)^{-1}U$ is $-\lambda^2(\lambda - 1) = 0$ so that $\lambda_1 = \lambda_2 = 0$, $\lambda_3 = 1$, and $r[(D + L)^{-1}U] = 1$. Thus *the Gauss-Seidel iterates diverge*. ∎

Remark 1. It can be further shown that the *rate* of convergence in the two methods depends on the values of $r[D^{-1}(L + U)]$ and $r[(D + L)^{-1}U]$. The smaller the value of r, the faster will be the rate of convergence.†

Remark 2. Let $|A|$ denote the max-row sum norm of A (see equation 6.7.11 on page 441):

$$|A| = \max_{1 \le i \le n} \sum_{j=1}^{n} |a_{ij}| \tag{9}$$

_____

†A precise definition can be given for the term *rate of convergence*. Intuitively, the rate of convergence is a measure of how many iterates it takes to obtain a certain degree of accuracy.

Using Gershgorin's circle theorem (see Problem 18), it is not difficult to prove that, for any $n \times n$ matrix A,

$$r(A) \le |A| \tag{10}$$

The following result then follows directly from Theorem 2.

THEOREM 3   Let A, D, L, and U be as in Theorem 2. Then:

i. The sequence of Jacobi iterates converges if

$$|D^{-1}(L + U)| < 1 \tag{11}$$

ii. The sequence of Gauss-Seidel iterates converges if

$$|(D + L)^{-1}U| < 1 \tag{12}$$

∎

Remark. Each of the conditions (i) and (ii) is sufficient, not necessary.

EXAMPLE 8   **A Matrix for Which the Gauss-Seidel Iterates Converge**   Let $A = \begin{pmatrix} 1 & 0 & \frac{1}{2} \\ \frac{1}{2} & 1 & 1 \\ \frac{1}{2} & 0 & 1 \end{pmatrix}$.

Then A is not strictly diagonally dominant, but we can still show that the Gauss-Seidel iterates will converge. For $D + L = \begin{pmatrix} 1 & 0 & 0 \\ \frac{1}{2} & 1 & 0 \\ \frac{1}{2} & 0 & 1 \end{pmatrix}$,

$$(D + L)^{-1} = \begin{pmatrix} 1 & 0 & 0 \\ -\frac{1}{2} & 1 & 0 \\ -\frac{1}{2} & 0 & 1 \end{pmatrix}$$

and

$$(D + L)^{-1}U = \begin{pmatrix} 1 & 0 & 0 \\ -\frac{1}{2} & 1 & 0 \\ -\frac{1}{2} & 0 & 1 \end{pmatrix}\begin{pmatrix} 0 & 0 & \frac{1}{2} \\ 0 & 0 & 1 \\ 0 & 0 & 0 \end{pmatrix} = \begin{pmatrix} 0 & 0 & \frac{1}{2} \\ 0 & 0 & \frac{3}{4} \\ 0 & 0 & -\frac{1}{4} \end{pmatrix}$$

Thus $|(D + L)^{-1}U| = \frac{3}{4} < 1$.   ∎

Error Analysis—Or When Do We Stop? In solving problems by iteration, we always have the question of determining when to stop. There are two ways to make this decision. First, we can agree to stop after a fixed number of iterations, say, 10 or 20. But since we do not know how many iterations it will take to get a reasonably accurate answer, this method is not very useful.

A better device is to stop when the relative error ϵ_r is sufficiently small. Remember:

$$\epsilon_r = \left| \frac{x^* - x}{x} \right| \tag{13}$$

where x is the exact solution and x^* is the approximation. Of course, we cannot compute ϵ_r exactly since we do not know the exact answer x. (If we did, we would not have any problem in the first place.) We can, however, for many numerical schemes, estimate the relative error in an iteration scheme by the formula

$$\epsilon_r^{(n)} = \left| \frac{x^{(n)} - x^{(n-1)}}{x^{(n)}} \right| \tag{14}$$

Formula (14) can be explained in the following way: If we know that the scheme converges, then the iterate $x^{(n)}$ is getting closer and closer to the "correct" answer x. Thus the absolute error $\epsilon_a = |x^{(n)} - x|$ is approaching zero. But then, since $x^{(n)} \approx x$, we have $|x^{(n)} - x^{(n-1)}| \approx |x^{(n)} - x|$. This means that formula (14) approximates the true relative error

$$\epsilon_r = \left| \frac{x^{(n)} - x}{x} \right|$$

Thus we can agree to stop when $\epsilon_r^{(n)}$ is smaller than some agreed upon value ϵ. Typically we have $\epsilon = 0.1, 0.01, 0.001$, or some similar value.

In Example 1 suppose we agree to iterate until the estimated relative error $\epsilon_r^{(n)}$ in the computation of x_1 is less than 0.01. From Table 7.4 we get Table 7.5.

Table 7.5

Iterate	$x_1^{(n)}$	$\left| x_1^{(n)} - x_1^{(n-1)} \right|$	$\epsilon_r^{(n)} = \left| \dfrac{x_1^{(n)} - x_1^{(n-1)}}{x_1^{(n)}} \right|$
0	0		
1	−1.6886	1.6886	1
2	1.6552	3.3438	2.0202
3	2.9507	1.2955	0.43905
4	2.3158	0.6349	0.27416
5	2.3684	0.0526	0.02221
6	2.5617	0.1933	0.07546
7	2.5063	0.0554	0.02210
8	2.4808	0.0255	0.01028
9	2.5037	0.0229	0.00915
10	2.5034	0.0003	0.00012

Here we would stop after the ninth iterate. This would give us the estimate $x_1 \approx 2.5037$. Since $x_1 = 2.5$, the true relative error is $0.0037/2.5 = 0.00148$. It is apparent that this method gives us only a crude measure of the relative error. It is easy, however, to compute the approximations $\epsilon_r^{(n)}$; and, under conditions of convergence, they do provide a reasonable measure of how close we are getting to the right answer.

Focus on . . .

Carl Gustav Jacob Jacobi, 1804–1851

Carl Gustav Jacob Jacobi *(Historical Pictures Service, Chicago)*

The son of a prosperous banker, Carl Gustav Jacob Jacobi was born in Potsdam, Germany, in 1804. He was educated at the University of Berlin, where he received his doctorate in 1825. In 1827 he was appointed Extraordinary Professor of Mathematics at the University of Königsberg. Jacobi taught at Königsberg until 1842, when he returned to Berlin under a pension from the Prussian government. He remained in Berlin until his death in 1851.

A prolific writer of mathematical treatises, Jacobi was best known in his time for his results in the theory of elliptic functions. Today, however, he is most remembered for his work on determinants. He was one of the two most creative developers of determinant theory, the other being Cauchy. In 1829 Jacobi published a paper on algebra that contained the notation for the Jacobian that we use today in multivariable calculus. In 1841 he published an extensive treatise titled *De determinantibus functionalibus,* which was devoted to results about the Jacobian. Jacobi showed the relationship between the Jacobian of functions of several variables and the derivative of a function of one variable. He also showed that n functions of n variables are linearly independent if and only if their Jacobian is not identically zero.

In addition to being a fine mathematician, Jacobi was considered the greatest teacher of mathematics of his generation. He inspired and influenced an astonishing number of students. To dissuade his students from mastering great amounts of mathematics before setting off to do their own research, Jacobi often remarked, "Your father would never have married, and you would not be born, if he had insisted on knowing all the girls in the world before marrying one."

Jacobi believed strongly in research in pure mathematics and frequently defended it against the claim that research should always be applicable to something. He once said, "The real end of science is the honor of the human mind."

PROBLEMS   7.3

In Problems 1–6 determine whether the given matrix is strictly diagonally dominant.

1. $\begin{pmatrix} 2 & 1 \\ 1 & 2 \end{pmatrix}$

2. $\begin{pmatrix} 3 & 3 \\ 4 & 5 \end{pmatrix}$

3. $\begin{pmatrix} 1 & \frac{1}{2} & \frac{1}{2} \\ \frac{1}{2} & 1 & \frac{1}{2} \\ \frac{1}{2} & \frac{1}{2} & 1 \end{pmatrix}$

4. $\begin{pmatrix} 1 & \frac{1}{2} & \frac{1}{3} \\ \frac{1}{2} & 1 & -\frac{1}{3} \\ -\frac{1}{2} & -\frac{1}{3} & 1 \end{pmatrix}$

5. $\begin{pmatrix} 3 & -2 & 0 \\ 1 & -4 & 2 \\ -3 & 1 & -5 \end{pmatrix}$

6. $\begin{pmatrix} 6 & -2 & 3 \\ -3 & 5 & 2 \\ -2 & -4 & -7 \end{pmatrix}$

In Problems 7–12 solve the given system by using the Jacobi method and the Gauss-Seidel method. Carry out your computations until the estimated relative error $\epsilon_r^{(n)}$ is smaller than the number given in parentheses. Start with all initial approximations equal to zero and use five significant figures.

7. $\begin{aligned} 2x_1 - x_2 &= 7 \\ 3x_1 + 5x_2 &= 4 \end{aligned}$ (0.01)

8. $\begin{aligned} 3.3x_1 - 2.7x_2 &= -0.6 \\ -4.2x_1 + 8.3x_2 &= 11.95 \end{aligned}$ (0.001)

9. $\begin{aligned} 3x_1 - x_2 + x_3 &= 4 \\ 2x_1 + 5x_2 + 2x_3 &= -5 \\ x_1 + 2x_2 + 4x_3 &= 20 \end{aligned}$ (0.01)

10. $\begin{aligned} 3.8x_1 + 1.6x_2 + 0.9x_3 &= 3.72 \\ -0.7x_1 + 5.4x_2 + 1.6x_3 &= 3.16 \\ 1.5x_1 + 1.1x_2 - 3.2x_3 &= 43.78 \end{aligned}$ (0.001)

11. $\begin{aligned} 5.2x_1 + 3.1x_2 - 1.6x_3 &= 1.64 \\ 1.7x_1 + 2.4x_2 + 0.3x_3 &= 20.42 \\ -6.3x_1 - 3.7x_2 - 12.6x_3 &= 0.27 \end{aligned}$ (0.001)

12. $\begin{aligned} -3.1x_1 + 1.9x_2 - 0.77x_3 &= -12.806 \\ 0.9x_1 - 2.4x_2 + 1.06x_3 &= 12.165 \\ 7.6x_1 - 3.9x_2 + 16.5x_3 &= 27.931 \end{aligned}$ (0.0001)

13. Consider the system

$$x_1 + \tfrac{1}{2}x_2 + \tfrac{1}{2}x_3 = 2$$
$$\tfrac{1}{2}x_1 + x_2 + \tfrac{1}{2}x_3 = 2$$
$$\tfrac{1}{2}x_1 + \tfrac{1}{2}x_2 + x_3 = 2$$

    a.   Show that the matrix of the system is not strictly diagonally dominant.

    b.   Starting with $x_1^{(0)} = x_2^{(0)} = x_3^{(0)} = 0.8$, show that the Jacobi iterates oscillate back and forth between the values 0.8 and 1.2. That is, show that the sequence of Jacobi iterates diverges.

    c.   Show that the Gauss-Seidel iterates converge to the solution $x_1 = x_2 = x_3 = 1$ by computing eight iterates and rounding to five significant figures.

  *d.   Explain the results of parts (b) and (c) in light of Theorem 2.

*14. Let A be an $n \times n$ matrix with $\det A \neq 0$. Show that it is always possible to rearrange the rows of A so that the diagonal components of A are all nonzero.

15. Let A be a diagonal matrix with $\det A \neq 0$. Show that

$$r[D^{-1}(L + U)] = r[(D + L)^{-1}U] = 0.$$

16. Let A be an invertible upper or lower triangular matrix. Show that the sequences of Jacobi and Gauss-Seidel iterates always converge.

17. Let $A = \begin{pmatrix} a & b \\ c & d \end{pmatrix}$. Show that both the Jacobi and the Gauss-Seidel iterates converge if and only if $|bc/ad| < 1$. This shows that it is impossible to find an example of a 2×2 system where one sequence of iterates converges while the other does not.

*18. Use Gershgorin's circle theorem (Theorem 6.8.3) to show that $r(A) \leq |A|$, where

$$|A| = \max_{1 \leq i \leq n} \sum_{j=1}^{n} |a_{ij}|.$$

*19. Use part (i) of Theorem 3 to show that if A is strictly diagonally dominant, then the Jacobi iterates converge.

7.4   COMPUTING EIGENVALUES AND EIGENVECTORS

As we have seen, the computation of eigenvalues and eigenvectors for a given matrix A is important for a variety of applications. It is tempting to estimate eigenvalues by first finding the characteristic polynomial $p(\lambda) = \det (A - \lambda I)$ and then estimating, directly, the roots of $p(\lambda)$. There are two problems with this approach. First, polynomials are often ill-conditioned; that is, a small round-off error in the coefficients of the polynomial can lead to large errors in the roots. Second, even if the coefficients of $p(\lambda)$ are exact, it is still difficult to find all the roots of a polynomial. For these reasons a number of techniques have been devised for computing eigenvalues and eigenvectors directly. The first of these is used to compute the eigenvalue of largest absolute value.

DEFINITION 1   **Dominant Eigenvalue and Eigenvector** Let $\lambda_1, \lambda_2, \ldots, \lambda_n$ be the eigenvalues of A. Then the eigenvalue λ_1 is **dominant** if

$$|\lambda_1| > |\lambda_i| \qquad \text{for } i = 2, \ldots, n \tag{1}$$

If $\mathbf{v}_1$ is an eigenvector of A corresponding to λ_1, then $\mathbf{v}_1$ is called a **dominant eigenvector.**

EXAMPLE 1   **Determining the Dominant Eigenvalue**   If the eigenvalues of A are $-4, -2, 1, 3$, then -4 is dominant.   ∎

EXAMPLE 2   **The Case of No Dominant Eigenvalue**   If the eigenvalues of A are $-5, 3, 5$, then A has no dominant eigenvalue since $|-5| = |5|$.   ∎

We now describe a method, called the **power method,** for computing the dominant eigenvalue and eigenvector of a matrix.

The Power Method

Let $\lambda_1, \lambda_2, \ldots, \lambda_n$ be the eigenvalues of A and suppose that

$$|\lambda_1| > |\lambda_2| \geq |\lambda_3| \geq \cdots \geq |\lambda_n| \tag{2}$$

That is, λ_1 is the dominant eigenvalue. Suppose further that A is diagonalizable; that is, A has n linearly independent eigenvectors $\mathbf{u}_1, \mathbf{u}_2, \ldots, \mathbf{u}_n$ where $\mathbf{u}_1$ corresponds to λ_1, and so on. Let $\mathbf{x}_0$ be a vector in $\mathbb{R}^n$. There are constants $c_1, c_2, \ldots,$ c_n such that

$$\mathbf{x}_0 = c_1\mathbf{u}_1 + c_2\mathbf{u}_2 + \cdots + c_n\mathbf{u}_n \tag{3}$$

We assume that $c_1 \neq 0$. Define a sequence of iterates by the formula

$$\boxed{\mathbf{x}_{n+1} = A\mathbf{x}_n} \tag{4}$$

Then

$$\mathbf{x}_1 = A\mathbf{x}_0 = c_1 A\mathbf{u}_1 + c_2 A\mathbf{u}_2 + \cdots + c_n A\mathbf{u}_n$$
$$= c_1\lambda_1\mathbf{u}_1 + c_2\lambda_2\mathbf{u}_2 + \cdots + c_n\lambda_n\mathbf{u}_n$$

Continuing to multiply by powers of A, we find that

$$\mathbf{x}_2 = A\mathbf{x}_1 = A^2\mathbf{x}_0 = A(A\mathbf{x}_0) = c_1\lambda_1 A\mathbf{u}_1 + c_2\lambda_2 A\mathbf{u}_2 + \cdots + c_n\lambda_n A\mathbf{u}_n$$
$$= c_1\lambda_1^2\mathbf{u}_1 + c_2\lambda_2^2\mathbf{u}_2 + \cdots + c_n\lambda_n^2\mathbf{u}_n$$
$$\vdots$$

$$\mathbf{x}_k = A^k\mathbf{x}_0 = c_1\lambda_1^k\mathbf{u}_1 + c_2\lambda_2^k\mathbf{u}_2 + \cdots + c_n\lambda_n^k\mathbf{u}_n \tag{5}$$

or

$$\mathbf{x}_k = A^k\mathbf{x}_0 = \lambda_1^k\left[c_1\mathbf{u}_1 + c_2\left(\frac{\lambda_2}{\lambda_1}\right)^k\mathbf{u}_2 + \cdots + c_n\left(\frac{\lambda_n}{\lambda_1}\right)^k\mathbf{u}_n\right] \tag{6}$$

Since $|\lambda_i| < |\lambda_1|$ for $i = 2, 3, \ldots, n$, we see that $|\lambda_i/\lambda_1|^k$ approaches zero as k increases. Therefore we may write

$$\boxed{\mathbf{x}_k = A^k\mathbf{x}_0 \approx \lambda_1^k c_1\mathbf{u}_1} \tag{7}$$

Suppose $\mathbf{u}_1 = \begin{pmatrix} a_1 \\ a_2 \\ \vdots \\ a_n \end{pmatrix}$. Then $\lambda_1^k c_1\mathbf{u}_1 = \begin{pmatrix} \lambda_1^k c_1 a_1 \\ \lambda_1^k c_1 a_2 \\ \vdots \quad \vdots \\ \lambda_1^k c_1 a_n \end{pmatrix}$.

Now let a_j be nonzero. Then we form the quotient

$$\boxed{\alpha_j^{(k+1)} = \frac{j\text{th component of } A^{k+1}\mathbf{x}_0}{j\text{th component of } A^k\mathbf{x}_0} \approx \frac{\lambda_1^{k+1}c_1 a_j}{\lambda_1^k c_1 a_j} = \lambda_1} \tag{8}$$

This gives us a method for computing λ_1. We simply look at the ratio of the jth components of $\mathbf{x}_{k+1}$ and $\mathbf{x}_k$ and let k get large. Moreover, once we have found λ_1, we also know an eigenvector corresponding to λ_1—for, by equation (7),

$$\boxed{\mathbf{x}_k \approx \lambda_1^k c_1 \mathbf{u}_1} \tag{9}$$

is an eigenvector corresponding to λ_1 since $\lambda_1^k c_1$ is a scalar and $\mathbf{u}_1$ is an eigenvector.

WARNING   The power method described above will work only if A has a dominant eigenvalue.†   □

EXAMPLE 3   **Finding the Dominant Eigenvalue and Eigenvector by the Power Method**
Use the power method to find the dominant eigenvalue and eigenvector of

$$A = \begin{pmatrix} -4 & -5 \\ 1 & 2 \end{pmatrix}.$$

Solution   Here $\mathbf{x}_0$ is arbitrary, so we choose a simple value for it: $\mathbf{x}_0 = \begin{pmatrix} 1 \\ 1 \end{pmatrix}$. Then

$$\mathbf{x}_1 = A\mathbf{x}_0 = \begin{pmatrix} -4 & -5 \\ 1 & 2 \end{pmatrix}\begin{pmatrix} 1 \\ 1 \end{pmatrix} = \begin{pmatrix} -9 \\ 3 \end{pmatrix} \qquad \alpha_1^{(1)} = -\tfrac{9}{1} = -9 \qquad \alpha_2^{(1)} = \tfrac{3}{1} = 3$$

$$\mathbf{x}_2 = A\mathbf{x}_1 = \begin{pmatrix} -4 & -5 \\ 1 & 2 \end{pmatrix}\begin{pmatrix} -9 \\ 3 \end{pmatrix} = \begin{pmatrix} 21 \\ -3 \end{pmatrix} \qquad \alpha_1^{(2)} = -\tfrac{21}{9} = -2.3333 \qquad \alpha_2^{(2)} = -\tfrac{3}{3} = -1$$

Continuing in this fashion, we obtain Table 7.6. All results are rounded to five significant figures.

Table 7.6

Iterate	$\mathbf{x}_k$ (as a row vector)	$\alpha_1^{(k)}$	$\alpha_2^{(k)}$
0	(1, 1)	—	—
1	(−9, 3)	−9	3
2	(21, −3)	−2.3333	−1
3	(−69, 15)	−3.2857	−5
4	(201, −39)	−2.9130	−2.6
5	(−609, 123)	−3.0299	−3.1538
6	(1821, −363)	−2.9901	−2.9512
7	(−5469, 1095)	−3.0033	−3.0165
8	(16401, −3279)	−2.9989	−2.9945
9	(−49209, 9843)	−3.0004	−3.0018

---

† It can be shown that the power method will work even when A is not diagonalizable. In that case, however, convergence is at a slower rate.

It appears that $\alpha_1^{(k)}$ and $\alpha_2^{(k)}$ are converging to -3—which, as is easily verified, is the dominant eigenvalue of A (the other one is $\lambda_2 = 1$). Moreover, we see that $\mathbf{v}_1 = \begin{pmatrix} -49209 \\ 9843 \end{pmatrix}$ is approximately equal to an eigenvector of A. To simplify this vector, we "normalize" it by dividing through by its largest component (in absolute value) $-49{,}209$ to obtain $\mathbf{v}_1' = \begin{pmatrix} 1 \\ -0.20002 \end{pmatrix} \approx \begin{pmatrix} 1 \\ -\frac{1}{5} \end{pmatrix}$. It is easily verified that $\begin{pmatrix} 1 \\ -\frac{1}{5} \end{pmatrix}$ is an eigenvector of A corresponding to the eigenvalue $\lambda_1 = -3$. ∎

The Power Method with Scaling

In the last example we saw that the iterates grew very rapidly in size. To prevent this, we do what we did to complete the problem: We *normalize* or *scale* the vector $\mathbf{x}_k$ by dividing it by its largest component (in absolute value). If we call the new scaled iterate $\mathbf{x}_k'$, then

$$\boxed{\mathbf{x}_{k+1} = A\mathbf{x}_k'} \tag{10}$$

This new method is called the **power method with scaling.** It will give us an eigenvector $\mathbf{u}$ with largest component 1 and we can then find the dominant eigenvalue by solving the equation $A\mathbf{u} = \lambda_1\mathbf{u}$ for λ_1.

EXAMPLE 4 **Finding the Dominant Eigenvalue and Eigenvector by the Power Method with Scaling**   Redo Example 3 by using the power method with scaling.

Solution   If $\mathbf{x}_0 = \begin{pmatrix} 1 \\ 1 \end{pmatrix}$, then $\mathbf{x}_1 = \begin{pmatrix} -9 \\ 3 \end{pmatrix}$ as before, and

$$\mathbf{x}_1' = -\frac{1}{9}\begin{pmatrix} -9 \\ 3 \end{pmatrix} = \begin{pmatrix} 1 \\ -\frac{1}{3} \end{pmatrix}$$

Then

$$\mathbf{x}_2 = A\mathbf{x}_1' = \begin{pmatrix} -4 & -5 \\ 1 & 2 \end{pmatrix}\begin{pmatrix} 1 \\ -\frac{1}{3} \end{pmatrix} = \begin{pmatrix} -\frac{7}{3} \\ \frac{1}{3} \end{pmatrix}$$

so that

$$\mathbf{x}_2' = -\frac{3}{7}\begin{pmatrix} -\frac{7}{3} \\ \frac{1}{3} \end{pmatrix} = \begin{pmatrix} 1 \\ -\frac{1}{7} \end{pmatrix} = \begin{pmatrix} 1 \\ -0.14286 \end{pmatrix}$$

We carry out further iterations in Table 7.7.

As before, we can conclude that $\mathbf{v} = \begin{pmatrix} 1 \\ -0.2 \end{pmatrix}$ is an eigenvector of A corresponding to λ_1. Then $A\mathbf{v} = \lambda_1\mathbf{v}$ or $\begin{pmatrix} -4 & -5 \\ 1 & 2 \end{pmatrix}\begin{pmatrix} 1 \\ -0.2 \end{pmatrix} = \begin{pmatrix} \lambda_1 \\ -0.2\lambda_1 \end{pmatrix}$, which yields $\begin{pmatrix} -3 \\ 0.6 \end{pmatrix} = \begin{pmatrix} \lambda_1 \\ -0.2\lambda_1 \end{pmatrix}$. Hence $\lambda_1 = -3$.

Table 7.7

Iterate	$\mathbf{x}_k$	$\mathbf{x}_k'$ (normalized)
0	(1, 1)	(1, 1)
1	(−9, 3)	(1, −0.33333)
2	(−2.3333, 0.33333)	(1, −0.14286)
3	(−3.2857, 0.71428)	(1, −0.21739)
4	(−2.9131, 0.56522)	(1, −0.19403)
5	(−3.0299, 0.61194)	(1, −0.20197)
6	(−2.9902, 0.59606)	(1, −0.19934)
7	(−3.0033, 0.60132)	(1, −0.20022)
8	(−2.9989, 0.59956)	(1, −0.19993)
9	(−3.0004, 0.60014)	(1, −0.20002)

Note. In Table 7.7 the first component of $\mathbf{x}_k$ tends to -3. This must be the case since, for k large,

$$\mathbf{x}_k = A\mathbf{x}_{k-1}' \approx \lambda_1 \mathbf{x}_{k-1}'$$

But the first component of $\mathbf{x}_{k-1}'$ is 1. Hence the first component of $\mathbf{x}_k \approx \lambda_1$. This means that we can find λ_1 directly from the table.   ∎

Deflation

The power method has the obvious drawback of giving us only the dominant eigenvalue. There are many ways to compute other eigenvalues. We examine one method here. First, we need the following result.

THEOREM 1   Let $\lambda_1, \lambda_2, \ldots, \lambda_n$ be the eigenvalues of A. Let λ_1 be the dominant eigenvalue with eigenvector $\mathbf{u}_1$. Let $\mathbf{v}$ be a column vector such that $\mathbf{u}_1 \cdot \mathbf{v} = 1$. If the matrix B is given by

$$B = A - \lambda_1 \mathbf{u}_1 \mathbf{v}^t \qquad\qquad (11)$$

then the eigenvalues of B are $\{0, \lambda_2, \lambda_3, \ldots, \lambda_n\}$.† The proof of this theorem can be found in the referenced book by Blum (page 239).   ∎

---

† Note that since $\mathbf{u}_1$ is an $n \times 1$ matrix (a column vector) and $\mathbf{v}$ is also an $n \times 1$ matrix, then $\mathbf{v}^t$ is a $1 \times n$ matrix and $\mathbf{u}_1\mathbf{v}^t$ is an $n \times n$ matrix.

The process of finding the second and subsequent eigenvalues by finding the eigenvalues of B by the power method with scaling is called **deflation.**

EXAMPLE 5 **Finding All Eigenvalues by the Power Method with Scaling**   In Example 3 we had $A = \begin{pmatrix} -4 & -5 \\ 1 & 2 \end{pmatrix}$, $\lambda_1 = -3$, and $\mathbf{u}_1 = \begin{pmatrix} 1 \\ -\frac{1}{5} \end{pmatrix}$. If $\mathbf{v} = \begin{pmatrix} \frac{1}{2} \\ -\frac{5}{2} \end{pmatrix}$, then $\mathbf{u}_1 \cdot \mathbf{v} = 1$ and

$$B = A - \lambda_1 \mathbf{u}_1 \mathbf{v}^t = \begin{pmatrix} -4 & -5 \\ 1 & 2 \end{pmatrix} + 3 \begin{pmatrix} 1 \\ -\frac{1}{5} \end{pmatrix} (\tfrac{1}{2}, -\tfrac{5}{2})$$

$$= \begin{pmatrix} -4 & -5 \\ 1 & 2 \end{pmatrix} + 3 \begin{pmatrix} \frac{1}{2} & -\frac{5}{2} \\ -\frac{1}{10} & \frac{1}{2} \end{pmatrix} = \begin{pmatrix} -4 & -5 \\ 1 & 2 \end{pmatrix} + \begin{pmatrix} \frac{3}{2} & -\frac{15}{2} \\ -\frac{3}{10} & \frac{3}{2} \end{pmatrix} = \begin{pmatrix} -\frac{5}{2} & -\frac{25}{2} \\ \frac{7}{10} & \frac{7}{2} \end{pmatrix}$$

We find that $\det B = 0$; hence zero is an eigenvalue of B, as expected. The other eigenvalue of B is the second eigenvalue of A. We compute this by the power method with scaling. Starting with $x_0 = \begin{pmatrix} 1 \\ 1 \end{pmatrix}$, we obtain

$$\mathbf{x}_1 = B\mathbf{x}_0 = \begin{pmatrix} -2.5 & -12.5 \\ 0.7 & 3.5 \end{pmatrix} \begin{pmatrix} 1 \\ 1 \end{pmatrix} = \begin{pmatrix} -15 \\ 4.2 \end{pmatrix}$$

Then

$$\mathbf{x}_1' = \begin{pmatrix} 1 \\ -0.28 \end{pmatrix}$$

and

$$\mathbf{x}_2 = B\mathbf{x}_1 = \begin{pmatrix} -2.5 & -12.5 \\ 0.7 & 3.5 \end{pmatrix} \begin{pmatrix} 1 \\ -0.28 \end{pmatrix} = \begin{pmatrix} 1 \\ -0.28 \end{pmatrix}$$

Thus, without further ado, we see that $\begin{pmatrix} 1 \\ -0.28 \end{pmatrix}$ is an eigenvector of B corresponding to the eigenvalue $\lambda_2 = 1$. Therefore the eigenvalues of $\begin{pmatrix} -4 & -5 \\ 1 & 2 \end{pmatrix}$ are -3 and 1.   ∎

Note. It is not a coincidence that $B\mathbf{x}_0$ is an eigenvector of B if B is a 2×2 matrix—this is *always* the case. (See Problem 14 for a suggestion as to why this is so.)

EXAMPLE 6 **Finding All Eigenvalues by the Power Method with Deflation and Scaling** Compute the eigenvalues of $A = \begin{pmatrix} 4 & -1 & 1 \\ -1 & 3 & -2 \\ 1 & -2 & 3 \end{pmatrix}$ by the power method with scaling and deflation.

Solution   Let $\mathbf{x}_0 = \begin{pmatrix} 1 \\ 1 \\ 1 \end{pmatrix}$. Then $\mathbf{x}_1 = A\mathbf{x}_0 = \begin{pmatrix} 4 \\ 0 \\ 2 \end{pmatrix}$ and $\mathbf{x}_1' = \begin{pmatrix} 1 \\ 0 \\ 0.5 \end{pmatrix}$. Similarly,

$$\mathbf{x}_2 = A\mathbf{x}_1 = \begin{pmatrix} 4.5 \\ -2 \\ 2.5 \end{pmatrix} \text{ and } \mathbf{x}_2' = \begin{pmatrix} 1 \\ -0.44444 \\ 0.55556 \end{pmatrix}.$$ Continuing in this manner, we obtain the values in Table 7.8.

Table 7.8

Iterate	$\mathbf{x}_k$	$\mathbf{x}_k'$	α_k = 1st Component of $\mathbf{x}_k$
0	(1, 1, 1)	(1, 1, 1)	1
1	(4, 0, 2)	(1, 0, 0.5)	4
2	(4.5, −2, 2.5)	(1, −0.44444, 0.55556)	4.5
3	(5, −3.4444, 3.5556)	(1, −0.68888, 0.71112)	5
4	(5.4, −4.4889, 4.5111)	(1, −0.83128, 0.83539)	5.4
5	(5.6667, −5.1646, 5.1687)	(1, −0.91139, 0.91212)	5.6667
6	(5.8235, −5.5584, 5.5591)	(1, −0.95448, 0.95460)	5.8235
7	(5.9091, −5.7726, 5.7728)	(1, −0.97690, 0.97693)	5.9091
8	(5.9538, −5.8846, 5.8846)	(1, −0.98838, 0.98838)	5.9538
9	(5.9768, −5.9419, 5.9419)	(1, −0.99416, 0.99416)	5.9768
10	(5.9883, −5.9708, 5.9708)	(1, −0.99708, 0.99708)	5.9883

It appears that the α_k's are converging to $\lambda_1 = 6$ with corresponding eigenvector $\mathbf{u}_1 = \begin{pmatrix} 1 \\ -1 \\ 1 \end{pmatrix}$. This is easily verified. Next we find a vector $\mathbf{v}$ such that $\mathbf{u}_1 \cdot \mathbf{v} = 1$. One obvious choice is $\mathbf{v} = \begin{pmatrix} \frac{1}{3} \\ -\frac{1}{3} \\ \frac{1}{3} \end{pmatrix}$. Then

$$\mathbf{u}_1\mathbf{v}^t = \begin{pmatrix} 1 \\ -1 \\ 1 \end{pmatrix} (\tfrac{1}{3}, -\tfrac{1}{3}, \tfrac{1}{3}) = \begin{pmatrix} \frac{1}{3} & -\frac{1}{3} & \frac{1}{3} \\ -\frac{1}{3} & \frac{1}{3} & -\frac{1}{3} \\ \frac{1}{3} & -\frac{1}{3} & \frac{1}{3} \end{pmatrix}$$

so that

$$B = A - \lambda_1\mathbf{u}_1\mathbf{v}^t = \begin{pmatrix} 4 & -1 & 1 \\ -1 & 3 & -2 \\ 1 & -2 & 3 \end{pmatrix} - 6\begin{pmatrix} \frac{1}{3} & -\frac{1}{3} & \frac{1}{3} \\ -\frac{1}{3} & \frac{1}{3} & -\frac{1}{3} \\ \frac{1}{3} & -\frac{1}{3} & \frac{1}{3} \end{pmatrix}$$

$$= \begin{pmatrix} 4 & -1 & 1 \\ -1 & 3 & -2 \\ 1 & -2 & 3 \end{pmatrix} - \begin{pmatrix} 2 & -2 & 2 \\ -2 & 2 & -2 \\ 2 & -2 & 2 \end{pmatrix} = \begin{pmatrix} 2 & 1 & -1 \\ 1 & 1 & 0 \\ -1 & 0 & 1 \end{pmatrix}$$

We see that $\det B = 0$; hence zero is an eigenvalue of B. To find the dominant eigenvalue of B, we again use the power method with scaling. The results are tabulated in Table 7.9.

Table 7.9

Iterate	$\mathbf{x}_k$	$\mathbf{x}_k'$	$\alpha_k = $ 1st Component of $\mathbf{x}_k$
0	$(1, 1, 1)$	$(1, 1, 1)$	1
1	$(2, 2, 0)$	$(1, 1, 0)$	2
2	$(3, 2, -1)$	$(1, 0.66667, -0.33333)$	3
3	$(3, 1.6667, -1.3333)$	$(1, 0.55557, -0.44443)$	3
4	$(3, 1.5556, -1.4444)$	$(1, 0.51853, -0.48147)$	3
5	$(3, 1.5185, -1.4815)$	$(1, 0.50617, -0.49383)$	3
6	$(3, 1.5062, -1.4938)$	$(1, 0.50207, -0.49793)$	3
7	$(3, 1.5021, -1.4979)$	$(1, 0.50070, -0.49930)$	3

Now it seems that the iterates are converging to $\lambda_2 = 3$ and $\mathbf{u}_2 = (1, \frac{1}{2}, -\frac{1}{2})$. Again this is easily verified. Although $\mathbf{u}_2$ is an eigenvector of both A and B, this is not always the case. (In Example 5, for instance, $\begin{pmatrix} 1 \\ -0.28 \end{pmatrix}$ was an eigenvector of B but not of A.)

Finally, we use deflation again to find the last eigenvalue of B (and therefore of A). We set $\mathbf{v}_1 = \begin{pmatrix} 1 \\ 0 \\ 0 \end{pmatrix}$. Then $\mathbf{u}_2 \cdot \mathbf{v}_1 = 1$ and $\mathbf{u}_2 \mathbf{v}_1' = \begin{pmatrix} 1 \\ \frac{1}{2} \\ -\frac{1}{2} \end{pmatrix} (1, 0, 0) = \begin{pmatrix} 1 & 0 & 0 \\ \frac{1}{2} & 0 & 0 \\ -\frac{1}{2} & 0 & 0 \end{pmatrix}$ and $C = B - \lambda_2 \mathbf{u}_2 \mathbf{v}_1' = \begin{pmatrix} 2 & 1 & -1 \\ 1 & 1 & 0 \\ -1 & 0 & 1 \end{pmatrix} - \begin{pmatrix} 3 & 0 & 0 \\ \frac{3}{2} & 0 & 0 \\ -\frac{3}{2} & 0 & 0 \end{pmatrix} = \begin{pmatrix} -1 & 1 & -1 \\ -\frac{1}{2} & 1 & 0 \\ \frac{1}{2} & 0 & 1 \end{pmatrix}$. We omit the iteration, which shows that the dominant eigenvalue of C is $\lambda_3 = 1$. Thus the eigenvalues of A are 6, 3, and 1. ∎

The power method together with deflation provides a reasonable way to find the eigenvalues of A if no two eigenvalues of A have the same absolute value and if each approximation to an eigenvalue is a good one. If, for example, λ_1 is inaccurate, then the computation of λ_2 by deflation could be a good deal more inaccurate.

There are many other ways to compute, numerically, the eigenvalues of a square matrix. One method that works fairly well on a symmetric matrix is called **Jacobi's method.** The idea is to compute a sequence of orthogonal matrices whose diagonal components approach the eigenvalues of A. Many of the references listed in Section 7.1 discuss that method. Finally, we note that the decision "when to stop" can be made, as in the last section, by computing approximate values of the relative error $\epsilon_r^{(n)}$.

PROBLEMS 7.4

In Problems 1–6 estimate the dominant eigenvalue and eigenvector of A by using the power method with scaling.

1. $\begin{pmatrix} -2 & -2 \\ -5 & 1 \end{pmatrix}$
2. $\begin{pmatrix} 8 & 3 \\ -3 & -2 \end{pmatrix}$
3. $\begin{pmatrix} -22.3 & -32 \\ 12 & 17.7 \end{pmatrix}$

4. $\begin{pmatrix} 1 & -1 & 4 \\ 3 & 2 & -1 \\ 2 & 1 & -1 \end{pmatrix}$
5. $\begin{pmatrix} 3 & 2 & 4 \\ 2 & 0 & 2 \\ 4 & 2 & 3 \end{pmatrix}$
6. $\begin{pmatrix} 5 & 4 & 2 \\ 4 & 5 & 2 \\ 2 & 2 & 2 \end{pmatrix}$

7. Use the power method to estimate the dominant eigenvalue of $A = \begin{pmatrix} 1 & 7 \\ 6 & 3 \end{pmatrix}$:

   a. Rounding to five significant figures, continue the iterations until the estimated relative error $\epsilon_r^{(n)} < 0.001$.

   b. Compute the dominant eigenvalue exactly. What is the exact value of ϵ_r?

8. For the matrix $A = \begin{pmatrix} -16.32 & 13 \\ 8 & 4.79 \end{pmatrix}$, follow the steps of Problem 7. Use six significant figures.

9. Show that the iterates of the power method fail to converge for the matrix $A = \begin{pmatrix} -3 & 5 \\ -2 & 3 \end{pmatrix}$. Explain why.

10. Do the same for the matrix $A = \begin{pmatrix} 2 & -1 \\ 5 & -2 \end{pmatrix}$.

In Problems 11–13 use deflation to find the other eigenvalues.

11. For the matrix of Problem 1   **12.** For the matrix of Problem 3

13. For the matrix of Problem 4

14. Let $A = \begin{pmatrix} a & 0 \\ 0 & b \end{pmatrix}$. Show that, for any 2-vector $\mathbf{x}_0$, $B\mathbf{x}_0$ is an eigenvector of B, where B is defined by equation (11).

SUMMARY

Consider the system

$$
\begin{aligned}
a_{11}x_1 + a_{12}x_2 + \cdots + a_{1n}x_n &= b_1 \\
a_{21}x_1 + a_{22}x_2 + \cdots + a_{2n}x_n &= b_2 \\
&\ \ \vdots \\
a_{m1}x_1 + a_{m2}x_2 + \cdots + a_{mn}x_n &= b_m
\end{aligned}
\qquad (*)
$$

- To solve the system (∗) by **Gaussian elimination with partial pivoting,** perform the following modification of Gaussian elimination: (pp. 475, 476)

  Step 1. Write the system in augmented-matrix form. From the first column with nonzero components (called the **pivot column**), select the component with the *largest absolute value.* This component is called the **pivot.**

  Step 2. Rearrange the rows to move the pivot to the top.

  Step 3. Divide the first row by the pivot.

  Step 4. Add multiples of the first row to the other rows to make all the other components in the pivot column equal to zero.

  Step 5. Cover the first row and perform Steps 1–4 on the resulting *submatrix.*

  Step 6. Continue in this manner until the matrix is in row echelon form.

  Step 7. Use **back substitution** to find the solution (if any) to the system.

- The **Jacobi method** for solving system (∗) is an iterative method in which it is assumed that $\det A \neq 0$ where A is the $n \times n$ coefficient matrix of (∗).   (pp. 481, 482)

  Step 1. Rewrite system (∗) so that, in the ith equation, x_i is written in terms of the other variables:

  Step 2. Arbitrarily choose an initial approximation to the solution: $x_1^{(0)}, x_2^{(0)}, \ldots, x_n^{(0)}$. If no other information is available, choose $x_1^{(0)} = x_2^{(0)} = \cdots = x_n^{(0)} = 0$.

  Step 3. Substitute these initial values into the right-hand side of the rewritten system obtained in Step 1 to obtain the new approximation $x_1^{(1)}, x_2^{(1)}, \ldots, x_n^{(1)}$.

  Step 4. Use the values computed in Step 3 to compute $x_1^{(2)}, x_2^{(2)}, \ldots, x_n^{(2)}$ and continue in this fashion to generate the sequences $\{x_1^{(k)}, x_2^{(k)}, \ldots, x_n^{(k)}\}$.

- The **Gauss-Seidel** iterative method is like the Jacobi method except that the newest values of $x_1^{(k)}, x_2^{(k)}, \ldots, x_n^{(k)}$ are used at each step in the iteration.   (p. 483)

- The matrix A is **strictly diagonally dominant** if, in every row, the absolute value of the diagonal component is greater than the sum of the absolute values of the off-diagonal components. That is,   (p. 485)

$$|a_{ii}| > |a_{i1}| + |a_{i2}| + \cdots + |a_{i,i-1}| + |a_{i,i+1}| + \cdots + |a_{in}| = \sum_{\substack{j=1 \\ j \neq i}}^{n} |a_{ij}|$$

  for $i = 1, 2, \ldots, n$

- If A is strictly diagonally dominant, then both the Jacobi and the Gauss-Seidel iterations converge to the unique solution to $A\mathbf{x} = \mathbf{b}$ for any vector $\mathbf{b}$.   (p. 486)

- Let $r(A)$ denote the absolute value of the eigenvalue of A with largest absolute value. Then, referring to the system $A\mathbf{x} = \mathbf{b}$ with $\det A \neq 0$:   (p. 486)

The Jacobi iterates will converge if and only if

$$r[D^{-1}(L + U)] < 1$$

The Gauss-Seidel iterates will converge if and only if

$$r[(D + L)^{-1}U] < 1$$

Here L, D, and U are the **lower triangular,** the **diagonal,** and the **upper triangular** parts of A and

$$A = L + D + V$$

- **Dominant Eigenvalue and Eigenvector**·Let $\lambda_1, \lambda_2, \ldots, \lambda_n$ be the eigenvalues of A. Then the eigenvalue λ_1 is **dominant** if   (p. 492)

$$|\lambda_1| > |\lambda_i| \qquad \text{for } i = 2, \ldots, n \qquad (1)$$

If $\mathbf{v}_1$ is an eigenvector of A corresponding to λ_1, then $\mathbf{v}_1$ is called a **dominant eigenvector.**
- The **Power Method for Computing the Dominant Eigenvalue**   (pp. 493, 494)
  Let $\lambda_1, \lambda_2, \ldots, \lambda_n$ be the eigenvalues of A and suppose that

$$|\lambda_1| > |\lambda_2| \geq |\lambda_3| \geq \cdots \geq |\lambda_n| \qquad (2)$$

That is, λ_1 is the dominant eigenvalue. Suppose further that A is diagonalizable; that is, A has n linearly independent eigenvectors $\mathbf{u}_1, \mathbf{u}_2, \ldots, \mathbf{u}_n$. Let $\mathbf{x}_0$ be a vector in $\mathbb{R}^n$. Then the sequence of iterates

$$\mathbf{x}_{n+1} = A\mathbf{x}_n$$

converges to an eigenvector corresponding to λ_1. Also

$$\alpha_j^{(n+1)} = \frac{j\text{th component of } A^{n+1}\mathbf{x}_0}{j\text{th component of } A^n\mathbf{x}_0} \qquad \text{converges to } \lambda_1$$

- The **power method with scaling** is a modification of the power method. At each step in the iteration divide $\mathbf{x}_k$ by its largest component to obtain the new iterate $\mathbf{x}'_n$. Then $\mathbf{x}_{n+1} = A\mathbf{x}'_n$.   (p. 495)
- **Deflation**·Obtain the dominant eigenvalue of A by the power method.   (pp. 496, 497)
  Let $\lambda_1, \lambda_2, \ldots, \lambda_n$ be the eigenvalues of A. Let λ_1 be the dominant eigenvalue with eigenvector $\mathbf{u}_1$. Let $\mathbf{v}$ be a column vector such that $\mathbf{u}_1 \cdot \mathbf{v} = 1$. If the matrix B is given by

$$B = A - \lambda_1 \mathbf{u}_1 \mathbf{v}^t$$

then the eigenvalues of B are $\{0, \lambda_2, \lambda_3, \ldots, \lambda_n\}$.
  Now obtain the second largest eigenvalue of A, λ_2, by applying the power method to B. Continue to obtain all eigenvalues of A, assuming that no two eigenvalues have the same absolute value.

REVIEW EXERCISES

In Exercises 1–4 the number x and an approximation x^* are given. Find the absolute and relative errors ϵ_a and ϵ_r.

1. $x = -7$; $x^* = -6.98$ **2.** $x = 1000$; $x^* = 1.002 \times 10^3$

3. $x = \frac{1}{75}$; $x^* = 1 \times 10^{-2}$ **4.** $x = 37539$; $x^* = 3.7 \times 10^4$

5. Reduce the matrix $\begin{pmatrix} 2 & -4 & 6 \\ 1 & -3 & 5 \\ -4 & 9 & -13 \end{pmatrix}$ to row echelon form.

In Exercises 6 and 7 solve the given system by Gaussian elimination with partial pivoting. Round to six significant digits at every step.

6. $3.6x_1 + 8.2x_2 - 6.4x_3 = 1.26$
$\quad -4.5x_1 - 5.9x_2 + 0.3x_3 = 2.57$
$\quad 0.7x_1 + 3.6x_2 - 4.8x_3 = 2.15$

7. $1.3x_1 - 9.6x_2 + 5.35x_3 = 0.515$
$\quad -12x_1 - 15x_2 + 3.8x_3 = -71.966$
$\quad 1.06x_1 - 22.2x_2 + 9.93x_3 = 1.809$

In Exercises 8 and 9 determine whether the given matrix is strictly diagonally dominant.

8. $\begin{pmatrix} -1 & \frac{1}{2} & \frac{1}{3} \\ -\frac{5}{6} & 1 & 0 \\ 2 & \frac{3}{2} & -4 \end{pmatrix}$

9. $\begin{pmatrix} -1 & \frac{1}{3} & -\frac{1}{3} \\ -\frac{5}{6} & 1 & \frac{1}{6} \\ 2 & \frac{3}{2} & -4 \end{pmatrix}$

In Exercises 10 and 11 solve the given system by using the Jacobi and Gauss-Seidel methods. Carry out the iterations until the estimated relative error $\epsilon_r^{(n)}$ is smaller than the number given in parentheses. Use six significant figures in all computations.

10. $2.7x_1 - 0.9x_2 + 1.3x_3 = 6.98$
       $-0.3x_1 + x_2 + 0.4x_3 = -2.77$ (0.001)
         $4x_1 - 3.3x_2 + 9.6x_3 = 21.79$

11. $42.31x_1 + 8.62x_2 + 19.4x_3 = -2.2502$
       $-4.73x_1 + 80.4x_2 - 37.2x_3 = 3.5402$ (0.0001)
         $8.37x_1 + 30.9x_2 - 57.4x_3 = -24.0858$

In Exercises 12–14 estimate the dominant eigenvalue and eigenvector by using the power method with scaling.

12. $\begin{pmatrix} 8 & -2 \\ 4 & 2 \end{pmatrix}$

13. $\begin{pmatrix} -6 & 3 \\ 6 & 1 \end{pmatrix}$

14. $\begin{pmatrix} 1 & -1 & 0 \\ -1 & 2 & -1 \\ 0 & -1 & 1 \end{pmatrix}$

15. Use deflation to find the second eigenvalue of the matrix of Exercise 13.

Appendix 1
Mathematical Induction

Mathematical induction is the name given to an elementary logical principle that can be used to prove a certain type of mathematical statement. Typically, we use mathematical induction to prove that a certain statement or equation holds for every positive integer. For example, we may need to prove that $2^n > n$ for all integers $n \geq 1$. To do this, we proceed in two steps:

Step 1. We prove that the statement is true for some integer N (usually $N = 1$).

Step 2. We *assume* that the statement is true for an integer k and then *prove* that it is true for the integer $k + 1$.

If we can complete these two steps, then we shall have demonstrated the validity of the statement for *all* positive integers greater than or equal to N. To convince you of this fact, we reason as follows: Since the statement is true for N [by step (1)], it is true for the integer $N + 1$ [by step (2)]. Then it is also true for the integer $(N + 1) + 1 = N + 2$ [again by step (2)], and so on. We now demonstrate the procedure with some examples.

EXAMPLE 1   Prove that the sum of the first n positive integers is equal to $n(n + 1)/2$.

Solution   We are asked to show that

$$1 + 2 + 3 + \cdots + n = \frac{n(n + 1)}{2} \tag{1}$$

You may first wish to try a few examples to illustrate that formula (1) really works.

For example,

$$1 + 2 + 3 + 4 + 5 + 6 + 7 + 8 + 9 + 10 = \frac{10(11)}{2} = 55$$

Step 1. If $n = 1$, then the sum of the first 1 integer is 1. But $(1)(1 + 1)/2 = 1$ so equation (1) holds in the case $n = 1$.

Step 2. Assume that (1) holds for $n = k$; that is,

$$1 + 2 + 3 + \cdots + k = \frac{k(k + 1)}{2}$$

We must now show that it holds for $n = k + 1$. That is, we must show that

$$1 + 2 + 3 + \cdots + k + (k + 1) = \frac{(k + 1)(k + 2)}{2}$$

But

$$1 + 2 + 3 + \cdots + k + (k + 1) = \overbrace{(1 + 2 + 3 + \cdots + k)}^{= \, k(k + 1)/2 \text{ by assumption}} + (k + 1)$$

$$= \frac{k(k + 1)}{2} + (k + 1)$$

$$= \frac{k(k + 1) + 2(k + 1)}{2}$$

$$= \frac{k^2 + 3k + 2}{2}$$

$$= \frac{(k + 1)(k + 2)}{2}$$

and the proof is complete. ∎

Where the Difficulty Lies

Mathematical induction is sometimes difficult at first sight because of Step 2. Step 1 is usually easy to carry out. In Example 1, for instance, we inserted the value $n = 1$ on both sides of equation (1) and verified that $1 = 1(1 + 1)/2$. Step 2 was much more difficult. Let us look at it again.

Induction
hypothesis

We *assumed* that equation (1) was valid for $n = k$. We did not prove it. That assumption is called the **induction hypothesis.** We then used the induction hypothesis to show that equation (1) holds for $n = k + 1$. Perhaps this will be clearer if we look at a particular value for k, say, $k = 10$. Then we have

Assumption

$1 + 2 + 3 + 4 + 5 + 6 + 7 + 8 + 9 + 10$

$$= \frac{10(10 + 1)}{2} = \frac{10(11)}{2} = 55 \qquad (2)$$

To prove

$1 + 2 + 3 + 4 + 5 + 6 + 7 + 8 + 9 + 10 + 11$

$$= \frac{11(11 + 1)}{2} = \frac{11(12)}{2} = 66 \qquad (3)$$

The actual proof

$(1 + 2 + 3 + 4 + 5 + 6 + 7 + 8 + 9 + 10) + 11$

By the induction
hypothesis (2)
↓

$$= \frac{10(11)}{2} + 11 = \frac{10(11)}{2} + \frac{2(11)}{2}$$

$$= \frac{11(10 + 2)}{2} = \frac{11(12)}{2}$$

which is equation (3). Thus, *if* (2) is true, then (3) is true.

The beauty of the method of mathematical induction is that we do not have to prove each case separately, as we did in this illustration. Rather, we prove it for a first case, *assume* it for a general case, and then prove it for the general case plus 1. Two steps take care of an infinite number of cases. It's really quite a remarkable idea.

EXAMPLE 2   Prove that the sum of the squares of the first n positive integers is $n(n + 1)(2n + 1)/6$.

Solution   We must prove that

$$1^2 + 2^2 + 3^2 + \cdots + n^2 = \frac{n(n + 1)(2n + 1)}{6} \qquad (4)$$

Step 1. Since $\dfrac{1(1 + 1)(2 \cdot 1 + 1)}{6} = 1 = 1^2$, equation (4) is valid for $n = 1$.

Step 2. Suppose that equation (4) holds for $n = k$; that is

Induction
hypothesis $1^2 + 2^2 + 3^2 + \cdots + k^2 = \dfrac{k(k + 1)(2k + 1)}{6}$

Then to prove that (4) is true for $n = k + 1$, we have

$$1^2 + 2^2 + 3^2 + \cdots + k^2 + (k + 1)^2 = (1^2 + 2^2 + 3^2 \cdots + k^2) + (k + 1)^2$$

$$= \underbrace{\frac{k(k + 1)(2k + 1)}{6}}_{\substack{\text{Induction} \\ \text{hypothesis}}} + (k + 1)^2$$

$$= \frac{k(k + 1)(2k + 1) + 6(k + 1)^2}{6}$$

$$= \frac{k + 1}{6}[k(2k + 1) + 6(k + 1)]$$

$$= \frac{k + 1}{6}[2k^2 + 7k + 6]$$

$$= \frac{k + 1}{6}[(k + 2)(2k + 3)]$$

$$= \frac{(k + 1)(k + 2)[2(k + 1) + 1]}{6}$$

which is equation (4) for $n = k + 1$, and the proof is complete. Again you may wish to experiment with this formula. For example,

$$1^2 + 2^2 + 3^2 + 4^2 + 5^2 + 6^2 + 7^2 = \frac{7(7 + 1)(2 \cdot 7 + 1)}{6}$$

$$= \frac{7 \cdot 8 \cdot 15}{6} = 140 \qquad \blacksquare$$

EXAMPLE 3 For $a \neq 1$, use mathematical induction to prove the formula for the sum of a geometric progression:

$$1 + a + a^2 + \cdots + a^n = \frac{1 - a^{n+1}}{1 - a} \qquad (5)$$

Solution **Step 1.** If $n = 0$ (the first integer in this case), then

$$\frac{1 - a^{0+1}}{1 - a} = \frac{1 - a}{1 - a} = 1$$

Thus equation (5) holds for $n = 0$. (We use $n = 0$ instead of $n = 1$ since $a^0 = 1$ is the first term.)

Step 2. Assume that (5) holds for $n = k$; that is,

$$\underset{\substack{\text{Induction} \\ \text{hypothesis}}}{} 1 + a + a^2 + \cdots + a^k = \frac{1 - a^{k+1}}{1 - a}$$

Then

$$1 + a + a^2 + \cdots + a^k + a^{k+1} = (1 + a + a^2 + \cdots + a^k) + a^{k+1}$$

Induction hypothesis
↓

$$= \frac{1 - a^{k+1}}{1 - a} + a^{k+1}$$

$$= \frac{1 - a^{k+1} + (1 - a)a^{k+1}}{1 - a} = \frac{1 - a^{k+2}}{1 - a}$$

so that equation (5) also holds for $n = k + 1$, and the proof is complete.   ■

EXAMPLE 4   Use mathematical induction to prove that $2n + n^3$ is divisible by 3 for every positive integer n.

Solution   **Step 1.** If $n = 1$, then $2n + n^3 = 2 \cdot 1 + 1^3 = 2 + 1 = 3$, which is divisible by 3

Step 2. Assume that $2k + k^3$ is divisible by 3.   Induction hypothesis
This means that $\dfrac{2k + k^3}{3} = m$, an integer. Then, expanding $(k + 1)^3$, we obtain

$$2(k + 1) + (k + 1)^3 = 2k + 2 + (k^3 + 3k^2 + 3k + 1)$$
$$= k^3 + 2k + 3k^2 + 3k + 3$$
$$= k^3 + 2k + 3(k^2 + k + 1)$$

Then

$$\frac{2(k + 1) + (k + 1)^3}{3} = \frac{k^3 + 2k}{3} + \frac{3(k^2 + k + 1)}{3}$$
$$= m + k^2 + k + 1 = \text{an integer}$$

Thus, $2(k + 1) + (k + 1)^3$ is divisible by 3. This shows that the statement is true for $n = k + 1$.   ■

EXAMPLE 5   Let $A_1, A_2, \ldots, A_k$ be k invertible $n \times n$ matrices. Show that

$$(A_1 A_2 \cdots A_m)^{-1} = A_m^{-1} A_{m-1}^{-1} \cdots A_2^{-1} A_1^{-1} \tag{6}$$

For $m = 2$, we have $(A_1 A_2)^{-1} = A_2^{-1} A_1^{-1}$ by Theorem 1.9.3. Thus equation (6) holds for $m = 2$. We assume it is true for $m = k$ and prove it for $m = k + 1$. Let $B = A_1 A_2 \cdots A_k$. Then

$$(A_1 A_2 \cdots A_k A_{k+1})^{-1} = (B A_{k+1})^{-1} = A_{k+1}^{-1} B^{-1} \tag{7}$$

But, by the induction assumption,

$$B^{-1} = (A_1 A_2 \cdots A_k)^{-1} = A_k^{-1} A_{k-1}^{-1} \cdots A_2^{-1} A_1^{-1} \tag{8}$$

Substituting (8) into (7) completes the proof.   ■

Focus on . . .

Mathematical Induction

The first mathematician to give a formal proof by the explicit use of mathematical induction was the Italian clergyman Franciscus Maurolicus (1494–1575), who was the abbot of Messina in Sicily, and is considered the greatest geometer of the sixteenth century. In his *Arithmetic*, published in 1575, Maurolicus used mathematical induction to prove, among other things, that, for every positive integer n,

$$1 + 3 + 5 + \cdots + (2n - 1) = n^2$$

You are asked to prove this in Problem 4.

The induction proofs of Maurolicus were given in a sketchy style that is difficult to follow. A clearer exposition of the method was given by the French mathematician Blaise Pascal (1623–1662). In his *Traité du Triangle Arithmétique*, published in 1662, Pascal proved a formula for the sum of binomial coefficients. He used his formula to develop what is today called the *Pascal triangle*.

Although the method of mathematical induction was used formally in 1575, the term *mathematical induction* was not used until 1838. In that year, one of the originators of set theory, Augustus de Morgan (1806–1871), published an article in the *Penny Cyclopedia* (London) entitled "Induction (Mathematics)." At the end of that article, he used the term we use today. However, the term did not enjoy widespread use until the 20th century.

PROBLEMS   A1

In Problems 1–20 prove the given formula using mathematical induction. Unless otherwise stated, assume that n is a positive or nonnegative integer.

1. $2 + 4 + 6 + \cdots + 2n = n(n + 1)$

2. $1 + 4 + 7 + \cdots + (3n - 2) = \dfrac{n(3n - 1)}{2}$

3. $2 + 5 + 8 + \cdots + (3n - 1) = \dfrac{n(3n + 1)}{2}$

4. $1 + 3 + 5 + \cdots + (2n - 1) = n^2$

5. $2^n > n$

6. $2^n < n!$ for $n = 4, 5, 6, \ldots$, where

$$n! = 1 \cdot 2 \cdot 3 \cdots (n - 1) \cdot n$$

7. $1 + 2 + 4 + 8 + \cdots + 2^n = 2^{n+1} - 1$

8. $1 + 3 + 9 + 27 + \cdots + 3^n = \dfrac{3^{n+1} - 1}{2}$

9. $1 + \dfrac{1}{2} + \dfrac{1}{4} + \cdots + \dfrac{1}{2^n} = 2 - \dfrac{1}{2^n}$

10. $1 - \dfrac{1}{3} + \dfrac{1}{9} - \cdots + \left(-\dfrac{1}{3}\right)^n = \dfrac{3}{4}\left[1 - \left(-\dfrac{1}{3}\right)^{n+1}\right]$

11. $1^3 + 2^3 + 3^3 + \cdots + n^3 = \dfrac{n^2(n+1)^2}{4}$

12. $1 \cdot 2 + 2 \cdot 3 + 3 \cdot 4 + \cdots + n(n+1) = \dfrac{n(n+1)(n+2)}{3}$

13. $1 \cdot 2 + 3 \cdot 4 + 5 \cdot 6 + \cdots + (2n-1)(2n) = \dfrac{n(n+1)(4n-1)}{3}$

14. $\dfrac{1}{2^2 - 1} + \dfrac{1}{3^2 - 1} + \dfrac{1}{4^2 - 1} + \cdots + \dfrac{1}{(n+1)^2 - 1} = \dfrac{3}{4} - \dfrac{1}{2(n+1)} - \dfrac{1}{2(n+2)}$

15. $n + n^2$ is even

16. $n < \dfrac{n^2 - n}{12} + 2$ if $n > 10$

17. $n(n^2 + 5)$ is divisible by 6

*18. $3n^5 + 5n^3 + 7n$ is divisible by 15

*19. $x^n - 1$ is divisible by $x - 1$

*20. $x^n - y^n$ is divisible by $x - y$

*21. Give a formal proof that $(ab)^n = a^n b^n$ for every positive integer n.

22. Assuming that every polynomial has at least one root, prove that a polynomial of degree n has exactly n roots (counting multiplicities).

23. Given that $\det AB = \det A \det B$, prove that $\det A_1 A_2 \cdots A_m = \det A_1 \det A_2 \cdots \det A_m$, where $A_1, \ldots, A_m$ are $n \times n$ matrices.

24. If $A_1, A_2, \ldots, A_k$ are $m \times n$ matrices, show that $(A_1 + A_2 + \cdots + A_k)^t = A_1^t + A_2^t + \cdots + A_k^t$. You may assume that $(A + B)^t = A^t + B^t$.

25. Prove that there are exactly 2^n subsets of a set containing n elements.

26. Prove that if $2k - 1$ is an even integer for some integer k, then $2(k + 1) - 1 = 2k + 2 - 1 = 2k + 1$ is also an even integer. What, if anything, can you conclude by the proof?

27. What is wrong with the following proof that each horse in a set of n horses has the same color as every other horse in the set?

Step 1. It is true for $n = 1$ since there is only one horse in the set and it obviously has the same color as itself.

Step 2. Suppose it is true for $n = k$. That is, each horse in a set containing k horses is the same color as every other horse in the set. Let $h_1, h_2, \ldots, h_k, h_{k+1}$ denote the $k + 1$ horses in a set S. Let $S_1 = \{h_1, h_2, \ldots, h_k\}$ and $S_2 = \{h_2, h_3, \ldots, h_k, h_{k+1}\}$. Then both S_1 and S_2 contain k horses, so the horses in each set are of the same color. We write $h_i = h_j$ to indicate that horse i has the same color as horse j. Then we have

$$h_1 = h_2 = h_3 = \cdots = h_k$$

and

$$h_2 = h_3 = h_4 = \cdots = h_k = h_{k+1}$$

This means that

$$h_1 = h_2 = h_3 = \cdots = h_k = h_{k+1}$$

so all the horses in S have the same color. This proves the statement in the case $n = k + 1$, so the statement is true for all n.

Appendix 2
Complex Numbers

In Chapter 6 we encountered the problem of finding the roots of the polynomial

$$\lambda^2 + b\lambda + c = 0 \tag{1}$$

To find the roots, we use the quadratic formula to obtain

$$\lambda = \frac{-b \pm \sqrt{b^2 - 4c}}{2} \tag{2}$$

If $b^2 - 4c > 0$, there are two real roots. If $b^2 - 4c = 0$, we obtain the single root (of multiplicity 2) $\lambda = -b/2$. To deal with the case $b^2 - 4c < 0$, we introduce the **imaginary number:**[†]

$$\boxed{i = \sqrt{-1}} \tag{3}$$

[†] You should not be troubled by the term "imaginary." It's just a name. The British mathematician Alfred North Whitehead, in the chapter on imaginary numbers in his *Introduction to Mathematics,* wrote:

At this point it may be useful to observe that a certain type of intellect is always worrying itself and others by discussion as to the applicability of technical terms. Are the incommensurable numbers properly called numbers? Are the positive and negative numbers really numbers? Are the imaginary numbers imaginary, and are they numbers?—are types of such futile questions. Now, it cannot be too clearly understood that, in science, technical terms are names arbitrarily assigned, like Christian names to children. There can be no question of the names being right or wrong. They may be judicious or injudicious; for they can sometimes be so arranged as to be easy to remember, or so as to suggest relevant and important ideas. But the essential principle involved was quite clearly enunciated in Wonderland to Alice by Humpty Dumpty, when he told her, apropos of his use of words, "I pay them extra and make them mean what I like." So we will not bother as to whether imaginary numbers are imaginary, or as to whether they are numbers, but will take the phrase as the arbitrary name of a certain mathematical idea, which we will now endeavour to make plain.

Then, for $b^2 - 4c < 0$,

$$\sqrt{b^2 - 4c} = \sqrt{(4c - b^2)(-1)} = \sqrt{4c - b^2}\, i$$

and the two roots of (1) are given by

$$\lambda_1 = -\frac{b}{2} + \frac{\sqrt{4c - b^2}}{2}\, i \quad \text{and} \quad \lambda_2 = -\frac{b}{2} - \frac{\sqrt{4c - b^2}}{2}\, i$$

EXAMPLE 1   Find the roots of the quadratic equation $\lambda^2 + 2\lambda + 5 = 0$.

Solution   We have $b = 2$, $c = 5$, and $b^2 - 4c = -16$. Thus $\sqrt{b^2 - 4c} = \sqrt{-16} = \sqrt{16}\sqrt{-1} = 4i$ and the roots are

$$\lambda_1 = \frac{-2 + 4i}{2} = -1 + 2i \quad \text{and} \quad \lambda_2 = -1 - 2i \qquad \blacksquare$$

DEFINITION 1   A **complex number** is a number of the form

$$\boxed{z = \alpha + i\beta} \qquad (4)$$

where α and β are real numbers. α is called the **real part** of z and is denoted by Re z. β is called the **imaginary part** of z and is denoted by Im z. Representation (4) is sometimes called the **Cartesian form** of the complex number z.

Remark. If $\beta = 0$ in equation (4), then $z = \alpha$ is a real number. In this context we can regard the set of real numbers as a subset of the set of complex numbers.

EXAMPLE 2   In Example 1, Re $\lambda_1 = -1$ and Im $\lambda_1 = 2$. $\blacksquare$

We can add and multiply complex numbers by using standard rules of algebra.

EXAMPLE 3   Let $z = 2 + 3i$ and $w = 5 - 4i$. Calculate **(i)** $z + w$, **(ii)** $3w - 5z$, and **(iii)** zw.

Solution   **i.** $z + w = (2 + 3i) + (5 - 4i) = (2 + 5) + (3 - 4)i = 7 - i$.

ii. $3w = 3(5 - 4i) = 15 - 12i$; $5z = 10 + 15i$; and $3w - 5z = (15 - 12i) - (10 + 15i) = (15 - 10) + i(-12 - 15) = 5 - 27i$.

iii. $zw = (2 + 3i)(5 - 4i) = (2)(5) + 2(-4i) + (3i)(5) + (3i)(-4i) = 10 - 8i + 15i - 12i^2 = 10 + 7i + 12 = 22 + 7i$. Here we used the fact that $i^2 = -1$. $\blacksquare$

We can plot a complex number z in the xy-plane by plotting Re z along the x-axis and Im z along the y-axis. Thus each complex number can be thought of as a point in the xy-plane. With this representation the xy-plane is called the **complex plane.** Some representative points are plotted in Figure A.1.

Complex plane

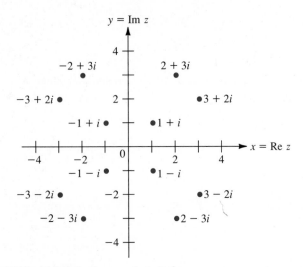

Figure A.1   Twelve points in the complex plane

Conjugate

If $z = \alpha + i\beta$, then we define the **conjugate** of z, denoted by $\bar{z}$, by

$$\boxed{\bar{z} = \alpha - i\beta} \qquad (5)$$

Figure A.2 depicts a representative value of z and $\bar{z}$.

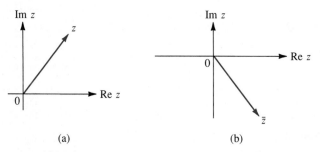

(a)   (b)

Figure A.2 $\bar{z}$ is obtained by reflecting z about the real axis

EXAMPLE 4   Compute the conjugate of **(i)** $1 + i$, **(ii)** $3 - 4i$, **(iii)** $-7 + 5i$, and **(iv)** -3.

Solution   **(i)** $\overline{1 + i} = 1 - i$; **(ii)** $\overline{3 - 4i} = 3 + 4i$; **(iii)** $\overline{-7 + 5i} = -7 - 5i$; **(iv)** $\overline{-3} = -3$.

∎

It is not difficult to show (see Problem 35) that

$$\bar{z} = z \qquad \text{if and only if } z \text{ is real} \tag{6}$$

Pure imaginary number   If $z = \beta i$ with β real, then z is said to be **pure imaginary.** We can then show (see Problem 36) that

$$\bar{z} = -z \qquad \text{if and only if } z \text{ is pure imaginary} \tag{7}$$

Let $p_n(x) = a_0 + a_1 x + a_2 x^2 + \cdots + a_n x^n$ be a polynomial with real coefficients. Then it can be shown (see Problem 41) that the complex roots of the equation $p_n(x) = 0$ occur in complex conjugate pairs. That is, if z is a root of $p_n(x) = 0$, then so is $\bar{z}$. We saw this fact illustrated in Example 1 in the case $n = 2$.

Magnitude   For $z = \alpha + i\beta$, we define the **magnitude** of z, denoted by $|z|$, by

$$\text{Magnitude of } z = |z| = \sqrt{\alpha^2 + \beta^2} \tag{8}$$

Argument   and we define the **argument** of z, denoted by arg z, as the angle θ between the line $0z$ and the positive $x =$ axis. By definition

$$-\pi < \arg z \le \pi$$

From Figure A.3 we see that $r = |z|$ is the distance from z to the origin. If $\alpha > 0$, then

$$\theta = \arg z = \tan^{-1} \frac{\beta}{\alpha}$$

Recall that $\tan^{-1} x$ always takes values in the interval $\left(-\dfrac{\pi}{2}, \dfrac{\pi}{2}\right)$. If $\alpha = 0$ and

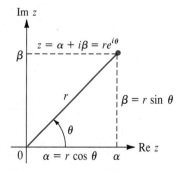

Figure A.3   If $z = \alpha + i\beta$, then $\alpha = r \cos \theta$ and $\beta = r \sin \theta$

$\beta > 0$, then $\theta = \arg z = \dfrac{\pi}{2}$. If $\alpha = 0$ and $\beta < 0$, then $\theta = \arg z = -\dfrac{\pi}{2}$. If $\alpha < 0$ and $\beta > 0$, then θ is in the second quadrant and is given by

$$\theta = \arg z = \pi - \tan^{-1}\frac{\beta}{\alpha}$$

Finally, if $\alpha < 0$ and $\beta < 0$, then θ is the third quadrant and

$$\theta = \arg z = -\pi + \tan^{-1}\frac{\beta}{\alpha}.$$

In sum, we have

Argument of z

Let $z = \alpha + \beta i$. Then

$$\arg z = \tan\frac{\beta}{\alpha} \quad \text{if } \alpha > 0$$

$$\arg z = \frac{\pi}{2} \quad \text{if } \alpha = 0 \text{ and } \beta > 0$$

$$\arg z = -\frac{\pi}{2} \quad \text{if } \alpha = 0 \text{ and } \beta < 0 \tag{9}$$

$$\arg z = \pi - \tan^{-1}\frac{\beta}{\alpha} \quad \text{if } \alpha < 0 \text{ and } \beta > 0$$

$$\arg z = -\pi + \tan^{-1}\frac{\beta}{\alpha} \quad \text{if } \alpha < 0 \text{ and } \beta < 0 \tag{10}$$

$\arg 0$ is not defined.

From Figure A.4 we see that

$$|\bar{z}| = |z| \tag{11}$$

and

$$\arg \bar{z} = -\arg z \tag{12}$$

We can use $|z|$ and $\arg z$ to describe what is often a more convenient way to represent complex numbers.† From Figure A.3 it is evident that if $z = \alpha + i\beta$,

---

†Those of you who have studied polar coordinates in a calculus course will find this representation very familiar.

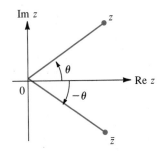

Figure A.4 arg $\bar{z}$ = −arg z

$r = |z|$, and θ = arg z, then

$$\alpha = r \cos \theta \quad \text{and} \quad \beta = r \sin \theta \tag{13}$$

We shall see at the end of this appendix that

$$e^{i\theta} = \cos \theta + i \sin \theta \tag{14}$$

Since cos (−θ) = cos θ and sin (−θ) = −sin θ, we also have

$$e^{-i\theta} = \cos (-\theta) + i \sin (-\theta) = \cos \theta - i \sin \theta \tag{14$'$}$$

Formula (14) is called **Euler's formula.**† Using Euler's formula and equation (13), we have

$$z = \alpha + i\beta = r \cos \theta + ir \sin \theta = r(\cos \theta + i \sin \theta)$$

or

$$z = re^{i\theta} \tag{15}$$

Polar form   Representation (15) is called the **polar form** of the complex number z.

EXAMPLE 5   Determine the polar forms of the following complex numbers: **(i)** 1, **(ii)** −1, **(iii)** i, **(iv)** 1 + i, **(v)** −1 − $\sqrt{3}i$, and **(vi)** −2 + 7i.

Solution   The six points are plotted in Figure A.5.

---

† Named for the great Swiss mathematician Leonhard Euler (1707–1783).

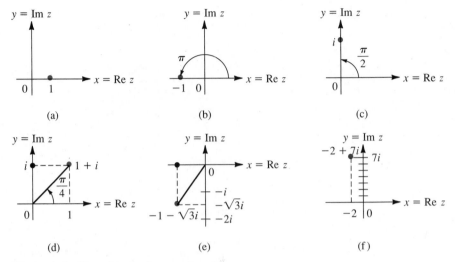

Figure A.5   Six points in the complex plane

i. From Figure A.5a it is clear that arg $1 = 0$. Since Re $1 = 1$, we see that, in polar form, $1 = 1e^{i0} = 1e^0 = 1$.

ii. Since $\arg(-1) = \pi$ (Figure A.5b) and $|-1| = 1$, we have
$$-1 = 1e^{\pi i} = e^{i\pi}$$

iii. From Figure A.5c we see that arg $i = \pi/2$. Since $|i| = \sqrt{0^2 + 1^2} = 1$, it follows that
$$i = e^{i\pi/2}$$

iv. $\arg(1 + i) = \tan^{-1}(1/1) = (\pi/4)$ and $|1 + i| = \sqrt{1^2 + 1^2} = \sqrt{2}$, so that
$$1 + i = \sqrt{2}e^{i\pi/4}$$

v. Here $\tan^{-1}(\beta/\alpha) = \tan^{-1}\sqrt{3} = \pi/3$. However arg z is in the third quadrant so $\theta = \pi/3 - \pi = -2\pi/3$. Also, $|-1 - \sqrt{3}| = \sqrt{1^2 + (\sqrt{3})^2} = \sqrt{1 + 3} = 2$ so that
$$-1 - \sqrt{3} = 2e^{4\pi i/3}$$

vi. To compute this we need a calculator. We find that, in radians,
$$\arg z = \tan^{-1}(-\tfrac{7}{2}) = \tan^{-1}(-3.5) \approx -1.2925.$$

But $\tan^{-1} x$ is defined as a number in the interval $(-\pi/2, \pi/2)$. Since, from Figure A.5f, θ is in the second quadrant, we see that arg $z = \tan^{-1}(-3.5) + \pi \approx 1.8491$. Next, we see that
$$|-2 + 7i| = \sqrt{(-2)^2 + 7^2} = \sqrt{53}.$$

Hence
$$-2 + 7i \approx \sqrt{53}e^{1.8491i}$$

EXAMPLE 6   Convert the following complex numbers from polar to Cartesian form:
(i) $2e^{i\pi/3}$; (ii) $4e^{3\pi i/2}$.

Solution   i. $e^{i\pi/3} = \cos \pi/3 + i \sin \pi/3 = \frac{1}{2} + (\sqrt{3}/2)i$. Thus $2e^{i\pi/3} = 1 + \sqrt{3}i$.

ii. $e^{3\pi i/2} = \cos 3\pi/2 + i \sin 3\pi/2 = 0 + i(-1) = -i$. Thus $4e^{3\pi i/2} = -4i$. ■

If $\theta = \arg z$, then, by equation (12), $\arg \bar{z} = -\theta$. Thus, since $|\bar{z}| = |z|$;

$$\boxed{\text{If } z = re^{i\theta}, \quad \text{then } \bar{z} = re^{-i\theta}} \tag{16}$$

Suppose we write a complex number in its polar form $z = re^{i\theta}$. Then

$$z^n = (re^{i\theta})^n = r^n(e^{i\theta})^n = r^n e^{in\theta} = r^n(\cos n\theta + i \sin n\theta) \tag{17}$$

Formula (17) is useful for a variety of computations. In particular, when $r = |z| = 1$, we obtain the **De Moivre formula.**†

$$\boxed{\begin{array}{c} \textbf{De Moivre Formula} \\ (\cos \theta + i \sin \theta)^n = \cos n\theta + i \sin n\theta \end{array}} \tag{18}$$

EXAMPLE 7   Compute $(1 + i)^5$.

Solution   In Example 5(iv) we showed that $1 + i = \sqrt{2}e^{\pi i/4}$. Then

$$(1 + i)^5 = (\sqrt{2}e^{\pi i/4})^5 = (\sqrt{2})^5 e^{5\pi i/4} = 4\sqrt{2}\left(\cos \frac{5\pi}{4} + i \sin \frac{5\pi}{4}\right)$$

$$= 4\sqrt{2}\left(-\frac{1}{\sqrt{2}} - \frac{1}{\sqrt{2}}i\right) = -4 - 4i$$

This can be checked by direct calculation. If the direct calculation seems no more difficult, then try to compute $(1 + i)^{20}$ directly. Proceeding as above, we obtain

$$(1 + i)^{20} = (\sqrt{2})^{20} e^{20\pi i/4} = 2^{10}(\cos 5\pi + i \sin 5\pi)$$
$$= 2^{10}(-1 + 0) = -1024 \qquad\qquad ■$$

Proof of Euler's   We shall show that
Formula

$$e^{i\theta} = \cos \theta + i \sin \theta \tag{19}$$

---

† Abraham De Moivre (1667–1754) was a French mathematician well known for his work in probability theory, infinite series, and trigonometry. He was so highly regarded that Newton often told those who came to him with questions on mathematics, "Go to M. De Moivre; he knows these things better than I do."

by using power series. If these are unfamiliar to you, skip the proof. We have

$$e^x = 1 + x + \frac{x^2}{2!} + \frac{x^3}{3!} + \cdots \dagger \tag{20}$$

$$\sin x = x - \frac{x^3}{3!} + \frac{x^5}{5!} - \cdots \tag{21}$$

$$\cos x = 1 - \frac{x^2}{2!} + \frac{x^4}{4!} - \cdots \tag{22}$$

Then

$$e^{i\theta} = 1 + (i\theta) + \frac{(i\theta)^2}{2!} + \frac{(i\theta)^3}{3!} + \frac{(i\theta)^4}{4!} + \frac{(i\theta)^5}{5!} + \cdots \tag{23}$$

Now $i^2 = -1$, $i^3 = -i$, $i^4 = 1$, $i^5 = i$, and so on. Thus (23) can be written

$$e^{i\theta} = 1 + i\theta - \frac{\theta^2}{2!} - \frac{i\theta^3}{3!} + \frac{\theta^4}{4!} + \frac{i\theta^5}{5!} - \cdots$$

$$= \left(1 - \frac{\theta^2}{2!} + \frac{\theta^4}{4!} - \cdots\right) + i\left(\theta - \frac{\theta^3}{3!} + \frac{\theta^5}{5!} - \cdots\right)$$

$$= \cos\theta + i\sin\theta$$

This completes the proof. ∎

PROBLEMS   A2

In Problems 1–5 perform the indicated operation.

1. $(2 - 3i) + (7 - 4i)$ **2.** $3(4 + i) - 5(-3 + 6i)$

3. $(1 + i)(1 - i)$ **4.** $(2 - 3i)(4 + 7i)$

5. $(-3 + 2i)(7 + 3i)$

In Problems 6–15 convert the complex number to its polar form.

6. $5i$ **7.** $5 + 5i$ **8.** $-2 - 2i$ **9.** $3 - 3i$

10. $2 + 2\sqrt{3}i$ **11.** $3\sqrt{3} + 3i$ **12.** $1 - \sqrt{3}i$ **13.** $4\sqrt{3} - 4i$

14. $-6\sqrt{3} - 6i$ **15.** $-1 - \sqrt{3}i$

In Problems 16–25 convert from polar to Cartesian form.

16. $e^{3\pi i}$ **17.** $2e^{-7\pi i}$ **18.** $\frac{1}{2}e^{3\pi i/4}$ **19.** $\frac{1}{2}e^{-3\pi i/4}$

20. $6e^{\pi i/6}$ **21.** $4e^{5\pi i/6}$ **22.** $4e^{-5\pi i/6}$ **23.** $3e^{-2\pi i/3}$

24. $\sqrt{3}e^{23\pi i/4}$ **25.** e^i

---

† Although we shall not prove it here, this series expansion is also valid when x is a complex number.

In Problems 26–34 compute the conjugate of the given number.

26. $3 - 4i$ **27.** $4 + 6i$ **28.** $-3 + 8i$

29. $-7i$ **30.** 16 **31.** $2e^{\pi i/7}$

32. $4e^{3\pi i/5}$ **33.** $3e^{-4\pi i/11}$ **34.** $e^{0.012i}$

35. Show that $z = \alpha + i\beta$ is real if and only if $z = \bar{z}$. [*Hint:* If $z = \bar{z}$, show that $\beta = 0$.]

36. Show that $z = \alpha + i\beta$ is pure imaginary if and only if $z = -\bar{z}$. [*Hint:* If $z = -\bar{z}$, show that $\alpha = 0$.]

37. For any complex number z, show that $z\bar{z} = |z|^2$.

38. Show that the circle of radius 1 centered at the origin (the *unit circle*) is the set of points in the complex plane that satisfy $|z| = 1$.

39. For any complex number z_0 and positive real number a, describe $\{z: |z - z_0| = a\}$.

40. Describe $\{z: |z - z_0| \le a\}$, where z_0 and a are as in Problem 39.

*41. Let $p(\lambda) = \lambda^n + a_{n-1}\lambda^{n-1} + a_{n-2}\lambda^{n-2} + \cdots + a_1\lambda + a_0$ with $a_0, a_1, \ldots, a_{n-1}$ real numbers. Show that if $p(z) = 0$, then $p(\bar{z}) = 0$. That is: *The roots of polynomials with real coefficients occur in complex conjugate pairs.*

42. Derive expressions for $\cos 4\theta$ and $\sin 4\theta$ by comparing the De Moivre formula and the expansion of $(\cos \theta + i \sin \theta)^4$.

43. Prove De Moivre's formula by mathematical induction. [*Hint:* Recall the trigonometric identities $\cos (x + y) = \cos x \cos y - \sin x \sin y$ and $\sin (x + y) = \sin x \cos y + \cos x \sin y$.]

Answers to Odd-Numbered Problems

Chapter 1

Problems 1.2, page 5

1. $x = -\frac{13}{5}$, $y = -\frac{11}{5}$; det $= -10$

3. no solutions; det $= 0$

5. $x = \frac{11}{2}$, $y = -30$; det $= -2$

7. infinite number of solutions; $y = \frac{2}{3}x$, where x is arbitrary; det $= 0$

9. $x = -1$, $y = 2$; det $= -1$

11. det $= a^2 - b^2$; if $a^2 - b^2 \neq 0$ (i.e., if $a \neq \pm b$), then $x = y = c/(a + b)$. If $a^2 - b^2 = 0$, then $a = \pm b$. If $a \neq 0$ and $a = b$, then there is an infinite number of solutions given by $y = c/a - x$. If $a \neq 0$ and $a = -b$, then there are no solutions (unless $c = 0$ in which case $x = y$ is a solution).

13. det $= -2ab$; so unique solution if both a and b are nonzero

15. $a = b = 0$ and $c \neq 0$ or $d \neq 0$

17. no point of intersection

19. The lines are coincident. Any point of the form $(x, (4x - 10)/6)$ is a point of intersection.

21. $(\frac{67}{45}, \frac{2}{15})$ 23. $\sqrt{13}/13$

25. $\sqrt{61}/5$ 27. $\sqrt{5}$

29. Since the slope of the given line L is $-\dfrac{a}{b}$, the slope of $L_\perp$ is $\dfrac{b}{a}$. The equation of a line $L_\perp$ perpendicular to L and passing through (x_1, y_1) is given by $\dfrac{y - y_1}{x - x_1} = \dfrac{b}{a}$, or $bx - ay = bx_1 - ay_1$. The unique point of intersection of L and $L_\perp$ is found to be $(x_0, y_0) =$
$$\left(\frac{ac - aby_1 + b^2 x_1}{a^2 + b^2}, \frac{bc - abx_1 + a^2 y_1}{a^2 + b^2} \right).$$
Then d is the distance between (x_0, y_0) and (x_1, y_1) and, after some algebra,
$$d^2 = \frac{1}{(a^2 + b^2)^2} \times (a^2 c^2 -$$
$$2a^2 bcy_1 + a^2 b^2 y_1^2 - 2a^3 cx_1 + 2a^3 bx_1 y_1 + a^4 x_1^2 + c^2 b^2 - 2ab^2 cx_1 + a^2 b^2 x_1^2 - 2b^3 cy_1 + 2ab^3 x_1 y_1 + b^4 y_1^2) =$$
$$\frac{a^2 + b^2}{(a^2 + b^2)^2} (c^2 - 2abcy_1 + b^2 y_1^2 - 2acx_1 + 2abx_1 y_1 + a^2 x_1^2) =$$
$$\frac{1}{(a^2 + b^2)} (ax_1 + by_1 - c)^2. \text{ Thus}$$
$$d = \frac{|ax_1 + by_1 - c|}{\sqrt{a^2 + b^2}}.$$

31. If det $A = 0$, then $a_{11}a_{22} -$
$a_{12}a_{21} = 0$. So $-a_{11}a_{22} = -a_{12}a_{21}$
and $-\dfrac{a_{11}}{a_{12}} = -\dfrac{a_{21}}{a_{22}}$ if $a_{12}a_{22} \neq 0$.
Thus the two lines are parallel since
they have equal slopes. If $a_{12} = 0$
and $a_{22} = 0$, then the lines are parallel because they are both vertical.

33. The unique solution can be found to
be $x = \dfrac{a_{22}b_1 - a_{12}b_2}{a_{11}a_{22} - a_{12}a_{21}}$ and $y = $
$\dfrac{a_{11}b_2 - a_{21}b_1}{a_{11}a_{22} - a_{12}a_{21}}$.

35. Let $x = $ no. of cups; $y = $ no. of saucers; solutions are $(x, 240 - \frac{3}{2}x)$.
[There is a finite number of solutions because x and y must be integers.]

37. 32 sodas, 128 milk shakes

Problems 1.3, page 21

Note: Where there were an infinite number of solutions, we wrote the solutions
with the last variable chosen arbitrarily.
The solutions can be written in other
ways as well.

1. $(2, -3, 1)$

3. $(3 + \frac{2}{9}x_3, \frac{8}{9}x_3, x_3)$, x_3 arbitrary.

5. $(-9, 30, 14)$

7. no solution

9. $(-\frac{4}{5}x_3, \frac{9}{5}x_3, x_3)$, x_3 arbitrary

11. $(-1, \frac{5}{2} + \frac{1}{2}x_3, x_3)$, x_3 arbitrary

13. no solution

15. $(\frac{20}{13} - \frac{4}{13}x_4, -\frac{28}{13} + \frac{3}{13}x_4, -\frac{45}{13} +$
$\frac{9}{13}x_4, x_4)$, x_4 arbitrary

17. $(18 - 4x_4, -\frac{15}{2} + 2x_4, -31 +$
$7x_4, x_4)$, x_4 arbitrary

19. no solution

21. row echelon form

23. reduced row echelon form

25. neither

27. reduced row echelon form

29. neither

31. row echelon form: $\begin{pmatrix} 1 & -6 \\ 0 & 1 \end{pmatrix}$; reduced row echelon form: $\begin{pmatrix} 1 & 0 \\ 0 & 1 \end{pmatrix}$

33. row echelon form:
$$\begin{pmatrix} 1 & -2 & 4 \\ 0 & 1 & -\frac{4}{11} \\ 0 & 0 & 1 \end{pmatrix}$$
reduced row echelon form:
$$\begin{pmatrix} 1 & 0 & 0 \\ 0 & 1 & 0 \\ 0 & 0 & 1 \end{pmatrix}$$

35. row echelon form:
$$\begin{pmatrix} 1 & -\frac{7}{2} \\ 0 & 1 \\ 0 & 0 \end{pmatrix}$$
reduced row echelon form:
$$\begin{pmatrix} 1 & 0 \\ 0 & 1 \\ 0 & 0 \end{pmatrix}$$

37. $x_1 = 30{,}000 - 5x_3$
$x_2 = x_3 - 5000$
$5000 \leq x_3 \leq 6000$; no

39. no unique solution (2 equations in 3
unknowns); if 200 shares of
McDonald's, then 100 shares of
Hilton and 300 shares of Eastern

41. The row echelon form of the augmented matrix representing this system is
$$\left(\begin{array}{ccc|c} 1 & -\frac{1}{2} & \frac{3}{2} & a/2 \\ 0 & 1 & -\frac{19}{5} & \frac{2}{5}(b - \frac{3}{2}a) \\ 0 & 0 & 0 & -2a + 3b + c \end{array}\right)$$
which is inconsistent if $-2a + 3b +$
$c \neq 0$ or $c \neq 2a - 3b$.

43. $a_{11}a_{22}a_{33} + a_{12}a_{23}a_{31} + a_{13}a_{32}a_{21}$
$-a_{13}a_{22}a_{31} - a_{12}a_{21}a_{33}$
$- a_{11}a_{32}a_{23} \neq 0$

45. $(1.90081, 4.19411, -11.34852)$

Problems 1.4, page 25

1. $(0, 0)$ **3.** $(0, 0, 0)$

5. $(\frac{4}{6}x_3, \frac{5}{6}x_3, x_3)$, x_3 arbitrary

7. $(0, 0)$

9. $(-4x_4, 2x_4, 7x_4, x_4)$, x_4 arbitrary.

11. $(0, 0)$ **13.** $(0, 0, 0)$

15. $k = \frac{95}{11}$

Problems 1.5, page 34

1. $\begin{pmatrix} 2 \\ -3 \\ 11 \end{pmatrix}$ **3.** $\begin{pmatrix} -4 \\ 0 \\ 4 \end{pmatrix}$

5. $\begin{pmatrix} -31 \\ 22 \\ -27 \end{pmatrix}$ **7.** $\begin{pmatrix} 0 \\ 0 \\ 0 \end{pmatrix}$

9. $\begin{pmatrix} -11 \\ 11 \\ -10 \end{pmatrix}$ **11.** $(1, 2, 5, 7)$

13. $(-8, 12, 4, 20)$

15. $(8, -5, 7, -1)$

17. $(7, 2, 4, 11)$

19. $(-11, 9, 18, 18)$

21. $\mathbf{a} + \mathbf{0} = \begin{pmatrix} a_1 \\ a_2 \\ \vdots \\ a_n \end{pmatrix} + \begin{pmatrix} 0 \\ 0 \\ \vdots \\ 0 \end{pmatrix}$

$= \begin{pmatrix} a_1 + 0 \\ a_2 + 0 \\ \vdots \\ a_n + 0 \end{pmatrix} = \begin{pmatrix} a_1 \\ a_2 \\ \vdots \\ a_n \end{pmatrix} = \mathbf{a}$

$0\mathbf{a} = 0 \begin{pmatrix} a_1 \\ a_2 \\ \vdots \\ a_n \end{pmatrix} = \begin{pmatrix} 0a_1 \\ 0a_2 \\ \vdots \\ 0a_n \end{pmatrix} = \begin{pmatrix} 0 \\ 0 \\ \vdots \\ 0 \end{pmatrix} = \mathbf{0}$

23. $\mathbf{a} + \mathbf{b} = \begin{pmatrix} a_1 + b_1 \\ a_2 + b_2 \\ \vdots \\ a_n + b_n \end{pmatrix}$,

$\alpha(\mathbf{a} + \mathbf{b}) = \begin{pmatrix} \alpha(a_1 + b_1) \\ \alpha(a_2 + b_2) \\ \vdots \\ \alpha(a_n + b_n) \end{pmatrix}$

$= \begin{pmatrix} \alpha a_1 + \alpha b_1 \\ \alpha a_2 + \alpha b_2 \\ \vdots \\ \alpha a_n + \alpha b_n \end{pmatrix}$

$= \begin{pmatrix} \alpha a_1 \\ \alpha a_2 \\ \vdots \\ \alpha a_n \end{pmatrix} + \begin{pmatrix} \alpha b_1 \\ \alpha b_2 \\ \vdots \\ \alpha b_n \end{pmatrix}$

$= \alpha \begin{pmatrix} a_1 \\ a_2 \\ \vdots \\ a_n \end{pmatrix} + \alpha \begin{pmatrix} b_1 \\ b_2 \\ \vdots \\ b_n \end{pmatrix}$

$= \alpha \mathbf{a} + \alpha \mathbf{b}$

$(\alpha + \beta)\mathbf{a} = \begin{pmatrix} (\alpha + \beta)a_1 \\ (\alpha + \beta)a_2 \\ \vdots \\ (\alpha + \beta)a_n \end{pmatrix}$

$= \begin{pmatrix} \alpha a_1 + \beta a_1 \\ \alpha a_2 + \beta a_2 \\ \vdots \\ \alpha a_n + \beta a_n \end{pmatrix}$

$= \begin{pmatrix} \alpha a_1 \\ \alpha a_2 \\ \vdots \\ \alpha a_n \end{pmatrix} + \begin{pmatrix} \beta a_1 \\ \beta a_2 \\ \vdots \\ \beta a_n \end{pmatrix}$

$= \alpha \begin{pmatrix} a_1 \\ a_2 \\ \vdots \\ a_n \end{pmatrix} + \beta \begin{pmatrix} a_1 \\ a_2 \\ \vdots \\ a_n \end{pmatrix}$

$= \alpha \mathbf{a} + \beta \mathbf{a}$

$(\alpha\beta)\mathbf{a} = \begin{pmatrix} \alpha\beta a_1 \\ \alpha\beta a_2 \\ \vdots \\ \alpha\beta a_n \end{pmatrix} = \alpha \begin{pmatrix} \beta a_1 \\ \beta a_2 \\ \vdots \\ \beta a_n \end{pmatrix}$

$= \alpha \left[\beta \begin{pmatrix} a_1 \\ a_2 \\ \vdots \\ a_n \end{pmatrix} \right] = \alpha(\beta\mathbf{a})$

25. $\mathbf{d} + \mathbf{e}$ represents the combined demand of the two factories for each of the four raw materials needed to produce one unit of each factory's product; **2d** represents the demand of factory 1 for each of the four raw materials needed to produce two units of its product.

27. $\mathbf{w} = \begin{pmatrix} 3 \\ 0 \\ 5 \end{pmatrix}$

Problems 1.6, page 41

1. $\begin{pmatrix} 3 & 9 \\ 6 & 15 \\ -3 & 6 \end{pmatrix}$ **3.** $\begin{pmatrix} 2 & 2 \\ -2 & -1 \\ 6 & -1 \end{pmatrix}$

5. $\begin{pmatrix} 0 & 0 \\ 0 & 0 \\ 0 & 0 \end{pmatrix}$ **7.** $\begin{pmatrix} -2 & 4 \\ 7 & 15 \\ -15 & 10 \end{pmatrix}$

9. $\begin{pmatrix} 4 & 10 \\ 17 & 22 \\ -9 & 1 \end{pmatrix}$ **11.** $\begin{pmatrix} 0 & 6 \\ 5 & 14 \\ -9 & 9 \end{pmatrix}$

13. $\begin{pmatrix} 1 & -5 & 0 \\ -3 & 4 & -5 \\ -14 & 13 & -1 \end{pmatrix}$

15. $\begin{pmatrix} 1 & 1 & 5 \\ 9 & 5 & 10 \\ 7 & -7 & 3 \end{pmatrix}$

17. $\begin{pmatrix} -1 & -1 & -1 \\ -3 & -3 & -10 \\ -7 & 3 & 5 \end{pmatrix}$

19. $\begin{pmatrix} -1 & -1 & -5 \\ -9 & -5 & -10 \\ -7 & 7 & -3 \end{pmatrix}$

25. $\begin{pmatrix} 0 & 1 & 0 & 1 & 0 \\ 1 & 0 & 1 & 1 & 0 \\ 0 & 1 & 0 & 0 & 1 \\ 1 & 1 & 0 & 0 & 0 \\ 0 & 0 & 1 & 0 & 0 \end{pmatrix}$

Problems 1.7, page 55

1. -14 **3.** 1 **5.** $ac + bd$

7. 51 **9.** $a = 0$ **11.** 4

13. 28

15. $\begin{pmatrix} 8 & 20 \\ -4 & 11 \end{pmatrix}$ **17.** $\begin{pmatrix} -3 & -3 \\ 1 & 3 \end{pmatrix}$

19. $\begin{pmatrix} 13 & 35 & 18 \\ 20 & 26 & 20 \end{pmatrix}$

21. $\begin{pmatrix} 19 & -17 & 34 \\ 8 & -12 & 20 \\ -8 & -11 & 7 \end{pmatrix}$

23. $\begin{pmatrix} 18 & 15 & 35 \\ 9 & 21 & 13 \\ 10 & 9 & 9 \end{pmatrix}$ **25.** $(7 \quad 16)$

27. $\begin{pmatrix} 3 & -2 & 1 \\ 4 & 0 & 6 \\ 5 & 1 & 9 \end{pmatrix}$ **29.** $\begin{pmatrix} a & b & c \\ d & e & f \\ g & h & j \end{pmatrix}$

31. If $D = a_{11}a_{22} - a_{12}a_{21}$, then
$$\begin{pmatrix} b_{11} & b_{12} \\ b_{21} & b_{22} \end{pmatrix} = \begin{pmatrix} a_{22}/D & -a_{12}/D \\ -a_{21}/D & a_{11}/D \end{pmatrix}$$

33. a. 3 in group 1, 4 in group 2, 5 in group 3

b. $\begin{pmatrix} 2 & 1 & 1 & 0 & 0 \\ 1 & 1 & 0 & 1 & 0 \\ 1 & 0 & 2 & 0 & 1 \end{pmatrix}$

35. orthogonal

37. orthogonal **39.** orthogonal

41. all α and β that satisfy $5\alpha + 4\beta = 25$ ($\beta = (25 - 5\alpha)/4$, α arbitrary)

43. a. (2, 3, 5, 1)

b. $\begin{pmatrix} 1 \\ \frac{3}{2} \\ \frac{1}{2} \\ 2 \end{pmatrix}$ **c.** 11

45. a. $\begin{pmatrix} 80{,}000 & 45{,}000 & 40{,}000 \\ 50 & 20 & 10 \end{pmatrix}$

b. $\begin{pmatrix} 1 \\ 3 \\ 1 \end{pmatrix}$ **c.** money: 255,000; shares: 120

47. $\begin{pmatrix} 0 & -8 \\ 32 & 32 \end{pmatrix}$ **49.** $\begin{pmatrix} 11 & 38 \\ 57 & 106 \end{pmatrix}$

51. $A^2 = \begin{pmatrix} 0 & 0 & 1 & 0 & 0 \\ 0 & 0 & 0 & 1 & 0 \\ 0 & 0 & 0 & 0 & 1 \\ 0 & 0 & 0 & 0 & 0 \\ 0 & 0 & 0 & 0 & 0 \end{pmatrix}$

$$A^3 = \begin{pmatrix} 0 & 0 & 0 & 1 & 0 \\ 0 & 0 & 0 & 0 & 1 \\ 0 & 0 & 0 & 0 & 0 \\ 0 & 0 & 0 & 0 & 0 \\ 0 & 0 & 0 & 0 & 0 \end{pmatrix}$$

$$A^4 = \begin{pmatrix} 0 & 0 & 0 & 0 & 1 \\ 0 & 0 & 0 & 0 & 0 \\ 0 & 0 & 0 & 0 & 0 \\ 0 & 0 & 0 & 0 & 0 \\ 0 & 0 & 0 & 0 & 0 \end{pmatrix}$$

$$A^5 = \begin{pmatrix} 0 & 0 & 0 & 0 & 0 \\ 0 & 0 & 0 & 0 & 0 \\ 0 & 0 & 0 & 0 & 0 \\ 0 & 0 & 0 & 0 & 0 \\ 0 & 0 & 0 & 0 & 0 \end{pmatrix}$$

53. $PQ = \begin{pmatrix} \frac{11}{90} & \frac{41}{90} & \frac{19}{45} \\ \frac{11}{120} & \frac{71}{120} & \frac{19}{60} \\ \frac{1}{5} & \frac{1}{5} & \frac{3}{5} \end{pmatrix}$; all entries

are nonnegative and $\frac{11}{90} + \frac{41}{90} + \frac{19}{45} =$
$\frac{11}{120} + \frac{71}{120} + \frac{19}{60} = \frac{1}{5} + \frac{1}{5} + \frac{3}{5} = 1$.

55. Let $P = (p_{ij})$ and $Q = (q_{ij})$ be
$k \times k$ probability matrices. Let
$PQ = C = (c_{ij})$. The sum of the
elements in the mth row of PQ is
$c_{m1} + c_{m2} + c_{m3} + \cdots$
$+ c_{mk} = p_{m1}q_{11} + p_{m2}q_{21}$
$+ p_{m3}q_{31} + \cdots + p_{mk}q_{k1}$
$+ p_{m1}q_{12} + p_{m2}q_{22} + p_{m3}q_{32}$
$+ \cdots + p_{mk}q_{k2} + p_{m1}q_{13}$
$+ p_{m2}q_{23} + p_{m3}p_{33} + \cdots$
$+ p_{mk}q_{k3}$
$\vdots$
$+ p_{m1}q_{1k} + p_{m2}q_{2k} + p_{m3}q_{3k}$
$+ \cdots + p_{mk}q_{kk}$

(The elements in parentheses are
those of a row of Q, whose sum
is 1.)

$\downarrow$

$= p_{m1}(q_{11} + q_{12} + q_{13} + \cdots$
$+ q_{1k}) + p_{m2}(q_{21} + q_{22}$
$+ q_{23} + \cdots + q_{2k})$
$+ p_{m3}(q_{31} + q_{32} + q_{33} + \cdots$
$+ q_{3k}) + \cdots + p_{mk}(q_{k1} + q_{k2}$
$+ q_{k3} + \cdots + q_{kk})$
$= p_{m1}(1) + p_{m2}(1) + p_{m3}(1)$
$+ \cdots + p_{mk}(1) = 1.$

57. a. player 2 > player 4 > player 1 >
player 3
b. score = number of games won
plus one-half the number of
games that were won by each
player that this given player
beat

59. $A(B + C)$

$$= \begin{pmatrix} 1 & 2 & 4 \\ 3 & -1 & 0 \end{pmatrix} \begin{pmatrix} 1 & 9 \\ 2 & 11 \\ 10 & 1 \end{pmatrix}$$

$$= \begin{pmatrix} 45 & 35 \\ 1 & 16 \end{pmatrix}$$

$$AB + AC = \begin{pmatrix} 24 & 15 \\ 7 & 17 \end{pmatrix}$$
$$+ \begin{pmatrix} 21 & 20 \\ -6 & -1 \end{pmatrix}$$
$$= \begin{pmatrix} 45 & 35 \\ 1 & 16 \end{pmatrix}$$

61. 36 **63.** 9840

65. $\frac{13}{3} + \frac{15}{4} + \frac{17}{5} = \frac{689}{60}$

67. $(1^2 + 2^2 + 3^2)(2^3 + 3^3 + 4^3) = 1386$

69. $\displaystyle\sum_{k=0}^{5} (-3)^k$ **71.** $\displaystyle\sum_{k=1}^{n} k^{1/k}$

73. $\displaystyle\sum_{k=0}^{9} \frac{(-1)^{k+1}}{a^k}$

75. $\displaystyle\sum_{k=2}^{7} k^2 \cdot 2k = \sum_{k=2}^{7} 2k^3$

77. $\displaystyle\sum_{i=1}^{3}\sum_{j=1}^{2} a_{ij}$ **79.** $\displaystyle\sum_{k=1}^{5} a_{3k}b_{k2}$

81. $\displaystyle\sum_{k=M}^{N} (a_k + b_k) = (a_M + b_M) +$
$(a_{M+1} + b_{M+1}) + (a_{M+2} +$
$b_{M+2}) + \cdots + (a_N + b_N)$
$= (a_M + a_{M+1} + a_{M+2} + \cdots + a_N) +$
$(b_M + b_{M+1} + b_{M+2} + \cdots + b_N)$
$= \displaystyle\sum_{k=M}^{N} a_k + \sum_{k=M}^{N} b_k$

83. $\displaystyle\sum_{k=M}^{N} a_k = a_M + a_{M+1} + \cdots + a_m +$

$$a_{m+1} + \cdots + a_N = (a_M + a_{M+1} + \cdots + a_m) + (a_{m+1} + a_{m+2} + \cdots +$$

$$a_N) = \sum_{k=M}^{m} a_k + \sum_{k=m+1}^{N} a_k$$

Problems 1.8, page 63

1. $\begin{pmatrix} 2 & -1 \\ 4 & 5 \end{pmatrix} \begin{pmatrix} x_1 \\ x_2 \end{pmatrix} = \begin{pmatrix} 3 \\ 7 \end{pmatrix}$

3. $\begin{pmatrix} 3 & 6 & -7 \\ 2 & -1 & 3 \end{pmatrix} \begin{pmatrix} x_1 \\ x_2 \\ x_3 \end{pmatrix} = \begin{pmatrix} 0 \\ 1 \end{pmatrix}$

5. $\begin{pmatrix} 0 & 1 & -1 \\ 1 & 0 & 1 \\ 3 & 2 & 0 \end{pmatrix} \begin{pmatrix} x_1 \\ x_2 \\ x_3 \end{pmatrix} = \begin{pmatrix} 7 \\ 2 \\ -5 \end{pmatrix}$

7. $\begin{aligned} x_1 + x_2 - x_3 &= 7 \\ 4x_1 - x_2 + 5x_3 &= 4 \\ 6x_1 + x_2 + 3x_3 &= 20 \end{aligned}$

9. $\begin{aligned} 2x_1 \quad\quad + x_3 &= 2 \\ -3x_1 + 4x_2 \quad\quad &= 3 \\ 5x_2 + 6x_3 &= 5 \end{aligned}$

11. $\begin{aligned} x_1 \quad\quad\quad\quad &= 2 \\ x_2 \quad\quad\quad &= 3 \\ x_3 \quad &= -5 \\ x_4 &= 6 \end{aligned}$

13. $\begin{aligned} 6x_1 + 2x_2 + x_3 &= 2 \\ -2x_1 + 3x_2 + x_3 &= 4 \\ 0x_1 + 0x_2 + 0x_3 &= 2 \end{aligned}$

15. $\begin{aligned} 7x_1 + 2x_2 &= 1 \\ 3x_1 + x_2 &= 2 \\ 6x_1 + 9x_2 &= 3 \end{aligned}$

17. The simplest solution to the nonhomogeneous equation is obtained by setting $x_2 = 0$. Then the general solution is $(2, 0) + x_2(3, 1)$; x_2 arbitrary.

19. If $x_3 = 0$, one nonhomogeneous solution is $(2, 0, 0)$ and the general solution is $(2, 0, 0) + x_3(-\frac{1}{3}, -\frac{4}{3}, 1)$; x_3 arbitrary.

21. If $x_3 = x_4 = 0$, one nonhomogeneous solution is $(-1, 4, 0, 0)$ and the general solution is $(-1, 4, 0, 0) + x_3(-3, 4, 1, 0) + x_4(5, -7, 0, 1)$.

23. $(c_1 y_1 + c_2 y_2)''$

$$+ a(x)(c_1 y_1 + c_2 y_2)' + b(x)(c_1 y_1 + c_2 y_2)$$
$$= c_1 y_1'' + c_2 y_2'' + a(x)c_1 y_1' + a(x)c_2 y_2' + b(x)c_1 y_1 + b(x)c_2 y_2$$
$$= c_1(y_1'' + a(x)y_1' + b(x)y_1) + c_2(y_2'' + a(x)y_2' + b(x)y_2)$$
$$= c_1 \cdot 0 + c_2 \cdot 0 = 0 \text{ since } y_1 \text{ and } y_2 \text{ solve (7)}.$$

Problems 1.9, page 80

1. $\begin{pmatrix} 2 & -1 \\ -3 & 2 \end{pmatrix}$ **3.** $\begin{pmatrix} 0 & 1 \\ 1 & 0 \end{pmatrix}$

5. not invertible

7. $\begin{pmatrix} \frac{1}{3} & -\frac{1}{3} & -\frac{1}{3} \\ 0 & \frac{1}{2} & 1 \\ 0 & 0 & -1 \end{pmatrix}$

9. not invertible

11. not invertible

13. $\begin{pmatrix} \frac{7}{3} & -\frac{1}{3} & -\frac{1}{3} & -\frac{2}{3} \\ \frac{4}{9} & -\frac{1}{9} & -\frac{4}{9} & \frac{1}{9} \\ -\frac{1}{9} & -\frac{2}{9} & \frac{1}{9} & \frac{2}{9} \\ -\frac{5}{3} & \frac{2}{3} & \frac{2}{3} & \frac{1}{3} \end{pmatrix}$

15. $\begin{pmatrix} 0 & 1 & 0 & 2 \\ 1 & -1 & -2 & 2 \\ 0 & 1 & 3 & -3 \\ -2 & 2 & 3 & -2 \end{pmatrix}$

17. $(A_1 A_2 \cdots A_m)^{-1} = A_m^{-1} A_{m-1}^{-1} \cdots A_2^{-1} A_1^{-1}$ since $(A_m^{-1} A_{m-1}^{-1} \cdots A_2^{-1} A_1^{-1})(A_1 A_2 \cdots A_{m-1} A_m) = (A_m^{-1} A_{m-1}^{-1} \cdots A_2^{-1})(A_1^{-1} A_1) \times A_2 \cdots A_{m-1} A_m = (A_m^{-1} A_{m-1}^{-1} \cdots A_2^{-1}) \times (A_2 \cdots A_{m-1} A_m) = \cdots = I.$

19. $A^{-1} = \dfrac{1}{a_{11} a_{22} - a_{21} a_{12}} \times$
$$\begin{pmatrix} a_{22} & -a_{12} \\ -a_{21} & a_{11} \end{pmatrix}. \text{ If } A = \pm I, \text{ then}$$
$A^{-1} = A$. If $a_{11} = -a_{22}$ and $a_{21} a_{12} = 1 - a_{11}^2$, then $a_{11} a_{22} - a_{21} a_{12} = -a_{11}^2 - (1 - a_{11}^2) = -1$. Thus

$$A^{-1} = \begin{pmatrix} -a_{22} & a_{12} \\ a_{21} & -a_{11} \end{pmatrix}$$
$$= \begin{pmatrix} a_{11} & a_{12} \\ a_{21} & a_{22} \end{pmatrix} = A$$

21. The system $B\mathbf{x} = \mathbf{0}$ has an infinite number of solutions (by Theorem 1.4.1). But if $B\mathbf{x} = \mathbf{0}$, then $AB\mathbf{x} = \mathbf{0}$. Thus, from Theorem 6 [parts (i) and (ii)], AB is not invertible.

23. $\begin{pmatrix} \sin\theta & \cos\theta & 0 \\ \cos\theta & -\sin\theta & 0 \\ 0 & 0 & 1 \end{pmatrix}$ is its own inverse (since $\sin^2\theta + \cos^2\theta = 1$).

25. If the ith diagonal component is 0, then in the row reduction of A the ith row is zero so that, by the statement in step 3(b) on page 69, A is not invertible. Otherwise, if

$$A = \text{diag}(a_1, a_2, \ldots, a_n)$$

then

$$A^{-1} = \text{diag}\left(\frac{1}{a_1}, \frac{1}{a_2}, \ldots, \frac{1}{a_n}\right)$$

27. $\begin{pmatrix} \frac{1}{2} & -\frac{1}{6} & \frac{7}{30} \\ 0 & \frac{1}{3} & -\frac{4}{15} \\ 0 & 0 & \frac{1}{5} \end{pmatrix}$

29. We prove the result in the case A is upper triangular. The proof in the lower triangular case is similar. Consider the homogeneous system

$$\begin{pmatrix} a_{11} & a_{12} & a_{13} & \cdots & a_{1,n-1} & a_{1n} \\ 0 & a_{22} & a_{23} & \cdots & a_{2,n-1} & a_{2n} \\ \vdots & \vdots & \vdots & & \vdots & \vdots \\ 0 & 0 & 0 & \cdots & a_{n-1,n-1} & a_{n-1,n} \\ 0 & 0 & 0 & \cdots & 0 & a_{nn} \end{pmatrix}$$

$$\times \begin{pmatrix} x_1 \\ x_2 \\ \vdots \\ x_{n-1} \\ x_n \end{pmatrix} = \begin{pmatrix} 0 \\ 0 \\ \vdots \\ 0 \\ 0 \end{pmatrix}$$

Suppose that $a_{11}, a_{22}, \ldots, a_{nn}$ are all nonzero. The last equation in the homogeneous system is $a_{nn}x_n = 0$, and since $a_{nn} \neq 0$, $x_n = 0$. The next-to-the-last equation is

$$a_{n-1,n-1}x_{n-1} + a_{n-1,n}x_n = 0$$

and $a_{n-1,n-1} \neq 0$, $x_n = 0$ implies that $x_{n-1} = 0$. Similarly, we con-

clude that $x_1 = x_2 = \cdots = x_{n-1} = x_n = 0$, so the only solution to the homogeneous system is the trivial solution. By Theorem 6 [parts (i) and (ii)], A is invertible. Conversely, suppose one of the diagonal components, say, a_{11}, is equal to 0. Then the homogeneous system $A\mathbf{x} = \mathbf{0}$ has the solution

$$\mathbf{x} = \begin{pmatrix} 1 \\ 0 \\ \vdots \\ 0 \end{pmatrix}$$

[If $a_{jj} = 0$ with $j \neq 1$, then choose $\mathbf{x}$ to be the vector with 1 in the jth position and 0 everywhere else.] Using Theorem 6 again, we conclude that A is not invertible.

31. any nonzero multiple of $(1, 2)$

33. 3 chairs and 2 tables

35. 4 units of A and 5 units of B

37. a. $A = \begin{pmatrix} 0.293 & 0 & 0 \\ 0.014 & 0.207 & 0.017 \\ 0.044 & 0.010 & 0.216 \end{pmatrix}$;

$I - A$

$= \begin{pmatrix} 0.707 & 0 & 0 \\ -0.014 & 0.793 & -0.017 \\ -0.044 & -0.010 & 0.784 \end{pmatrix}$

b. $\begin{pmatrix} 18,689 \\ 22,598 \\ 3,615 \end{pmatrix}$

39. $\begin{pmatrix} 1 & \frac{1}{2} \\ 0 & 1 \end{pmatrix}$; yes

41. $\begin{pmatrix} 1 & \frac{2}{3} & \frac{1}{3} \\ 0 & 1 & 1 \\ 0 & 0 & 1 \end{pmatrix}$; yes

43. $\begin{pmatrix} 1 & -\frac{1}{2} & 2 \\ 0 & 1 & -14 \\ 0 & 0 & 0 \end{pmatrix}$; no

45. $\begin{pmatrix} 1 & 0 & 2 & 3 \\ 0 & 1 & 2 & 7 \\ 0 & 0 & 1 & \frac{10}{7} \\ 0 & 0 & 0 & 0 \end{pmatrix}$; no

Problems 1.10, page 85

1. $\begin{pmatrix} -1 & 6 \\ 4 & 5 \end{pmatrix}$ **3.** $\begin{pmatrix} 2 & -1 & 1 \\ 3 & 2 & 4 \end{pmatrix}$

5. $\begin{pmatrix} 1 & -1 & 1 \\ 2 & 0 & 5 \\ 3 & 4 & 5 \end{pmatrix}$ **7.** $\begin{pmatrix} 1 & 0 \\ 0 & 1 \\ 1 & 0 \\ 0 & 1 \end{pmatrix}$

9. $\begin{pmatrix} a & d & g \\ b & e & h \\ c & f & j \end{pmatrix}$

11. $[(A + B)^t]_{ij} = (A + B)_{ji} = a_{ji} + b_{ji} = (A^t)_{ij} + (B^t)_{ij}$. Thus the ijth component of $(A + B)^t$ equals the ijth component of A^t plus the ijth component of B^t.

13. $(A + B)^t = A^t + B^t = A + B$

15. If A is $m \times n$, then A^t is $n \times m$ and AA^t is $m \times m$. Also, $(AA^t)^t = (A^t)^t A^t = AA^t$.

17. If A is upper triangular and $B = A^t$, then $b_{ij} = a_{ji} = 0$ if $j > i$. Thus B is lower triangular.

19. $(A + B)^t = A^t + B^t$
$= -A - B = -(A + B)$

21. $(AB)^t = B^t A^t = (-B)(-A)$
$= (-1)^2 BA = BA$

23. $[\frac{1}{2}(A - A^t)]^t = \frac{1}{2}(A^t - (A^t)^t)$
$\qquad = \frac{1}{2}(A^t - A)$
$\qquad = -[\frac{1}{2}(A - A^t)]$

25. (ii) tells us that $a_{11}a_{21} + a_{12}a_{22} = 0$. So

$AA^t = \begin{pmatrix} a_{11}^2 + a_{12}^2 & a_{11}a_{21} + a_{12}a_{22} \\ a_{11}a_{21} + a_{12}a_{22} & a_{21}^2 + a_{22}^2 \end{pmatrix}$

$\quad = \begin{pmatrix} 1 & 0 \\ 0 & 1 \end{pmatrix}$

Then, from Theorem 1.9.8, we see that $A^t = A^{-1}$.

27. $\begin{pmatrix} 2 & -3 \\ -1 & 2 \end{pmatrix}$

29. $\begin{pmatrix} \frac{13}{8} & -\frac{15}{8} & \frac{5}{4} \\ -\frac{1}{2} & \frac{1}{2} & 0 \\ -\frac{1}{8} & \frac{3}{8} & -\frac{1}{4} \end{pmatrix}$

Problems 1.11, page 93

1. yes, $R_1 \rightleftarrows R_2$

3. no [two operations are used: $R_1 \rightleftarrows R_2$ followed by $R_2 \to R_2 + R_1$]

5. no [two operations are used: $R_1 \to 3R_1$ and $R_2 \to 3R_2$]

7. no [two operations are used: $R_1 \rightleftarrows R_3$ followed by $R_1 \rightleftarrows R_2$]

9. yes, $R_2 \to R_2 + 2R_1$

11. no [two operations are used: $R_2 \to R_2 + R_1$ and $R_4 \to R_4 + R_3$]

13. $\begin{pmatrix} 1 & 0 & 0 \\ 0 & 4 & 0 \\ 0 & 0 & 1 \end{pmatrix}$

15. $\begin{pmatrix} 1 & -3 & 0 \\ 0 & 1 & 0 \\ 0 & 0 & 1 \end{pmatrix}$

17. $\begin{pmatrix} 0 & 0 & 1 \\ 0 & 1 & 0 \\ 1 & 0 & 0 \end{pmatrix}$ **19.** $\begin{pmatrix} 1 & 0 & 0 \\ 0 & 1 & 1 \\ 0 & 0 & 1 \end{pmatrix}$

21. $\begin{pmatrix} 1 & 0 \\ 0 & -2 \end{pmatrix}$ **23.** $\begin{pmatrix} 1 & 2 \\ 0 & 1 \end{pmatrix}$

25. $\begin{pmatrix} 0 & 0 & 1 \\ 0 & 1 & 0 \\ 1 & 0 & 0 \end{pmatrix}$ **27.** $\begin{pmatrix} -1 & 0 & 0 \\ 0 & 1 & 0 \\ 0 & 0 & 1 \end{pmatrix}$

29. $\begin{pmatrix} 1 & 0 & 0 \\ 0 & 1 & 0 \\ -5 & 0 & 1 \end{pmatrix}$

31. $\begin{pmatrix} 0 & 1 \\ 1 & 0 \end{pmatrix}$ **33.** $\begin{pmatrix} 1 & 0 \\ 0 & \frac{1}{4} \end{pmatrix}$

35. $\begin{pmatrix} 1 & 2 & 0 \\ 0 & 1 & 0 \\ 0 & 0 & 1 \end{pmatrix}$

37. $\begin{pmatrix} 1 & 0 & 0 \\ 0 & -2 & 0 \\ 0 & 0 & 1 \end{pmatrix}$

39. $\begin{pmatrix} 1 & 0 & 0 & -5 \\ 0 & 1 & 0 & 0 \\ 0 & 0 & 1 & 0 \\ 0 & 0 & 0 & 1 \end{pmatrix}$

41. $\begin{pmatrix} 2 & 0 \\ 0 & 1 \end{pmatrix}\begin{pmatrix} 1 & 0 \\ 3 & 1 \end{pmatrix}\begin{pmatrix} 1 & 0 \\ 0 & \frac{1}{2} \end{pmatrix}\begin{pmatrix} 1 & \frac{1}{2} \\ 0 & 1 \end{pmatrix}$

43. $\begin{pmatrix} 1 & 0 & 0 \\ 0 & 1 & 0 \\ 5 & 0 & 1 \end{pmatrix}\begin{pmatrix} 1 & 0 & 0 \\ 0 & 2 & 0 \\ 0 & 0 & 1 \end{pmatrix}\begin{pmatrix} 1 & 1 & 0 \\ 0 & 1 & 0 \\ 0 & 0 & 1 \end{pmatrix}\begin{pmatrix} 1 & 0 & 0 \\ 0 & 1 & 0 \\ 0 & 0 & -4 \end{pmatrix}\begin{pmatrix} 1 & 0 & -\frac{1}{2} \\ 0 & 1 & 0 \\ 0 & 0 & 1 \end{pmatrix}\begin{pmatrix} 1 & 0 & 0 \\ 0 & 1 & \frac{3}{2} \\ 0 & 0 & 1 \end{pmatrix}$

45. $\begin{pmatrix} 0 & 0 & 1 \\ 0 & 1 & 0 \\ 1 & 0 & 0 \end{pmatrix}\begin{pmatrix} 1 & 0 & 0 \\ 0 & 1 & 0 \\ 0 & -1 & 1 \end{pmatrix}\begin{pmatrix} 1 & 0 & 0 \\ 0 & 1 & 0 \\ 0 & 0 & -1 \end{pmatrix}\begin{pmatrix} 1 & 0 & 1 \\ 0 & 1 & 0 \\ 0 & 0 & 1 \end{pmatrix}\begin{pmatrix} 1 & 0 & 0 \\ 0 & 1 & -1 \\ 0 & 0 & 1 \end{pmatrix}$

47. $\begin{pmatrix} 2 & 0 & 0 & 0 \\ 0 & 1 & 0 & 0 \\ 0 & 0 & 1 & 0 \\ 0 & 0 & 0 & 1 \end{pmatrix}\begin{pmatrix} 1 & 0 & 0 & 0 \\ 0 & 3 & 0 & 0 \\ 0 & 0 & 1 & 0 \\ 0 & 0 & 0 & 1 \end{pmatrix}\begin{pmatrix} 1 & 0 & 0 & 0 \\ 0 & 1 & 0 & 0 \\ 0 & 0 & -4 & 0 \\ 0 & 0 & 0 & 1 \end{pmatrix}\begin{pmatrix} 1 & 0 & 0 & 0 \\ 0 & 1 & 0 & 0 \\ 0 & 0 & 1 & 0 \\ 0 & 0 & 0 & 5 \end{pmatrix}$

49. $\begin{pmatrix} a & 0 \\ 0 & 1 \end{pmatrix}\begin{pmatrix} 1 & 0 \\ 0 & c \end{pmatrix}\begin{pmatrix} 1 & b/a \\ 0 & 1 \end{pmatrix}$; the first two matrices are elementary because $a \neq 0$ and $c \neq 0$

51. The 2×2 and 3×3 cases are the results of Problems 49 and 50. In the answer to Problem 1.9.29 we proved this result. Another proof can be given by showing, as in Problems 49 and 50, that A can be written as the product of elementary matrices. The key step is to reduce A to I by noting that the only times we divide, we divide by the numbers on the diagonal, which are nonzero by assumption.

53. A^t is upper triangular, so $(A^t)^{-1}$ is upper triangular by the result of Problem 52. But $(A^t)^{-1} = (A^{-1})^t$, so $(A^{-1})^t$ is upper triangular, which means that $A^{-1} = [(A^{-1})^t]^t$ is lower triangular.

55. Let $B = A_{ij}$ and $D = A_{ij}A$. Then the krth component d_{kr} of D is given by

$$d_{kr} = \sum_{l=1}^{n} b_{kl}a_{lr} \qquad (*)$$

If $k \neq j$, the kth row of B is the kth row of the identity, so $b_{kl} = 1$ if $l = k$ and 0 otherwise. Thus

$$d_{kr} = b_{kk}a_{kr} = a_{kr} \quad \text{if } k \neq j$$

If $k = j$, then

$$b_{jl} = \begin{cases} 1, & \text{if } l = j \\ c, & \text{if } l = i \\ 0, & \text{otherwise} \end{cases}$$

and (*) becomes

$$a_{jr} = b_{jj}a_{jr} + b_{ji}a_{ir}$$
$$= a_{jr} + ca_{jr}$$

Thus each component in the jth row of $A_{ij}A$ is the sum of the corresponding component in the jth row of A and c times the corresponding component in the ith row of A.

57. $\begin{pmatrix} 1 & 0 \\ 2 & 1 \end{pmatrix}\begin{pmatrix} 1 & 2 \\ 0 & 0 \end{pmatrix}$

59. $\begin{pmatrix} 0 & 1 \\ 1 & 0 \end{pmatrix}\begin{pmatrix} 1 & 0 \\ 0 & 0 \end{pmatrix}$

61. $\begin{pmatrix} 1 & 0 & 0 \\ 0 & 1 & 0 \\ 1 & 0 & 1 \end{pmatrix}\begin{pmatrix} 1 & 0 & 0 \\ 0 & -3 & 0 \\ 0 & 0 & 1 \end{pmatrix}$
$\times \begin{pmatrix} 1 & 0 & 0 \\ 0 & 1 & 0 \\ 0 & 3 & 1 \end{pmatrix}\begin{pmatrix} 1 & -3 & 3 \\ 0 & 1 & -\frac{1}{3} \\ 0 & 0 & 0 \end{pmatrix}$

Problems 1.12, page 102

1. $\begin{pmatrix} 0 & 1 & 1 & 0 \\ 0 & 0 & 0 & 0 \\ 1 & 1 & 0 & 0 \\ 1 & 0 & 0 & 0 \end{pmatrix}$
3. $\begin{pmatrix} 0 & 0 & 1 & 0 & 0 \\ 1 & 0 & 0 & 1 & 0 \\ 0 & 1 & 0 & 1 & 0 \\ 1 & 1 & 1 & 0 & 1 \\ 0 & 1 & 0 & 0 & 0 \end{pmatrix}$

5.

7.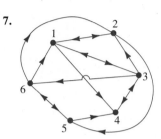

9. There is one 2-chain between 12, 14, 21, 22, 25, 32, 33, 34, 35, 41, 43, 51, 54; two 2-chains between 23, 31, 42, 44; one 3-chain between 12, 13, 14, 15, 21, 23, 35, 51, 52, 55; two 3-chain between 11, 31, 34, 45, 53; three 3-chains between 22, 24, 32, 33, 42, 43, 44; four 3-chains between 41; one 4-chain between 15, 51, 53; two 4-chains between 11, 14, 35; three 4-chains between 12, 13, 25, 45, 52, 54; four 4-chains between 22, 23, 24, 33; five 4-chains between 31; six 4-chains between 21, 32, 34, 41, 44; seven 4-chains between 43; eight 4-chains between 42.

11. Let $B = A + A^2$. Then $(B)_{ij} = (A)_{ij} + (A^2)_{ij}$. But $(A)_{ij}$ is the number of one-step links between vertices i and j, and $(A^2)_{ij}$ is the number of two-step links between vertices i and j. So $(B)_{ij}$ is the total number of one-step or two-step links between vertices i and j.

Chapter 1—Review, page 107

1. $(\frac{4}{7}, \frac{10}{7})$ **3.** no solution

5. $(0, 0, 0)$ **7.** $(-\frac{1}{2}, 0, \frac{5}{2})$

9. $(\frac{1}{3}x_3, \frac{7}{3}x_3, x_3)$, x_3 arbitrary

11. no solution

13. $(0, 0, 0, 0)$

15. $\begin{pmatrix} -6 & 3 \\ 0 & 12 \\ 6 & 9 \end{pmatrix}$

17. $\begin{pmatrix} 16 & 2 & 3 \\ -20 & 10 & -1 \\ -36 & 8 & 16 \end{pmatrix}$

19. $\begin{pmatrix} 17 & 39 & 41 \\ 14 & 20 & 42 \end{pmatrix}$

21. $\begin{pmatrix} 9 & 10 \\ 30 & 32 \end{pmatrix}$

23. reduced row echelon form

25. neither

27. row echelon form:

$$\begin{pmatrix} 1 & 4 & -1 \\ 0 & 1 & \frac{5}{4} \end{pmatrix}$$

reduced row echelon form:

$$\begin{pmatrix} 1 & 0 & -6 \\ 0 & 1 & \frac{5}{4} \end{pmatrix}$$

29. $\begin{pmatrix} 1 & \frac{3}{2} \\ 0 & 1 \end{pmatrix}$; inverse is

$$\begin{pmatrix} \frac{4}{11} & -\frac{3}{11} \\ \frac{1}{11} & \frac{2}{11} \end{pmatrix}$$

31. $\begin{pmatrix} 1 & 2 & 0 \\ 0 & 1 & \frac{1}{3} \\ 0 & 0 & 1 \end{pmatrix}$; inverse is

$$\begin{pmatrix} -\frac{1}{4} & \frac{1}{4} & \frac{1}{4} \\ \frac{5}{8} & -\frac{1}{8} & -\frac{1}{8} \\ \frac{1}{8} & -\frac{5}{8} & \frac{3}{8} \end{pmatrix}$$

33. $\begin{pmatrix} 1 & 0 & 2 \\ 0 & 1 & 1 \\ 0 & 0 & 1 \end{pmatrix}$; inverse is

$$\begin{pmatrix} \frac{5}{6} & \frac{2}{3} & -2 \\ \frac{1}{3} & \frac{2}{3} & -1 \\ -\frac{1}{6} & -\frac{1}{3} & 1 \end{pmatrix}$$

35. $\begin{pmatrix} 1 & 2 & 0 \\ 2 & 1 & -1 \\ 3 & 1 & 1 \end{pmatrix} \begin{pmatrix} x_1 \\ x_2 \\ x_3 \end{pmatrix} = \begin{pmatrix} 3 \\ -1 \\ 7 \end{pmatrix}$; A^{-1} is given in Exercise 31;

$$x_1 = \tfrac{3}{4}, \quad x_2 = \tfrac{9}{8}, \quad x_3 = \tfrac{29}{8}$$

37. $\begin{pmatrix} 2 & -1 \\ 3 & 0 \\ 1 & 2 \end{pmatrix}$; neither

39. $\begin{pmatrix} 2 & 3 & 1 \\ 3 & -6 & -5 \\ 1 & -5 & 9 \end{pmatrix}$; symmetric

41. $\begin{pmatrix} 1 & -1 & 4 & 6 \\ -1 & 2 & 5 & 7 \\ 4 & 5 & 3 & -8 \\ 6 & 7 & -8 & 9 \end{pmatrix}$; symmetric

43. $\begin{pmatrix} 1 & 0 & 0 \\ 0 & -2 & 0 \\ 0 & 0 & 1 \end{pmatrix}$

45. $\begin{pmatrix} 1 & 0 & 0 \\ 0 & 1 & 0 \\ -5 & 0 & 1 \end{pmatrix}$

47. $\begin{pmatrix} 1 & 0 & 0 \\ 0 & 1 & \frac{1}{5} \\ 0 & 0 & 1 \end{pmatrix}$

49. $\begin{pmatrix} 0 & 1 & 0 \\ 1 & 0 & 0 \\ 0 & 0 & 1 \end{pmatrix}$

51. $\begin{pmatrix} 2 & 0 \\ 0 & 1 \end{pmatrix} \begin{pmatrix} 1 & 0 \\ -1 & 1 \end{pmatrix} \times \begin{pmatrix} 1 & 0 \\ 0 & \frac{1}{2} \end{pmatrix} \begin{pmatrix} 1 & -\frac{1}{2} \\ 0 & 1 \end{pmatrix}$

53. $\begin{pmatrix} 2 & 0 \\ 0 & 1 \end{pmatrix} \begin{pmatrix} 1 & 0 \\ -4 & 1 \end{pmatrix} \begin{pmatrix} 1 & -\frac{1}{2} \\ 0 & 0 \end{pmatrix}$

55. $\begin{pmatrix} 0 & 1 & 0 & 0 \\ 0 & 0 & 1 & 0 \\ 1 & 0 & 0 & 1 \\ 0 & 1 & 0 & 0 \end{pmatrix}$

57.

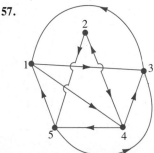

Chapter 2

Problems 2.1, page 118

1. -10　**3.** 47　**5.** 4　**7.** 56

9. 274

11.
Let $A = \begin{pmatrix} a_{11} & 0 & 0 & \cdots & 0 \\ 0 & a_{22} & 0 & \cdots & 0 \\ 0 & 0 & a_{33} & \cdots & 0 \\ \vdots & \vdots & \vdots & & \vdots \\ 0 & 0 & 0 & \cdots & a_{nn} \end{pmatrix}$

and

$$B = \begin{pmatrix} b_{11} & 0 & 0 & \cdots & 0 \\ 0 & b_{22} & 0 & \cdots & 0 \\ 0 & 0 & b_{33} & \cdots & 0 \\ \vdots & \vdots & \vdots & & \vdots \\ 0 & 0 & 0 & \cdots & b_{nn} \end{pmatrix}$$

Then $\det A = a_{11}a_{22}a_{33}\cdots a_{nn}$, $\det B = b_{11}b_{22}b_{33}\cdots b_{nn}$,

$$AB = \begin{pmatrix} a_{11}b_{11} & 0 & 0 & \cdots & 0 \\ 0 & a_{22}b_{22} & 0 & \cdots & 0 \\ 0 & 0 & a_{33}b_{33} & \cdots & 0 \\ \vdots & \vdots & & & \vdots \\ 0 & 0 & 0 & \cdots & a_{nn}b_{nn} \end{pmatrix}$$

and

$$\begin{aligned}
\det AB &= (a_{11}b_{11})(a_{22}b_{22}) \\
&\quad \times (a_{33}b_{33})\cdots(a_{nn}b_{nn}) \\
&= (a_{11}a_{22}a_{33}\cdots a_{nn}) \\
&\quad \times (b_{11}b_{22}b_{33}\cdots b_{nn}) \\
&= \det A \det B
\end{aligned}$$

13. Almost any example will work. For instance, $\det \begin{pmatrix} 1 & 0 \\ 0 & 1 \end{pmatrix} = 1$, but

$$\det \begin{pmatrix} 1 & 0 \\ 0 & 0 \end{pmatrix} + \det \begin{pmatrix} 0 & 0 \\ 0 & 1 \end{pmatrix} = 0 + 0 \neq 1.$$

As another example, let $A = \begin{pmatrix} 1 & 2 \\ 3 & 4 \end{pmatrix}$ and $B = \begin{pmatrix} 5 & 6 \\ 7 & 8 \end{pmatrix}$; then

$(A + B) = \begin{pmatrix} 6 & 8 \\ 10 & 12 \end{pmatrix}$, $\det A = -2$,

$\det B = -2$, and $\det (A + B) = -8 \neq \det A + \det B$.

15. Let $A = \begin{pmatrix} a_{11} & 0 & \cdots & 0 \\ a_{21} & a_{22} & \cdots & 0 \\ \vdots & \vdots & & \vdots \\ a_{n1} & a_{n2} & \cdots & a_{nn} \end{pmatrix}$

Then, continually expanding in the first row, we obtain

$$\det A = a_{11} \begin{vmatrix} a_{22} & 0 & \cdots & 0 \\ a_{32} & a_{33} & \cdots & 0 \\ \vdots & \vdots & & \vdots \\ a_{n2} & a_{n3} & \cdots & a_{nn} \end{vmatrix}$$

$$= a_{11}a_{22} \begin{vmatrix} a_{33} & 0 & \cdots & 0 \\ a_{43} & a_{44} & \cdots & 0 \\ \vdots & \vdots & & \vdots \\ a_{n3} & a_{n4} & \cdots & a_{nn} \end{vmatrix}$$

$$= \cdots = a_{11}a_{22}a_{33} \cdots$$

$$\times a_{n-2} \begin{vmatrix} a_{n-1,n-1} & 0 \\ a_{n,n-1} & a_{nn} \end{vmatrix}$$

$$= a_{11}a_{22}a_{33} \cdots$$

$$\times a_{n-2,n-2}a_{n-1,n-1}a_{nn}$$

17. Suppose that $A = (0, c)$. Then the situation is as depicted in the figure. The base of the parallelogram is $\overline{AC} = \sqrt{b^2 + d^2}$. The slope of AC is $\dfrac{d}{b}$ so the slope of $BQ' = -\dfrac{b}{d}$ and the equation of the line passing through 0 and Q is $y = -\dfrac{b}{d}x$. The equation of line AC is $y = c + \dfrac{d}{b}x$. Q is the point of intersection of these two lines and is

$$\left(\frac{d(-bc)}{b^2 + d^2}, \frac{-b(-bc)}{b^2 + d^2} \right)$$

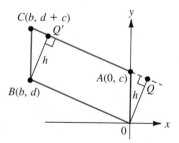

$C(b, d + c)$
Q'
h
$B(b, d)$
$A(0, c)$
h
Q
y
0
x

$$= \left(\frac{d \det A}{b^2 + d^2}, -\frac{b \det A}{b^2 + d^2} \right)$$

Then $h = \overline{0Q} = \dfrac{|\det A|}{\sqrt{b^2 + d^2}}$, so

base $\times$ height $= \overline{AC} \times \overline{0Q} = |\det A|$.

A similar proof works in the case $A = (a, 0)$.

Problems 2.2, page 130

1. 28 **3.** 2 **5.** 32 **7.** -36

9. -260 **11.** -183 **13.** 24

15. -296 **17.** 138

19. $abcde$ **21.** -8 **23.** 16

25. -16 **27.** -16

29. proof by induction: true for $n = 2$ since

$$\begin{vmatrix} 1 + x_1 & x_2 \\ x_1 & 1 + x_2 \end{vmatrix}$$
$$= (1 + x_1)(1 + x_2) - x_1x_2$$
$$= 1 + x_1 + x_2$$

Assume true for $n = k$. That is,

$$\begin{vmatrix} 1 + x_1 & x_2 & x_3 & \cdots & x_k \\ x_1 & 1 + x_2 & x_3 & \cdots & x_k \\ x_1 & x_2 & 1 + x_3 & \cdots & x_k \\ \vdots & \vdots & \vdots & & \vdots \\ x_1 & x_2 & x_3 & \cdots & 1 + x_k \end{vmatrix}$$
$$= 1 + x_1 + x_2 + \cdots + x_k$$

Then, for $n = k + 1$,

$$\begin{vmatrix} 1 + x_1 & x_2 & x_3 & \cdots & x_k & x_{k+1} \\ x_1 & 1 + x_2 & x_3 & \cdots & x_k & x_{k+1} \\ x_1 & x_2 & 1 + x_3 & \cdots & x_k & x_{k+1} \\ \vdots & \vdots & \vdots & & \vdots & \vdots \\ x_1 & x_2 & x_3 & \cdots & x_k & 1 + x_{k+1} \end{vmatrix}$$

(using Property 3 in the first column)

$$= \begin{vmatrix} 1 & x_2 & x_3 & \cdots & x_k & x_{k+1} \\ 0 & 1 + x_2 & x_3 & \cdots & x_k & x_{k+1} \\ 0 & x_2 & 1 + x_3 & \cdots & x_k & x_{k+1} \\ \vdots & \vdots & \vdots & & \vdots & \vdots \\ 0 & x_2 & x_3 & \cdots & x_k & 1 + x_{k+1} \end{vmatrix} \quad ①$$

$$+ \begin{vmatrix} x_1 & x_2 & x_3 & \cdots & x_k & x_{k+1} \\ x_1 & 1+x_2 & x_3 & \cdots & x_k & x_{k+1} \\ x_1 & x_2 & 1+x_3 & \cdots & x_k & x_{k+1} \\ \vdots & \vdots & \vdots & & \vdots & \vdots \\ x_1 & x_2 & x_3 & \cdots & x_k & 1+x_{k+1} \end{vmatrix} \;\; ②$$

But, expanding det ① in its first column, we have

$$\det ① = \begin{vmatrix} 1+x_2 & x_3 & \cdots & x_k & x_{k+1} \\ x_2 & 1+x_3 & \cdots & x_k & x_{k+1} \\ \vdots & \vdots & & \vdots & \vdots \\ x_2 & x_3 & \cdots & x_k & 1+x_{k+1} \end{vmatrix}$$

$$= 1 + x_2 + x_3 + \cdots + x_{k+1}$$

by the induction assumption (since ① is a $k \times k$ determinant). To evaluate det ②, subtract the first row from all other rows:

$$\det ② = \begin{vmatrix} x_1 & x_2 & x_3 & \cdots & x_k & x_{k+1} \\ 0 & 1 & 0 & \cdots & 0 & 0 \\ 0 & 0 & 1 & \cdots & 0 & 0 \\ \vdots & \vdots & \vdots & & \vdots & \vdots \\ 0 & 0 & 0 & \cdots & 0 & 1 \end{vmatrix} = x_1$$

Adding det ① and det ② completes the proof.

31. If n is odd, det $A = -\det A$, so that $2 \det A = 0$ and $\det A = 0$.

33. $\dfrac{1}{2} \begin{vmatrix} 1 & x_1 & y_1 \\ 1 & x_2 & y_2 \\ 1 & x_3 & y_3 \end{vmatrix}$

$$= \frac{1}{2} \begin{vmatrix} x_2 - x_1 & x_3 - x_1 \\ y_2 - y_1 & y_3 - y_1 \end{vmatrix}$$

Look at the figures below.

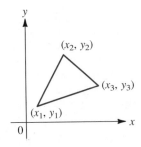

The area A of the triangle is half the area of the parallelogram generated by the vectors $\mathbf{u}_1$ and $\mathbf{u}_2$ which, by the result of Problem 2.1.16, is given by

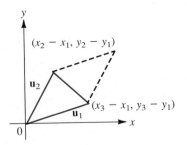

$$A = \pm \frac{1}{2} \begin{vmatrix} x_2 - x_1 & x_3 - x_1 \\ y_2 - y_1 & y_3 - y_1 \end{vmatrix}$$

35. $D_3 = \begin{vmatrix} 1 & 1 & 1 \\ a_1 & a_2 & a_3 \\ a_1^2 & a_2^2 & a_3^2 \end{vmatrix}$

$$= \begin{vmatrix} 1 & 0 & 0 \\ a_1 & a_2 - a_1 & a_3 - a_1 \\ a_1^2 & a_2^2 - a_1^2 & a_3^2 - a_1^2 \end{vmatrix}$$

$$= \begin{vmatrix} a_2 - a_1 & a_3 - a_1 \\ (a_2 + a_1)(a_2 - a_1) & (a_3 - a_1)(a_3 - a_1) \end{vmatrix}$$

$$= (a_2 - a_1)(a_3 - a_1) \times \begin{vmatrix} 1 & 1 \\ a_2 + a_1 & a_3 + a_1 \end{vmatrix}$$

$$= (a_2 - a_1)(a_3 - a_1) \times (a_3 - a_2)$$

37. a. $D_n = \begin{vmatrix} 1 & 1 & \cdots & 1 \\ a_1 & a_2 & \cdots & a_n \\ a_1^2 & a_2^2 & \cdots & a_n^2 \\ \vdots & \vdots & & \vdots \\ a_1^{n-1} & a_2^{n-1} & \cdots & a_n^{n-1} \end{vmatrix}$

b. We prove this by induction. The result is true for $n = 3$ by the result of Problem 35. We assume it true for $n = k$. Now

$D_{k+1} =$

$$\begin{vmatrix} 1 & 1 & \cdots & 1 & 1 \\ a_1 & a_2 & \cdots & a_k & a_{k+1} \\ a_1^2 & a_2^2 & \cdots & a_k^2 & a_{k+1}^2 \\ \vdots & \vdots & & \vdots & \vdots \\ a_1^{k-1} & a_2^{k-1} & \cdots & a_k^{k-1} & a_{k+1}^{k-1} \\ a_1^k & a_2^k & \cdots & a_k^k & a_{k+1}^k \end{vmatrix}$$

We subtract the first column from each of the other k columns:

$D_{k+1} =$

$$\begin{vmatrix} 1 & 0 & \cdots \\ a_1 & a_2 - a_1 & \cdots \\ a_1^2 & a_2^2 - a_1^2 & \cdots \\ \vdots & \vdots & \\ a_1^{k-1} & a_2^{k-1} - a_1^{k-1} & \cdots \\ a_1^k & a_2^k - a_1^k & \cdots \end{vmatrix}$$

$$\begin{matrix} 0 & 0 \\ a_k - a_1 & a_{k+1} - a_1 \\ a_k^2 - a_1^2 & a_{k+1}^2 - a_1^2 \\ \vdots & \vdots \\ a_k^{k-1} - a_1^{k-1} & a_{k+1}^{k-1} - a_1^{k-1} \\ a_k^k - a_1^k & a_{k+1}^k - a_1^k \end{matrix}$$

$$= \begin{vmatrix} a_2 - a_1 & a_3 - a_1 \\ a_2^2 - a_1^2 & a_3^2 - a_1^2 \\ \vdots & \vdots \\ a_2^{k-1} - a_1^{k-1} & a_3^{k-1} - a_1^{k-1} \\ a_2^k - a_1^k & a_3^k - a_1^k \end{vmatrix}$$

$$\begin{matrix} \cdots & a_k - a_1 & a_{k+1} - a_1 \\ \cdots & a_k^2 - a_1^2 & a_{k+1}^2 - a_1^2 \\ & \vdots & \vdots \\ \cdots & a_k^{k-1} - a_1^{k-1} & a_{k+1}^{k-1} - a_1^{k-1} \\ \cdots & a_k^k - a_1^k & a_{k+1}^k - a_1^k \end{matrix}$$

Now $a_2^k - a_1^k = (a_2 - a_1) \times (a_2^{k-1} + a_2^{k-2}a_1 + a_2^{k-3}a_1^2 + \cdots + a_2^2 a_1^{k-3} + a_2 a_1^{k-2} + a_1^{k-1})$, and $a_2^{k-1} - a_1^{k-1} = (a_2 - a_1) \times (a_2^{k-2} + a_2^{k-3}a_1 + \cdots + a_2^2 a_1^{k-4} + a_2 a_1^{k-3} + a_1^{k-2})$. Note that if the terms in the second factor of the last expression are multiplied by a_1 and then subtracted from the second factor of $a_2^k - a_1^k$, only the term

a_2^{k-1} remains. Thus we (i) expand the last determinant obtained above in the first row, (ii) factor $a_{j-1} - a_1$ from the jth column for $1 \le j \le k$, and (iii) multiply the lth row by a_1 and subtract it from the $(l+1)$st row, for $l = k-1, k-2, \ldots, 3, 2$ in succession. This yields $D_{k+1} = (a_2 - a_1)(a_3 - a_1) \cdots (a_{k+1} - a_1)$

$$\times \begin{vmatrix} 1 & 1 & \cdots & 1 & 1 \\ a_2 & a_3 & \cdots & a_k & a_{k+1} \\ a_2^2 & a_3^2 & \cdots & a_k^2 & a_{k+1}^2 \\ \vdots & \vdots & & \vdots & \vdots \\ a_2^{k-1} & a_3^{k-1} & \cdots & a_k^{k-1} & a_{k+1}^{k-1} \\ a_2^k & a_3^k & \cdots & a_k^k & a_{k+1}^k \end{vmatrix}$$

$$= \prod_{j=2}^{k+1} (a_j - a_1) \prod_{\substack{i=2 \\ j>i}}^{k+1} (a_j - a_i)$$

(from the induction assumption since the last determinant is $k \times k$)

$$= \prod_{\substack{i=1 \\ j>i}}^{k+1} (a_j - a_i)$$

This completes the proof.

39. a. $A^2 = \begin{pmatrix} 0 & 0 \\ 0 & 0 \end{pmatrix}$; $k = 2$

b. $A^3 = \begin{pmatrix} 0 & 0 & 0 \\ 0 & 0 & 0 \\ 0 & 0 & 0 \end{pmatrix}$; $k = 3$

41. $\det A^2 = \det A \det A = \det A$. If $\det A \ne 0$, then $\det A = 1$. The answer is 0 or 1.

Problems 2.3, page 140

1. $a_{1k}A_{1k}$ is the only term in the expansion in the first column of A involving the component a_{1k}. But

$$a_{1k}A_{1k} = a_{1k}(-1)^{1+k}|M_{1k}|$$

If we expand $|M_{1k}|$ about its lth column for $l \ne k$, a term in the expansion takes the form a_{il} (cofactor of a_{il} in M_{1k}). But this is the only occurrence of a_{il} in the expansion of

M_{1k} since the other terms have the form a_{jl} (cofactor of a_{jl} in M_{1k}), which deletes the column corresponding to the lth column of A, and a_{il} is in the lth column. Therefore $a_{1k}A_{1k} = (-1)^{1+k}a_{1k}a_{il} \cdot$ (cofactor of a_{il} in m_{1k}).

3. Expand $|A|$ about its kth column. A term is $a_{ik}A_{ik}$ and this is the only occurrence of a_{ij} in the expansion of $|A|$. Now $A_{ik} = (-1)^{i+k}|M_{ik}|$, and if this is expanded in the lth column (for $l \neq k$), the only term in the expansion containing a_{jl} is $a_{jl} \cdot$ (cofactor of a_{jl} in M_{ik}) for the same reason as in Problem 1. Thus the only occurrence of $a_{ij}a_{jl}$ is $(-1)^{i+k}a_{ik}a_{jl} \cdot$ (cofactor of a_{jl} in M_{ik}).

5. -6

7. EB is the matrix obtained by multiplying the ith row of B by c and adding it to the jth row. By Property 7,

$$\det EB = \det B = $$
$$1 \det B = \det E \det B$$

since, from (16), $\det E = 1$.

Problems 2.4, page 146

1. $\begin{pmatrix} \frac{1}{2} & -\frac{1}{2} \\ -\frac{1}{4} & \frac{3}{4} \end{pmatrix}$ 3. $\begin{pmatrix} 0 & 1 \\ 1 & 0 \end{pmatrix}$

5. $\begin{pmatrix} \frac{1}{3} & -\frac{1}{4} & -\frac{1}{6} \\ 0 & \frac{1}{4} & \frac{1}{2} \\ 0 & \frac{1}{4} & -\frac{1}{2} \end{pmatrix}$

7. $\begin{pmatrix} 0 & 1 & -1 \\ 2 & -2 & -1 \\ -1 & 1 & 1 \end{pmatrix}$

9. not invertible

11. $\begin{pmatrix} \frac{7}{3} & -\frac{1}{3} & -\frac{1}{3} & -\frac{2}{3} \\ \frac{4}{9} & -\frac{1}{9} & -\frac{4}{9} & \frac{1}{9} \\ -\frac{1}{9} & -\frac{2}{9} & \frac{1}{9} & \frac{2}{9} \\ -\frac{5}{3} & \frac{2}{3} & \frac{2}{3} & \frac{1}{3} \end{pmatrix}$

13. Follows from the fact that $\det A^t = \det A$.

15. $A^{-1} = \begin{pmatrix} \frac{1}{14} & \frac{1}{14} & \frac{9}{28} \\ -\frac{5}{7} & \frac{2}{7} & -\frac{3}{14} \\ \frac{1}{14} & \frac{1}{14} & -\frac{5}{28} \end{pmatrix}$,

$\det A = -28, \quad \det A^{-1} = -\frac{1}{28}$

17. no inverse if α is any real number

Problems 2.5, page 150

1. $x_1 = -5, x_2 = 3$

3. $x_1 = 2, x_2 = 5, x_3 = -3$

5. $x_1 = \frac{45}{13}, x_2 = -\frac{11}{13}, x_3 = \frac{23}{13}$

7. $x_1 = \frac{3}{2}, x_2 = \frac{3}{2}, x_3 = \frac{1}{2}$

9. $x_1 = \frac{21}{29}, x_2 = \frac{171}{29}, x_3 = -\frac{284}{29}, x_4 = -\frac{182}{29}$

Chapter 2—Review, page 152

1. -4 3. 24 5. 60 7. 34

9. $\begin{pmatrix} -\frac{1}{11} & \frac{4}{11} \\ \frac{2}{11} & \frac{3}{11} \end{pmatrix}$

11. not invertible

13. $\begin{pmatrix} \frac{1}{11} & \frac{1}{14} & 0 & \frac{3}{11} \\ \frac{9}{11} & -\frac{2}{11} & 0 & -\frac{6}{11} \\ \frac{3}{11} & \frac{3}{11} & 0 & -\frac{2}{11} \\ \frac{1}{22} & \frac{1}{22} & -\frac{1}{2} & \frac{3}{22} \end{pmatrix}$

15. $x_1 = \frac{11}{7}, x_2 = \frac{1}{7}$

17. $x_1 = \frac{1}{4}, x_2 = \frac{5}{4}, x_3 = -\frac{3}{4}$

Chapter 3

Problems 3.1, page 163

1. $|\mathbf{v}| = 4\sqrt{2}, \theta = \pi/4$

3. $|\mathbf{v}| = 4\sqrt{2}, \theta = 7\pi/4$

5. $|\mathbf{v}| = 2, \theta = \pi/6$

7. $|\mathbf{v}| = 2, \theta = 2\pi/3$

9. $|\mathbf{v}| = 2, \theta = 4\pi/3$

11. $|\mathbf{v}| = \sqrt{89}, \theta = \pi + \tan^{-1}\left(-\frac{8}{5}\right) \approx 2.13$ (in the second quadrant)

13. a. $(6, 9)$

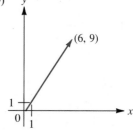

b. $(-3, 7)$

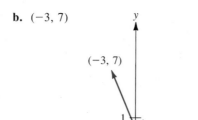

$(-3, 7)$

c. $(-7, 1)$

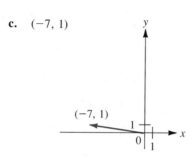

$(-7, 1)$

d. $(39, -22)$

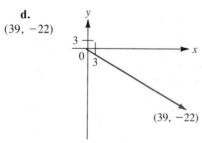

$(39, -22)$

15. $|\mathbf{i}| = |(1, 0)|$
$$= \sqrt{1^2 + 0^2} = \sqrt{1} = 1;$$
$$|\mathbf{j}| = |(0, 1)|$$
$$= \sqrt{0^2 + 1^2} = \sqrt{1} = 1$$

17. $|\mathbf{u}| = \sqrt{\left(\dfrac{a}{\sqrt{a^2 + b^2}}\right)^2 + \left(\dfrac{b}{\sqrt{a^2 + b^2}}\right)^2}$

$$= \sqrt{\dfrac{a^2}{a^2 + b^2} + \dfrac{b^2}{a^2 + b^2}} = 1$$

Direction of $|\mathbf{u}| =$

$$\tan^{-1} \dfrac{\dfrac{b}{\sqrt{a^2 + b^2}}}{\dfrac{a}{\sqrt{a^2 + b^2}}} =$$

$$\tan^{-1}\left(\dfrac{b}{a}\right) = \text{direction of } \mathbf{v}$$

19. $(1/\sqrt{2})\mathbf{i} - (1/\sqrt{2})\mathbf{j}$

21. $(1/\sqrt{2})\mathbf{i} + (1/\sqrt{2})\mathbf{j}$ if $a > 0$;
$-(1/\sqrt{2})\mathbf{i} - (1/\sqrt{2})\mathbf{j}$ if $a < 0$

23. $\sin\theta = -3/\sqrt{13}$,
$\cos\theta = 2/\sqrt{13}$

25. $-(1/\sqrt{2})\mathbf{i} - (1/\sqrt{2})\mathbf{j}$

27. $\frac{3}{5}\mathbf{i} - \frac{4}{5}\mathbf{j}$

29. **a.** $(1/\sqrt{2})\mathbf{i} - (1/\sqrt{2})\mathbf{j}$
   b. $(7/\sqrt{193})\mathbf{i} - (12/\sqrt{193})\mathbf{j}$
   c. $-(2/\sqrt{53})\mathbf{i} + (7/\sqrt{53})\mathbf{j}$

31. $\overrightarrow{PQ}$ is a representation of
$(c + a - c)\mathbf{i} + (d + b - d)\mathbf{j} =$
$a\mathbf{i} + b\mathbf{j}$. Thus $\overrightarrow{PQ}$ and (a, b) are
representations of the same vector.

33. $4\mathbf{i} + 4\sqrt{3}\mathbf{j}$ **35.** $-3\mathbf{i} + 3\sqrt{3}\mathbf{j}$

37. **i.** Suppose $\mathbf{u} = \alpha\mathbf{v}$ where $\alpha > 0$.
   Then $|\mathbf{u} + \mathbf{v}| = |\alpha\mathbf{v} + \mathbf{v}| =$
   $|(\alpha + 1)\mathbf{v}| = |\alpha + 1|\ |\mathbf{v}| =$
   $(\alpha + 1)|\mathbf{v}|$ (since $\alpha + 1 > 0$) $=$
   $\alpha|\mathbf{v}| + |\mathbf{v}| = |\alpha\mathbf{v}| + |\mathbf{v}| =$
   $|\mathbf{u}| + |\mathbf{v}|$.

   ii. Conversely, suppose
   $\mathbf{u} = (a, b)$, $\mathbf{v} = (c, d)$, and
   $|\mathbf{u} + \mathbf{v}| = |\mathbf{u}| + |\mathbf{v}|$. Then
   $\mathbf{u} + \mathbf{v} = (a + c, b + d)$, and
   $|\mathbf{u} + \mathbf{v}|^2 = (|\mathbf{u}| + |\mathbf{v}|)^2 =$
   $|\mathbf{u}|^2 + 2|\mathbf{u}|\ |\mathbf{v}| + |\mathbf{v}|^2$, which
   implies that

   $$(a + c)^2 + (b + d)^2 = a^2 + b^2$$
   $$+ 2\sqrt{(a^2 + b^2)(c^2 + d^2)}$$
   $$+ c^2 + d^2$$

   and, after multiplying through
   and canceling like terms, we
   obtain $ac + bd =$
   $\sqrt{a^2c^2 + a^2d^2 + b^2c^2 + b^2d^2}$.
   Then, squaring both sides and
   again canceling like terms, we
   have $2abcd = a^2d^2 + b^2c^2$, or
   $(ad - bc)^2 = a^2d^2 - 2abcd +$
   $b^2c^2 = 0$ so that $ad = bc$. If
   $d \neq 0$, then $a = \dfrac{b}{d}c$, $b = \dfrac{b}{d}d$,

   and $|\mathbf{u}| = \left|\dfrac{b}{d}\right| |\mathbf{v}|$. Then

$$\left|\frac{b+d}{d}\right||\mathbf{v}| =$$

$$\left|\frac{b+d}{d}(c, d)\right| = \left|\left(\frac{b}{d}c + c,\right.\right.$$

$$\left.\left.\frac{b}{d}d + d\right)\right| = |(a + c, b + d)| =$$

$$|\mathbf{u} + \mathbf{v}| = |\mathbf{u}| + |\mathbf{v}| = \left|\frac{b}{d}\mathbf{v}\right| +$$

$$|\mathbf{v}| = \left(\left|\frac{b}{d}\right| + 1\right)|\mathbf{v}| =$$

$$\frac{|b| + |d|}{|d|}|\mathbf{v}|. \text{ Thus } \left|\frac{b+d}{d}\right| =$$

$$\frac{|b| + |d|}{|d|} \text{ so that } |b + d| = |b| +$$

$|d|$. Since b and d are real numbers, this implies that b and d have the same sign so that $\dfrac{b}{d}$ is positive. Thus if $\alpha =$

$\left|\dfrac{b}{d}\right| = \dfrac{b}{d}$, then $\mathbf{u} = \alpha\mathbf{v}$. If $d =$

0, then $c \neq 0$ and $\mathbf{u} = \dfrac{a}{c}\mathbf{v}$ by

the same reasoning as above

where $\dfrac{a}{c} > 0$. Thus $\mathbf{u}$ is a positive scalar multiple of $\mathbf{v}$.

Problems 3.2, page 170

1. 0; 0 **3.** 0; 0 **5.** 20; $\frac{20}{29}$

7. -22; $-22/5\sqrt{53}$

9. $\mathbf{u} \cdot \mathbf{v} = \alpha\beta - \beta\alpha = 0$

11. parallel

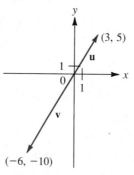

13. neither

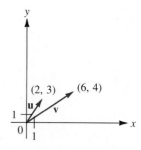

15. orthogonal

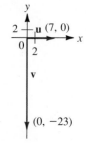

17. a. $-\frac{3}{4}$ **b.** $\frac{4}{3}$ **c.** $\frac{1}{7}$
d. $(-96 + \sqrt{7500})/78 \approx -0.12$

19. If $\mathbf{u}$ and $\mathbf{v}$ have opposite directions, then $\theta_{\mathbf{u}} = \theta_{\mathbf{v}} + \pi$. Thus $\cos\theta_{\mathbf{u}} = \cos(\theta_{\mathbf{v}} + \pi) = -\cos\theta_{\mathbf{v}}$. This implies that $\cos\theta_{\mathbf{u}} = \frac{3}{5} =$

$$-\frac{1}{\sqrt{1 + \alpha^2}} = -\cos\theta_{\mathbf{v}}, \text{ or}$$

$\sqrt{1 + \alpha^2} = -\frac{5}{3} < 0$, which is impossible since $\sqrt{1 + \alpha^2} = |\mathbf{v}| \geq 0$.

21. $\frac{3}{2}\mathbf{i} + \frac{3}{2}\mathbf{j}$ **23.** 0 **25.** $-\frac{2}{13}\mathbf{i} + \frac{3}{13}\mathbf{j}$

27. $[(\alpha + \beta)/2]\mathbf{i} + [(\alpha + \beta)/2]\mathbf{j}$

29. $[(\alpha - \beta)/2]\mathbf{i} + [(\alpha - \beta)/2]\mathbf{j}$

31. $a_1a_2 + b_1b_2 \geq 0$

33. $\text{Proj}_{\overrightarrow{PQ}}\overrightarrow{RS} = \frac{51}{25}\mathbf{i} + \frac{68}{25}\mathbf{j}$;
$\text{Proj}_{\overrightarrow{RS}}\overrightarrow{PQ} = -\frac{17}{26}\mathbf{i} + \frac{85}{26}\mathbf{j}$

35. i. If $\mathbf{u} = \alpha\mathbf{v}$, then $\mathbf{u} \cdot \mathbf{v} = \alpha\mathbf{v} \cdot \mathbf{v} = \alpha|\mathbf{v}|^2$, $|\mathbf{u}| = |\alpha| \, |\mathbf{v}|^2$ so

that $\cos\varphi = \dfrac{\alpha|\mathbf{v}|^2}{|\alpha| \, |\mathbf{v}| \, |\mathbf{v}|} = \dfrac{\alpha}{|\alpha|} =$

± 1.

ii. Suppose that $\mathbf{u}$ and $\mathbf{v}$ are parallel. Then, if $\mathbf{u} = (a, b)$ and $\mathbf{v} = (c, d)$, we have $1 =$

$$\cos^2\varphi = \frac{|\mathbf{u} \cdot \mathbf{v}|^2}{|\mathbf{u}|^2|\mathbf{v}|^2} =$$

$$\frac{(ac + bd)^2}{(a^2 + b^2)(c^2 + d^2)}. \text{ Multiply-}$$

ing through and simplifying, we obtain $0 = a^2d^2 - 2abcd + b^2c^2$. Thus $ad = bc$. If $a \neq 0$,

then $d = \left(\dfrac{c}{a}\right)b$ and $c = \left(\dfrac{c}{a}\right)a$

so that $\mathbf{v} = \left(\dfrac{c}{a}\right)\mathbf{u}$. If $a = 0$,

then $b \neq 0$ and $\mathbf{v} = \dfrac{d}{b}\mathbf{u}$.

37. The line $ax + by + c = 0$ has slope $-\dfrac{a}{b}$. A vector parallel to the line is

$\mathbf{u} = \mathbf{i} - \dfrac{a}{b}\mathbf{j}$, and

$\mathbf{u} \cdot \mathbf{v} = 1 \cdot a - \dfrac{a}{b} \cdot b = 0.$

39. $52/5\sqrt{113} \approx 0.9783$;
$61/\sqrt{34}\sqrt{113} \approx 0.9841$;
$-27/5\sqrt{34} \approx -0.9261$

41. If either $a_1 = a_2 = 0$ or $b_1 = b_2 = 0$, both sides of the inequality are zero. If at least one of a_1 and $a_2 \neq 0$ and at least one of b_1 and $b_2 \neq 0$, let $\mathbf{u} = a_1\mathbf{i} + a_2\mathbf{j}$, $\mathbf{v} = b_1\mathbf{i} + b_2\mathbf{j}$. Then $\mathbf{u} \neq 0$, $\mathbf{v} \neq 0$, $|\mathbf{u}|\,|\mathbf{v}| \neq 0$,

and $\left|\dfrac{\mathbf{u} \cdot \mathbf{v}}{|\mathbf{u}|\,|\mathbf{v}|}\right| = |\cos \varphi| \leq 1.$ Thus

$|a_1b_1 + a_2b_2| = |\mathbf{u} \cdot \mathbf{v}| \leq |\mathbf{u}|\,|\mathbf{v}| = \sqrt{a_1^2 + a_2^2}\sqrt{b_1^2 + b_2^2}$. Equality holds when $|\cos \varphi| = 1$, which is true if and only if $\mathbf{u}$ and $\mathbf{v}$ are parallel.

43. $\sqrt{5}$

45. Let $A = \begin{pmatrix} a_1 & b_1 \\ a_2 & b_2 \end{pmatrix}$. Then $A^t =$

$\begin{pmatrix} a_1 & a_2 \\ b_1 & b_2 \end{pmatrix}$. Let $\mathbf{u} = (a_1, a_2)$ and

$\mathbf{v} = (b_1, b_2)$. Then $\mathbf{u} \cdot \mathbf{v} = 0$, $|\mathbf{u}| = 1$ and $|\mathbf{v}| = 1$. $A^tA =$

$\begin{pmatrix} a_1^2 + b_1^2 & a_1a_2 + b_1b_2 \\ a_2a_1 + b_2b_1 & a_2^2 + b_2^2 \end{pmatrix}$

$= \begin{pmatrix} |\mathbf{u}|^2 & \mathbf{u} \cdot \mathbf{v} \\ \mathbf{u} \cdot \mathbf{v} & |\mathbf{v}|^2 \end{pmatrix} = \begin{pmatrix} 1 & 0 \\ 0 & 1 \end{pmatrix}.$

Similarly, $AA^t = I$. Hence A is invertible and $A^{-1} = A^t$.

Problems 3.3, page 182

1. $\sqrt{40}$ **3.** 6

5. 3; $-1, 0, 0$

7. $\sqrt{5}$; $1/\sqrt{5}, 0, 2/\sqrt{5}$

9. $\sqrt{3}$; $1/\sqrt{3}, 1/\sqrt{3}, -1/\sqrt{3}$

11. $\sqrt{3}$; $1/\sqrt{3}, -1/\sqrt{3}, -1/\sqrt{3}$

13. $\sqrt{3}$; $-1/\sqrt{3}, -1/\sqrt{3}, 1/\sqrt{3}$

15. $\sqrt{78}$; $2/\sqrt{78}, 5/\sqrt{78}, -7/\sqrt{78}$

17. $\sqrt{29}$; $-2/\sqrt{29}, -3/\sqrt{29}, -4/\sqrt{29}$

19. $4\sqrt{3}\mathbf{i} + 4\sqrt{3}\mathbf{j} + 4\sqrt{3}\mathbf{k}$

21. $(1/\sqrt{26})\mathbf{i} - (3/\sqrt{26})\mathbf{j} + (4/\sqrt{26})\mathbf{k}$

23. $R = (-3, y, z)$, y, z arbitrary; this set of points constitutes a plane parallel to the yz-plane.

25. $\left|\dfrac{\mathbf{u} \cdot \mathbf{v}}{|\mathbf{u}|\,|\mathbf{v}|}\right| = |\cos \varphi| \leq 1.$ Thus

$|\mathbf{u} \cdot \mathbf{v}| \leq |\mathbf{u}|\,|\mathbf{v}|.$ Then

$$|\mathbf{u} + \mathbf{v}|^2 = (\mathbf{u} + \mathbf{v}) \cdot (\mathbf{u} + \mathbf{v})$$
$$= |\mathbf{u}|^2 + 2\mathbf{u} \cdot \mathbf{v} + |\mathbf{v}|^2$$
$$\leq |\mathbf{u}|^2 + 2|\mathbf{u}|\,|\mathbf{v}| + |\mathbf{v}|^2$$
$$= (|\mathbf{u}| + |\mathbf{v}|)^2$$

27. $-6\mathbf{j} + 9\mathbf{k}$ **29.** $8\mathbf{i} - 14\mathbf{j} + 9\mathbf{k}$

31. $16\mathbf{i} + 29\mathbf{j} + 42\mathbf{k}$ **33.** $\sqrt{59}$

35. $\cos^{-1}(35/\sqrt{29}\sqrt{59})$
$\approx \cos^{-1}(0.8461)$
≈ 0.5621
$\approx 32.21°$

37. $\frac{25}{29}\mathbf{u} = \frac{50}{29}\mathbf{i} - \frac{75}{29}\mathbf{j} + \frac{100}{29}\mathbf{k}$

39. Since the line segments PS and SR are perpendicular (in Figure 3.28), the triangle PSR is a right triangle and

$$\overline{PR}^2 = \overline{PS}^2 + \overline{SR}^2 \qquad \text{(i)}$$

But triangle PRQ is also a right triangle so that

$$\overline{PQ}^2 = \overline{PR}^2 + \overline{RQ}^2 \qquad \text{(ii)}$$

So, combining (i) and (ii), we get

$$\overline{PQ}^2 = \overline{PS}^2 + \overline{SR}^2 + \overline{RQ}^2 \quad \text{(iii)}$$

Since the x and z coordinates of P and S are equal,

$$\overline{PS}^2 = (y_2 - y_1)^2 \qquad \text{(iv)}$$

Similarly,

$$\overline{RS}^2 = (x_2 - x_1)^2 \qquad \text{(v)}$$

and $\qquad \overline{RQ}^2 = (z_2 - z_1)^2 \qquad \text{(vi)}$

Thus, using (iv), (v), and (vi) in (iii) yields

$$\overline{PQ}^2 = (x_2 - x_1)^2$$
$$+ (y_2 - y_1)^2$$
$$+ (z_2 - z_1)^2$$

41. i. If $\mathbf{v} = \alpha\mathbf{u}$, then $\cos \varphi =$
$$\frac{\mathbf{u} \cdot \mathbf{v}}{|\mathbf{u}|\,|\mathbf{v}|} = \frac{\alpha|\mathbf{u}|^2}{|\alpha|\,|\mathbf{u}|^2} = \pm 1. \text{ If } \mathbf{u}$$

and $\mathbf{v}$ are parallel, then $\dfrac{\mathbf{u}}{|\mathbf{u}|} =$

$\pm \dfrac{\mathbf{v}}{|\mathbf{v}|}$ so that

$$\mathbf{v} = \pm \frac{|\mathbf{v}|}{|\mathbf{u}|}\mathbf{u} = \alpha\mathbf{u}$$

ii. If $\mathbf{u} \cdot \mathbf{v} = 0$, then $\cos \varphi = 0$
and $\varphi = \dfrac{\pi}{2}$. If $\varphi = \dfrac{\pi}{2}$, then

$$\mathbf{u} \cdot \mathbf{v} = |\mathbf{u}|\,|\mathbf{v}| \cos \varphi = 0$$

Problems 3.4, page 192

1. $-6\mathbf{i} - 3\mathbf{j}$ **3.** $-\mathbf{i} - \mathbf{j} + \mathbf{k}$

5. $12\mathbf{i} + 8\mathbf{j} - 21\mathbf{k}$

7. $(bc - ad)\mathbf{j}$

9. $-5\mathbf{i} - \mathbf{j} + 7\mathbf{k}$ **11.** $\mathbf{0}$

13. $42\mathbf{i} + 6\mathbf{j}$

15. $-9\mathbf{i} + 39\mathbf{j} + 61\mathbf{k}$

17. $-4\mathbf{i} + 8\mathbf{k}$ **19.** $\mathbf{0}$

21. $\pm[-(9/\sqrt{181})\mathbf{i} - (6/\sqrt{181})\mathbf{j} + (8/\sqrt{181})\mathbf{k}]$

23. $\sqrt{30}/\sqrt{6}\sqrt{29} \approx 0.415$

25. $5\sqrt{5}$ **27.** $\sqrt{523}$

29. $\sqrt{a^2b^2 + a^2c^2 + b^2c^2}$

31. Let $\mathbf{u} = a_1\mathbf{i} + b_1\mathbf{j} + c_1\mathbf{k}$ and $\mathbf{v} =$

$a_2\mathbf{i} + b_2\mathbf{j} + c_2\mathbf{k}$. Then $\mathbf{u} \times \mathbf{v} =$
$(b_1c_2 - c_1b_2)\mathbf{i} + (c_1a_2 - a_1c_2)\mathbf{j} +$
$(a_1b_2 - b_1a_2)\mathbf{k}$ so that $|\mathbf{u} \times \mathbf{v}|^2 =$
$(b_1c_2 - c_1b_2)^2 + (c_1a_2 - a_1c_2)^2 +$
$(a_1b_2 - b_1a_2)^2 = b_1^2c_2^2 -$
$2b_1c_2c_1b_2 + c_1^2b_2^2 + c_1^2a_2^2 -$
$2c_1a_2a_1c_2 + a_1^2c_2^2 + a_1^2b_2^2 -$
$2a_1b_2b_1a_2 + b_1^2a_2^2$. This is equal to
$|\mathbf{u}|^2|\mathbf{v}|^2 - (\mathbf{u} \cdot \mathbf{v})^2 = (a_1^2 + b_1^2 + c_1^2)$
$(a_2^2 + b_2^2 + c_2^2) - (a_1a_2 + b_1b_2 + c_1c_2)^2$.

33. Let $\mathbf{u} = a_1\mathbf{i} + b_1\mathbf{j} + c_1\mathbf{k}$, $\mathbf{v} = a_2\mathbf{i} + b_2\mathbf{j} + c_2\mathbf{k}$, and $\mathbf{w} = a_3\mathbf{i} + b_3\mathbf{j} + c_3\mathbf{k}$. Then

$(\mathbf{u} \times \mathbf{v}) \cdot \mathbf{w}$
$$= [(b_1c_2 - c_1b_2)\mathbf{i}$$
$$+ (c_1a_2 - a_1c_2)\mathbf{j}$$
$$+ (a_1b_2 - b_1a_2)\mathbf{k}]$$
$$\cdot [a_3\mathbf{i} + b_3\mathbf{j} + c_3\mathbf{k}]$$
$$= b_1c_2a_3 - c_1b_2a_3$$
$$+ c_1a_2b_3 - a_1c_2b_3$$
$$+ a_1b_2c_3 - b_1a_2c_3$$

and

$\mathbf{u} \cdot (\mathbf{v} \times \mathbf{w})$
$$= [a_1\mathbf{i} + b_1\mathbf{j} + c_1\mathbf{k}]$$
$$\times [(b_2c_3 - c_2b_3)\mathbf{i}$$
$$+ (c_2a_3 - a_2c_3)\mathbf{j}$$
$$+ (a_2b_3 - b_2a_3)\mathbf{k}]$$
$$= a_1b_2c_3 - a_1c_2b_3$$
$$+ b_1c_2a_3 - b_1a_2c_3$$
$$+ c_1a_2b_3 - c_1b_2a_3$$

35. If $\mathbf{u}$ and $\mathbf{v}$ are parallel and neither is $\mathbf{0}$, then $\mathbf{v} = t\mathbf{u}$ for some constant t. Then if $\mathbf{u} = a\mathbf{i} + b\mathbf{j} + c\mathbf{k}$,

$$\mathbf{u} \times \mathbf{v} = \begin{vmatrix} \mathbf{i} & \mathbf{j} & \mathbf{k} \\ a & b & c \\ ta & tb & tc \end{vmatrix}$$

$= 0$ by Property 6 on page 124

37. 23

39. This problem will rely heavily on Property 3 of the determinants, which states that if the ith column (or row) of a determinant consists of a pair of elements, the determinant can be rewritten as a sum of two determinants whose columns (or rows) are identical, except for the ith column (or row). The first determinant contains one of the elements of the pair, while the other member of each pair of elements of the ith column (or row) appears in the second determinant. Also note that the volume generated by $\mathbf{u}$, $\mathbf{v}$, $\mathbf{w}$ is given by

$$\text{Volume} = \begin{vmatrix} u_1 & u_2 & u_3 \\ v_1 & v_2 & v_3 \\ w_1 & w_2 & w_3 \end{vmatrix}$$

Let

$$A = \begin{pmatrix} a_{11} & a_{12} & a_{13} \\ a_{21} & a_{22} & a_{23} \\ a_{31} & a_{32} & a_{33} \end{pmatrix}$$

$$\mathbf{u}_1 = A\mathbf{u}, \quad \mathbf{v}_1 = A\mathbf{v}, \quad \mathbf{w}_1 = A\mathbf{w}$$

Thus

$$\mathbf{u}_1 = \begin{pmatrix} a_{11} & a_{12} & a_{13} \\ a_{21} & a_{22} & a_{23} \\ a_{31} & a_{32} & a_{33} \end{pmatrix} \begin{pmatrix} u_1 \\ u_2 \\ u_3 \end{pmatrix}$$

$$= \begin{pmatrix} a_{11}u_1 + a_{12}u_2 + a_{13}u_3 \\ a_{21}u_1 + a_{22}u_2 + a_{23}u_3 \\ a_{31}u_1 + a_{32}u_2 + a_{33}u_3 \end{pmatrix}$$

Similarly,

$$\mathbf{v}_1 = \begin{pmatrix} a_{11}v_1 + a_{12}v_2 + a_{13}v_3 \\ a_{21}v_1 + a_{22}v_2 + a_{23}v_3 \\ a_{31}v_1 + a_{32}v_2 + a_{33}v_3 \end{pmatrix}$$

and

$$\mathbf{w}_1 = \begin{pmatrix} a_{11}w_1 + a_{12}w_2 + a_{13}w_3 \\ a_{21}w_1 + a_{22}w_2 + a_{23}w_3 \\ a_{31}w_1 + a_{32}w_2 + a_{33}w_3 \end{pmatrix}$$

By Problem 36, the volume generated by $\mathbf{u}_1$, $\mathbf{v}_1$, $\mathbf{w}_1$ is

$V =$

$$\begin{vmatrix} a_{11}u_1 + a_{12}u_2 + a_{13}u_3 \\ a_{11}v_1 + a_{12}v_2 + a_{13}v_3 \\ a_{11}w_1 + a_{12}w_2 + a_{13}w_3 \end{vmatrix}$$

$$\begin{vmatrix} a_{21}u_1 + a_{22}u_2 + a_{23}u_3 \\ a_{21}v_1 + a_{22}v_2 + a_{23}v_3 \\ a_{21}w_1 + a_{22}w_2 + a_{23}w_3 \\ a_{31}u_1 + a_{32}u_2 + a_{33}u_3 \\ a_{31}v_1 + a_{32}v_2 + a_{33}v_3 \\ a_{31}w_1 + a_{32}w_2 + a_{33}w_3 \end{vmatrix}$$

By expansion, it can be verified that

$$V = \begin{vmatrix} a_{11} & a_{12} & a_{13} \\ a_{21} & a_{22} & a_{23} \\ a_{31} & a_{32} & a_{33} \end{vmatrix} \begin{vmatrix} u_1 & u_2 & u_3 \\ v_1 & v_2 & v_3 \\ w_1 & w_2 & w_3 \end{vmatrix}$$

$= (\det A)(\text{volume generated by } \mathbf{u},$ $\mathbf{v}, \mathbf{w})$, if $\det A \geq 0$, otherwise use $-\det A$.

41. Let $\mathbf{u} = (u_1, u_2, u_3)$, $\mathbf{v} = (v_1, v_2, v_3)$, and $\mathbf{w} = (w_1, w_2, w_3)$. Then

$$\mathbf{v} \times \mathbf{w} = \begin{vmatrix} \mathbf{i} & \mathbf{j} & \mathbf{k} \\ v_1 & v_2 & v_3 \\ w_1 & w_2 & w_3 \end{vmatrix}$$

$= (v_2w_3 - v_3w_2, \ v_3w_1 - v_1w_3,$
    $v_1w_2 - v_2w_1);$

$$\mathbf{u} \times (\mathbf{v} \times \mathbf{w}) = \begin{vmatrix} \mathbf{i} \\ u_1 \\ v_2w_3 - v_3w_2 \end{vmatrix}$$

$$\begin{vmatrix} \mathbf{j} & \mathbf{k} \\ u_2 & u_3 \\ v_3w_1 - v_1w_3 & v_1w_2 - v_2w_1 \end{vmatrix}$$

$= (-u_3v_3w_1 + u_3v_1w_3 + u_2v_1w_2 -$
$u_2v_2w_1, u_3v_2w_3 - u_3v_3w_2 -$
$u_1v_1w_2 + u_1v_2w_1, u_1v_3w_1 -$
$u_1v_1w_3 - u_2v_2w_3 + u_2v_3w_2)(*)$
$(\mathbf{u} \cdot \mathbf{w})\mathbf{v} = (u_1w_1 + u_2w_2 + u_3w_3)(v_1,$
$v_2, v_3) = (u_1v_1w_1 + u_2v_1w_2 +$
$u_3v_1w_3, u_1v_2w_1 + u_2v_2w_2 +$
$u_3v_2w_3, u_1v_3w_1 + u_2v_3w_2 + u_3v_3w_3)$
$-(\mathbf{u} \cdot \mathbf{v})\mathbf{w} = -(u_1v_1 + u_2v_2 +$
$u_3v_3)(w_1, w_2, w_3) = (-u_1v_1w_1 -$
$u_2v_2w_1 - u_3v_3w_1, -u_1v_1w_2 -$
$u_2v_2w_2 - u_3v_3w_2, -u_1v_1w_3 -$
$u_2v_2w_3 - u_3v_3w_3).$

If we add the last two vectors, we obtain the vector (*).

Problems 3.5, page 202

In the answers to Problems 1–5 we assume that the first point is P and the second point is Q. The vector equations are of the form $\overrightarrow{QR} = \overrightarrow{QP} + t\mathbf{v}$. Only $\mathbf{v}$ is given in the answers.

1. $\mathbf{v} = -\mathbf{i} + \mathbf{j} - 4\mathbf{k}$; $x = 2 - t$, $y = 1 + t$, $z = 3 - 4t$;
$(x - 2)/(-1) = y - 1 = (z - 3)/(-4)$

3. $\mathbf{v} = -\mathbf{j} - 2\mathbf{k}$; $x = -4$, $y = 1 - t$, $z = 3 - 2t$;
$x = -4$ and $z = 1 + 2y$

5. $\mathbf{v} = 2\mathbf{i} - 2\mathbf{k}$; $x = 1 + 2t$, $y = 2$, $z = 3 - 2t$;
$y = 2$ and $x = 4 - z$

In Problems 7–11 $\mathbf{v}$ is already given.

7. $x = 2 + 2t$, $y = 2 - t$, $z = 1 - t$;
$(x - 2)/2 = (y - 2)/(-1) = (z - 1)/(-1)$

9. $x = -1$, $y = -2 - 3t$, $z = 5 + 7t$; $x = -1$ and $7y + 3z = 1$

11. $x = a + dt$, $y = b + et$, $z = c$;
$(x - a)/d = (y - b)/e$ and $z = c$

13. $\mathbf{v} = 3\mathbf{i} + 6\mathbf{j} + 2\mathbf{k}$; $x = 4 + 3t$, $y = 1 + 6t$, $z = -6 + 2t$;
$(x - 4)/3 = (y - 1)/6 = (z + 6)/2$

15. The vector $\mathbf{v}_1 = a_1\mathbf{i} + b_1\mathbf{j} + c_1\mathbf{k}$ is parallel to L_1, while the vector $\mathbf{v}_2 = a_2\mathbf{i} + b_2\mathbf{j} + c_2\mathbf{k}$ is parallel to L_2. Thus $L_1 \perp L_2$ if $\mathbf{v}_1 \perp \mathbf{v}_2$ or $\mathbf{v}_1 \cdot \mathbf{v}_2 = 0$. But $\mathbf{v}_1 \cdot \mathbf{v}_2 = a_1a_2 + b_1b_2 + c_1c_2$.

17. $3\mathbf{i} + 6\mathbf{j} + 9\mathbf{k} = 3(\mathbf{i} + 2\mathbf{j} + 3\mathbf{k})$, so the direction vectors of the lines are parallel. Note that they are not coincident since, for example, the point $(1, -3, -3)$ is on L_1 but not on L_2.

19. If they had a point in common, we would have
$$2 - t = 1 + s$$
$$1 + t = -2s$$
$$-2t = 3 + 2s$$

The unique solution of the first two of these equations is $s = -2$, $t = 3$; but this pair does not satisfy the third equation.

21. **a.** $(\sqrt{186}/3)(t = \frac{1}{3})$
b. $\sqrt{1518/11} = \sqrt{138/11}$, $(t = -\frac{4}{11})$
c. $\sqrt{30}/2$ $(t = -\frac{3}{2})$

23. $(x + 4)/26 = (y - 7)/1 = (z - 3)/37$

25. $(x - 4)/(-4) = (y - 6)/16 = z/24$

27. 3   29. $y = 0$ (xz-plane)

31. $x + y = 3$ 33. $y + z = 5$

35. $-3x - 4y + z = 45$

37. $2x - 7y - 8z = -20$

39. $-12x - 21y + 22z = 63$

41. $2x + y = 7$ 43. coincident

45. none of these

47. $(x, y, z) = (-1, -3, 0) + t(1, 2, 1)$

49. $(x, y, z) = (-11/4, 3/2, 0) + t(9, 16, 2)$

51. $13/\sqrt{69}$ 53. $19/\sqrt{35}$

55. $\cos^{-1}(9/\sqrt{3}\sqrt{29})$
$= \cos^{-1}(0.9649)$
≈ 0.2657
$\approx 15.23°$

57. $\cos^{-1}(20/\sqrt{294}\sqrt{6})$
$= \cos^{-1}(\frac{20}{42})$
≈ 1.074
$\approx 61.56°$

59. $\mathbf{n} = \mathbf{v} \times \mathbf{w}$ is orthogonal to $\mathbf{v}$ and $\mathbf{w}$. If $\mathbf{u} \cdot (\mathbf{v} \times \mathbf{w}) = 0$, then $\mathbf{u} \perp \mathbf{n}$, which means that $\mathbf{u}$ lies in the plane determined by $\mathbf{v}$ and $\mathbf{w}$.

61. coplanar; $29x - y + 11z = 0$

63. not coplanar; $\mathbf{u} \cdot (\mathbf{v} \times \mathbf{w}) = -9$.

Chapter 3—Review, page 208

1. $|\mathbf{v}| = 3\sqrt{2}$, $\theta = \pi/4$

3. $|\mathbf{v}| = 4$, $\theta = 5\pi/3$

5. $|\mathbf{v}| = 12\sqrt{2}$, $\theta = 5\pi/4$

7. $2\mathbf{i} + 2\mathbf{j}$

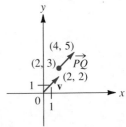

9. $4\mathbf{i} + 2\mathbf{j}$

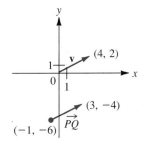

11. a. $(10, 5)$, **b.** $(5, -3)$,
c. $(-31, 12)$

13. $(1/\sqrt{2})\mathbf{i} + (1/\sqrt{2})\mathbf{j}$

15. $(2/\sqrt{29})\mathbf{i} + (5/\sqrt{29})\mathbf{j}$ **17.** $\frac{3}{5}\mathbf{i} + \frac{4}{5}\mathbf{j}$

19. $(1/\sqrt{2})\mathbf{i} - (1/\sqrt{2})\mathbf{j}$ if $a > 0$ and
$-(1/\sqrt{2})\mathbf{i} + (1/\sqrt{2})\mathbf{j}$ if $a < 0$

21. $-(5/\sqrt{29})\mathbf{i} - (2/\sqrt{29})\mathbf{j}$

23. $-(10/\sqrt{149})\mathbf{i} + (7/\sqrt{149})\mathbf{j}$

25. $\mathbf{j}$ **27.** $-\frac{7}{2}\sqrt{3}\mathbf{i} + \frac{7}{2}\mathbf{j}$ **29.** $0; 0$

31. $-14, -14/\sqrt{5}\sqrt{41}$

33. neither

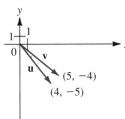

35. parallel

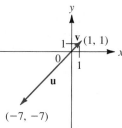

37. parallel

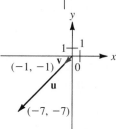

39. $7\mathbf{i} + 7\mathbf{j}$

41. $\frac{15}{13}\mathbf{i} + \frac{10}{13}\mathbf{j}$ **43.** $-\frac{3}{2}\mathbf{i} - \frac{7}{2}\mathbf{j}$

45. $\text{Proj}_{\overrightarrow{RS}}\overrightarrow{PQ} = -\frac{99}{25}\mathbf{i} + \frac{132}{25}\mathbf{j};$
$\text{Proj}_{\overrightarrow{PQ}}\overrightarrow{RS} = -\frac{33}{82}\mathbf{i} - \frac{297}{82}\mathbf{j}$

47. $\sqrt{216}$

49. $\sqrt{130}; 0, 3/\sqrt{130}, 11/\sqrt{130}$

51. $\sqrt{53}; -4/\sqrt{53}, 1/\sqrt{53}, 6/\sqrt{53}$

53. $(2/\sqrt{6})\mathbf{i} - (1/\sqrt{6})\mathbf{j} + (1/\sqrt{6})\mathbf{k}$

55. $\mathbf{i} - 14\mathbf{j} + 20\mathbf{k}$

57. $\frac{26}{21}\mathbf{i} - \frac{52}{21}\mathbf{j} + \frac{13}{21}\mathbf{k}$ **59.** 22

61. $\cos^{-1}(-9/\sqrt{798}) \approx 1.895 \approx 108.6°$

63. $-7\mathbf{i} - 7\mathbf{k}$

65. $-26\mathbf{i} - 8\mathbf{j} + 7\mathbf{k}$

67. $\sqrt{2065}$

69. $\overrightarrow{OR} = (-4\mathbf{i} + \mathbf{j}) + t(7\mathbf{i} - \mathbf{j} + 7\mathbf{k});$
$x = -4 + 7t, y = 1 - t, z = 7t;$
$$(x + 4)/7 = (y - 1)/(-1) = z/7$$

71. $\overrightarrow{OR} = \mathbf{i} - 2\mathbf{j} - 3\mathbf{k} + t(5\mathbf{i} - 3\mathbf{j} + 2\mathbf{k});$
$x = 1 + 5t, y = -2 - 3t, z = -3 + 2t;$
$$(x - 1)/5 = (y + 2)/(-3)$$
$$= (z + 3)/2$$

73. $\sqrt{165}/3$ **75.** $x + z = -1$

77. $2x - 3y + 5z = 19$

79. $x = \frac{1}{2} - \frac{9}{2}t, y = \frac{7}{2} - \frac{11}{2}t, z = t$

81. $x = \frac{4}{3} - \frac{5}{2}t, y = -4 - \frac{7}{2}t, z = t$

83. $\cos^{-1}|-1/\sqrt{207}|$
$= \cos^{-1} 1/\sqrt{207}$
$\approx 1.501 \approx 86.01°$

Chapter 4

Problems 4.2, page 217

1. yes

3. no; (iv); also (vi) does not hold if
$\alpha < 0$

5. yes **7.** yes

9. no (i), (iii), (iv), (vi) do not hold

11. yes **13.** yes

15. no; (i), (iii), (iv), (vi) do not hold

17. yes **19.** yes

21. Suppose that $\mathbf{0}$ and $\mathbf{0}'$ are additive identities. Then, by the definition of additive identity, $\mathbf{0} = \mathbf{0} + \mathbf{0}'$ and $\mathbf{0}' = \mathbf{0}' + \mathbf{0} = \mathbf{0} + \mathbf{0}'$. Thus $\mathbf{0} = \mathbf{0}'$.

23. For $\mathbf{x}$, $\mathbf{y}$ in V define $\mathbf{z}$ by $\mathbf{z} = -\mathbf{x} + \mathbf{y}$. $\mathbf{z}$ exists since every $\mathbf{x}$ has an additive inverse $-\mathbf{x}$ and V is closed under addition. Then $\mathbf{x} + \mathbf{z} = \mathbf{x} + (-\mathbf{x} + \mathbf{y}) = (\mathbf{x} - \mathbf{x}) + \mathbf{y} = \mathbf{0} + \mathbf{y} = \mathbf{y}$. Assume there exist $\mathbf{z}$ and $\mathbf{z}'$ such that $\mathbf{x} + \mathbf{z} = \mathbf{y}$ and $\mathbf{x} + \mathbf{z}' = \mathbf{y}$. Then $\mathbf{z} = -\mathbf{x} + \mathbf{y} = \mathbf{z}'$. So $\mathbf{z}$ is unique.

25. Let y_1 and y_2 be solutions to the equation. Then

$$y_1'' + a(x)y_1' + b(x)y_1(x) = 0$$

and

$$y_2'' + a(x)y_2' + b(x)y_2(x) = 0$$

Then

$$(y_1 + y_2)'' + a(x)(y_1 + y_2)' + b(x)(y_1 + y_2)$$
$$= [y_1'' + a(x)y_1' + b(x)y_1]$$
$$+ [y_2'' + a(x)y_2' + b(x)y_2]$$
$$= 0 + 0 = 0$$

so $y_1 + y_2$ is a solution. Similarly, $(\alpha y_1)'' + a(x)(\alpha y_1') + b(x)(\alpha y) = \alpha[y_1'' + a(x)y_1' + b(x)y_1] = \alpha \cdot 0 = 0$, so αy_1 is also a solution. Thus the closure rules hold. Since $-y_1 = (-1)y$ is also a solution, we have the additive inverse. The other axioms follow easily.

Problems 4.3, page 222

1. no; because $\alpha(x, y) \notin H$ if $\alpha < 0$

3. yes **5.** yes **7.** yes

9. yes **11.** yes **13.** yes

15. no; the zero polynomial $\notin H$

17. no; the function $f(x) \equiv 0 \notin V$

19. yes

21. a. If $A_1, A_2 \in H_1$, then

$(A_1 + A_2)_{11} = (A_1)_{11} + (A_2)_{11} = 0 + 0 = 0$, and $(\alpha A_1)_{11} = \alpha(A_1)_{11} = \alpha 0 = 0$ so that H_1 is a subspace. If $A_1, A_2 \in H_2$, then

$$A_1 = \begin{pmatrix} -b_1 & a_1 \\ a_1 & b_1 \end{pmatrix}$$

$$A_2 = \begin{pmatrix} -b_2 & a_2 \\ a_2 & b_2 \end{pmatrix}$$

$A_1 + A_2$
$$= \begin{pmatrix} -(b_1 + b_2) & (a_1 + a_2) \\ (a_1 + a_2) & (b_1 + b_2) \end{pmatrix}$$
$$= \begin{pmatrix} -c & d \\ d & c \end{pmatrix} \in H_2 \text{ Also,}$$

$$\alpha A_1 = \begin{pmatrix} -\alpha b_1 & \alpha a_1 \\ \alpha a_1 & \alpha b_1 \end{pmatrix}$$
$$= \begin{pmatrix} -c & d \\ d & c \end{pmatrix} \in H_2$$

and so H_2 is also a subspace.

b. $H = H_1 \cap H_2 = \left\{ A \in M_{22} \colon A = \begin{pmatrix} 0 & a \\ a & 0 \end{pmatrix} \text{ for some scalar } a \right\}$. If $A_1, A_2 \in H$, then $A_1 = \begin{pmatrix} 0 & a_1 \\ a_1 & 0 \end{pmatrix}$, $A_2 = \begin{pmatrix} 0 & a_2 \\ a_2 & 0 \end{pmatrix}$, $A_1 + A_2 = \begin{pmatrix} 0 & a_1 + a_2 \\ a_1 + a_2 & 0 \end{pmatrix} = \begin{pmatrix} 0 & b \\ b & 0 \end{pmatrix} \in H$ and $\alpha A_1 = \begin{pmatrix} 0 & \alpha a_1 \\ \alpha a_1 & 0 \end{pmatrix} = \begin{pmatrix} 0 & c \\ c & 0 \end{pmatrix} \in H.$

23. If $\mathbf{x}_1$, $\mathbf{x}_2 \in H$, then $A(\mathbf{x}_1 + \mathbf{x}_2) = A\mathbf{x}_1 + A\mathbf{x}_2 = 0 + 0 = 0$, so $\mathbf{x}_1 + \mathbf{x}_2 \in H$. Also, $A(\alpha \mathbf{x}_1) = \alpha A\mathbf{x}_1 = \alpha 0 = 0$ so $\alpha \mathbf{x}_1 \in H$ and H is a subspace.

25. Let $\mathbf{u} = (x_1, y_1, z_1, w_1)$ and $\mathbf{v} = (x_2, y_2, z_2, w_2) \in H$. Then $\mathbf{u} + \mathbf{v} = (x_1 + x_2, y_1 + y_2, z_1 + z_2, w_1 + w_2)$ and $a(x_1 + x_2) + b(y_1 + y_2) +$

$c(z_1 + z_2) + d(w_1 + w_2) = (ax_1 + by_1 + cz_1 + dw_1) + (ax_2 + by_2 + cz_2 + dw_2) = 0 + 0 = 0$, so $\mathbf{u} + \mathbf{v} \in H$. Similarly, $\alpha\mathbf{u} = (\alpha x_1, \alpha y_1, \alpha z_1, \alpha w_1)$ and $a(\alpha x_1) + b(\alpha y_1) + c(\alpha z_1) + d(\alpha w_1) = \alpha(ax_1 + by_1 + cz_1 + dw_1) = \alpha 0 = 0$, so $\alpha\mathbf{u} \in H$. Thus H is a subspace.

27. Let $\mathbf{x}, \mathbf{y} \in H$. Then $\mathbf{x} = \mathbf{u}_1 + \mathbf{v}_1$ and $\mathbf{y} = \mathbf{u}_2 + \mathbf{v}_2$, where $\mathbf{u}_1, \mathbf{u}_2 \in H_1$ and $\mathbf{v}_1, \mathbf{v}_2 \in H_2$. Then $\mathbf{x} + \mathbf{y} = (\mathbf{u}_1 + \mathbf{v}_1) + (\mathbf{u}_2 + \mathbf{v}_2) = (\mathbf{u}_1 + \mathbf{u}_2) + (\mathbf{v}_1 + \mathbf{v}_2)$. Since H_1 and H_2 are subspaces, $\mathbf{u}_1 + \mathbf{u}_2 \in H_1$ and $\mathbf{v}_1 + \mathbf{v}_2 \in H_2$, so $\mathbf{x} + \mathbf{y} \in H$. Similarly, $\alpha\mathbf{x} = \alpha(\mathbf{u}_1 + \mathbf{v}_1) = \alpha\mathbf{u}_1 + \alpha\mathbf{v}_1$. But $\alpha\mathbf{u}_1 \in H_1$ and $\alpha\mathbf{v}_1 \in H_2$, so $\alpha\mathbf{x} \in H$ and H is a subspace.

29. Let $\mathbf{v}_1 = \begin{pmatrix} x_1 \\ y_1 \end{pmatrix}$ and $\mathbf{v}_2 = \begin{pmatrix} x_2 \\ y_2 \end{pmatrix}$. $\mathbf{v}_1$ is not a multiple of $\mathbf{v}_2$ since the vectors are not collinear. Let $A = \begin{pmatrix} x_1 & x_2 \\ y_1 & y_2 \end{pmatrix}$. Then $\det A = x_1 y_2 - x_2 y_1$. If $\det A = 0$, then $x_1 y_2 = x_2 y_1$ or $x_1/x_2 = y_1/y_2$ (if $x_2 = 0$ or $y_2 = 0$, a similar conclusion can be drawn). Let $c = x_1/x_2 = y_1/y_2$. Then $x_1 = cx_2$ and $y_1 = cy_2$, so $\mathbf{v}_1 = c\mathbf{v}_2$ contradicting what was stated above. Thus $\det A \neq 0$. Let $\mathbf{v} = \begin{pmatrix} x \\ y \end{pmatrix}$ be any other vector in $\mathbb{R}^2$. We wish to find scalars a and b such that $\mathbf{v} = a\mathbf{v}_1 + b\mathbf{v}_2$, or

$$\begin{pmatrix} x \\ y \end{pmatrix} = a\begin{pmatrix} x_1 \\ y_1 \end{pmatrix} + b\begin{pmatrix} x_2 \\ y_2 \end{pmatrix}$$
$$= \begin{pmatrix} ax_1 + bx_2 \\ ay_1 + by_2 \end{pmatrix}$$

or

$$\begin{pmatrix} x_1 & x_2 \\ y_1 & y_2 \end{pmatrix}\begin{pmatrix} a \\ b \end{pmatrix} = \begin{pmatrix} x \\ y \end{pmatrix}$$

or

$$A\begin{pmatrix} a \\ b \end{pmatrix} = \begin{pmatrix} x \\ y \end{pmatrix}$$

Since $\det A \neq 0$, this system has the unique solution $\begin{pmatrix} a \\ b \end{pmatrix} = A^{-1}\begin{pmatrix} x \\ y \end{pmatrix}$. Thus $\mathbf{v} \in H$, which shows that $\mathbb{R}^2 \subset H$. But since $H \subset \mathbb{R}^2$, we have

$$H = \mathbb{R}^2$$

Problems 4.4, page 228

1. yes

3. no; for example, $\begin{pmatrix} 1 \\ 2 \end{pmatrix} \notin$ span of the three vectors $\left[\text{each is a multiple of } \begin{pmatrix} 1 \\ 1 \end{pmatrix} \right]$

5. yes 7. yes

9. no; for example, $x \notin$ span $\{1 - x, 3 - x^2\}$

11. yes 13. yes

15. Let $p_i(x) = a_{0i} + a_{1i}x + \cdots + a_{ni}x^n$ for $i = 1, 2, \ldots, n$ and define the vector

$$\mathbf{a}_i = (a_{0i}, a_{1i}, \ldots, a_{ni})$$

Choose a vector $\mathbf{b} = (b_0, \ldots, b_n)$ such that $\mathbf{a}_i \cdot \mathbf{b} = 0$ for $i = 1, 2, \ldots, m$. If $m < n + 1$, this last homogeneous system of equations always has a nontrivial solution $\mathbf{b}$ (by Theorem 1.4.1). Suppose that $\mathbf{b}$ can be written as $\mathbf{b} = \alpha_1\mathbf{a}_1 + \alpha_2\mathbf{a}_2 + \cdots + \alpha_m\mathbf{a}_m$. Then $\mathbf{b} \cdot \mathbf{b} = \alpha_1\mathbf{a}_1 \cdot \mathbf{b} + \alpha_2\mathbf{a}_2 \cdot \mathbf{b} + \cdots + \alpha_m\mathbf{a}_m \cdot \mathbf{b} = 0$. But $\mathbf{b} \neq 0$. This contradiction shows that $\mathbf{b}$ cannot be written as a linear combination of the $\mathbf{a}_i$'s if $m < n + 1$. Let $q(x) = b_0 + b_1x + \cdots + b_nx^n$. Then $q(x)$ cannot be written as a linear combination of the p_i's and therefore the p_i's do not span. We are left to conclude that $m \geq n + 1$.

17. If $p(x) \in P$, then $p(x) = a_0 + a_1x + \cdots + a_kx^k$ for some integer k. Thus $p(x)$ can be written as a linear combination of $1, x, x^2, \ldots, x^k$.

19. $\alpha\mathbf{v}_1 + \beta\mathbf{v}_2 = (\alpha + \beta c)\mathbf{v}_1$. Let $\mathbf{v}_1 = (x_1, y_1, z_1)$. Then any vector $\mathbf{v} = (x, y, z)$ in span $\{\mathbf{v}_1, \mathbf{v}_2\}$ can be written as $(x, y, z) = ((\alpha + \beta c)x_1, (\alpha + \beta c)y_1, (\alpha + \beta c)z_1)$, or

$$x = (\alpha + \beta c)x_1$$
$$y = (\alpha + \beta c)y_1$$
$$z = (\alpha + \beta c)z_1$$

which, from Section 3.5, is the equation of a line passing through $(0, 0, 0)$ with direction (x_1, y_1, z_1).

21. Let $\mathbf{v} \in V$. Then there are scalars $\alpha_1, \alpha_2, \ldots, \alpha_n$ such that $\mathbf{v} = \alpha_1\mathbf{v}_1 + \alpha_2\mathbf{v}_2 + \cdots + \alpha_n\mathbf{v}_n = \alpha_1\mathbf{v}_1 + \alpha_2\mathbf{v}_2 + \cdots + \alpha_n\mathbf{v}_n + 0\mathbf{v}_{n+1}$. The last equation shows that $\mathbf{v}_1, \mathbf{v}_2, \ldots, \mathbf{v}_{n+1}$ span V.

23. Let $\mathbf{v}_i = (v_{i1}, v_{i2}, \ldots, v_{in})$ and $\mathbf{u}_i = (u_{i1}, u_{i2}, \ldots, u_{in})$. Also, let

$$\mathbf{w}_j = \begin{pmatrix} v_{1j} \\ v_{2j} \\ \vdots \\ v_{nj} \end{pmatrix}, \quad \mathbf{z}_j = \begin{pmatrix} u_{1j} \\ u_{2j} \\ \vdots \\ u_{nj} \end{pmatrix}$$

and $A = (a_{ij})$. Then, writing out the components in the given inequalities, we have

(*) $v_{ij} = a_{i1}u_{1j} + a_{i2}u_{2j} + \cdots + a_{in}u_{nj}$

or

$$\begin{array}{c} \text{given} \\ \det A \neq 0 \\ \downarrow \end{array}$$
$$\mathbf{w}_j = A\mathbf{z}_j; \text{ so } \mathbf{z}_j = A^{-1}\mathbf{w}_j$$

Let $A^{-1} = B = (b_{ij})$. Then the expression for $\mathbf{z}_j$ can be written

$$u_{ij} = b_{i1}v_{1j} + b_{i2}v_{2j} + \cdots + b_{in}v_{nj}$$

We see that this is similar to the expression (*) with u's and v's interchanged. Thus

$$\mathbf{u}_i = \sum_{k=1}^{n} b_{ik}\mathbf{v}_k \quad \text{for } i = 1, 2, \ldots, n$$

so $\mathbf{u}_i \in$ span $\{\mathbf{v}_1, \mathbf{v}_2, \ldots, \mathbf{v}_n\}$ for

$i = 1, 2, \ldots, n$. Since it was given that $\mathbf{v}_i \in$ span $\{\mathbf{u}_1, \mathbf{u}_2, \ldots, \mathbf{u}_n\}$, we conclude that the spans are equal.

Problems 4.5, page 240

1. independent
3. dependent;

$$-2\begin{pmatrix} 2 \\ -1 \\ 4 \end{pmatrix} + \begin{pmatrix} 4 \\ -2 \\ 8 \end{pmatrix} = \begin{pmatrix} 0 \\ 0 \\ 0 \end{pmatrix}$$

5. dependent (from Theorem 2)
7. independent
9. independent
11. independent
13. independent
15. independent
17. dependent
19. independent
21. independent
23. $ad - bc = 0$ **25.** $\alpha = -\frac{13}{2}$
27. System (7) can be written as

(*) $$c_1\begin{pmatrix} a_{11} \\ a_{21} \\ \vdots \\ a_{m1} \end{pmatrix} + c_2\begin{pmatrix} a_{12} \\ a_{22} \\ \vdots \\ a_{m2} \end{pmatrix} + \cdots$$
$$+ c_n\begin{pmatrix} a_{1n} \\ a_{2n} \\ \vdots \\ a_{mn} \end{pmatrix} = \begin{pmatrix} 0 \\ 0 \\ \vdots \\ 0 \end{pmatrix}$$

If (7) has even one nontrivial solution, then the columns of A are linearly dependent. If the columns of A are dependent, then there are number $c_1, c_2, \ldots, c_n$ not all zero such that (*) holds.

29. If $\mathbf{0} = c_1\mathbf{v}_1 + c_2\mathbf{v}_2 + \cdots + c_k\mathbf{v}_k$, then $\mathbf{0} = c_1\mathbf{v}_1 + c_2\mathbf{v}_2 + \cdots + c_k\mathbf{v}_k + 0\mathbf{v}_{k+1} + 0\mathbf{v}_{k+2} + \cdots + 0\mathbf{v}_n$. Since $\mathbf{v}_1, \mathbf{v}_2, \ldots, \mathbf{v}_n$ are independent, we have $c_1 = c_2 = \cdots = c_k = 0$.

31. If $c_1\mathbf{v}_1 + c_2\mathbf{v}_2 + c_3\mathbf{v}_3 = \mathbf{0}$, then
$0 = 0 \cdot \mathbf{v}_1 = (c_1\mathbf{v}_1 + c_2\mathbf{v}_2 + c_3\mathbf{v}_3) \cdot$
$\mathbf{v}_1 = c_1(\mathbf{v}_1 \cdot \mathbf{v}_1) + c_2(\mathbf{v}_2 \cdot \mathbf{v}_1) +$
$c_3(\mathbf{v}_3 \cdot \mathbf{v}_1) = c_1\mathbf{v}_1 \cdot \mathbf{v}_1 + c_20 +$
$c_30 = c_1\mathbf{v}_1 \cdot \mathbf{v}_1$. Since $\mathbf{v}_1 \neq \mathbf{0}$,
$\mathbf{v}_1 \cdot \mathbf{v}_1 \neq 0$, so we must have $c_1 =$
0. A similar computation shows that
$c_2 = c_3 = 0$.

33. $x_2 \begin{pmatrix} -1 \\ 1 \\ 0 \end{pmatrix} + x_3 \begin{pmatrix} -1 \\ 0 \\ 1 \end{pmatrix}$

35. $x_3 \begin{pmatrix} 13 \\ -6 \\ 1 \end{pmatrix}$

37. $x_2 \begin{pmatrix} -2 \\ 1 \\ 0 \\ 0 \end{pmatrix} + x_3 \begin{pmatrix} 3 \\ 0 \\ 1 \\ 0 \end{pmatrix}$
$+ x_4 \begin{pmatrix} -5 \\ 0 \\ 0 \\ 1 \end{pmatrix}$

39. Any nonzero $\mathbf{u}$ will lead to a similar
result. For example, let $\mathbf{u} =$
$(1, 0, 0)$. Then $\mathbf{x} = (0, 1, 0)$ and
$\mathbf{y} = (0, 0, 1)$ are in H.
$\mathbf{w} = \mathbf{x} \times \mathbf{y} = (1, 0, 0) = \mathbf{u}$.
  e. H is the plane orthogonal to $\mathbf{u}$
  and $\mathbf{w}$ is orthogonal to this
  plane and so it must be parallel
  to $\mathbf{u}$.

41. Let $p(x) = a_0 + a_1x + a_2x^2$ and
$q(x) = b_0 + b_1x + b_2x^2$ be two
polynomials in P_2. Let $r(x) = c_0 +$
$c_1x + c_2x^2$ be a third polynomial in
P_2. If p and q span P_2, then there
are scalars α and β such that $r =$
$\alpha p + \beta q$; that is,

$$c_0 = \alpha a_0 + \beta b_0$$
$$c_1 = \alpha a_1 + \beta b_1$$
$$c_2 = \alpha a_2 + \beta b_2$$

This "overdetermined" system of
three equations in two unknowns
will have a solution if and only if

the third equation is a linear combination of the first two. Since c_0, c_1,
and c_2 are arbitrary, this will rarely
be the case. Thus p and q cannot
span P_2. To see this in another
way, suppose that $\begin{pmatrix} \alpha \\ \beta \end{pmatrix}$ is a solution
to the system. Then $\begin{pmatrix} c_0 \\ c_1 \end{pmatrix} =$
$\begin{pmatrix} a_0 & b_0 \\ a_1 & b_1 \end{pmatrix}\begin{pmatrix} \alpha \\ \beta \end{pmatrix}$ or $\begin{pmatrix} \alpha \\ \beta \end{pmatrix} =$
$\begin{pmatrix} a_0 & b_0 \\ a_1 & b_1 \end{pmatrix}^{-1}\begin{pmatrix} c_0 \\ c_1 \end{pmatrix}$. But, also,
$\begin{pmatrix} \alpha \\ \beta \end{pmatrix} = \begin{pmatrix} a_1 & b_1 \\ a_2 & b_2 \end{pmatrix}^{-1}\begin{pmatrix} c_1 \\ c_2 \end{pmatrix}$. In general, these two expressions for $\begin{pmatrix} \alpha \\ \beta \end{pmatrix}$
will not be equal, which means that
in general the system does not have
a solution.

43. Let $S = \{\mathbf{v}_1, \mathbf{v}_2, \ldots, \mathbf{v}_k\}$ be a subset of the linearly independent set
$T = \{\mathbf{v}_1, \mathbf{v}_2, \ldots, \mathbf{v}_n\}$ with $k < n$.
Suppose S is dependent. Then there
exist scalars not all zero with
$c_1\mathbf{v}_1 + c_2\mathbf{v}_2 + \cdots + c_k\mathbf{v}_k = \mathbf{0}$. This
shows that T is dependent, a contra-
diction. Simply form a linear combination of the vectors in T, using c_i
whenever $\mathbf{v}_i \in S$ and 0 otherwise.

45. Write the matrices as $A^{(1)}$, $A^{(2)}$,
$\ldots$, $A^{(mn+1)}$. Suppose that $A^{(i)} =$
$(a_{jk}^{(i)})$. Consider the system

$$a_{11}^{(1)}\alpha_1 + a_{11}^{(2)}\alpha_2 + \cdots$$
$$\vdots$$
$$a_{mn}^{(1)}\alpha_1 + a_{mn}^{(2)}\alpha_2 + \cdots$$
$$+ a_{11}^{(mn+1)}\alpha_{mn+1} = 0$$
$$\vdots$$
$$+ a_{mn}^{(mn+1)}\alpha_{mn+1} = 0$$

This is a homogeneous system with
mn equations and $mn + 1$ unknowns. Therefore it has a nonzero
solution, and there are scalars α_1,
$\alpha_2, \ldots, \alpha_{mn+1}$ not all zero such
that $\alpha_1 A^{(1)} + \alpha_2 A^{(2)} + \cdots +$

$\alpha_{mn+1}A^{mn+1} = 0$ (the $m \times n$ zero matrix).

47. Assume that $1, x, \ldots, x^k$ are linearly independent. Suppose that $c_0 + c_1 x + \cdots + c_k x^k + c_{k+1} x^{k+1} = 0$.

If $c_{k+1} \neq 0$, then $x^{k+1} = -\dfrac{c_0}{c_{k+1}} - \dfrac{c_1}{c_{k+1}} x - \cdots - \dfrac{c_k}{c_{k+1}} x^k$, which is clearly impossible. Thus $c_{k+1} = 0$. But then $c_0 + c_1 x + \cdots + c_k x^k = 0$, which implies that $c_0 = c_1 = \cdots = c_k = 0$ since $1, x, x^2, \ldots, x^k$ are linearly independent. Thus $1, x, x^2, \ldots, x^k, x^{k+1}$ are also linearly independent, and this completes the induction proof (see Appendix 1).

49. There are scalars $\alpha_1, \alpha_2, \ldots, \alpha_n$ not all zero such that $\alpha_1 \mathbf{v}_1 + \alpha_2 \mathbf{v}_2 + \cdots + \alpha_n \mathbf{v}_n = \mathbf{0}$. Let k be the largest integer for which $\alpha_k \neq 0$ (k may equal n). Then the equation reads $\alpha_1 \mathbf{v}_1 + \alpha_2 \mathbf{v}_2 + \cdots + \alpha_k \mathbf{v}_k = \mathbf{0}$ so that

$$\mathbf{v}_k = -\frac{\alpha_1}{\alpha_k} \mathbf{v}_1 - \frac{\alpha_2}{\alpha_k} \mathbf{v}_2 - \cdots$$
$$-\frac{\alpha_{k-1}}{\alpha_k} \mathbf{v}_{k-1}$$

51. Suppose f and g are dependent. Then $g = cf$ and $g' = cf'$ for some constant c and

$$
w(f, g)(x) = \begin{vmatrix} f(x) & g(x) \\ f'(x) & g'(x) \end{vmatrix}
$$
$$
= \begin{vmatrix} f(x) & cf(x) \\ f'(x) & cf'(x) \end{vmatrix}
$$
$$
= cf(x)f'(x) - cf(x)f'(x) = 0
$$

53. Suppose that $c_1(\mathbf{u} + \mathbf{v}) + c_2(\mathbf{u} + \mathbf{w}) + c_3(\mathbf{v} + \mathbf{w}) = \mathbf{0}$. Then $(c_1 + c_2)\mathbf{u} + (c_1 + c_3)\mathbf{v} + (c_2 + c_3)\mathbf{w} = \mathbf{0}$. Since $\mathbf{u}$, $\mathbf{v}$, and $\mathbf{w}$ are linearly independent,

$$
\begin{aligned}
c_1 + c_2 & = 0 \\
c_1 \quad\;\; + c_3 & = 0 \\
c_2 + c_3 & = 0
\end{aligned}
$$

The determinant of this homogeneous system is

$$
\begin{vmatrix} 1 & 1 & 0 \\ 1 & 0 & 1 \\ 0 & 1 & 1 \end{vmatrix} = -2 \neq 0
$$

so the only solution is $c_1 = c_2 = c_3 = 0$ and the three vectors are independent.

55. $\begin{vmatrix} 1 & 1 & 1 \\ a & b & c \\ a^2 & b^2 & c^2 \end{vmatrix} =$

$(b - a)(c - a)(c - b)$, according to the result of Problem 2.2.35 on page 132.

57. $\begin{pmatrix} 2 \\ 1 \\ 2 \end{pmatrix}, \begin{pmatrix} -1 \\ 3 \\ 4 \end{pmatrix}, \begin{pmatrix} 1 \\ 2 \\ 2 \end{pmatrix}$. (There are many choices for the third vector.)

59. a. By the result of Problem 3.5.59 on page 205, $\mathbf{u} \times (\mathbf{v} \times \mathbf{w}) = 0$. From Theorem 2 part (vi) on page 185,

$$\mathbf{v} \cdot (\mathbf{v} \times \mathbf{w}) = \mathbf{w} \cdot (\mathbf{v} \times \mathbf{w}) = 0$$

Let

$$
\mathbf{a} = \begin{pmatrix} a \\ b \\ c \end{pmatrix} = \mathbf{v} \times \mathbf{w}. \text{ Then}
$$

$$\mathbf{a} \cdot \mathbf{u} = \mathbf{a} \cdot \mathbf{v} = \mathbf{a} \cdot \mathbf{w} = 0$$

and writing out the terms in each scalar product gives the desired result.

b. Think of the system in (a) as a homogeneous system of three equations in the three unknowns a, b, and c. Since the system has nontrivial solutions, its determinant is 0.

c. This follows from (a) since we have

$$a\mathbf{u} + b\mathbf{v} + c\mathbf{w} = \mathbf{0}$$

but a, b, and c are not all equal to 0.

Problems 4.6, page 251

1. no; doesn't span

3. no; dependent

5. no; doesn't span

7. yes 9. yes

11. $\left\{ \begin{pmatrix} 0 \\ 1 \\ -1 \end{pmatrix}, \begin{pmatrix} 1 \\ 0 \\ 2 \end{pmatrix} \right\}$ 13. $\left\{ \begin{pmatrix} 2 \\ 3 \\ 4 \end{pmatrix} \right\}$

15. Since dim $\mathbb{R}^2 = 2$, a proper subspace H must have dimension 1. Let $\{(x_0, y_0)\}$ be a basis for H. If $(x, y) \in H$, then $(x, y) = c(x_0, y_0)$ for some number c. This means that

$$x = cx_0, \; y = cy_0 \text{ or } c = \frac{x}{x_0} = \frac{y}{y_0}$$

and $y = \left(\dfrac{y_0}{x_0} \right) x$, which is the equation of a straight line through the origin with slope $\dfrac{y_0}{x_0}$ if $x_0 \neq 0$. If $x_0 = 0$, then the line is the y-axis.

17. Let $\{\mathbf{v}_1, \mathbf{v}_2, \ldots, \mathbf{v}_{n-1}\}$ be a basis for H. Let $\mathbf{v}_n$ be another vector in H. Then the vectors $\mathbf{v}_1, \mathbf{v}_2, \ldots, \mathbf{v}_{n-1}, \mathbf{v}_n$ are linearly dependent. Let A be the matrix whose rows are $\mathbf{v}_1, \mathbf{v}_2, \ldots, \mathbf{v}_{n-1}, \mathbf{v}_n$. Then det $A = 0$, and the equation $A\mathbf{a} = \mathbf{0}$ has a nontrivial solution

$$\mathbf{a} = \begin{pmatrix} a_1 \\ a_2 \\ \vdots \\ a_n \end{pmatrix}$$

This means that $\mathbf{v}_i \cdot \mathbf{a} = 0$ for $i = 1, 2, \ldots, n$. In particular, if $\mathbf{v}_n = (x_1, x_2, \ldots, x_n)$, then $\mathbf{v}_n \cdot \mathbf{a} = \mathbf{0}$, or $a_1 x_1 + a_2 x_2 + \cdots + a_n x_n = 0$. Since $\mathbf{v}_n$ was an arbitrary vector in H, this proves the result.

19. $\left\{ \begin{pmatrix} 1 \\ 1 \end{pmatrix} \right\}$ 21. $\left\{ \begin{pmatrix} -2 \\ -3 \\ 1 \end{pmatrix} \right\}$

23. $\left\{ \begin{pmatrix} 3 \\ 1 \\ 0 \end{pmatrix}, \begin{pmatrix} -2 \\ 0 \\ 1 \end{pmatrix} \right\}$ 25. n

27. V has a basis $\{\mathbf{u}_1, \mathbf{u}_2, \ldots, \mathbf{u}_n\}$, so there exist scalars such that

$$\begin{aligned} \mathbf{v}_1 &= a_{11}\mathbf{u}_1 + a_{12}\mathbf{u}_2 + \cdots + a_{1n}\mathbf{u}_n \\ \mathbf{v}_2 &= a_{21}\mathbf{u}_1 + a_{22}\mathbf{u}_2 + \cdots + a_{2n}\mathbf{u}_n \\ &\;\;\vdots \qquad \vdots \qquad \vdots \qquad \vdots \\ \mathbf{v}_m &= a_{m1}\mathbf{u}_1 + a_{m2}\mathbf{u}_2 + \cdots + a_{mn}\mathbf{u}_n \end{aligned}$$

Let $\mathbf{a}_i = (a_{i1}, a_{i2}, \ldots, a_{in})$. The $\mathbf{a}_i$'s are m linearly independent vectors in $\mathbb{R}^n$ (otherwise the $\mathbf{v}_i$'s wouldn't be independent). Expand the $\mathbf{a}_i$'s into a basis $\mathbf{a}_1, \mathbf{a}_2, \ldots, \mathbf{a}_m, \mathbf{a}_{m+1}, \ldots, \mathbf{a}_n$ for $\mathbb{R}^n$ by adding $n - m$ linearly independent vectors to the set. Then if $\mathbf{a}_k = (a_{k1}, a_{k2}, \ldots, a_{kn})$ for $m < k \leq n$, define $\mathbf{v}_k = a_{k1}\mathbf{u}_1 + a_{k2}\mathbf{u}_2 + \cdots + a_{kn}\mathbf{u}_n$ for $k = m + 1, m + 2, \ldots, n$. Since

$$\det \begin{pmatrix} a_{11} & a_{12} & \cdots & a_{1n} \\ a_{21} & a_{22} & \cdots & a_{2n} \\ \vdots & \vdots & & \vdots \\ a_{n1} & a_{n2} & \cdots & a_{nn} \end{pmatrix} \neq 0$$

the set $\{\mathbf{v}_1, \mathbf{v}_2, \ldots, \mathbf{v}_n\}$ forms a basis for V since it consists of n linearly independent vectors in V with dim $V = n$.

29. If the vectors are independent, then they form a basis and dim $V = n$. If not, then by Problem 4.5.49, at least one of them can be written as a linear combination of the ones that precede it. Throw this vector out. Proceed in this manner until m linearly independent vectors remain. These must still span V by the manner in which they were chosen. Thus dim $V = m < n$. In either event dim $V \leq n$.

31. **a.** See Problem 4.3.27.
    b. Let $\{\mathbf{v}_1, \ldots, \mathbf{v}_n\}$ be a basis for H, and let $\{\mathbf{u}_1, \mathbf{u}_2, \ldots, \mathbf{u}_m\}$ be a basis for K. Clearly

$B = \{v_1, v_2, \ldots, v_n, u_1, u_2, \ldots, u_m\}$ span $H + K$. Suppose that $\alpha_1 v_1 + \alpha_2 v_2 + \cdots + \alpha_n v_n + \beta_1 u_1 + \beta_2 u_2 + \cdots + \beta_m u_m = 0$ where not all of the coefficients are zero. Let $h = \alpha_1 v_1 + \alpha_2 v_2 + \cdots + \alpha_n v_n$ and $k = \beta_1 u_1 + \beta_2 u_2 + \cdots + \beta_m u_m$. Then by linear independence, neither h nor k is the zero vector. Also, $h \in H$ and $k \in K$. But then $h + k = 0$ or $h = -k \in K$. Thus $0 \neq h \in H \cap K$, which contradicts the fact that $H \cap K = \{0\}$. Hence all the α's and β's are zero, which implies that the vectors in B are linearly independent. Thus B is a basis for $H + K$ and $\dim(H + K) = \dim H + \dim K$.

33. i. If $\dim \operatorname{span} \{v_1, v_2\} = 1$, then choose a basis $\{v\}$ for span $\{v_1, v_2\}$. Then $v_1 = \alpha v$, and $v_2 = \beta v$. If $v_1 = 0$, then v_1 is a single point lying on v_2. If $v_1 \neq 0$, then $\alpha \neq 0$ so that

$$v_2 = \beta v = \frac{\beta}{\alpha}(\alpha v) = \frac{\beta}{\alpha} v_1.$$ In

either case the vectors are collinear.

ii. If the vectors are collinear, then $v_2 = c v_1$ for some scalar c so that v_1 is a basis for span $\{v_1, v_2\}$ and $\dim \operatorname{span} \{v_1, v_2\} = 1$.

35. If they are not linearly independent, then, as in the answer to Problem 29, $\dim V < n$. Since $\dim V = n$, the vectors must be independent, and therefore constitute a basis for V.

37. $B_1 = \left\{ \begin{pmatrix} 1 \\ 0 \\ 1 \\ 0 \end{pmatrix}, \begin{pmatrix} 0 \\ 1 \\ 0 \\ 1 \end{pmatrix}, \begin{pmatrix} 1 \\ 0 \\ 0 \\ 0 \end{pmatrix}, \begin{pmatrix} 0 \\ 1 \\ 0 \\ 0 \end{pmatrix} \right\}$

$B_2 = \left\{ \begin{pmatrix} 1 \\ 0 \\ 1 \\ 0 \end{pmatrix}, \begin{pmatrix} 0 \\ 1 \\ 0 \\ 1 \end{pmatrix}, \begin{pmatrix} 0 \\ 0 \\ 1 \\ 0 \end{pmatrix}, \begin{pmatrix} 0 \\ 0 \\ 0 \\ 1 \end{pmatrix} \right\}$

There are infinitely many other choices.

Problems 4.7, page 265

1. $\rho = 2, \nu = 0$ **3.** $\rho = 1, \nu = 2$

5. $\rho = 2, \nu = 1$ **7.** $\rho = 2, \nu = 2$

9. $\rho = 2, \nu = 0$ **11.** $\rho = 2, \nu = 2$

13. $\rho = 3, \nu = 1$ **15.** $\rho = 2, \nu = 1$

17. range basis $= \left\{ \begin{pmatrix} 1 \\ 3 \\ 5 \end{pmatrix}, \begin{pmatrix} -1 \\ 1 \\ -1 \end{pmatrix} \right\}$; these are the first two columns of A.

null space basis $= \left\{ \begin{pmatrix} -3 \\ 1 \\ 2 \end{pmatrix} \right\}$

19. range basis $= \left\{ \begin{pmatrix} 1 \\ 0 \\ 1 \end{pmatrix}, \begin{pmatrix} -1 \\ 1 \\ 0 \end{pmatrix}, \begin{pmatrix} 3 \\ 3 \\ 5 \end{pmatrix} \right\}$; these are the first three (linearly independent) columns of A.

null space basis $= \left\{ \begin{pmatrix} -6 \\ -4 \\ 1 \\ 0 \end{pmatrix} \right\}$

21. range basis $= \left\{ \begin{pmatrix} 1 \\ -2 \\ 2 \\ 3 \end{pmatrix} \right\}$

null space basis $=$

$\left\{ \begin{pmatrix} 1 \\ 1 \\ 0 \\ 0 \end{pmatrix}, \begin{pmatrix} 0 \\ 2 \\ 1 \\ 0 \end{pmatrix}, \begin{pmatrix} 0 \\ 3 \\ 0 \\ 1 \end{pmatrix} \right\}$

23. $\left\{ \begin{pmatrix} 1 \\ 4 \\ -2 \end{pmatrix}, \begin{pmatrix} 0 \\ 1 \\ -\frac{6}{7} \end{pmatrix} \right\}$

25. $\{(1, 0, 0, \frac{1}{2}), (0, 1, -1, \frac{3}{2}), (0, 0, 0, 1)\}$

27. no **29.** yes

31. If c_i denotes the ith column of D, then

$$c_i = d_i \begin{pmatrix} 0 \\ 0 \\ \vdots \\ 1 \\ 0 \\ \vdots \\ 0 \end{pmatrix} \leftarrow i\text{th position.}$$

Thus the c_i's are linearly independent when $d_i \neq 0$, and the number of linearly independent columns is the rank.

33. $\rho(A^t) =$ dimension of column space of $A^t =$ dimension of row space of $A =$ dimension of column space of A (by Theorem 3) $= \rho(A)$.

35. **i.** Let $H = $ range of A and let $\{v_1, v_2, \ldots, v_k\}$ be a basis for H. Since B is invertible, $\ker B = \{0\}$, which means that $\{Bv_1, Bv_2, \ldots, Bv_k\}$ is a linearly independent set in $\mathbb{R}^m$, and is therefore a basis for range BA. Then $\rho(BA) = k = \rho(A)$.

ii. Since C is invertible, range of $C = \mathbb{R}^n$. Let $h \in H$; then there is an $x \in \mathbb{R}^n$ such that $Ax = h$. Since range of $C = \mathbb{R}^n$, there is a $y \in \mathbb{R}^n$ such that $Cy = x$. Then $ACy = h$. Thus $H \subset$ range of AC. If $v \in$ range of AC, there is a u in $\mathbb{R}^n$ such that $ACu = v$. But then $v = A(Cu)$ so that $v \in$ range of $A = H$. Hence range of $AC \subset H$ so that range of $AC = H$ and $\rho(A) = \rho(AC)$.

37. Since $\rho(A) = 5$, the five rows of A are linearly independent. Thus the five rows of (A, b) are linearly independent and $\rho(A, b) = 5$.

39. By Problem 35, $\rho(A) = \rho(AD) = \rho(C(AD)) = \rho(B)$.

41. **i.** If there is an $x \neq 0$ such that $Ax = 0$, then $A(\alpha x) = \alpha Ax = 0$ for every $\alpha \in \mathbb{R}$ so that $v(A) = \dim \ker A \geq 1$, and $\rho(A) = n - v(A) \leq n - 1 < n$.

ii. If $\rho(A) < n$, then $v(A) = n - \rho(A) > 0$ so that there is an $x \neq 0$ such that $Ax = 0$.

Problems 4.8, page 269

1. 1 **3.** 2 **5.** 2 **7.** 3 **9.** 3

11. Let $|M|$ be a subdeterminant of order $k + 1$. Expand $|M|$ in its first row. In doing so we obtain $k + 1$ determinants of order k. Since each such determinant is zero by assumption, $|M| = 0$.

Problems 4.9, page 279

1. $\dfrac{x + y}{2}\begin{pmatrix} 1 \\ 1 \end{pmatrix} + \dfrac{x - y}{2}\begin{pmatrix} 1 \\ -1 \end{pmatrix} = \begin{pmatrix} x \\ y \end{pmatrix}$

3. $\dfrac{4x + 3y}{41}\begin{pmatrix} 5 \\ 7 \end{pmatrix} + \dfrac{7x - 5y}{41}\begin{pmatrix} 3 \\ -4 \end{pmatrix} = \begin{pmatrix} x \\ y \end{pmatrix}$

5. $\dfrac{dx - by}{ad - bc}\begin{pmatrix} a \\ c \end{pmatrix} + \dfrac{-cx + ay}{ad - bc}\begin{pmatrix} b \\ d \end{pmatrix} = \begin{pmatrix} x \\ y \end{pmatrix}$

7. $(x - y)\begin{pmatrix} 1 \\ 0 \\ 0 \end{pmatrix} + (y - z)\begin{pmatrix} 1 \\ 1 \\ 0 \end{pmatrix}$

$+ z\begin{pmatrix} 1 \\ 1 \\ 1 \end{pmatrix} = \begin{pmatrix} x \\ y \\ z \end{pmatrix}$

9. $\dfrac{6x - 11y + 10z}{31}\begin{pmatrix} 2 \\ 1 \\ 3 \end{pmatrix}$

$+ \dfrac{2x + 17y - 7z}{31}\begin{pmatrix} -1 \\ 4 \\ 5 \end{pmatrix}$

$$+ \frac{7x + 13y - 9z}{31} \begin{pmatrix} 3 \\ -2 \\ -4 \end{pmatrix}$$

$$= \begin{pmatrix} x \\ y \\ z \end{pmatrix}$$

11. $a_0 + a_1 x + a_2 x^2$
$$= (a_0 + a_1 + a_2)1$$
$$+ a_1(x - 1) + a_2(x^2 - 1)$$

13. $a_0 + a_1 x + a_2 x_2$
$$= \frac{(a_0 + a_1 + a_2)}{2}(x + 1)$$
$$+ \frac{(a_1 - a_0 - a_2)}{2}(x - 1)$$
$$+ a_2(x^2 - 1)$$

15. $2(x^3 + x^2) - 5(x^2 + x)$
$$+ 10(x + 1) - 16(1)$$

17. $(\mathbf{x})_{B_2} = \begin{pmatrix} -\frac{1}{3} \\ 0 \end{pmatrix}$

19. $(\mathbf{x})_{B_2} = \begin{pmatrix} \frac{86}{33} \\ -\frac{20}{11} \\ \frac{7}{11} \end{pmatrix}$

21. independent

23. dependent

25. independent

27. independent

29. If they were linearly independent, they would span P_n. But $1 \in P_n$ and $1 \notin \text{span}\{p_1, p_2, \ldots, p_{n+1}\}$ since the constant term in each polynomial is 0.

31. If they were linearly independent, they would span M_{mn}. But the matrix $A = (a_{ij})$, where $a_{11} = 1$ and $a_{ij} = 0$ otherwise is not in the span of $A_1, A_2, \ldots, A_{mn+1}$ since a linear combination of matrices with a 0 in the 1, 1 position also has a 0 in the 1, 1 position.

33. $\begin{pmatrix} \cos\theta & -\sin\theta \\ \sin\theta & \cos\theta \end{pmatrix} \begin{pmatrix} 1 \\ 0 \end{pmatrix} = \begin{pmatrix} \cos\theta \\ \sin\theta \end{pmatrix};$
$\begin{pmatrix} \cos\theta & -\sin\theta \\ \sin\theta & \cos\theta \end{pmatrix} \begin{pmatrix} 0 \\ 1 \end{pmatrix} = \begin{pmatrix} -\sin\theta \\ \cos\theta \end{pmatrix}.$

A^{-1} is obtained by rotating through an angle of $-\theta$. Thus

$$A^{-1} = \begin{pmatrix} \cos(-\theta) & -\sin(-\theta) \\ \sin(-\theta) & \cos(-\theta) \end{pmatrix}$$
$$= \begin{pmatrix} \cos\theta & \sin\theta \\ -\sin\theta & \cos\theta \end{pmatrix}$$

Alternatively, since

$$A = \begin{pmatrix} \cos\theta & -\sin\theta \\ \sin\theta & \cos\theta \end{pmatrix},$$

$$A^{-1} = \begin{pmatrix} \cos\theta & \sin\theta \\ -\sin\theta & \cos\theta \end{pmatrix}$$

35. $A^{-1} =$
$$\begin{pmatrix} \cos(\pi/4) & \sin(\pi/4) \\ -\sin(\pi/4) & \cos(\pi/4) \end{pmatrix}$$
$$= \begin{pmatrix} 1/\sqrt{2} & 1/\sqrt{2} \\ -1/\sqrt{2} & 1/\sqrt{2} \end{pmatrix}$$

(from Problem 33), so

$$A^{-1} \begin{pmatrix} 2 \\ -7 \end{pmatrix} = \begin{pmatrix} -5/\sqrt{2} \\ -9/\sqrt{2} \end{pmatrix}$$

37. Since C is invertible, the columns of C are linearly independent. That is, $\{\mathbf{c}_1, \mathbf{c}_2, \ldots, \mathbf{c}_n\}$ are n linearly independent vectors in V, which are therefore a basis for V since $\dim V = n$.

39. If $CA = I$, then $(\mathbf{x})_{B_1} = I(\mathbf{x})_{B_1} = CA(\mathbf{x})_{B_1}$. Conversely, suppose that $(\mathbf{x})_{B_1} = CA(\mathbf{x})_{B_1}$. Let $B_1 = \{\mathbf{v}_1, \mathbf{v}_2, \ldots, \mathbf{v}_n\}$. Then

$$(\mathbf{v}_1)_{B_1} = \begin{pmatrix} 1 \\ 0 \\ \vdots \\ 0 \end{pmatrix} = CA \begin{pmatrix} 1 \\ 0 \\ \vdots \\ 0 \end{pmatrix}$$

Let

$$CA = \begin{pmatrix} r_{11} & r_{12} & \cdots & r_{1n} \\ r_{21} & r_{22} & \cdots & r_{2n} \\ \vdots & \vdots & & \vdots \\ r_{n1} & r_{n2} & \cdots & r_{nn} \end{pmatrix}$$

Then $CA \begin{pmatrix} 1 \\ 0 \\ \vdots \\ 0 \end{pmatrix}$ = the first column of

$CA = \begin{pmatrix} r_{11} \\ r_{22} \\ \vdots \\ r_{n1} \end{pmatrix} = \begin{pmatrix} 1 \\ 0 \\ \vdots \\ 0 \end{pmatrix}$. Similarly, the

second column of $CA = \begin{pmatrix} 0 \\ 1 \\ 0 \\ \vdots \\ 0 \end{pmatrix}$ since

$(\mathbf{v}_2)_{B_1} = \begin{pmatrix} 0 \\ 1 \\ 0 \\ \vdots \\ 0 \end{pmatrix}$. Continuing in this

manner, we see that $CA = I$.

Problems 4.10, page 294

1. $\begin{pmatrix} 1/\sqrt{2} \\ 1/\sqrt{2} \end{pmatrix}, \begin{pmatrix} -1/\sqrt{2} \\ 1/\sqrt{2} \end{pmatrix}$

3. **i.** If $a = b = 0$, $\{(1, 0), (0, 1)\}$
    ii. If $a = 0$, $b \neq 0$, $\{(1, 0)\}$
    iii. If $a \neq 0$, $b = 0$, $\{(0, 1)\}$
    iv. If $a \neq 0$, $b \neq 0$, $\{(b/\sqrt{a^2 + b^2}, -a/\sqrt{a^2 + b^2})\}$

5. $\{(1/\sqrt{5}, 0, 2/\sqrt{5}), (2/\sqrt{30}, 5/\sqrt{30}, -1/\sqrt{30})\}$

7. $\{(2/\sqrt{29}, 3/\sqrt{29}, 4/\sqrt{29})\}$

9. $\{(1/\sqrt{5}, 0, 0, 2/\sqrt{5}), (2/\sqrt{30}, 5/\sqrt{30}, 0, -1/\sqrt{30}), (-2/\sqrt{10}, 1/\sqrt{10}, 2/\sqrt{10}, 1/\sqrt{10})\}$

11. $\{(a/\sqrt{a^2 + b^2 + c^2}, b/\sqrt{a^2 + b^2 + c^2}, c/\sqrt{a^2 + b^2 + c^2})\}$

13. $\{(-7/\sqrt{66}, -1/\sqrt{66}, 4/\sqrt{66})\}$

15. $Q^t = \begin{pmatrix} \frac{2}{3} & \frac{1}{3} & -\frac{2}{3} \\ \frac{1}{3} & \frac{2}{3} & \frac{2}{3} \\ \frac{2}{3} & -\frac{2}{3} & \frac{1}{3} \end{pmatrix}$ and $Q^t Q =$
$I = QQ^t$

17. $PQ = \dfrac{1}{3\sqrt{2}} \times \begin{pmatrix} 1 - \sqrt{8} & -1 - \sqrt{8} \\ 1 + \sqrt{8} & 1 - \sqrt{8} \end{pmatrix}$

$(PQ)^t = \dfrac{1}{3\sqrt{2}}$
$\times \begin{pmatrix} 1 - \sqrt{8} & 1 + \sqrt{8} \\ -1 - \sqrt{8} & 1 - \sqrt{8} \end{pmatrix}$

$(PQ)(PQ)^t = \dfrac{1}{18} \begin{pmatrix} 18 & 0 \\ 0 & 18 \end{pmatrix} = I$

19. $I = Q^{-1}Q = Q^t Q$. But $\det (Q^t Q) = \det Q^t \det Q = \det Q \det Q = (\det Q)^2$. Since

$$1 = \det I = \det Q^t Q$$
$$= (\det Q)^2$$

we have

$$\det Q = \pm 1$$

21. If $\mathbf{v}_i = \mathbf{0}$, then $0\mathbf{v}_1 + 0\mathbf{v}_2 + \cdots + 0\mathbf{v}_{i-1} + \mathbf{v}_i + 0\mathbf{v}_{i+1} + \cdots + 0\mathbf{v}_n = \mathbf{0}$, which implies that the $\mathbf{v}_i$'s are linearly dependent. Thus $\mathbf{v}_i \neq \mathbf{0}$ for $i = 1, 2, \ldots, n$.

23. **a.** $\mathbf{0}$

b. $\dfrac{1}{\sqrt{a^2 + b^2}} \begin{pmatrix} a \\ b \end{pmatrix}$

c. $\mathbf{v} = \begin{pmatrix} a \\ b \end{pmatrix} + \begin{pmatrix} 0 \\ 0 \end{pmatrix}$

25. **a.** $\dfrac{1}{49} \begin{pmatrix} -186 \\ 75 \\ 118 \end{pmatrix}$

b. $\dfrac{1}{7} \begin{pmatrix} 3 \\ -2 \\ 6 \end{pmatrix}$

c. $\mathbf{v} = \dfrac{1}{49} \begin{pmatrix} -186 \\ 75 \\ 118 \end{pmatrix}$
$+ \dfrac{13}{49} \begin{pmatrix} 3 \\ -2 \\ 6 \end{pmatrix}$

27. **a.** $\dfrac{1}{5} \begin{pmatrix} 1 \\ -3 \\ 4 \\ 17 \end{pmatrix}$

b. $\dfrac{1}{\sqrt{15}}\begin{pmatrix}2\\-1\\3\\-1\end{pmatrix}$

c. $\dfrac{1}{5}\begin{pmatrix}1\\-3\\4\\17\end{pmatrix} + \dfrac{2}{5}\begin{pmatrix}2\\-1\\3\\-1\end{pmatrix}$

29. $|\mathbf{u}_1 - \mathbf{u}_2|^2 = (\mathbf{u}_1 - \mathbf{u}_2) \cdot (\mathbf{u}_1 - \mathbf{u}_2) = \mathbf{u}_1 \cdot \mathbf{u}_1 - \mathbf{u}_2 \cdot \mathbf{u}_1 - \mathbf{u}_1 \cdot \mathbf{u}_2 + \mathbf{u}_2 \cdot \mathbf{u}_2 = 1 - 0 - 0 + 1 = 2$ since $\mathbf{u}_1, \mathbf{u}_2$ are orthonormal.

31. $a^2 + b^2 = 1$.

33. $|\mathbf{u} + \mathbf{v}|^2 = (|\mathbf{u}| + |\mathbf{v}|)^2$. This means that $(\mathbf{u} + \mathbf{v}) \cdot (\mathbf{u} + \mathbf{v}) = |\mathbf{u}|^2 + 2\mathbf{u} \cdot \mathbf{v} + |\mathbf{v}|^2 = (|\mathbf{u}| + |\mathbf{v}|)^2 = |\mathbf{u}|^2 + 2|\mathbf{u}| \, |\mathbf{v}| + |\mathbf{v}|^2$. Thus $\mathbf{u} \cdot \mathbf{v} = |\mathbf{u}| \, |\mathbf{v}|$, which can occur only if $\mathbf{u} = \lambda\mathbf{v}$; that is, $\mathbf{u}$ and $\mathbf{v}$ are linearly dependent.

35. We prove this by mathematical induction. If $k = 2$, this is the result of Problem 33. We assume it is true for $k = n$ and prove it for $k = n + 1$. Suppose that $|\mathbf{x}_1 + \mathbf{x}_2 + \cdots + \mathbf{x}_n + \mathbf{x}_{n+1}| = |\mathbf{x}_1| + |\mathbf{x}_2| + \cdots + |\mathbf{x}_n| + |\mathbf{x}_{n+1}|$. (*) This implies that $|\mathbf{x}_1 + \mathbf{x}_2 + \cdots + \mathbf{x}_n| = |\mathbf{x}_1| + |\mathbf{x}_2| + \cdots + |\mathbf{x}_n|^{(\checkmark)}$, for if this is not true, then, by the triangle inequality,

$|\mathbf{x}_1 + \mathbf{x}_2 + \cdots + \mathbf{x}_n|$
$\qquad\qquad < |\mathbf{x}_1| + |\mathbf{x}_2| + \cdots + |\mathbf{x}_n|$

But then

$|\mathbf{x}_1 + \mathbf{x}_2 + \cdots + \mathbf{x}_n + \mathbf{x}_{n+1}|$
$\le |\mathbf{x}_1 + \mathbf{x}_2 + \cdots + \mathbf{x}_n|$
$\quad + |\mathbf{x}_{n+1}| < |\mathbf{x}_1|$
$\quad + |\mathbf{x}_2| + \cdots + |\mathbf{x}_n|$
$\quad + |\mathbf{x}_{n+1}|$

which contradicts (*). Thus, by the induction assumption, dim span $\{\mathbf{x}_1, \mathbf{x}_2, \ldots, \mathbf{x}_n\} = 1$. Let $\mathbf{u} = \mathbf{x}_1 + \mathbf{x}_2 + \cdots + \mathbf{x}_n$. By (*) and $(\checkmark)$ $|\mathbf{u} + \mathbf{x}_{n+1}| = |\mathbf{u}| + |\mathbf{x}_{n+1}|$ so that by

Problem 33, $\mathbf{x}_{n+1} = \lambda\mathbf{u}$ for some number λ. That is, $\mathbf{x}_{n+1} \in$ span $\{\mathbf{x}_1, \mathbf{x}_2, \ldots, \mathbf{x}_n\}$ so that dim span $\{\mathbf{x}_1, \mathbf{x}_2, \ldots, \mathbf{x}_n, \mathbf{x}_{n+1}\} = 1$ also. Thus the result is true for $k = n + 1$, and the proof is complete.

37. $(H^{\perp})^{\perp} = \{\mathbf{v} \in \mathbb{R}^n; \mathbf{v} \cdot \mathbf{k} = 0$ for every $\mathbf{k} \in H^{\perp}\}$. Let $\mathbf{x} \in H$; then $\mathbf{x} \cdot \mathbf{k} = 0$ for every $\mathbf{k} \in H^{\perp}$ so that $\mathbf{x} \in (H^{\perp})^{\perp}$, which shows that $H \subseteq (H^{\perp})^{\perp}$. Conversely, if $\mathbf{v} \in (H^{\perp})^{\perp}$, then $\mathbf{v} \cdot \mathbf{k} = 0$ for every $\mathbf{k} \in H^{\perp}$. But $\mathbf{v} = \mathbf{h}' + \mathbf{k}'$ where $\mathbf{h}' \in H$ and $\mathbf{k}' \in H^{\perp}$. Then $0 = \mathbf{v} \cdot \mathbf{k} = \mathbf{h}' \cdot \mathbf{k} + \mathbf{k}' \cdot \mathbf{k} = 0 + \mathbf{k}' \cdot \mathbf{k}$. Thus $\mathbf{k}' \cdot \mathbf{k} = 0$ for every $\mathbf{k} \in H$, which means, in particular, that $\mathbf{k}' \cdot \mathbf{k}' = 0$. Thus $\mathbf{k}' = \mathbf{0}$ and $\mathbf{v} = \mathbf{h}' \in H$. Thus $(H^{\perp})^{\perp} \subset H$ and, together with $H \subset (H^{\perp})^{\perp}$, shows that $(H^{\perp})^{\perp} = H$.

39. Let $\mathbf{k} \in H_2^{\perp}$. Then $\mathbf{k} \cdot \mathbf{h} = 0$ for every $\mathbf{h} \in H_2$. Since $H_1 \subset H_2$, this shows that $\mathbf{k} \cdot \mathbf{h} = 0$ for every $\mathbf{h} \in H_1$. That is, $\mathbf{k} \in H_1^{\perp}$. Thus $H_2^{\perp} \subset H_1^{\perp}$.

Problems 4.11, page 305

1. $y = \dfrac{408}{126} - \dfrac{57}{126}x \approx 3.24 - 0.45x$

3. $y = \dfrac{162}{84} - \dfrac{10}{84}x \approx 1.93 - 0.12x$

5. $y = \dfrac{13,536}{5184} + \dfrac{10,800}{5184}x + \dfrac{1,584}{5,184}x^2 \approx 2.61 + 2.08x + 0.31x^2$
This is the equation of the parabola passing through the three points.

7. The argument here closely parallels the arguments given for linear and quadratic approximations.

9. This is a generalization of Problem 7.

11. $y \approx 108.71 + 4.906x - 0.00973x^2$

Problems 4.12, page 315

1. **i.** $(A, A) = a_{11}^2 + a_{22}^2 + \cdots + a_{nn}^2 \ge 0$.

  ii. $(A, A) = 0$ implies that $a_{ii}^2 = 0$ for $i = 1, 2, \ldots, n$ so that

$A = 0$. If $A = 0$, then $(A, A) = 0$.

iii. $(A, B + C) = a_{11}(b_{11} + c_{11}) + \cdots + a_{nn}(b_{nn} + c_{nn}) = a_{11}b_{11} + a_{11}c_{11} + \cdots + a_{nn}b_{nn} + a_{nn}c_{nn} = (a_{11}b_{11} + \cdots + a_{nn}b_{nn}) + (a_{11}c_{11} + \cdots + a_{nn}c_{nn}) = (A, B) + (A, C)$

iv. Similarly, $(A + B, C) = (A, C) + (B, C)$

v. $(A, B) = (B, A) = (\overline{B, A})$ since all components are real and $a_{ii}b_{ii} = b_{ii}a_{ii}$

vi. $(\alpha A, B) = (\alpha a_{11})b_{11} + \cdots + (\alpha a_{nn})b_{nn} = \alpha[a_{11}b_{11} + \cdots + a_{nn}b_{nn}] = \alpha(A, B)$

vii. $(A, \alpha B) = (\overline{\alpha B, A}) = (\alpha B, A) = \alpha(B, A) = \alpha(A, B) = \alpha(A, B)$

3. Let E_i be the $n \times n$ matrix with a 1 in the i, i position and 0 everywhere else. It is easy to see that $\{E_1, E_2, \ldots, E_n\}$ is an orthonormal basis for D_n.

5. $\left\{ \left(\dfrac{1}{\sqrt{2}}, \dfrac{i}{\sqrt{2}} \right), \left(\dfrac{i}{\sqrt{2}}, \dfrac{1}{\sqrt{2}} \right) \right\}$

7. $\left\{ \dfrac{1}{\sqrt{2}}, \sqrt{\dfrac{3}{2}}x, \sqrt{\dfrac{5}{8}}(3x^2 - 1) \right\}$

9. First note that if $A = (a_{ij})$ and $B^t = (b_{ji})$, then

$$(AB^t)_{ij} = \sum_{k=1}^{n} a_{ik}b_{jk}$$

so that

$$\text{tr}(AB^t) = \sum_{i=1}^{n}\sum_{j=1}^{n} a_{ij}b_{ij}$$

i. $(A, A) = \text{tr}(AA^t)$

$$= \sum_{i=1}^{n}\left(\sum_{j=1}^{n} a_{ij}^2 \right) \geq 0$$

ii. $(A, A) = 0$ implies that $a_{ij}^2 = 0$ for every i and j so that $A = 0$. If $A = 0$, then $A^t = 0$ and $AA^t = 0$ so that $\text{tr}(AA^t) = 0$.

iii. $(A, B + C) = \text{tr}[A(B + C)^t] + \text{tr}[A(B^t + C^t)] = \text{tr}(AB^t +$

$AC^t) = \text{tr}(AB^t) + \text{tr}(AC^t) = (A, B) + (A, C)$

iv. Similarly, $(A + B, C) = (A, C) + (B, C)$

v. $(A, B) = \sum_{i=1}^{n}\sum_{j=1}^{n} a_{ij}b_{ij}$

$= \text{tr}(BA^t) = (B, A)$

vi. $(\alpha A, B) = \text{tr}(\alpha AB^t) = \alpha \text{tr}(AB^t) = \alpha(A, B)$

vii. $(A, \alpha B) = (\alpha B, A) = \alpha(B, A) = \alpha(A, B)$

11. $\begin{pmatrix} 1 & 0 \\ 0 & 0 \end{pmatrix}, \begin{pmatrix} 0 & 1 \\ 0 & 0 \end{pmatrix}, \begin{pmatrix} 0 & 0 \\ 1 & 0 \end{pmatrix}, \begin{pmatrix} 0 & 0 \\ 0 & 1 \end{pmatrix}$

13. a. i. $(p, p) = p(a)^2 + p(b)^2 + p(c)^2 \geq 0$

ii. $(p, p) = 0$ implies that $p(a) = p(b) = p(c) = 0$. But a quadratic can have at most two roots. Thus $p(x) = 0$ for all x. Conversely, if $p \equiv 0$, then $p(a) = p(b) = p(c) = 0$, so $(p, p) = 0$.

iii. $(p, q + r)$

$= p(a)(q(a) + r(a))$
$\quad + p(b)(q(b) + r(b))$
$\quad + p(c)(q(c) + r(c))$
$= [p(a)q(a) + p(b)q(b)$
$\quad + p(c)q(c)]$
$\quad + [p(a)r(a) + p(b)r(b)$
$\quad + p(c)r(c)]$
$= (p, q) + (p, r)$

iv. Similarly, $(p + q, r) = (p, r) + (q, r)$

v. $(p, q) = p(a)q(a) + (p(b)q(b)$
$\quad + p(c)q(c)$
$= q(a)p(a) + q(b)p(b)$
$\quad + q(c)p(c)$
$= (q, p)$

vi. $(\alpha p, q) = [\alpha p(a)]q(a)$
$\quad + [\alpha p(b)]q(b)$
$\quad + [\alpha p(c)]q(c)$

$$= \alpha[p(a)q(a)$$
$$+ p(b)q(b)$$
$$+ p(c)q(c)]$$
$$= \alpha(p, q)$$

vii. $(p, \alpha q) = (\alpha p, q) = \alpha(p, q) = \alpha(p, q)$

b. No, since (**ii**) is violated. For example, let $a = 1$, $b = -1$, and $p(x) = (x - 1)(x + 1) = x^2 - 1 \neq 0$. Then $p(a) = p(b) = 0$ so that $(p, p) = 0$ even though $p \neq 0$. In fact, for any polynomial q, we have $(p, q) = 0$.

15. $\sqrt{31}$

17. $0 \leq \left(\left(\dfrac{\mathbf{u}}{|\mathbf{u}|} - \dfrac{\mathbf{v}}{|\mathbf{v}|}\right), \left(\dfrac{\mathbf{u}}{|\mathbf{u}|} - \dfrac{\mathbf{v}}{|\mathbf{v}|}\right)\right)$

$$= \dfrac{(\mathbf{u}, \mathbf{u})}{|\mathbf{u}|^2} - \dfrac{(\mathbf{u}, \mathbf{v})}{|\mathbf{u}| \, |\mathbf{v}|} - \dfrac{(\mathbf{v}, \mathbf{u})}{|\mathbf{u}| \, |\mathbf{v}|}$$
$$+ \dfrac{(\mathbf{v}, \mathbf{v})}{|\mathbf{v}|^2}$$

$$= \dfrac{|\mathbf{u}|^2}{|\mathbf{u}|^2} - \left[\dfrac{(\mathbf{u}, \mathbf{v}) + (\overline{\mathbf{u}, \mathbf{v}})}{|\mathbf{u}| \, |\mathbf{v}|}\right]$$
$$+ \dfrac{|\mathbf{v}|^2}{|\mathbf{v}|^2}$$

Now if $z = a + bi$, then $z + \bar{z} = (a + bi) + (a - bi) = 2a = 2 \operatorname{Re} z$ (and $z - \bar{z} = 2bi = 2i Imz$). Thus $(\mathbf{u}, \mathbf{v}) + (\overline{\mathbf{u}, \mathbf{v}}) = 2 \operatorname{Re}(\mathbf{u}, \mathbf{v})$, and we have $2 - \dfrac{2 \operatorname{Re}(\mathbf{u}, \mathbf{v})}{|\mathbf{u}| \, |\mathbf{v}|} \geq 0$ or

$\dfrac{\operatorname{Re}(\mathbf{u}, \mathbf{v})}{|\mathbf{u}| \, |\mathbf{v}|} \leq 1$. Let λ be a real number. Then $0 \leq ((\lambda\mathbf{u} + (\mathbf{u}, \mathbf{v})\mathbf{v}), (\lambda\mathbf{u} + (\mathbf{u}, \mathbf{v})\mathbf{v})) = \lambda^2|\mathbf{u}|^2 + |(\mathbf{u}, \mathbf{v})|^2|\mathbf{v}|^2 + \lambda(\overline{\mathbf{u}, \mathbf{v}})(\mathbf{u}, \mathbf{v}) + \lambda(\mathbf{u}, \mathbf{v})(\mathbf{v}, \mathbf{u}) = $ (since λ is real) $\lambda^2|\mathbf{u}|^2 + 2\lambda|(\mathbf{u}, \mathbf{v})|^2 + |(\mathbf{u}, \mathbf{v})|^2|\mathbf{v}|^2$. The last line is a quadratic equation in λ. If we have $a\lambda^2 + b\lambda + c \geq 0$, then the equation $a\lambda^2 + b\lambda + c = 0$ can have at most one real root, and therefore $b^2 - 4ac \leq 0$. Thus

$$4((|\mathbf{u}, \mathbf{v})|^2)^2 - 4|\mathbf{u}|^2|(\mathbf{u}, \mathbf{v})|^2|\mathbf{v}|^2 \leq 0$$

or $|(\mathbf{u}, \mathbf{v})|^2 \leq |\mathbf{u}|^2|\mathbf{v}|^2$

and $|(\mathbf{u}, \mathbf{v})| \leq |\mathbf{u}| \, |\mathbf{v}|$.

19. $H^{\perp} = \operatorname{span} \{(-15x^2 + 16x - 3), (20x^3 - 30x^2 + 12x - 1)\}$

21. $1 + 2x + 3x^2 - x^3$
$$= \dfrac{30x^2 + 52x + 19}{20}$$
$$+ \dfrac{(-20x^3 + 30x^2 - 12x + 1)}{20}$$

23. $\dfrac{2}{\pi} + \sqrt{3}\left(\dfrac{2}{\pi} - \dfrac{8}{\pi^2}\right)\sqrt{3}(2x - 1)$
$$+ \sqrt{5}\left(\dfrac{2}{\pi} + \dfrac{24}{\pi^2} - \dfrac{96}{\pi^3}\right)$$
$$\times \sqrt{5}(6x^2 - 6x + 1)$$
$$\approx -0.8346x^2 - 0.2091x$$
$$+ 1.0194$$

25. $A^* = \begin{pmatrix} \dfrac{1}{\sqrt{2}} & \dfrac{1}{\sqrt{2}} \\ -\dfrac{1}{2} - \dfrac{i}{2} & \dfrac{1}{2} + \dfrac{i}{2} \end{pmatrix}$

Verify that $A^*A = I$.

27. We check the seven conditions on page 307.

i. $(f, f) = \displaystyle\int_a^b f\bar{f} = \int_a^b f_1^2 + f_2^2 \geq 0$ since $f_1^2 \geq 0$ and $f_2^2 \geq 0$

ii. follows from (i)

iii. $(f, g + h) = \displaystyle\int_a^b f(\overline{g + h})$
$$= \int_a^b f\bar{g} + f\bar{h} = (f, g) + (f, h)$$

iv. Similar to (iii)

v. $(f, g) = \displaystyle\int_a^b f\bar{g} = \int_a^b \bar{g}f$
$$= \overline{(g, f)} = \overline{\int_a^b g\bar{f}} = \int_a^b \overline{g\bar{f}} = \int_a^b \bar{g}f$$

vi. $(\alpha f, g) = \displaystyle\int_a^b \alpha f\bar{g} = \alpha \int_a^b f\bar{g}$

vii. $(f, \alpha g) = \int_a^b f(\bar{\alpha}\bar{g}) = \int_a^b f\bar{\alpha}\bar{g}$

$= \bar{\alpha} \int_a^b f\bar{g} = \bar{\alpha}(f, g)$

29. $\sqrt{\pi}$

Problems 4.13, page 322

1. Let L be a linearly independent set in V. Let S be the collection of all linearly independent subsets of V, partially ordered by inclusion such that every set in S contains L. The proof then follows as in the proof of Theorem 2.

3. The result is true for $n = 2$. Assume it is true for $n = k$. Consider the $k + 1$ sets $A_1, A_2, \ldots, A_k, A_{k+1}$ in a chain. The first k sets form a chain, and, by the induction assumption, one of them contains the other $k - 1$ sets. Call this set A_i. Then either $A_i \subseteq A_{k+1}$ or $A_{k+1} \subseteq A_i$. In either case we have found a set that contains the other k sets and the result is true for $n = k + 1$. This completes the induction proof.

Chapter 4—Review, page 327

1. yes; dimension 2; basis $\{(1, 0, 1), (0, 1, 2)\}$

3. yes; dimension 3; basis $\{(1, 0, 0, -1), (0, 1, 0, -1), (0, 0, 1, -1)\}$

5. yes; dimension $[n(n + 1)]/2$; basis $\{(E_{ij}: j \geq i\}$, where E_{ij} is the matrix with 1 in the i, j position and 0 everywhere else

7. no; for example, $(x^5 - 2x) + (-x^5 + x^2) = x^2 - 2x$, which is not a polynomial of degree 5, so the set is not closed under addition.

9. no; for example, $\begin{pmatrix} 1 & 1 \\ 0 & 2 \\ 3 & 1 \end{pmatrix} +$

$\begin{pmatrix} 2 & 1 \\ -1 & 2 \\ 1 & 0 \end{pmatrix} = \begin{pmatrix} 3 & 2 \\ -1 & 4 \\ 4 & 1 \end{pmatrix}$, which does not satisfy $a_{12} = 1$.

11. independent

13. dependent

15. independent

17. independent

19. independent

21. dimension 2; basis $\{(2, 0, 1), (0, 4, 3)\}$

23. dimension 3; basis $\{(1, 0, 3, 0), (0, 1, -1, 0), (0, 0, 1, 1)\}$

25. dimension 4; basis $\{D_1, D_2, D_3, D_4\}$, where D_i is the matrix with a 1 in the i, i position and 0 everywhere else

27. range $A = \text{span} \left\{ \begin{pmatrix} 1 \\ -2 \end{pmatrix} \right\}$; ker $A =$ span $\left\{ \begin{pmatrix} 2 \\ 1 \end{pmatrix} \right\}$; $\rho(A) = \nu(A) = 1$

29. range $A = \mathbb{R}^3$; ker $A = \{0\}$; $\rho(A) = 3$, $\nu(A) = 0$

31. range $A = \text{span} \left\{ \begin{pmatrix} 2 \\ -1 \\ 4 \end{pmatrix}, \begin{pmatrix} 3 \\ 2 \\ 6 \end{pmatrix} \right\}$; ker $A = \{0\}$; $\rho(T) = 2$, $\nu(T) = 0$

33. $\frac{3}{4} \begin{pmatrix} 1 \\ 2 \end{pmatrix} - \frac{5}{4} \begin{pmatrix} -1 \\ 2 \end{pmatrix} = \begin{pmatrix} 2 \\ -1 \end{pmatrix}$

35. $1(1 + x^2) + 0(1 + x) + 3(1) = 4 + x^2$

37. 2

39. $\left\{ \frac{1}{\sqrt{13}} \begin{pmatrix} 2 \\ 3 \end{pmatrix}, \frac{1}{\sqrt{13}} \begin{pmatrix} -3 \\ 2 \end{pmatrix} \right\}$

41. $\begin{pmatrix} 1/\sqrt{3} \\ 1/\sqrt{3} \\ 1/\sqrt{3} \end{pmatrix}$

43. a. $\begin{pmatrix} \frac{4}{3} \\ -\frac{1}{3} \\ \frac{5}{3} \end{pmatrix}$ **b.** $\begin{pmatrix} -1/\sqrt{3} \\ 1/\sqrt{3} \\ 1/\sqrt{3} \end{pmatrix}$

c. $\begin{pmatrix} \frac{4}{3} \\ -\frac{1}{3} \\ \frac{5}{3} \end{pmatrix} + \begin{pmatrix} -\frac{7}{3} \\ \frac{7}{3} \\ \frac{7}{3} \end{pmatrix}$

45. a. $\begin{pmatrix} \frac{1}{2} \\ \frac{1}{2} \\ \frac{1}{2} \\ \frac{1}{2} \end{pmatrix}$

b. $\left\{ \begin{pmatrix} 1/\sqrt{2} \\ 0 \\ -1/\sqrt{2} \\ 0 \end{pmatrix}, \begin{pmatrix} 0 \\ 1/\sqrt{2} \\ 0 \\ -1/\sqrt{2} \end{pmatrix} \right\}$

c. $\begin{pmatrix} \frac{1}{2} \\ \frac{1}{2} \\ \frac{1}{2} \\ \frac{1}{2} \end{pmatrix} + \begin{pmatrix} \frac{1}{2} \\ -\frac{1}{2} \\ -\frac{1}{2} \\ \frac{1}{2} \end{pmatrix}$

47. $\dfrac{e^2 - 1}{2} + 3(x - 1) + (\frac{15}{4}e^2 - \frac{105}{4})$
$\times (x^2 - 2x + \frac{2}{3}) \approx 1.167 + 0.0821x + 1.459x^2$

49. $y = \frac{7}{6}x^2 + \frac{3}{2}x - \frac{8}{3}$

Chapter 5

Problems 5.1, page 336

1. linear   **3.** linear

5. not linear, since

$$T\left(\alpha \begin{pmatrix} x \\ y \\ z \end{pmatrix} \right) = T \begin{pmatrix} \alpha x \\ \alpha y \\ \alpha z \end{pmatrix} = \begin{pmatrix} 1 \\ \alpha z \end{pmatrix}$$

while

$$\alpha T \begin{pmatrix} x \\ y \\ z \end{pmatrix} = \alpha \begin{pmatrix} 1 \\ z \end{pmatrix} = \begin{pmatrix} \alpha \\ \alpha z \end{pmatrix}$$

7. linear

9. not linear, since

$$T\left(\alpha \begin{pmatrix} x \\ y \end{pmatrix} \right) = T \begin{pmatrix} \alpha x \\ \alpha y \end{pmatrix}$$
$$= (\alpha x)(\alpha y)$$
$$= \alpha^2 xy$$

while $\alpha T \begin{pmatrix} x \\ y \end{pmatrix} = \alpha xy$

11. linear

13. not linear, since

$$T\left(\alpha \begin{pmatrix} x \\ y \\ z \\ w \end{pmatrix} \right) = \alpha^2 T \begin{pmatrix} x \\ y \\ z \\ w \end{pmatrix}$$

$$\neq \alpha T \begin{pmatrix} x \\ y \\ z \\ w \end{pmatrix}$$

if $\alpha \neq 1$ or 0

15. not linear, since

$$T(A + B) = (A + B)^t(A + B)$$
$$= (A^t + B^t)(A + B)$$
$$= A^tA + A^tB$$
$$\qquad + B^tA + B^tB$$

But

$$T(A) + T(B) = A^tA + B^tB$$
$$\neq T(A + B)$$

unless $A^tB + B^tA = 0$.

17. not linear, since $T(\alpha D) = (\alpha D)^2 = \alpha^2 D^2 \neq \alpha T(D) = \alpha D^2$ unless $\alpha = 1$ or 0.

19. linear   **21.** linear

23. not linear, since $T(f + g) = (f + g)^2 \neq f^2 + g^2 = T(f) + T(g)$

25. linear   **27.** linear

29. not linear, since

$$T(\alpha A) = \det (\alpha A)$$
$$= \alpha^n \det A$$
$$\neq \alpha \det A$$
$$= \alpha T(A)$$

unless $\alpha = 0$ or 1. [det $\alpha A = \alpha^n$ det A by Problem 2.2.28.] Also, in general

$$\det (A + B) \neq \det A + \det B$$

31. a. $\begin{pmatrix} -14 \\ 4 \\ 26 \end{pmatrix}$ **b.** $\begin{pmatrix} -31 \\ -6 \\ 26 \end{pmatrix}$

33. It rotates a vector counterclockwise around the z-axis through an angle of θ in a plane parallel to the xy-plane.

35. Suppose $\alpha < 0$. Then $T[(\alpha - \alpha)\mathbf{x}] = T(0\mathbf{x}) = 0T\mathbf{x} = \mathbf{0}$. Thus $T[(\alpha - \alpha)\mathbf{x}] = T(0\mathbf{x}) = 0T\mathbf{x} = \mathbf{0}$ and $T(\alpha\mathbf{x}) + T(-\alpha\mathbf{x}) = T((\alpha - \alpha)\mathbf{x}) = \mathbf{0}$. But $-\alpha > 0$ so that $T(-\alpha\mathbf{x}) = -\alpha T\mathbf{x}$. Therefore $T(\alpha\mathbf{x}) - \alpha T\mathbf{x} = \mathbf{0}$, or $T(\alpha\mathbf{x}) = \alpha T\mathbf{x}$ for $\alpha < 0$ as well.

37. $T(\mathbf{x} - \mathbf{y}) = T\mathbf{x} + T(-\mathbf{y}) = T\mathbf{x} + T[(-1)\mathbf{y}] = T\mathbf{x} + (-1)T\mathbf{y} = T\mathbf{x} - T\mathbf{y}$.

41. $T(\mathbf{v}_1 + \mathbf{v}_2) = (\mathbf{v}_1 + \mathbf{v}_2, \mathbf{u}_1)\mathbf{u}_1$
$\qquad\qquad + (\mathbf{v}_1 + \mathbf{v}_2, \mathbf{u}_2)\mathbf{u}_2$
$\qquad + \cdots + (\mathbf{v}_1 + \mathbf{v}_2, \mathbf{u}_n)\mathbf{u}_n$
$= (\mathbf{v}_1, \mathbf{u}_1)\mathbf{u}_1 + (\mathbf{v}_2, \mathbf{u}_1)\mathbf{u}_1 + (\mathbf{v}_1, \mathbf{u}_2)\mathbf{u}_2$
$+ (\mathbf{v}_2, \mathbf{u}_2)\mathbf{u}_2 + \cdots + (\mathbf{v}_1, \mathbf{u}_n)\mathbf{u}_n + (\mathbf{v}_2, \mathbf{u}_n)\mathbf{u}_n$
$= (\mathbf{v}_1, \mathbf{u}_1)\mathbf{u}_1 + (\mathbf{v}_1, \mathbf{u}_2)\mathbf{u}_2 + \cdots + (\mathbf{v}_1, \mathbf{u}_n)\mathbf{u}_n$
$+ (\mathbf{v}_2, \mathbf{u}_1)\mathbf{u}_1 + (\mathbf{v}_2, \mathbf{u}_2)\mathbf{u}_2 + \cdots + (\mathbf{v}_2, \mathbf{u}_n)\mathbf{u}_n$
$= T\mathbf{v}_1 + T\mathbf{v}_2$

$T(\alpha\mathbf{v}) = (\alpha\mathbf{v}, \mathbf{u}_1)\mathbf{u}_1 + (\alpha\mathbf{v}, \mathbf{u}_2)\mathbf{u}_2 + \cdots + (\alpha\mathbf{v}, \mathbf{u}_n)\mathbf{u}_n$
$= \alpha[(\mathbf{v}, \mathbf{u}_1)\mathbf{u}_1 + (\mathbf{v}, \mathbf{u}_2)\mathbf{u}_2 + \cdots + (\mathbf{v}, \mathbf{u}_n)\mathbf{u}_n] = \alpha T\mathbf{v}$

Problems 5.2, page 344

1. kernel $= \{(0, y) \colon y \in \mathbb{R}\}$, i.e., the y-axis; range $= \{(x, 0) \colon x \in \mathbb{R}\}$, i.e., the x-axis; $\rho(T) = \nu(T) = 1$

3. kernel $= \{(x, -x) \colon x \in \mathbb{R}\}$—this is the line $x + y = 0$; range $= \mathbb{R}$; $\rho(T) = \nu(T) = 1$.

5. kernel $= \left\{\begin{pmatrix} 0 & 0 \\ 0 & 0 \end{pmatrix}\right\}$; range $= M_{22}$; $\rho(T) = 4$, $\nu(T) = 0$

7. kernel $= \{A \colon A^t = -A\} = \{A \colon A$ is skew-symmetric$\}$; range $= \{A \colon A$ is symmetric$\}$; $\rho(T) = (n^2 + n)/2$; $\nu(T) = (n^2 - n)/2$

9. kernel $= \{f \in C[0, 1] \colon f(\tfrac{1}{2}) = 0\}$; range $= \mathbb{R}$; $\rho(T) = 1$; the kernel is an infinite dimensional space so that $\nu(T) = \infty$. For example, the linearly independent functions $x - \tfrac{1}{2}$,

$(x - \tfrac{1}{2})^2$, $(x - \tfrac{1}{2})^3$, $(x - \tfrac{1}{2})^4$, $\ldots$, $(x - \tfrac{1}{2})^n$, $\ldots$ all satisfy $f(\tfrac{1}{2}) = 0$.

11. If $\mathbf{v} \in V$, then $\mathbf{v} = c_1\mathbf{v}_1 + c_2\mathbf{v}_2 + \cdots + c_n\mathbf{v}_n$ so that $T\mathbf{v} = T(c_1\mathbf{v}_1 + c_2\mathbf{v}_2 + \cdots + c_n\mathbf{v}_n) = c_1T\mathbf{v}_1 + c_2T\mathbf{v}_2 + \cdots + c_nT\mathbf{v}_n) = c_1\mathbf{0} + c_2\mathbf{0} + \cdots + c_n\mathbf{0} = \mathbf{0}$. Thus $T\mathbf{v} = \mathbf{0}$ for every $\mathbf{v} \in V$, and is therefore the zero transformation.

13. The range of T is a subspace of $\mathbb{R}^3$ and, by Example 4.6.9, the subspaces of $\mathbb{R}^3$ are $\{\mathbf{0}\}$, $\mathbb{R}^3$, and lines and planes passing through the origin.

15. $T\mathbf{x} = A\mathbf{x}$, where $A = \begin{pmatrix} 0 & a \\ b & c \end{pmatrix}$, a, b, c real

17. $T\mathbf{x} = A\mathbf{x}$, where $A = \begin{pmatrix} 2 & -1 & 1 \\ 2 & -1 & 1 \\ 2 & -1 & 1 \end{pmatrix}$

19. i. If $A \in \ker T$, then $A - A^t = 0$, or $A = A^t$.

  ii. If $A \in$ range of T, then there is a matrix B such that $B - B^t = A$. Then $A^t = (B - B^t)^t = B^t - (B^t)^t = B^t - B = -A$ so that A is skew-symmetric.

21. Let $T_{ij}(\mathbf{u}_i) = \mathbf{w}_j$ and $T_{ij}(\mathbf{u}_k) = \mathbf{0}$ if $k \neq i$. These form a basis for $L(V, W)$, so dim $L(V, W) = $ nm.

23. False. Let S and $T \colon \mathbb{R}^2 \to \mathbb{R}^2$ be given by $S(\mathbf{x}) = A\mathbf{x}$ and $T(\mathbf{x}) = B\mathbf{x}$, where $A = \begin{pmatrix} 0 & 1 \\ 0 & 0 \end{pmatrix}$ and $B = \begin{pmatrix} 1 & 0 \\ 0 & 0 \end{pmatrix}$. Then $ST(\mathbf{x}) = AB\mathbf{x} = \begin{pmatrix} 0 & 0 \\ 0 & 0 \end{pmatrix}\mathbf{x} = \mathbf{0}$. However, $TS(\mathbf{x})$ is not the zero transformation because $BA = \begin{pmatrix} 0 & 1 \\ 0 & 0 \end{pmatrix} \neq$ the zero matrix.

Problems 5.3, page 363

1. $\begin{pmatrix} 1 & -2 \\ -1 & 1 \end{pmatrix}$; $\ker T = \{\mathbf{0}\}$; range $T = \mathbb{R}^2$; $\nu(T) = 0$, $\rho(T) = 2$

3. $\begin{pmatrix} 1 & -1 & 1 \\ -2 & 2 & -2 \end{pmatrix}$; range $T =$

span $\left\{ \begin{pmatrix} 1 \\ -2 \end{pmatrix} \right\}$; ker $T =$

span $\left\{ \begin{pmatrix} 1 \\ 1 \\ 0 \end{pmatrix}, \begin{pmatrix} 0 \\ 1 \\ 1 \end{pmatrix} \right\}$; $\rho(T) = 1$,

$\nu(T) = 2$

5. $\begin{pmatrix} 1 & -1 & 2 \\ 3 & 1 & 4 \\ 5 & -1 & 8 \end{pmatrix}$; range $T =$

span $\left\{ \begin{pmatrix} 1 \\ 3 \\ 5 \end{pmatrix}, \begin{pmatrix} -1 \\ 1 \\ -1 \end{pmatrix} \right\}$; ker $T =$

span $\left\{ \begin{pmatrix} -3 \\ 1 \\ 2 \end{pmatrix} \right\}$; $\rho(T) = 2$,

$\nu(T) = 1$

7. $\begin{pmatrix} 1 & -1 & 2 & 3 \\ 0 & 1 & 4 & 3 \\ 1 & 0 & 6 & 6 \end{pmatrix}$; range $T =$

span $\left\{ \begin{pmatrix} 1 \\ 0 \\ 1 \end{pmatrix}, \begin{pmatrix} -1 \\ 1 \\ 0 \end{pmatrix} \right\}$; ker $T =$

span $\left\{ \begin{pmatrix} -6 \\ -4 \\ 1 \\ 0 \end{pmatrix}, \begin{pmatrix} -6 \\ -3 \\ 0 \\ 1 \end{pmatrix} \right\}$;

$\rho(T) = 2$, $\nu(T) = 2$

9. $\begin{pmatrix} \frac{5}{4} & -\frac{13}{4} \\ \frac{5}{4} & \frac{3}{4} \end{pmatrix}$; range $T = \mathbb{R}^2$; ker $T =$

$\{\mathbf{0}\}$; $\rho(T) = 2$, $\nu(T) = 0$

11. $\begin{pmatrix} 3 & \frac{7}{5} & \frac{16}{5} \\ 0 & \frac{4}{5} & \frac{2}{5} \end{pmatrix}$; range $T = \mathbb{R}^2$;

$(\ker T)_{B_1} = $ span $\left\{ \begin{pmatrix} 5 \\ 3 \\ -6 \end{pmatrix} \right\}$; $\rho(T) =$

2, $\nu(T) = 1$

13. $\begin{pmatrix} 0 & 1 & 0 \\ 0 & -1 & 0 \\ 0 & 0 & 0 \\ 1 & 0 & 0 \end{pmatrix}$; range $T =$

span $\{1 - x, x^3\}$; ker $T =$ span $\{x^2\}$;
$\rho(T) = 2$, $\nu(T) = 1$

15. $(0, 0, 1, 0)$; range $T = \mathbb{R}$; ker $T =$
span $\{1, x, x^3\}$; $\rho(T) = 1$, $\nu(T) = 3$

17. $\begin{pmatrix} 1 & -1 & 2 & 3 \\ 0 & 1 & 4 & 3 \\ 1 & 0 & 6 & 5 \end{pmatrix}$; range $T =$

span $\{1 + x^2, -1 + x, 3 + 3x + 5x^2\} = P_2$; ker $T =$ span $\{x^2 - 4x - 6\}$; $\rho(T) = 3$, $\nu(T) = 1$

19. $\begin{pmatrix} 1 & 1 & 1 & 1 \\ 1 & 1 & 1 & 0 \\ 1 & 1 & 0 & 0 \\ 1 & 0 & 0 & 0 \end{pmatrix}$; range $T = M_{22}$;

ker $T = \left\{ \begin{pmatrix} 0 & 0 \\ 0 & 0 \end{pmatrix} \right\}$; $\rho(T) = 4$,

$\nu(T) = 0$

21. $\begin{pmatrix} 0 & 1 & 0 & 0 & 0 \\ 0 & 0 & 2 & 0 & 0 \\ 0 & 0 & 0 & 3 & 0 \\ 0 & 0 & 0 & 0 & 4 \end{pmatrix}$; range $D = P_3$;

ker $D = \mathbb{R}$; $\rho(D) = 4$, $\nu(D) = 1$

23. $\begin{pmatrix} 0 & 1 & 0 & 0 & \cdots & 0 \\ 0 & 0 & 2 & 0 & \cdots & 0 \\ 0 & 0 & 0 & 3 & \cdots & 0 \\ \vdots & \vdots & \vdots & \vdots & & \vdots \\ 0 & 0 & 0 & 0 & \cdots & n \end{pmatrix}$;

range $D = P_{n-1}$; ker $D = \mathbb{R}$;
$\rho(D) = n$, $\nu(D) = 1$

25. $\begin{pmatrix} 2 & 0 & 2 & 0 & 0 \\ 0 & 3 & 0 & 6 & 0 \\ 0 & 0 & 4 & 0 & 12 \\ 0 & 0 & 0 & 5 & 0 \\ 0 & 0 & 0 & 0 & 6 \end{pmatrix}$;

range $T = P_4$; ker $T = \{\mathbf{0}\}$; $\rho(T) = 5$, $\nu(T) = 0$

27. $A_T = $ diag $(b_0, b_1, b_2, \ldots, b_n)$,

where $b_j = \sum_{i=1}^{j+1} \dfrac{j!}{(j + 1 - i)!}$; range

$T = P_n$; ker $T = \{\mathbf{0}\}$; $\rho(T) = n + 1$, $\nu(T) = 0$

29. $\begin{pmatrix} 1 & 0 & 0 \\ 0 & 1 & 0 \\ 0 & 0 & 1 \end{pmatrix}$; range $T = P_2$; ker $T = \{0\}$, $\rho(T) = 3$, $\nu(T) = 0$

31. For example, in M_{34},

$$A_T = \begin{pmatrix} 1 & 0 & 0 & 0 & 0 & 0 & 0 & 0 & 0 & 0 & 0 & 0 \\ 0 & 0 & 0 & 0 & 1 & 0 & 0 & 0 & 0 & 0 & 0 & 0 \\ 0 & 0 & 0 & 0 & 0 & 0 & 0 & 0 & 1 & 0 & 0 & 0 \\ 0 & 1 & 0 & 0 & 0 & 0 & 0 & 0 & 0 & 0 & 0 & 0 \\ 0 & 0 & 0 & 0 & 0 & 1 & 0 & 0 & 0 & 0 & 0 & 0 \\ 0 & 0 & 0 & 0 & 0 & 0 & 0 & 0 & 0 & 1 & 0 & 0 \\ 0 & 0 & 1 & 0 & 0 & 0 & 0 & 0 & 0 & 0 & 0 & 0 \\ 0 & 0 & 0 & 0 & 0 & 0 & 1 & 0 & 0 & 0 & 0 & 0 \\ 0 & 0 & 0 & 0 & 0 & 0 & 0 & 0 & 0 & 0 & 1 & 0 \\ 0 & 0 & 0 & 1 & 0 & 0 & 0 & 0 & 0 & 0 & 0 & 0 \\ 0 & 0 & 0 & 0 & 0 & 0 & 0 & 1 & 0 & 0 & 0 & 0 \\ 0 & 0 & 0 & 0 & 0 & 0 & 0 & 0 & 0 & 0 & 0 & 1 \end{pmatrix}$$

In general, $A_T = (a_{ij})$ where

$$a_{ij} = \begin{cases} 1, & \text{if } i = km + l, \\ & \text{and } j = (l - 1)n + k + 1 \\ & \text{for } k = 1, 2, \ldots, n - 1 \\ & \text{and } l = 1, 2, \ldots, m \\ 0, & \text{otherwise} \end{cases}$$

33. $\begin{pmatrix} 0 & 0 & 0 \\ 0 & 0 & -1 \\ 0 & 1 & 0 \end{pmatrix}$; range $D =$ span $\{\sin x, \cos x\}$; ker $D = \mathbb{R}$; $\rho(D) = 2$, $\nu(D) = 1$

35. $\begin{pmatrix} \frac{1}{2} & -i/2 \\ i/2 & \frac{1}{2} \end{pmatrix}$

37. Let B_1 and B_2 be bases for V and W, respectively. We have $(Tv)_{B_2} = A_T(\mathbf{v})_{B_1}$ for every $\mathbf{v} \in V$. Then $\mathbf{v} \in$ ker T if and only if $T\mathbf{v} = \mathbf{0}$ if and only if $A_T(\mathbf{v})_{B_1} = (\mathbf{0})_{B_2}$ if and only if $(\mathbf{v})_{B_1} \in$ ker A_T. Thus nullity of $T = N_{A_T}$ so that $\nu(T) = \nu(A_T)$. If $\mathbf{w} \in$ range T, then $T\mathbf{v} = \mathbf{w}$ for some $\mathbf{v} \in V$ so that $A_T(\mathbf{v})_{B_1} = (T\mathbf{v})_{B_2} = (\mathbf{w})_{B_2}$. This means that $(\mathbf{w})_{B_2} \in R_{A_T}$. Thus $R_{A_T} =$ range T so that $\rho(T) = \rho(A_T)$. Since $\nu(A_T) + \rho(A_T) = n$

from Theorem 4.7.5, we see that $\nu(T) + \rho(T) = n$ also.

39. compression along the y-axis with $c = \frac{1}{4}$

41. shear along the x-axis with $c = 2$

43. shear along the y-axis with $c = \frac{1}{2}$

45. reflection about the line $y = x$

47. $\begin{pmatrix} \frac{1}{4} & 0 \\ 0 & 1 \end{pmatrix}$;

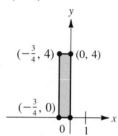

49. $\begin{pmatrix} 1 & 0 \\ 3 & 1 \end{pmatrix}$;

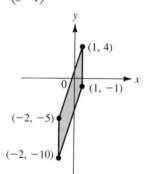

51. $\begin{pmatrix} 1 & \frac{1}{5} \\ 0 & 1 \end{pmatrix}$;

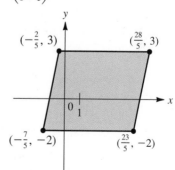

53. $\begin{pmatrix} -1 & 0 \\ 0 & 1 \end{pmatrix}$;

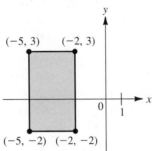

55. $\begin{pmatrix} 0 & 1 \\ 1 & 0 \end{pmatrix}$;

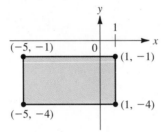

57. $\begin{pmatrix} 3 & 0 \\ 0 & 1 \end{pmatrix}\begin{pmatrix} 1 & 0 \\ -1 & 1 \end{pmatrix}\begin{pmatrix} 1 & 0 \\ 0 & \frac{14}{3} \end{pmatrix}\begin{pmatrix} 1 & \frac{2}{3} \\ 0 & 1 \end{pmatrix}$

59. $\begin{pmatrix} 3 & 0 \\ 0 & 1 \end{pmatrix}\begin{pmatrix} 1 & 0 \\ 4 & 1 \end{pmatrix}\begin{pmatrix} 1 & 0 \\ 0 & -1 \end{pmatrix}\begin{pmatrix} 1 & 0 \\ 0 & 6 \end{pmatrix}\begin{pmatrix} 1 & 2 \\ 0 & 1 \end{pmatrix}$

61. $\begin{pmatrix} 0 & 1 \\ 1 & 0 \end{pmatrix}\begin{pmatrix} 5 & 0 \\ 0 & 1 \end{pmatrix}\begin{pmatrix} 1 & 0 \\ 0 & -1 \end{pmatrix}\begin{pmatrix} 1 & 0 \\ 0 & 2 \end{pmatrix}\begin{pmatrix} 1 & \frac{7}{5} \\ 0 & 1 \end{pmatrix}$

63. $\begin{pmatrix} -1 & 0 \\ 0 & 1 \end{pmatrix}\begin{pmatrix} 1 & 0 \\ 6 & 1 \end{pmatrix}\begin{pmatrix} 1 & 0 \\ 0 & 62 \end{pmatrix}\begin{pmatrix} 1 & -10 \\ 0 & 1 \end{pmatrix}$

Problems 5.4, page 373

1. Since $(\alpha A)^t = \alpha A^t$ and $(A + B)^t = A^t + B^t$, T is linear. $A^t = 0$ if and only if $A = 0$ so ker $T = \{0\}$ and T is $1 - 1$. For any matrix A, $(A^t)^t = A$, so T is onto.

3. i. If T is an isomorphism, then $Tx = A_T x = 0$ if and only if $x = 0$. Thus, by the Summing Up Theorem, det $A_T \neq 0$.

ii. If det $A_T \neq 0$, then $A_T x = 0$ has only the trivial solution. Thus T is $1 - 1$, and since V and W are finite dimensional, T is also onto.

5. $m = [n(n + 1)]/2 = \dim \{A: A$ is $n \times n$ and symmetric$\}$.

7. Define $T: P_4 \to W$ by $Tp = xp$. $Tp = 0$ implies $p(x) = 0$; that is, p is the zero polynomial. Thus T is $1 - 1$, and since $\dim W = 5$, T is also onto.

9. $mn = pq$

11. The proof of Theorem 6 proves the assertion with the understanding that the scalars $c_1, c_2, \ldots, c_n$ are complex numbers.

13. $T(A_1 + A_2) = (A_1 + A_2)B = A_1B + A_2B = TA_1 + TA_2$; $T(\alpha A) = (\alpha A)B = \alpha(AB) = \alpha TA$. Thus T is linear. Suppose $TA = 0$. Then $AB = 0$. Since B is invertible, we can multiply on the left by B^{-1} to obtain $A = ABB^{-1} = 0B^{-1} = 0$ or $A = 0$. Thus T is $1 - 1$, and since $\dim M_{nn} = n^2 < \infty$, T is an isomorphism.

15. Choose $\mathbf{h} \in H$. Then $\text{Proj}_H \mathbf{h} = \mathbf{h}$ so that T is onto. If $H = V$, then T is also $1 - 1$.

17. Since T is an isomorphism, ker $T =$ ker $A = \{\mathbf{0}\}$ so that, by the Summing Up Theorem, A is invertible. If $\mathbf{x} = T^{-1}\mathbf{y}$, then $T\mathbf{x} = A\mathbf{x} = \mathbf{y}$ so that $\mathbf{x} = A^{-1}\mathbf{y}$ since A^{-1} exists. Thus $T^{-1}\mathbf{y} = A^{-1}\mathbf{y}$ for every $\mathbf{y} \in \mathbb{R}^n$.

19. For $z = a + ib \in \mathbb{C}$, define $Tz = (a, b) \in \mathbb{R}^2$. Then $T(z_1 + z_2) = T((a_1 + a_2) + i(b_1 + b_2)) = (a_1 + a_2, b_1 + b_2) = (a_1, b_1) + (a_2, b_2) = Tz_1 + Tz_2$. If $\alpha \in \mathbb{R}$, then $T(\alpha z) = T(\alpha(a + ib)) = T(\alpha a + i\alpha b) = (\alpha a, \alpha b) = \alpha(a, b) = \alpha Tz$. Thus T is linear. Finally, if $T(z) = (0, 0)$, then clearly $z = a + ib = 0 + i0 = 0$.

Thus T is $1-1$ and because dim $\mathbb{C}$ (over the reals) $=$ dim $\mathbb{R}^2 = 2$, T is an isomorphism.

Problems 5.5, page 379

1. $T\mathbf{x} \cdot T\mathbf{y}$

$$= \begin{pmatrix} x_1 \sin\theta + x_2 \cos\theta \\ x_1 \cos\theta - x_2 \sin\theta \\ x_3 \end{pmatrix}$$

$$\cdot \begin{pmatrix} y_1 \sin\theta + y_2 \cos\theta \\ y_1 \cos\theta - y_2 \sin\theta \\ y_3 \end{pmatrix}$$

$$= x_1 y_1 (\sin^2\theta + \cos^2\theta)$$
$$+ x_2 y_2 (\sin^2\theta + \cos^2\theta)$$
$$+ x_3 y_3$$
$$= x_1 y_1 + x_2 y_2 + x_3 y_3$$
$$= \mathbf{x} \cdot \mathbf{y}$$

(all other terms in the scalar product drop out).

3. Using Theorem 1, $T\mathbf{x} \cdot T\mathbf{y} =$
$(AB\mathbf{x}) \cdot (AB\mathbf{y}) = \mathbf{x} \cdot (AB)^t(AB\mathbf{y}) =$
$\mathbf{x} \cdot (B^t A^t)(AB)\mathbf{y} =$
$\mathbf{x} \cdot (B^{-1}A^{-1}AB)\mathbf{y} = \mathbf{x} \cdot \mathbf{y}$

5. Same proof as for Theorem 2 except replace $\mathbf{x} \cdot \mathbf{y}$ with $(\mathbf{x}, \mathbf{y})$ and $T\mathbf{x} \cdot T\mathbf{y}$ with $(T\mathbf{x}, T\mathbf{y})$.

7. $T\mathbf{x} = \alpha\mathbf{x}$ where α is a scalar and $\alpha \neq 0$ or 1.

9. $T\mathbf{x} \cdot T\mathbf{y} = \mathbf{x} \cdot \mathbf{y} = A\mathbf{x} \cdot A\mathbf{y}$ and $A^t = A^{-1}$ so that $A = (A^{-1})^t$. Then
$\mathbf{x} \cdot \mathbf{y} = \mathbf{x} \cdot (I\mathbf{y}) = \mathbf{x} \cdot (A^{-1})^t A^{-1}\mathbf{y} =$
$A^{-1}\mathbf{x} \cdot A^{-1}\mathbf{y} = S\mathbf{x} \cdot S\mathbf{y}$ so that $S\mathbf{x} = A^{-1}\mathbf{y}$ is an isometry.

11. $T(a_0 + a_1 x + a_2 x^2 + a_3 x^3) =$
$(a_0/\sqrt{2} - (\sqrt{5}/2\sqrt{2})a_2,$
$\sqrt{(3/2)}a_1 - (3\sqrt{7}/2\sqrt{2})a_3,$
$(3\sqrt{5}/2\sqrt{2})a_2, (5\sqrt{7}/2\sqrt{2})a_3)$

13. $T\begin{pmatrix} a & b \\ c & d \end{pmatrix} = (a/\sqrt{2} - (\sqrt{5}/2\sqrt{2})c,$
$\sqrt{(3/2)}b - (3\sqrt{7}/2\sqrt{2})d,$
$(3\sqrt{5}/2\sqrt{2})c, (5\sqrt{7}/2\sqrt{2})d)$

15. $A^* = \begin{pmatrix} 1-i & 3 \\ -4-2i & 6+3i \end{pmatrix}$

17. If A is hermitian, then $A^* = A$. In particular, the diagonal components of A do not move when we take the transpose so that $\overline{a_{ii}} = a_{ii}$, which means that a_{ii} is real.

19. Let $A^* = B = (b_{ij})$ and let $\mathbf{c}_i$ be the ith column of A. Then $AB = I = (\delta_{ij})$ where

$$\delta_{ij} = \begin{cases} 1, & \text{if } i = j \\ 0, & \text{if } i \neq j \end{cases}$$

But $\delta_{ij} = \displaystyle\sum_{k=1}^{n} a_{ik} b_{kj} = \sum_{k=1}^{n} a_{ik}\overline{a_{jk}} = $
$\mathbf{c}_i \cdot \mathbf{c}_j = \delta_{ij}$.

21. Since the ith component of $A\mathbf{x}$ is

$$\sum_{j=1}^{n} a_{ij} x_j, \text{ we have } (A\mathbf{x}, \mathbf{y}) =$$

$$\sum_{i=1}^{n}\sum_{j=1}^{n} a_{ij} x_j \overline{y_i}. \text{ Similarly, if } A^* =$$

$B = (b_{ij})$, $(\mathbf{x}, A^*\mathbf{y}) =$

$$\sum_{i=1}^{n}\sum_{j=1}^{n} x_j \overline{b_{ji} y_i} = \sum_{j=1}^{n}\sum_{i=1}^{n} x_j \overline{b_{ji}} y_i =$$

$$\sum_{j=1}^{n}\sum_{i=1}^{n} x_j a_{ij} \overline{y_i} = \sum_{i=1}^{n}\sum_{j=1}^{n} a_{ij} x_j \overline{y_i} =$$

$(A\mathbf{x}, \mathbf{y})$.

Chapter 5—Review, page 383

1. linear

3. not linear, since $T(\alpha(x, y)) =$
$T(\alpha x, \alpha y) = \alpha x / \alpha y = x/y =$
$T(x, y) \neq \alpha T(x, y)$ unless $\alpha = 1$.

5. not linear, since $T(p_1 + p_2) = 1 + p_1 + p_2$, but
$$Tp_1 + Tp_2 = (1 + p_1) + (1 + p_2)$$
$$= 2 + p_1 + p_2.$$

7. $\ker T = \left\{ \begin{pmatrix} 0 \\ 0 \end{pmatrix} \right\}$; range $T = \mathbb{R}^2$;
$\rho(T) = 2$; $\nu(T) = 0$

9. $\ker T = \text{span}\left\{ \begin{pmatrix} 0 \\ 0 \\ 1 \end{pmatrix} \right\}$; range $T = \mathbb{R}^2$; $\rho(T) = 2$; $\nu(T) = 1$

11. $\ker T = \left\{ \begin{pmatrix} 0 & 0 \\ 0 & 0 \end{pmatrix} \right\}$; range $T = M_{22}$;
$\rho(T) = 4$; $\nu(T) = 0$

13. $\begin{pmatrix} 0 & 0 \\ 0 & -1 \end{pmatrix}$; range $T = \text{span} \left\{ \begin{pmatrix} 0 \\ 1 \end{pmatrix} \right\}$;
$\ker T = \text{span} \left\{ \begin{pmatrix} 1 \\ 0 \end{pmatrix} \right\}$; $\rho(T) = \nu(T) = 1$

15. $\begin{pmatrix} 1 & 0 & -2 & 0 \\ 0 & 2 & 0 & 3 \end{pmatrix}$; range $T = \mathbb{R}^2$;
$\ker T = \text{span} \left\{ \begin{pmatrix} 2 \\ 0 \\ 1 \\ 0 \end{pmatrix}, \begin{pmatrix} 0 \\ -3 \\ 0 \\ 2 \end{pmatrix} \right\}$;
$\rho(A) = \nu(A) = 2$

17. $\begin{pmatrix} -1 & 1 & 0 & 0 \\ 0 & 2 & 0 & 0 \\ 0 & 0 & -1 & 1 \\ 0 & 0 & 0 & 2 \end{pmatrix}$; range $T = M_{22}$;
$\ker T = \{\mathbf{0}\}$; $\rho(T) = 4$, $\nu(T) = 0$

19. expansion along the x-axis with $c = 3$

21. shear along the y-axis with $c = -2$

23. $\begin{pmatrix} 3 & 0 \\ 0 & 1 \end{pmatrix}$;

25. $\begin{pmatrix} 0 & 1 \\ 1 & 0 \end{pmatrix}$;

27. $\begin{pmatrix} 1 & 0 \\ -2 & 1 \end{pmatrix} \begin{pmatrix} 1 & 0 \\ 0 & 8 \end{pmatrix} \begin{pmatrix} 1 & 3 \\ 0 & 1 \end{pmatrix}$

29. $\begin{pmatrix} 0 & 1 \\ 1 & 0 \end{pmatrix} \begin{pmatrix} 1 & 0 \\ -6 & 1 \end{pmatrix} \begin{pmatrix} 1 & 0 \\ 0 & 22 \end{pmatrix} \begin{pmatrix} 1 & 3 \\ 0 & 1 \end{pmatrix}$

31. $T(a_0 + a_1 x + a_2 x^2) = \begin{pmatrix} a_0 \\ a_1 \\ a_2 \end{pmatrix}$

Chapter 6

Problems 6.1, page 398

1. $-4, 3$; $E_{-4} = \text{span} \left\{ \begin{pmatrix} 1 \\ 1 \end{pmatrix} \right\}$;
$E_3 = \text{span} \left\{ \begin{pmatrix} 2 \\ -5 \end{pmatrix} \right\}$

3. $i, -i$; $E_i = \text{span} \left\{ \begin{pmatrix} 2 + i \\ 5 \end{pmatrix} \right\}$;
$E_{-i} = \text{span} \left\{ \begin{pmatrix} 2 - i \\ 5 \end{pmatrix} \right\}$

5. $-3, -3$; $E_{-3} = \text{span} \left\{ \begin{pmatrix} 1 \\ 0 \end{pmatrix} \right\}$;
geom. mult. is 1

7. $0, 1, 3$; $E_0 = \text{span} \left\{ \begin{pmatrix} 1 \\ 1 \\ 1 \end{pmatrix} \right\}$;
$E_1 = \text{span} \left\{ \begin{pmatrix} -1 \\ 0 \\ 1 \end{pmatrix} \right\}$;
$E_3 = \text{span} \left\{ \begin{pmatrix} 1 \\ -2 \\ 1 \end{pmatrix} \right\}$

9. $1, 1, 10$;
$E_1 = \text{span} \left\{ \begin{pmatrix} 1 \\ 0 \\ -2 \end{pmatrix}, \begin{pmatrix} 0 \\ 1 \\ -2 \end{pmatrix} \right\}$;
$E_{10} = \text{span} \left\{ \begin{pmatrix} 2 \\ 2 \\ 1 \end{pmatrix} \right\}$;
geom. mult. of 1 is 2

11. $1, 1, 1;\ E_1 = \text{span}\left\{\begin{pmatrix} 1 \\ 1 \\ 1 \end{pmatrix}\right\};$

geom. mult. is 1 (alg. mult. is 3)

13. $-1, i, -i;$

$E_{-1} = \text{span}\left\{\begin{pmatrix} 0 \\ -1 \\ 1 \end{pmatrix}\right\};$

$E_i = \text{span}\left\{\begin{pmatrix} 1 + i \\ 1 \\ 1 \end{pmatrix}\right\};$

$E_{-i} = \text{span}\left\{\begin{pmatrix} 1 - i \\ 1 \\ 1 \end{pmatrix}\right\}$

15. $1, 2, 2;$

$E_1 = \text{span}\left\{\begin{pmatrix} 4 \\ 1 \\ -3 \end{pmatrix}\right\};$

$E_2 = \text{span}\left\{\begin{pmatrix} 3 \\ 1 \\ -2 \end{pmatrix}\right\};$

geom. mult. of 2 is 1

17. $a, a, a, a;\ E_a = \mathbb{R}^4;$ geom. mult. of a = alg. mult. of a = 4

19. $a, a, a, a;$

$E_a = \text{span}\left\{\begin{pmatrix} 1 \\ 0 \\ 0 \\ 0 \end{pmatrix}, \begin{pmatrix} 0 \\ 0 \\ 0 \\ 1 \end{pmatrix}\right\};$

alg. mult. of a = 4; geom. mult. of a = 2.

21. Eigenvalues are $a \pm ib$. Then

$[A - (a + ib)I]\left\{\begin{pmatrix} 1 \\ i \end{pmatrix}\right\} =$

$\begin{pmatrix} -ib & b \\ -b & -ib \end{pmatrix}\begin{pmatrix} 1 \\ i \end{pmatrix} = \begin{pmatrix} 0 \\ 0 \end{pmatrix}$

Similarly,

$[A - (a - ib)I]\begin{pmatrix} 1 \\ -i \end{pmatrix} = \begin{pmatrix} 0 \\ 0 \end{pmatrix}$

23. Let $\beta_1, \beta_2, \ldots, \beta_m$ be the eigenvalues of αA. Then, for each i, there is a vector $\mathbf{v}_i \neq \mathbf{0}$ such that $(\alpha A)\mathbf{v}_i = \beta_i \mathbf{v}_i$. Thus $(\alpha A - \beta_i I)\mathbf{v} = \mathbf{0}$, or $\alpha\left(A - \dfrac{\beta_i}{\alpha}I\right)\mathbf{v} = \mathbf{0}$, which implies that $\det\left(A - \dfrac{\beta_i}{\alpha}I\right) = 0.$

Thus $\dfrac{\beta_i}{\alpha}$ is an eigenvalue of A and $\dfrac{\beta_i}{\alpha} = \lambda_j$ for some j and $\beta_i = \alpha\lambda_j$. Thus each eigenvalue of αA is of the form $\alpha\lambda_j$. Conversely, if $\mu_i = \alpha\lambda_i$, choose a vector $\mathbf{v}_i$ such that $A\mathbf{v}_i = \lambda_i\mathbf{v}_i$. Then $(\alpha A)\mathbf{v}_i = \alpha\lambda_i\mathbf{v}_i = \mu_i\mathbf{v}_i$ so that μ_i is an eigenvalue of αA.

25. $\det(A - \lambda_i I) = 0$ and $\det A^{-1} \neq 0$ because $(A^{-1})^{-1} = A$ exists. Thus $0 = \det(A - \lambda_i I)\det A^{-1} = \det[(A - \lambda_i I)A^{-1}] = \det(I - \lambda_i A^{-1}) = \det \lambda_i \left(\dfrac{1}{\lambda_i}I - A^{-1}\right)$ $\lambda_i^n \det\left(\dfrac{1}{\lambda_i}I - A^{-1}\right)$. In the last step we used the fact that $\det \alpha A = \alpha^n \det A$ if A is an $n \times n$ matrix (see Problem 2.2.28), and $\lambda_i \neq 0$ by Theorem 6, parts (i) and (x). Thus

$\det\left(A^{-1} - \dfrac{1}{\lambda_i}I\right) =$

$(-1)^n \det\left(\dfrac{1}{\lambda_i}I - A^{-1}\right) = 0$

and $\dfrac{1}{\lambda_i}$ is an eigenvalue of A^{-1}. Conversely, if μ_i is an eigenvalue of A^{-1}, then since $(A^{-1})^{-1} = A$, $\dfrac{1}{\mu_i}$ is an eigenvalue of A so that $\dfrac{1}{\mu_i} = \lambda_j$ for some j, or $\mu_i = \dfrac{1}{\lambda_j}$. Thus all eigenvalues of A^{-1} are of the form $\dfrac{1}{\lambda_i}$. Alternatively, if $A\mathbf{v} =$

Chapter 6

$\lambda\mathbf{v}$, then $\mathbf{v} = A^{-1}\lambda\mathbf{v}$ so that $A^{-1}\mathbf{v} = \dfrac{1}{\lambda}\mathbf{v}$ and $\dfrac{1}{\lambda}$ is an eigenvalue of A^{-1}.

27. Since $\det(A - \lambda_i I) = 0$, we see that $\det(A^2 - \lambda_i^2 I) = \det(A - \lambda_i I)(A + \lambda_i I) = \det(A - \lambda_i I)\det(A + \lambda_i I) = 0$. Thus λ_i^2 is an eigenvalue of A^2. Conversely, if μ_i is an eigenvalue of A^2, then $0 = \det(A^2 - \mu_i I) = \det[(A - \sqrt{\mu_i}I)(A + \sqrt{\mu_i}I)]$ so that either $\det(A - \sqrt{\mu_i}I)$ or $\det(A + \sqrt{\mu_i}I) = 0$. In either case $\pm\sqrt{\mu_i}$ is an eigenvalue of A so that $\mu_i = (\pm\sqrt{\mu_i})^2 = (\pm\lambda_j)^2 = \lambda_j^2$ for some j.

29. From Problem 28, $a_i A^i \mathbf{v} = a_i \lambda^i \mathbf{v}$ so that $p(A)\mathbf{v} = \displaystyle\sum_{i=1}^{n}(a_i A^i)\mathbf{v} =$

$$\sum_{i=1}^{n}(a_i A^i \mathbf{v}) = \sum_{i=1}^{n}a_i \lambda^i \mathbf{v} =$$

$$\left(\sum a_i \lambda^i\right)\mathbf{v} = p(\lambda)\mathbf{v}.$$

31. If A is upper triangular, then so is $A - \lambda I$ so that, by Theorem 2.1.1, $\det(A - \lambda I) = (a_{ii} - \lambda)(a_{22} - \lambda)\cdots(a_{nn} - \lambda) = 0$ when $\lambda = a_{ii}$ for some $i = 1, 2, \dots, n$.

33. $A\mathbf{v} = \lambda\mathbf{v}$ where $\mathbf{v} \neq 0$. Then $\overline{A\mathbf{v}} = \overline{\lambda\mathbf{v}}$, which implies that $\overline{A}\,\overline{\mathbf{v}} = \overline{\lambda}\,\overline{\mathbf{v}}$. But if A is real, then $\overline{A} = A$. Thus $A\overline{\mathbf{v}} = \overline{\lambda}\,\overline{\mathbf{v}}$ and $\mathbf{v} \neq \mathbf{0}$ so that $\overline{\lambda}$ is an eigenvalue of A with eigenvector $\overline{\mathbf{v}}$. Here we have used the easily verified fact that $\overline{A\mathbf{v}} = \overline{A}\,\overline{\mathbf{v}}$.

Problems 6.2, page 404

1.

n	$p_{j,n}$	$p_{a,n}$	T_n	$p_{j,n}/p_{a,n}$	T_n/T_{n-1}
0	0	12	12	0	—
1	36	7	43	5.14	3.58
2	21	19	40	1.11	0.930
5	104	45	149	2.31	—
10	600	291	891	2.06	—
19	16,090	7737	23827	2.08	—
20	23,170	11140	34310	2.08	1.44

Note that the eigenvalues are 1.44 and -0.836. The corresponding eigenvectors are $\begin{pmatrix} 2.09 \\ 1 \end{pmatrix}$ and $\begin{pmatrix} -3.57 \\ 1 \end{pmatrix}$.

3.

n	$p_{j,n}$	$p_{a,n}$	T_n	$p_{j,n}/p_{a,n}$	T_n/T_{n-1}
0	0	20	20	0	—
1	80	16	96	5	4.8
2	64	69	133	0.928	1.39
5	1092	498	1590	2.19	—
10	42,412	22,807	65,219	1.86	—
19	3.69×10^7	1.95×10^7	5.64×10^7	1.89	—
20	7.82×10^7	4.14×10^7	11.96×10^7	1.89	2.12

The eigenvalues are 2.12 and -1.32 with corresponding eigenvectors $\begin{pmatrix} 1.89 \\ 1 \end{pmatrix}$ and $\begin{pmatrix} -3.03 \\ 1 \end{pmatrix}$.

5. From equation (9), $p_n \approx a_1 \lambda_1^n \mathbf{v}_i$

for n large. If $\mathbf{v}_1 = \begin{pmatrix} x \\ y \end{pmatrix}$, then

$$\frac{p_{j,n}}{p_{a,n}} \approx \frac{a_1 \lambda_1^n x}{a_1 \lambda_1^n y} = \frac{x}{y}; \text{ but}$$

$$\begin{pmatrix} -\lambda_1 & k \\ \alpha & \beta - \lambda_1 \end{pmatrix} \begin{pmatrix} x \\ y \end{pmatrix} = \begin{pmatrix} 0 \\ 0 \end{pmatrix} \text{ so that}$$

$-\lambda_1 x + ky = 0$ and $\dfrac{x}{y} = \dfrac{k}{\lambda_1}$. Thus

$$\frac{p_{j,n}}{p_{a,n}} \approx \frac{x}{y} = \frac{k}{\lambda_1} \text{ for } n \text{ large.}$$

Problems 6.3, page 412

1. yes; $C = \begin{pmatrix} 1 & 2 \\ 1 & -5 \end{pmatrix}$,

$$C^{-1}AC = \begin{pmatrix} -4 & 0 \\ 0 & 3 \end{pmatrix}$$

3. yes; $C = \begin{pmatrix} 1 & 1 \\ 2-i & 2+i \end{pmatrix}$;

$$C^{-1}AC = \begin{pmatrix} i & 0 \\ 0 & -i \end{pmatrix}$$

5. yes;

$$C = \begin{pmatrix} 2 & 2 \\ -1+3i & -1-3i \end{pmatrix};$$

$$C^{-1}AC = \begin{pmatrix} 2+3i & 0 \\ 0 & 2-3i \end{pmatrix}$$

7. yes; $C = \begin{pmatrix} 3 & 1 & 1 \\ 2 & 3 & 0 \\ 1 & 1 & 1 \end{pmatrix}$;

$$C^{-1}AC = \begin{pmatrix} 1 & 0 & 0 \\ 0 & 2 & 0 \\ 0 & 0 & -1 \end{pmatrix}$$

9. yes; $C = \begin{pmatrix} 0 & 0 & 1 \\ 1 & 1 & 0 \\ 0 & 2 & 0 \end{pmatrix}$;

$$C^{-1}AC = \begin{pmatrix} 0 & 0 & 0 \\ 0 & 2 & 0 \\ 0 & 0 & 3 \end{pmatrix}$$

11. $C = \begin{pmatrix} 1 & 0 & 2 \\ 3 & -2 & 1 \\ 0 & 1 & 2 \end{pmatrix}$;

$$C^{-1}AC = \begin{pmatrix} 1 & 0 & 0 \\ 0 & 1 & 0 \\ 0 & 0 & 2 \end{pmatrix}$$

13. No, since 1 is an eigenvalue of algebraic multiplicity 3 and geometric multiplicity 1

15. yes;

$$C = \begin{pmatrix} 0 & -1 & 1 & 1 \\ 0 & 1 & 1 & 1 \\ 1 & 0 & 1 & -1 \\ 0 & 1 & -1 & 1 \end{pmatrix};$$

$$C^{-1}AC = \begin{pmatrix} 2 & 0 & 0 & 0 \\ 0 & 2 & 0 & 0 \\ 0 & 0 & 4 & 0 \\ 0 & 0 & 0 & 6 \end{pmatrix}$$

17. $B = C^{-1}AC$ so $B^n =$
$(C^{-1}AC)(C^{-1}AC)\cdots(C^{-1}AC) =$
$C^{-1}A(CC^{-1})A(CC^{-1})\cdots AC =$
$C^{-1}AIA\cdots IAC = C^{-1}AA\cdots AC =$
$C^{-1}A^nC.$

19. $\begin{pmatrix} 1 & 0 \\ 0 & 1 \end{pmatrix}$

21. If $D = C^{-1}AC$, then, as in Problem 17, $D^n = (C^{-1}AC)$
$\times (C^{-1}AC)\cdots(C^{-1}AC) = C^{-1}A^nC$
so that $A^n = CD^nC^{-1}$.

23. Clearly A has c as an eigenvalue of algebraic multiplicity n. Thus if A is diagonalizable, there must be an invertible matrix E such that
$E^{-1}AE = \operatorname{diag}(c, c, \ldots, c) = cI$
so that $A = E(cI)E^{-1} = cEIE^{-1} = cI.$

25. If A and B have distinct eigenvalues, then both have n linearly independent eigenvectors, and we have
$D_1 = C_1^{-1}AC_1$ and $D_2 = C_2^{-1}BC_2$.
  i. If A and B have the same eigenvectors, then $C_1 = C_2 = C$
  and $AB = (CD_1C^{-1})$
  $\times (CD_2C^{-1}) = CD_1D_2C^{-1}$
  $= CD_2D_1C^{-1} = (CD_2C^{-1})$
  $\times (CD_1C^{-1}) = BA$ (since diagonal matrices of the same order always commute).

ii. If $BA = AB$, let $\mathbf{x}$ be an eigen-vector of B corresponding to λ. Then $BA\mathbf{x} = AB\mathbf{x} = A(\lambda\mathbf{x}) = \lambda A\mathbf{x}$ so that $\mathbf{y} = A\mathbf{x}$ is an eigenvector of B corresponding to λ. Thus $A\mathbf{x}$ and $\mathbf{x}$ are line-arly dependent so that there is a scalar μ with $A\mathbf{x} = \mu\mathbf{x}$. But this shows that $\mathbf{x}$ is also an eigenvector of A. Thus every eigenvector of B is an eigen-vector of A. A similar argu-ment shows that every eigen-vector of A is an eigenvector of B.

Problems 6.4, page 419

1. $Q = \begin{pmatrix} 2/\sqrt{5} & 1/\sqrt{5} \\ 1/\sqrt{5} & -2/\sqrt{5} \end{pmatrix}$,

$D = \begin{pmatrix} 5 & 0 \\ 0 & -5 \end{pmatrix}$

3. $Q = \begin{pmatrix} 1/\sqrt{2} & 1/\sqrt{2} \\ 1/\sqrt{2} & -1/\sqrt{2} \end{pmatrix}$,

$D = \begin{pmatrix} 0 & 0 \\ 0 & 2 \end{pmatrix}$

5. $Q = \begin{pmatrix} 1/\sqrt{2} & \frac{1}{2} & \frac{1}{2} \\ -1/\sqrt{2} & \frac{1}{2} & \frac{1}{2} \\ 0 & 1/\sqrt{2} & -1/\sqrt{2} \end{pmatrix}$.

$D = \begin{pmatrix} -3 & 0 & 0 \\ 0 & 1+2\sqrt{2} & 0 \\ 0 & 0 & 1-2\sqrt{2} \end{pmatrix}$

7. $Q = \begin{pmatrix} -\frac{2}{3} & \frac{1}{3} & \frac{2}{3} \\ \frac{2}{3} & \frac{2}{3} & \frac{1}{3} \\ \frac{1}{3} & -\frac{2}{3} & \frac{2}{3} \end{pmatrix}$,

$D = \begin{pmatrix} 0 & 0 & 0 \\ 0 & 3 & 0 \\ 0 & 0 & 6 \end{pmatrix}$

9. Let $\mathbf{u}$ be an eigenvector correspond-ing to λ with $|\mathbf{u}| = 1$. Then $Q\mathbf{u} = \lambda\mathbf{u}$ and $1 = |\mathbf{u}| = |Q^{-1}Q\mathbf{u}| = |\lambda Q^{-1}\mathbf{u}| = |\lambda Q^t\mathbf{u}| = |\lambda Q\mathbf{u}| =$
(since Q is symmetric)
$|\lambda^2\mathbf{u}| = \lambda^2|\mathbf{u}| = \lambda^2$. Thus $\lambda^2 = 1$ and $\lambda = \pm 1$.

11. $1 = \det I = \det(Q^{-1}Q) = \det(Q^tQ) = (\det Q^t)(\det Q) = (\det Q)^2$ — since $\det A^t = \det A$ for any matrix A. Thus

$\det Q = \pm 1$ and $\begin{pmatrix} a & c \\ b & d \end{pmatrix} = Q^t = $

$Q^{-1} = \begin{pmatrix} \dfrac{d}{\det Q} & -\dfrac{b}{\det Q} \\ -\dfrac{c}{\det Q} & \dfrac{a}{\det Q} \end{pmatrix}$.

If $\det Q = 1$, then $c = -b$. If $\det Q = -1$, then $c = b$.

13. If the 2×2 matrix A has orthogo-nal eigenvectors, then A is or-thogonally diagonalizable, which means that A is symmetric by Theo-rem 4.

15. Let λ be an eigenvalue of A with eigenvector $\mathbf{v}$ and suppose that $A^* = A$. Then $\lambda(\mathbf{v}, \mathbf{v}) = (\lambda\mathbf{v}, \mathbf{v}) = (A\mathbf{v}, \mathbf{v}) = (\mathbf{v}, A^*\mathbf{v}) = (\mathbf{v}, A\mathbf{v}) = (\mathbf{v}, \lambda\mathbf{v}) = \overline{\lambda}(\mathbf{v}, \mathbf{v})$. Since $\mathbf{v} \neq 0$, this means that $\lambda = \overline{\lambda}$ so that λ is real.

17. Use Problem 16 after showing that to every eigenvalue of algebraic multiplicity k there correspond k or-thonormal eigenvectors. Let Q be obtained exactly as in the proof of Theorem 3. Recall that $(\mathbf{u}, \mathbf{v}) = \mathbf{u}_1 \cdot \overline{\mathbf{v}}_1 + \cdots + \mathbf{u}_n \cdot \overline{\mathbf{v}}_n$. $Q^t = \overline{Q}^{-1}$ and A is similar to $Q^tA\overline{Q}$ or $\overline{Q}^tAQ$. $|Q^tA\overline{Q} - I| = |A - I|$; $\overline{Q}^tAQ = (\overline{Q}^tA)Q$

$= \begin{pmatrix} \overline{\mathbf{u}}_1^tA \\ \overline{\mathbf{u}}_2^tA \\ \vdots \\ \overline{\mathbf{u}}_n^tA \end{pmatrix}(\mathbf{u}_1, \mathbf{u}_2, \ldots, \mathbf{u}_n)$

$= \begin{pmatrix} \overline{\mathbf{u}}_1^t \\ \overline{\mathbf{u}}_1^t \\ \vdots \\ \overline{\mathbf{u}}_n^t \end{pmatrix}(A^*\mathbf{u}_1, A^*\mathbf{u}_2, \ldots, A^*\mathbf{u}_n)$.

Now $(\overline{\mathbf{u}}_1^t, A^*\mathbf{u}_1) = (\overline{\mathbf{u}}_1^t, A\mathbf{u}_1) = (\overline{\mathbf{u}}_1^t, \lambda_1\mathbf{u}_1) = \overline{\lambda}_1(\overline{\mathbf{u}}_1^t, \mathbf{u}_1) = \overline{\lambda}_1 = \lambda_1$ (by Problem 15) and since $(\overline{\mathbf{u}}_1^t, \mathbf{u}_1) = \overline{\mathbf{u}}_1^t \cdot \overline{\mathbf{u}}_1 = 1 = \overline{\mathbf{u}}_1^t \cdot \mathbf{u}_1$. Then $\overline{Q}_tAQ =$

$$\begin{pmatrix} \lambda_1 & \bar{\mathbf{u}}_1' A \mathbf{u}_2 & \cdots & \bar{\mathbf{u}}_1' A \mathbf{u}_n \\ 0 & \bar{\mathbf{u}}_2' A \mathbf{u}_2 & \cdots & \bar{\mathbf{u}}_2' A \mathbf{u}_n \\ \vdots & \vdots & & \vdots \\ 0 & \bar{\mathbf{u}}_n' A \mathbf{u}_2 & \cdots & \bar{\mathbf{u}}_n' A \mathbf{u}_n \end{pmatrix}$$

$\bar{\mathbf{u}}_1' A \mathbf{u}_j = A \bar{\mathbf{u}}_1' \cdot \bar{\mathbf{u}}_j = \overline{A \mathbf{u}_1'} \cdot \mathbf{u}_j = 0$
if $j \neq 1$. Now $\overline{(Q^t A Q)^t} =$
$\overline{Q^t A^t (Q^t)^t} = \overline{Q^t A^t Q} = \bar{Q}^t \bar{A}^t Q =$
$\bar{Q}^t A Q$, since $\bar{A}^t = A^* = A$. Thus
$\bar{Q}^t A Q$ is hermitian, which means
that the zeros in the first row of
$\bar{Q}^t A Q$ must match the zeros in the
first column. The rest of the proof
follows, as in the proof of Theorem
3, with Q^t replaced by $\bar{Q}^t$.

19. $U = \dfrac{1}{\sqrt{3}} \begin{pmatrix} -1 + i & 1 \\ 1 & 1 + i \end{pmatrix}$;

$U^* A U = \begin{pmatrix} -1 & 0 \\ 0 & 8 \end{pmatrix}$

Problems 6.5, page 429

1. $\begin{pmatrix} 3 & -1 \\ -1 & 0 \end{pmatrix} \begin{pmatrix} x \\ y \end{pmatrix} \cdot \begin{pmatrix} x \\ y \end{pmatrix} = 5$;

$Q = \begin{pmatrix} \dfrac{2}{\sqrt{26 - 6\sqrt{13}}} & \dfrac{2}{\sqrt{26 + 6\sqrt{13}}} \\ \dfrac{3 - \sqrt{13}}{\sqrt{26 - 6\sqrt{13}}} & \dfrac{3 + \sqrt{13}}{\sqrt{26 + 6\sqrt{13}}} \end{pmatrix} =$

$\begin{pmatrix} 0.9571 & 0.2898 \\ -0.2898 & 0.9571 \end{pmatrix}$;

$\dfrac{x'^2}{\left(\dfrac{10}{\sqrt{13} + 3}\right)} - \dfrac{y'^2}{\left(\dfrac{10}{\sqrt{13} - 3}\right)} = 1$;

hyperbola; $\theta = 5.989 = 343°$

3. $\begin{pmatrix} 4 & 2 \\ 2 & -1 \end{pmatrix} \begin{pmatrix} x \\ y \end{pmatrix} \cdot \begin{pmatrix} x \\ y \end{pmatrix} = 9$;

$Q = \begin{pmatrix} \dfrac{5 + \sqrt{41}}{\sqrt{82 + 10\sqrt{41}}} & \dfrac{5 - \sqrt{41}}{\sqrt{82 - 10\sqrt{41}}} \\ \dfrac{4}{\sqrt{82 + 10\sqrt{41}}} & \dfrac{4}{\sqrt{82 - 10\sqrt{41}}} \end{pmatrix} =$

$\begin{pmatrix} 0.9436 & -0.3310 \\ 0.3310 & 0.9436 \end{pmatrix}$;

$\dfrac{x'^2}{\left(\dfrac{18}{\sqrt{41} + 3}\right)} - \dfrac{y'^2}{\left(\dfrac{18}{\sqrt{41} - 3}\right)} = 1$;

hyperbola; $\theta \approx 0.3374 \approx 19.33°$

5. $\begin{pmatrix} 0 & \frac{1}{2} \\ \frac{1}{2} & 0 \end{pmatrix} \begin{pmatrix} x \\ y \end{pmatrix} \cdot \begin{pmatrix} x \\ y \end{pmatrix} = a > 0$;

$Q = \begin{pmatrix} 1/\sqrt{2} & 1/\sqrt{2} \\ -1/\sqrt{2} & 1/\sqrt{2} \end{pmatrix}$;

$\dfrac{x'^2}{2a} - \dfrac{y'^2}{2a} = 1$; hyperbola; $\theta = 7\pi/4 = 315°$.

7. Same as Problem 5 except that now
we have a hyperbola with the roles
of x' and y' reversed; since $a < 0$,
we have

$$\dfrac{y'^2}{(-2a)} - \dfrac{x'^2}{(-2a)} = 1$$

9. $\begin{pmatrix} -1 & 1 \\ 1 & -1 \end{pmatrix} \begin{pmatrix} x \\ y \end{pmatrix} \cdot \begin{pmatrix} x \\ y \end{pmatrix} = 0$;

$Q = \begin{pmatrix} 1/\sqrt{2} & -1/\sqrt{2} \\ 1/\sqrt{2} & 1/\sqrt{2} \end{pmatrix}$;

$y'^2 = 0$, which is the equation of a
straight line through the origin; $\theta = \pi/4 = 45°$.

11. $\begin{pmatrix} 3 & -3 \\ -3 & 5 \end{pmatrix} \begin{pmatrix} x \\ y \end{pmatrix} \cdot \begin{pmatrix} x \\ y \end{pmatrix} = 36$;

$Q = \begin{pmatrix} \dfrac{1 + \sqrt{10}}{\sqrt{20 + 2\sqrt{10}}} & \dfrac{1 - \sqrt{10}}{\sqrt{20 - 2\sqrt{10}}} \\ \dfrac{3}{\sqrt{20 + 2\sqrt{10}}} & \dfrac{3}{\sqrt{20 - 2\sqrt{10}}} \end{pmatrix} =$

$\begin{pmatrix} 0.8112 & -0.5847 \\ 0.5847 & 0.8112 \end{pmatrix}$;

$\dfrac{x'^2}{\left(\dfrac{36}{4 - \sqrt{10}}\right)} + \dfrac{y'^2}{\left(\dfrac{36}{4 + \sqrt{10}}\right)} = 1$;

ellipse; $\theta \approx 0.6245 \approx 35.78°$

13. $\begin{pmatrix} 6 & \frac{5}{2} \\ \frac{5}{2} & -6 \end{pmatrix} \begin{pmatrix} x \\ y \end{pmatrix} \cdot \begin{pmatrix} x \\ y \end{pmatrix} = -7$;

$Q = \begin{pmatrix} 5/\sqrt{26} & -1/\sqrt{26} \\ 1/\sqrt{26} & 5/\sqrt{26} \end{pmatrix}$;

$$\frac{y'^2}{(14/13)} - \frac{x'^2}{(14/13)} = 1; \text{ hyperbola;}$$
$$\theta \approx 0.197 \approx 11.31°$$

15. $\begin{pmatrix} 1 & -1 & -1 \\ -1 & 1 & -1 \\ -1 & -1 & 1 \end{pmatrix}\begin{pmatrix} x \\ y \\ z \end{pmatrix} \cdot \begin{pmatrix} x \\ y \\ z \end{pmatrix};$

$$Q = \begin{pmatrix} 1/\sqrt{3} & 1/\sqrt{2} & 1/\sqrt{6} \\ 1/\sqrt{3} & -1/\sqrt{2} & 1/\sqrt{6} \\ 1/\sqrt{3} & 0 & -2/\sqrt{6} \end{pmatrix};$$
$$-x'^2 + 2y'^2 + 2z'^2$$

17. $\begin{pmatrix} 3 & 2 & 2 \\ 2 & 2 & 0 \\ 2 & 0 & 4 \end{pmatrix}\begin{pmatrix} x \\ y \\ z \end{pmatrix} \cdot \begin{pmatrix} x \\ y \\ z \end{pmatrix};$

$$Q = \begin{pmatrix} -\frac{2}{3} & \frac{1}{3} & \frac{2}{3} \\ \frac{2}{3} & \frac{2}{3} & \frac{1}{3} \\ \frac{1}{3} & -\frac{2}{3} & \frac{2}{3} \end{pmatrix};$$
$$3y'^2 + 6z'^2$$

19. $\begin{pmatrix} 1 & 1 & 2 & \frac{7}{2} \\ 1 & 1 & 3 & -1 \\ 2 & 3 & 3 & 0 \\ \frac{7}{2} & -1 & 0 & 1 \end{pmatrix}$

21. $\begin{pmatrix} 3 & -\frac{7}{2} & \frac{1}{2} & -1 & \frac{3}{2} \\ -\frac{7}{2} & -2 & -\frac{1}{2} & \frac{1}{2} & 0 \\ \frac{1}{2} & -\frac{1}{2} & 3 & -2 & -\frac{5}{2} \\ -1 & \frac{1}{2} & -2 & -6 & \frac{1}{2} \\ \frac{3}{2} & 0 & -\frac{5}{2} & \frac{1}{2} & -1 \end{pmatrix}$

23. $\begin{pmatrix} \cos\theta & -\sin\theta \\ \sin\theta & \cos\theta \end{pmatrix}\begin{pmatrix} x \\ y \end{pmatrix} =$
$$\begin{pmatrix} x\cos\theta & -y\sin\theta \\ x\sin\theta & +y\cos\theta \end{pmatrix} = \begin{pmatrix} x' \\ y' \end{pmatrix}.$$

Then the quadratic equation $ax'^2 + bx'y' + cy'^2$ becomes $a(x\cos\theta - y\sin\theta)^2 + b(x\cos\theta - y\sin\theta) \times (x\sin\theta + y\cos\theta) + c(x\sin\theta + y\cos\theta)^2$; the cross product term is $-2axy(\sin\theta\cos\theta + bxy \times [\cos^2\theta - \sin^2\theta] + 2cxy \times \sin\theta\cos\theta = xy[-a\sin 2\theta + b\cos 2\theta + c\sin 2\theta] = 0$ so that $(c - a)\sin 2\theta + b\cos 2\theta = 0$ and
$$\frac{a - c}{b} = \frac{\cos 2\theta}{\sin 2\theta} = \cot 2\theta.$$

25. Suppose that $ax^2 + bxy + cy^2$ is converted to $a'x'^2 + b'x'y' + c'y'^2$ by a rotation. Let

$$A = \begin{pmatrix} a & \frac{b}{2} \\ \frac{b}{2} & c \end{pmatrix} \text{ and } A' = \begin{pmatrix} a' & \frac{b'}{2} \\ \frac{b'}{2} & c' \end{pmatrix}.$$

There is an orthogonal matrix Q such that $A = QA'Q^t$ and Q is also a rotation matrix. Thus $\det A = \det Q \det A' \det Q^t = \det QQ^t \det A' = \det A'$ since $QQ^t = I$. But $\det A = ac - \frac{b^2}{4}$ and $\det A' = a'c' - \frac{b'^2}{4}$. Finally, since A and A' are similar, they have the same eigenvalues. But the sum of the eigenvalues of A is $a + c$, while the sum of the eigenvalues of A' is $a' + c'$. Thus $a + c = a' + c'$.

27. Let $\lambda_1, \lambda_2, \ldots, \lambda_n$ be the eigenvalues of A. Then, removing the cross product terms, we have $F(\mathbf{x}) = F'(\mathbf{x}') = \lambda_1 x_1'^2 + \lambda_2 x_2'^2 + \cdots + \lambda_n x_2'^2$ where $\mathbf{x}' = Q^t\mathbf{x}$. If $\lambda_i \geq 0$ for $i = 1, 2, \ldots, n$, then $F'(\mathbf{x}') \geq 0$. If $F'(\mathbf{x}') \geq 0$, then $\lambda_i \geq 0$ since, if not, there is a λ_j with $\lambda_j < 0$. Let $\mathbf{x}^*$ be the vector with 0's in every position except the jth and a 1 in the jth position. Then $H(\mathbf{x}^*) = \lambda_j < 0$, which is a contradiction.

29. negative definite

31. positive definite

33. indefinite

35. negative definite

37. i. If $\det A \neq 0$, then neither λ_1 nor λ_2 is zero. Thus, with $d = 0$, equation (22) becomes $\lambda_1 x'^2 + \lambda_2 y'^2 = 0$. If now both λ_1 and λ_2 are positive or negative, then the equation is satisfied only when $x' = 0$ and

$y' = 0$. These are the equations of two straight lines. If λ_1 and λ_2 have opposite signs, then the equations become $x' = \pm\sqrt{\dfrac{\lambda_2}{-\lambda_1}}\,y'$, which are again the equations of two straight lines. If $\det A = 0$, then one of λ_1 or λ_2 is zero, and the equation becomes $x' = 0$ or $y' = 0$, each of which is the equation of a single straight line.

Problems 6.6, page 437

1. no **3.** no **5.** yes **7.** no

9. yes **11.** no **13.** yes

15. I

17. $\begin{pmatrix} 1 & 0 \\ -1 & -\frac{1}{7} \end{pmatrix}$; $J = \begin{pmatrix} -3 & 1 \\ 0 & -3 \end{pmatrix}$

19. a. Let $\mathbf{x} \in \mathbb{C}^3$ be a fixed vector that is not an eigenvector. Since the geometric multiplicity of the eigenvalue λ is one, $\mathbf{x}$ is not a multiple of $\mathbf{v}_1$, where $\mathbf{v}_1$ is an eigenvector. Then $\mathbf{x}$ and $\mathbf{v}_1$ are linearly independent. Let $\mathbf{w} = c_1\mathbf{v}_1 + c_2\mathbf{x}$. Assume that $\mathbf{w} = (A - \lambda I)\mathbf{x}$. Then $A\mathbf{x} - \lambda\mathbf{x} = c_1\mathbf{v}_1 + c_2\mathbf{x}$. Let $B = A - (c_2 + \lambda)I$ so that $B\mathbf{x} = c_1\mathbf{v}_1$. Assume that $c_2 \neq 0$. Then $\lambda + c_2$ is not an eigenvalue since λ is the only eigenvalue. We have $\det B = \det[A - (\lambda + c_2)I] \neq 0$. Then B^{-1} exists, $\mathbf{x} = c_1 B^{-1}\mathbf{v}_1$, and $\lambda\mathbf{x} = c_1 B^{-1}\lambda\mathbf{v}_1 = c_1 B^{-1} A\mathbf{v}_1$. Since $A = B + (c_2 + \lambda)I$, $\lambda\mathbf{x} = c_1[B^{-1}B + (c_2 + \lambda)B^{-1}]\mathbf{v}_1 = c_1\mathbf{v}_1 + c_1(c_2 + \lambda)B^{-1}\mathbf{v}_1 = c_1\mathbf{v}_1 + (c_2 + \lambda)B^{-1}B\mathbf{x} = c_1\mathbf{v}_1 + c_2\mathbf{x} + \lambda\mathbf{x}$. Thus $c_1\mathbf{v}_1 + c_2\mathbf{x} = \mathbf{0}$ and $c_1 = c_2 = 0$ because $\mathbf{v}_1$ and $\mathbf{x}$ are linearly independent. This contradicts our previous assumption that $c_2 \neq 0$ and $\mathbf{w} = (A - \lambda I)\mathbf{x} = c_1\mathbf{v}_1$.

Let $c_1\mathbf{x} = \mathbf{v}_2$; then $(A - \lambda I)\mathbf{v}_2 = \mathbf{v}_1$.

b. Let $\mathbf{y} \in \mathbb{C}^3$ with $\mathbf{y}$ not an eigenvector of A; $\mathbf{y}$ can be chosen linearly independent of $\mathbf{v}_2$ (it is already independent of $\mathbf{v}_1$) so that $\mathbf{z} = d_1\mathbf{v}_2 + d_2\mathbf{y}$ is not an eigenvector. Write $\mathbf{z}$ as $\mathbf{z} = (A - \lambda I)\mathbf{y}$. Then $A\mathbf{y} - \lambda\mathbf{y} = d_1\mathbf{v}_2 + d_2\mathbf{y}$. Let $D = A - (d_2 + \lambda)I$ so that $D\mathbf{y} = d_1\mathbf{v}_2$ since $A\mathbf{y} - (\lambda I)\mathbf{y} - d_2\mathbf{y} = d_1\mathbf{v}_2$. Assume that $d_2 \neq 0$; then $d_2 + \lambda$ is not an eigenvalue. Clearly $\det D \neq 0$, D^{-1} exists, $\mathbf{y} = d_1 D^{-1}\mathbf{v}_2$, and $\lambda\mathbf{y} = d_1 D^{-1}\lambda\mathbf{v}_2$. Then $\lambda\mathbf{v}_2 = \mathbf{v}_1 - A\mathbf{v}_2$; $\lambda\mathbf{y} = d_1 D^{-1}(\mathbf{v}_1 - A\mathbf{v}_2) = d_1 D^{-1}\mathbf{v}_1 - d_1 D^{-1} A\mathbf{v}_2$. $A = D + (d_2 + \lambda)I$. $\lambda\mathbf{y} = d_1 D^{-1}\mathbf{v}_1 - d_1\mathbf{v}_2 - d_1 d_2 D^{-1}\mathbf{v}_2 - d_1 D^{-1}\lambda\mathbf{v}_2 = d_1 D^{-1}\mathbf{v}_1 - d_1 D^{-1} A\mathbf{v}_2 - d_1\mathbf{v}_2 - d_1 d_2 D^{-1}\mathbf{v}_2 = d_1 D^{-1}\lambda\mathbf{v}_2 - d_1(\mathbf{v}_2 + d_2 D^{-1}\mathbf{v}_2) = \lambda\mathbf{y} - d_1(I + d_2 D^{-1})\mathbf{v}_2$. So $0 = d_1(I + d_2 D^{-1})\mathbf{v}_2$. $d_1 \neq 0$, otherwise $\lambda + d_2$ would be an eigenvalue and $\mathbf{y}$ an eigenvector. Then $(d_2 DD^{-1} + D)\mathbf{v}_2 = D\mathbf{0} = \mathbf{0}$. $d_2\mathbf{v}_2 + [A - (d_2 + \lambda)I]\mathbf{v}_2 = \mathbf{0}$. $d_2\mathbf{v}_2 + (A - \lambda I)\mathbf{v}_2 - d_2\mathbf{v}_2 = \mathbf{0}$, or $(A - \lambda I)\mathbf{v}_2 = \mathbf{0}$, contrary to the result of part (a). Thus $d_2 = 0$ and $(A - I\lambda)\mathbf{y} = d_1\mathbf{v}_2$. Let $\mathbf{y} = d_1\mathbf{v}_3$; then $(A - I\lambda)\mathbf{v}_3 = \mathbf{v}_2$.

c. Let $C = (\mathbf{v}_1, \mathbf{v}_2, \mathbf{v}_3)$, where $\mathbf{v}_1$, $\mathbf{v}_2$, $\mathbf{v}_3$ are as above and linearly independent; then C^{-1} exists. $AC = A(\mathbf{v}_1, \mathbf{v}_2, \mathbf{v}_3) = (A\mathbf{v}_1, A\mathbf{v}_2, A\mathbf{v}_3) = (\lambda\mathbf{v}_1, \mathbf{v}_1 + \lambda\mathbf{v}_2, \mathbf{v}_2 + \lambda\mathbf{v}_3);$

$$CJ = (\mathbf{v}_1, \mathbf{v}_2, \mathbf{v}_3)\begin{pmatrix} \lambda & 1 & 0 \\ 0 & \lambda & 1 \\ 0 & 0 & \lambda \end{pmatrix} =$$

$(\lambda\mathbf{v}, \mathbf{v}_1 + \lambda\mathbf{v}_2, \mathbf{v}_2 + \lambda\mathbf{v}_3) = AC$; so $J = C^{-1}AC$.

21. $C = \begin{pmatrix} 1 & 1 & 0 \\ 0 & -1 & -2 \\ -1 & 0 & 3 \end{pmatrix}$;

$J = \begin{pmatrix} 0 & 1 & 0 \\ 0 & 0 & 1 \\ 0 & 0 & 0 \end{pmatrix}$

23. If $m = n$, then $A^m = 0$ by definition of index of nilpotency. If $m > n$, then $A^m = A^{m-n}A^n = A^{m-n}0 = 0$.

25. $\begin{pmatrix} \lambda_1 & 0 & 0 & 0 \\ 0 & \lambda_2 & 0 & 0 \\ 0 & 0 & \lambda_3 & 0 \\ 0 & 0 & 0 & \lambda_4 \end{pmatrix}$

$\begin{pmatrix} \lambda_1 & 1 & 0 & 0 \\ 0 & \lambda_1 & 0 & 0 \\ 0 & 0 & \lambda_2 & 0 \\ 0 & 0 & 0 & \lambda_3 \end{pmatrix}$

$\begin{pmatrix} \lambda_1 & 1 & 0 & 0 \\ 0 & \lambda_1 & 1 & 0 \\ 0 & 0 & \lambda_1 & 0 \\ 0 & 0 & 0 & \lambda_2 \end{pmatrix}$

$\begin{pmatrix} \lambda_1 & 1 & 0 & 0 \\ 0 & \lambda_1 & 0 & 0 \\ 0 & 0 & \lambda_2 & 1 \\ 0 & 0 & 0 & \lambda_2 \end{pmatrix}$

$\begin{pmatrix} \lambda_1 & 1 & 0 & 0 \\ 0 & \lambda_1 & 1 & 0 \\ 0 & 0 & \lambda_1 & 1 \\ 0 & 0 & 0 & \lambda_1 \end{pmatrix}$

Here the λ_i's are not necessarily distinct. Also, the blocks may be permuted on the diagonal.

27. $\begin{pmatrix} 3 & 0 & 0 & 0 \\ 0 & 3 & 0 & 0 \\ 0 & 0 & 3 & 0 \\ 0 & 0 & 0 & -4 \end{pmatrix}$

$\begin{pmatrix} 3 & 1 & 0 & 0 \\ 0 & 3 & 0 & 0 \\ 0 & 0 & 3 & 0 \\ 0 & 0 & 0 & -4 \end{pmatrix}$

$\begin{pmatrix} 3 & 1 & 0 & 0 \\ 0 & 3 & 1 & 0 \\ 0 & 0 & 3 & 0 \\ 0 & 0 & 0 & -4 \end{pmatrix}$

The Jordan blocks may be permuted along the diagonal.

29. $\begin{pmatrix} 4 & 0 & 0 & 0 & 0 \\ 0 & 4 & 0 & 0 & 0 \\ 0 & 0 & 4 & 0 & 0 \\ 0 & 0 & 0 & -3 & 0 \\ 0 & 0 & 0 & 0 & -3 \end{pmatrix}$

$\begin{pmatrix} 4 & 1 & 0 & 0 & 0 \\ 0 & 4 & 0 & 0 & 0 \\ 0 & 0 & 4 & 0 & 0 \\ 0 & 0 & 0 & -3 & 0 \\ 0 & 0 & 0 & 0 & -3 \end{pmatrix}$

$\begin{pmatrix} 4 & 1 & 0 & 0 & 0 \\ 0 & 4 & 1 & 0 & 0 \\ 0 & 0 & 4 & 0 & 0 \\ 0 & 0 & 0 & -3 & 0 \\ 0 & 0 & 0 & 0 & -3 \end{pmatrix}$

$\begin{pmatrix} 4 & 1 & 0 & 0 & 0 \\ 0 & 4 & 1 & 0 & 0 \\ 0 & 0 & 4 & 0 & 0 \\ 0 & 0 & 0 & -3 & 1 \\ 0 & 0 & 0 & 0 & -3 \end{pmatrix}$

$\begin{pmatrix} 4 & 1 & 0 & 0 & 0 \\ 0 & 4 & 0 & 0 & 0 \\ 0 & 0 & 4 & 0 & 0 \\ 0 & 0 & 0 & -3 & 1 \\ 0 & 0 & 0 & 0 & -3 \end{pmatrix}$

$\begin{pmatrix} 4 & 0 & 0 & 0 & 0 \\ 0 & 4 & 0 & 0 & 0 \\ 0 & 0 & 4 & 0 & 0 \\ 0 & 0 & 0 & -3 & 1 \\ 0 & 0 & 0 & 0 & -3 \end{pmatrix}$

The Jordan blocks may be permuted along the diagonal.

31.

$$\begin{pmatrix} -7 & 0 & 0 & 0 & 0 \\ 0 & -7 & 0 & 0 & 0 \\ 0 & 0 & -7 & 0 & 0 \\ 0 & 0 & 0 & -7 & 0 \\ 0 & 0 & 0 & 0 & -7 \end{pmatrix}$$

$$\begin{pmatrix} -7 & 1 & 0 & 0 & 0 \\ 0 & -7 & 0 & 0 & 0 \\ 0 & 0 & -7 & 0 & 0 \\ 0 & 0 & 0 & -7 & 0 \\ 0 & 0 & 0 & 0 & -7 \end{pmatrix}$$

$$\begin{pmatrix} -7 & 1 & 0 & 0 & 0 \\ 0 & -7 & 1 & 0 & 0 \\ 0 & 0 & -7 & 0 & 0 \\ 0 & 0 & 0 & -7 & 0 \\ 0 & 0 & 0 & 0 & -7 \end{pmatrix}$$

$$\begin{pmatrix} -7 & 1 & 0 & 0 & 0 \\ 0 & -7 & 1 & 0 & 0 \\ 0 & 0 & -7 & 1 & 0 \\ 0 & 0 & 0 & -7 & 0 \\ 0 & 0 & 0 & 0 & -7 \end{pmatrix}$$

$$\begin{pmatrix} -7 & 1 & 0 & 0 & 0 \\ 0 & -7 & 1 & 0 & 0 \\ 0 & 0 & -7 & 1 & 0 \\ 0 & 0 & 0 & -7 & 1 \\ 0 & 0 & 0 & 0 & -7 \end{pmatrix}$$

$$\begin{pmatrix} -7 & 1 & 0 & 0 & 0 \\ 0 & -7 & 0 & 0 & 0 \\ 0 & 0 & -7 & 1 & 0 \\ 0 & 0 & 0 & -7 & 0 \\ 0 & 0 & 0 & 0 & -7 \end{pmatrix}$$

$$\begin{pmatrix} -7 & 1 & 0 & 0 & 0 \\ 0 & -7 & 1 & 0 & 0 \\ 0 & 0 & -7 & 0 & 0 \\ 0 & 0 & 0 & -7 & 1 \\ 0 & 0 & 0 & 0 & -7 \end{pmatrix}$$

The Jordan blocks may be permuted along the diagonal.

Problems 6.7, page 448

1. $\dfrac{1}{7}\begin{pmatrix} 5e^{-4t} + 2e^{3t} & 2e^{-4t} - 2e^{3t} \\ 5e^{-4t} - 5e^{3t} & 2e^{-4t} + 5e^{3t} \end{pmatrix}$

3. $\begin{pmatrix} 2\sin t + \cos t & -\sin t \\ 5\sin t & -2\sin t + \cos t \end{pmatrix}$

5. $e^{-3t}\begin{pmatrix} 1 - 7t & -7t \\ 7t & 1 + 7t \end{pmatrix}$

7. $e^{-5t}\begin{pmatrix} 1 - 7t & 7t \\ -7t & 1 + 7t \end{pmatrix}$

9. $\begin{pmatrix} 4e^t - 3e^{2t} + 6te^{2t} & -12e^t + 12e^{2t} - 6te^{2t} & 6te^{2t} \\ e^t - e^{2t} + 2te^{2t} & -3e^t + 4e^{2t} - 2te^{2t} & 2te^{2t} \\ -3e^t + 3e^{2t} - 4te^{2t} & 9e^t - 9e^{2t} + 4te^{2t} & -4te^{2t} + e^{2t} \end{pmatrix}$

11. $\mathbf{x}(t) =$

$-\dfrac{1}{3}\begin{pmatrix} -e^t - 2e^{4t} & -e^t + e^{4t} \\ -2e^t + 2e^{4t} & -2e^t - e^{4t} \end{pmatrix}$

$\times \begin{pmatrix} x_1(0) \\ x_2(0) \end{pmatrix}$, which leads to

$x_1(t) = \frac{1}{3}[(x_1(0) + x_2(0))e^t + (2x_1(0) - x_2(0))e^{4t}]$

$= \frac{1}{3}[(x_1(0) + x_2(0)) + (2x_1(0) - x_2(0))e^{3t}]e^t$

If $2x_1(0) < x_2(0)$, then the first population will be extinct when $x_1(0) + x_2(0) = [x_2(0) - 2x_1(0)]e^{3t}$, or

$$t = \frac{1}{3}\ln\left(\frac{x_1(0) + x_2(0)}{x_2(0) - 2x_1(0)}\right)$$

13. $\begin{pmatrix} x_1 \\ x_2 \end{pmatrix}' = \dfrac{1}{1000}\begin{pmatrix} -30 & 10 \\ 30 & -30 \end{pmatrix}\begin{pmatrix} x_1 \\ x_2 \end{pmatrix};$

$\begin{pmatrix} x_1 \\ x_2 \end{pmatrix} = \begin{pmatrix} 500\,(e^{\alpha t} + e^{\beta t}) \\ 500\sqrt{3}\,(e^{\alpha t} - e^{\beta t}) \end{pmatrix}$

where $\alpha = -0.03 + \sqrt{0.0003} \approx -0.0127$ and $\beta = -0.03 - \sqrt{0.0003} \approx -0.0473$.

15. a. $\begin{pmatrix} x_1' \\ x_2' \end{pmatrix} = \begin{pmatrix} 0 & 1 \\ -b & -a \end{pmatrix}\begin{pmatrix} x_1 \\ x_2 \end{pmatrix}$

b. $\det\begin{pmatrix} -\lambda & 1 \\ -b & -a-\lambda \end{pmatrix} = \lambda^2 + a\lambda + b$ so that $p(\lambda) = \lambda^2 + a\lambda + b = 0$.

17. $(1 + 5t)e^{-3t}$ **19.** $\frac{8}{7}e^{5t} + \frac{13}{7}e^{-2t}$

21. By Problem 20, $N_3^k = 0$ for $k \geq 3$.

Thus $e^{N_3 t} = I + N_3 t + N_3^2 \dfrac{t^2}{2}$

$$= \begin{pmatrix} 1 & 0 & 0 \\ 0 & 1 & 0 \\ 0 & 0 & 1 \end{pmatrix} + \begin{pmatrix} 0 & t & 0 \\ 0 & 0 & t \\ 0 & 0 & 0 \end{pmatrix} +$$

$$\begin{pmatrix} 0 & 0 & t^2/2 \\ 0 & 0 & 0 \\ 0 & 0 & 0 \end{pmatrix} = \begin{pmatrix} 1 & t & t^2/2 \\ 0 & 1 & t \\ 0 & 0 & 1 \end{pmatrix}.$$

23. From 6.6.20, $C = \begin{pmatrix} 1 & 1 & 0 \\ 1 & 2 & 1 \\ 0 & 1 & 0 \end{pmatrix}$ and

$$J = \begin{pmatrix} -1 & 1 & 0 \\ 0 & -1 & 1 \\ 0 & 0 & -1 \end{pmatrix} \text{ so that}$$

$e^{At} = Ce^{Jt}C^{-1}$

$$= \begin{pmatrix} 1 & 1 & 0 \\ 1 & 2 & 1 \\ 0 & 1 & 0 \end{pmatrix}e^{-t}\begin{pmatrix} 1 & t & t^2/2 \\ 0 & 1 & t \\ 0 & 0 & 1 \end{pmatrix}\begin{pmatrix} 1 & 0 & -1 \\ 0 & 0 & 1 \\ -1 & 1 & -1 \end{pmatrix}$$

$$= e^{-t}\begin{pmatrix} 1 - t - t^2/2 & t + t^2/2 & -t^2/2 \\ -2t - t^2/2 & 1 + 2t + t^2/2 & -t - t^2/2 \\ -t & t & 1 - t \end{pmatrix}$$

25. $e^{\lambda t}\begin{pmatrix} 1 & t & t^2/2 & t^3/6 \\ 0 & 1 & t & t^2/2 \\ 0 & 0 & 1 & t \\ 0 & 0 & 0 & 1 \end{pmatrix}$

27. $\begin{pmatrix} e^{-4t} & te^{-4t} & (t^2/2)e^{-4t} & 0 \\ 0 & e^{-4t} & te^{-4t} & 0 \\ 0 & 0 & e^{-4t} & 0 \\ 0 & 0 & 0 & e^{3t} \end{pmatrix}$

Problems 6.8, page 457

1. a. $p(\lambda) = \lambda^2 + \lambda - 12 = 0$;
  b. $p(A) = A^2 + A - 12I$

$$= \begin{pmatrix} 14 & 2 \\ 5 & 11 \end{pmatrix}$$

$$+ \begin{pmatrix} -2 & -2 \\ -5 & 1 \end{pmatrix} + \begin{pmatrix} -12 & 0 \\ 0 & -12 \end{pmatrix}$$

$$= \begin{pmatrix} 0 & 0 \\ 0 & 0 \end{pmatrix}$$

  c. $A^{-1} = \dfrac{1}{12}\begin{pmatrix} -1 & -2 \\ -5 & 2 \end{pmatrix}$

3. a. $p(\lambda) = -\lambda^3 + 4\lambda^2 - 3\lambda$;
  b. $p(A) = -A^3 + 4A^2 - 3A$

$$= -\begin{pmatrix} 5 & -9 & 4 \\ -9 & 18 & -9 \\ 4 & -9 & 5 \end{pmatrix}$$

$$+ \begin{pmatrix} 8 & -12 & 4 \\ -12 & 24 & -12 \\ 4 & -12 & 8 \end{pmatrix}$$

$$- \begin{pmatrix} 3 & -3 & 0 \\ -3 & 6 & -3 \\ 0 & -3 & 3 \end{pmatrix}$$

$$= \begin{pmatrix} 0 & 0 & 0 \\ 0 & 0 & 0 \\ 0 & 0 & 0 \end{pmatrix}$$

  c. A^{-1} does not exist.

5. a. $p(\lambda) = -\lambda^3 + 3\lambda^2 - 3\lambda + 1 = 0$
  b. $p(A) = -A^3 + 3A^2 - 3A + I$

$$= -\begin{pmatrix} 1 & -3 & 3 \\ 3 & -8 & 6 \\ 6 & -15 & 10 \end{pmatrix}$$

$$+ \begin{pmatrix} 0 & 0 & 3 \\ 3 & -9 & 9 \\ 9 & -24 & 18 \end{pmatrix}$$

$$- \begin{pmatrix} 0 & 3 & 0 \\ 0 & 0 & 3 \\ 3 & -9 & 9 \end{pmatrix}$$

$$+ \begin{pmatrix} 1 & 0 & 0 \\ 0 & 1 & 0 \\ 0 & 0 & 1 \end{pmatrix}$$

$$= \begin{pmatrix} 0 & 0 & 0 \\ 0 & 0 & 0 \\ 0 & 0 & 0 \end{pmatrix}$$

  c. $A^{-1} = \begin{pmatrix} 3 & -3 & 1 \\ 1 & 0 & 0 \\ 0 & 1 & 0 \end{pmatrix}$

7. a. $p(\lambda) = -\lambda^3 + 6\lambda^2 + 18\lambda + 9 = 0$
  b. $p(A) = -A^3 + 6A^2 + 18A + 9I$

$$= -\begin{pmatrix} 63 & 54 & 108 \\ 180 & 189 & 324 \\ 168 & 204 & 315 \end{pmatrix}$$

$$+ \begin{pmatrix} 18 & 72 & 54 \\ 108 & 162 & 216 \\ 150 & 114 & 252 \end{pmatrix}$$

$$+ \begin{pmatrix} 36 & -18 & 54 \\ 72 & -18 & 108 \\ 18 & 90 & 54 \end{pmatrix}$$

$$+ \begin{pmatrix} 9 & 0 & 0 \\ 0 & 9 & 0 \\ 0 & 0 & 9 \end{pmatrix}$$

$$= \begin{pmatrix} 0 & 0 & 0 \\ 0 & 0 & 0 \\ 0 & 0 & 0 \end{pmatrix}$$

c. $A^{-1} = \dfrac{1}{9}\begin{pmatrix} -27 & 18 & -9 \\ -6 & 3 & 0 \\ 19 & -11 & 6 \end{pmatrix}$

9. a. $p(\lambda) = (a - \lambda)^4$

b. $p(A) = (aI - A)^4$

$$= \begin{pmatrix} 0 & -b & 0 & 0 \\ 0 & 0 & -c & 0 \\ 0 & 0 & 0 & -d \\ 0 & 0 & 0 & 0 \end{pmatrix}^4$$

$$= \begin{pmatrix} 0 & 0 & 0 & 0 \\ 0 & 0 & 0 & 0 \\ 0 & 0 & 0 & 0 \\ 0 & 0 & 0 & 0 \end{pmatrix}$$

c.

$$A^{-1} = \begin{pmatrix} 1/a & -b/a^2 & cb/a^3 & -bcd/a^4 \\ 0 & 1/a & -c/a^2 & cd/a^3 \\ 0 & 0 & 1/a & -d/a^2 \\ 0 & 0 & 0 & 1/a \end{pmatrix}$$

11. $|\lambda| \le 7$ and $\text{Re } \lambda \ge \frac{13}{6}$

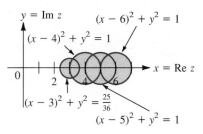

$y = \text{Im } z$ $(x - 6)^2 + y^2 = 1$

$(x - 4)^2 + y^2 = 1$

$x = \text{Re } z$

$(x - 3)^2 + y^2 = \frac{25}{36}$

$(x - 5)^2 + y^2 = 1$

13. $|\lambda| \le 10.5$ and $-10.5 \le \text{Re } \lambda \le 5.75$

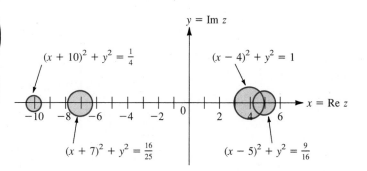

$(x + 10)^2 + y^2 = \frac{1}{4}$ $(x - 4)^2 + y^2 = 1$

$y = \text{Im } z$

$x = \text{Re } z$

$(x + 7)^2 + y^2 = \frac{16}{25}$ $(x - 5)^2 + y^2 = \frac{9}{16}$

15. Since A is symmetric, the eigenvalues of A are real. Then, by Gershgorin's theorem, $\lambda = \text{Re } \lambda \ge 4 - (2 + 1 + \frac{1}{4}) = \frac{3}{4}$.

17. a. $F(\lambda) = B_0C_0 + B_0C_1\lambda + B_1C_0\lambda + B_1C_1\lambda^2$

b. $P(A)Q(A) = (B_0 + B_1A) \times (C_0 + C_1A) = B_0C_0 + B_0C_1A + B_1AC_0 + B_1AC_1A$
$F(A) = B_0C_0 + B_0C_1A + B_1C_0A + B_1C_1A^2$
$F(A) = P(A)Q(A)$ if and only if $C_0A = AC_0$ (in the third term) and $AC_1A = C_1A^2$ in the fourth term.

19. $\det A = \lambda_1\lambda_2\cdots\lambda_n$. If $\det A = 0$, then $\lambda_i = 0$ for some i. But $|\lambda_i - a_{ii}| \le r_i$ so that $|0 - a_{ii}| = |a_{ii}| \le r_i$, which is impossible since A is strictly diagonally dominant. Thus $\lambda_i \ne 0$ for $i = 1, 2, \ldots, n$ and $\det A \ne 0$.

Chapter 6—Review, page 463

1. $4, -2$;

$$E_4 = \text{span}\left\{\begin{pmatrix} 1 \\ 1 \end{pmatrix}\right\};$$

$$E_{-2} = \text{span}\left\{\begin{pmatrix} 2 \\ 1 \end{pmatrix}\right\}$$

3. $1, 7, -5;$

$$E_1 = \text{span}\left\{\begin{pmatrix} -6 \\ 3 \\ 4 \end{pmatrix}\right\};$$

$$E_7 = \text{span}\left\{\begin{pmatrix} 0 \\ 3 \\ 1 \end{pmatrix}\right\};$$

$$E_{-5} = \text{span}\left\{\begin{pmatrix} 0 \\ 0 \\ 1 \end{pmatrix}\right\}$$

5. $1, 3, 3 + \sqrt{2}i, 3 - \sqrt{2}i;$

$$E_1 = \text{span}\left\{\begin{pmatrix} 1 \\ 2 \\ 0 \\ 0 \end{pmatrix}\right\};$$

$$E_3 = \text{span}\left\{\begin{pmatrix} 1 \\ 1 \\ 0 \\ 0 \end{pmatrix}\right\};$$

$$E_{3+\sqrt{2}i} = \text{span}\left\{\begin{pmatrix} 0 \\ 0 \\ -1 \\ \sqrt{2}i \end{pmatrix}\right\};$$

$$E_{3-\sqrt{2}i} = \text{span}\left\{\begin{pmatrix} 0 \\ 0 \\ 1 \\ \sqrt{2}i \end{pmatrix}\right\}$$

7. $C = \begin{pmatrix} -3 & 1 \\ 4 & -1 \end{pmatrix};\ C^{-1}AC = \begin{pmatrix} 2 & 0 \\ 0 & -3 \end{pmatrix}$

9. $C = \begin{pmatrix} 0 & -1-i & -1+i \\ 1 & 1 & 1 \\ -1 & 1 & 1 \end{pmatrix};$

$$C^{-1}AC = \begin{pmatrix} -1 & 0 & 0 \\ 0 & i & 0 \\ 0 & 0 & -i \end{pmatrix}$$

11. not diagonalizable

13. $Q = \begin{pmatrix} 1/\sqrt{2} & 0 & 1/\sqrt{2} \\ 1/\sqrt{2} & 0 & -1/\sqrt{2} \\ 0 & 1 & 0 \end{pmatrix};$

$$Q^tAQ = \begin{pmatrix} 4 & 0 & 0 \\ 0 & -3 & 0 \\ 0 & 0 & 0 \end{pmatrix}$$

15. $C = \begin{pmatrix} 1 & 1 & 1 & -1 \\ -1 & 0 & 0 & 0 \\ 0 & 1 & 0 & 0 \\ -1 & -1 & -1 & 2 \end{pmatrix};$

$$C^{-1}AC = \begin{pmatrix} -1 & 0 & 0 & 0 \\ 0 & -1 & 0 & 0 \\ 0 & 0 & 3 & 0 \\ 0 & 0 & 0 & 3 \end{pmatrix}$$

17. $\dfrac{x'^2}{8/(3+\sqrt{2})} + \dfrac{y'^2}{8/(3-\sqrt{2})} = 1$: ellipse

19. $\dfrac{y'^2}{10/(\sqrt{13}+3)} - \dfrac{x'^2}{10/(\sqrt{13}-3)} = 1$: hyperbola

21. $4x'^2 - 3y'^2$

23. $C = \begin{pmatrix} 2 & -1 \\ 1 & 0 \end{pmatrix};$

$$C^{-1}AC = \begin{pmatrix} -2 & 1 \\ 0 & -2 \end{pmatrix}$$

25. $\begin{pmatrix} -e^t + 2e^{-t} & 2e^t - 2e^{-t} \\ -e^t + e^{-t} & 2e^t - e^{-t} \end{pmatrix}$

27.
$$e^{-t}\begin{pmatrix} \cos 2t - \sin 2t & -2\sin 2t \\ \sin 2t & \cos 2t + \sin 2t \end{pmatrix}$$

29. $|\lambda| \leq 5$ and $-5 \leq \text{Re}\ \lambda \leq \frac{14}{3}$

Chapter 7

Problems 7.1, page 473

1. 0.33333333×10^0

3. -0.35×10^{-4}

5. 0.77777777×10^0

7. 0.77272727×10^1

9. -0.18833333×10^2

11. 0.23705963×10^9

13. 0.83742×10^{-20}

15. $\epsilon_a = 0.1$, $\epsilon_r = 0.0002$

17. $\epsilon_a = 0.005$, $\epsilon_r = 0.04$

19. $\epsilon_a = 0.00333\ldots$,
    $\epsilon_r \approx 0.57143 \times 10^{-3}$

21. $\epsilon_a = 1$, $\epsilon_r \approx 0.1419144 \times 10^{-4}$

23. There are three different operations:
    (1) dividing row i by a_{ii}, (2) multiplying row i by a_{ji}, $j > i$, and subtracting it from row j; (3) the back substitution. Operation (1) requires $\sum_{k=1}^{n} k = \frac{n(n+1)}{2}$ multiplications. Operation (2) requires

$$\sum_{k=1}^{n-1} k(k+1) = \sum_{k=1}^{n-1} k^2 + \sum_{k=1}^{n-1} k$$

$$= \frac{(n-1)n(2n-1)}{6} + \frac{(n-1)n}{2}$$

$$= \frac{n^3 - n}{3}$$

multiplications and additions. Operation (3) requires

$$\sum_{k=1}^{n-1} k = \frac{(n-1)n}{2} = \frac{n^2 - n}{2}$$

multiplications and additions. Adding these fractions together gives the desired results.

25. There are three different operations:
    (1) dividing row i by a_{ii}; (2) multiplying row i by a_{ji}, $j > i$ and subtracting it from row j; (3) keeping track of the n diagonal elements and multiplying them together at the end. Operation (1) requires $\sum_{k=1}^{n-1} k = \frac{n(n-1)}{2}$ multiplications. Operation (2) requires $\sum_{k=1}^{n-1} k^2 = \frac{n(n-1)(2n-1)}{6}$ multiplications. Operation (3) requires $n-1$ multiplications. The sum is $\frac{n-1}{6}[3n + n(2n-1) + 6] = \frac{1}{6}(n-1)(2n^2 + 2n + 6) = \frac{1}{6}(2n^3 + 4n - 6) = \frac{n^3}{3} + \frac{2}{3}n - 1$ multiplications. A similar computation yields the number of additions given in Table 7.1.

27. 7545 microseconds $= 7.545 \times 10^{-3}$ seconds

29. mqn multiplications and $mq(n-1)$ additions

Problems 7.2, page 480

1. $x_1 = 1.6$, $x_2 = -0.800002$ (actual value is -0.8), $x_3 = -3.7$

3. $x_1 = -0.000001$, $x_2 = -2.61001$, $x_3 = 4.3$ Exact solution is $(0, -2.61, 4.3)$.

5. **a.** with pivoting: $x_1 = 5.99$, $x_2 = -2$, $x_3 = 3.99$

b. without pivoting: $x_1 = 6$, $x_2 = -2$, and $x_3 = 4$ (Yes, sometimes it's better to follow the simplest path. In Problem 6 pivoting gives much more accurate answers.) The relative errors with pivoting are $\frac{1}{600} = 0.0017$, 0, and $\frac{1}{400} = 0.0025$.

7. A solution with rounding to 3 significant figures is $x_1 = 1050$ and $x_2 = -1000$. The exact solution is $x_1 = \frac{15650}{13} \approx 1204$ and $x_2 = -\frac{15000}{13} \approx -1154$. The relative errors are $0.1279 \approx 13\%$ and $0.1334 \approx 13\%$

Problems 7.3, page 491

1. yes **3.** no **5.** yes

7. Jacobi: $x_1 = 2.9757$, $x_2 = -0.9919$ (8 iterations) Gauss–Seidel: $x_1 = 3.0041$, $x_2 = -1.0025$ (5 iterations) Exact solution is $(3, -1)$.

9. Jacobi: $x_1 = -1.9999$, $x_2 = -2.988$, $x_3 = 7.0146$ (7 iterations) Gauss–Seidel: $x_1 = -1.9974$, $x_2 = -2.9993$, $x_3 = 6.999$ (6 iterations) Exact solution is $(-2, -3, 7)$.

11. Jacobi: $x_1 = -8.2864$, $x_2 = 14.386$, $x_3 = -0.10296$ (13 iterations) Gauss–Seidel: $x_1 = -8.2989$, $x_2 = 14.3995$, $x_3 = -0.10039$ (7 iterations) Exact solution is $(-8.3, 14.4, -0.1)$.

13. a. $|a_{ii}| = 1$ and
$$|a_{12}| + |a_{13}| = |a_{21}| + |a_{23}|$$
$$= |a_{31}| + |a_{32}|$$
$$= 1$$
thus $|a_{ii}| = \sum_{\substack{j=1 \\ j\neq i}}^{3} |a_{ij}|$ so that

$$a_{ii} \not> \sum_{\substack{j=1 \\ j\neq 1}}^{3} |a_{ij}|$$

b. *Jacobi*

n	$x_1^{(n)}$	$x_2^{(n)}$	$x_3^{(n)}$
0	0.8	0.8	0.8
1	1.2	1.2	1.2
2	0.8	0.8	0.8
3	1.2	1.2	1.2
4	0.8	0.8	0.8
5	1.2	1.2	1.2

c. *Gauss–Seidel*

n	$x_1^{(n)}$	$x_2^{(n)}$	$x_3^{(n)}$
0	0.8	0.8	0.8
1	1.2	1	0.9
2	1.05	1.025	0.9625
3	1.0063	1.0156	0.98906
4	0.99766	1.0066	0.99785
5	0.99775	1.0022	1
6	0.99889	1.0006	1.0003
7	0.99958	1.0001	1.0002
8	0.99988	0.99998	1.0001

d. i. $D^{-1}(L + U) = \begin{pmatrix} 0 & \frac{1}{2} & \frac{1}{2} \\ \frac{1}{2} & 0 & \frac{1}{2} \\ \frac{1}{2} & \frac{1}{2} & 0 \end{pmatrix}$,

which has the eigenvalue 1, so, by Theorem 2(i), the Jacobi iterates fail to converge.

ii. $(D + L)^{-1}U = \begin{pmatrix} 0 & \frac{1}{2} & \frac{1}{2} \\ 0 & -\frac{1}{4} & \frac{1}{4} \\ 0 & -\frac{1}{8} & -\frac{3}{8} \end{pmatrix}$,

whose characteristic polynomial is $-\lambda(\lambda^2 + \frac{5}{8}\lambda + \frac{1}{8}) = 0$ with roots 0, $(-5 \pm \sqrt{7}i)/16$.

$$|(-5 + \sqrt{7}i)/16|$$
$$= |(-5 - \sqrt{7}i)/16|$$
$$= \sqrt{(\tfrac{5}{16})^2 + (\sqrt{7}/16)^2}$$
$$\approx 0.3536.$$

Thus $r[(D + L)^{-1}U] \approx 0.35 < 1$ and, by Theorem 2(ii), the Gauss–Seidel iterates converge.

15. If A is diagonal, then $L = U = 0$. Then $r[D^{-1}(L + U)] = r(0) = 0$, and $r[(D + L)^{-1}U] = r(D0) = r(0) = 0$.

17. $D = \begin{pmatrix} a & 0 \\ 0 & d \end{pmatrix}$, $L = \begin{pmatrix} 0 & 0 \\ c & 0 \end{pmatrix}$, and

$U = \begin{pmatrix} 0 & b \\ 0 & 0 \end{pmatrix}$ so that

$$D^{-1}(L + U) = \begin{pmatrix} 0 & \dfrac{b}{a} \\ \dfrac{c}{d} & 0 \end{pmatrix} \text{ and}$$

$$(D + L)^{-1}U = \begin{pmatrix} 0 & \dfrac{b}{a} \\ 0 & -\dfrac{bc}{ad} \end{pmatrix}.$$

The eigenvalues of $D^{-1}(L + U)$ are $\pm\sqrt{\dfrac{bc}{ad}}$, and the eigenvalues of $(D + L)^{-1}U$ are 0 and $-\dfrac{bc}{ad}$. These are all less than one in absolute value if and only if $\left|\dfrac{bc}{ad}\right| < 1$.

19. $D^{-1}(L + U)$

$$= \begin{pmatrix} \dfrac{1}{a_{11}} & 0 & \cdots & 0 \\ 0 & \dfrac{1}{a_{22}} & \cdots & 0 \\ \vdots & \vdots & & \vdots \\ 0 & 0 & \cdots & \dfrac{1}{a_{nn}} \end{pmatrix}$$

$$\times \begin{pmatrix} 0 & a_{12} & a_{13} & \cdots & a_{1n} \\ a_{21} & 0 & a_{23} & \cdots & a_{2n} \\ \vdots & \vdots & \vdots & & \vdots \\ a_{n1} & a_{n2} & a_{n3} & \cdots & 0 \end{pmatrix}$$

$$= \begin{pmatrix} 0 & \dfrac{a_{12}}{a_{11}} & \dfrac{a_{13}}{a_{11}} & \cdots & \dfrac{a_{1n}}{a_{11}} \\ \dfrac{a_{21}}{a_{22}} & 0 & \dfrac{a_{23}}{a_{22}} & \cdots & \dfrac{a_{2n}}{a_{22}} \\ \vdots & \vdots & \vdots & & \vdots \\ \dfrac{a_{n1}}{a_{nn}} & \dfrac{a_{n2}}{a_{nn}} & \dfrac{a_{n3}}{a_{nn}} & \cdots & 0 \end{pmatrix}$$

Since A is strictly diagonally dominant, $|a_{11}| > |a_{12}| + |a_{13}| + \cdots + |a_{1n}|$. Thus $\left|\dfrac{a_{12}}{a_{11}}\right| + \left|\dfrac{a_{13}}{a_{11}}\right| + \cdots + \left|\dfrac{a_{1n}}{a_{11}}\right| < 1$. A similar fact holds in every row of $D^{-1}(L + U)$. Thus, by Gershgorin's theorem, all the eigenvalues of $D^{-1}(L + U)$ lie in circles centered at the origin with radii less than 1. This means that if λ is an eigenvalue of $D^{-1}(L + U)$, then $|\lambda| < 1$, which implies that $r[D^{-1}(L + U)] < 1$.

Problems 7.4, page 500

1. $-4, \begin{pmatrix} 1 \\ 1 \end{pmatrix}$ **3.** $-6.3, \begin{pmatrix} 1 \\ -0.5 \end{pmatrix}$

5. $8, \begin{pmatrix} 1 \\ 0.5 \\ 1 \end{pmatrix}$

7. a. $\lambda \approx 8.5536$ (without scaling); $\epsilon_r \approx 0.00016374$
   b. $\lambda = 2 + \sqrt{43} = 8.5574$. The actual relative error is 0.000444.

9. Starting with $x_0 = \begin{pmatrix} 1 \\ 1 \end{pmatrix}$ and without scaling, we obtain $\begin{pmatrix} 1 \\ 1 \end{pmatrix}, \begin{pmatrix} 2 \\ 1 \end{pmatrix}$, $\begin{pmatrix} -1 \\ -1 \end{pmatrix}, \begin{pmatrix} -2 \\ -1 \end{pmatrix}, \begin{pmatrix} 1 \\ 1 \end{pmatrix}, \begin{pmatrix} 2 \\ 1 \end{pmatrix}, \ldots$
That is, the method does not converge. Note that the eigenvalues of A are $\pm i$, both of which have absolute value 1. That is, A does not have a dominant eigenvalue.

11. The second eigenvalue is $\lambda_2 = 3$.

13. The other eigenvalues are -2 and 1.

Chapter 7—Review, page 502

1. $\epsilon_a = 0.02$, $\epsilon_r = 0.002857$

3. $\epsilon_a = \frac{1}{300} = 0.003333\ldots$, $\epsilon_r = 0.25$

5. $\begin{pmatrix} 1 & -2 & 3 \\ 0 & 1 & -2 \\ 0 & 0 & 1 \end{pmatrix}$

7. $x_1 = 7.11004$, $x_2 = -2.39005$, $x_3 = -5.92009$. Exact solution is $(7.11, -2.39, -5.92)$.

9. no

11. Jacobi: $x_1 = -0.34008$, $x_2 = 0.260018$, $x_3 = 0.509983$ (12 iterates) Gauss–Seidel: $x_1 = -0.340001$, $x_2 = 0.26$, $x_3 = 0.51$ (7 iterates) Exact solution is $(-0.34, 0.26, 0.51)$.

13. -8, $\begin{pmatrix} 1 \\ -\frac{2}{3} \end{pmatrix}$ 15. $\lambda_2 = 3$

Appendix 1, page A-6

1. First, is it true for $n = 1$? $2 = 1(1 + 1)$, so it is. Now assume it is true for $n = k$. Then $2 + 4 + 6 + \cdots + 2k = k(k + 1)$. We must now show that it is true for $n = k + 1$; that is, we must show that $2 + 4 + 6 + \cdots + 2k + 2(k + 1) = (k + 1)[(k + 1) + 1]$. We know that $2 + 4 + 6 + \cdots + 2k + 2(k + 1) = k(k + 1) + 2(k + 1)$ (induction hypothesis) $= (k + 2)(k + 1) = (k + 1)[(k + 1) + 1]$

3. First, is it true for $n = 1$? $2 = \dfrac{1(3 \cdot 1 + 1)}{2}$, so it is. Now assume it is true for $n = k$. Then $2 + 5 + 8 + \cdots + (3k - 1) = \dfrac{k(3k + 1)}{2}$.
We must now show it is true for $n = k + 1$; that is, we must show that

$$2 + 5 + 8 + \cdots + (3k - 1) + (3k + 2)$$
$$= \frac{(k + 1)[3(k + 1) + 1]}{2}$$
$$= \frac{(k + 1)(3k + 4)}{2}$$
$$= \frac{3k^2 + 7k + 4}{2}$$

Add $3k + 2$ to both sides of the equation in the induction hypothesis and get

$$2 + 5 + 8 + \cdots + (3k - 1) + (3k + 2)$$
$$= \frac{k(3k + 1)}{2} + (3k + 2)$$
$$= \frac{3k^2 + k}{2} + \frac{6k + 4}{2}$$
$$= \frac{3k^2 + 7k + 4}{2}$$

5. Is it true for $n = 1$? Yes, $2^1 > 1$. Is it true for $n = 2$? Yes, $2^2 > 2$. Now assume it is true for $n = k$; that is, $2^k > k$. We must now prove it true for $n = k + 1$, or that $2^{k+1} > k + 1$. Multiply both sides of the induction hypothesis by 2 to obtain $2^{k+1} > 2k$. Now $2k = k + k > k + 1$ if $k > 1$, which it is. Therefore $2^{k+1} > k + 1$.

7. Is it true for $n = 1$? Yes, since $1 + 2 = 2^2 - 1$. Now assume it is true for $n = k$; that is, $1 + 2 + 4 + \cdots + 2^k = 2^{k+1} - 1$. We must now prove that it is true for $n = k + 1$, or that $1 + 2 + 4 + \cdots + 2^k + 2^{k+1} = 2^{k+2} - 1$. Add 2^{k+1} to both sides of the induction hypothesis and obtain

$$1 + 2 + 4 + \cdots + 2^k + 2^{k+1} = 2^{k+1} - 1 + 2^{k+1}$$
$$= 2 \cdot 2^{k+1} - 1$$
$$= 2^{k+2} - 1$$

9. Is it true for $n = 1$? Yes, since $1 + \frac{1}{2} = 2 - \dfrac{1}{2^1}$. Now assume it is true for $n = k$; that is,

$$1 + \frac{1}{2} + \frac{1}{4} + \cdots + \frac{1}{2^k} = 2 - \frac{1}{2^k}$$

We must now prove for $n = k + 1$; that is,

$$1 + \frac{1}{2} + \frac{1}{4} + \cdots + \frac{1}{2^k} + \frac{1}{2^{k+1}} =$$
$$2 - \frac{1}{2^{k+1}}$$

Add $\dfrac{1}{2^{k+1}}$ to both sides of the induction hypothesis and obtain

$$1 + \frac{1}{2} + \frac{1}{4} + \cdots + \frac{1}{2^k} + \frac{1}{2^{k+1}}$$

$$= 2 - \frac{1}{2^k} + \frac{1}{2^{k+1}}$$

$$= 2 - \frac{2}{2^{k+1}} + \frac{1}{2^{k+1}}$$

$$= 2 - \frac{1}{2^{k+1}}$$

11. Is it true for $n = 1$? Yes, $1^3 = \dfrac{1^2(1+1)^2}{4}$. Now assume it is true for $n = k$; that is,

$$1^3 + 2^3 + 3^3 + \cdots + k^3 = \frac{k^2(k+1)^2}{4}$$

We must prove that it is true for $n = k + 1$; that is,

$$1^3 + 2^3 + 3^3 + \cdots + k^3 + (k+1)^3$$

$$= \frac{(k+1)^2[(k+1)+1]^2}{4}$$

$$= \frac{k^4 + 6k^3 + 13k^2 + 12k + 4}{4}$$

Add $(k+1)^3$ to both sides of the induction hypothesis.

$$1^3 + 2^3 + \cdots + k^3 + (k+1)^3$$

$$= \frac{k^2(k+1)^2}{4} + (k+1)^3$$

$$= \frac{k^4 + 2k^3 + k^2}{4} + k^3 + 3k^2 + 3k + 1$$

$$= \frac{k^4 + 2k^3 + k^2}{4} + \frac{4k^3 + 12k^2 + 12k + 4}{4}$$

$$= \frac{k^4 + 6k^3 + 13k^2 + 12k + 4}{4}$$

13. Is it true for $n = 1$? Yes, $1 \cdot 2 = \dfrac{1 \cdot (1+1) \cdot (4-1)}{3}$ Assume it is true for $n = k$; that is,

$$1 \cdot 2 + 3 \cdot 4 + \cdots + (2k-1)(2k) =$$
$$\frac{k(k+1)(4k-1)}{3}$$

Now prove for $n = k + 1$; that is,

$$1 \cdot 2 + 3 \cdot 4 + (2k-1)(2k) + (2k+1)(2k+2)$$

$$= \frac{(k+1)(k+2)(4k+3)}{3}$$

$$= \frac{4k^3 + 15k^2 + 17k + 6}{3}$$

Add $[2(k+1) - 1][2(k+1] = (2k+1)(2k+2)$ to both sides of the induction hypothesis. Obtain

$$1 \cdot 2 + 3 \cdot 4 + \cdots + (2k-1)(2k) + (2k+1)(2k+2)$$

$$= \frac{k(k+1)(4k-1)}{3} + (2k+1)(2k+2)$$

$$= \frac{4k^3 + 3k^2 - k}{3} + 4k^2 + 6k + 2$$

$$= \frac{4k^3 + 3k^2 - k}{3} + \frac{12k^2 + 18k + 6}{3}$$

$$= \frac{4k^3 + 15k^2 + 17k + 6}{3}$$

Several of the following use the fact that if an integer m divides evenly into an integer a and divides evenly into an integer b, then m divides evenly into $a + b$.

15. Is it true for $n = 1$? Yes, since $1^2 + 1 = 2$ is even. Assume that $k^2 + k$ is even. Now prove true for k; that is, we now must prove that $(k+1)^2 + (k+1)$ is even. But

$$(k+1)^2 + (k+1) = k^2 + 2k + 1 + k + 1$$
$$= (k^2 + k) + (2k + 2)$$

Now 2 divides evenly into $k^2 + k$ by the induction hypothesis. It is evident that 2 divides evenly into $2k$ and that 2 divides evenly into 2. Therefore 2 divides evenly into $k^2 + k + 2k + 2$, meaning that the number is even.

17. Is it true for $n = 1$? Yes, because $1(1^2 + 5) = 6$ is divisible by 6.

Now assume it is true for k; that is, $k(k^2 + 5)$ is divisible by 6. We now must prove that $(k + 1)[(k + 1)^2 + 5]$ is divisible by 6.

$$(k + 1)[(k + 1)^2 + 5]$$
$$= (k + 1)(k^2 + 2k + 6)$$
$$= (k + 1)(k^2 + 5 + 2k + 1)$$
$$= k(k^2 + 5) + (k^2 + 5) + k(2k + 1) + (2k + 1)$$
$$= k(k^2 + 5) + 3(k^2 + k) + 6$$

Now $k(k^2 + 5)$ is divisible by 6 by the induction hypothesis; $3(k^2 + k)$ is clearly divisible by 3 and is even by Problem 15, so it is divisible by 6; and certainly 6 is divisible by 6, so the expression given is divisible by 6.

19. The problem is true if $n = 1$ since $x^1 - 1$ is divisible by $x - 1$. Now assume that $x^k - 1$ is divisible by $x - 1$. We have to prove that $x^{k+1} - 1$ is divisible by $x - 1$. Now

$$x^{k+1} - 1 = x^k x - 1 = x^k x - x + x - 1$$
$$= x(x^k - 1) + (x - 1).$$

The first term is divisible by $x - 1$ by the induction hypothesis, and the second term in the sum is divisible by $x - 1$, so the expression is divisible by $x - 1$.

21. If $n = 1$, $(ab)^1 = a^1 b^1 = ab$, so it is true. Now assume that $n = k$; that is, $(ab)^k = a^k b^k$. We must prove for $k + 1$; that is, $(ab)^{k+1} = a^{k+1} b^{k+1}$. Now

$$(ab)^{k+1} = (ab)^k(ab) = a^k b^k ab$$
$$= a^k ab^k b \text{ (since multiplication is commutative)}$$
$$= a^{k+1} b^{k+1}$$

23. From Theorem 2.2.4, $\det A_1 A_2 = \det A_1 \det A_2$, so the result holds for $n = 2$. Assume that it holds for $n = k$. Then

$$\det A_1 A_2 \cdots A_k A_{k+1}$$
$$= \det A_1 A_2 \cdots A_k \det A_{k+1}$$
(using the result for $n = 2$)
$$= (\det A_1 \det A_2 \cdots \det A_k) \det A_{k+1}$$
(using the result for $n = k$)
$$= \det A_1 \det A_2 \cdots \det A_k \det A_{k+1}$$
which is the result for $n = k + 1$.

25. $n = 1$; there are exactly two subsets of a set with one element: the set itself and the empty set. Now assume there are exactly 2^k subsets of a set with k elements. Now consider a set A with $k + 1$ elements. Remove one, call it a_{k+1}. The remaining elements form a set of k elements. This set has 2^k subsets. The other subsets of A are found by taking the union of each of the subsets of $A - \{a_{k+1}\}$ and $\{a_{k+1}\}$; there are 2^k of these also. In other words, A has 2^k subsets containing the element a_{k+1} and 2^k subsets not containing a_{k+1}, for a total of $2^k + 2^k = 2^{k+1}$ subsets.

27. It is not true for $n = 2$. In that case S_1 and S_2 are disjoint, and therefore you cannot say that $h_1 = h_2$.

Appendix 2, page A-17

1. $9 - 7i$ 3. 2 5. $-27 + 5i$
7. $5\sqrt{2}e^{i(\pi/4)}$
9. $3\sqrt{2}e^{i(7\pi/4)} = 3\sqrt{2}e^{-i(\pi/4)}$
11. $6e^{(\pi/6)i}$
13. $8e^{i(11\pi/6)} = 8e^{-i(\pi/6)}$
15. $2e^{i(4\pi/3)} = 2e^{-i(2\pi/3)}$
17. -2 19. $-\sqrt{2}/4 - i(\sqrt{2}/4)$
21. $-2\sqrt{3} + 2i$
23. $-\frac{3}{2} - \frac{3}{2}\sqrt{3}i$
25. $\cos 1 + i \sin 1 \approx 0.5403 + 0.8415i$
27. $4 - 6i$ 29. $7i$
31. $2e^{-i(\pi/7)}$ 33. $3e^{i(4\pi/11)}$
35. If $z = \bar{z}$, then $\alpha + i\beta = \alpha - i\beta$, or $i\beta = -i\beta$, which is possible if and

only if $\beta = 0$ so that z is real. If z is real, then $z = \alpha = \bar{z}$.

37. $z\bar{z} = (\alpha + i\beta)(\alpha - i\beta) = \alpha^2 - (i^2\beta^2) = \alpha^2 + \beta^2 = |z|^2$

39. The locus of points on a circle in the complex plane centered at z_0 with radius a. If $z_0 = x_0 + iy_0$, then in x and y coordinates this is the circle whose equation is $(x - x_0)^2 + (y - y_0)^2 = a^2$.

41. Suppose that $p(z) = z^n + a_{n-1}z^{n-1} + \cdots + a_1z + a_0 = 0$. Then
$$\overline{z^n + a_{n-1}z^{n-1} + \cdots + a_1z + a_0} = \bar{0}$$
$$= 0 = \bar{z}^n + \overline{a_{n-1}z^{n-1}} + \cdots + \overline{a_1z + a_0}$$
$$= \bar{z}^n + a_{n-1}\bar{z}^{n-1} + \cdots + a_1\bar{z} + a_0$$
(since the a_i's are real) $= \bar{z}^n + a_{n-1}\bar{z}^{n-1} + \cdots + a_1\bar{z} + a_0 = p(\bar{z}) = 0$. Here we have used the

fact that for any integer k, $\overline{z^k} = \bar{z}^k$. This follows easily if we write z in polar form. If $z = re^{i\theta}$, then $z^n = r^ne^{in\theta}$, $\overline{z^n} = r^ne^{-in\theta}$, $\bar{z} = re^{-i\theta}$, and $\bar{z}^n = r^ne^{-in\theta} = \overline{z^n}$.

43. Since $(\cos\theta + i\sin\theta)^1 = \cos 1 \cdot \theta + i\sin 1 \cdot \theta$, DeMoivre's formula holds for $n = 1$. Assume it holds for $n = k$. That is, $(\cos\theta + i\sin\theta)^k = \cos k\theta + i\sin k\theta$. Then $(\cos\theta + \sin\theta)^{k+1} = (\cos\theta + i\sin\theta)^k(\cos\theta + i\sin\theta) = (\cos k\theta + i\sin k\theta) \times (\cos\theta + i\sin\theta) = [\cos k\theta\cos\theta - \sin k\theta\sin\theta] + i[\sin k\theta\cos\theta + \cos k\theta\sin\theta] = \cos(k\theta + \theta) + i\sin(k\theta + \theta) = \cos(k + 1)\theta + i\sin(k + 1)\theta$, which is DeMoivre's formula for $n = k + 1$.

Index

DATE DUE